EVS35—International Electric Vehicle Symposium and Exhibition (Oslo, Norway)

EVS35—International Electric Vehicle Symposium and Exhibition (Oslo, Norway)

Guest Editor

Joeri Van Mierlo

MDPI

Basel • Beijing • Wuhan • Barcelona • Belgrade • Novi Sad • Cluj • Manchester

Guest Editor
Joeri Van Mierlo
MOBI—Electromobility
Research Centre
Vrije Universiteit Brussel
Brussels
Belgium

Editorial Office
MDPI AG
Grosspeteranlage 5
4052 Basel, Switzerland

This is a reprint of the Special Issue, published open access by the journal *World Electric Vehicle Journal* (ISSN 2032-6653), freely accessible at: https://www.mdpi.com/journal/wevj/special_issues/EVS35_International_Electric_Vehicle_Symposium_Exhibition.

For citation purposes, cite each article independently as indicated on the article page online and as indicated below:

Lastname, A.A.; Lastname, B.B. Article Title. *Journal Name* **Year**, *Volume Number*, Page Range.

ISBN 978-3-7258-3941-4 (Hbk)
ISBN 978-3-7258-3942-1 (PDF)
https://doi.org/10.3390/books978-3-7258-3942-1

Contents

About the Editor

Joeri Van Mierlo

Joeri Van Mierlo is a key player in the Electromobility scene. He is a professor at the Vrije Universiteit Brussels, one of the top universities in this field. Prof. Dr. ir. Joeri Van Mierlo leads the MOBI – Electromobility Research Centre and is head of the ETEC - Electric Engineering and Energy Technology department. Prof. Van Mierlo was a visiting professor at Chalmers University of Technology, Sweden (2012). He is an expert in the field of Electric and Hybrid vehicles (batteries, power converters, energy management simulations), as well as in the environmental and economic comparison of vehicles with different drive trains and fuels (LCA, TCO). Prof. Van Mierlo served as Vice President of AVERE (2013–2019), the European Electric Vehicle Association and as Vice President (2015–2019) of its Belgian section, EV-Belgium. He is appointed as a member of the KVAB (Royal Flemish Academy of Belgium for Science and the Arts). He is an active member of the EARPA (European Automotive Research Partner Association) and a member of EGVIA for 2Zero (European Green Vehicles Initiative Association for the 2Zero partnership). He is a member of the Core Lab Committee of Flanders Make. He is the editor-in-chief of the World Electric Vehicle Journal and the author of more than 500 scientific publications. "Highly Cited Researcher" award (2023 and 2024) from Clarivate - Web of Science.

Preface

This Special Issue gathers a selection of the best papers from EVS35, the 35th International Electric Vehicle Symposium and Exhibition, which was held from 11 to 15 June 2022 in Oslo, Norway.

The International Electric Vehicle Symposium (EVS) is the world's longest, largest, and highest-spec event in the field of electric vehicles, covering areas including pure electric vehicles, hybrid vehicles, fuel cell vehicles, and core components. With the support of the World Electric Vehicle Association, the Symposium is hosted by three regional professional organizations under the World Electric Vehicle Association in North America (Electric Drive Transportation Association, EDTA), Europe (The European Association for Electromobility, E-Mobility Europe), and Asia (Electric Vehicle Association of Asia Pacific, EVAAP) in turn. EVS already has a long history of more than 50 years since its birth in Phoenix, Arizona, USA, in 1969.

The theme of EVS35 was The Electric Future is Here. The topics that are included are technical advances in research findings in hybrid electric vehicles, pure electric vehicles, fuel cell vehicles, infrastructure construction and service research, marketing and policy support research, etc.

The authors of the best papers presented at EVS35 were invited to further extend their EVS35 paper, including their most recent research findings. After a second thorough round of peer reviews, these papers were published in this Special Issue of the *World Electric Vehicle Journal* (WEVJ), the official journal of the World Electric Vehicle Association (WEVA).

Joeri Van Mierlo
Guest Editor

World Electric Vehicle Journal

MDPI

Article

Perspectives on Modeling Energy and Mobility Transitions for Stakeholders: A Dutch Case

Younjung Choi *, Marcus Vinicius Pereira Pessoa and G. Maarten Bonnema

Systems Engineering and Multidisciplinary Design, Department of Design, Production, and Management, Faculty of Engineering Technology, University of Twente, Drienerlolaan 5, 7522 NB Enschede, The Netherlands; m.v.pereirapessoa@utwente.nl (M.V.P.P.); g.m.bonnema@utwente.nl (G.M.B.)
* Correspondence: y.choi@utwente.nl; Tel.: +31-53-4891878

Abstract: We address the value of engaging stakeholders in energy and mobility transitions by using models. As a communication medium, models can facilitate the collaborative exploration of a future between modeling researchers and stakeholders. Developing models to engage stakeholders requires an understanding of state-of-the-art models and the usability of models from the stakeholder perspective. We employ mixed methods in our research. We present the overview of models that have been proposed to make sense of the transitions in the scientific literature through a systematic literature mapping (n = 105). We interviewed 10 stakeholders based in The Netherlands to elaborate on use cases in which models can benefit stakeholders in practice and the characteristics of usable models. We conclude our research by elaborating on two challenges of model design that modeling research can consider to engage stakeholders. First, we argue that understanding the epistemic requirements of both modeling researchers and stakeholders that models can simultaneously meet is crucial (e.g., questions addressed using models and assumptions). Second, we seek technical solutions for producing models in a time-wise manner and developing interfaces that allow models distant in formalism and represented phenomena to communicate in tandem. Our research creates awareness of the model design aspect by considering its usability.

Keywords: energy transition; sustainable mobility; modeling; transition models; stakeholder engagement; learning; usability of transition models; epistemic requirement; model integration

Citation: Choi, Y.; Pessoa, M.V.P.; Bonnema, G.M. Perspectives on Modeling Energy and Mobility Transitions for Stakeholders: A Dutch Case. *World Electr. Veh. J.* **2023**, *14*, 178. https://doi.org/10.3390/wevj14070178

Academic Editor: Joeri Van Mierlo

Received: 30 April 2023
Revised: 16 June 2023
Accepted: 24 June 2023
Published: 6 July 2023

1. Introduction

Achieving sustainable energy and mobility systems requires stakeholder decisions to be made and actions to be taken over time [1–4]. Accordingly, various models with different formalisms and represented phenomena have been introduced to support these heterogeneous decision-making processes. These include policymaking frameworks that articulate relationships between policies and their environmental impacts [5], energy audit models that explore industrial measures for improving energy efficiency [6–8], and quantitative models that simulate future scenarios to enhance policymaking consistency and validate underlying theories [9,10].

1.1. Engaging Stakeholders in the Transitions by Using Models

Engaging stakeholders is recognized as a crucial element for achieving societal transformation [11–13]. Models can serve as a valuable medium for collaborative research between stakeholders and researchers about systems transitions. However, it is vital to investigate how adaptable models are for engagement [9].

Researchers working on energy or mobility system modeling have focused on enhancing modeling practices by better representing transitioning systems. This involves understanding modeling methodologies [14–17], the characteristics of systems (e.g., techno-economic details, heterogeneity of actors, emergent behaviors, dynamics of transition scenarios) [17–20], and geographic, temporal, and sectoral resolutions [19,20]. Despite this progress, research on the usability aspect of models from the stakeholder perspective is still in its early stages.

Recent studies have addressed concerns regarding the usability of models from the stakeholder perspective. First, the studies highlighted the need for models to better represent human behavioral aspects (e.g., cultural dimensions) and balance price policies (e.g., regulatory policies) with non-price policies (e.g., public awareness) [21,22]. Second, the studies revealed that improving the comprehensibility of models through visual aids and instructions could enhance their usability [22,23]. Third, they showed that transparent communication regarding involved assumptions, databases, and modeling frameworks could also improve model usage [21–24]. Last, the articles emphasized the applicability of models in specific use cases, such as integrating energy models into transport planning and using them for policymaking and collaboration between municipalities [22,23].

Modeling is an epistemic activity that helps us understand complex phenomena [25,26]. In energy and mobility transitions, engaging stakeholders through modeling can support decision-making and facilitate collaborative learning between researchers and stakeholders [27–29]. However, we must broaden our understanding of two aspects. First, we must understand models that are diverse in formalism in the represented phenomena relevant to the transitions. By understanding the diversity of models and each modeling approach's strength(s), researchers can employ suitable modeling methods by balancing the use context in practice and the research environment. Second, we need to consider characteristics that can ensure the usability and effective use of models in practice.

Muratori et al. [30] reviewed approximately 30 models that were developed to project the future of integrated energy and mobility systems and categorized the models by specific purposes (e.g., models for estimating vehicle choices). However, as stakeholders' needs may exceed these categories, we expanded our exploration of models in our research. While the usability of models from a stakeholder perspective has mainly been discussed for energy system models, we incorporated user perspective knowledge to identify common stakeholder perspectives for future research. We consider that doing so would help researchers to better understand the needs and viewpoints of stakeholders and thereby develop more valuable and relevant models for exploring integrated energy and mobility systems.

1.2. Research Questions

To address the knowledge gap, we investigated the following research question:

"What are the key considerations and approaches for effectively engaging stakeholders in energy and mobility transitions using models?"

To this end, we answered three sub-research questions. The sub-research questions and the associated research aims and objectives are presented in Table 1.

Demarcation

Our research aims to support energy and mobility system transitions in The Netherlands. Therefore, we focused our interviews on practitioners working in the Dutch energy and mobility sectors.

Table 1. Research questions, objectives, and methods.

	RsQ1	RsQ2	RsQ3
Research question	*"What models have been proposed in the scientific literature to understand energy and mobility transitions?"*	*"What traits of models are supportive of stakeholder engagement?"*	*"How can models be designed to engage stakeholders in the transitions effectively?"*
Aim	Providing state-of-the-art models used for understanding the transitions, which can support modeling researchers to employ models and modeling approaches that are suitable for engaging stakeholders	Providing knowledge about model usability, which can enhance the effectiveness of stakeholder engagement by using models	Providing insights to bridge the gap between available models and stakeholders' needs, which can support modeling researchers to design models for effective engagements
Research objective	Identifying state-of-the-art models presented in scientific literature considering forms, represented phenomena, and utilities	Identifying traits of models that can enhance their usability from the stakeholder perspective	Identifying approaches for the model design
Method	Systematic literature mapping	Stakeholder interview	Synthesis
Section	Section 2	Section 3	Section 4

1.3. Research Methodology

This paper utilized a three-step methodology (Table 1). First, a systematic literature mapping was conducted to review models in the scientific literature that pertained to energy and mobility transitions, thereby addressing RsQ1 in Section 2. Second, stakeholders from the local and regional government, businesses, and innovation management involved in energy and mobility transitions were interviewed to gather ideas for use cases of models in practice and the traits of usable models necessary to answer RsQ2 in Section 3. Third, the reviewed models were evaluated from the stakeholder perspective to determine how to design models that better engage stakeholders. Finally, the evaluation and reflection results were synthesized to answer RsQ3 in Sections 3 and 4.

2. Models for Understanding the Transitions

In this section, we explain the methodology used for the literature mapping in Section 2.1 and present six grouped models we reviewed in detail in Section 2.2.

2.1. Method: Systematic Literature Mapping

We utilized systematic literature mapping techniques described in systems engineering, software engineering, and environmental management research papers [31–33].

2.1.1. Literature Acquisition Procedure

To acquire the scientific literature that presented the models in question, we used the PICO search tool to determine keywords and specify a search query [33–36]. Our population of interest was scientific articles, while our intervention involved producing models for decision support and simulation to explain or predict the transition phenomena. We did not make comparisons or aim to favor any one model. Our search string was "(model OR simulation OR 'decision support system') AND (mobility OR transport OR vehicle) AND ('energy transition' OR decarbonization OR sustainab*)," which we applied to Scopus and Web of Science databases for articles published from 2016 to 2021. We only gathered scientific papers whose titles involved the search string. We retrieved 215 articles but only used 105 that met our inclusion criteria of being written in English and having free full-text availability. We excluded papers such as conference summaries and those whose titles merely included homonyms of our search terms (e.g., suspended solids transport).

2.1.2. Literature Mapping Protocol

We used an iterative process to map the retrieved literature due to the diversity of the reviewed models and their potential uses. We initially attempted deductive reasoning using existing frameworks (e.g., the Transition Management framework presented by [3]) by assuming that models would be used for governance activities, such as vision development. The results of this approach were of limited value, as not all models were designed for governance activities. We then tried to develop a new framework, but the deterministic process was unsuitable for identifying the various models. Ultimately, we used inductive reasoning and divided the literature into qualitative and quantitative models based on their primary functions and utilities (e.g., system articulation and sustainable business models) (Figure 1). We further categorized the models based on their representations, as explained in Section 2.2.

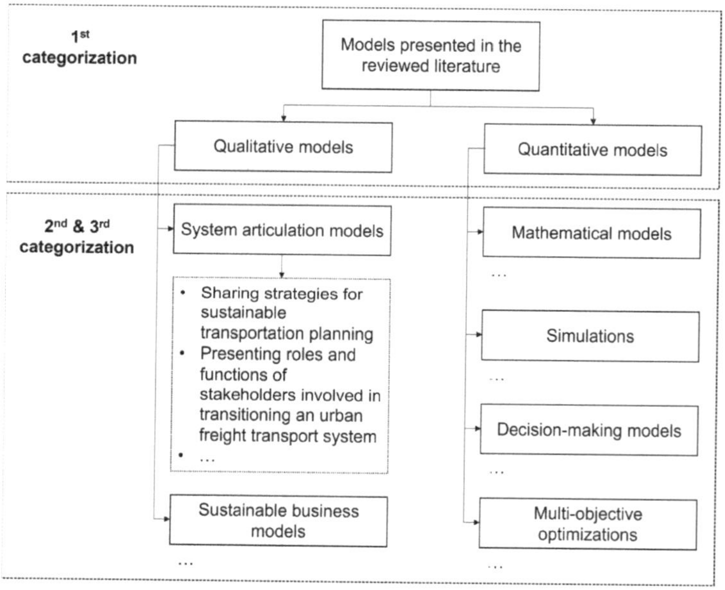

Figure 1. The literature mapping protocol.

2.2. Results

The literature mapping informed the following six types of models:

- Qualitative articulation models (Section 2.2.1);
- Sustainable business models (Section 2.2.2);
- Mathematical models (Section 2.2.3);
- Simulations (Section 2.2.4);
- Decision-making models (Section 2.2.5); and
- Multi-objective optimizations (Section 2.2.6).

Table 2 represents the main functions of six types of models and answerable question(s) by utilizing the models.

Table 2. Categorized models.

Model	Main Function(s)	Answerable Question(s) by Utilizing the Models *
Qualitative articulation models (n = 22)	Sharing both cutting-edge and underexplored knowledge	What are the functions and responsibilities of the stakeholders engaged in transitioning the urban freight transport system?
Sustainable business models (n = 18)	Suggesting viable business models that align with sustainability in the field of mobility	What are the strategies for managing electric vehicles at the end of their lifespan?
Mathematical models (n = 23)	Depicting human attitudes and behaviors	What methods can be used to forecast the adoption behavior of emerging technologies and services?
	Depicting the performance of an organizational activity	How can we evaluate the effectiveness of a policymaking tool?
	Depicting the effects of an organizational activity	What methods can be used to measure the environmental sustainability of an urban mobility design?
Simulations (n = 24)	Anticipating the future of the area by comprehending alterations in human conduct, technological advancement, market trends, and policy execution	How does implementing an international policy target affect a national economy?
	Examining the interaction of a hypothetically designed system with existing systems through analysis	What materials are appropriate for manufacturing batteries for electric vehicles?
	Assessing the influence of decision-making principles employed by practitioners	How can sustainable mobility transition scenarios be effectively generated?
Decision-making models (n = 6)	Developing decision-making criteria for a mobility system maintenance or development project	How are the decision-making criteria shared and applied in regional end-of-life vehicle management?
Multi-objective optimizations (n = 12)	Optimizing the operation of a dynamic system while balancing multiple objectives	How can a logistics company determine the most efficient route for its vehicle?

* We formulated the questions by only reviewing the literature that was presented as examples. Thus, the capability of each type of model is not limited to answering the written questions.

2.2.1. Qualitative Articulation Models

The initial literature group introduced qualitative articulation models that visually and descriptively presented concepts and knowledge relevant to energy and mobility transitions (Table 3). The literature mainly offered these qualitative articulation models to increase stakeholder awareness and understanding of lesser-known expertise and contribute to field-specific academic research practice.

2.2.2. Sustainable Business Models

The second literature group identified the enablers of implementing sustainable business models in the mobility sector through business modeling (Table 4).

Table 3. Knowledge disseminated through qualitative articulation models.

Shared Knowledge through the Qualitative Articulation Models	Reviewed Articles
Strategies for sustainable transportation planning (process), including modeling approaches ($n = 9$)	[37–45]
The functions and responsibilities of stakeholders involved in transitioning an urban freight transport system ($n = 1$)	[46]
Urban spatial planning aimed at addressing mobility issues such as traffic congestion ($n = 2$)	[47,48]
Technical solutions for adapting an urban transportation system to fit the unique characteristics of a city, such as a tourism-based economy ($n = 2$)	[49,50]
The current state of academic knowledge and practices related to emerging research topics, such as end-of-life vehicle management and modeling techniques for electric vehicle batteries ($n = 8$)	[51–58]

Table 4. Knowledge disseminated through the sustainable business models.

Shared Knowledge through the Sustainable Business Models	Reviewed Articles
Factors that facilitate sustainable urban mobility, such as the endorsement of celebrities ($n = 4$)	[59–62]
Ways to promote shared vehicle usage, such as offering user incentives ($n = 3$)	[63–65]
Ways to facilitate the management of end-of-life electric vehicles and batteries, such as fostering cross-sectoral collaboration ($n = 3$)	[66–68]
Revising business models to suit local contexts and mobility-related industries, such as biofuel transportation, while ensuring sustainability ($n = 8$)	[67,69–75]

2.2.3. Mathematical Models

The third literature group consisted of mathematical models representing the critical phenomena relevant to transitioning energy and mobility systems in mathematical equations. These models allowed authors to contribute niche knowledge about the phenomena and communicate with stakeholders such as policymakers, urban and traffic planners, designers, and private businesses. Generally, the models represented human behavior and attitude, organizational activity performance, and the impact of organizational activity (Table 5).

Table 5. Phenomena represented in mathematical models.

<table>
<tr><th colspan="2">Displayed Phenomena</th><th>Reviewed Articles</th></tr>
<tr><td rowspan="3">Human attitude and behavior ($n = 10$)</td><td>Phenomena related to the choice of travel mode, such as the influence of social norms, emotions, and expert opinions</td><td>[76–79]</td></tr>
<tr><td>Emerging technology and service adoption by capturing decision-making episodes and consumer knowledge, particularly in the context of electric vehicles</td><td>[80–84]</td></tr>
<tr><td>Developing trust in emerging mobility concepts</td><td>[85]</td></tr>
<tr><td rowspan="2">The performance of organizational activity ($n = 5$)</td><td>The effectiveness of a decision support system or a policymaking tool</td><td rowspan="2">[86–90]</td></tr>
<tr><td>Other possible factors that could impact the performance of sectoral mobility practices, including aspects such as organizational innovation, that may not be commonly considered or well-known</td></tr>
<tr><td rowspan="3">The effects of organizational activity ($n = 8$)</td><td>The environmental sustainability of technological solutions, such as urban mobility designs and electric vehicles</td><td>[91–93]</td></tr>
<tr><td>The economic impact of last-mile delivery and sectoral transportation activities when redesigning logistics chains</td><td>[94,95]</td></tr>
<tr><td>The sustainability of urban mobility through an integrated assessment approach</td><td>[96–98]</td></tr>
</table>

2.2.4. Simulations

The fourth literature group focused on simulating mathematical models to achieve three objectives: Projecting the future of a locale, analyzing hypothetically designed systems, and testing practitioners' decision-making principles (Table 6). The simulation results were then used to provide recommendations for policymaking and business practices based on changes in human behaviors, technology development, market dynamics, and policy implementation.

Table 6. Objectives of simulation contents.

Objectives of Simulations	Simulation Contents	Reviewed Articles
Anticipating the future of the area by comprehending alterations in human conduct, technological advancement, market trends, and policy execution (n = 11)	The evaluation of the effectiveness of implementing international policies, such as the EU's decarbonization target, concerning future economic and technological mobility advancements	[99,100]
	The effects of applying international policies on the economies of individual nations	[101,102]
	Examining the sustainability of a city through the lens of demographic changes, land use, travel behaviors, and technological advancements	[103–105]
	The influence of social media on public perception of sustainable mobility	[106]
	The evolution of the mobility sector due to drivers such as advances in Information and Communication Technology (ICT) and changes in user behavior	[107–109]
Examining the interaction of a hypothetically designed system with existing systems through analysis (n = 9)	Investigating potential materials for the production of electric vehicle batteries and other vehicle components Examining sustainable practices for operating shared autonomous vehicles and developing charging and swapping stations	[110–115]
	The effectiveness of a connected vehicle system, taking into account factors such as safety, vehicle diversity, and technology market readiness	[116,117]
	The impact of user incentives on the performance of a bike-sharing system	[118]
Assessing the influence of decision-making principles employed by practitioners (n = 4) (e.g., principles applied to vehicle routing problem-solving and sustainable mobility scenario generation)		[119–122]

2.2.5. Decision-Making Models

In transportation projects, stakeholders such as citizens, local governments, and academics may have varying expert knowledge and preferences. Therefore, the authors used modeling techniques to capture stakeholders' priorities. This approach aimed to develop decision-making criteria models that are coherent and reflect the priorities of all stakeholders. The authors applied this modeling technique to transportation development projects and end-of-life vehicle management [123–128].

2.2.6. Multi-Objective Optimizations

The transportation sector is constantly changing due to the adoption of new technologies, such as electric vehicles, and the integration of electricity and gas systems. As a result, the system performance evaluation model is evolving, emphasizing societal impacts such as greenhouse gas emissions. Stakeholders, including logistics and transport providers, urban planners, supply chain managers, transport infrastructure managers, and policymakers, must decide on the best strategy for operating dynamic systems that address multiple objectives, including cost efficiency and environmental impact. To support multi-objective consideration, the authors formulated mathematical equations to describe operational systems, such as the supply chain, and the operation's objectives, such as profitability, to determine the optimal solution (Table 7).

Table 7. Multi-objective problems addressed by the reviewed operation research.

Stakeholders	Operation Problem	Reviewed Articles
Logistic and transport service providers	Multi-objective vehicle routing objectives are the amount of energy consumed, the quality of a transported good, etc.	[129–134]
Transportation infrastructure managers	Managing a transportation infrastructure considering emergent problems due to rapid urbanization or energy transition and attempting to fulfill objectives such as cost minimization and environmental friendliness	[135–139]

We examined six distinct types of models that varied in their forms, representing phenomena, and utilities. In the next section, we discuss how to design such models to engage stakeholders in the transition process effectively.

3. Stakeholder Perspective on the Usability of Models

This section addresses our second and third research questions: "What traits of models are supportive of stakeholder engagement?" and "How can models be designed to engage stakeholders in the transitions effectively?" We conducted semi-structured interviews with practitioners in the Dutch energy and mobility sectors to gain insight into the stakeholder perspective on model usability. Section 3.1 provides a description of our interview methodology. Section 3.2 examines use cases in which models can be beneficial in practice and discusses the traits of usable models addressed by the interviewees.

3.1. Method

We utilized a qualitative research method involving semi-structured interviews to gain a stakeholder perspective on model usability. The interviews comprised four parts, as shown in Table 8. Our goal was to gain insight into tasks the interviewees perform so that we can understand the circumstances in which energy and mobility transition models can be helpful in practice. We also gained insight into the interviewees' perspectives on the strengths and weaknesses of models.

3.1.1. Data Collection

We used quota sampling to select interviewees with diverse perspectives from provincial and municipal levels of government and industry (Table 9). We invited 10 practitioners who were accessible to the authors and willing to participate in the research using convenience sampling rather than a randomized group of people [140]. The interviewees had diverse types of experience from relevant stakeholder groups in energy and mobility transitions, representing the government, the power grid operation, and the industry that provides energy and mobility solutions. The areas of expertise ranged from energy and electric vehicle charging infrastructure, sustainability, and mobility program management to supporting policymaking. Each interview lasted about one hour, and all sessions took place online from February 2022 to April 2022 due to COVID-19 complications. During the discussions, the practitioners shared their experiences of using models and outputs for policy planning, communicating with stakeholders, and gaining a better understanding of the development of products and business models. In addition, they provided thoughts on improving the utility of energy and mobility transition models. All interviews were conducted in English, recorded, transcribed, and anonymized.

Table 8. Semi-structured interview protocol applied in this research.

Interview Part	Objectives	Asked Questions/Activities per Part
Introduction	Letting interviewees acclimatize to the interviewer's research project and the objectives of interview	A short presentation on the research background and research interest in understanding how to produce models for stakeholders and supportive user scenarios
Understanding interviewees and supportive models to them from multiple perspectives: Working organizations, conducting tasks in the organizations, and individual voices on the transitions	Understanding the tasks conducted by interviewees in their organizations	Q1: "What are your usual tasks in your organization?" Q2: "Can you explain the energy and transport transition projects you are responsible for?"
	Exploring the circumstances in which interviewees make complex decisions wherein energy and mobility transition models can potentially be useful	Q3: "What kinds of decisions do you (have to) make about energy and transport transitions?" Q4: "Do you experience any dilemmas during such decision-making processes?"
Understanding strategies for designing useful models from user experience	Understanding whether interviewees are directly engaged in using models	Q5: "When working on energy and transport transition projects, have you or your organization ever used computer software/tools/games?" Q5-1-1 (if the answer to Q5 was "Yes."): "What software/tools/games did you use?"
	Understanding the effectiveness of using models and/or content generated from models	Q5-1-2: "What support did you receive?" Q5-1-3: "What were the strengths and weaknesses of the software/tools/games?"
Finalization	Concluding interviews	A statement of gratitude for participating into the interview

Table 9. The information on the interviewees based in The Netherlands.

Sector	Job Description	Number of Interviewees
Provincial government	Regional energy network system design Stakeholder communication for regional energy system planning Regional electric vehicle charging infrastructure management	3
Municipal government	Local sustainability program guidance Local sustainable mobility program management	2
Knowledge management	Power grid management	1
Business	Electric vehicle technology development Electric vehicle charging infrastructure Flexibility solution development Sustainability solution development	4

3.1.2. Data Analysis

To analyze the data gathered from the interviews, we utilized the ATLAS.ti tool and adopted the QUAGOL qualitative data analysis approach [141]. We used Vivo coding to gain an overview of the interviews and summarize the main content. Then, we collected the most frequently occurring keywords in each interview and employed keyword searching to obtain the results [142]. Table 10 presents a summary of the interview results.

Table 10. A summary of the interview results.

Discussed Content	Summary
Tasks performed by the interviewees, which could be supported by using models (Q1 to Q4): Section 3.2.1	Governmental officers: • Identifying synergies and dilemmas that would be emerged as a result of multiple local initiatives by industry and government sectors (e.g., unexpectedly high cost for acquiring an overarching system, such as the electricity grid and its upkeep) • Managing regional electric mobility infrastructure to achieve local and national governments objectives simultaneously
	Knowledge management (power grid): • Developing the proof of concept of a technical solution in the context of customer engagement
	Businesses • Supporting product design for the transitions in the context of engineering optimization and communication • Exploring future circumstances in which products will be sold (e.g., customer preference) and the consequences of introducing products to a future market (e.g., the number of required public charging stations) • Asset management: Electric charging stations
The interviewees' experience with models (Q5)	Interacting with models directly or only utilizing the outputs of models: Models were generated by either internal employees (e.g., engineers, data analysts) or external personnel (e.g., universities, consultants)
Functions of the models used (Q5-1-1): Sections 3.2.1 and 3.2.2	Governmental officers: • Presenting locations that require existing electric charging stations to be updated. • Indicating traffic flows and predicting electric charging demands • Estimating the number of electric charging stations demanded in future Knowledge management (power grid): • Estimating the impacts on the power grid and computing the effects of applying diverse smart charging profiles Businesses: • Supporting the product design (e.g., optimization) • Estimating the carbon footprint • Optimizing flexibility solutions
Strengths of the models used (Q5-1-2): Sections 3.2.1 and 3.2.2	• Presenting the status of electric charger usage: Supporting the creation of new business opportunities • Easy-to-change parameters • Possibility of using in-house data • Enabling the exploration of the impacts of flexibility solutions by applying diverse scenarios • Supporting decisions over the number of electric charging stations, which helps communication between stakeholders (e.g., provincial and municipal governments, and charging point operators)
Weaknesses of the models used (Q5-1-3): Section 3.2.2	• Limited representation of the real world (e.g., a lack of realistic illustration of human behaviors) • Transparency of models (e.g., codes, assumptions) • Unnecessarily detailed information (e.g., indicating lots of correlations) • Extensive development processes • Less comprehensive definition of a key concept (e.g., mobility) • Lack of compliance between Dutch and European systems

3.2. Results

The interviews yielded two types of results. First, in Section 3.2.1, we discuss two use cases where stakeholders can be supported by using models. Second, in Section 3.2.2, we explore the characteristics of models that are essential for effective usage.

3.2.1. Use Cases of Models Supportive to Practitioners

We identified two situations where stakeholders could effectively use models to understand energy and mobility transitions in practice. The first situation is when organizations use models to gain insight into future energy and mobility systems, such as understanding the customer segment of a future mobility market. The second situation is when stakeholders collaboratively design local and regional infrastructure across sectors, such as provinces, municipalities, industries, and civic groups.

Organizational Learning to Understand Future Systems

The interviews revealed three specific examples of organizational learning. First, models that indicate the state of organizational progress in the transitions could offer valuable information. For instance, practitioners in electric vehicle charging innovation development and marketing communications discussed the evaluation of flexibility solutions for preventing grid congestion and ensuring the sustainability of product delivery and employee mobility patterns. A consultant experienced in flexibility solutions stressed the importance of identifying the societal values provided by the solutions, such as energy independence: "If you work on models, the models should also integrate external costs that are indirect effects of benefit".

Second, the governmental officers indicated a preference for models that can support the design of energy and mobility infrastructure. We found that designing models while considering the different responsibilities of local and regional governments is crucial. Models that can simulate the adaptation strategies of local energy and mobility infrastructure in accordance with future projections (e.g., the trend of electric vehicle purchasing) and regulations (e.g., the EU's zero-emission policies) appear to be relevant to the local government level, according to a municipality officer: "As a city, [...] we don't have any influences about car manufacturers. [...] How are we coping with the grid capacity shortage? How are we combining it with the other mobility transition programs?" On the other hand, models that can present changes in regional energy infrastructure potentially made by such local initiatives and industries would be helpful for regional governments. A regional energy infrastructure planner highlighted the challenge of assessing proposals for developing energy supply systems that could accommodate renewable energy parks and electric vehicles. The proposals submitted by companies responsible for designing and implementing these systems often presented the best-case scenarios from the companies' point of view. However, as a province, it was crucial to consider the values associated with these system development solutions from the public's perspective, including factors such as land use, environmental impact, and potential effects on energy prices resulting from their implementation. Such reflections required a thorough review of alternatives, but generating other options and comparing them to the company's solutions was limited due to the lack of resources, such as a "design tool" as mentioned by the interviewee.

Third, the interviewees highlighted models that allow stakeholders to spot future business opportunities. For example, a government charging infrastructure practitioner noted that models providing the expected demand for electric vehicle charging points, capacity, and installation locations could facilitate tendering between municipalities and charging point operators. An industry practitioner in charging point management emphasized the usefulness of a model that can help identify potential business cases required in the near future. This proactive approach aims to prevent user disappointment resulting from the current lack of available charging poles at charging stations due to limited availability.

Collaborative Infrastructure Design with a Diverse Set of Stakeholders

The interviewees revealed that models could also support the collaborative design of regional energy systems among heterogeneous stakeholder groups. A provincial officer mentioned that planning a regional energy system change was challenging due to inadequate communication between government and industry. To deliver the energy to end-users, the province had to understand the type of energy required by the industry ahead of time. According to the officer in question, there was a different understanding between the types of energy carriers (e.g., hydrogen, green gas, electricity) to be provided to the industrial area and the kind of energy (e.g., electricity, heat) required by the industrial processes such as production. However, communication was hindered due to the confidentiality of future industry plans. To address this, a provincial government officer devised an idea to set up a "task force" that included a diverse range of stakeholders. Using models, stakeholders could test their energy supply and demand plans and receive feedback. Doing so could help to avoid mismatches between supply and demand and identify better energy-sourcing options.

Models for Organizational Learning and Collaborative Infrastructure Design

Both use cases discussed earlier could be supported using computational simulations. This is because the use cases involve predicting the future of the business environment and regional infrastructure, analyzing the impacts of business solutions on multiple scales, and testing the implications of various choices. These are the primary functions of simulation (Section 2.2.4). In addition, developing sustainable business models is crucial to facilitate changes in business practices.

Collaborative infrastructure design would demand decision-making models and optimizations, as the collaborative process involves multiple stakeholders from different sectors who may not hold homogenous decision-making criteria. Therefore, decision-making modeling can help structure the shared decision-making process (Section 2.2.5). Furthermore, since the collaborative process is participatory, exploring optimum design decisions that meet diverse requirements served by optimizations is an essential activity (Section 2.2.6).

Developing the models discussed above, including sustainable business models, decision-making models, simulations, and optimizations, would ultimately require a detailed articulation of the systems on which the models are founded (Section 2.2.1). Additionally, transforming qualitative expressions into mathematical formulas is critical for computational prediction and optimization (Section 2.2.3). Finally, it follows that applying a range of diverse models while considering their interconnectedness would be crucial to allow modeling researchers to produce models for effective engagement (Figure 2).

In addition, we consider that the term "model" risks creating confusion when engaging stakeholders because some people lack modeling knowledge. We noted that interviewees used atypical terms, such as "digital twin". A municipal officer said: "Working with a model-based [. . .] or an agent-based model is really abstract for me or vague".

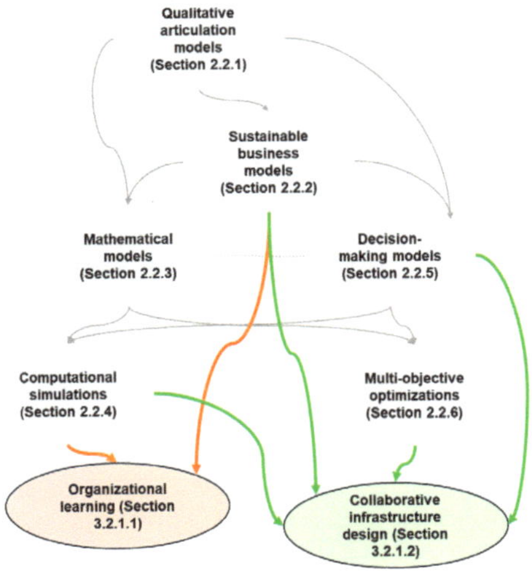

Figure 2. Required models for the two use cases.

3.2.2. Supportive Traits of Models to Engage Stakeholders

Based on the interviews, we found that models with specific characteristics could be better used. These uses include:

- Considering stakeholder perspectives while selecting phenomena to be modeled, including key concepts and assumptions;
- Providing insights into the near future within a short amount of time;
- Conveying balanced information involving reliability and usability;
- Ensuring transparent communication of involved assumptions; and
- Enabling communication between other models.

To evaluate the six grouped models in Section 2.2 in terms of these characteristics, we created Table 11.

Table 11. Assessing the degree to which the reviewed models possess the essential traits to engage stakeholders.

Required Traits of Models	Considering Stakeholder Perspectives While Selecting Phenomena	Providing a Near-Future Projection	Balancing Reliability against Usability, and Communicating Assumptions Transparently	Enabling Real-Time Communication between Models
Qualitative articulation models	-	-	▲	-
Sustainable business models	••	-	▲	-
Mathematical models	-	○	▲	•
Simulations	-	○	▲	•
Decision-making models	••	-	▲	-
Multi-objective optimizations	-	-	▲	○
Rating scale	(-) We barely observed models providing the feature. (○) We observed a few models that partially provide the feature. (•) We observed less than half of the models providing the feature. (••) We observed more than half of the models providing the feature. (▲) The examination required subjective judgment.			

Considering Stakeholder Perspectives While Selecting Phenomena to Be Modeled

Perspectives may differ between the modeling researchers who create the models and the stakeholders who utilize them for energy transitions. This can result in differences in the phenomena the two groups consider and the interpretation of concepts, including their reasoning procedures. To ensure that models are adopted and used appropriately, it is crucial to represent relevant phenomena for researchers and stakeholders. For example, modeled city traffic may be incomplete if pedestrian and passenger behaviors are not considered from the perspective of municipal government officers. Additionally, the interpretation of a concept can vary among researchers and stakeholders. For instance, an energy solution business practitioner found an oversimplified definition of "mobility" in a model for quantifying a carbon footprint that considered an equivalence between commuting via personal vehicles and public transportation.

We examined whether the models reviewed in this study incorporated the perspectives of model users. Sustainable business models were produced by considering stakeholders' interests through qualitative research, such as organizing workshops to co-create locally adapted business models [70,71,73,74] (Section 2.2.2). Decision-making models captured stakeholders' decision-making procedures using customized research methods such as the Fuzzy Analytic Hierarchy and Interval Analytic Hierarchy Processes [125–127] (Section 2.2.5). In some mathematical models, stakeholders appeared to be involved as research subjects whose behaviors were observed and measured rather than direct perspective providers. However, most models did not significantly engage stakeholders in the model production process.

Providing Insights into the near Future within a Short Amount of Time

We determined the importance of quickly producing models that project the near future. For example, an interviewee responsible for electric vehicle charging infrastructure stressed the need for prompt future projections (one to three years) to facilitate electric vehicle charging point tendering. A power grid management practitioner also indicated: "I think 2040, 2050 is far ahead, far in the future. We already, or at least certain areas, need outcome". Thus, the interviewee in question recommended that model producers communicate results iteratively and not wait until the modeling is complete: "I am also open to more or less giving a quicker indication. [...] for five days with an accuracy degree [...] then first you have an indication and [...] then you do some more in-depth analysis [...]".

The articles reviewed in Sections 2.2.1–2.2.4 showcased various modeling approaches for future predictions. Simulations were primarily utilized for long-term future projections spanning decades, such as exploring the consequences of international policy targets on national economies by specific years such as 2030 and 2050 [99–102]. However, there was a lack of modeling methods that could provide near-future projections in a cost-efficient manner.

Conveying Balanced Information between Reliability and Usability

We noted that models could be designed to better focus on presenting the essential information primarily by balancing reliability and usability rather than providing overly detailed results. For example, a provincial officer warned against getting bogged down in detail by recalling a model that outlined an intensely frequent prediction of the amount of energy generated: "I don't need from all these companies on the minutes [...] but I need to have an idea about what they use on the everyday level. [...] You have to be reliable but not on the very detail".

An objective evaluation concerning whether models succeeded at balancing between reliability and usability was challenging to achieve because the evaluation appeared to rely on the subjective judgment of a stakeholder adopting models depending on the context of model usage, personal preference, etc.

Attaching Transparent Communication of Involved Assumptions

The interviews highlighted the importance of a clear description of the assumptions made when constructing models to increase their credibility. An interviewee working for a new product design research team used metaphorical expressions, namely "white box" and a "grey box", and discussed the transparency of model elements (e.g., codes). In addition, a provincial officer recalled a modeling result in 2019 that projected that the demand for electric vehicle fast charging stations in 2025 would be 5000, but in 2021 the projected demand for 2025 had increased to 6000. The officer emphasized the importance of communicating assumptions such as parameter settings to ensure the reliability of modeling results. However, it was challenging to evaluate objectively whether the communication of assumptions was transparent.

Enabling Communication between Other Models

During stakeholder interviews, it was suggested that artifacts capable of letting distant models at different geographical scales and with different content communicate with each other, preferably in real-time, would be beneficial. For example, a municipal civil servant mentioned integrating models representing electric vehicle charging demand, grid capacity, and traffic into a single, high-level model: "Well, we're combining these models. [...] they put it in our heads, and we hope to say wise words. That's how we do it at this moment". Likewise, a practitioner from an energy flexibility solution business preferred a model that could communicate the impacts of intertwined system changes in national and European energy systems (e.g., standardization).

Similar concepts of integrated models were also found in the literature, such as frameworks for assessing the sustainability of urban mobility by involving multiple models [39,43–45]. Mathematical models and simulations were also used to understand the relationships between long-term policy targets and national economic and technological systems [96,98,104,117]. However, the technical measures needed to enable the interfaces were barely discussed.

4. Discussion

Stakeholders' decision-making affects the transitions of energy and mobility systems. Modeling research has developed models to support these decisions. Effective use of models by stakeholders requires access to the models and an understanding of the characteristics of models that could enhance usability. Understanding the various types of models that exist allows us to employ suitable modeling approaches tailored to use contexts. In addition, understanding the essential characteristics of usable models is crucial for effective utilization. We add to the understanding of both aspects by addressing the last sub-research question, "How can models be designed to engage stakeholders in the transitions effectively"?

4.1. Interpretations and Implications

4.1.1. The Models Covering the Transitions in the Representations

The overview of six grouped models representing energy and mobility transitions revealed that authors developed and used them for different epistemic purposes (e.g., disseminating cutting-edge knowledge to stakeholders and translating complex phenomena into mathematical formulas). It may be the case that the models will effectively address specific questions they are designed for. We, therefore, consider understanding the epistemic objectives and requirements of stakeholders to be an essential aspect of a preparation stage. For instance, decision-making models would primarily suit situations where heterogeneous stakeholder groups must make shared decisions. On the other hand, extensive computational simulations might not be the best tool for stakeholders needing only a brief overview of emerging technologies and adoption behavior research.

This epistemology aspect did not appear to be widely addressed in the transition modeling research. To engage stakeholders effectively, it appears to be crucial to diagnose the questions that need to be answered and to design models that meet the requirements for epistemic activities. To achieve this, exploring how to determine the requirements based on an understanding of inquiry in the context of transitions and how to relate the requirements to selecting appropriate modeling approaches can be helpful.

4.1.2. Approaches for Designing the Models for Effective Stakeholder Engagement

To ensure that stakeholders use models effectively, we interviewed Dutch stakeholders who had experience using models to make decisions. We highlighted two issues to be addressed in order to contribute to effective transition model design. The first issue is representing shared phenomena between modeling researchers and stakeholders while simultaneously acknowledging potentially different epistemic requirements (Section 4.1.1). The second issue is implementing technical measures to produce models cost-efficiently and developing interface artifacts that can make models communicate simultaneously. These two problems are distinct yet related.

We regard the first issue in transition model design as needing a mechanism to facilitate the formulation of shared perception about the state of transitions between modeling researchers and stakeholders and the elaboration of epistemic requirements. We saw the diversity of phenomena represented in the models in Section 2.2. The transitions of such diverse phenomena already amplify the complexity of transition modeling. Conceiving the transitions necessitates individuals to make assumptions, hypotheses, etc. (Section 3.2). If modeling researchers are unaware of this heterogeneity, their models may not be easily accepted by stakeholders. The mechanism can be realized as a form of pre-communication in an early stage of model development.

It could involve illustrating explored phenomena by using models and contemplating the epistemic purposes and requirements of both researchers and stakeholders, which can then support the identification of utilizable types of models (Section 2.2). Transparent and credible models may be produced by outlining models that have the potential to be developed and explaining the cognitive processes involved in the analysis of transitions. During this process, modeling researchers may also communicate the scientific theories and methods used for transcribing the phenomena into mathematical formulas and quantification (Section 4.1.2). In addition, we observed the difficulty of examining some factors of model usability (e.g., balanced reliability and usability) (Section 3.2.2). Modeling research may benefit from tools that measure model usability. Nevertheless, we acknowledge that the concepts discussed here must be rigorously examined scientifically and articulated to avoid encouraging speculation.

Regarding the second problem of model design, providing technical solutions for producing future predictions cost-efficiently and developing interfaces between distant models, the research does not provide sufficient evidence to offer specific recommendations for addressing this problem. However, one suggestion may involve experts from other disciplines collaborating with modeling researchers to improve productivity and streamline the interface development process.

4.2. Limitations of the Research and Suggestions for Future Research

In this section, we discuss research limitations and suggest future research directions accordingly.

4.2.1. Systematic Literature Mapping: The Limited Scope of the Reviewed Scientific Literature and Subjectivity Intervention in the Mapping Mechanism

Our research aimed to offer an impartial survey of existing models by utilizing an iterative mapping approach and reviewing a broad range of scientific literature. Our approach sought to avoid bias toward specific models or scientific disciplines. Nevertheless, we suggest using diverse keywords when formulating search strings to better reflect the diversity of scientific disciplines and associated themes in transition research (e.g., energy justice) (Section 2.1). Additionally, we encountered challenges during the literature mapping process due to our literature categorization approach and taxonomy of models. The associated semantic issues appeared to impact how we identified and analyzed the models (e.g., simulations as design space explorations versus technical system models). As a result, we recommend mapping literature using multiple perspectives more effectively to determine the progress of modeling research and its utility.

4.2.2. Stakeholder Interview: Limited Generalizability

Our exploratory research for understanding model usability from the stakeholder perspective is valuable in raising awareness among the scientific community. Through the interviews, we found that using models can benefit stakeholders' learning about transitions concerning their organizational performance and collaborative infrastructure design. As discussed in Section 3.2.2, our research acknowledges the characteristics of usable models articulated by the previous study (Section 1.1). Nonetheless, our sample size of 10 practitioners from the Dutch energy and mobility sectors may limit the generalizability of our findings (Section 3.1). Moreover, we faced challenges in harmonizing our interviewees' diverse needs and use cases. Thus, we strongly suggest continuing the model user research beyond exploratory research and expanding model usability by employing rigorous scientific approaches.

4.2.3. The need to Validate the Findings with Modeling Researchers

In summary, our research aimed to facilitate stakeholder engagement by using models and support modeling researchers' modeling practices for engagement, as presented in Table 1. However, we recognize that our research findings were not validated by other modeling researchers, limiting our evaluation's objectivity. Therefore, we recommend further research to investigate the perspective of modeling researchers on engaging stakeholders in their modeling practice. We can develop a more balanced and effective engagement practice by considering both parties' viewpoints.

5. Conclusions

This research aimed to understand the key considerations and approaches for effectively engaging stakeholders in energy and mobility transitions using models. We answer our sub-research questions and the main research question as follows.

RsQ1: *"What models have been proposed in the scientific literature to understand energy and mobility transitions?"* In total, we explored six types of models through systematic literature mapping: Qualitative articulation models, sustainable business models, mathematical models, simulations, decision-making models, and multi-objective optimizations. Each type of model held a set of unique forms, functions, and questions to be answered. Moreover, the purpose of presenting each type of model differed (e.g., increasing stakeholder topic awareness by using qualitative articulation models versus providing recommendations to policymaking through simulating futures).

RsQ2: *"What traits of models are supportive of stakeholder engagement?"* From our interviews, we identified two instances in which models can be useful for stakeholders. First, we consider that models can support internal organizational learning. Stakeholders can adjust models to help them understand the progress being made by organizational solutions toward achieving transitions, (re)design local and regional infrastructure adaptable to local transition initiatives and national targets, and find business opportunities (e.g., installing electric charging stations). Second, models can facilitate collaborative infrastructure design across sectors and governments and mediate conversation. The scientific rigor of a model is undoubtedly crucial. To engage stakeholders, the usability of a model also appears to be an essential concern. We thereby identified five strategies: Considering stakeholders' perspectives when selecting phenomena to be modeled, providing near-future projections, balancing the reliability and usability of a model, transparent communication concerning assumptions, and enabling communication between models.

RsQ3: *"How can models be designed to engage stakeholders in the transitions effectively?"* Designing models that engage stakeholders effectively can be initiated by understanding their needs and the use contexts. Stakeholders' needs and use contexts vary, such as the required completeness and forms: Some stakeholders would require models to facilitate complex future scenario simulations, while others would better appreciate the acquisition of straightforward information. Involving stakeholders in the early process of model development and articulating epistemic requirements, as well as suitable forms of models, can be useful in ensuring effective model design.

Regarding the main research question, "What are the key considerations and approaches for effectively engaging stakeholders in energy and mobility transitions using models?", we would first stress the importance of pre-communication with stakeholders at the early stage of model development. Communicating the collaboratively explored phenomena transparently by using models and revealing the associated thought processes of both researchers and stakeholders (e.g., through assumptions or hypotheses) appears to be essential. Furthermore, modeling researchers may cooperate with researchers or practitioners in other disciplines to enhance a model's productivity or develop interface artifacts that enable models to communicate better and generate coherent results.

Future research can increase the understanding of diverse models by reviewing broader scientific disciplines relevant to the transitions and applying multiple modes of observing models. We look forward to continuing the investigation into the model usability aspect by interacting with stakeholders that are more diverse and larger in number. Finally, understanding the perspective of modeling researchers is essential to maintain the viability of stakeholder engagement. Thus, we suggest examining the challenges that the modeling researchers may experience concerning the expansion of modeling research practice toward engagement.

Our essential scientific contribution is to provide the perspective of model design to engage stakeholders in the transitions by using models. Regarding the research outcomes, we will further investigate the pre-communication concept that may facilitate the acquisition of both epistemic requirements of modeling researchers and stakeholders and a shared transition perception mechanism as part of the model design process. First, we should rigorously examine the validity of the concept of pre-communication.

Author Contributions: Conceptualization, Y.C. and M.V.P.P.; Methodology, Y.C., M.V.P.P. and G.M.B.; Analysis, Y.C.; Investigation, Y.C.; Writing—original draft preparation, Y.C.; Writing—review and editing, M.V.P.P. and G.M.B.; Visualization, Y.C.; Supervision, M.V.P.P. and G.M.B.; Funding acquisition, G.M.B. All authors have read and agreed to the published version of the manuscript.

Funding: This work was conducted within the Dutch research consortium NEON and funded by the Dutch Research Council (Project No. 17628).

Data Availability Statement: Our research utilized two types of data. One was the scientific literature in the body text according to their usage. The second data was acquired from the semi-structured interviews. We anonymized the interview data in line with the ethical approval of the University of Twente. Therefore, providing the original transcripts of interviews requires permission from the interviewees.

Acknowledgments: We highly appreciate researcher and educator colleagues who provided invaluable support for the research practicality in the Departments of Design, Production, and Management and Civil Engineering, University of Twente, and the research consortium NEON.

Conflicts of Interest: The authors declare no conflict of interest. The funders had no role in the study's design; in the collection, analyses, or interpretation of data; in the writing of the manuscript; or in the decision to publish the results.

References

1. Biresselioglu, M.E.; Demirbag Kaplan, M.; Yilmaz, B.K. Electric Mobility in Europe: A Comprehensive Review of Motivators and Barriers in Decision Making Processes. *Transp. Res. Part A Policy Pract.* **2018**, *109*, 1–13. [CrossRef]
2. Geels, F.W.; Sovacool, B.K.; Schwanen, T.; Sorrell, S. The Socio-Technical Dynamics of Low-Carbon Transitions. *Joule* **2017**, *1*, 463–479. [CrossRef]
3. Loorbach, D. Transition Management for Sustainable Development: A Prescriptive, Complexity-Based Governance Framework. *Governance* **2010**, *23*, 161–183. [CrossRef]
4. van der Schoor, T.; van Lente, H.; Scholtens, B.; Peine, A. Challenging Obduracy: How Local Communities Transform the Energy System. *Energy Res. Soc. Sci.* **2016**, *13*, 94–105. [CrossRef]
5. Meyar-Naimi, H.; Vaez-Zadeh, S. Sustainable Development Based Energy Policy Making Frameworks, a Critical Review. *Energy Policy* **2012**, *43*, 351–361. [CrossRef]
6. Dongellini, M.; Marinosci, C.; Morini, G.L. Energy Audit of an Industrial Site: A Case Study. *Energy Procedia* **2014**, *45*, 424–433. [CrossRef]
7. Thollander, P.; Karlsson, M.; Rohdin, P.; Wollin, J.; Rosenqvist, J. Design, Monitoring, and Evaluation of Industrial Energy Policy Programs. In *Introduction to Industrial Energy Efficiency*; Elsevier: Amsterdam, The Netherlands, 2020; pp. 325–348. ISBN 978-0-12-817247-6.
8. Zanardo, R.P.; Siluk, J.C.M.; de Souza Savian, F.; Schneider, P.S. Energy Audit Model Based on a Performance Evaluation System. *Energy* **2018**, *154*, 544–552. [CrossRef]
9. Holtz, G.; Alkemade, F.; de Haan, F.; Köhler, J.; Trutnevyte, E.; Luthe, T.; Halbe, J.; Papachristos, G.; Chappin, E.; Kwakkel, J.; et al. Prospects of Modelling Societal Transitions: Position Paper of an Emerging Community. *Environ. Innov. Soc. Transit.* **2015**, *17*, 41–58. [CrossRef]
10. Hoekstra, A.; Steinbuch, M.; Verbong, G. Creating Agent-Based Energy Transition Management Models That Can Uncover Profitable Pathways to Climate Change Mitigation. *Complexity* **2017**, *2017*, 1–23. [CrossRef]
11. Heaslip, E.; Fahy, F. Developing Transdisciplinary Approaches to Community Energy Transitions: An Island Case Study. *Energy Res. Soc. Sci.* **2018**, *45*, 153–163. [CrossRef]
12. Hoppe, T.; De Vries, G. Social Innovation and the Energy Transition. *Sustainability* **2019**, *11*, 141. [CrossRef]
13. Köhler, J.; Geels, F.W.; Kern, F.; Markard, J.; Onsongo, E.; Wieczorek, A.; Alkemade, F.; Avelino, F.; Bergek, A.; Boons, F.; et al. An Agenda for Sustainability Transitions Research: State of the Art and Future Directions. *Environ. Innov. Soc. Transit.* **2019**, *31*, 1–32. [CrossRef]
14. Hansen, P.; Liu, X.; Morrison, G.M. Agent-Based Modelling and Socio-Technical Energy Transitions: A Systematic Literature Review. *Energy Res. Soc. Sci.* **2019**, *49*, 41–52. [CrossRef]
15. Johannsen, R.M.; Prina, M.G.; Østergaard, P.A.; Mathiesen, B.V.; Sparber, W. Municipal Energy System Modelling–A Practical Comparison of Optimisation and Simulation Approaches. *Energy* **2023**, *269*, 126803. [CrossRef]
16. Pfenninger, S.; Hirth, L.; Schlecht, I.; Schmid, E.; Wiese, F.; Brown, T.; Davis, C.; Gidden, M.; Heinrichs, H.; Heuberger, C.; et al. Opening the Black Box of Energy Modelling: Strategies and Lessons Learned. *Energy Strategy Rev.* **2018**, *19*, 63–71. [CrossRef]
17. Savvidis, G.; Siala, K.; Weissbart, C.; Schmidt, L.; Borggrefe, F.; Kumar, S.; Pittel, K.; Madlener, R.; Hufendiek, K. The Gap between Energy Policy Challenges and Model Capabilities. *Energy Policy* **2019**, *125*, 503–520. [CrossRef]
18. Li, F.G.N.; Trutnevyte, E.; Strachan, N. A Review of Socio-Technical Energy Transition (STET) Models. *Technol. Forecast. Soc. Change* **2015**, *100*, 290–305. [CrossRef]
19. Muratori, M.; Jadun, P.; Bush, B.; Bielen, D.; Vimmerstedt, L.; Gonder, J.; Gearhart, C.; Arent, D. Future Integrated Mobility-Energy Systems: A Modeling Perspective. *Renew. Sustain. Energy Rev.* **2020**, *119*, 109541. [CrossRef]
20. Pfenninger, S.; Hawkes, A.; Keirstead, J. Energy Systems Modeling for Twenty-First Century Energy Challenges. *Renew. Sustain. Energy Rev.* **2014**, *33*, 74–86. [CrossRef]
21. Laes, E.; Couder, J. Probing the Usefulness of Technology-Rich Bottom-up Models in Energy and Climate Policies: Lessons Learned from the Forum Project. *Futures* **2014**, *63*, 123–133. [CrossRef]
22. Süsser, D.; Gaschnig, H.; Ceglarz, A.; Stavrakas, V.; Flamos, A.; Lilliestam, J. Better Suited or Just More Complex? On the Fit between User Needs and Modeller-Driven Improvements of Energy System Models. *Energy* **2022**, *239*, 121909. [CrossRef]
23. Ben Amer, S.; Gregg, J.S.; Sperling, K.; Drysdale, D. Too Complicated and Impractical? An Exploratory Study on the Role of Energy System Models in Municipal Decision-Making Processes in Denmark. *Energy Res. Soc. Sci.* **2020**, *70*, 101673. [CrossRef]
24. Iyer, G.; Edmonds, J. Interpreting Energy Scenarios. *Nat. Energy* **2018**, *3*, 357–358. [CrossRef]
25. Passmore, C.; Gouvea, J.S.; Giere, R. Models in Science and in Learning Science: Focusing Scientific Practice on Sense-Making. In *International Handbook of Research in History, Philosophy and Science Teaching*; Matthews, M.R., Ed.; Springer: Dordrecht, The Netherlands, 2014; pp. 1171–1202. ISBN 978-94-007-7653-1.
26. Upmeier zu Belzen, A.; Engelschalt, P.; Krüger, D. Modeling as Scientific Reasoning—The Role of Abductive Reasoning for Modeling Competence. *Educ. Sci.* **2021**, *11*, 495. [CrossRef]
27. Clark, W.C.; Tomich, T.P.; Van Noordwijk, M.; Guston, D.; Catacutan, D.; Dickson, N.M.; McNie, E. Boundary Work for Sustainable Development: Natural Resource Management at the Consultative Group on International Agricultural Research (CGIAR). *Proc. Natl. Acad. Sci. USA* **2016**, *113*, 4615–4622. [CrossRef]

28. Cuppen, E.; Nikolic, I.; Kwakkel, J.; Quist, J. Participatory Multi-Modelling as the Creation of a Boundary Object Ecology: The Case of Future Energy Infrastructures in the Rotterdam Port Industrial Cluster. *Sustain. Sci.* **2021**, *16*, 901–918. [CrossRef]
29. Lang, D.J.; Wiek, A.; Bergmann, M.; Stauffacher, M.; Martens, P.; Moll, P.; Swilling, M.; Thomas, C.J. Transdisciplinary Research in Sustainability Science: Practice, Principles, and Challenges. *Sustain. Sci.* **2012**, *7*, 25–43. [CrossRef]
30. Muratori, M.; Jadun, P.; Bush, B.; Hoehne, C.; Vimmerstedt, L.; Yip, A.; Gonder, J.; Winkler, E.; Gearhart, C.; Arent, D. Exploring the Future Energy-Mobility Nexus: The Transportation Energy & Mobility Pathway Options (TEMPO) Model. *Transp. Res. Part D Transp. Environ.* **2021**, *98*, 102967. [CrossRef]
31. Axelsson, J. A Systematic Mapping of the Research Literature on System-of-Systems Engineering. In Proceedings of the 2015 10th System of Systems Engineering Conference (SoSE), San Antonio, TX, USA, 17–20 May 2015; pp. 18–23.
32. Mumtaz, H.; Singh, P.; Blincoe, K. A Systematic Mapping Study on Architectural Smells Detection. *J. Syst. Softw.* **2021**, *173*, 110885. [CrossRef]
33. Petersen, K.; Vakkalanka, S.; Kuzniarz, L. Guidelines for Conducting Systematic Mapping Studies in Software Engineering: An Update. *Inf. Softw. Technol.* **2015**, *64*, 1–18. [CrossRef]
34. Methley, A.M.; Campbell, S.; Chew-Graham, C.; McNally, R.; Cheraghi-Sohi, S. PICO, PICOS and SPIDER: A Comparison Study of Specificity and Sensitivity in Three Search Tools for Qualitative Systematic Reviews. *BMC Health Serv. Res.* **2014**, *14*, 579. [CrossRef]
35. Nishikawa-Pacher, A. Research Questions with PICO: A Universal Mnemonic. *Publications* **2022**, *10*, 21. [CrossRef]
36. Schardt, C.; Adams, M.B.; Owens, T.; Keitz, S.; Fontelo, P. Utilization of the PICO Framework to Improve Searching PubMed for Clinical Questions. *BMC Med. Inf. Decis. Mak.* **2007**, *7*, 16. [CrossRef] [PubMed]
37. Borodulina, S.; Pantina, T. Model of Sustainable Economic Development in the Context of Inland Water Transport Management. In *International Scientific Conference Energy Management of Municipal Facilities and Sustainable Energy Technologies EMMFT 2019*; Murgul, V., Pukhkal, V., Eds.; Advances in Intelligent Systems and Computing; Springer International Publishing: Cham, Switzerland, 2021; Volume 1258, pp. 806–819. ISBN 978-3-030-57449-9.
38. Bravos, G.; Loucopoulos, P.; Dimitrakopoulos, G.; Anagnostopoulos, D.; Kiousi, A. Enabling Smart Objects in Cities Towards Urban Sustainable Mobility-as-a-Service: A Capability–Driven Modeling Approach. In *Smart Objects and Technologies for Social Good*; Gaggi, O., Manzoni, P., Palazzi, C., Bujari, A., Marquez-Barja, J.M., Eds.; Lecture Notes of the Institute for Computer Sciences, Social Informatics and Telecommunications Engineering; Springer International Publishing: Cham, Switzerland, 2017; Volume 195, pp. 342–352. ISBN 978-3-319-61948-4.
39. Kaszubowski, D. A Method for the Evaluation of Urban Freight Transport Models as a Tool for Improving the Delivery of Sustainable Urban Transport Policy. *Sustainability* **2019**, *11*, 23. [CrossRef]
40. Lovelace, R.; Parkin, J.; Cohen, T. Open Access Transport Models: A Leverage Point in Sustainable Transport Planning. *Transp. Policy* **2020**, *97*, 47–54. [CrossRef]
41. Robèrt, K.-H.; Borén, S.; Ny, H.; Broman, G. A Strategic Approach to Sustainable Transport System Development-Part 1: Attempting a Generic Community Planning Process Model. *J. Clean. Prod.* **2017**, *140*, 53–61. [CrossRef]
42. Semin, V.G.; Grigoreva, S.V.; Dmitrieva, T.V.; Ilyina, E.A. A Process Model of Risk Management in the System of Management of Strategic Sustainability of Cargo Motor Transport Enterprises. In Proceedings of the 2016 Institution of Electrical and Electronics Engineers (IEEE) Conference on Quality Management, Transport and Information Security, Information Technologies (IT MQ IS), Nalchik, Russia, 4–11 October 2016; pp. 172–175.
43. Strulak-Wójcikiewicz, R.; Lemke, J. Concept of a Simulation Model for Assessing the Sustainable Development of Urban Transport. *Transp. Res. Procedia* **2019**, *39*, 502–513. [CrossRef]
44. Torrisi, V.; Ignaccolo, M.; Inturri, G. Innovative Transport Systems to Promote Sustainable Mobility: Developing the Model Architecture of a Traffic Control and Supervisor System. In *Computational Science and Its Applications–International Conference on Computational Science and Its Applications (ICCSA) 2018*; Gervasi, O., Murgante, B., Misra, S., Stankova, E., Torre, C.M., Rocha, A.M.A.C., Taniar, D., Apduhan, B.O., Tarantino, E., Ryu, Y., Eds.; Lecture Notes in Computer Science; Springer International Publishing: Cham, Switzerland, 2018; Volume 10962, pp. 622–638. ISBN 978-3-319-95167-6.
45. Wolff, M.G.d.C.; Caldas, M.A.F. A Model for the Evaluation of Brazilian Road Transport: A Sustainable Perspective. *J. Adv. Transp.* **2018**, *2018*, 1–12. [CrossRef]
46. Cassiano, D.R.; Bertoncini, B.V.; de Oliveira, L.K. A Conceptual Model Based on the Activity System and Transportation System for Sustainable Urban Freight Transport. *Sustainability* **2021**, *13*, 5642. [CrossRef]
47. Faron, A. Spatial Development City Models in Transport Sustainble Mobility Issue. In Proceedings of the 18th International Multidisciplinary Scientific GeoConference (SGEM 2018), Albena, Bulgaria, 2–8 July 2018.
48. Yamu, C.; van Nes, A. Fractal Urban Models and Their Potential for Sustainable Mobility. In Proceedings of the 12th International Space Syntax Symposium, Beijing, China, 8–13 July 2019; p. 15.
49. Roda, M.; Giorgi, D.; Joime, G.P.; Anniballi, L.; London, M.; Paschero, M.; Mascioli, F.M.F. An Integrated Methodology Model for Smart Mobility System Applied to Sustainable Tourism. In Proceedings of the 2017 Institution of Electrical and Electonics Engineers (IEEE) 3rd International Forum on Research and Technologies for Society and Industry (RTSI), Modena, Italy, 11–13 September 2017; pp. 1–6.
50. Udoji, C.U.; Szpytko, J. The Concept of a Sustainable Urban Transport System Model for Developing Countries on the Example of Lagos. *J. KONBiN* **2020**, *50*, 29–38. [CrossRef]

51. Araujo-Morera, J.; Verdejo, R.; López-Manchado, M.A.; Hernández Santana, M. Sustainable Mobility: The Route of Tires through the Circular Economy Model. *Waste Manag.* **2021**, *126*, 309–322. [CrossRef]
52. Bischoff, J.; Maciejewski, M. Current and Future Dynamic Passenger Transport Services—Modeling, Simulation, and Optimization in a Sustainable Transport System. In *Sustainable Transportation and Smart Logistics*; Elsevier: Amsterdam, The Netherlands, 2019; pp. 337–360. ISBN 978-0-12-814242-4.
53. Briggs, I.; Murtagh, M.; Kee, R.; McCulloug, G.; Douglas, R. Sustainable Non-Automotive Vehicles: The Simulation Challenges. *Renew. Sustain. Energy Rev.* **2017**, *68*, 840–851. [CrossRef]
54. Li, Y.; Yang, J.; Song, J. Design Structure Model and Renewable Energy Technology for Rechargeable Battery towards Greener and More Sustainable Electric Vehicle. *Renew. Sustain. Energy Rev.* **2017**, *74*, 19–25. [CrossRef]
55. Lopez-Arboleda, E.; Sarmiento, A.T.; Cardenas, L.M. Systematic Review of Integrated Sustainable Transportation Models for Electric Passenger Vehicle Diffusion. *Sustainability* **2019**, *11*, 2513. [CrossRef]
56. Mor, N.; Sood, H.; Goyal, T. An Approach Driven Critical Review on the Use of Accident Prediction Models for Sustainable Transport System. *J. Discret. Math. Sci. Cryptogr.* **2020**, *23*, 313–320. [CrossRef]
57. Muller, M.; Park, S.; Lee, R.; Fusco, B.; Correia, G.H. de A. Review of Whole System Simulation Methodologies for Assessing Mobility as a Service (MaaS) as an Enabler for Sustainable Urban Mobility. *Sustainability* **2021**, *13*, 5591. [CrossRef]
58. Tomasov, M.; Kajanova, M.; Bracinik, P.; Motyka, D. Overview of Battery Models for Sustainable Power and Transport Applications. *Transp. Res. Procedia* **2019**, *40*, 548–555. [CrossRef]
59. Bellini, F.; Dulskaia, I.; Savastano, M.; D'Ascenzo, F. Business Models Innovation for Sustainable Urban Mobility in Small and Medium-Sized European Cities. *Manag. Marketing. Chall. Knowl. Soc.* **2019**, *14*, 266–277. [CrossRef]
60. Hanna, P.; Kantenbacher, J.; Cohen, S.; Gössling, S. Role Model Advocacy for Sustainable Transport. *Transp. Res. Part D Transp. Environ.* **2018**, *61*, 373–382. [CrossRef]
61. Schnee, R.; Chrenko, D.; Rodet-Kroichvili, N.; Neugebauer, P. Examination of Charging Infrastructure for Electric Vehicles Based on Components of Sustainable Business Models. In Proceedings of the 2020 Institution of Electrical and Electronics Engineers (IEEE) Vehicle Power and Propulsion Conference (VPPC), Gijon, Spain, 18 November–16 December 2020; pp. 1–6.
62. Valsecchi Ribeiro de Souza, J.; Marotti de Mello, A.; Marx, R. When Is an Innovative Urban Mobility Business Model Sustainable? A Literature Review and Analysis. *Sustainability* **2019**, *11*, 1761. [CrossRef]
63. Santos, G. Sustainability and Shared Mobility Models. *Sustainability* **2018**, *10*, 3194. [CrossRef]
64. Signorile, P.; Larosa, V.; Spiru, A. Mobility as a Service: A New Model for Sustainable Mobility in Tourism. *WHATT* **2018**, *10*, 185–200. [CrossRef]
65. Vinci, G.; Musarra, M. Digital Services for New Model of Sustainable Mobility. In *Digitally Supported Innovation: A Multi-Disciplinary View on Enterprise, Public. Sector and User Innovation*; Springer International Publishing: Berlin/Heidelberg, Germany, 2016; Volume 18, pp. 235–246. ISBN 978-3-319-40265-9.
66. Al-Quradaghi, S.; Kucukvar, M.; Onat, N. Towards a Sustainable Management of End-of-Life Vehicles in Qatar: A Closed-Loop Circular Economy Model. In Proceedings of the International Conference on Industrial Engineering and Operations Management, Washington, DC, USA, 27–29 September 2018; Volume 2018, pp. 647–654.
67. Jiao, N.; Evans, S. Business Models for Sustainability: The Case of Repurposing a Second-Life for Electric Vehicle Batteries. In *Sustainable Design and Manufacturing 2017*; Campana, G., Howlett, R.J., Setchi, R., Cimatti, B., Eds.; Smart Innovation, Systems and Technologies; Springer International Publishing: Cham, Switzerland, 2017; Volume 68, pp. 537–545. ISBN 978-3-319-57077-8.
68. Reinhardt, R.; Christodoulou, I.; Gassó-Domingo, S.; Amante García, B. Towards Sustainable Business Models for Electric Vehicle Battery Second Use: A Critical Review. *J. Environ. Manag.* **2019**, *245*, 432–446. [CrossRef] [PubMed]
69. Bignami, D.F.; Rogel, L. Testing a New Model for a Sustainable Mobility in the City of Milan: The Condominium Car Sharing. In *Electric Vehicle Sharing Services for Smarter Cities*; Bignami, D.F., Colorni Vitale, A., Lué, A., Nocerino, R., Rossi, M., Savaresi, S.M., Eds.; Research for Development; Springer International Publishing: Cham, Switzerland, 2017; pp. 79–93. ISBN 978-3-319-61963-7.
70. Flodén, J.; Williamsson, J. Business Models for Sustainable Biofuel Transport: The Potential for Intermodal Transport. *J. Clean. Prod.* **2016**, *113*, 426–437. [CrossRef]
71. Heydkamp, C. User Types for Sustainable Mobility Incentive Models. In Proceedings of the 30th International Electric Vehicle Symposium (EVS30), Stuttgart, Germany, 9–11 October 2017.
72. Jiao, N.; Evans, S. Business Models for Sustainability: The Case of Second-Life Electric Vehicle Batteries. *Procedia CIRP* **2016**, *40*, 250–255. [CrossRef]
73. Pardo-Bosch, F.; Pujadas, P.; Morton, C.; Cervera, C. Sustainable Deployment of an Electric Vehicle Public Charging Infrastructure Network from a City Business Model Perspective. *Sustain. Cities Soc.* **2021**, *71*, 102957. [CrossRef]
74. Pronello, C.; Camusso, C. Users' Needs and Business Models for a Sustainable Mobility Information Network in the Alpine Space. *Transp. Res. Procedia* **2017**, *25*, 3590–3605. [CrossRef]
75. Reinhardt, R.; Christodoulou, I.; García, B.A.; Gassó-Domingo, S. Sustainable Business Model Archetypes for the Electric Vehicle Battery Second Use Industry: Towards a Conceptual Framework. *J. Clean. Prod.* **2020**, *254*, 119994. [CrossRef]
76. Liu, Y.; Sheng, H.; Mundorf, N.; Redding, C.; Ye, Y. Integrating Norm Activation Model and Theory of Planned Behavior to Understand Sustainable Transport Behavior: Evidence from China. *Int. J. Environ. Res. Public Health* **2017**, *14*, 1593. [CrossRef]
77. Manca, S.; Altoè, G.; Schultz, P.W.; Fornara, F. The Persuasive Route to Sustainable Mobility: Elaboration Likelihood Model and Emotions Predict Implicit Attitudes. *Environ. Behav.* **2020**, *52*, 830–860. [CrossRef]

78. Davidson, R. The Algorithmic Governance of Connected Autonomous Vehicles: Data-Driven Decision Support Systems and Smart Sustainable Urban Mobility Behaviors. *Contemp. Read. Law Soc. Justice* **2020**, *12*, 16. [CrossRef]
79. Youssef, Z.; Alshuwaikhat, H.; Reza, I. Modeling the Modal Shift towards a More Sustainable Transport by Stated Preference in Riyadh, Saudi Arabia. *Sustainability* **2021**, *13*, 337. [CrossRef]
80. Tran, T.D.; Ovtracht, N. Promoting Sustainable Mobility by Modelling Bike Sharing Usage in Lyon. *IOP Conf. Ser. Earth Environ. Sci.* **2018**, *143*, 012070. [CrossRef]
81. Xie, Y.; Danaf, M.; Lima Azevedo, C.; Akkinepally, A.P.; Atasoy, B.; Jeong, K.; Seshadri, R.; Ben-Akiva, M. Behavioral Modeling of On-Demand Mobility Services: General Framework and Application to Sustainable Travel Incentives. *Transportation* **2019**, *46*, 2017–2039. [CrossRef]
82. Johnson, A. The Smart Automation of Urban Governance: Sustainable Mobilities and Data-Driven Decision Support Systems. *Contemp. Read. Law Soc. Justice* **2020**, *12*, 51. [CrossRef]
83. Ortega, J.; Tóth, J.; Péter, T.; Moslem, S. An Integrated Model of Park-And-Ride Facilities for Sustainable Urban Mobility. *Sustainability* **2020**, *12*, 4631. [CrossRef]
84. Kaur, N.; Sahdev, S.L.; Bhutani, R.S. Analyzing Adoption of Electric Vehicles in India for Sustainable Growth Through Application of Technology Acceptance Model. In Proceedings of the 2021 International Conference on Innovative Practices in Technology and Management (ICIPTM), Noida, India, 17–19 February 2021; pp. 255–260.
85. Szaruga, E.; Skąpska, E.; Załoga, E.; Matwiejczuk, W. Trust and Distress Prediction in Modal Shift Potential of Long-Distance Road Freight in Containers: Modeling Approach in Transport Services for Sustainability. *Sustainability* **2018**, *10*, 2370. [CrossRef]
86. Chai, G.; Bartlett, R.; Kelly, G.; Yang, L.; Oh, E. Model for Benchmarking a Pavement Maintenance Budget for Sustainable Road Transport Infrastructure. In *Geo-China 2016*; American Society of Civil Engineers: Shandong, China, 2016; pp. 180–187.
87. García-Melero, G.; Sainz-González, R.; Coto-Millán, P.; Valencia-Vásquez, A. Sustainable Mobility Policy Analysis Using Hybrid Choice Models: Is It the Right Choice? *Sustainability* **2021**, *13*, 2993. [CrossRef]
88. Haritonova, L. The Neural Modeling for the Assessment of Hazardous Hydro Meteorological Phenomena Including the Case of the Providing a Sustainable Work (Reliability) on Transport. In *International Scientific Conference Energy Management of Municipal Transportation Facilities and Transport EMMFT 2017*; Murgul, V., Popovic, Z., Eds.; Advances in Intelligent Systems and Computing; Springer International Publishing: Cham, Switzerland, 2018; Volume 692, pp. 700–709. ISBN 978-3-319-70986-4.
89. Kumar, K.M.; Rahman, A.A.; Jayaraman; Rahim, S.A. Conceptualization of a Research Model for Sustainable Logistics Practices and Logistics Transport Performance. *J. Manag.* **2017**, *51*, 1–12. [CrossRef]
90. Revunova, S.; Vlasenko, V.; Bukreev, A. Modeling the Sustainable Development of Innovation in Transport Construction Based on the Communication Approach. *IOP Conf. Ser. Earth Environ. Sci.* **2017**, *90*, 012164. [CrossRef]
91. Oliveira, C.d.M.C.d.; Wolff, M.G.d.C. Sustainable Urban Mobility in Rio de Janeiro: A Model to Quantify Greenhouse Gas Emissions and Purpose of Practical Application. *Braz. J. Oper. Prod. Manag.* **2020**, *17*, e2020993. [CrossRef]
92. D'Ambra, L.; Crisci, A.; Meccariello, G.; Della Ragione, L.; Palma, R. Evaluation of the Social and Economic Impact of Carbon Dioxide (CO_2) Emissions on Sustainable Mobility Using Cumulative Ordinal Models: Trend Odds Model. *Socio-Econ. Plan. Sci.* **2021**, *75*, 100817. [CrossRef]
93. Zeng, D.; Dong, Y.; Cao, H.; Li, Y.; Wang, J.; Li, Z.; Hauschild, M.Z. Are the Electric Vehicles More Sustainable than the Conventional Ones? Influences of the Assumptions and Modeling Approaches in the Case of Typical Cars in China. *Resour. Conserv. Recycl.* **2021**, *167*, 105210. [CrossRef]
94. Kin, B.; Spoor, J.; Verlinde, S.; Macharis, C.; Van Woensel, T. Modelling Alternative Distribution Set-Ups for Fragmented Last Mile Transport: Towards More Efficient and Sustainable Urban Freight Transport. *Case Stud. Transp. Policy* **2018**, *6*, 125–132. [CrossRef]
95. Auzina-Emsina, A.; Pocs, R. Impact of Transport and Storage Sector on Sustainable Development: Evaluation Using Input-Output Model. In Proceedings of the 10th International Multi-Conference on Complexity, Informatics and Cybernetics (IMCIC 2019), Orlando, FL, USA, 12–15 March 2019; p. 6.
96. Nadi, P.A.; Murad, A. Modelling Sustainable Urban Transport Performance in the Jakarta City Region: A GIS Approach. *Sustainability* **2019**, *11*, 1879. [CrossRef]
97. Rivera-González, L.; Bolonio, D.; Mazadiego, L.F.; Naranjo-Silva, S.; Escobar-Segovia, K. Long-Term Forecast of Energy and Fuels Demand Towards a Sustainable Road Transport Sector in Ecuador (2016–2035): A LEAP Model Application. *Sustainability* **2020**, *12*, 472. [CrossRef]
98. Illahi, U. Transport Sustainability Performance Evaluation Using a Multi-Stage Multi-Tool Hybrid Model. *Eur. Transp. /Trasp. Eur.* **2021**, *81*, 1–15. [CrossRef]
99. Halim, R.; Kirstein, L.; Merk, O.; Martinez, L. Decarbonization Pathways for International Maritime Transport: A Model-Based Policy Impact Assessment. *Sustainability* **2018**, *10*, 2243. [CrossRef]
100. Siskos, P.; Zazias, G.; Petropoulos, A.; Evangelopoulou, S.; Capros, P. Implications of Delaying Transport Decarbonisation in the EU: A Systems Analysis Using the PRIMES Model. *Energy Policy* **2018**, *121*, 48–60. [CrossRef]
101. Zhang, H.; Chen, W.; Huang, W. TIMES Modelling of Transport Sector in China and USA: Comparisons from a Decarbonization Perspective. *Appl. Energy* **2016**, *162*, 1505–1514. [CrossRef]
102. Charalampidis, I.; Karkatsoulis, P.; Capros, P. A Regional Economy-Energy-Transport Model of the EU for Assessing Decarbonization in Transport. *Energies* **2019**, *12*, 3128. [CrossRef]

103. López-Iglesias, E.; Peón, D.; Rodríguez-Álvarez, J. Mobility Innovations for Sustainability and Cohesion of Rural Areas: A Transport Model and Public Investment Analysis for Valdeorras (Galicia, Spain). *J. Clean. Prod.* **2018**, *172*, 3520–3534. [CrossRef]
104. Pautasso, E.; Osella, M.; Caroleo, B. Addressing the Sustainability Issue in Smart Cities: A Comprehensive Model for Evaluating the Impacts of Electric Vehicle Diffusion. *Systems* **2019**, *7*, 29. [CrossRef]
105. Zhang, R.; Zhang, J.; Long, Y.; Wu, W.; Liu, J.; Jiang, Y. Long-Term Implications of Electric Vehicle Penetration in Urban Decarbonization Scenarios: An Integrated Land Use–Transport–Energy Model. *Sustain. Cities Soc.* **2021**, *68*, 102800. [CrossRef]
106. Borowski, E.; Chen, Y.; Mahmassani, H. Social Media Effects on Sustainable Mobility Opinion Diffusion: Model Framework and Implications for Behavior Change. *Travel. Behav. Soc.* **2020**, *19*, 170–183. [CrossRef]
107. Shin, J.; Lim, T.; Kim, M.; Choi, J. Can Next-Generation Vehicles Sustainably Survive in the Automobile Market? Evidence from Ex-Ante Market Simulation and Segmentation. *Sustainability* **2018**, *10*, 607. [CrossRef]
108. Moallemi, E.A.; Köhler, J. Coping with Uncertainties of Sustainability Transitions Using Exploratory Modelling: The Case of the MATISSE Model and the UK's Mobility Sector. *Environ. Innov. Soc. Transit.* **2019**, *33*, 61–83. [CrossRef]
109. Lee, J.; Kim, J.; Kim, H.; Hwang, J. Sustainability of Ride-Hailing Services in China's Mobility Market: A Simulation Model of Socio-Technical System Transition. *Telemat. Inform.* **2020**, *53*, 101435. [CrossRef]
110. Choi, H.; Lim, N.; Lee, S.J.; Park, J. Feasibility Study for Sustainable Use of Lithium-Ion Batteries Considering Different Positive Electrode Active Materials under Various Driving Cycles by Using Cell to Electric Vehicle (EV) Simulation. *Sustainability* **2020**, *12*, 9764. [CrossRef]
111. Dlugosch, O.; Brandt, T.; Neumann, D. Combining Analytics and Simulation Methods to Assess the Impact of Shared, Autonomous Electric Vehicles on Sustainable Urban Mobility. *Inf. Manag.* **2020**, *59*, 103285. [CrossRef]
112. Fournier, G.; Boos, A.; Wörner, R.; Jaroudi, I.; Morozova, I.; Nemoto, E.H. Substituting Individual Mobility by Mobility on Demand Using Autonomous Vehicles-a Sustainable Assessment Simulation of Berlin and Stuttgart. *Int. J. Automot. Technol. Manag.* **2020**, *20*, 369–407. [CrossRef]
113. Ginigeme, O.; Fabregas, A. Model Based Systems Engineering High Level Design of a Sustainable Electric Vehicle Charging and Swapping Station Using Discrete Event Simulation. In Proceedings of the 2018 Annual Institution of Electrical and Eletronics Engineers (IEEE) International Systems Conference (SysCon), Vancouver, BC, Canada, 23–26 April 2018; pp. 1–6.
114. Moriwaki, K. On Sustainable Vehicle Management—A Simulation Study. In Proceedings of the 2017 11th Asian Control Conference (ASCC), Gold Coast, Australia, 17–20 December 2017; pp. 485–488.
115. Palconit, E.V.; Abundo, M.L.S. Electric Ferry Ecosystem for Sustainable Inter-Island Transport in the Philippines: A Prospective Simulation for Davao City–Samal Island Route. *Int. J. Sustain. Energy* **2019**, *38*, 368–381. [CrossRef]
116. Li, W.; Wu, G.; Barth, M.J.; Zhang, Y. Safety, Mobility and Environmental Sustainability of Eco-Approach and Departure Application at Signalized Intersections: A Simulation Study. In Proceedings of the 2016 Institute of Electrical and Electronic Engineers (IEEE) Intelligent Vehicles Symposium (IV), Gotenburg, Sweden, 19–22 June 2016; pp. 1109–1114.
117. Shi, Y.; Li, Y.; Cai, Q.; Zhang, H.; Wu, D. How Does Heterogeneity Affect Freeway Safety? A Simulation-Based Exploration Considering Sustainable Intelligent Connected Vehicles. *Sustainability* **2020**, *12*, 8941. [CrossRef]
118. Santiago, A.L.; Iglesias, C.A.; Carrera, Á. Improving Sustainable Mobility with a Variable Incentive Model for Bike-Sharing Systems Based on Agent-Based Social Simulation. In *Advances in Practical Applications of Agents, Multi-Agent. Systems, and Trustworthiness. The PAAMS Collection*; Demazeau, Y., Holvoet, T., Corchado, J.M., Costantini, S., Eds.; Lecture Notes in Computer Science; Springer International Publishing: Cham, Switzerland, 2020; Volume 12092, pp. 158–170. ISBN 978-3-030-49777-4.
119. Awasthi, A.; Omrani, H. A Scenario Simulation Approach for Sustainable Mobility Project Evaluation Based on Fuzzy Cognitive Maps. *Int. J. Model. Simul.* **2018**, *38*, 1–11. [CrossRef]
120. Lee, K.; Chae, J.; Song, B.; Choi, D. A Model for Sustainable Courier Services: Vehicle Routing with Exclusive Lanes. *Sustainability* **2020**, *12*, 1077. [CrossRef]
121. Li, X.; Kan, H.; Hua, X.; Wang, W. Simulation-Based Electric Vehicle Sustainable Routing with Time-Dependent Stochastic Information. *Sustainability* **2020**, *12*, 2464. [CrossRef]
122. Mitropoulos, L.K.; Prevedouros, P.D.; Yu, X.A.; Nathanail, E.G. A Fuzzy and a Monte Carlo Simulation Approach to Assess Sustainability and Rank Vehicles in Urban Environment. *Transp. Res. Procedia* **2017**, *24*, 296–303. [CrossRef]
123. Aghamohagheghi, M.; Hashemi, S.M.; Tavakkoli-Moghaddam, R. An Advanced Decision Support Framework to Assess Sustainable Transport Projects Using a New Uncertainty Modeling Tool: Interval-Valued Pythagorean Trapezoidal Fuzzy Numbers. *Iran. J. Fuzzy Syst.* **2021**, *21*, 53–73. [CrossRef]
124. Ahmed, S.; Ahmed, S.; Shumon, M.R.H.; Falatoonitoosi, E.; Quader, M.A. A Comparative Decision-Making Model for Sustainable End-of-Life Vehicle Management Alternative Selection Using AHP and Extent Analysis Method on Fuzzy AHP. *Int. J. Sustain. Dev. World Ecol.* **2016**, *23*, 83–97. [CrossRef]
125. Duleba, S.; Moslem, S. Sustainable Urban Transport Development with Stakeholder Participation, an AHP-Kendall Model: A Case Study for Mersin. *Sustainability* **2018**, *10*, 3647. [CrossRef]
126. Ghorbanzadeh, O.; Moslem, S.; Blaschke, T.; Duleba, S. Sustainable Urban Transport Planning Considering Different Stakeholder Groups by an Interval-AHP Decision Support Model. *Sustainability* **2018**, *11*, 9. [CrossRef]
127. Moslem, S.; Duleba, S. Sustainable Urban Transport Development by Applying a Fuzzy-AHP Model: A Case Study from Mersin, Turkey. *Urban. Sci.* **2019**, *3*, 55. [CrossRef]

128. Rivero Gutiérrez, L.; De Vicente Oliva, M.A.; Romero-Ania, A. Managing Sustainable Urban Public Transport Systems: An AHP Multicriteria Decision Model. *Sustainability* **2021**, *13*, 4614. [CrossRef]
129. Abdullahi, H.; Reyes-Rubiano, L.; Ouelhadj, D.; Faulin, J.; Juan, A.A. Modelling and Multi-Criteria Analysis of the Sustainability Dimensions for the Green Vehicle Routing Problem. *Eur. J. Oper. Res.* **2021**, *292*, 143–154. [CrossRef]
130. Fathi, A.; Saen, R.F. A Novel Bidirectional Network Data Envelopment Analysis Model for Evaluating Sustainability of Distributive Supply Chains of Transport Companies. *J. Clean. Prod.* **2018**, *184*, 696–708. [CrossRef]
131. Gharehyakheh, A.; Krejci, C.C.; Cantu, J.; Rogers, K.J. A Multi-Objective Model for Sustainable Perishable Food Distribution Considering the Impact of Temperature on Vehicle Emissions and Product Shelf Life. *Sustainability* **2020**, *12*, 6668. [CrossRef]
132. Li, Y.; Lim, M.K.; Xiong, W. An Optimization Model of Vehicle Routing Problem for Logistics Based on Sustainable Development Theory. In *Recent. Advances in Intelligent Manufacturing*; Wang, S., Price, M., Lim, M.K., Jin, Y., Luo, Y., Chen, R., Eds.; Communications in Computer and Information Science; Springer: Singapore, 2018; Volume 923, pp. 179–190. ISBN 9789811323959.
133. Vale, C.; Ribeiro, I.M. Intermodal Routing Model for Sustainable Transport Through Multi-Objective Optimization. In *Intelligent Transport Systems, from Research and Development to the Market Uptake*; Ferreira, J.C., Martins, A.L., Monteiro, V., Eds.; Lecture Notes of the Institute for Computer Sciences, Social Informatics and Telecommunications Engineering; Springer International Publishing: Cham, Switzerland, 2019; Volume 267, pp. 144–154. ISBN 978-3-030-14756-3.
134. Velarde, J.M.; García, S.; López, M.; Bueno-Solano, A. Implementation of a Mathematical Model to Improve Sustainability in the Handling of Transport Costs in a Distribution Network. *Sustainability* **2019**, *12*, 63. [CrossRef]
135. Bakó, A.; Gáspár, L. Development of a Sustainable Optimization Model for the Rehabilitation of Transport Infrastructure. *APH* **2018**, *15*, 11–33. [CrossRef]
136. Cao, Z.; Wang, J.; Zhao, Q.; Han, Y.; Li, Y. Decarbonization Scheduling Strategy Optimization for Electricity-Gas System Considering Electric Vehicles and Refined Operation Model of Power-to-Gas. *IEEE Access* **2021**, *9*, 5716–5733. [CrossRef]
137. Chen, W.; Xu, M.; Xing, Q.; Cui, L.; Jiao, L. A Fuzzy Demand-Profit Model for the Sustainable Development of Electric Vehicles in China from the Perspective of Three-Level Service Chain. *Sustainability* **2020**, *12*, 6389. [CrossRef]
138. Ma, X.; Chen, X.; Li, X.; Ding, C.; Wang, Y. Sustainable Station-Level Planning: An Integrated Transport and Land Use Design Model for Transit-Oriented Development. *J. Clean. Prod.* **2018**, *170*, 1052–1063. [CrossRef]
139. Qiao, W.; Sun, Y.; Wang, D. Towards Sustainable Transport: A Transit Signal Priority Model Based on Emission and Delay Reduction. In Proceedings of the 17th COTA International Conference of Transportation Professionals, Shanghai, China, 7–9 July 2017; American Society of Civil Engineers: Reston, VA, USA, 2018; pp. 3280–3289. [CrossRef]
140. Robinson, O.C. Sampling in Interview-Based Qualitative Research: A Theoretical and Practical Guide. *Qual. Res. Psychol.* **2014**, *11*, 25–41. [CrossRef]
141. Dierckx de Casterlé, B.; Gastmans, C.; Bryon, E.; Denier, Y. QUAGOL: A Guide for Qualitative Data Analysis. *Int. J. Nurs. Stud.* **2012**, *49*, 360–371. [CrossRef] [PubMed]
142. Saldaña, J. Coding and Analysis Strategies. In *The Oxford Handbook of Qualitative Research*; Leavy, P., Ed.; Oxford University Press: Oxford, UK, 2014; ISBN 978-0-19-981175-5.

World Electric Vehicle Journal

MDPI

Article

Investigation and Development of Textile Lightweight Bodies for Urban Logistic Vehicles

Fabian Edel [1,*], Corinna Pulkus [2], Sarah Kim [2], Juergen Erhardt [3] and Sven Kuijpens [2]

1 Fraunhofer-Institute for Industrial Engineering IAO, 70569 Stuttgart, Germany
2 German Institutes of Textile and Fiber Research Denkendorf (DITF), 73770 Denkendorf, Germany
3 Erhardt GmbH, 70499 Stuttgart, Germany
* Correspondence: fabian.edel@iao.fraunhofer.de; Tel.: +49-152-28835285

Abstract: The urban logistic sector is struggling with the following two problems: a particularly high level of traffic and air pollution. For those reasons, small electric vehicles are the perfect technical solution to cope sustainably with the increasing parcel volume in urban areas. To compensate the weight disadvantage of electric drive, a textile lightweight solution for the body of logistics vehicles is examined in this study. Therefore, various textiles and mounting solutions are examined and developed, as well as tested by multiple testing procedures. The result of this work is an ideal combination of a textile and suitable mounting method for a textile lightweight body for last—mile logistic vehicles.

Keywords: textile; lightweight; urban logistic vehicles; freight transport; materials

Citation: Edel, F.; Pulkus, C.; Kim, S.; Erhardt, J.; Kuijpens, S. Investigation and Development of Textile Lightweight Bodies for Urban Logistic Vehicles. *World Electr. Veh. J.* **2023**, *14*, 121. https://doi.org/10.3390/wevj14050121

Academic Editor: Joeri Van Mierlo

Received: 13 March 2023
Revised: 4 April 2023
Accepted: 27 April 2023
Published: 4 May 2023

1. Introduction

Nowadays, inner—city logistics is struggling with two significant problems: first, an increasing volume of parcels has been recorded for the last few years. Second, urban areas are facing air pollution due to the constant increase in traffic [1–4]. The transportation sector is responsible for the emissions of greenhouse gases and air pollution such as carbon monoxide, unburned hydrocarbons and particulate matter [5]. One effective solution to solve these problems is the use of small electric—driven logistic vehicles for the last—mile in urban areas. Electric driven vehicles are one of the most promising technical solutions because they do not produce local emissions and, depending on the way the electricity is generated, have a huge potential to reduce emissions [6]. From a technical point of view these vehicles represent an ideal application field as they have plannable routes and pre—defined depots [7]. Additionally, electric—driven logistic vehicles are smaller and lighter compared to conventional logistic vehicles. Additionally, the combination of EVs and a lightweight design can reduce environmental impacts [8]. However, the electrification and reduction of the logistic vehicle size as well as the high weight of the battery lead to a reduction in payload which is already smaller for small logistic vehicles. However, the payload is the most important variable in the logistic business.

The main target of this work is to replace conventional parts of small electric logistic vehicles by lightweight components, to compensate the losses. In general, the mass of electric vehicles (EV) is not a new problem. Years ago, EVs were based on conventional vehicle structures, except for the EV components, which led to a high vehicle mass. For example, the BMW i3 was not based on a conventional structure; it had a new lightweight design for electric vehicles [9]. A number of lightweight materials such as aluminium alloys, high strength steel, plastics and composites are integrated in the design of EVs [10].

In terms of the components, it is to improve the weight and additionally to optimize them with technologies from other industries. The cross—industry innovation approach is a methodical approach that explores information across industries. The transfer of knowledge, technologies and know—how from outside the sector is the base of the cross—industry innovation [11]. For this work, the principle of the cross —industry innovation method was used to identify usable technologies, especially textiles and mounting systems. This work specifically focuses on the use of technical textiles in vehicle exteriors, which are investigated. Previous studies have already presented the general feasibility of textiles in the exterior of the vehicles (passenger vehicles). Within this work, textiles for exteriors are transferred into the logistics area [12]. In the case of logistic vehicles, the vehicle body of the cargo space offers a great potential to reduce the weight.

2. State of the Art in Science and Technology

Textiles are already being used in the exterior of logistic vehicles in the form of truck tarpaulins. Truck tarpaulins, which are polyester carrier fabrics with a PVC (polyvinyl chloride) coating on both sides are commercially available in weights of 670 to 900 g/m^2 [13–15]. They are considered to be highly tear—resistant and serve as a mobile advertising space. In the roof area, translucent PVC tarpaulins are preferred. Horizontal surfaces often suffer, in addition to their own weight, from the weight of snow mass, which can lead to damage. Tarpaulins are attached to the frame of truck bodies using piping, straps, eyelets or similar. However, additional fastenings lead to additional weight. For logistic vehicles with a body made of truck tarpaulins, there is a risk of theft due to cutting into the tarpaulin. Currently, segments made of PVC with attached wires that trigger an alarm when cut represent the solution to this problem [16]. This can also be used to upgrade existing tarpaulins. In the calculation of the permitted total mass of a truck and smaller sized logistic vehicles, saved weight means lower energy consumption and thus the possibility of a higher payload.

Sandwich structures are mainly used for logistic vehicle bodies. Sandwich structures are made up of two face layers and a core material. The core materials usually consist of lightweight foam, or a honeycomb structure enclosed by two solid, force—absorbing surfaces. While the top surface absorbs compressive stress, the core provides cohesion and dissipates shear forces. Often, further layers for heat—insulating or sound—insulating effects are added. The lightweight construction of sandwich structures combines high stiffness and compressive strength at low weight.

The materials used are subject to constant change, driven by increasing requirements in terms of lightweight design, safety, rigidity, reliability, environmental compatibility and economy. The general target from the point of view of lightweight construction is to meet all the requirements of a vehicle components with minimum mass demand. Automotive developers are already presenting approaches in lightweight concept vehicles in which the vehicle exterior is completely covered with a textile. At the beginning of 2019, the simulation of the "Futurelight Camper", a trailer with a frame made of carbon fibre, was presented by the outdoor company The North Face in collaboration with the BMW Group subsidiary [17]. A waterproof and breathable membrane made of a nanofiber structure covers the frame. It is possible to adjust weight, stretchability, air permeability and robustness in the existing "Futurelight" textile during manufacturing [18]. Details on the realization have not been provided yet.

A very light variant of a sandwich panel structure can be made of polyester resin panels with a glass fibre reinforcement and a PET foam core. With a thickness of about 14 mm, such a panel has a price of about 45 €/m^2 and weighs 4.35 kg/m^2 [19]. The weight of a commercial truck tarpaulin without a frame starts at about 670 g/m^2.

The potential of outdoor textiles to resist UV radiation and intense climatic fluctuations has been proven for quite some time in many fields. They are used in the field of architectural textiles, boat construction, air sports, clothing, tents, garden furniture and sunshades.

3. Methodical Approach

To investigate the usage of textile in urban logistic vehicles, a methodological approach has been developed consisting of four individual steps. This approach was developed independent of materials or technologies. This offers the opportunity to transfer the method to other materials or mounting systems and supports the development of further exterior components for light electric last mile delivery vehicles through a methodical sequence of steps to be carried out. The first step is the identification of basic requirements. Based on the defined requirements, textiles are researched that are already used in various areas and industries in the exterior, followed by the selection of the most prospective ones. Basic tests of the textiles help to verify the compliance with the requirements. Mounting possibilities for textiles at the construction of the vehicle body have also been researched, selected and tested. The results were merged and transferred into early prototypes. Using these early prototypes, further tests such as weathering, air flow (wind tunnel) as well as cargo securing, and the load resistance of the lateral surface were executed. After all tests were completed, one textile and one mounting mechanism were assembled in a prototype. Figure 1 shows the approach.

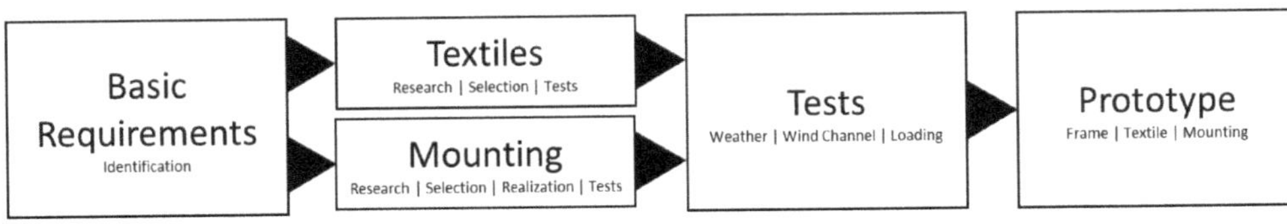

Figure 1. Methodical approach for examining textile bodies for urban logistic vehicles.

Basic Requirements: The identification of the basic requirements consists of the definition of the application area, the components, and legal requirements. The first step is the definition of the application area, in this case the vehicle category such as passenger transportation, logistic transportation, etc. If the application area is defined, the components, which should be replaced by textiles, have to be defined. If the components are defined, research about the valid legal requirements is necessary to obtain information about the legal framework and vehicle registration. To obtain the information, research of literature and laws was necessary. The research project was located in Germany; therefore, the German laws were focused on for the research of laws. In addition, further technical, emotional, and aesthetic requirements are included which is important for the product development later.

Textiles: To find suitable textiles for the use case, eligible materials are researched for the application, based on a material and availability research. This involves searching the use case industries but also in external industries. Based on the technical specification, researched textiles are shortlisted. These textiles are subject to comprehensive laboratory testing such as artificial weathering, soiling and cleaning behaviour, the determination of water and oil mark or burst properties. The three most convincing textiles, which meet the requirements for use in logistic bodies are selected to execute further tests [20].

Mounting: Mounting techniques must be researched and reviewed to find the most suitable joining technique for the special use case. Afterwards, they are systematically evaluated whether they can be transferred to the application. The individual mounting techniques must be tested under laboratory conditions.

Tests: Once the textiles and mounting mechanism are selected and tested under laboratory conditions, pre—prototypes of the construction are created and tested as a system. In this use case, three different tests with a prototype are performed:

Wind tunnel test: In a wind tunnel test, the effect of the flow around the pre—prototype is tested. With the airflow, the tensile and compressive force on the textile and the deformation of the textiles (fluttering) measured.

Load test: A drop test is conducted to check the mechanical resilience of the lateral surface (textile) and mountings. The test simulates an impact of the package on the lateral surface of the body when driving along curves by "falling" onto the textile surface with the same force. Accordingly, a calculation of the impact force is performed to generate consistent values of a curve run.

Weathering test: The textile is tested in an endurance test by being exposed to all climatic conditions for 12 months. This shall examine the weather compatibility of the textiles and mounting systems.

Figure 2 shows the several principles of testing.

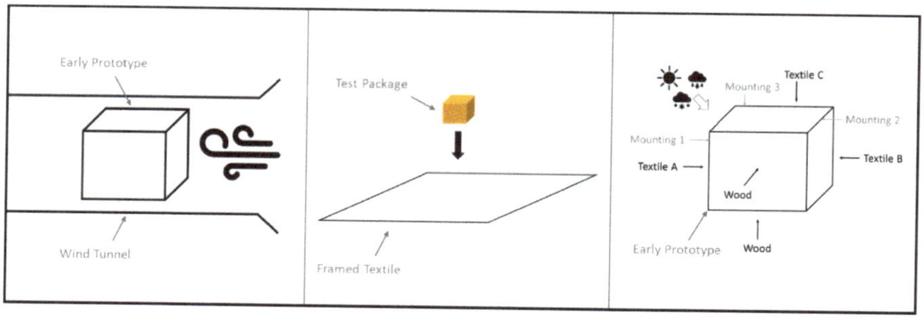

Figure 2. Test procedures of the wind tunnel test (**left**), load test (**middle**) and weathering test (**right**).

Prototype: After implementing and evaluating the tests, a prototype of the logistics body has been built. The prototype contains both the most appropriate textile as well as the most appropriate mounting mechanism, selected by taking the basic requirements into account.

This methodical approach was applied in this work to generate knowledge about the usage of textiles for vehicle bodies for urban logistic vehicles.

4. Results

According to the methodical approach developed in this project, the research work was conducted, and the results presented in this chapter were obtained.

4.1. Basic Requirements

The application areas in this project were predefined through the project partner. The usage of textiles in the vehicle exterior should be analysed in logistic transportation and in passenger transportation. This paper shows only the results of logistic transportation; the methodical approach is the same for both application areas. The results of passenger transportation may follow in another scientific paperwork. For the logistic transportation, the focus was set on an urban logistic vehicle (<3.5 t).

To define the component which should be replaced by textiles, a workshop with relevant project partners was conducted. In this meeting a lot of "textile" ideas came up. At urban logistic vehicles, the most prospective assembly group for a "textilization" was the vehicle body. The experts in the project have seen the highest potential to reduce weight in that assembly group. The whole sides of the body can be designed by textiles. Additionally, mechanical elements such as roller blinds and sliding doors should be analysed in that work. For a higher resistance of the textiles, the inside of the textiles should be protected by a textile net. This could be necessary if the textiles get impacted by moving packages inside of the cargo hold. Summarised, the following components should be implemented:

- Vehicle body with textile sides
- Inner textile net for a higher resistance
- Roller blinds and sliding doors.

After defining the components, the legal requirements for the construction of urban logistic vehicle bodies must be researched. Before this step, the daily use—case of an urban logistic vehicle was elaborated for a same understanding of the research object. There exist a few legal (technical) requirements (for the defined user region Germany) for the construction of a logistic vehicle body. The most relevant for this project were:

- no sharp, pointed or protruding edges [21]

- closable, also while driving [22]
- width of 2.55 m and height of 4 m must not be exceeded [23]
- lashing possibilities must be installed [24].

In addition to the legal technical requirements, further technical, emotional, and aesthetical requirements also were considered (Table 1).

Table 1. Technical, emotional and aesthetical requirements.

Technical Requirements	Emotional Requirements	Aesthetical Requirements
Weather—resistance Theft—resistant No acoustic disturbances Wear and tear—resistant Dirt—resistant Weather—resistance	Roller blind and sliding door with high usability Textile visually and haptically high—quality and appealing Individuality for recognition value	Valence Modern design Corporate design

A catalogue of requirements for further research and development of the textiles and mounting techniques was drawn up based on the basic requirements identified.

4.2. Textiles

Based on the basic requirements identified in chapter 4.1, specific technical requirements for the textile were derived and defined. These technical requirements include, for example, a water tightness of a 1500 mm water column, an air impermeability of >5 L/m^2/s, an UV resistance of 5 years, climate resistance at temperatures from −10 °C to +70 °C, no wetting of oil and water, dirt repellence and a weight that is well below 520 g/m^2, (equivalent to the weight of conventional PVC truck tarpaulin) [13–15]. In a first step, textiles already available on the market from various outdoor sectors such as construction textiles, tents, boat building, air sports, sun protection and outdoor clothing with potential suitability for this project were researched and compared with the defined technical requirements.

In addition, suitable test methods were identified to model the behaviour of the textiles in an actual application, so that an assessment could be made for their suitability for the usage of textile in the vehicle exterior. These are water and oil repellence—the determination of oil and water marks, scratch resistance—the determination of the tendency for thread pulling and snagging, colour fastness—the determination of colour fastness after artificial weathering with UV irradiation, bursting properties before and after artificial weathering with UV irradiation and soiling and cleaning behaviours. Figure 3 shows an overview of the test methods.

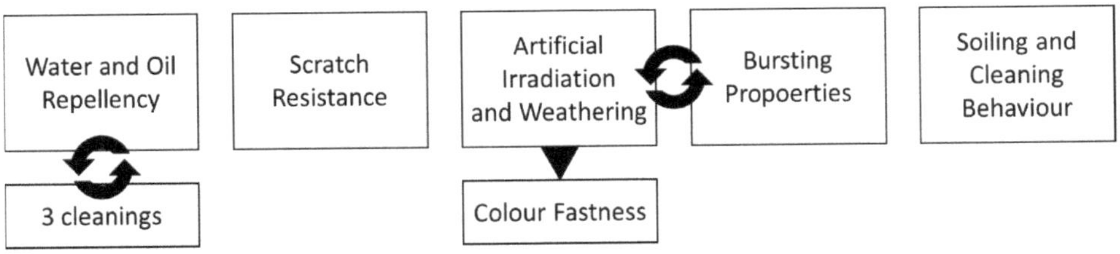

Figure 3. Overview of test methods for textile behaviour.

Water repellency and oil repellency: For use in vehicle exteriors, cleaning is considered a necessity. The repellence of oil— and water—containing liquids should therefore not lose its effect. The determination of the oil repellence (DIN EN ISO 14419) and water repellence ("3M—Water repellency") were tested on the textiles before and after three cleanings with a cleaning agent typical for the automotive sector (car shampoo concentrate R527 from Förch) and a commercially available all—purpose cleaner (neutral cleaner from Frosch). In addition, the "colour stain test", a test developed at the DITF, was performed, in which the wetting behaviour is evaluated with a dye solution. For the application in automotive exteriors, the test is intended to reveal no wetting of the textile.

Scratch resistance: The Orbitor Pilling and Snagging Tester 516 was selected for testing the pilling tendency and snagging. The corresponding standard is BS 8479. This test was selected to produce results similar to the striping or snagging of a bush with thorns and spines when the vehicle is parked. The determination of the thread pulling tendency and the snagging proved the robustness of the textile under the influence of sharp objects, which can cause damage to the surface of coatings, holes or thread pullers. As a comparison and confirmation, a test was performed with branches with spines of a rose plant, which were pulled over the textiles.

Artificial irradiation and weathering: Artificial irradiation and weathering was selected to mimic time—accelerated aging. The influence of weather and solar radiation varies greatly depending on the climatic zone. Artificial aging was performed according to the climatic zone "Central Europe" with treatment cycles set up at the DITF. The test includes temperature changes from −10 to +70 °C, humidity changes and irradiation with global radiation. For the artificial irradiation and weathering with UV fluorescent lamps, a UV—global tester "Weiss UV200" was used. The procedure was carried out in accordance with the standard DIN EN ISO 4892-3. Artificial weathering can deviate from the behaviour under practical conditions during long—term outdoor weathering. The effect of photodegradation was then determined with colour fastness and the determination of the bursting strength.

Colour Fastness: The effect of weathering on the colour can cause the complete loss of it. The loss of the colour means at the same time the loss of the aesthetics expected by the customer. The colour before and after weathering and irradiation was measured on the spectrophotometer "X—rite colour i7" according to the CIELab system under daylight. The corresponding standard is DIN EN 20105—A02.

Bursting properties: The determination of bursting strength was chosen to gain an insight into the possible change in strength in combination with the artificial weathering and irradiation, as well as to gain knowledge of the deformation and recovery of the textiles as a basis for further tests in a wind tunnel. Due to the driving wind, outward pressure on the textiles is expected. The determination of the bursting properties is carried out according to DIN EN 13938-2 with the TruBurst Model 140 using a pneumatic method before and after artificial irradiation and weathering.

Soiling and cleaning behaviour: The aim was to replicate soiling in road traffic. Since the soil composition and the soil application of existing standards and test methods do not correspond to the actual application, an own soil was compiled, consisting of 9.59 g "Leverkusen Standard Soil 40" and 1.71 g brake dust dissolved in 4 L of water. The outside of textiles was brought into direct contact with the soil solution and the water was evaporated completely, depositing the soil on the textile. The textiles were visually assessed.

Subsequent cleaning was carried out with the same cleaning agent used for the oil and water mark. The cleaning tests examine whether the textiles can be cleaned without leaving residues. Figure 4 shows some impressions of the laboratory tests.

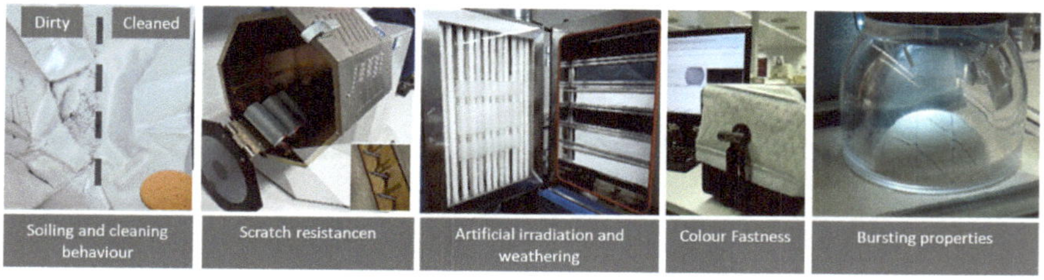

Figure 4. Impressions of the different textile laboratory tests.

Textile A (Laminate) is a three—layer laminate consisting of a polyamide fabric on the outside, a polyurethane membrane in the middle layer and an inner layer of a knitted polyester and elastane fabric. The textile is impregnated oil and water repellent with a fluorocarbon finish. Applications of this textile are motorcycle clothing or backpacks [25].

Textile B (Aircraft) is a woven polyester fabric used as an aircraft covering. The fabric, with a weight of 90 to 210 g/m^2, is applied and tensioned by gluing and shrinking by means of heat. The warp and weft density are identical; therefore, the woven textile has no preferred direction. The textile is non—combustible (DIN EN ISO 6941), UV—resistant and air—impermeable. Furthermore, the textile is re—shrinkable. A loss of tension caused by snow or rain load can thus be easily retightened with hot air. The woven textile which is produced in Germany is available in different translucencies. For the roof, a translucent material is preferred to increase the brightness inside the vehicles body [26].

Textile C (Kite) is very light with only 53 g/m^2 and is used for kites. It is a double ripstop polyester woven fabric with very low stretch. Furthermore, the fabric is air—impermeable and water repellent [27].

The three textiles were implemented in a pre—prototype with different textiles on each side of the body in order to be able to carry out further tests (see Section 4.4). Figure 5 shows an overview of the test—results of the selected textiles.

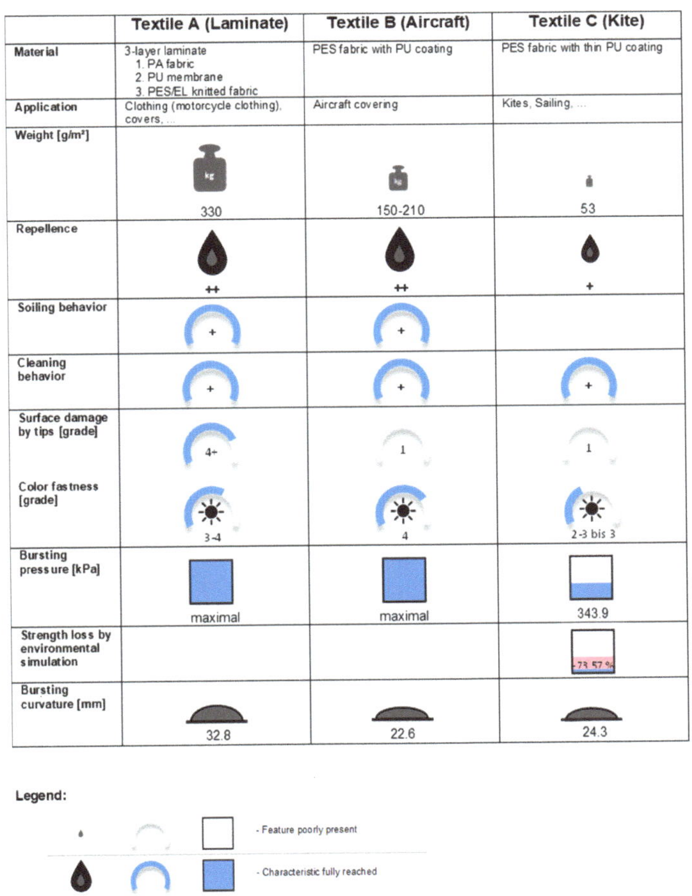

Figure 5. Overview of the test—results of the textiles.

4.3. Mounting

A conventional vehicle body consists of sandwich panels with aluminium profiling. Unlike textiles, the sandwich panels have a load—bearing and stabilizing function. Thus, the frame construction for a textile covering must be built differently. Several frame constructions were manufactured and tested for the project as individual elements from commercially available vehicle construction extruded profiles.

To find out whether it is possible to cover the large surfaces of the box structure or whether additional struts are required in the structure, test frames were manufactured, covered, and then tested in a load test (see Section 4.4.3). In addition, various joining techniques such as sewing, gluing, and riveting were examined.

It has been shown that a two—layer structure with an additional inner fabric net for load securing and the protection of the outer textiles is most suitable. For the inner textile, a polyester net fabric coated with PVC with very high tear resistance was chosen. The fabric can be processed both by welding and sewing. Due to its very low price (about 3 €/m^2 [28]), it is economically justifiable to replace it in case of damage. For long durability, reinforcements have been sewn on the locations that are prone to a lot of stress due to friction. In addition, piping was applied to distribute forces over the entire surface in the event of an impact. The piping rails were attached to the frame interior with blind rivets, thus pulling in the textile and bringing it to tension.

As a result of the tests, the techniques selected as most appropriate were clamping by frame components, a solution of a piping system and gluing the textile onto the frame. Furthermore, textile B was selected as the outer skin with which good tension can be achieved due to its shrinkage behaviour of 15%. First, the textile was made up to the required size and then the PVC piping was glued on using a dispersion hotmelt adhesive and a hot air blower. Since textiles might lose tension over time due to weathering, it is essential to be able to tension them again. The rail system used has several tensioning grooves to achieve this. The textile is pre—tensioned by means of the piping rails and then finally tensioned with a hot air dryer at a maximum of 180 °C. The textile is then re—tensioned. To further improve the durability and stability of the frames and textiles, additional struts were built in. To keep the weight as low as possible, flexible PVC square profiles were installed, which are both light and stable. The mounting was carried out with blind rivets. Figure 6 shows the results of the mounting research.

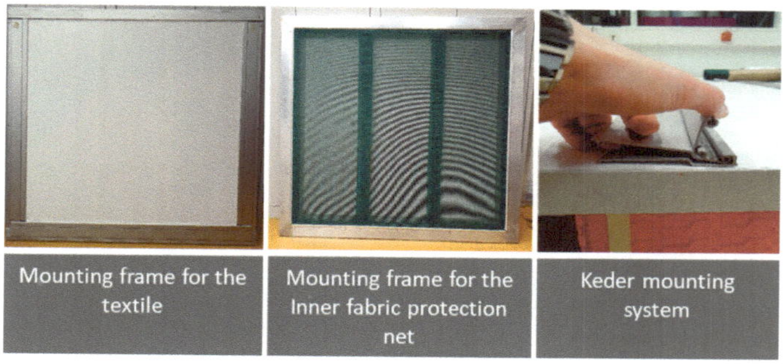

Figure 6. Results of the mounting techniques research (**left**: outside, **middle**: inside) and the piping system (**right**).

4.4. Tests

After the research, selection and testing of promising textiles and mounting techniques, several early prototypes were built to test the whole system with the assembled components. An early prototype is defined as an easy— and fast—to—build but incomplete type of a prototype, as it focusses on those aspects that are necessary for a defined test. Unlike a commonly known prototype, an early prototype only includes properties which are necessary to realize a prototype at an explicit moment for the development of new technologies and to reach the defined testing goals [29].

4.4.1. Wind Tunnel Test

For the wind tunnel and weathering test, an early prototype named "the cube", was built. All sides of the cube have the maximum frame size that a final vehicle body would have (1500 mm). The cube was equipped with the three identified textiles (Section 4.2) and three different mounting techniques (Section 4.3). The front and the bottom side are made of wood. The rear side is made of ordinary textile with two zippers for the entry of the inside of the cube. Figure 7 shows the textile and mounting distribution at the cube.

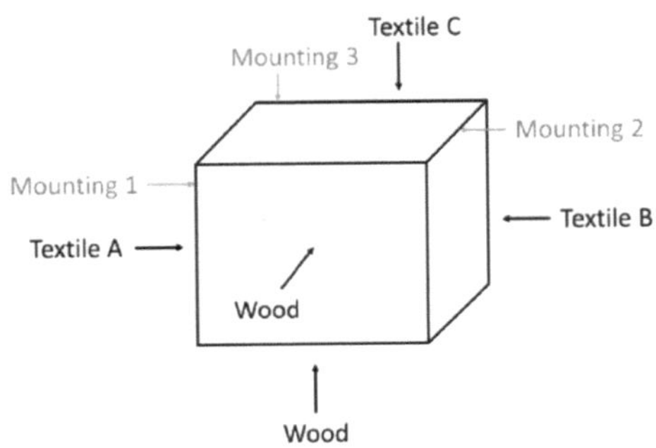

Textile A	Laminate
Textile B	Aircraft
Textile C	Kite
Mounting 1	Piping system
Mounting 2	Gluing, Adhesion
Mounting 3	Clamping by frame components

Figure 7. Textile and mounting distribution at the system early—prototype, the cube.

At the wind tunnel test, the effect of the flow around the cube was tested. For a simulated and continuous air flow, a gust wind tunnel was used. By means of the air flow, the tensile and compressive force on the textile and the deformation of the textiles (fluttering) was measured. The tensile and compressive force is measured by a force gauge, the deformation of the textile by a camera with a slow—motion function. Seven different test runs have been completed to obtain all data for each textile in two flow—around positions of the cube. In position A, the cube was flown around to simulate straight ahead driving (wood/textiles = 90° to the flow). In position B the cube was flown around to simulate cornering (wood/textiles = 45° to the flow). Figure 8 shows the different positions.

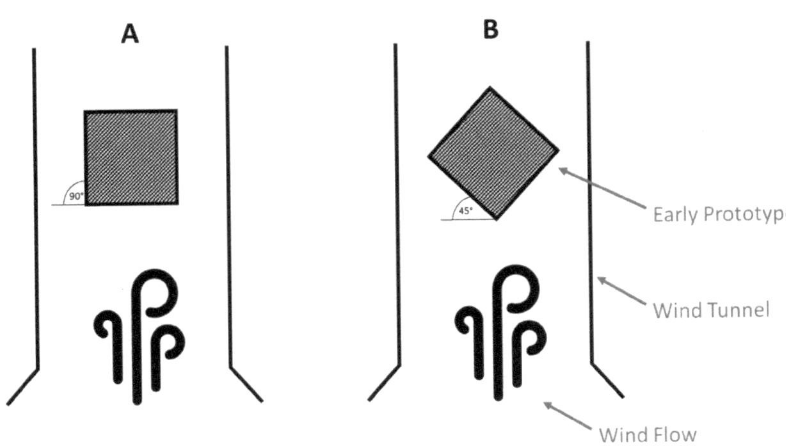

Figure 8. Flow around position of the cube at the wind tunnel.

The measuring went from 0 m/s to 16.67 m/s. At the wind speed of 8.33 m/s (30 km/h), 13.89 m/s (50 km/h) and 16.67 m/s (60 km/h), the flow was constant for 30 s to obtain consistent conditions for predefined wind speeds.

The results of the wind tunnel test show that textile B (Aircraft) has the best overall values. The fluttering (deformation) and the tensile force show the smallest amplitudes. The laminate (textile A), which has an elastic portion, already showed the first signs of material fatigue after stringing. It performed worst in the wind tunnel test due to strong fluttering, high tensile force and high noise development. Figure 9 shows an example of the measured data at the wind tunnel test (Textile B, aircraft).

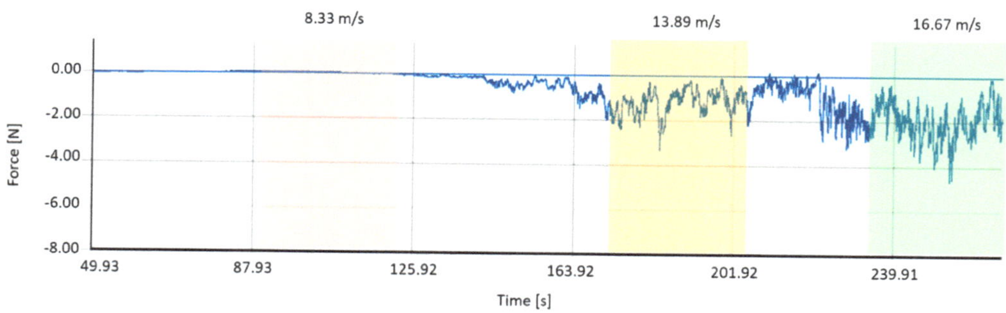

Figure 9. Example of the measured data at the wind tunnel test.

4.4.2. Weathering Test

For the weathering test the cube was exposed for an endurance test to the weather. The test period went on for 12 months in all climatic conditions of central Europe, i.e., sun, wind/storm, rain and snow (48.74047623855775, 9.096374837039678). This examined the weather compatibility of the textiles and mounting systems.

The following findings were made after 12 months:

- Textile C/kite does not resist the weather. Cracks appeared. It became porous.
- Textile A/3—layer—laminate lost some tension.
- The adhesive bonding of Textile B/aircraft came off.

The results after 12 months showed that the textile B/aircraft and the piping system had the best withering properties for the use case as a vehicle body.

4.4.3. Load Test

For the load test, an early prototype was built to check the mechanical resilience of the textile and net at the lateral surface and the mountings. The test simulates an impact of the package on the lateral surface of the vehicle body when driving along curves. Testing this scenario with a vehicle while driving along a curve was too opulent, so a falling test was constructed to simulate a package bumping into the lateral surface from a defined height h, which consists of a net and a textile. Figure 10 shows the technical concept of the load test.

To reach the same force acting in the lateral surface at the falling test ($F_{eff} = F_{imp}$), the effective force and the impact force were calculated. The aim of the calculation is to find out the height h; the package has to fall down to reach the same velocity $\left(v_{pack} = v_{Fall}\right)$ as a package can have during a bump into the side wall inside the vehicle body.

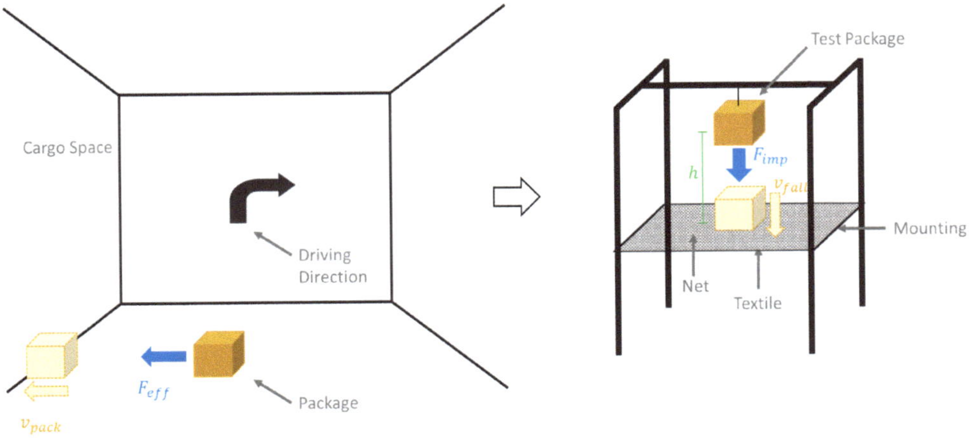

Figure 10. Technical concept of the load test.

First, the effective force F_{eff} had to be calculated, which is determined from the difference between the centripetal force F_a $[N]$ during cornering and the frictional force F_{fric}:

$$F_{eff} = F_a - F_{fric} \tag{1}$$

The centripetal force F_a is the product of the mass of the package m_p [kg] in the delivery vehicle and the centripetal acceleration a_c $\left[\frac{m}{s2}\right]$. The centripetal acceleration is the quotient of the truck velocity v_{truck} $\left[\frac{m}{s}\right]$ in the square divided by the radius of the curve r $[m]$:

$$F_a = m_p * a_c = m_p * \frac{(v_{truck})^2}{r} \tag{2}$$

The frictional force F_{fric} $[N]$ is the product of the mass of the package m_p [kg], the coefficient of friction μ_f [-] (cardboard—wood) and the acceleration of gravity g $\left[\frac{m}{s^2}\right]$:

$$F_{fric} = m_p * \mu_f * g \tag{3}$$

The height h $[m]$ can be calculated from the equation of the force of the impact on the lateral surface F_{imp} $[N]$ with the the acceleration due to gravity g $\left[\frac{m}{s^2}\right]$ and the velocity of the parcel at the moment of impact v_{fall}, which takes the same value as in the sideways motion:

$$F_{imp} = m_{p,fall} * a_{fall} = m_{p,\,fall} * \frac{{v_{pack}}^2}{2 * h} \tag{4}$$

$$with\ v_{pack} = v_{fall}\ and\ a_{fall} = g$$

The conversion of Equation (4) according to the variable h $[m]$ results in the following equation with the variables of the mass of the package falling $m_{p,\,fall}$ $[kg]$, the maximum loading areas with b $[m]$, the acceleration of gravity g $\left[\frac{m}{s^2}\right]$ and the effective force F_{eff} which result from Equation (1):

$$h = \frac{v_{fall^2}}{2 * g} = \frac{(a_{ges} * t)^2}{2 * g} = \frac{\left(\frac{F_{eff}}{m_{p,fall}} * \sqrt{\frac{2*b}{a_{ges}}}\right)^2}{2 * g} = \frac{\left(\frac{F_{eff}}{m_{p,fall}} * \sqrt{\frac{2*b}{\frac{F_{eff}}{m_{p,fall}}}}\right)^2}{2 * g} \tag{5}$$

The falling test enables repetitive consistent conditions. For the special use—case in this work, the following data were used to calculate the height h. All data are based on a literature research or predefined vehicle data [30–32]. The mass of the package was reduced from 31.5 kg to 10 kg for better handling during the tests.

Data for the calculation of the fall height:

$m_{p,truck} = 31.5$ kg
$m_{p,fall} = 10$ kg
$v_{truck} = 5.6\ \frac{m}{s}$
$r = 5\ m$
$\mu_f = 0.5\ (Cartridge - Wood\ /\ cartridge - cartridge)$
$g = 9.81\ \frac{m}{s^2}$
$b = 1.5$

By inserting the data above into the Formulas (1)–(5), a fall height $h = 0.45$ m is calculated. In addition to the calculated height, the maximum height $h_{max} = 1.45$ m given from the technical conditions and an absolute maximum test—force f_{edel} (person who throws the package) has also been tested. The point of impact (middle or at the edge) and the impact position (flat, edge or corner) of the package can be chosen variably. The testing of the fall test was recorded by two cameras which were filming the drift punch of the textiles in front of a reference scale. One bar corresponds to 2 cm. Figure 11 shows an example.

Figure 11. Example of recording the drift punch of a package in front of the reference scale.

The results for the height $h = 0.45$ m are very low. For all textiles, there was no drift punch existing (0 cm). For the maximum technical height $h_{max} = 1.45$ m, clear drift punches exist in a range from 1.5 to 3 (0.03–0.06 m) without any damage of the net and textiles. For the maximum test—force f_{edel}, the drift punches receive values from 5–7 (0.1–0.14 cm) without any damages. Table 2 shows the results.

Table 2. Results of the falling test.

	Textile A + Piping System	Textile B + Piping System	Textile C + Piping System
$h = 0.45$ m	0	0	0
$h_{max} = 1.45$ m	2	3	1.5
f_{edel}	5	7.5	5

The fall test showed the chosen textiles, safety net and mounting technique working excellent for the use as a vehicle body.

4.5. Prototype

At the end of the scientific work, a prototype of the vehicle body has been built. For a customer approach, because of the transportability, the prototype was built at a trailer to make the technologies presentable. The final prototype consists of one final selected textile B (aircraft) as the outer skin for protection against water and dirt and as an advertising medium and an inner textile for load securing and the protection of the outer textiles from slipping cargo (polyester net coated with PVC). The textile is attached to the frame with a new type of rail and piping system. The opening systems are one flap at the back and two sliding doors at the sides. The whole textile system is about 15–20 percent lighter than a conventional vehicle body (depends on the size of the vehicle body), which is a valuable contribution in the field of urban logistic. In comparison to a conventional vehicle body of this size, a payload increases about 80 kg (conventional 3.5 t truck: about 2 t vehicle, 400 kg vehicle body and 1 t payload) is achieved. The focus in this work was the selection of lightweight textiles and mounting techniques according to demand. The lightweight construction potential was also considered, which can be further increased by suitable frame construction methods. Besides the reduction of weight, other important aspects are the possibility of a rapid replacement or the repair of the textiles in the event of damage. Figure 12 shows the constructed prototype of the vehicle body.

Figure 12. Final prototype of the textile vehicle body at a trailer.

5. Conclusions

The work in this paper shows which textiles and mounting techniques can be used for vehicle bodies at urban logistic vehicles to reduce unladen weight for logistics.

- For the definition of the perfect textile setup, basic requirements were defined and textiles as well as mounting techniques were researched and tested.
- Three textiles and mounting techniques showed good results for the presented use case of a vehicle body.
- The selected textiles and mounting techniques were subjected to further tests such as weathering, loading and durability in wind (wind tunnel). At these tests, one textile, a woven polyester fabric with a polyurethane (PU) coating used for light aircraft covering, with polyvinylchloride (PVC) coated polyester mesh fabrics on the inside for load security and a piping mounting system proved to be suitable.
- At the end, a prototype was built to obtain information about the market viability for urban logistic vehicles under 3.5 t of total weight.

The results showed a weight reduction potential of about 15–20% in comparison to conventional materials such as sandwich constructions (depends also on the size of the vehicle body). The project showed that the weight saving potential for vehicles < 2.6 t could be way higher, because the savings ratio for textiles is higher than for sandwich bodies for example. The smaller the vehicle, the higher the potential weight saving. A second very big advantage of the textile vehicle body is the opportunity to distribute a decentralized construction kit. This is not possible with panels from conventional construction, as transport is more circumstantial. All the results and findings are of enormous added value and technical progress to promote electromobility in urban areas for the last—mile logistic industry. One disadvantage of the textile exterior is the lack of load—bearing capacity.

A corresponding clamping frame must always be provided. Ideas for a frame made of material lighter than aluminium are being planned and give further space for research questions. Further research work also consists of the implementation of a prototype of the body on a chassis and a practical application to gain further experience about the suitability of textiles in urban logistic vehicles.

Author Contributions: Conceptualization, F.E.; methodology, F.E., C.P., S.K. (Sarah Kim) and J.E.; investigation, F.E., C.P., S.K. (Sarah Kim), J.E. and S.K. (Sven Kuijpens); writing—original draft, F.E.; writing—review and editing, C.P. and S.K. (Sarah Kim); visualization, F.E. All authors have read and agreed to the published version of the manuscript.

Funding: This research was funded by the Ministry of Economy, Labour and Housing Baden—Wuerttemberg, Germany, grant number 3—4332.62.

Data Availability Statement: The data in this article were generated within a project and are not publicly available.

Acknowledgments: The work and the results presented are part of the project TexEx: identification, evaluation, and demonstration of meaningful applications of textiles in vehicle exteriors founded by the Ministry of Economy, Labour and Housing Baden—Wuerttemberg, Germany.

Conflicts of Interest: Juergen Erhardt is an employee of Erhardt GmbH. The paper reflects the views of the scientists, and not the company.

References

1. Winkler, R. Innenstadtlogistik Mit Lastenrädern am Beispiel der KEP—Branche, München, May. 2019. Available online: https://www.adac.de/-/media/pdf/vek/fachinformationen/radverkehr/innenstadtlogistik-mit-lastenraedern-adac-sp.pdf (accessed on 2 May 2023).
2. Klauenberg, J.; Dr, K.J.; Hanna, J.; Michael, K.; Dag, R.; Daniel, R.; Schröder. Die Veränderung des Gewerblichen Lieferverkehrs und Dessen Auswirkungen auf die Städtische Logistik, Berlin/Dortmund, Nov. 2020. Available online: https://bmdv.bund.de/SharedDocs/DE/Anlage/G/staedtische-logistik-bericht-veraenderungen-lieferverkehr.html (accessed on 2 May 2023).
3. Pöhler, D.; Platt, U. Luftverschmutzung in den Städten. *Ruperto Carola* **2016**, *9*, 50–59. [CrossRef]
4. Wohlgemut, L.; Graff, A. Feinstaub in Städten-Welche Probleme Gibt es Weiltweit?: Fine Particulate Matter in Cities—Global Problems to Face. Dessau-Roßlau. 2013. Available online: https://www.umweltbundesamt.de/sites/default/files/medien/360/publikationen/feinstaub_in_staedten_weltweit_s_11-17.pdf (accessed on 2 May 2023).
5. Zhou, W.; Cleaver, C.J.; Dunant, C.F.; Allwood, J.M.; Lin, J. Cost, range anxiety and future electricity supply: A review of how today's technology trends may influence the future uptake of BEVs. *Renew. Sustain. Energy Rev.* **2023**, *173*, 113074. [CrossRef]
6. Quak, H.; Nesterova, N.; van Rooijen, T. Possibilities and Barriers for Using Electric-powered Vehicles in City Logistics Practice. *Transp. Res. Procedia* **2016**, *12*, 157–169. [CrossRef]
7. Lesemann, M.; Welfers, T.; Mohrmann, B.; Eckstein, L. Konzeption und Aufbau eines elektrischen Lieferfahrzeugs. *ATZ-Automob. Z.* **2014**, *116*, 34–41. [CrossRef]
8. Del Pero, F.; Delogu, M.; Pierini, M. The effect of lightweighting in automotive LCA perspective: Estimation of mass-induced fuel consumption reduction for gasoline turbocharged vehicles. *J. Clean. Prod.* **2017**, *154*, 566–577. [CrossRef]
9. Delogu, M.; Zanchi, L.; Dattilo, C.A.; Pierini, M. Innovative composites and hybrid materials for electric vehicles lightweight design in a sustainability perspective. *Mater. Today Commun.* **2017**, *13*, 192–209. [CrossRef]
10. Bubna, P.; Wiseman, M. *Impact of Light-Weight Design on Manufacturing Cost-A Review of BMW i3 and Toyota Corolla Body Components*; SAE Technical Paper Series; SAE International: Warrendale, PA, USA, 2016.
11. Kerl, A. *Management von Multi-Cross-Industry Innovation: Wirkungsabschatzung, Organisationale Strukturen … und Gestaltungshinweise*; GABLER; Springer Fachmedien Wiesbaden: Wiesbaden, Germany, 2018.
12. Edel, F.; Schäffner, P.; Kern, M. Lightweight components for light electric vehicles based in textile exterior. In Proceedings of the 32nd Electric Vehicle Symposium (EVS32), Fraunhofer IAO, Lyon, France, 19–22 May 2019.
13. Giese, S. Produktkatalog: Friedrich Kabrinke Raumausstatter und Polstereibedarf. 2021. Available online: https://kabrinke.de/download/Kabrinke-Produktkatalog-2021.pdf (accessed on 2 May 2023).
14. SCAR B1 LKW Plane-PVC Beschichtet-500 g-Planenlager.de, 18.08 €, SCAR B1 LKW Plane-PVC Beschichtet-500 g-Planenlager.de, 18.08 €. Available online: https://www.planenlager.de/lkw-plane-pvc-schwer-entflammbar/ (accessed on 28 April 2022).
15. Anhängerplanen–Bensch Planen. Available online: https://www.bensch-planen.de/planenkonfektion/anhaengerplanen/ (accessed on 28 April 2022).
16. Idw/Hochschule Bremerhaven. Alarmsystem für Planenauflieger. Available online: https://www.gruender-mv.de/2017/07/04/alarmsystem-fuer-planenauflieger/ (accessed on 28 April 2022).

17. BMW Group PressClub, Designworks Collaborates with The North Face to Imagine New Camper Concept. Available online: https://www.press.bmwgroup.com/global/article/detail/T0289592EN/designworks-collaborates-with-the-north-face-to-imagine-new-camper-concept?language=en (accessed on 28 April 2022).
18. The North Face, Futurelight Technologie. Available online: https://www.thenorthface.de/innovation/technologies/futurelight.html (accessed on 28 April 2022).
19. Erhardt, J. Erhardt Nutzfahrzeugaufbauten. Available online: http://www.erhardt-fahrzeugbau.de/page/home/index.html (accessed on 28 April 2022).
20. Pulkus, C. Identifikation und Evaluation von Textilien für das Fahrzeugexterieur. Bachelor's Thesis, Textil & Design, Hochschule Reutlingen, Reutlingen, Germany, 2020.
21. European Union. *Regulation No 26—Uniform Provisions Concerning the Approval of Vehicles with Regard to Their External Projections: UN/ECE No 26*; European Union: Brussels, Belgium, 2007.
22. European Union. *Regulation No 11—Uniform Provisions Concerning the Approval of Vehicles with Regard to Door Latches and Door retention Component: UN/ECE No 11*; European Union: Brussels, Belgium, 2010.
23. Bundesministerium der Justiz. *Straßenverkehrs-Ordnung: StVO*; Bundesministerium der Justiz: Berlin, Germany, 2013.
24. European Parliament. *Directive 2014/47/EU on the Technical Roadside Inspection of the Roadworthiness of Commercial Vehicles Circulating in the Union and Repealing Directive 2000/30/EC: 2014/47/EU*; European Parliament, Council of the European Union: Berlin, Germany, 2014.
25. Axel Suijker Textil. Stratos-Cordura®3-Lagen Laminat. Available online: https://www.aktivstoffe.de/stratos-cordura.html (accessed on 29 April 2022).
26. Lanitz-Aviation. Produkte-Lanitz Aviation. Available online: https://www.lanitz-aviation.com/ (accessed on 3 April 2022).
27. Extremtextil, e.K. Double Ripstop Kite-Polyester. Available online: https://www.extremtextil.de/double-ripstop-kite-polyester-high-tenacity-pu-beschichtet-53g-qm.html (accessed on 29 April 2022).
28. Mehler Texnologies. Scrim Compact-Mehler Texnologies. Available online: https://www.mehler-texnologies.com/produkte-archive/scrim-compact/ (accessed on 29 April 2022).
29. Braun, F.; Edel, F.; Ardilio, A. Enhancing Driver's Experience through Emotion Sensitive Lighting Interaction. In Proceedings of the International Conference on Usability and User Experience 2022, New York, NY, USA, 24–28 July 2022.
30. *DIN EN 12195-1:2021-01*; Load Restraining on Road Vehicles-Safety—Part 1: Calculation of Securing Forces. European Parliament, Council of the European Union: Berlin, Germany, 2014.
31. Mitschke, M. *Dynamik der Kraftfahrzeuge*, 5th ed.; Springer Vieweg: Wiesbaden, Germany, 2014.
32. Deutsche Post DHL Group. Paket | DHL: Prices and Products. Available online: https://www.dhl.de/en/privatkunden/pakete-versenden/deutschlandweit-versenden/paket.html (accessed on 26 April 2022).

World Electric Vehicle Journal

MDPI

Communication

Performance Comparison of Si IGBT and SiC MOSFET Power Module Driving IPMSM or IM under WLTC

Hirokatsu Umegami *, Toshikazu Harada and Ken Nakahara

Research & Development Center, ROHM Co., Ltd., Kyoto 615-8585, Japan
* Correspondence: hirokatsu.umegami@dsn.rohm.co.jp

Abstract: The cumulative inverter losses and power consumption of a silicon insulated gate bipolar transistor (Si IGBT) and three types of silicon carbide metal-oxide-semiconductor field-effect transistors (SiC MOSFETs) were evaluated on an electric motor test bench under a worldwide harmonized light vehicles test cycle (WLTC). SiC MOSFETs showed higher performance than Si IGBT regardless of the motor type and test vehicles. In the case of driving an interior permanent magnet synchronous motor (IPMSM), the latest 4th generation SiC MOSFET (SiC-4G) in ROHM has the lowest inverter loss and energy consumption compared with the other generations. In the case of driving an induction motor (IM), on the other hand, the 2nd generation SiC MOSFET (SiC-2G) in ROHM has the best energy consumption despite the fact that the inverter losses of SiC-2G are slightly larger than the loss of SiC-4G. The latest or later generation power device does not necessarily contribute to better performance in a total system by simply replacing early power devices.

Keywords: inverter; powertrain; efficiency; energy consumption; EV (electric vehicle)

Citation: Umegami, H.; Harada, T.; Nakahara, K. Performance Comparison of Si IGBT and SiC MOSFET Power Module Driving IPMSM or IM under WLTC. *World Electr. Veh. J.* **2023**, *14*, 112. https://doi.org/10.3390/wevj14040112

Academic Editor: Joeri Van Mierlo

Received: 29 March 2023
Revised: 13 April 2023
Accepted: 14 April 2023
Published: 17 April 2023

1. Introduction

Fossil fuel-powered vehicles cannot stop emitting green-house-gas, particle matter, etc. wherever they go. Vehicle electrification is one of the worldwide trends to reduce air pollution, energy consumption, and the environmental load. Automotive electrification can reduce fossil fuel dependence and the exhaust emission of passenger cars and road freight vehicles [1–3].

Electric vehicles (EVs) can convert energy into driving force more efficiently than a combustion engine. However, EVs can only run a much shorter distance than petrol cars because the energy density of a battery is smaller than fuel. In order to extend driving distance, it is necessary to improve electricity conversion efficiency from a battery to a motor. xEV, referring to all kinds of EVs, has several electric power components, e.g., a motor, a traction inverter, a dc–dc converter, and a battery charger. Semiconductors used in power conversion circuits are silicon metal oxide semiconductor field effect transistors (Si MOSFETs), silicon insulated gate bipolar transistors (Si IGBTs), and silicon carbide metal oxide semiconductor field effect transistors (SiC MOSFETs). Si IGBTs are mainly used for xEV's traction inverters because of their high withstand voltage, low loss in the high current range, and their continuing performance evolution. Meanwhile, SiC MOSFETs are steadily catching up to Si IGBTs in terms of performance and are increasingly adopted in commercial traction inverters. The SiC MOSFET has excellent characteristics: low transient loss, low conduction loss, high-speed operation, and high-temperature operation. These features are effective in terms of energy saving and achieving a reduction in size and weight.

Technical innovation and improvement often occur on an individual technology basis. In the power device field, new material is selected based on the performance of physical properties for a next generation power device [4–7]. Device structures are devised to unlock the material potential for closing or exceeding the ideal performance [4,8–11]. However, there are not too many reports of actual use in a power converter or a system except for simple circuits to evaluate the tolerance and reliability of materials and device structures [12–15]. In the power conversion field, there reports of, e.g., increasing the power density and downsizing the total volume of a power converter [16–18], improving circuit configurations or control methods [19,20], analyzing losses of each electronic components in a circuit [21,22], etc. The evaluations focus on only a power converter, although there have been reports comparing performance among several power devices [23–27]. In the xEV system field close to electricity, the main contents are the loss distribution and junction temperature of power devices not including mechanical parts [28–31]. On the other hand, in the field close to the system or machines, the main topic is far from electricity, such as the difference between the energy consumption of standardized drive cycle and real use [32], comparison among commercial cars [33], and motor optimization and analysis [34]. Few reports evaluating the transmission of electricity to machines compare performances with changing drive cycles or motors [35,36] but not power devices.

ROHM released a commercial planar, trench, and improved-trench SiC MOSFET [37–40]. The structure arrangement certainly improves the static characteristics of conduction resistance, switching loss, short-circuit withstand time, etc. It is, however, not certain whether a system that combines individually optimized devices will perform as intended, because complex systems cannot be built up without mutual interactions. We are interested in how the difference in the types of power devices influences the performance of a power converter or a total system.

In this paper, a type of Si IGBTs and three types of ROHM SiC MOSFETs are applied to a three-phase, two-level inverter assuming application to EVs. An interior permanent magnet synchronous motor (IPMSM) and an induction motor (IM) are employed as a traction motor in order to verify whether there is a difference depending on the motor type. An electric motor test bench gives a traction motor road load according to vehicle specifications. The evaluation method mostly follows the worldwide harmonized light duty driving test procedure (WLTP). Performance indexes are integrated inverter loss, energy consumption, and energy ration over a driving test cycle.

2. Materials and Methods

2.1. System Configuration Outline

Figure 1 shows the overall experimental system configuration for evaluating power modules. The device under test (DUT) includes components of a traction inverter and 4 types of power module, namely Infineon Si IGBT (FF450R12ME4), Rohm 2nd generation planar SiC MOSFET (SiC-2G: BSM400D12P2G003), Rohm 3rd generation trench SiC MOSFET (SiC-3G: BSM400D12P2G002), and Rohm 4th generation improved trench SiC MOSFET (SiC-4G: the same module packaging as the others). The power modules are selected under the condition of the same module packaging, voltage rating, and current capacity. The specifications of the traction motors of IPMSM and IM are the same as the dynamometer of the motor test bench: 12,000 rpm/300 Nm/100 kW. The dynamometer imitates the preset road load. The controller (PE-Expert4 of Myway Plus) generates a gate drive signal following the worldwide harmonized light vehicles test cycle (WLTC). There are two-line liquid cooling systems for heat dissipation of the motor and the power modules. The coolant is an ethylene glycol-based water solution.

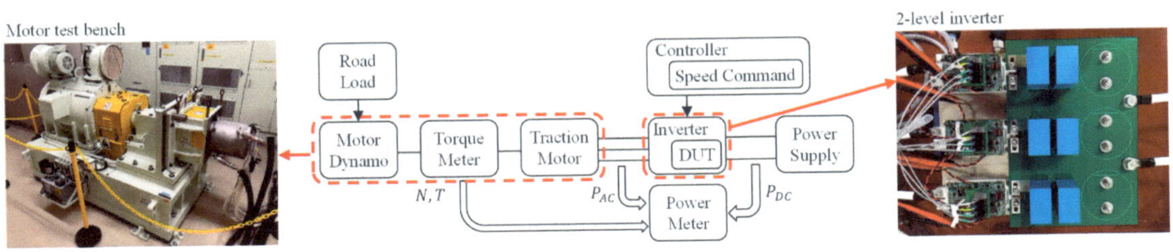

Figure 1. System configuration diagram.

Input and output electric power of the inverter are measured and integrated by the power meter (PW6001 of Hioki) with the current sensors (CT6876 of Hioki) and motor power is calculated with number of rotations and torque measured by the torque meter (T40B of HBM). Table 1 presents the equipment details.

Table 1. Equipment detail.

	Maker	Model	Specifications
Power Supply	SINFONIA TECHNOLOGY	-	850 V/500 A/100 kW
DUT	INFINEON	Si IGBT FF450R12ME4 [41]	1200 V/450 A
	ROHM	SiC-2G BSM400D12P2G [42] SiC-3G BSM400D12P3G [43] SiC-4G prototype	1200 V/400 A
Torque Meter	HBM	T40B	20,000 rpm/500 Nm
Traction Motor	MOTION SYSTEM TECH	IPMSM IM	850 V/500 A/100kW 12,000 rpm/300 Nm/100 kW
Dynamometer	SINFONIA TECHNOLOGY	-	12,000 rpm/300 Nm/100 kW
Controller	MYWAY PLUS	PE-EXPERT4	-
Power Meter	HIOKI	PW6001	w/CT6876

2.2. Experimental Conditions

The evaluation cycle shown in Figure 2 illustrates the pre-set reference vehicle speed to control motor speed by the controller. This is one dynamic segment of the WLTC Class 3b Shortened Type 1 test that excludes the extra high-speed phase in Japan. The original Shortened Type 1 test has two dynamic segments and two continuous speed segments [44]. We did not conduct a reproduction of the battery profile, and battery voltage drop depended on current in this study for simplifying the following analyses.

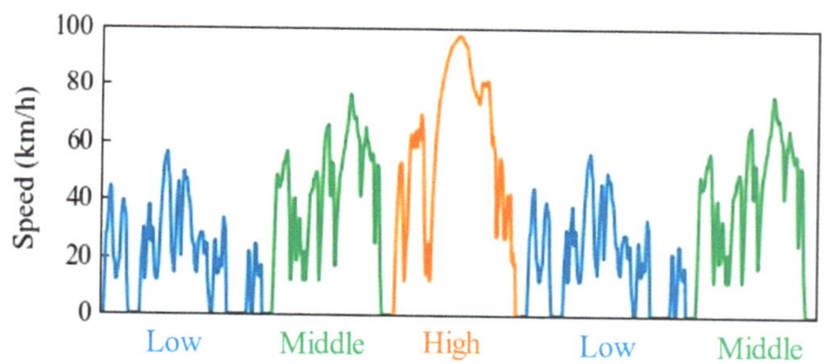

Figure 2. Evaluation cycle of one dynamic segment in WLTC Class 3b in Japan.

The motor dynamo determines the road load as follows:

$$T_{R/L} = J \cdot \dot{\omega} + \left(0.5 C_d A \rho v^2 + \mu m g\right) \cdot r/G \quad (1)$$

where $T_{R/L}$ is the road load torque; J is the vehicle inertia automatically calculated with set vehicle parameters in the motor dynamo system; $\dot{\omega}$ is the angular acceleration; C_d is the aerodynamic drag coefficient; A is the frontal area; ρ is the air density; v is the vehicle speed; μ is the rolling resistance coefficient; m is the vehicle mass; g is the gravity constant; r is the tire radius; G is the gear ratio.

We choose two vehicles for comparative evaluation. One is the 2017 Nissan Leaf G which is one of the major EVs in Japan. The other is the 2022 BMW i4 eDrive 40 whose required current is the maximum in other vehicles for which we set available parameters. Table 2 presents vehicle parameters for calculating road load [45].

Table 2. Parameters of test vehicles.

Parameter	Symbol	Unit	2017 Nissan Leaf G	2022 BMW i4 eDrive 40
Tire Radius	r	m	0.323	0.356
Gear Ratio	G	-	8.193	8.774
Aerodynamic Drag Coefficient	C_d	-	0.28	0.24
Frontal Area	A	m^2	2.48	2.41
Vehicle Mass	m	kg	1646	2251
Air Density	ρ	kg/m^3	1.189	
Rolling Resistance Coefficient	μ	-	0.011	
Gravity Constant	g	m/s^2	9.8	

In this paper, we set the inverter input voltage as 800 V for high voltage use evaluation although the battery voltage of the original vehicles is around 400 V. The gate resistance values of each type of power module are decided such that the peak voltage of the switching surge is approximately 1100 V when the test current is approximately 800 A in the double pulse test. The on-/off- gate resistances of Si IGBT, SiC-2G, SiC-3G, and SiC-4G are 0.0/6.8 Ω, 1.2/2.2 Ω, 1.2/2.2 Ω, and 2.7/6.8 Ω. The gate driver circuit (BSMGD2G17D24-EVK001) can output the recommended gate voltage for the inverter use of SiC power modules, while negative voltage for Si IGBT is limited to −4 V. The on-/off-gate voltages of Si IGBT, SiC-2G, SiC-3G, and SiC-4G are 15/−4 V, 18/−4 V, 18/−2 V, and 18/−2 V. The switching frequency and the dead time of the inverter are set 10 kHz and 2 us. The list of parameters is shown in Table 3.

Table 3. Parameters of driven DUT.

Parameter	Unit	IGBT	SiC-2G	SiC-3G	SiC-4G
Inverter Input Voltage	V	800			
Inverter Input Capacitor	μF	960 (320 μF × 3: 947C321K122CDMS) 150 (25 μF × 6: B32778G1256K000)			
Snubber Capacitor	nF	4050 (1350 nF × 3: EVSM1D72J2-142H16)			
Discharge Resistance	MΩ	0.73 (2.2 MΩ × 3-parallel)			
Gate Resistance (ON/OFF)	Ω	0.0/6.8	1.2/2.2	1.2/2.2	2.7/6.8
Gate Voltage (ON/OFF)	V	15/−4	18/−4	18/−2	18/−2
Switching Frequency	kHz	10			
Dead Time	us	2			

2.3. Evaluation Index

The original energy consumption requires complex calculation procedures [41]. However, in this paper, we employed a simplified calculation as follows:

$$EC = EC_{DC}/D \tag{2}$$

where EC is the energy consumption per 100 km; EC_{DC} is the total integrated inverter input energy; D is the total driving distance standardized by 100 km.

$$EC_{DC} = \sum P_{DC} \cdot \Delta\tau_1 \tag{3}$$

where P_{DC} is the inverter input power; $\Delta\tau_1$ is 50 ms (refresh rate is 20 Hz) and is set by the accumulation function of the power meter, PW6001.

$$D = 2\pi r \cdot N_{total} / (4096 \cdot G) \tag{4}$$

where N_{total} is the total number of the a/b pulses converted from a resolver of the traction motor; G is the gear ratio.

The integrated inverter loss energy E_{loss_inv} is calculated as follows:

$$E_{loss\_inv} = \sum (P_{DC} - P_{AC}) \cdot \Delta\tau_1 \tag{5}$$

where P_{AC} is inverter output power.

The integrated motor loss energy E_{loss_mtr} is calculated as follows:

$$E_{loss\_mtr} = \sum P_{AC} \cdot \Delta\tau_1 - E_{trc} \tag{6}$$

where E_{trc} is the traction energy.

The traction energy E_{trc} is calculated as follows:

$$E_{trc} = \sum P_{MC} \cdot \Delta\tau_2 \tag{7}$$

where P_{MC} is the motor output; $\Delta\tau_2$ is approximately 200 milliseconds (refresh rate is 5 Hz) and is decided by the data accumulation speed between a PC and other measurement equipment.

The motor output P_{MC} is calculated as follows:

$$P_{MC} = 2\pi \cdot T \cdot N / 60 \tag{8}$$

where T is the torque and N is the number of rotations. Both parameters are measured using the torque meter.

3. Experimental Results and Discussion

In Figure 3, total energy consumption overall the evaluation cycle is compared in a combination of three categories, a type of power modules, traction motors and target vehicles. Each stacked bar of EC is color-coded by three categories, traction energy, motor loss energy, and inverter loss energy. Each data label on the bars means traction energy consumption (EC_{trc}) calculated with only traction energy, mechanical energy consumption (EC_{mc}) with traction energy and motor loss energy, and total energy consumption (EC_{ttl}) with energy of all 3 categories. EC_{ttl} and EC_{mc} are important values because the accuracy is guaranteed by the power meter. On the other hand, EC_{trc} is just a reference value because the torque meter does not guarantee the accuracy of transient response and the refresh rate of communication speed is slow. The ratio of the traction energy and the motor loss energy is not so accurate.

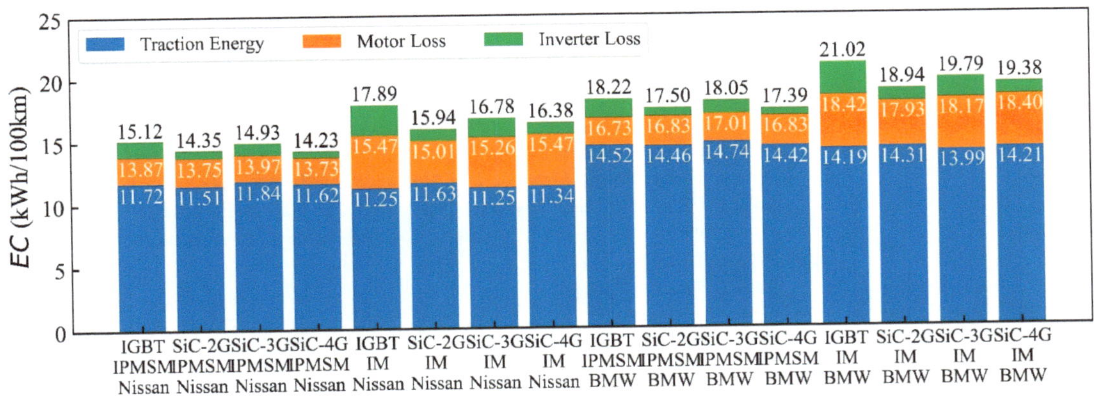

Figure 3. Energy consumption of each experimental case.

In the case of driving IPMSM, the good order of EC_{ttl} in the types of power modules is SiC-4G < SiC-2G < SiC-3G < Si IGBT regardless of the target vehicles. The order of EC_{mc} is SiC-4G ≈ SiC-2G < Si IGBT < SiC-3G in Nissan Leaf G and Si IGBT < SiC-4G ≈ SiC-2G < SiC-3G in BMW i4 eDrive 40. SiC MOSFETs much improve the total energy consumption, whereas motor loss can be increased since the traction energy should be uniquely determined by vehicle parameters.

In the case of driving IM, the order of EC_{ttl} is SiC-2G < SiC-4G < SiC-3G < Si IGBT and the order of EC_{mc} is SiC-2G < SiC-3G < SiC-4G ≈ Si IGBT regardless of the vehicles. All SiC MOSFETs demonstrated better performance than Si IGBT. SiC-2G of planar structure realizes the best energy saving although SiC-3G and -4G of the trench have better standalone device characteristics.

Pure inverter losses are shown in Figure 4. The lowest order of the loss is SiC-4G < SiC-2G < SiC-3G < Si IGBT regardless of the motor types and the target vehicles. The difference in material properties is very clear. The remarkable point is that SiC-2G of the planar can keep the loss lower than SiC-3G of the trench in power converter level.

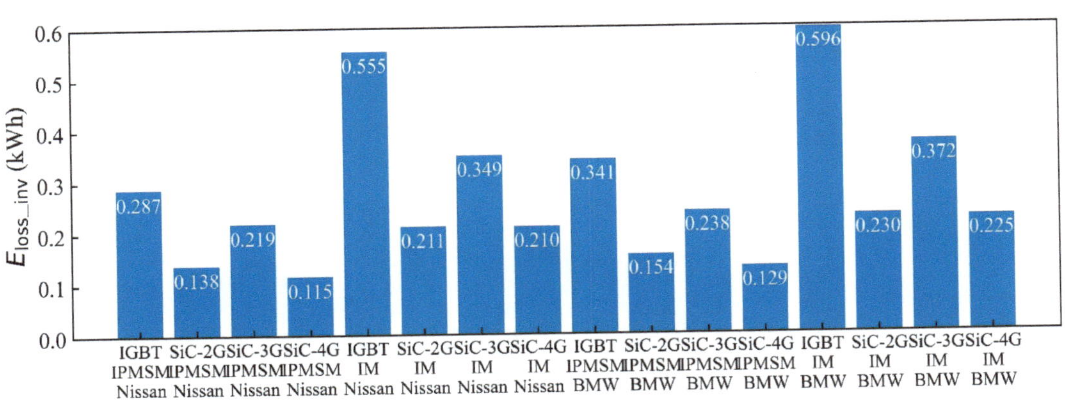

Figure 4. Inverter loss comparison between experimental cases.

Figure 5 shows the relationship of EC and power train loss of E_{loss_mtr} and E_{loss_inv}. The results in Figure 5a are calculated with the data of the overall evaluation cycle of one dynamic segment. Figure 5b–d show the data of the low, middle, and high phases of the cycle. The dashed line shows the Nissan Leaf G and the dot-dashed line the BMW i4 eDrive 40. The two lines are drawn by each simple average of all data. The slop of those lines signifies the impact of the powertrain loss on *EC*. In the low phase shown in Figure 5b, a short drive distance makes the slope of each line steep, and therefore the loss reduction effect is easily apparent as an improvement in energy consumption. On the other hand, in the high phase shown in Figure 5d, the loss has only a limited effect on energy consumption since the slope of each line is loose. The intercept of each line represents experimentally obtained ideal energy consumption ignoring motor and inverter loss. The intercept is approximately 11.56 kWh/100 km for the Nissan Leaf G, and approximately 14.41 kWh/100 km for the BMW i4 eDrive 40, as shown in Figure 5a. Therefore, the energy increase in traction purely due to the difference between the target vehicles is approximately 25%.

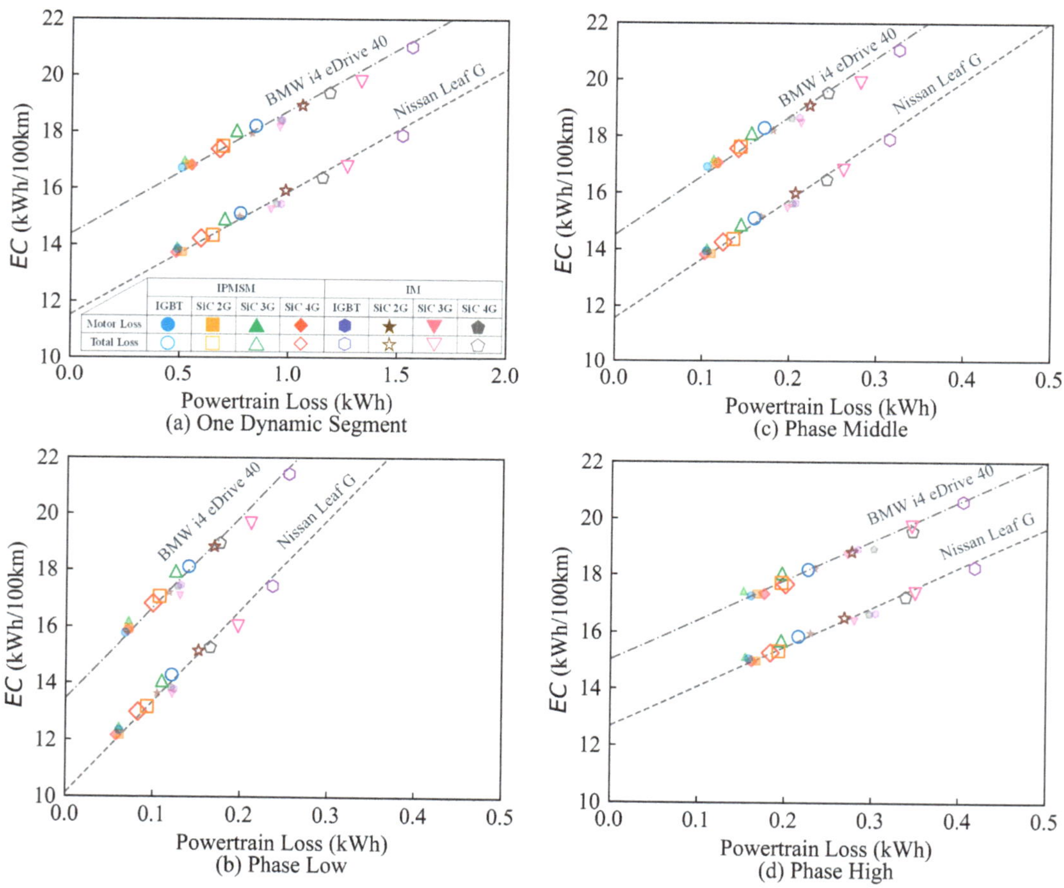

Figure 5. Energy consumption vs. powertrain loss of each phase.

The filled markers in Figure 3 represent motor loss. The motor loss increases in the order of low, middle, and high phase as drive distance increases. The markers are roughly overlapped if grouped by the target vehicles and the motor type except SiC-2G of IM. The open markers mean the total value of motor loss and inverter loss. The distance between the filled and the open marker of the same color and shape shows the amount of inverter loss and the amount of improvable energy consumption. Figure 5 helps to visually understand the ratio and the amount of motor and inverter loss, and the effect of each loss on energy consumption.

Figure 6 shows percentages of traction energy, motor loss energy, and inverter loss energy in each EC_{ttl}. The inverter loss ratio of SiC-4G is the lowest regardless of the motor types and the target vehicles.

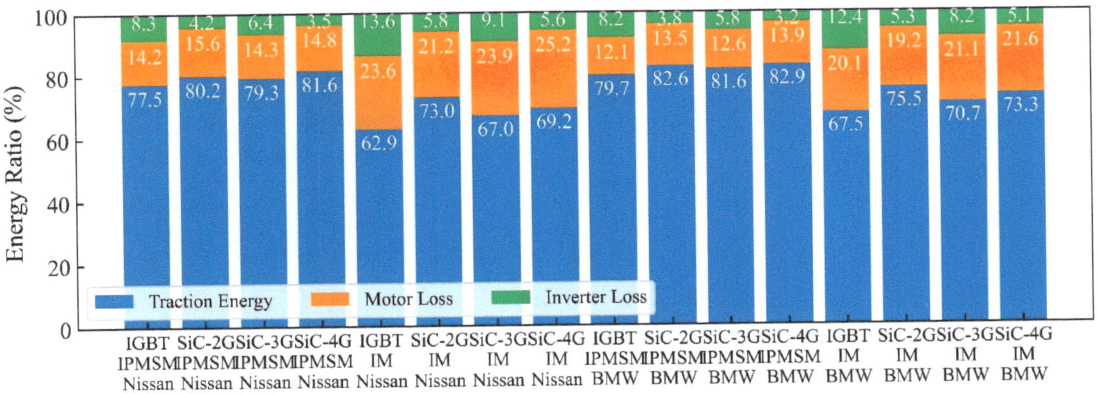

Figure 6. Energy consumption ratio of each experimental case.

4. Conclusions

In this paper, we investigated the effects of device materials and structures on the power converter and the EV system by just replacing power modules. Two types of vehicle parameter sets did not influence the trend of results except for increasing the required overall energy for the evaluation cycle of one dynamic segment of the Japanese WLTC. However, types of power modules and motors affected inverter loss and mechanical energy converted to traction. The good order in inverter loss and energy consumption was not equal to the individual performance of evaluated power modules, i.e., in the order of Si IGBT, SiC-2G (planar SiC MOSFET), SiC-3G (trench SiC MOSFET), and SiC-4G (improved trench SiC MOSFET). In addition, the type of power module can influence the motor loss.

SiC-2G demonstrated good balance in terms of inverter loss, motor loss, and energy consumption. SiC-4G had the lowest inverter loss and reduced energy consumption to 3% of the entire energy in the case of driving IPMSM. On the other hand, the mechanical energy of SiC-4G is the same as Si IGBT in the case of driving IM. Our future work aims to find what device parameters influence motor loss, mechanical energy, and energy consumption.

Author Contributions: Conceptualization, H.U. and K.N.; methodology, H.U.; software, H.U.; validation, H.U.; formal analysis, H.U.; investigation, H.U. and T.H.; resources, K.N.; data curation, H.U. and T.H.; writing—original draft, H.U.; visualization, H.U.; supervision, K.N.; project administration, K.N.; funding acquisition, K.N. All authors have read and agreed to the published version of the manuscript.

Funding: This research received no external funding.

Data Availability Statement: Not applicable.

Conflicts of Interest: The authors declare no conflict of interest.

References

1. Kazemzadeh, E.; Koengkan, M.; Fuinhas, A.J.; Mejdalani, A. Impact of Electrification of Road Transport on Premature Deaths from Outdoor Air Pollution: A Macroeconomic Evidence from 29 European Countries. *World Electr. Veh. J.* **2022**, *13*, 155. [CrossRef]
2. Kromer, A.M.; Bandivadekar, A.; Evans, C. Long-term greenhouse gas emission and petroleum reduction goals: Evolutionary pathways for the light-duty vehicle sector. *Energy* **2010**, *35*, 387–397. [CrossRef]
3. Hasselwander, S.; Galich, A.; Nieland, S. Impact of Climate Change on the Energy Consumption of Passenger Car Vehicles. *World Electr. Veh. J.* **2022**, *13*, 146.

4. Elasser, A.; Chow, P.T. Silicon carbide benefits and advantages for power electronics circuits and systems. *Proc. IEEE* **2022**, *90*, 969–986. [CrossRef]
5. Spaziani, L.; Lu, L. Silicon, GaN and SiC: There's room for all: An application space overview of device considerations. In Proceedings of the 2018 IEEE 30th International Symposium on Power Semiconductor Devices and ICs (ISPSD), Chicago, IL, USA, 13–17 May 2018.
6. Agarwal, K.A. An overview of SiC power devices. In Proceedings of the 2010 International Conference on Power, Control and Embedded Systems, Allahabad, India, 29 November–1 December 2010.
7. Kaminski, N.; Hilt, O. SiC and GaN Devices—Competition or Coexistence? In Proceedings of the 2012 7th International Conference on Integrated Power Electronics Systems (CIPS), Nuremberg, Germany, 6–8 March 2012.
8. Lee, S.J.; Chun, H.D.; Park, H.J.; Jung, K.Y.; Kang, G.E.; Sung, Y.M. Design of a novel SiC MOSFET structure for EV inverter efficiency improvement. In Proceedings of the 2014 IEEE 26th International Symposium on Power Semiconductor Devices & IC's (ISPSD), Waikoloa, HI, USA, 15–19 June 2014.
9. Zhang, L.; Dai, T.; Gammon, P.; Shah, V.; Mawby, P.; Antoniou, M. Comparison of a 3.3 kV SiC Hybrid-Channel Trench MOSFET and a Planar MOSFET. In Proceedings of the 2022 IEEE Workshop on Wide Bandgap Power Devices and Applications in Europe (WiPDA Europe), Coventry, UK, 18–20 September 2022.
10. Chaturvedi, M.; Dimitrijev, S.; Haasmann, D.; Moghadam, A.H.; Pande, P.; Jadli, U. Comparison of Commercial Planar and Trench SiC MOSFETs by Electrical Characterization of Performance-Degrading Near-Interface Traps. *IEEE Trans. Electron Devices* **2022**, *69*, 6225–6230. [CrossRef]
11. Tian, K.; Hallen, A.; Qi, J.; Nawaz, M.; Ma, S.; Wang, M.; Guo, S.; Elg, K. Comprehensive Characterization of the 4H-SiC Planar and Trench Gate MOSFETs From Cryogenic to High Temperature. *IEEE Trans. Electron Devices* **2019**, *66*, 4279–4286. [CrossRef]
12. Fukunaga, S.; Castellazzi, A.; Funaki, T. Development of reliable multi-chip power modules with parallel planar- and trench-gate SiC MOSFETs. In Proceedings of the 2022 IEEE 34th International Symposium on Power Semiconductor Devices and ICs (ISPSD), Vancouver, BC, Canada, 22–25 May 2022.
13. Bashar, E.; Agbo, N.; Wu, R.; Mendy, S.; Jahdi, S.; Jennings, M.; Withey, A.; Evans, S.; Davies, G.; Demitrove, J.; et al. A Review of Short Circuit Performance in 650 V Power Devices: SiC MOSFETs, Silicon Super-junction MOSFETs, SiC Cascode JFETs, Silicon MOSFETs and Silicon IGBTs. In Proceedings of the PCIM Europe 2022; International Exhibition and Conference for Power Electronics, Intelligent Motion, Renewable Energy and Energy Management, Nuremberg, Germany, 10–12 May 2022.
14. Cao, L.; Guo, Q.; Sheng, K. Comparative Evaluation of the Short circuit Capability of SiC Planar and Trench Power MOSFET. In Proceedings of the 2018 IEEE 2nd International Electrical and Energy Conference (CIEEC), Beijing, China, 4–6 November 2018.
15. Huang, W.; Deng, X.; Li, X.; Wen, Y.; Li, X.; Li, Z.; Zhang, B. Investigation of Surge Current Reliability of 1200 V Planar and Trench SiC MOSFET. In Proceedings of the 2020 IEEE 15th International Conference on Solid-State & Integrated Circuit Technology (ICSICT), Kunming, China, 3–6 November 2020.
16. Murakami, Y.; Tajima, Y.; Tanimoto, S. Air-cooled full-SiC high power density inverter unit. In Proceedings of the 2013 World Electric Vehicle Symposium and Exhibition (EVS27), Barcelona, Spain, 17–20 November 2013.
17. Yamaguchi, K. Design and evaluation of SiC-based high power density inverter, 70 kW/liter, 50 kW/kg. In Proceedings of the 2018 IEEE Applied Power Electronics Conference and Exposition (APEC), San Antonio, TX, USA, 4–8 March 2018.
18. Shi YWang, L.; Xie, R.; Li, H. Design and implementation of a 100 kW SiC filter-less PV inverter with 5 kW/kg power density and 99.2% CEC efficiency. In Proceedings of the 2018 IEEE Applied Power Electronics Conference and Exposition (APEC), San Antonio, TX, USA, 4–8 March 2018.
19. Sakasegawa, E.; Chishiki, R.; Sedutsu, R.; Soeda, T.; Haga, H.; Kennel, M.R. Comparison of Interleaved Boost Converter and Two-Phase Boost Converter Characteristics for Three-Level Inverters. *World Electr. Veh. J.* **2023**, *14*, 7. [CrossRef]
20. Deng, W.; Zhao, Y.; Wu, J. Energy Efficiency Improvement via Bus Voltage Control of Inverter for Electric Vehicles. *IEEE Trans. Veh. Technol.* **2017**, *66*, 1063–1073. [CrossRef]
21. Mazumder, S.K.; Acharya, K.; Jedraszczak, P. High-temperature all-SiC bidirectional DC/DC converter for plug-in-hybrid vehicle (PHEV). In Proceedings of the 2008 34th Annual Conference of IEEE Industrial Electronics, Orlando, FL, USA, 10–13 November 2008.
22. Everts, J.; Das, J.; Keybus DV, J.; Genoe, J.; Germain, M.; Driesen, J. A high-efficiency, high-frequency boost converter using enhancement mode GaN DHFETs on silicon. In Proceedings of the 2010 IEEE Energy Conversion Congress and Exposition, Atlanta, GA, USA, 12–16 September 2010.
23. Shang, F.; Arribas, P.A.; Krishnamurthy, M. A comprehensive evaluation of SiC devices in traction applications. In Proceedings of the 2014 IEEE Transportation Electrification Conference and Expo (ITEC), Dearborn, MI, USA, 15–18 June 2014.
24. Ozdemir, S.; Acar, F.; Selamogullari, S.U. Comparison of silicon carbide MOSFET and IGBT based electric vehicle traction inverters. In Proceedings of the 2015 International Conference on Electrical Engineering and Informatics (ICEEI), Denpasar, Indonesia, 10–11 August 2015.
25. Zhao, T.; Wang, J.; Huang, Q.A.; Agrwal, A. Comparisons of SiC MOSFET and Si IGBT Based Motor Drive Systems. In Proceedings of the 2007 IEEE Industry Applications Annual Meeting, New Orleans, LA, USA, 23–27 September 2007.
26. Wang, G.; Wang, F.; Magai, G.; Lei, Y.; Huang, A.; Das, M. Performance comparison of 1200 V 100A SiC MOSFET and 1200 V 100A silicon IGBT. In Proceedings of the 2013 IEEE Energy Conversion Congress and Exposition, Denver, CO, USA, 15–19 September 2013.
27. Zhang, L.; Yuan, X.; Wu, X.; Shi, C.; Zhang, J.; Zhang, Y. Performance Evaluation of High-Power SiC MOSFET Modules in Comparison to Si IGBT Modules. *IEEE Trans. Power Electron.* **2019**, *34*, 1181–1196. [CrossRef]

28. Allca-Pekarovic, A.; Kollmeyer, P.J.; Mahvelatishamsabadi, P.; Mirfakhrai, T.; Naghshtabrizi, P.; Emadi, A. Comparison of IGBT and SiC Inverter Loss for 400 V and 800 V DC Bus Electric Vehicle Drivetrains. In Proceedings of the 2020 IEEE Energy Conversion Congress and Exposition (ECCE), Detroit, MI, USA, 11–15 October 2020.
29. Menon, R.; Azeez, A.N.; Kadam, H.A.; Williamson, S.S. Energy loss analysis of traction inverter drive for different PWM techniques and drive cycles. In Proceedings of the 2018 IEEE International Conference on Industrial Electronics for Sustainable Energy Systems (IESES), Hamilton, New Zealand, 31 January–2 February 2018.
30. Amirpour, S.; Thiringer, T.; Hagstedt, D. Mission-Profile-Based Lifetime Study for SiC/IGBT Modules in a Propulsion Inverter. In Proceedings of the 2021 IEEE 19th International Power Electronics and Motion Control Conference (PEMC), Gliwice, Poland, 25–29 April 2021.
31. Nisch, A.; Heller, M.; Wondrak, W.; Bucher, A.; Hasenohr, C.; Kefer, K.; Lunz, B.; Pawellek, A.; Smit, A.; Gartner, M.; et al. Simulation and Measurement-Based Analysis of Efficiency Improvement of SiC MOSFETs in a Series-Production Ready 300 kW/400 V Automotive Traction Inverter. In Proceedings of the 2020 22nd European Conference on Power Electronics and Applications (EPE'20 ECCE Europe), Lyon, France, 7–11 September 2020.
32. Milligan, R.; Muneer, T.; Smith, I. A comparative range approach using the Real World Drive Cycles and the Battery Electric Vehicle. In Proceedings of the 2015 IEEE International Transportation Electrification Conference (ITEC), Chennai, India, 27–29 August 2015.
33. Charadsuksawat, A.; Laoonual, Y.; Chollacoop, N. Comparative Study of Hybrid Electric Vehicle and Conventional Vehicle Under New European Driving Cycle and Bangkok Driving Cycle. In Proceedings of the 2018 IEEE Transportation Electrification Conference and Expo, Asia-Pacific (ITEC Asia-Pacific), Bangkok, Thailand, 6–9 June 2018.
34. Chen, L.; Wang, J.; Lazari, P.; Chen, X. Optimizations of a permanent magnet machine targeting different driving cycles for electric vehicles. In Proceedings of the 2013 International Electric Machines & Drives Conference, Chicago, IL, USA, 12–15 May 2013.
35. Gunther, S.; Ulbrich, S.; Hofmann, W. Driving cycle-based design optimization of interior permanent magnet synchronous motor drives for electric vehicle application. In Proceedings of the 2014 International Symposium on Power Electronics, Electrical Drives, Automation and Motion, Ischia, Italy, 18–20 June 2014.
36. Patzak, A.; Bachheibl, F.; Baumgardt, A.; Dajaku, G.; Moros, O.; Gerling, D. Driving range evaluation of a multi-phase drive for low voltage high power electric vehicles. In Proceedings of the 2015 International Conference on Sustainable Mobility Applications, Renewables and Technology (SMART), Kuwait, Kuwait, 23–25 November 2015.
37. Trench-Structure SiC-MOSFET and Actual Products. Available online: https://techweb.rohm.com/product/power-device/sic/sic-basic/6574/ (accessed on 1 February 2023).
38. New Products Under Development 3rd Gen, SCI. MOSFET. Available online: https://www.rohm.com/documents/11303/2871848/3rd-gen-MOSFETs.pdf (accessed on 1 February 2023).
39. New 4th Generation SiC MOSFETs Featuring the Industry's Lowest on Resistance. Available online: https://www.rohm.com/news-detail?news-title=new-4th-gen-sic-mosfets&defaultGroupId=false (accessed on 1 February 2023).
40. Disruptive Technology: ROHM Generation 4, SIC MOSFET. Available online: https://www.techinsights.com/blog/disruptive-technology-rohm-generation-4-sic-mosfet (accessed on 1 February 2023).
41. Infineon IGBT Module FF450R12ME4. Available online: https://www.infineon.com/dgdl/Infineon-FF450R12ME4-DS-v03_01-EN.pdf?fileId=db3a30431a5c32f2011a769c4a506c4b (accessed on 1 February 2023).
42. Rohm Full-SiC Power Module BSM400D12P2G003. Available online: https://fscdn.rohm.com/en/products/databook/datasheet/discrete/sic/power_module/bsm400d12p2g003-e.pdf (accessed on 1 February 2023).
43. Rohm Full-SiC Power Module BSM400D12P3G002. Available online: https://fscdn.rohm.com/en/products/databook/datasheet/discrete/sic/power_module/bsm400d12p3g002-e.pdf (accessed on 1 February 2023).
44. United Nations Global Technical Regulation on Worldwide Harmonized Light Vehicles Test Procedures (ECE/TRANS/180/Add.15/Amend.6). Available online: https://unece.org/sites/default/files/2021-01/ECE-TRANS-180a15am6e.pdf (accessed on 1 February 2023).
45. EVSpecifications Specifications, News and Comparisons. Available online: https://www.evspecifications.com/ (accessed on 1 February 2023).

World Electric Vehicle Journal

Article

A Multi-Disciplinary Approach for the Electrical and Thermal Characterization of Battery Packs—Case Study for an Electric Race Car

Claudio Santarelli [1], Christopher Helbig [1], An Li [2,*], Benoit Honel [2], Thomas Nyhues [3] and Fabian Böhm [3]

1 Siemens Industry Software GmbH, Am Kabellager 9, 51063 Köln, Germany
2 Siemens Digital Industries Software, 19 bd Jules Carteret, 69007 Lyon, France
3 Ecurie Aix-Formula Student Team RWTH Aachen e.V., Templergraben 55, 52062 Aachen, Germany
* Correspondence: an.li@siemens.com

Abstract: A novel, multi-disciplinary approach is presented where experiments, system simulation and Computational Fluid Dynamics are combined for the electrical and thermal characterization of an air-cooled battery pack. As a case study, a Formula Student race car is considered and the procedure proposed consists of three steps: (1) experimental characterization of the battery cells under several thermal conditions; (2) thermal and electrical modeling of the battery stack with system simulation; (3) three-dimensional, time-dependent Conjugate Heat Transfer simulation of the whole battery pack to investigate the cooling performance of the chosen design, and to access fundamental quantities of the batteries, such as state of charge, temperature and ohmic heating. Future improvements of the current work are discussed, including the extension to a liquid-cooled design, battery aging consideration and model integration into a full vehicle system model.

Keywords: battery pack modeling; cooling; simulation; testing process; thermal management; cell characterization

Citation: Santarelli, C.; Helbig, C.; Li, A.; Honel, B.; Nyhues, T.; Böhm, F. A Multi-Disciplinary Approach for the Electrical and Thermal Characterization of Battery Packs—Case Study for an Electric Race Car. *World Electr. Veh. J.* **2023**, *14*, 102. https://doi.org/10.3390/wevj14040102

Academic Editor: Joeri Van Mierlo

Received: 23 February 2023
Revised: 22 March 2023
Accepted: 24 March 2023
Published: 10 April 2023

1. Introduction

The battery electric vehicle is one of the most promising solutions to the energy crisis and environmental pollution. Thanks to rapid progress in lithium-ion battery technology in recent years, more and more traditional internal combustion engine vehicles will be replaced by electric vehicles in the near future [1,2]. The transition towards battery electric mobility brings many challenges, of which Battery Thermal Management (BTM) is one of the most important, where several disciplines such as electronics, heat transfer and controls need to be managed together to guarantee a safe and efficient operation of the battery. BTM is important since the battery temperature has a great impact on its performance, lifetime and safety [3]. During operation, the temperature of the battery cells inside the battery pack should be firstly maintained within an optimal operating range (e.g., between 15 and 50 °C [4] or even between 15 and 35 °C [5]) to avoid adverse effects such as accelerated aging and thermal runaway. At low temperature, the battery internal resistance increases significantly which results in reduced energy and power densities [6]. Charging the battery cell at low temperature also increases the risk of lithium plating, which seriously reduces lifespan [7]. When operating at too high temperature, the overheating of the battery cells can cause thermal runaway, which may result in a destruction of the whole battery pack [8]. Secondly, the temperature gradient of the battery cells inside a battery pack should be kept as small as possible (e.g., within 5 °C [3,5]) to avoid performance loss and uneven cell aging. The imbalance caused by uneven cell aging affects not only the performance of the related cells but also that of the whole pack [9].

To efficiently design a battery thermal management system, a complete and detailed battery pack model is needed in order to gain detailed insights into the cell temperature evolution under different operating conditions. The battery pack model includes not only the cell models, but also the models of other battery pack components, such as the electrical connectors, the air or liquid cooling channels, and the pack casing. To represent battery electrical and thermal behavior, two main types of models are available in the literature:

- Electrochemical models, such as the well-known pseudo-two-dimensional (p2D) model [10] and the single particle model [11,12], allow detailed insight into different electrochemical processes inside the battery cell. However, it is not suitable for real-time applications and battery pack level simulation due to a high computational cost [13,14]. In addition, calibration of the electrochemical models is a challenging step which requires considerable experimental effort to measure or estimate the model parameters.
- The empirical equivalent circuit models are widely used in battery applications [13–15]. They are simple and commonly consist of basic electrical elements (e.g., voltage source, resistance, capacitance) or electrochemical impedance (e.g., Warburg impedance or constant phase element). Due to the simplicity in model structure and limited number of parameters, the equivalent circuit models run faster compared to the electrochemical models and hence are more suitable for real-time applications and battery pack level simulation. The equivalent circuit models require, at the same time, less experimental effort to calibrate the model parameters from test data. When they are properly parameterized, the equivalent circuit models can give accurate voltage and temperature estimations for the battery [13,15].

To simulate the battery pack model, the methods employed in the literature can be summarized into three categories: one-dimensional system simulation (or 1D simulation in short), three-dimensional computational fluid dynamics (CFD) simulation (or 3D simulation), and a combined approach using both 1D and 3D simulations. Compared to 3D simulation, 1D simulation is much faster and hence more suitable for real-time applications and aging simulation. Several works have shown system level battery pack simulations based on empirical equivalent circuit models [13,16–18]. Li et al. in [13] proposed a real-time compatible model to simulate the electro-thermal behavior of a battery pack with air cooling. This system level battery pack model has been used in the work of Ponchant et al. [16] for the software-in-the-loop and hardware-in-the-loop tests of the battery management system. Wang et al. in [17] designed a system level model to simulate a liquid cooled battery pack. This model includes not only the cell equivalent circuit model, but also models for other pack elements such as the cooling plates and the flow channels. Ramesh Babu et al. in [18] presented a liquid cooled battery pack model based on a system level approach. The model was used to simulate the average temperature of the battery pack. While 3D simulation is very time consuming, it can be more accurate and provides a higher level of detail compared to 1D simulation [18]. Li et al. in [19] developed a three-dimensional electrochemical-thermal model coupled with a fluid dynamics module to analyze in detail the temperature distribution inside an air-cooled battery pack. Buidin et al. in [4] used steady-state 3D simulations to determine the values of the heat transfer coefficient on the surface of the cylindrical battery cells at different pre-defined air flow and heat generation values. The heat transfer maps obtained via 3D simulations at different pre-defined conditions were used by the 1D simulation model to evaluate the thermal behavior of the battery pack in transient mode. Li et al. in [20] presented a similar work for a liquid-cooled battery pack combining 1D and 3D simulations via the use of heat transfer maps.

In this paper, a novel and holistic approach is presented that leverages experiments, 1D system simulation and 3D CFD simulation for the thermal and electrical characterization of cells and battery packs. Compared to similar works in the literature, the main contribution of our work is threefold. Firstly, this approach describes a complete workflow of the battery pack electro-thermal characterization, including the very first step, which is the cell characterization from experimental data. Commonly, many automotive manufacturers acquire cells from suppliers, but the parameter values to model the cell behavior are usually not provided by the cell suppliers. In our work, a standard workflow is used to calibrate the cell model parameters from experimental data. This calibration workflow can be easily applied to other battery cells. Secondly, our approach employs a co-simulation of the 1D and 3D models using the Functional Mockup Interface (FMI) standard [21]. Co-simulation allows for the evaluation of any operating condition. Hence it avoids the need to generate heat transfer maps for pre-defined operating conditions, as in [4,20]. Finally, serval important issues encountered during the battery pack design are rarely covered in similar works, such as battery aging, the conversion from an air-cooled design to a liquid-cooled design, and the integration of the battery pack model into a complete vehicle model. These aspects can be addressed with our simulation approach and will be discussed in this paper.

The case study chosen to present this concept is an electric Formula Student race car (as described in Section 2) and this approach consists of three steps. First, a single cell is experimentally investigated to identify the parameters of the equivalent circuit model (Section 3). Afterwards, a system simulation is set up to represent part of the battery pack as an ensemble of characterized cells (Section 4). Successively, the battery pack is exported by means of the FMI standard and embedded in a CFD Conjugate Heat Transfer (CHT) simulation, to simulate the cooling of the battery pack (Section 5). In this framework, fundamental quantities of the battery such as the state of charge, the temperature and the ohmic heating can be accessed for optimal dimensioning of the cooling system. Future improvements or extensions of the current work are discussed (Section 6), including the extension to a liquid-cooled design, the consideration of battery aging, and model integration into a full vehicle system model.

2. Case Study

The case study addresses the electric race car eace08 [22] shown in Figure 1a, designed and built by the RWTH Aachen's Formula Student team Ecurie Aix. The eace08 is a 4-wheel-drive car with up to 32 kW traction power per wheel hub motor, with an overall peak power of 80 kW (limited by the competition rules) and a recuperation capability of up to 60 kW. Four IGBT inverters supply the PMSM motors with power. Both systems are water cooled. The drive train is powered by a high-power Li-Ion battery pack with a capacity of roughly 6 kWh.

(a) (b)

Figure 1. Pictures of the eace08 on the Hockenheim Ring (**a**) and of the battery pack (**b**).

The energy is stored in 288 high power Li-ion pouch cells, and 24 of these cells in a 12s2p configuration form one stack, held together by a 3D-printed cage that also contains a secondary controller of the battery management system (BMS). A total of 12 of the described stacks are placed in a carbon-fiber-reinforced plastics container, together with the BMS master controller and periphery devices, such as a fuse, relays, and an insulation monitoring device. This yields an overall battery configuration of 144s2p and a theoretical voltage range of 432–600 V (upper limit defined by the rules). To accomplish the lightweight construction requirements, the battery cells are air cooled. Three fans at the back of the battery container induce a mass flow through the air vents on both side of the monocoque and the effect of the instantaneous car velocity is negligible with respect to the mass flow imposed by the fans. The air streams over the heat sinks are mounted on the cell tabs (Figure 1b) and leave the car at the rear end. The first generation of the air-cooling system as described above was developed as a conservative, oversized proof-of-concept design.

At the formula student competition, the cars compete in various dynamic disciplines, with the endurance race being the most prestigious but also the most challenging: over a track length of 22 km, the cars must drive as quickly and efficiently as possible. Not only the fastest overall time, but also the lowest net energy consumption will be rewarded with points. Efficient energy and thermal management of the battery is therefore necessary to get the most out of the race. The race car is designed to operate closely to the limits of its performance, considering lightweight construction, energy efficiency and thermal design. Since the system must maintain acceptable temperatures for 22 km, the cooling system can be even smaller than a conservative design approach would suggest: an improvement of the first proof-of-concept design is the pivotal motivation of the Batterie Aix project described here.

To optimize the BTM for the endurance race, proper knowledge of the battery's dynamic behavior in terms of thermal and electrical aspects is mandatory: hence, a multi-disciplinary approach for the investigation of the battery cooling was developed by Ecurie Aix in collaboration with Siemens Digital Industries Software, as described in the present document. The main motivations of this project are: (1) to gain deep insight into the behavior of the current battery pack/cooling system; (2) to derive trustworthy models that can be used in the design phase of the next generation's cooling system, e.g., towards alternative cooling strategies and/or improved overall BTM concepts; (3) to gain a better understanding of the electric-thermal properties of the system in different scenarios. The present paper focuses on the first goal of the project and the new approach is described below.

3. Cell Characterization

The battery cell used in the race car is a commercial $LiCoO_2$ pouch cell from MELASTA. The cell specification is summarized in Table 1.

Table 1. Specification of the battery cell.

Items	Specifications
Cathode chemistry	$LiCoO_2$
Nominal capacity	5.6 Ah
Nominal voltage	3.7 V
Lower/Upper cut-off voltage	3.0 V/4.2 V
Maximum charge current	56 A (<2 s)
Maximum continuous discharge current	84 A
Temperature range for both charge and discharge	0–60 °C

In our work, equivalent circuit models as shown in Figure 2 are chosen to model the battery cell behavior because they are simple, fast, suitable for battery pack level simulation, with a good compromise of accuracy/simplicity, and have been implemented in Simcenter Amesim [23], a multi-physical system simulation software from Siemens Digital Industries Software. A detailed description of these models can be found in [13].

The workflow to calibrate the equivalent circuit model parameters from the experimental data is summarized in Figure 2:

- Firstly, two types of current test profile have been designed to characterize the battery cell. The first type is a constant discharge and CC/CV (Constant Current/Constant Voltage) charge test as in Figure 3a. This profile allows verification of the cell capacity via coulomb counting, in order to detect eventual defected cells at an early stage of the experimental tests. It also allows collection of data along with significant cell temperature increase, which will be used to identify thermal parameters such as the cell specific heat (C_p) and the convective heat exchange coefficient (h_{cov}) of the cell surface in the thermal model in Figure 2. The second type is a HPPC (Hybrid Pulse Power Characterization), like the pulse test current profile from [13]. The pulse profile is adapted to the battery cell according to its specification in Table 1. Figure 3b–d show the detail of the pulse test profile. This pulse test profile tests the battery at different states of charge and different current amplitudes which allows identification of the parameters of the battery equivalent circuit models. At each tested state of charge, as shown in Figure 3d, the voltage and current data of the short-duration charge and discharge pulses is used to identify the instantaneous ohmic resistances (R_{ohm}) at different current amplitudes. The data of the long-duration charge and discharge pulses are used to identify the RC circuits of the electrical model (R_{diff} and C_{diff} in Figure 2). After each long-duration pulse, a rest period is imposed to stabilize the cell voltage. The stabilized cell voltage at the end of the rest period is considered to be the open circuit voltage (OCV) at the corresponding state of charge. More detail of the model parameter identification can be found in [13,23].

- Secondly, the two types of test profiles were applied experimentally to the battery cells at the Institute for Power Electronics and Electrical Drives (ISEA) of the RWTH Aachen. The battery cells were placed in a climate chamber and were connected to a Digatron battery test bench. Two battery cells were tested, as shown in Figure 2, to ensure the repeatability of the test results. The constant discharge and CC/CV charge test profile was applied to the cell at 25 °C of ambient temperature to evaluate the cell's real capacity. The pulse test profile was applied to the cell at 25 and 55 °C of ambient temperature. The tested temperatures cover the operating ambient temperature range of the battery pack for the competition in summer. For all the tests, the cell current, voltage, and temperature were measured. Figure 3 shows the experimental test results. The minor temperature variation of the pulse test at 25 °C in Figure 3b is neglectable for the electrical parameter identification. In this case, an average temperature is considered as 27 °C for this pulse test.
- Finally, the Battery Electro-thermal Identification Tool in Simcenter Amesim [23] was used to identify the electrical and thermal parameters of the equivalent circuit models from the experimental test data. The voltage and temperature estimation of the identified battery models are also shown in Figure 3.

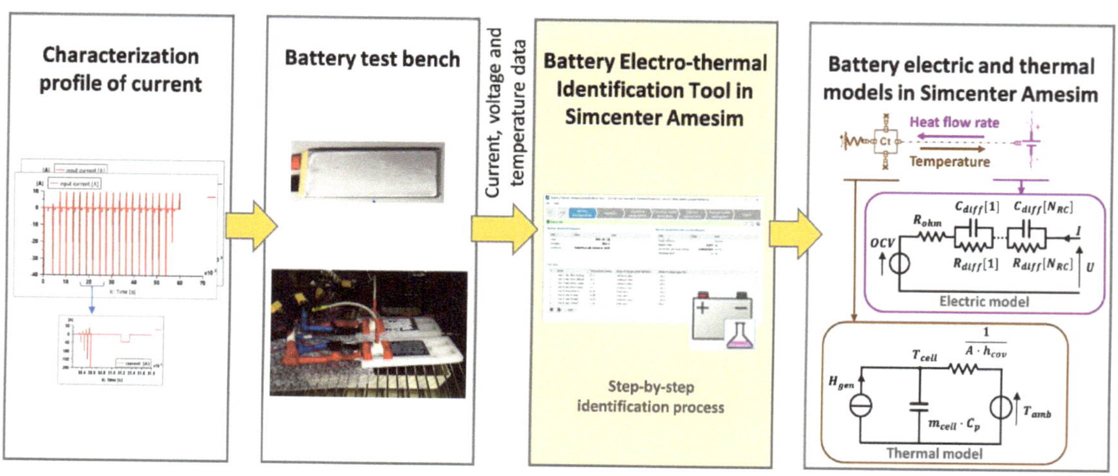

Figure 2. One-dimensional cell model parameterization workflow.

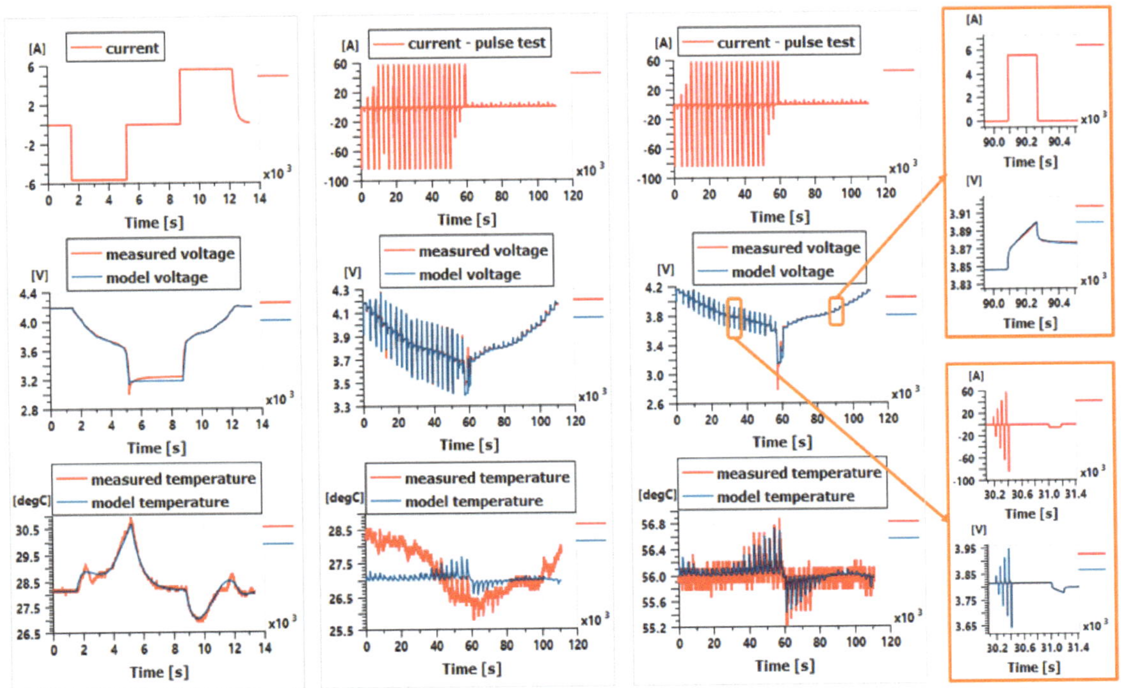

(a) (b) (c) (d)

Figure 3. Experimental test data and model estimations from the Battery Electro-thermal Identification Tool in Simcenter Amesim: (**a**) constant discharge and standard CC/CV charge test; (**b**) pulse test at 25 °C; (**c**) pulse test at 55 °C; (**d**) zoom of pulse test at 55 °C.

To validate the cell electrical and thermal models, a validation profile was applied to the cell via the test bench at different ambient temperatures. The validation profile includes a fast charge phase from about 0% State of Charge (SoC) and several dynamic driving cycles at different SoCs, as shown in Figure 4b,c. A battery cell simulation sketch in Simcenter Amesim, as in Figure 4a, is used to compare the model voltage and temperature estimations to the experimental data. Figure 4b,c show the comparison results, respectively, for 25 and 55 °C. A minor temperature difference can be seen at the beginning of the test at 25 °C, which is due to the ambient temperature variation in the climatic chamber, like the test at 25 °C in Figure 3b. At very low states of charge (<5% SoC), the battery electrical behavior becomes very nonlinear, which cannot be accurately represented by the equivalent circuit model. Otherwise, the identified models give accurate estimation of the battery volage and temperature behavior at these two temperatures. The identified cell electrical and thermal models are then validated.

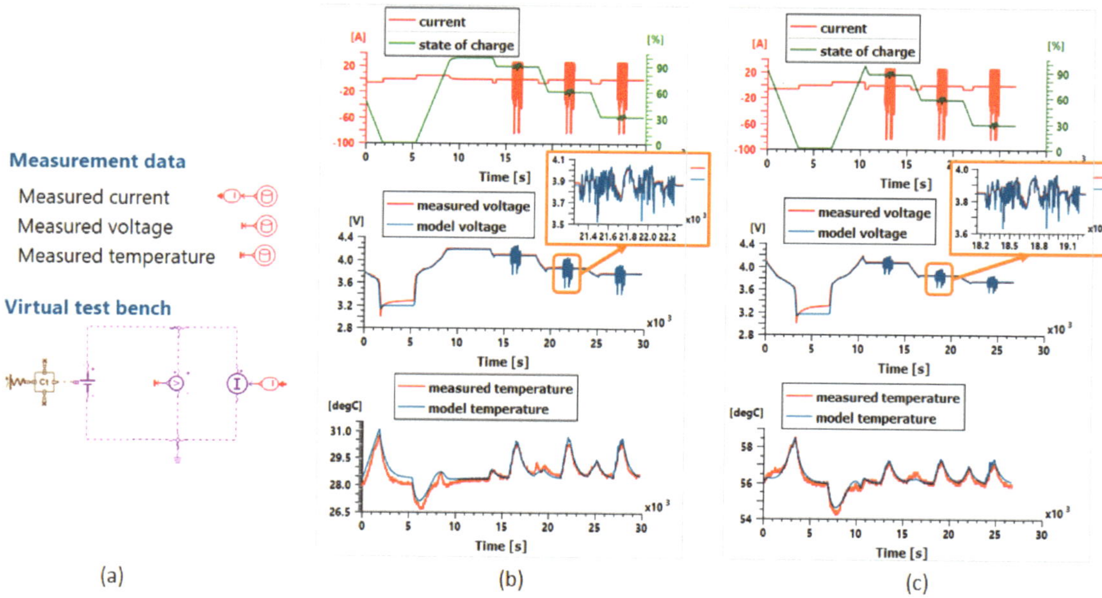

Figure 4. (**a**) Cell model validation in Simcenter Amesim; (**b**) and (**c**) comparison between model estimation and experimental data.

4. System Simulation Model

Based on the model of the single cell, electrically and thermally characterized as described in the previous section, the model of a cell stack consisting of several cells is built. To this end, and according to the 12s2p stack configuration in the race car described in Section 2, 24 instances of the cell model are electrically linked with each other to form a stack in the system simulation environment. It is noteworthy that building the stack model using any pre-calibrated cell is possible even before the cell characterization. In this case, the cell parameters of the pre-calibrated cell would be improved accordingly once the cell characterization is complete. This process of parallelization may be useful in view of a reduced project time.

In terms of a modular and efficient model construction, in Simcenter Amesim a cell pair of 1s2p configuration is first created as a so-called super-component, as portrayed in Figure 5. As suggested by the name, a super-component comprises several components combined into one, and in this way facilitates the re-use of model parts sharing the same setup. By means of exposed parameters and exposed variables, relevant "inner" quantities are defined as accessible outside the super-component. In this way, the super-components consisting of several individual components, such as the 1s2p cell configuration, simplify the visual appearance and ease the parametrization of complex models [23].

Figure 5 shows the super-component created and the electrical parallel connection of two single cells, which deliver their potential voltage through the electrical ports 1 and 3. The voltage of the cell pair can be extracted with the signal output of the voltmeter and processed further if needed. The thermal ports of both cells are connected to a node that enables the data exchange via the common thermal port 2. In this way, the heat flow rates due to the power losses generated in the cells are added and the simplifying assumption is that both cells have the same temperature, as two cells in parallel are thermally connected in this case.

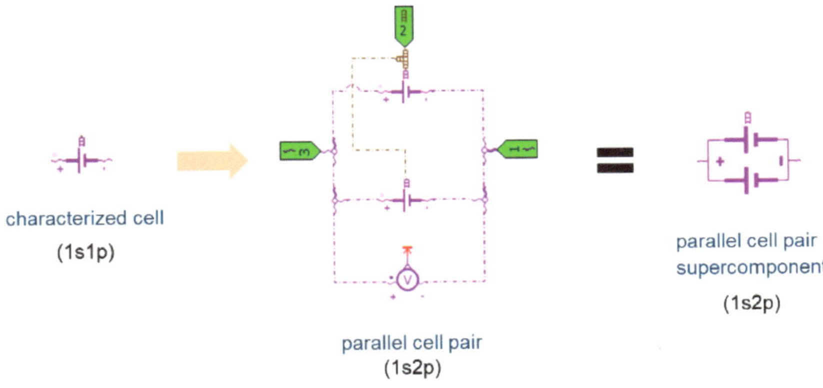

Figure 5. Super-component representing 1s2p cell pair.

The electrical stack configuration 12s2p is implemented by interconnecting 12 of the cell pairs (or 12 instances of the super-component) in series, as shown in the bottom part of Figure 6. The current through the stack is consequently divided between the two cells and combined accordingly at the nodes connecting the 1s2p elements. Thermal adapters are connected to the thermal ports of the cell pairs to enable the exchange of temperature as input and heat flow as output at the signal level. This exchange applies to each pair of cells and is ultimately used for co-simulation with the CFD model.

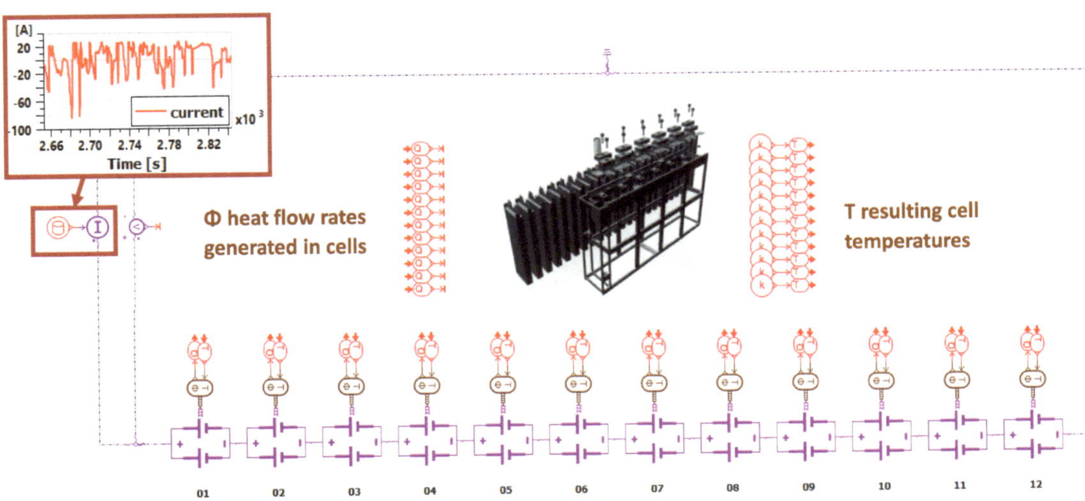

Figure 6. System simulation battery stack model.

To account for different situations, i.e., different current loads, and to ensure the re-usability of the model, the current profile is applied by means of a component (a current source) that points to a table file. To simulate different scenarios, the user only needs to swap the table file. This procedure is also applied to validate the system simulation model: current profiles from measurements are used to assess and verify the voltage response during charging and discharging, read from the voltage sensor connected in parallel with the current source. At this stage, the model is used in an open-loop fashion, meaning that it does not account for changing temperatures due to power losses in the cells, but rather prescribes constant cell temperatures. Once the model is validated, it can then be embedded in the CFD simulation, as described in the following.

Whereas the system simulation reflects the electrical setup of the stack, the thermal interconnections between the cells, and thus their orientations to each other, as well as their thermal properties, are modeled in CFD. Figure 7 shows the basic idea of the co-simulation between the electrical system simulation model and the thermal CFD model. Between the models, the temperatures are passed from the thermal model to the electrical model and, vice versa, the heat flow rates from the electrical to the thermal. The so-called watch parameters and watch variables are defined within the system simulation model, to possibly change system simulation model parameters and to access result variables from the CFD model.

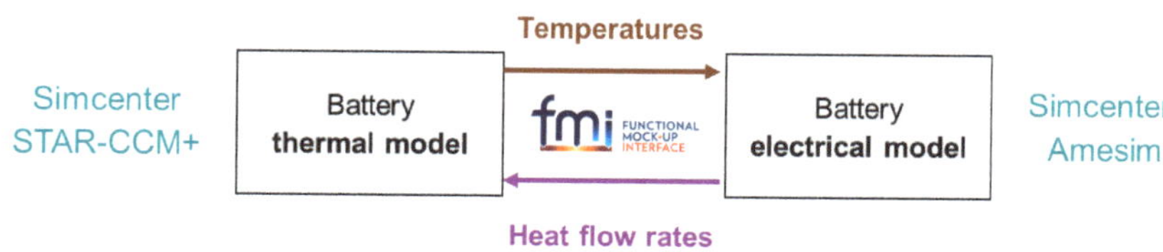

Figure 7. Co-simulation between battery thermal (CFD) and electrical (system simulation) model.

The implementation of the co-simulation was made by means of the FMI standard [21] that allows exchange of models between different simulation tools, since both Simcenter Amesim and Simcenter STAR-CCM+ support this standard. Specifically, the FMI standard is used in the co-simulation protocol according to the FMI 2.0 standard. A Functional Mockup Unit (FMU) is generated in Simcenter Amesim with three steps: adding an interface block to the model (shown in Figure 6); removing the dummy signal components; connecting it to the current source and thermal adapter components. Eventually, the FMU is imported into Simcenter STAR-CCM+ with the complete electrical characterization that was conducted during the experiments and system simulation.

5. CHT Simulation

In this section, the setup of the three-dimensional CHT simulation is explained. The software used is Simcenter STAR-CCM+, version 2021.1 [24], a multi-physics CFD-focused platform, based on a Finite-Volume approach for space discretization and a second-order temporal discretization of the transport equations, i.e., Navier-Stokes and energy equation.

Due to the constant air mass flow provided by the fans, the flow field can be seen as steady, and changes only in reaction to the temperature effects via heat transfer from the surrounding solid regions, e.g., the cooling fins. Hence, in the present case study, a multi-time scale workflow is employed, in which steady and transient solvers are combined. The transient behavior of the solid domains is calculated by using an algebraic multi-grid based iterative solver that is second order accurate in space and time. The fluid flow is captured by applying an eddy-viscosity based Reynolds Averaged Navier Stokes (RANS) solver, in which a predictor-corrector approach is used for the velocity-pressure coupling, based on a Semi-Implicit Method for Pressure Linked Equation (SIMPLE) algorithm, with second-order accuracy in space. The $\kappa\omega$-SST turbulence model [25] with a low-y^+ mesh approach is employed for an optimal modeling of wall turbulence.

The solid domains consist of the cooling fins, the printed circuit boards (PCBs), the cell tabs, the stack cages, and the battery cells, as portrayed in Figure 8. For each of these, components were modeled with corresponding material values (e.g., thermal conductivity and density). The voids of the stack cages are modeled as solid regions (with air material properties) for solver stability and reduction of calculation time: separate investigations are conducted to assess that this choice has only a marginal influence on the heat transfer, but great computational advantages. The stacks of the active material layers of the pouch cells are modeled as a homogenized body with orthotropic thermal conductivity λ, assuming $\lambda_{\parallel} = 20\ \mathrm{W{\cdot}m^{-1}{\cdot}K^{-1}}$ z-direction and $\lambda_{\perp} = 0.8\ \mathrm{W{\cdot}m^{-1}{\cdot}K^{-1}}$ in in x- and y-direction (see Figure 8). The battery cells are packed closely in the stack cages and are not in direct contact with the cooling fluid. Instead, the heat is conducted away through the cell tabs, which are located at the top end of the cells and connected to the cooling fins.

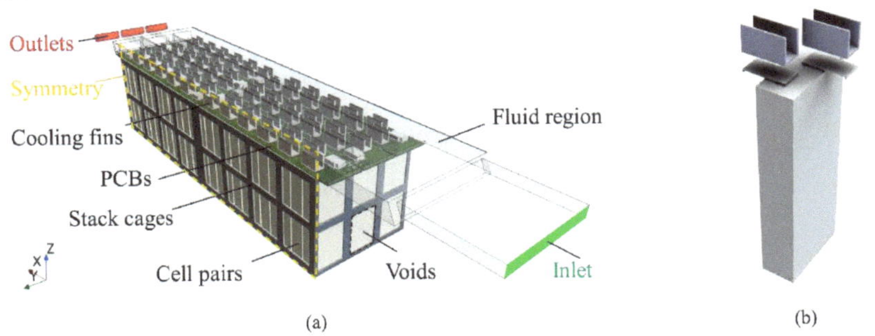

Figure 8. (**a**) Fluid and solid regions in the CFD model and (**b**) exploded view of the cell, tabs and the cooling fins.

The transient energy equation for the battery cell level can be written as:

$$\frac{\partial T_{cell}}{\partial t} = \frac{\lambda_{cell}}{\rho_{cell} c_{p,cell}} \nabla^2 T_{cell} + \dot{q}_{cell} \quad (1)$$

with $\dot{q}_{cell}$ the heat source calculated by the electrical cell model in the FMU, depending on the SoC, the cell temperature T_{cell} and the current I at the respective time step.

The symmetric architecture of the battery system allows the simulation setup to be reduced to six cell stacks, representing half of the battery system: a symmetry boundary condition is therefore applied to the symmetry plane, as displayed in Figure 8. On all other outer system boundaries apart from the inlet and the outlet, a convection boundary condition with a heat transfer coefficient of 6 W·m^{-2}·K^{-1} is applied, representing a natural convection in the battery container. The fans at the back of the battery container are modeled as a constant air mass flow of 0.0045 kg·s^{-1}. The driving heat transfer modes inside the battery container are conduction and convection and, due to the limited expected temperature below 60 °C, the influence of radiation is considered negligible and therefore neglected. For the fluid region, i.e., the cooling channel, the mesh consists of around 12 million cells, while all the solid regions together consist of around 11 million cells.

The small impact of the temperature change on the fluid flow and the different time scales between the solid and fluid continua in terms of propagation of temperature allow for the utilization of a steady-state fluid solver that is updated periodically, mimicking a transient behavior. For the multi-time scale workflow, a time step of 0.1 s is employed for the implicit unsteady solver of the solid continua, whilst the steady state solver of the fluid continua is updated every 60 s of simulation time by running 20 iterations, until converged, for the prevalent conditions of the solid continua. In the underlying case study, this multi-time scale workflow has proven to be computationally less expensive, with a negligible deviation compared to the fully time dependent simulation.

As described in Section 4, the system simulation model represents a complete cell stack, incorporating 24 of the previously parameterized battery cell models, consisting of two cell pairs. The system simulation model is now exported as a Standalone FMU, allowing Simcenter STAR-CCM+ (that acts as the master tool for the coupled electro-thermal simulation of the battery system) to call the FMU for the co-simulation, independently of the secondary controller tool (Simcenter Amesim [23]). For a FMU to be called by the CFD simulation, it must be embedded in Simcenter STAR-CCM+, specifying the exchanged variables, the exchange time step and optional watch parameters to access electrical quantities in the CFD simulation. Since six cell stacks are necessary in the CFD setup, the same FMU is imported six times and the inputs ($\dot{q}_{cell}$) and outputs (T_{cell}) are linked to the three-dimensional volumes of the respective cell pairs, allowing for the models to exchange data in a closed loop. The latter, with an exchange interval of 0.1 s, consists of the following steps: the mean temperature of each cell pair is evaluated in the CHT simulation at each (solid continua) time step, then passed to the FMUs; the latter evaluate the corresponding generated heat (calculated by the underlying electrical model in the FMU itself); finally, this value is imposed as a volume heat source for each cell pair in the CFD calculation. Each cell pair, as any other geometry part, is discretized in space in the three spatial dimensions, so that all physical quantities (e.g., temperature) can be evaluated in each discretization point (i.e., each mesh point) and an inhomogeneous profile on the surfaces and in the volumes can be observed. The current profile applied in the CFD simulation is a recorded profile measured during an endurance race.

The temperature of the battery pack and the velocity field inside the cooling channels at the end of the endurance race simulation can be seen in Figure 9a. The development of the average temperature, the voltage and the SoC over time in a selected cell pair (S1P12 = stack 1, cell pair 12) are shown in Figure 9b–d. In general, the results show the expected behavior of the electrical and thermal quantities for the applied current profile of the endurance race. The plateau that can be seen in electrical and thermal values from 725–975 s corresponds to the driver change phase, in which the car stands still and the current load is zero. Furthermore, the simulation results fit well with the target values that have been defined during the battery development. For an initial temperature for air and solids of 30 °C (which represents a canonical weather condition at Hockenheim in summer) the hottest cell of the battery pack reaches a maximum temperature of 59.2 °C at the end of the endurance race, which is slightly below the temperature limit defined by the competition rules [26]. The temperature spread inside the selected cell pair amounts to 7.9 °C and is below the limit of 10 °C that is linked to accelerated aging in a battery cell. According to the simulation, the car reaches the end of the endurance race with a remaining state of charge of 7%, which shows that the battery pack provides enough energy, but also points to a possible improvement of the energy management strategy.

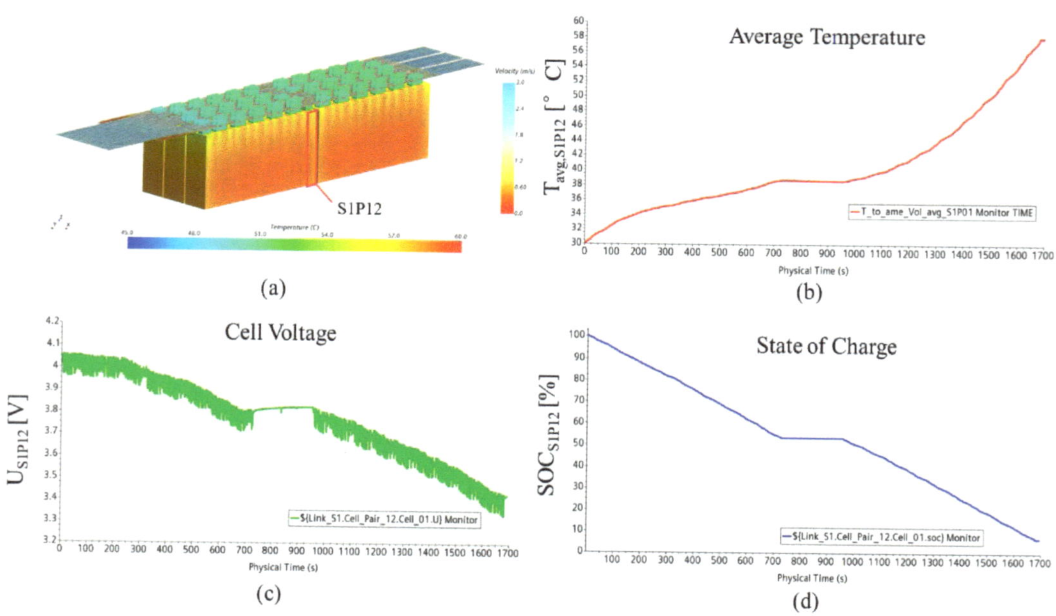

Figure 9. Velocity field of air region (**a**); cell pair temperature (**b**), SoC (**c**) and voltage (**d**) over time for one selected cell pair.

6. Discussion

In this section, several possible enhancements or extensions of the current work are briefly discussed.

6.1. Multi-Disciplinary Design Optimization

The first possible enhancement is a so-called Multi-disciplinary Design Optimization, employed with the objective of optimizing the performance of the cooling system. Focusing for example on possible design changes (such as number of fans—represented in the model by the value of the inlet mass flow and the shape of the air channel and cooling tabs), the SHERPA algorithm [27] can be employed to efficiently search for design optima. One possible optimization strategy could be to maximize the heat transfer between fluid and solid, i.e., increasing the cooling performance of the system, while trying to minimize the mass flow. This would allow for a reduction of the number of fans, reducing the weight of the car and the requirement for the electrical system. The two opposite optimization goals, i.e., increasing the heat transfer and reducing the mass flow, can be efficiently handled by the SHERPA algorithms within an optimization software such as Simcenter HEEDS, providing an optimal compromise between the optimization goals, and discovering the most efficient design parameter combinations (a so-called Pareto front).

6.2. Real-Time Application for Controlling

The second possible enhancement of the model is its utilization for real-time controlling of the battery cooling system. Simcenter Amesim provides the possibility of exporting the described model for real-time targets such as dSPACE devices [16], i.e., controller units that are often used as vehicle control units. By replacing the CFD-generated temperature values with instantaneously measured cell temperatures during the race, the presented model could improve the existing vehicle model with a detailed online electro-thermal battery model. This would enable, for example, a predictive race strategy, which can utilize the knowledge of the instantaneous thermal behavior to adjust the operating limits of the car. Another way to take advantage of the CFD model for real-time application is to perform a model reduction, which generates a reduced order model suitable for 1D system simulation. For example, the ROM builder of Simcenter Amesim can be used to train an artificial neural networks (ANN) model with the numerical simulation results from the CFD model. Similar model reduction to that of an ANN model has been reported in [19].

6.3. Evaluation of a Liquid-Cooled Design

For the air-cooled battery pack described previously, the simulation result in Figure 9a shows that the average battery temperature at the end of the driving scenario is higher than the standard limit usually defined between 40 and 45 °C. While this does yet not present any safety issue, working at such a high temperature level has a negative impact on the battery lifespan. In this case, a liquid-cooled battery pack can be considered to prevent the battery temperature from rising above the usual limit (e.g., 45 °C).

In order to find a functional liquid-cooled design for which the maximum battery temperature does not exceed the usual limit, a new battery stack model with a simplified thermal model for liquid cooling (Figure 10) is proposed based on the original model in Figure 6. In this new stack model, vertical cold plates are inserted between adjacent battery cell pairs. Each cold plate has a cooling channel thickness of 2 mm and a width equal to the cell height. The heat transfer coefficient for the simulation of the convective heat transfer between the cold plates and the battery cells is calculated by Simcenter Amesim based on a Nusselt correlation for rectangular pipes. Further simplifications include the neglection of the thermal resistance of the cold plates, as well as the contact resistances between the cold plates and the battery cells. As shown in Figure 10, several elements have been added compared to the original model in Figure 6:

- Thermal capacities (Ct) to represent the thermal masses of the connected batteries.
- Thermal resistances to represent the conductive heat exchange between adjacent battery cells.
- Hydraulic pipes to represent the cooling channels in which the liquid coolant flows. The inlet coolant flow is set to 30 kg/h at a temperature of 35 °C, which is the minimum temperature that can be reached by the coolant with ambient air at 30 °C.

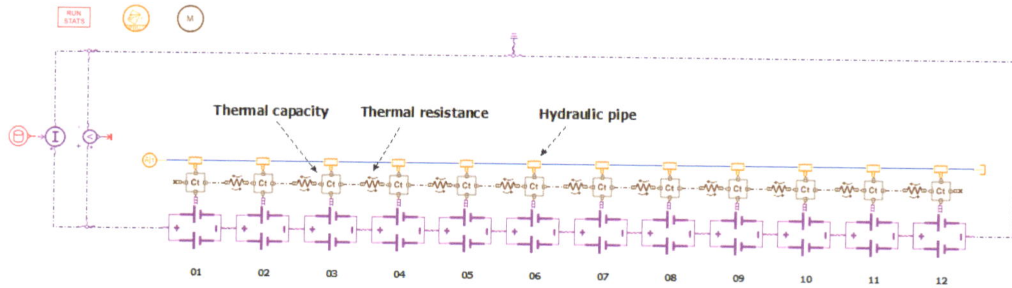

Figure 10. System simulation model in Simcenter Amesim of a liquid cooled 12S2P stack.

By running the model of Figure 10 using the same current profile of the endurance race, a significant improvement in battery temperature can be observed as shown in Figure 11.

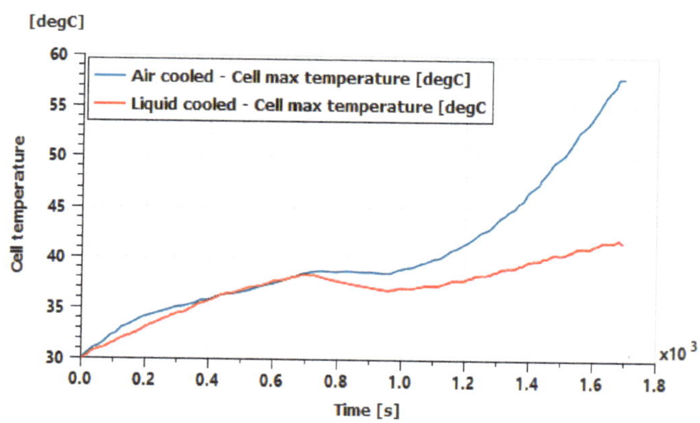

Figure 11. Simulation result of the battery maximum cell temperature with liquid cooling compared to that with air cooling.

This quick evaluation shows the possibilities of a liquid cooled design to limit the maximum cell temperature. However, this solution would have several drawbacks in the context of the Formula Student competition, since it would imply additional weight and space when considering the cold plates, the coolant, the pump and the radiator, compared to air cooling. It would consequently be difficult to consider this evolution without significant overall design changes.

6.4. Battery Aging Consideration

Battery aging is an important aspect to consider during battery pack design. The battery operating conditions (e.g., temperature, current and state of charge) have significant impact on battery lifespan. Different operating conditions result in different battery aging. To make sure the battery pack meets the expected lifespan, different scenarios of battery usage need to be studied. This can easily be performed with the system level model developed in Simcenter Amesim, such as that in Figure 6. In fact, a semi-empirical aging model [28,29] is available as an option for the battery components in the model of Figure 6. By using this empirical aging modeling option, different battery operating scenarios can be studied via simulation.

For example, Figure 12 shows the simulation results (original air-cooled setup) of the battery capacity loss for the following scenarios to study the impact of charge current on battery aging:

- Step 1: from 100% state of charge, the battery stack model is discharged with the current profile of the endurance race.
- Step 2: the battery stack is at rest during 200 s with increased cooling power for a quicker cooldown.
- Step 3: the battery stack is fast charged (2 C or 4 C) with CC/CV (constant current/constant voltage) protocol up to 100% state of charge.
- Step 4: the battery stack is at rest during 200 s.
- Step 5: repeat step 1 to step 4 for an additional 9 times.

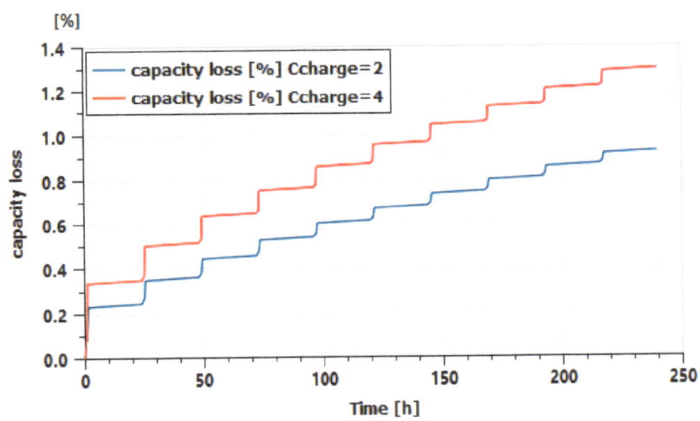

Figure 12. Simulated capacity loss for scenarios with two different charging current (2C or 4C) using the air cooled 12S2P stack model.

The results in Figure 12 show that higher charge current (4 C) has a stronger impact on battery aging. It is important to mention that this simulation was run using a pre-calibrated aging parameter set from the battery database in Simcenter Amesim. To achieve accurate estimation of the battery aging of a given battery, its empirical aging parameter values need to be identified from experimental aging data [30].

6.5. Model Integration into a Full Vehicle System

Integration of the battery pack model into a full vehicle model is one of the important steps in vehicle design. This enables, firstly, deep analysis of the interactions between different subsystems to optimally balance different vehicle KPIs (e.g., performance, safety, battery lifetime, and comfort). Secondly it extends the flexibility of the modeling approach by providing the boundary conditions resulting from the vehicle power demand. Finally, this type of full vehicle model is also widely used for control design. Figure 13 illustrates an example of a full vehicle system model including different subsystem models, such as the vehicle dynamic model, the electric motors, the inverter and the high voltage battery packs.

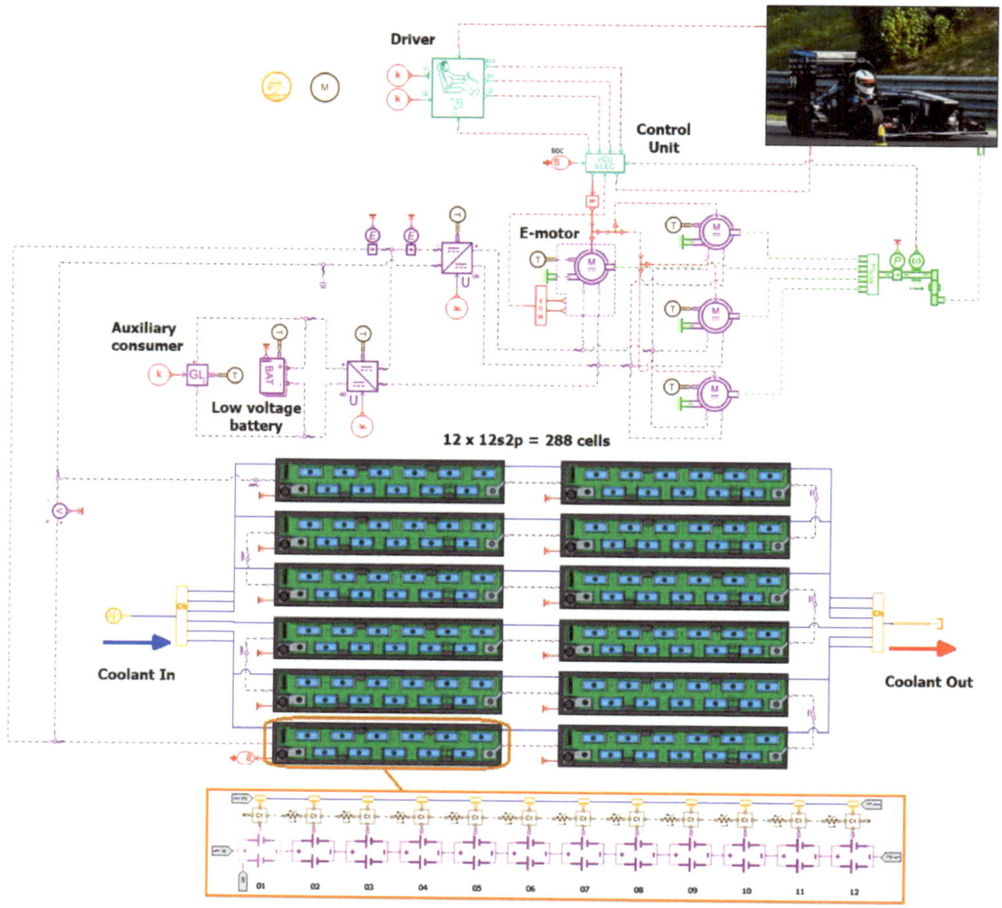

Figure 13. Full vehicle system simulation model in Simcenter Amesim.

7. Conclusions

In this paper a novel, multi-disciplinary approach is presented to characterize and simulate the BTM (Battery Thermal Management) of a battery pack for a race car, leveraging and combining experiments, systems and CFD simulation. First, a single cell is experimentally investigated to identify the equivalent circuit model under different temperatures. The cell model parameters are calibrated using a standard workflow from experimental data. This calibration workflow can be easily applied to other battery cells. By using the identified cell model, a system level battery stack model is set up and is then exported into a FMU by means of the FMI standard to ensure the scalability, re-usability, and portability of the model. Finally, the FMU is embedded in a CFD simulation where the cooling of the battery pack is simulated. This procedure allows a detailed CFD fluid-solid simulation that embeds experimental characterization, as well as the functional interconnection of cells at system level. In addition, the 1D and 3D co-simulation using the FMI standard highlights the advantages of the proposed approach to evaluate directly any operating conditions without the need to have a heat transfer map for pre-defined operating conditions, as in [4,20].

The same procedure can also be applied to model and simulate battery packs with different types of cooling strategies, e.g., with liquid cooling. In addition, if an empirical battery aging law is available, it can be added to the battery pack model in order to study the battery aging under different operating scenarios. Last but not least, the battery pack model can be integrated into a full vehicle system model to analyze the interactions with other subsystems, such as the electric motors and inverter.

As future research directions, the proposed approach can be adapted for other types of battery thermal management systems, such as those with heat pipe, phase change material, and immersion cooling. In addition, the cell thermal modeling can be further improved in the CFD model to distinguish the inner-cell material from the cell casings, as in [31].

Author Contributions: Conceptualization, methodology, software, formal analysis, resources, writing—original draft preparation, review and editing: C.S., C.H., A.L., B.H., T.N. and F.B. All authors have read and agreed to the published version of the manuscript.

Funding: This research received no external funding.

Institutional Review Board Statement: Not applicable.

Informed Consent Statement: Not applicable.

Data Availability Statement: Not applicable.

Acknowledgments: Ecurie Aix and Siemens Digital Industries Software kindly acknowledge the Institute for Power Electronics and Electrical Drives of the RWTH Aachen University for providing support to perform the tests in their laboratories in the framework of an ongoing and proficient cooperation between the team and the institute. Simulations were performed with computing resources granted by the RWTH Aachen University under project 0213.

Conflicts of Interest: The authors declare no conflict of interest. C.S., C.H., A.L. and B.H. are employees of Siemens. The paper reflects the views of the scientists, and not the company.

References

1. Houache, M.S.E.; Yim, C.-H.; Karkar, Z.; Abu-Lebdeh, Y. On the Current and Future Outlook of Battery Chemistries for Electric Vehicles—Mini Review. *Batteries* **2022**, *8*, 70. [CrossRef]
2. Un-Noor, F.; Padmanaban, S.; Mihet-Popa, L.; Mollah, M.N.; Hossain, E. A Comprehensive Study of Key Electric Vehicle (EV) Components, Technologies, Challenges, Impacts, and Future Direction of Development. *Energies* **2017**, *10*, 1217. [CrossRef]
3. Kim, J.; Oh, J.; Lee, H. Review on battery thermal management system for electric vehicles. *Appl. Therm. Eng.* **2019**, *149*, 192–212. [CrossRef]
4. Buidin, T.I.C.; Mariasiu, F. Modeling Approach of an Air-Based Battery Thermal Management System for an Electric Vehicle. *Appl. Sci.* **2021**, *11*, 7089. [CrossRef]
5. Wang, X.; Liu, S.; Zhang, Y.; Lv, S.; Ni, H.; Deng, Y.; Yuan, Y. A Review of the Power Battery Thermal Management System with Different Cooling, Heating and Coupling System. *Energies* **2022**, *15*, 1963. [CrossRef]
6. Jaguemont, J.; Boulon, L.; Dubé, Y. Characterization and Modeling of a Hybrid-Electric-Vehicle Lithium-Ion Battery Pack at Low Temperatures. *IEEE Trans. Veh. Technol.* **2016**, *65*, 1–14. [CrossRef]
7. Yang, X.-G.; Leng, Y.; Zhang, G.; Ge, S.; Wang, C.-Y. Modeling of lithium plating induced aging of lithium-ion batteries: Transition from linear to nonlinear aging. *J. Power Sources* **2017**, *360*, 28–40. [CrossRef]
8. Mallick, S.; Gayen, D. Thermal behaviour and thermal runaway propagation in lithium-ion battery systems—A critical review. *J. Energy Storage* **2023**, *62*, 106894. [CrossRef]
9. Omariba, Z.B.; Zhang, L.; Sun, D. Review of Battery Cell Balancing Methodologies for Optimizing Battery Pack Performance in Electric Vehicles. *IEEE Access* **2019**, *7*, 129335–129352. [CrossRef]
10. Doyle, M.; Fuller, T.F.; Newman, J. Modeling of Galvanostatic Charge and Discharge of the Lithium/Polymer/Insertion Cell. *J. Electrochem. Soc.* **1993**, *140*, 1526–1533. [CrossRef]
11. Petit, M.; Calas, E.; Bernard, J. A simplified electrochemical model for modelling Li-ion batteries comprising blend and bidispersed electrodes for high power applications. *J. Power Sources* **2020**, *479*, 228766. [CrossRef]
12. Edouard, C.; Petit, M.; Forgez, C.; Bernard, J.; Revel, R. Parameter sensitivity analysis of a simplified electrochemical and thermal model for Li-ion batteries aging. *J. Power Sources* **2016**, *325*, 482–494. [CrossRef]
13. Li, A.; Ponchant, M.; Sturm, J.; Jossen, A. Reduced-Order Electro-Thermal Battery Model Ready for Software-in-the-Loop and Hardware-in-the-Loop BMS Evaluation for an Electric Vehicle. *World Electr. Veh. J.* **2020**, *11*, 75. [CrossRef]
14. Tran, M.-K.; Mathew, M.; Janhunen, S.; Panchal, S.; Raahemifar, K.; Fraser, R.; Fowler, M. A comprehensive equivalent circuit model for lithium-ion batteries, incorporating the effects of state of health, state of charge, and temperature on model parameters. *J. Energy Storage* **2021**, *43*, 103252. [CrossRef]

15. Lacressonnière, F.; Varais, A.; Roboam, X.; Bru, E.; Mullins, T. Scale electro-thermal model of a lithium-ion battery for time-accelerated experiments in a hardware in the loop process. *J. Energy Storage* **2021**, *39*, 102576. [CrossRef]
16. Ponchant, M.; Li, A.; Beckers, C.; Paroha, M. Battery Management System Evaluation within a Complete Electric Vehicle Model with Software-in-the-Loop and Hardware-in-the-Loop Approaches. In Proceedings of the 2021 23rd European Conference on Power Electronics and Applications (EPE'21 ECCE Europe), Virtual Conference, 6–10 September 2021; pp. P.1–P.10.
17. Wang, J.; Xu, H.; Xu, X.; Pan, C. Design and simulation of liquid cooled system for power battery of PHEV. *IOP Conf. Ser. Mater. Sci. Eng.* **2017**, *231*, 012025. [CrossRef]
18. Babu, A.R.; Andric, J.; Minovski, B.; Sebben, S. System-Level Modeling and Thermal Simulations of Large Battery Packs for Electric Trucks. *Energies* **2021**, *14*, 4796. [CrossRef]
19. Li, A.; Yuen, A.C.Y.; Wang, W.; Chen, T.B.Y.; Lai, C.S.; Yang, W.; Wu, W.; Chan, Q.N.; Kook, S.; Yeoh, G.H. Integration of Computational Fluid Dynamics and Artificial Neural Network for Optimization Design of Battery Thermal Management System. *Batteries* **2022**, *8*, 69. [CrossRef]
20. Li, X.; Yao, M.; Zhu, L.; Yuan, X.; Shang, J.; Yang, B. Thermal Modeling of a Lithium-Ion Battery Pack in a Plug-in Hybrid Electric Vehicle. In Proceedings of the 2019 IEEE Energy Conversion Congress and Exposition (ECCE), Baltimore, MD, USA, 29 September–3 October 2019; pp. 2669–2673.
21. Blochwitz, T.; Otter, M.; Arnold, M.; Bausch, C.; Clauß, C.; Elmqvist, H.; Junghanns, A.; Mauss, J.; Monteiro, M.; Neidhold, T.; et al. The Functional Mockup Interface for Tool independent Exchange of Simulation Models. In Proceedings of the 8th International Modelica Conference, Dresden, Germany, 20–22 March 2011; pp. 105–114.
22. Ecurie Aix Formula Student Team RWTH Aachen e.V. eace08. Available online: https://www.ecurie-aix.de/eace082021 (accessed on 12 October 2022).
23. *Simcenter Amesim User Guide*; Version 2021.2; Siemens Digital Industries Software, Inc.: Plano, TX, USA, 2021.
24. *Simcenter STAR-CCM + Documentation*; Version 2021.1; Siemens Digital Industries Software, Inc.: Plano, TX, USA, 2021.
25. Menter, F.R. Two-Equation Eddy-Viscosity Turbulence Models for Engineering Applications. *AIAA J.* **1994**, *32*, 1598–1605. [CrossRef]
26. Formula Student Germany Rules 2022. Available online: https://www.formulastudent.de/fsg/rules/ (accessed on 12 October 2022).
27. Chase, N.; Rademacher, M.; Goodman, E.; Averill, R.; Sidhu, R. *A Benchmark Study of Optimization Search Algorithms*; Red Cedar Technology: East Lansing, MI, USA, 2010; pp. 1–15.
28. Wang, J.; Liu, P.; Hicks-Garner, J.; Sherman, E.; Soukiazian, S.; Verbrugge, M.; Tataria, H.; Musser, J.; Finamore, P. Cycle-life model for graphite-LiFePO4 cells. *J. Power Sources* **2011**, *196*, 3942–3948. [CrossRef]
29. Prada, E.; Petit, M. Aging modeling for advanced Li-ion battery pack sizing and management for HEV/EV through AMESim simulation platform. In Proceedings of the LMS Vehicle Users Conference, Saint-Quentin en Yvelines, France, 14 May 2013; pp. 29–30.
30. Mingant, R.; Petit, M.; Belaïd, S.; Bernard, J. Data-driven model development to predict the aging of a Li-ion battery pack in electric vehicles representative conditions. *J. Energy Storage* **2021**, *39*, 102592. [CrossRef]
31. Pirooz, A.; Van Mierlo, J.; Berecibar, M. 3D Thermal and 1D Electro-Thermal Model Coupling Framework for Lithium-Ion Battery Cells in Automotive Industry Platforms. In Proceedings of the 2021 IEEE Vehicle Power and Propulsion Conference (VPPC), Gijón, Spain, 25–28 October 2021; pp. 1–6.

World Electric Vehicle Journal

MDPI

Article

Second-Life Batteries Modeling for Performance Tracking in a Mobile Charging Station

Marwan Hassini [1,2,3,*], Eduardo Redondo-Iglesias [1,3] and Pascal Venet [2,3]

1 Univ Eiffel, Univ Lyon, ENTPE, LICIT-ECO7, F-69675 Lyon, France
2 Univ Lyon, Université Claude Bernard Lyon 1, INSA Lyon, Ecole Centrale de Lyon, CNRS, Ampère, UMR5005, 69622 Villeurbanne, France
3 Eco7/Ampère Joint Research Team for Energy Management and Storage for Transport, 69500 Bron, France
* Correspondence: marwan.hassini@univ-eiffel.fr

Abstract: Lithium-ion batteries are seen as a key element in reducing global greenhouse gas emissions from the transport and energy sectors. However, efforts are still needed to minimize their environmental impact. This article presents a path towards a circular economy and more sustainable batteries, thanks to their reuse in mobile charging stations for electric vehicles. This work presents the results of characterization tests and modeling of second life batteries. The presented characterization test and electrical models can be used as references to evaluate the performance of aged batteries after their first life. Detailed test procedures and data results are provided in an open-access data paper.

Keywords: second life battery; reuse; charging; sustainability; modeling; equivalent circuit model; open science

1. Introduction

Mitigating climate change is seen as a major challenge for the 21st century. According to the latest report from the Intergovernmental Panel on Climate Change, transport and electricity generation are responsible for 18% and 36% of the global greenhouse gas emissions, respectively [1]. To reduce this pollution, electrified vehicles and renewable energies are presented as interesting options. For these applications, lithium-ion batteries are key energy storage elements [1,2].

In 2021, 340 GWh of batteries have been produced, and these quantities are expected to increase significantly, possibly reaching 3500 GWh per year by 2030 [3]. In other words, to date, 10.8 kWh of batteries are currently being produced every second. Unfortunately, it is well known that their production is not yet sustainable. This means that none of the lithium-ion batteries produced can be manufactured without using natural resources and significant amounts of energy. Larcher and Tarascon pointed out that it takes over 400 kWh of energy to produce 1 kWh of lithium-ion battery with over 75% of this energy consumed during the material production phase [4]. In order to achieve sustainable energy production and transport, it is essential to develop solutions to reduce the negative impact of both the batteries already on the market as well as on the future generations of batteries.

The waste management hierarchy shown in Figure 1 establishes an order of preference for actions to reduce and manage waste [5]. Reduction, reuse, recycling, recovery and disposal are the actions that should be taken to reduce the environmental impact of a product. Reduction is the preferred action and consists of designing the product in order to limit the waste it will generate at the end of its life. In the context of batteries, reduction consists of minimizing the number of materials used in a battery, in other words, reducing the size of the battery. Electric vehicles powered by smaller batteries would be more economical and more environmentally-friendly [6,7].

Citation: Hassini, M.; Redondo-Iglesias, E.; Venet, P. Second-Life Batteries Modeling for Performance Tracking in a Mobile Charging Station. *World Electr. Veh. J.* **2023**, *14*, 94. https://doi.org/10.3390/wevj14040094

Academic Editor: Joeri Van Mierlo

Received: 27 February 2023
Revised: 24 March 2023
Accepted: 28 March 2023
Published: 3 April 2023

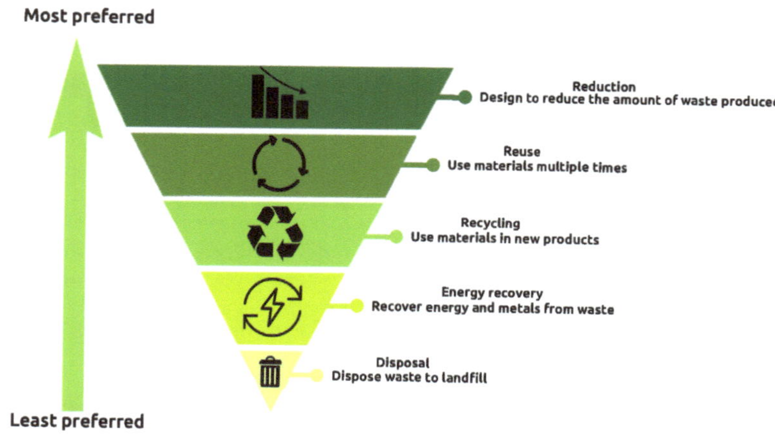

Figure 1. Waste Management Hierarchy.

However, such vehicles would have a shorter range and would be more dependent on charging infrastructure. To enable this shift, a reliable charging network needs to be developed. Mobile chargers are a solution to strengthen the existing infrastructure system. A mobile charging station can be defined as a charger capable of delivering energy to vehicles autonomously, i.e., without human assistance and without the need for an external power supply. This technology has the advantage of maximizing the utilization rate of the charging station as these chargers are able to move from one vehicle to another. They also provide a more flexible charging solution, as they are designed to be quickly deployed at any location, with the ability to adapt the energy embedded and the duration of stay to the needs of the drivers [8,9]. Figure 2 shows two autonomous mobile charging stations.

Figure 2. Mobile charging stations from Mob-Energy [10] and Volkswagen [11].

As the energy stored by these robot-like chargers is limited, they should be considered as a complementary technology to fixed chargers. To date, existing mobile charging stations can deliver up to 30 kWh at a maximum charging rate of 50 kW [12]. Powering these mobile charging stations with reused batteries is possible and would be a further step towards a circular economy. Reuse can be defined as the complete or partial reuse of the battery for the original purpose for which it was designed [13]. In the literature, reused batteries are commonly referred to as "second life batteries".

While there is a broad consensus in the battery community to recognize the environmental benefits of reuse, the economic interest of second life batteries is still being debated. Finding applications that could accept the constraints of reused batteries is still a challenge. To date, most studies have focused on stationary applications [14–17]. Since the volume required for these applications is limited, few authors have presented mobile applications as an interesting alternative [18–20]. In addition to the application choice, battery performance evolution needs to be carefully tracked as it has a significant impact on the economic interest of battery reuse [21,22].

Although research about second life batteries is gaining interest, two important limitations of the existing studies remain:

First, two definitions of battery reuse coexist in the literature. Most authors define second life batteries in terms of performance level. A threshold of 20 or 30% capacity loss is often considered as the boundary between first and second life [23–26]. Studies considering this definition answer the following research question: how does the performance of deeply aged batteries evolve? Other authors define reused batteries as batteries that have undergone a change of application [27,28]. This definition raises another research question: what is the effect of an application change on the evolution of battery performance? To date, there is a lack of experimental or modeling studies investigating the performance evolution of a battery that has undergone an application change. This is an important gap since the operating conditions can vary significantly from the first life to the second life.

Secondly, much of the existing research has examined batteries that have been aged in a laboratory. This method has been preferred because of the complexity of acquiring batteries after their electric vehicle life. Laboratory studies also allow control of operating conditions during the whole life of the battery. Nevertheless, in real-world situations users of second life batteries do not have access to the data during first-life use. The battery performance must be assessed to determine the battery's suitability for a second life. In recent years, few studies have assessed the performance of batteries extracted from electric vehicles [29–31]. Nevertheless, these papers are pioneer works focusing on experimental performance assessment. No work has yet been carried out on the electrical modeling of such batteries. Further studies should be conducted to get more insights into battery reuse after the automotive life.

1.1. Contributions

The authors aim to fill the gaps described above by presenting an experimental and modeling study on a battery extracted from an electric vehicle and reused in a mobile application. This article contributes to the existing research on second life batteries in various aspects.

1. Description of a new possible application for second life batteries

In this article, a new possible application for second life batteries is presented: the mobile charging station. To the best of the authors' knowledge, this is one of the first studies to present in detail a mobile application for reused batteries.

2. Experimental evaluation of second life battery performance

In this study, a characterization test is performed to evaluate the performance and calibrate the electrical models of the second life battery. The capacity of the battery cell is measured at three temperatures and the resistance of the battery cell is characterized at the same three temperatures using current at various charge/discharge rates and depths of discharge. This experimental work contributes to the existing literature on second life batteries as it is one of the first to evaluate the performance of a high-capacity prismatic cell extracted from real second life electric vehicle battery modules.

3. Modeling second life battery performance

This study presents two electrical models that can be used as a reference to evaluate and emulate second life battery performance. The influence of usage profile and temperature on model accuracy is also evaluated. This modeling work is central to estimating the energy stored in the mobile charging station, which is the first step in developing pricing, charging schedules, and energy management strategies. This is one of the first articles to model a second life battery cell extracted from an electric vehicle.

4. Open access publishing of the experimental data and software

Experimental data recorded during the characterization test are published as open-source along with the article. The open software used for data processing has also been shared. To the best of the authors' knowledge, this is one of the first articles to share experimental data on second life batteries and the software used to analyze these data [32–34].

1.2. Layout

The article is organized as follows: Section 2 presents the experimental evaluation of the second life battery performance. In Section 3, the electrical models are calibrated thanks to the experimental result and their accuracy is evaluated thanks to a mobile charging station and a WLTC profile in Section 4. Finally, conclusions are drawn and an outlook for future studies is given.

2. Second-Life Cell Performance Assessment

This section presents the experiments conducted to assess the performance of a second life battery cell. This performance assessment is necessary to tailor the reuse application to the capabilities of the battery. The data produced for this study are available in a public repository along with a detailed description of the cell and setup characteristics as well as the testing procedures and quality control [32].

2.1. Experimental Setup

In this study, a second life SAMSUNG SDI 94Ah cell with NMC positive electrode was tested at 0, 25 and 40 °C. The main characteristics of the cell are summarized in Table 1. This cell was extracted from a BMW i3 module purchased from the second life battery market. To ensure the representativeness of the study, the 12 cells of the module were tested and one cell with comparable performance to the others was selected.

Table 1. Samsung SDI 94Ah main characteristics. Data from [35].

Characteristics	Values
Format	Prismatic
Nominal capacity [Ah]	94
Positive electrode material	NMC
Negative electrode material	Graphite
Nominal voltage [V]	3.68
Maximal voltage [V]	4.15
Minimal voltage [V]	2.7
Specific energy [Wh/kg]	165
Size L × W × H [mm]	173 × 125 × 45
Weight [kg]	2.06

The experimental setup consisted of a Bitrode battery testing system and two climate chambers (Vötsch VT 3050 and Friocell 707).

Table 2 shows the characterization test protocol at a single temperature and Figure 3 is the voltage profile of the test. Step 1 is a capacity test, consisting of a sequence of three full charge/discharge cycles. Step 2 is an impedance measurement thanks to a sequence of current pulses at different depths of discharge (10, 20, 30, 40, 50, 60, 70, 80 and 90%) and current levels (0.3C, 0.5C, 0.8C, 1C and 1.3C). Step 3 is a pseudo-open circuit voltage measurement thanks to a full discharge/charge at C/20. Finally, step 4 is the validation cycle used to assess the accuracy of the models in mobile charging station usage. The validation profiles and results are described in detail in Sections 3 and 4.

Table 2. Reference characterization test.

Step	Test	Estimated Duration (h)
1	Capacity test	18
2	Impedance test	8
3	Pseudo OCV test	42
4	Mobile charging station cycles	12

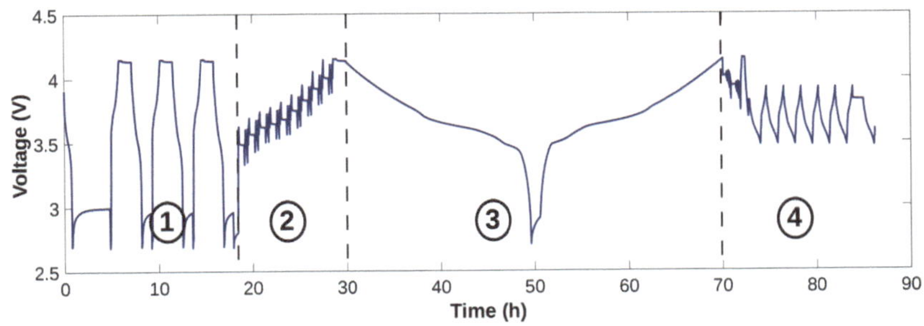

Figure 3. Voltage profile of the test. Step 1 is a capacity test, step 2 is the impedance measurement, step 3 is the low-current measurement and step 4 is the validation cycle.

For the characterization of a cell at 0, 25 and 40 °C, authors estimate that in 273 h (3 times 91 h) the climate chamber consumed 136.5 kWh and the cycler 94.5 kWh, giving a total consumption of 231 kWh. In France, at the time of the experiment, this electrical consumption generated about 14.7 kg CO_2 eq and cost 53 € [36]. In order to minimize the cost of battery research, the data has been made open access. The authors hope that this data will benefit future research and allow a reduction in the number of experiments conducted.

2.2. Capacity Measurement

The capacity is measured in step 1 of the characterization test shown in Figure 3. The third measurement of discharge capacity in step 1 is taken as the reference and the capacity is calculated according to Equation (A1). Figure 4 shows the influence of temperature on the discharge capacity.

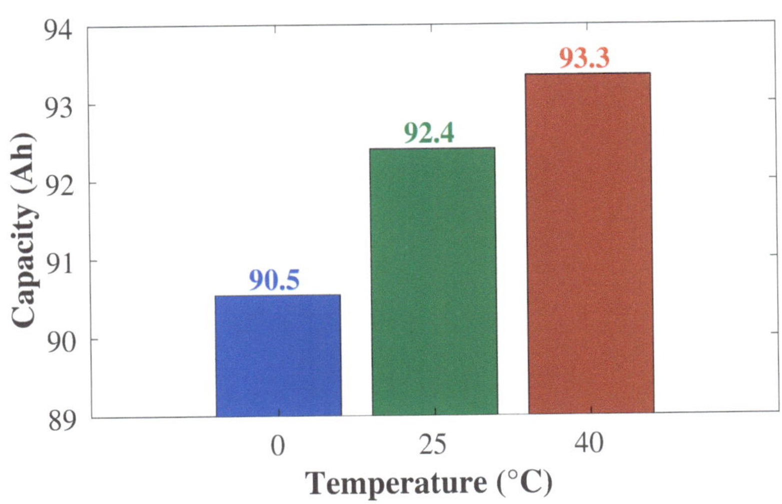

Figure 4. Discharge capacity measured at 0 °C (blue bar), 25 °C (green bar) and 40 °C (red bar).

The capacities listed in the datasheet were measured in discharge with a constant current-constant voltage profile and a 1C current. This method is similar to the one used in this work. A state of health "capacity" can therefore be calculated thanks to Equation (A2). Table 3 compares the capacities measured during step 1 at 0, 25 and 40 °C with the datasheet information.

2.3. Resistance Measurement

Resistance is determined during step 2 of the characterization test presented in Figure 3. Resistance is affected by temperature, sign and intensity of the current and the depth of discharge of the cell. The depth of discharge and resistance are calculated according to Equations (A3) and (A4).

Figure 5 shows the influence of temperature and depth of discharge over the discharge resistance. The measurement is made in discharge with a current rate of −1C and a 5 s resistance is considered.

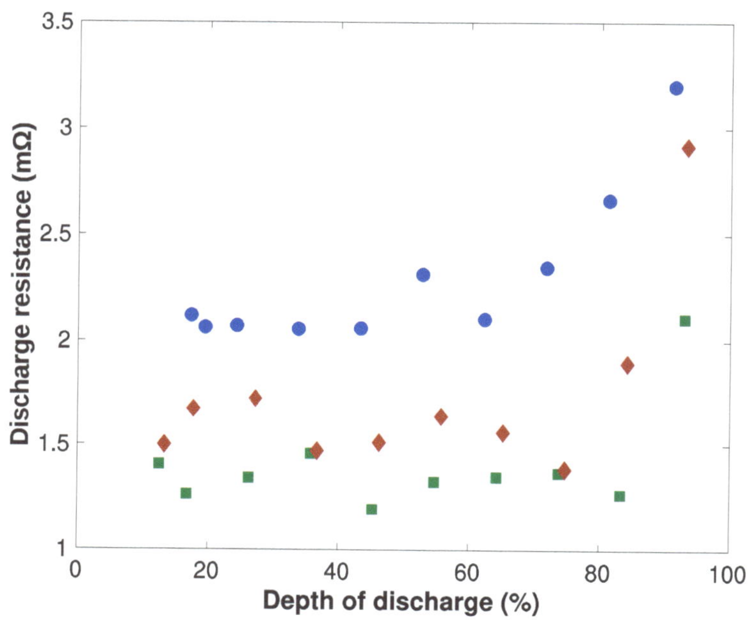

Figure 5. Discharge resistance of a second life cell at 0 (blue points), 25 (green squares) and 40 °C (red diamonds). Measurement is made at −1C and a 5 s resistance is considered.

State of health is an indicator of the evolution of the resistance since the beginning of the battery's life. It can be calculated using Equation (A5). Table 3 compares the resistances measured during step 2 at 0, 25 and 40 °C with the data sheet information.

2.4. Performance Comparison of a Fresh and Second Life Cell

This article evaluates the performance of a high-capacity prismatic cell taken from an electric vehicle. Table 3 compares the capacity and resistance measured on the second life cell with the data provided in the datasheet.

Table 3. Performance comparison between a fresh cell (datasheet) and a second life cell (experiments). "-" means that the state of health could not be calculated.

	Temperature	Fresh Cell	Second Life Cell	State of Health
Capacity (CC-CV, −1C)	0 °C	No data	90.5 Ah	-
	25 °C	95.2 Ah	92.4 Ah	97%
	40 °C	95.7 Ah	93.3 Ah	97%
Resistance (5 s, DoD 50%, −1C)	0 °C	No data	2.3 mΩ	-
	25 °C	0.75 mΩ	1.3 mΩ	27%
	40 °C	No data	1.6 mΩ	-

The result presented above shows through experimental results that a second life and a fresh cell can perform comparably. This result could be extended to the module level as the tested cell was selected as a representative element of the performance of the 11 other cells in the module. To explain this high level of performance, the authors hypothesize that the tested second life cell came from a crashed vehicle. This study highlights the pertinence of defining the second life battery based on an application change rather than a performance level. Batteries can reach the end of their automotive life with a performance level well above the commonly accepted threshold of 20% capacity loss.

3. Second-Life Cell Modeling

Accurate estimation of a lithium-ion cell's voltage behavior over time is paramount in most applications. In fact, the efficiency of thermal management, energy management, and balancing strategies depends heavily on the accuracy of the estimate [37–41]. In a large number of industrial applications where real-time control is required, electrical models are preferred over electrochemical [42]. Their ability to accurately emulate voltage with limited computational cost makes them suitable for embedded applications. In addition, they can be characterized using non-destructive methods [43–45].

This section presents the equivalent circuit models used in the study and the procedure used to calibrate and validate them. The work presented can be divided into three parts. First, two equivalent circuit models capable of emulating the voltage response of a lithium-ion cell are presented. Then, the parameters of the models at 0, 25 and 40 °C are identified. Finally, the accuracy of each model has been evaluated thanks to two usage profiles: a Worldwide Harmonized Light Vehicles Test Procedures cycle and a profile representative of a real mobile charging station usage. Data processing and modeling have been carried out thanks to two open software: DATTES and VEHLIB.

DATTES is a free and open-source Matlab/Octave set of tools for processing experimental data on energy storage systems [33]. This software is capable of processing data from many different battery cyclers and the most common battery techniques, including capacity and impedance measurements or OCV, ICA and EIS analysis. It allows users to process experimental data using a methodology carefully described in the documentation, or to customize the analysis method according to their needs.

VEHLIB is a free and open-source energy management simulation software for electrified vehicles developed in MATLAB/Simulink [34]. It is based on a block diagram description combined with principles from Bond Graph theory.

3.1. Model Definition

In this article, two equivalent circuit models have been chosen.

The first one is shown in Figure 6 and will be named "CPE model". It consists of an open circuit voltage (OCV) source in series with two elements: a resistance R_0 which models the ohmic behavior and a Constant Phase Element (CPE) associated with the other dynamic behavior.

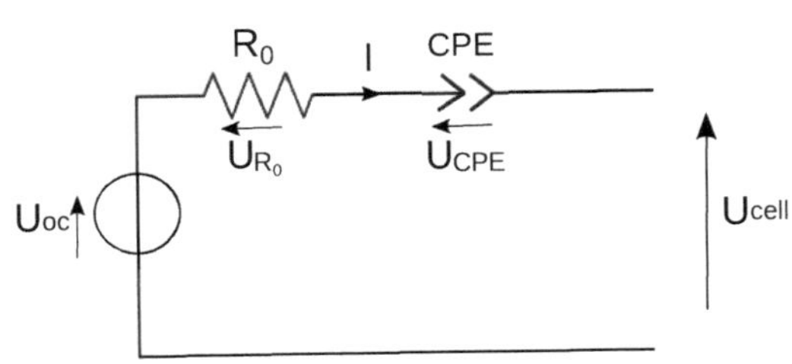

Figure 6. CPE model.

This model emulates the battery voltage response when a current is applied according to Equation (1).

$$U_{cell} = U_{oc} + U_{R0} + U_{CPE} \quad (1)$$

with

- U_{cell}: the voltage response of the cell (V)
- U_{oc}: the open circuit voltage (V)
- U_{R0}: the ohmic drop (V)
- U_{CPE}: the voltage dynamic modeled by the constant phase element (V)

The second model is shown in Figure 7 and will be named "RC model". It consists of a voltage source OCV in series with resistance R_0 which models the ohmic behavior, a first RC circuit for modeling the charge transfer and double layer processes and a second RC circuit associated with the diffusion behavior. To maximize the model accuracy during relaxation, adding more RC circuits would be necessary [46]. However, this would also dramatically increase the computational time for more advanced simulations using this model. This model emulates the battery voltage response when a current is applied according to Equation (2).

$$U_{cell} = U_{oc} + U_{R0} + U_{R1C1} + U_{R2C2} \quad (2)$$

with

- U_{cell}: the voltage response of the cell (V)
- U_{oc}: the open circuit voltage (V)
- U_{R0}: the ohmic drop (V)
- U_{RiCi}: the voltage dynamic modeled by i-th RC circuit (V)

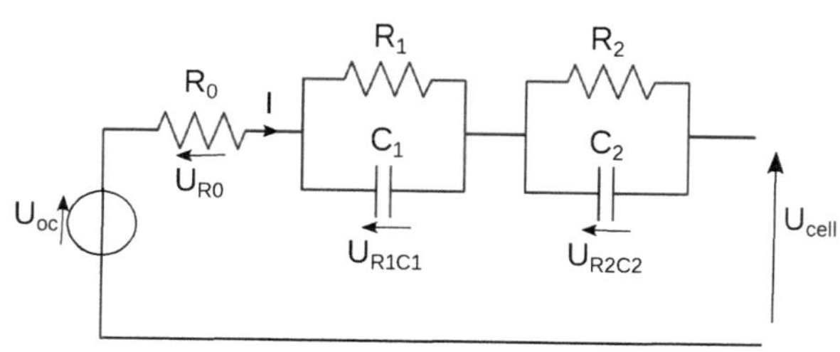

Figure 7. RC model.

For each temperature, the RC and CPE parameters of the models are calibrated in two steps according to the phenomena modeled. The static or thermodynamic behavior is modeled by the open circuit voltage, while the dynamic or polarization behavior is modeled by the parameters R0 + CPE or R0 + R1C1 + R2C2.

3.2. Static Behavior Identification

The static behavior of a battery is usually modeled by an open-circuit voltage source whose behavior evolves with the depth of discharge, temperature and state of health of the battery. The influence of aging is not considered in this study.

In this study, the open circuit voltage (OCV) is determined at C/20 thanks to step 3 of the test shown in Figure 3. It is calculated by averaging the low rate charge and discharge voltage curves, a technique known as pseudo open circuit voltage. In the datasheet, the open circuit voltage is measured using a galvanostatic intermittent titration technique at 1/3C. This method is slower but more accurate than pseudo-OCV because it is less subject to current polarization [47]. In this work, pseudo-OCV has been favored for its speed in producing a sufficient number of experimental points to accurately model the dependence of open circuit voltage on the depth of discharge. Figure 8 shows the influence of temperature and depth of discharge on the open circuit voltage.

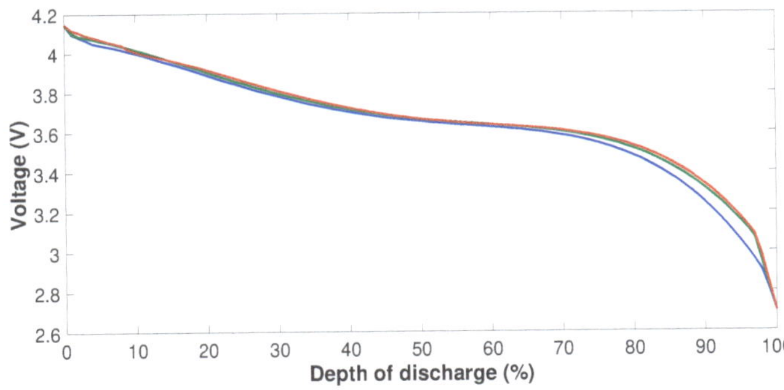

Figure 8. Experimental OCV-DoD relationship curves for second life battery at 0 °C (blue line), 25 °C (green line) and 40 °C (red line).

In order to model the influence of the depth of discharge on the open circuit voltage, 100 experimental points were considered, ranging from 0% to 100% depth of discharge. Linear interpolation is used between two experimental points for intermediate levels of discharge depth. This method was favored as it maximizes accuracy compared to modeling thanks to an equation.

3.3. Dynamic Behavior Identification

The dynamic behavior of a battery is modeled by the impedance parameters of the model. These parameters are influenced by the depth of discharge, temperature and state of health of the battery. The influence of aging is not considered in this study.

The resistance R_0 models the ohmic drop, it is determined first as it is the shortest dynamic phenomenon that will be modeled in this study. The influence of discharge depth and temperature on the resistance was determined experimentally in step 2 of the characterization test shown in Figure 3. To model the influence of discharge depth on R_0, 10 experimental points were considered from discharge depth 10% to 90% and at a current rate of -1C. A logarithmic interpolation is considered between two experimental points for intermediate levels of discharge depth. The influence of the depth of discharge on the 0 s resistance is modeled according to Equation (A6). The result of the resistance interpolation is shown in Figure 9.

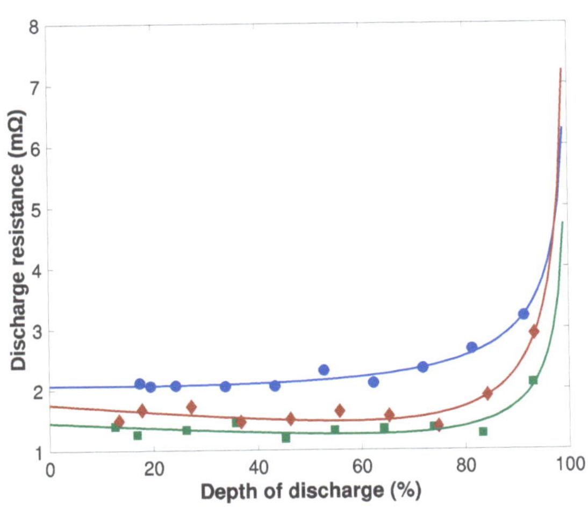

Figure 9. Experimental result and model response (lines) for 0 s discharge resistance versus depth of discharge and temperature 0 (blue), 25 (green) and 40 °C (red).

The parameters R_1, C_1, R_2, C_2, Q and α are intended to emulate the dynamic behavior of a mobile charging station. To model the influence of the depth of discharge on these parameters, 10 experimental points were considered, ranging from a depth of discharge of 10% to 90% and at a current rate of -1C. The equations used to model the influence of the depth of discharge on these parameters are presented in Appendix A and the values of models parameters are available in Appendix B.

4. Models Accuracy Assessment

The previous sections presented the identification of model parameters. It consisted in determining the parameters responsible for modeling the different battery dynamics separately thanks to the experimental data collected during the characterization test. In this section, the different parameters are collected to build the RC and CPE models and to validate their accuracy.

4.1. Accuracy on a WLTC Profile

To facilitate comparison with other modeling studies, a popular usage profile was used to validate the model: the worldwide harmonized light vehicles test procedures cycle. Wheel power demand data available for a Golf-GTE PHEV and previously published in [48] were used. In [48], the Golf-GTE PHEV chassis (mass, aerodynamics, rolling resistance, etc.) was used to evaluate the battery demand over the driving cycle. The VEHLIB library was used to perform the simulation [34,49]. Figure 10 shows the Worldwide Harmonized Light Vehicles Test Cycle profile used.

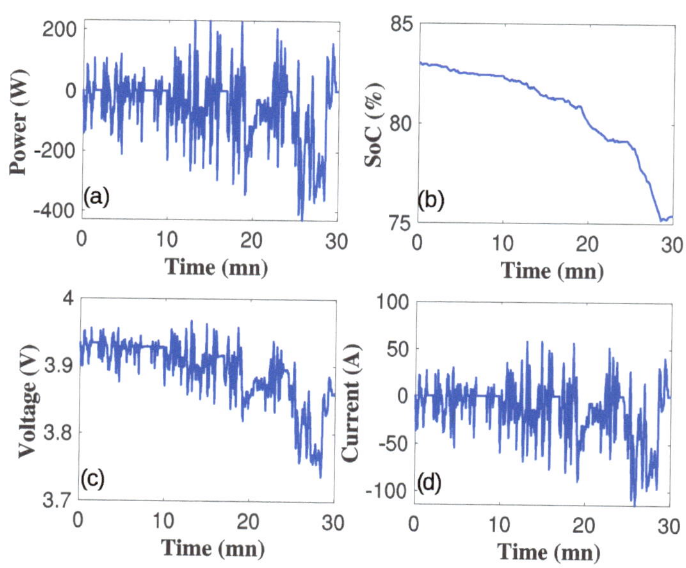

Figure 10. (**a**) Power, (**b**) state of charge, (**c**) voltage and (**d**) current evolution during a Worldwide Harmonized Light Vehicles Test Cycle profile.

4.1.1. Influence of the Model Choice over Models Accuracy

Figure 11 shows in blue line the second life cell voltage measured at 25 °C and compared to CPE model response in solid line and RC model in dashed line.

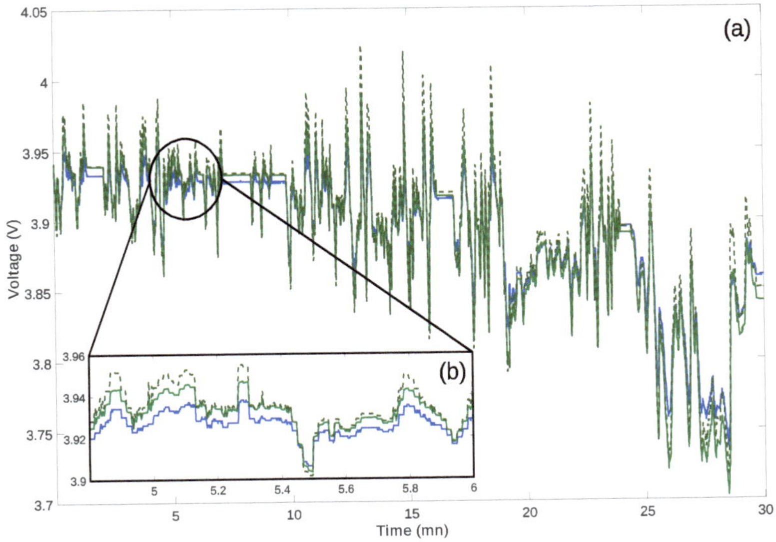

Figure 11. (**a**) Experimental data (blue line) compared to simulation results for the CPE model (solid line) and the RC model (dashed line) at 25 °C. (**b**) Zoom around 5 min.

A large part of the error comes from the dynamic behavior, more precisely from the interpolation of the resistance at the intermediate level of the depth of discharge. To improve the accuracy of the models, the parameters responsible for the voltage dynamics, especially at the end of charge and discharge, should be optimized. Efforts could also be made to more accurately account for the relaxation effect.

Figure 12 shows the error distribution functions of the models. For both the CPE model in the solid line and the RC model in the dashed line, the simulation error is less than ±7 mV, ±22 mV, and ±40 mV in 50, 90, and 99% of the profile, respectively.

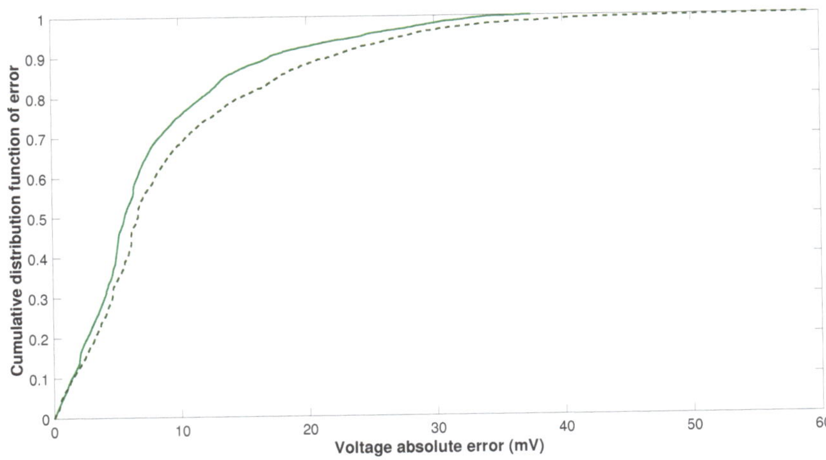

Figure 12. Error distribution functions for the CPE model (solid line) and the RC model (dashed line) emulating a cell voltage response used with a WLTC profile at 25 °C.

4.1.2. Influence of the Temperature over Models Accuracy

In this section, the influence of temperature on model accuracy is assessed thanks to a WLTC profile. Figure 13 presents the influence of temperature on the RC model accuracy. The error distribution function is plotted in blue for 0 °C, green for 25 °C and red for 40 °C.

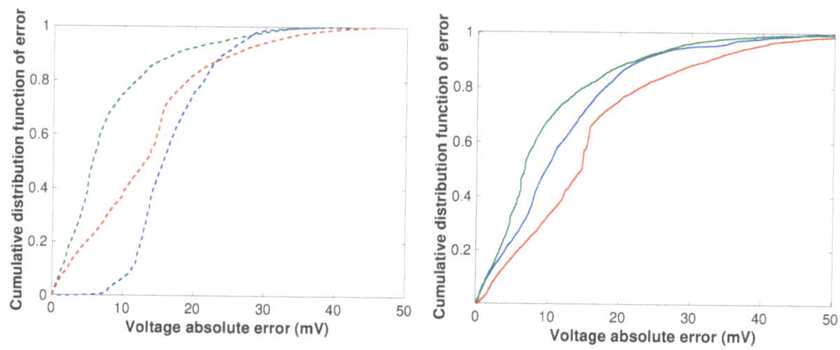

Figure 13. Influence of temperature over RC model ((**left**), dashed line) and CPE model ((**right**), solid line) accuracy. Blue lines are at 0 °C, green lines at 25 °C and red lines at 40 °C.

For the RC model, simulation error is lower than ±16 mV, ±25 mV and ±40 mV in, respectively, 50, 90 and 99% of the profile. While for the CPE model, simulation error is lower than ±15 mV, ±32 mV and ±50 mV in, respectively, 50, 90 and 99% of the profile.

4.2. *Accuracy on A Mobile Charging Station Profile*

A mobile charging station's usage profile inspired by real ones has been used to compare the model's capabilities to emulate the voltage response of the reused batteries. The real usage profiles were provided by the company Mob-Energy. Figure 14 shows the usage profile of the mobile charger with power levels given for a single lithium-ion cell.

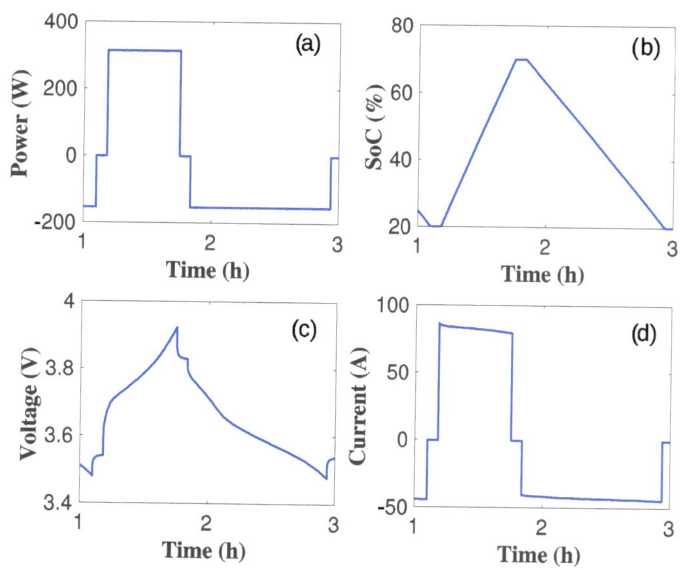

Figure 14. (**a**) Power, (**b**) state of charge, (**c**) voltage and (**d**) current evolution during a mobile charging station usage profile.

The first step of the profile consists of a high power discharge corresponding to the energy transfer from the charger to the vehicle. In this step, the discharge power reaches −150 W at the cell scale, which corresponds to a 7200 W discharge at the robot scale. Then the robot charger moves to grid charging, which corresponds to the low power discharge (almost 0 W). Finally, the mobile charger is charged from the grid. In this step, the charging power reaches 300 W at the cell scale, which corresponds to a 14,400 W charge at the robot scale. It can now move on to the next vehicle to charge. This pattern is repeated 12 times in a 24-h period. The voltage, state of charge and current responses of a cell are also shown.

Influence of the Model Choice over Models Accuracy

Figure 15 shows in blue line the second life cell voltage measured with a mobile charger profile at 25 °C. It is compared with the response of the CPE model in the solid line and the RC model in the dashed line. The cumulative distribution function of both models is also shown.

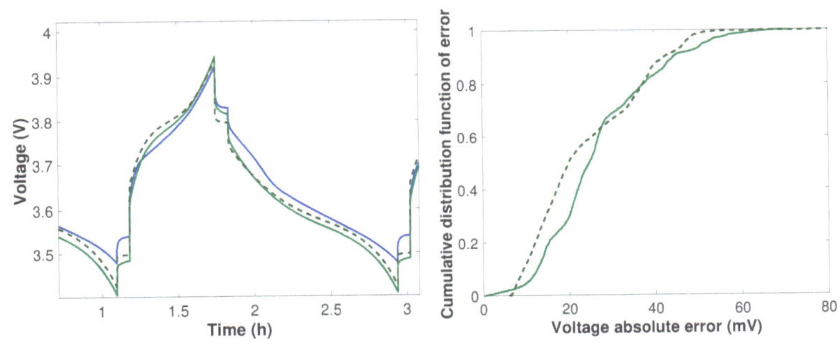

Figure 15. Influence of model choice over accuracy at 25 °C. Left, voltage profile; right error distribution (blue line = measurements, green solid line = CPE model, green dashed line = RC model).

For both the CPE and RC models, the simulation error is less than ±25 mV, ±45 mV and ±61 mV in 50, 90 and 99% of the profile, respectively. The accuracy of the models is lower in this usage profile, which can be explained by the wider range of the depth of discharge. The interpolation of the dynamic parameters in the middle of the depth of discharge is responsible for more than 75% of the model inaccuracy. To improve the accuracy of the models, more experimental data points should be collected and efforts should focus on interpolation techniques.

4.3. Computation Time and Error Assessment

Table 4 shows the accuracy and computation time of the models for different temperatures using a thirty-minute WLTC profile and an eleven-hour mobile charging station profile, respectively. The simulations were run on SIMULINK R2021a with an ode15s solver and a maximum step size of 100 ms. The computer used was a DELL Latitude 5290 laptop with an Intel® Core™ i7-8650U CPU and a clock speed of 1.90 GHz.

CPE and RC models have a comparable absolute average error, but simulation time is significantly in favor of the latter. Both models perform reasonably well, with accuracy levels around 1%, which is in the order of magnitude of models presented in the literature [50–53].

Models are more accurate with the WLTC profile compared to the mobile charging station one. This difference can be explained by the severity of the charger profile. The performance of the models is sufficient to estimate, accurately and with low computational load, the energy stored in the mobile charging station and its efficiency at a given time. These quantities are central for defining optimal pricing, scheduling, and energy management to maximize profits and minimize battery degradation [54–56].

Table 4. Accuracy and computation time for different temperatures using the mobile charging station and WLTC profiles.

			CPE Model	RC Model
WLTC	Accuracy (RMSE)	0 °C	17.5 mV	14.8 mV
		25 °C	10.5 mV	12.9 mV
		40 °C	15.9 mV	19.2 mV
	Computation time	0 °C	11.2 s	3 s
		25 °C	8.7 s	3.1 s
		40 °C	9.9 s	3.3 s
Charger			CPE model	RC model
	Accuracy (RMSE)	0 °C	32.6 mV	35.4 mV
		25 °C	29.5 mV	27.2 mV
	Computation time	0 °C	71.9 s	35.6 s
		25 °C	80.8 s	35.1 s

If a better estimation of the energy stored in the charging station at a given time is required, the accuracy of the models could be improved by adding parameters to the models. However, this strategy would certainly increase the computational time, so a trade-off between model accuracy and computational time must be made.

5. Conclusions

This article presents an experimental evaluation of the performance of a second life battery. This experimental work contributes to the existing literature on second life batteries as it is one of the first to evaluate the performance of a high-capacity prismatic cell extracted from an electric vehicle battery pack. The characterization of a cell extracted from an electric vehicle has shown that a second life and a fresh cell can perform comparably. This study highlights the relevance of defining second life batteries based on a change of application rather than a performance threshold between first and second life.

The modeling study conducted has demonstrated the ability of two models to accurately emulate the voltage response of a second life battery. The influence of usage profile and temperature on the accuracy of the models was also evaluated. This study contributes to the development of battery reuse by presenting a new possible application: mobile charging stations. This work is also the first step in the development of pricing, charging schedule, and energy management strategies for battery-powered mobile charging stations.

Finally, this study was conducted with an open-science approach. Characterization data are shared along with a detailed description of the test procedures [32]. Data processing has also been made thanks to two open software: DATTES and VEHLIB [33,34].

Author Contributions: Conceptualization: M.H., E.R.-I. and P.V.; Methodology: M.H., E.R.-I. and P.V.; Software: M.H. and E.R.-I.; Validation: M.H., E.R.-I. and P.V.; Data analysis: M.H.; Writing—original draft preparation: M.H.; Writing—review and editing: E.R.-I. and P.V.; Supervision: E.R.-I. and P.V.; Funding acquisition: E.R.-I. and P.V. All authors have read and agreed to the published version of the manuscript.

Funding: This research was lead as part of the project BAROM funded by Région Auvergne-Rhône-Alpes.

Data Availability Statement: Experimental data are shared on the a public depository at https://data.univ-gustave-eiffel.fr/dataverse/second_life_batteries (accessed on 27 February 2023). Code for data processing is shared at https://dattes.gitlab.io/page/examples/second_life/ (accessed on 27 February 2023).

Acknowledgments: Authors would like to thank Mob-Energy for the collaboration including sharing of a real mobile charging station usage profile. This article supports a project on the use of second life batteries in mobile charging station which would not have been possible without the help of Serge Pelissier, whom we would like to thank.

Conflicts of Interest: The authors declare no conflict of interest.

Abbreviations

The following abbreviations are used in this manuscript:

BOL	Beginning of Life
CC	Constant Current
CPE	Constant Phase Element
CV	Constant Voltage
DoD	Depth of discharge
EIS	Electrochemical Impedance Spectroscopy
ICA	Incremental Capacity Analysis
NMC	Nickel Manganese Cobalt
OCV	Open Circuit Voltage
RMSE	Root Mean Square Error
SoH	Stater of Health
WLTC	Worldwide Harmonized Light Vehicles Test Cycle

Appendix A. Equations

Appendix A.1. Capacity

$$Q(t) = \frac{1}{3600} \int_{t_0}^{t} i(t)dt \tag{A1}$$

with

- $Q(t)$: the discharge capacity measured (Ah)
- $t - t_0$: the duration of a discharge cycle (s)
- i: the discharge current applied to the cell (A)

Appendix A.2. State of Health Capacity

$$SoH_Q = 100 \times \frac{Q(t)}{Q_{BOL}} \tag{A2}$$

with

- SoH_Q: the state of health "capacity" (%)
- $Q(t)$: the capacity measured at a given time (Ah)
- Q_{BOL}: the capacity measured at the beginning of life (BOL) (Ah)

Appendix A.3. Depth of Discharge

$$DoD = DoD_{t0} - 100 \times \frac{1}{3600 \times Q_{BOL}} \int_{t_0}^{t} i(t)dt) \tag{A3}$$

with

- DoD: the depth of discharge of the cell (%)
- DoD_{t0}: the initial cell depth of discharge (%)
- Q_{BOL}: the capacity measured at the beginning of life (BOL) (Ah)
- $i(t)$: the current at a given time (A) (Current is negative in discharge)

Appendix A.4. Pulse Resistance

$$R_{\Delta t}(t) = \frac{\Delta V}{\Delta I} \tag{A4}$$

with

- $R_{\Delta t}(t)$: the resistance measured at a given time (Ω)
- Δt: the time after the pulse beginning (s)

- ΔV: the voltage variation after Δt (V)
- ΔI: the current pulse amplitude (A)

Appendix A.5. State of Health Resistance

$$SoH_R = 100 \times (1 - \frac{R(t) - R_{BOL}}{R_{BOL}}) \quad \text{(A5)}$$

with

- SoH_R: the state of health "resistance" (%)
- $R(t)$: the resistance measured at a given time (mΩ)
- R_{BOL}: the resistance measured at the beginning of life (BOL) (mΩ)

Appendix A.6. Resistance Dependence to Depth of Discharge

$$R(DoD) = r_2 \times log(DoD)^2 + r_1 \times log(DoD) + r_0 \quad \text{(A6)}$$

with

- $R(DoD)$: the resistance for a given depth of discharge (Ω)
- r_i: the polynomial coefficient of the model (Ω)

Appendix A.7. C1 Dependence to Depth of Discharge

$$C1(DoD) = c_{1,3} \times (DoD)^3 + c_{1,2} \times (DoD)^2 + c_{1,1} \times (DoD) + c_{1,0} \quad \text{(A7)}$$

with

- $C1(DoD)$: the resistance for a given depth of discharge (F)
- c_i: the polynomial coefficient of the model (F)

Appendix A.8. RC Impedance

$$Z_{RC}(jw) = \frac{1}{1 + RCjw} \quad \text{(A8)}$$

with

- $Z_{RC}(jw)$: the RC impedance (Ω)
- R: the resistance(Ω)
- R: the capacitor (F)

Appendix A.9. R1C1 Pulse Response Voltage

$$U_{R1C1} = I_{cell} \times exp(\frac{-t}{\tau_1}) \quad \text{(A9)}$$

with

- U_{R1C1}: the voltage polarization caused by R_1C_1 (V)
- τ_1: the polarization time constant of the fastest dynamic phenomena (s)

Appendix A.10. R2C2 Pulse Response Voltage

$$U_{R2C2} = I_{cell} \times exp(\frac{-t}{\tau_2}) \quad \text{(A10)}$$

with

- U_{R2C2}: the voltage polarization caused by R_2C_2 (V)
- τ_2: the polarization time constant of the fastest dynamic phenomena (s)

Appendix A.11. CPE Impedance

$$Z_{CPE}(jw) = \frac{1}{Q(jw)^{\alpha}} \tag{A11}$$

with

- $Z_{CPE}(jw)$: the CPE impedance (Ω)
- $Q(DoD)$: the main CPE element for a given depth of discharge ($\Omega^{-1}s^{\alpha}$)
- α: the depression factor of the CPE (-)

Appendix A.12. CPE Voltage

$$U_{CPE}(t) = I_{cell} \times \frac{(t-t_1)^{\alpha} - (t-t_2)^{\alpha}}{Q \times \Gamma(\alpha+1)} \tag{A12}$$

with

- U_{CPE}: the voltage polarization caused by the CPE element (V)
- t_1: the time at the beginning of the pulse (s)
- t_2: the time at the end of the pulse (s)
- α: the depression factor of the CPE (-)
- Q: the main CPE element ($\Omega^{-1}s^{\alpha}$)
- Γ: the Gamma function

Appendix A.13. Gamma Function

$$\Gamma(\alpha) = \int_0^{\infty} t^{\alpha-1} \times 10^{-t} dt \tag{A13}$$

with

- Γ: the Gamma function
- t: the time (s)
- α: the depression factor of the CPE (-)

Appendix B. Models Parameters

Table A1. OCV models parameters for DoD from 1% to 9%.

DoD	1	2	3	4	5	6	7	8	9
OCV	4.150 4.150 4.150	4.099 4.105 4.120	4.080 4.089 4.109	4.069 4.082 4.094	4.053 4.075 4.084	4.046 4.067 4.073	4.038 4.058 4.061	4.029 4.049 4.053	4.021 4.039 4.042

Table A2. OCV models parameters for DoD from 10% to 19%.

DoD	10	11	12	13	14	15	16	17	18	19
OCV	4.011 4.029 4.023	4.001 4.019 4.009	3.991 4.008 3.997	3.981 3.997 3.990	3.969 3.986 3.982	3.959 3.975 3.974	3.948 3.964 3.965	3.936 3.952 3.956	3.925 3.940 3.947	3.913 3.929 3.937

Table A3. OCV models parameters for DoD from 20% to 29%.

DoD	20	21	22	23	24	25	26	27	28	29
OCV	3.902 3.917 3.927	3.890 3.905 3.916	3.879 3.894 3.906	3.867 3.882 3.895	3.856 3.871 3.885	3.845 3.859 3.874	3.834 3.849 3.863	3.823 3.838 3.853	3.813 3.827 3.842	3.803 3.817 3.831

Table A4. OCV models parameters for DoD from 30% to 39%.

DoD	30	31	32	33	34	35	36	37	38	39
OCV	3.792 3.807 3.821	3.783 3.797 3.811	3.773 3.788 3.801	3.764 3.778 3.791	3.755 3.769 3.781	3.745 3.761 3.772	3.738 3.752 3.763	3.730 3.744 3.754	3.723 3.737 3.746	3.716 3.729 3.738

Table A5. OCV models parameters for DoD from 40% to 49%.

DoD	40	41	42	43	44	45	46	47	48	49
OCV	3.709 3.722 3.730	3.702 3.716 3.723	3.696 3.709 3.716	3.690 3.703 3.709	3.685 3.698 3.702	3.679 3.692 3.696	3.674 3.687 3.690	3.669 3.682 3.685	3.665 3.678 3.680	3.661 3.674 3.675

Table A6. OCV models parameters for DoD from 50% to 59%.

DoD	50	51	52	53	54	55	56	57	58	59
OCV	3.658 3.669 3.671	3.654 3.666 3.667	3.651 3.663 3.663	3.647 3.659 3.659	3.644 3.656 3.656	3.641 3.653 3.653	3.639 3.651 3.649	3.636 3.648 3.647	3.633 3.645 3.644	3.630 3.643 3.642

Table A7. OCV models parameters for DoD from 60% to 69%.

DoD	60	61	62	63	64	65	66	67	68	69
OCV	3.628 3.640 3.639	3.625 3.637 3.637	3.622 3.635 3.634	3.619 3.632 3.632	3.616 3.629 3.629	3.612 3.626 3.627	3.608 3.623 3.624	3.604 3.619 3.621	3.599 3.615 3.618	3.595 3.611 3.615

Table A8. OCV models parameters for DoD from 70% to 79%.

DoD	70	71	72	73	74	75	76	77	78	79
OCV	3.589 3.606 3.611	3.583 3.601 3.607	3.577 3.596 3.602	3.569 3.589 3.597	3.561 3.583 3.591	3.552 3.576 3.585	3.543 3.568 3.577	3.532 3.559 3.569	3.520 3.549 3.561	3.508 3.538 3.551

Table A9. OCV models parameters for DoD from 80% to 89%.

DoD	80	81	82	83	84	85	86	87	88	89
OCV	3.494 3.527 3.540	3.479 3.514 3.529	3.463 3.500 3.516	3.445 3.486 3.502	3.426 3.469 3.486	3.405 3.452 3.469	3.383 3.433 3.451	3.359 3.413 3.431	3.333 3.390 3.410	3.305 3.367 3.387

Table A10. OCV models parameters for DoD from 90% to 100%.

DoD	90	91	92	93	94	95	96	97	98	99	100
OCV	3.275 3.342 3.362	3.244 3.315 3.335	3.209 3.286 3.306	3.173 3.255 3.275	3.135 3.222 3.241	3.094 3.187 3.205	3.050 3.150 3.167	3.004 3.110 3.127	2.955 3.068 3.083	2.903 2.944 2.977	2.700 2.700 2.700

Table A11. R0 models parameters in discharge.

	R0 Discharge		
Temperature	$r_{0,2,dch}$	$r_{0,1,dch}$	$r_{0,0,dch}$
0 °C	5.746×10^{-4}	-4.720×10^{-3}	1.157×10^{-2}
25 °C	1.259×10^{-4}	8.614×10^{-4}	5.356×10^{-3}
40 °C	-1.501×10^{-4}	9.962×10^{-4}	-9.188×10^{-5}

Table A12. R0 models parameters in charge.

	R0 Charge		
Temperature	$r_{0,2,ch}$	$r_{0,1,ch}$	$r_{0,0,ch}$
0 °C	6.953×10^{-4}	-5.542×10^{-3}	1.311×10^{-2}
25 °C	-5.455×10^{-5}	2.307×10^{-2}	5.995×10^{-3}
40 °C	2.273×10^{-4}	-1.776×10^{-3}	5.070×10^{-3}

Table A13. R1 models parameters.

	R1		
Temperature	$r_{1,2}$	$r_{1,1}$	$r_{1,0}$
0 °C	2.037×10^{-4}	-1.697×10^{-3}	4.208×10^{-3}
25 °C	5.111×10^{-5}	-5.698×10^{-4}	1.847×10^{-4}
40 °C	6.204×10^{-5}	-4.265×10^{-4}	1.405×10^{-3}

Table A14. R2 models parameters.

	R2		
Temperature	$r_{2,2}$	$r_{2,1}$	$r_{2,0}$
0 °C	5.399×10^{-6}	-2.517×10^{-5}	2.366×10^{-5}
25 °C	2.813×10^{-6}	-1.355×10^{-5}	1.555×10^{-5}
40 °C	5.994×10^{-7}	-3.216×10^{-6}	6.739×10^{-6}

Table A15. C1 and C2 models parameters.

	C1				C2
Temperature	$c_{1,3}$	$c_{1,2}$	$c_{1,1}$	$c_{1,0}$	C2
0 °C	−3.094	547	-3.119×10^4	6.601×10^5	58,422
25 °C	−2.517	380.3	-1.765×10^4	4.053×10^5	62,841
40 °C	−1.610	178.7	−3902	1.703×10^5	95,467

Table A16. Q and α models parameters.

Temperature	Q	α
0 °C	3873	0.1
25 °C	4852	0.1
40 °C	5812	0.1

References

1. IPCC. *Climate Change 2022: Mitigation of Climate Change*; Contribution of Working Group III to the Sixth Assessment Report of the Intergovernmental Panel on Climate Change; IPCC: Geneva, Switzerland, 2022.
2. IEA. *Global EV Outlook 2021*; IEA: Paris, France, 2021.
3. IEA. *Global Supply Chains of EV Batteries*; IEA: Paris, France, 2022.
4. Larcher, D.; Tarascon, J.M. Towards greener and more sustainable batteries for electrical energy storage. *Nat. Chem.* **2015**, *7*, 19–29. [CrossRef] [PubMed]
5. European Union. Directive 2008/98/EC of the European Parliament and the Council of 19 November 2008 on Waste and Repealing Certain Directives. *Off. J. Eur. Union* **2008**, L 312/3. Available online: http://data.europa.eu/eli/dir/2008/98/oj (accessed on 26 February 2023).
6. Ellingsen, L.A.W.; Singh, B.; Strømman, A.H. The size and range effect: Lifecycle greenhouse gas emissions of electric vehicles. *Environ. Res. Lett.* **2016**, *11*, 054010. [CrossRef]
7. Redondo-Iglesias, E.; Vinot, E.; Venet, P.; Pelissier, S. Electric vehicle range and battery lifetime: A trade-off. In Proceedings of the Electric Vehicle Symposium (EVS32), Lyon, France, 21–22 May 2019.
8. Kong, P.Y. Autonomous Robot-Like Mobile Chargers for Electric Vehicles at Public Parking Facilities. *IEEE Trans. Smart Grid* **2019**, *10*, 5952–5963. [CrossRef]
9. Hassini, M.; Redondo-Iglesias, E.; Venet, P.; Gillet, S.; Zitouni, Y. Second-life batteries in a Mobile Charging Station: Model Based Performance Assessment. In Proceedings of the 35th International Electric Vehicle Symposium and Exhibition (EVS35), Oslo, Norway, 11–15 June 2022.

10. Mob-Energy. Mob-Energy. Available online: https://www.mob-energy.com/en/home (accessed on 27 March 2023).
11. Volkswagen. The Mobile Charging Robot—Presenting a Vision. Available online: https://www.volkswagen-newsroom.com/ (accessed on 27 March 2023).
12. Afshar, S.; Macedo, P.; Mohamed, F.; Disfani, V. Mobile charging stations for electric vehicles—A review. Renew. Sustain. Energy Rev. **2021**, *152*. [CrossRef]
13. European Commission. *Sustainable Batteries Regulation*; European Commission: Brussel, Belgium, 2020.
14. Braco, E.; San Martin, I.; Berrueta, A.; Sanchis, P.; Ursúa, A. Experimental assessment of cycling ageing of lithium-ion second life batteries from electric vehicles. *J. Energy Storage* **2020**, *32*, 101695. [CrossRef]
15. Abdel-Monem, M.; Hegazy, O.; Omar, N.; Trad, K.; De Breucker, S.; Van Den Bossche, P.; Van Mierlo, J. Design and analysis of generic energy management strategy for controlling second life battery systems in stationary applications. *Energies* **2016**, *9*, 889. [CrossRef]
16. Seger, P.V.; Rigo-Mariani, R.; Thivel, P.X.; Riu, D. A storage degradation model of Li-ion batteries to integrate ageing effects in the optimal management and design of an isolated microgrid. *Appl. Energy* **2023**, *333*. [CrossRef]
17. Etxandi-Santolaya, M.; Canals Casals, L.; Amante García, B.; Corchero, C. Circular Economy-Based Alternatives beyond Second-Life Applications: Maximizing the Electric Vehicle Battery First Life. *World Electr. Veh. J.* **2023**, *14*, 66. [CrossRef]
18. Hossain, E.; Murtaugh, D.; Mody, J.; Faruque, H.M.R.; Sunny, M.S.H.; Mohammad, N. A comprehensive review on second life batteries: Current state, manufacturing considerations, applications, impacts, barriers & potential solutions, business strategies, and policies. *IEEE Access* **2019**, *7*, 73215–73252. [CrossRef]
19. Haram, M.H.S.M.; Lee, J.W.; Ramasamy, G.; Ngu, E.E.; Thiagarajah, S.P.; Lee, Y.H. Feasibility of utilising second life EV batteries: Applications, lifespan, economics, environmental impact, assessment, and challenges. *Alex. Eng. J.* **2021**, *60*, 4517–4536. [CrossRef]
20. Montes, T.; Etxandi-Santolaya, M.; Eichman, J.; Ferreira, V.J.; Trilla, L.; Corchero, C. Procedure for Assessing the Suitability of Battery Second Life Applications after EV First Life. *Batteries* **2022**, *8*, 122. [CrossRef]
21. Martinez-Laserna, E.; Sarasketa-Zabala, E.; Sarria, I.V.; Stroe, D.I.; Swierczynski, M.; Warnecke, A.; Timmermans, J.-M.; Goutam, S.; Omar, N.; Rodriguez, P. Technical Viability of Battery Second Life: A Study From the Ageing Perspective. *IEEE Trans. Ind. Appl.* **2018**, *54*, 2703–2713. [CrossRef]
22. Debnath, U.K.; Ahmad, I.; Habibi, D. Quantifying economic benefits of second life batteries of gridable vehicles in the smart grid. *Int. J. Electr. Power Energy Syst.* **2014**, *63*, 577–587. [CrossRef]
23. Attia, P.; Bills, A.; Planella, F.B.; Dechent, P.; Dos Reis, G.; Dubarry, M.; Gasper, P.; Gilchrist, R.; Greenbank, S.; Howey, D.; et al. "Knees" in lithium-ion battery aging trajectories. *J. Electrochem. Soc.* **2022**, *169*, 060517. [CrossRef]
24. Dos Reis, G.; Strange, C.; Yadav, M.; Li, S. Lithium-ion battery data and where to find it. *Energy AI* **2021**, *5*, 100081. [CrossRef]
25. Coron, E.; Geniès, S.; Cugnet, M.; Thivel, P.X. High-Energy Li-Ion Cells: Impact of Electrode Ageing on Second Life Viability. *J. Electrochem. Soc.* **2021**, *168*, 100539. [CrossRef]
26. Seger, P.V.; Thivel, P.X.; Riu, D. A second life Li-ion battery ageing model with uncertainties: From cell to pack analysis. *J. Power Sources* **2022**, *541*, 231663. [CrossRef]
27. Michelini, E.; Höschele, P.; Ratz, F.; Stadlbauer, M.; Rom, W.; Ellersdorfer, C.; Moser, J. Potential and Most Promising Second-Life Applications for Automotive Lithium-Ion Batteries Considering Technical, Economic and Legal Aspects. *Energies* **2023**, *16*, 2830. [CrossRef]
28. Rallo, H.; Casals, L.C.; De La Torre, D.; Reinhardt, R.; Marchante, C.; Amante, B. Lithium-ion battery 2nd life used as a stationary energy storage system: Ageing and economic analysis in two real cases. *J. Clean. Prod.* **2020**, *272*, 122584. [CrossRef]
29. Quinard, H.; Redondo-Iglesias, E.; Pelissier, S.; Venet, P. Fast Electrical Characterizations of High-Energy Second Life Lithium-Ion Batteries for Embedded and Stationary Applications. *Batteries* **2019**, *5*, 33. [CrossRef]
30. Braco, E.; San Martín, I.; Berrueta, A.; Sanchis, P.; Ursúa, A. Experimental assessment of first-and second life electric vehicle batteries: Performance, capacity dispersion, and aging. *IEEE Trans. Ind. Appl.* **2021**, *57*, 4107–4117. [CrossRef]
31. Albuquerque, L.; Lacressonnière, F.; Roboam, X.; Forgez, C. Incremental Capacity Analysis as a diagnostic method applied to second life Li-ion batteries. In Proceedings of the Electrimacs 2022, Nancy, France, 16–19 May 2022.
32. Hassini, M. Dataset Second-life battery Characterization Test. Available online: https://data.univ-gustave-eiffel.fr/dataverse/second_life_batteries (accessed on 27 March 2023).
33. Redondo-Iglesias, E.; Hassini, M. Data Analysis Tools for Tests on Energy Storage. Available online: https://dattes.gitlab.io (accessed on 27 March 2023).
34. Jeanneret, B. VEHLIB. Available online: https://gitlab.univ-eiffel.fr/eco7/vehlib (accessed on 27 March 2023).
35. PushEV. Samsung SDI 94Ah battery cell full specifications. Available online: https://pushevs.com/2018/04/05/samsung-sdi-94-ah-battery-cell-full-specifications/ (accessed on 31 March 2023).
36. RTE, France's Transmission System Operator. eCO2mix—All of France's Electricity Data in Real Time. Available online: https://www.rte-france.com/en/eco2mix/ (accessed on 27 March 2023).
37. Damay, N.; Forgez, C.; Bichat, M.P.; Friedrich, G. Thermal modeling of large prismatic LiFePO4/graphite battery. Coupled thermal and heat generation models for characterization and simulation. *J. Power Sources* **2015**, *283*, 37–45. [CrossRef]
38. Lievre, A.; Sari, A.; Venet, P.; Hijazi, A.; Ouattara-Brigaudet, M.; Pelissier, S. Practical online estimation of lithium-ion battery apparent series resistance for mild hybrid vehicles. *IEEE Trans. Veh. Technol.* **2015**, *65*, 4505–4511. [CrossRef]

39. Shili, S.; Hijazi, A.; Sari, A.; Lin-Shi, X.; Venet, P. Balancing Circuit New Control for Supercapacitor Storage System Lifetime Maximization. *IEEE Trans. Power Electron.* **2017**, *32*, 4939–4948. [CrossRef]
40. Ndiaye, A.; Locment, F.; De Bernardinis, A.; Sechilariu, M.; Redondo-Iglesias, E. A Techno-Economic Analysis of Energy Storage Components of Microgrids for Improving Energy Management Strategies. *Energies* **2022**, *15*, 1556. [CrossRef]
41. Parthasarathy, C.; Laaksonen, H.; Redondo-Iglesias, E.; Pelissier, S. Aging aware adaptive control of Li-ion battery energy storage system for flexibility services provision. *J. Energy Storage* **2023**, *57*, 106268. [CrossRef]
42. Seaman, A.; Dao, T.S.; McPhee, J. A survey of mathematics-based equivalent-circuit and electrochemical battery models for hybrid and electric vehicle simulation. *J. Power Sources* **2014**, *256*, 410–423. [CrossRef]
43. Tran, M.K.; DaCosta, A.; Mevawalla, A.; Panchal, S.; Fowler, M. Comparative study of equivalent circuit models performance in four common lithium-ion batteries: LFP, NMC, LMO, NCA. *Batteries* **2021**, *7*, 51. [CrossRef]
44. Nikolian, A.; Jaguemont, J.; De Hoog, J.; Goutam, S.; Omar, N.; Van Den Bossche, P.; Van Mierlo, J. Complete cell-level lithium-ion electrical ECM model for different chemistries (NMC, LFP, LTO) and temperatures (−5 °C to 45 °C). Optimized modelling techniques. *Int. J. Electr. Power Energy Syst.* **2018**, *98*, 133–146. [CrossRef]
45. Sun, L.; Li, G.; You, F. Combined internal resistance and state-of-charge estimation of lithium-ion battery based on extended state observer. *Renew. Sustain. Energy Rev.* **2020**, *131*, 109994. [CrossRef]
46. Li, A.; Pelissier, S.; Venet, P.; Gyan, P. Fast Characterization Method for Modeling Battery Relaxation Voltage. *Batteries* **2016**, *2*, 7. [CrossRef]
47. Dubarry, M.; Baure, G. Perspective on Commercial Li-ion Battery Testing, Best Practices for Simple and Effective Protocols. *Electronics* **2020**, *9*, 152. [CrossRef]
48. Patil, T.D.; Vinot, E.; Ehrenberger, S.; Trigui, R.; Redondo-Iglesias, E. Sensitivity Analysis of Battery Aging for Model-Based PHEV Use Scenarios. *Energies* **2023**, *16*, 1749. [CrossRef]
49. Jeanneret, B.; Trigui, R.; Badin, F.; Harel, F.; Damemme, F. New Hybrid concept simulation tools, evaluation on the Toyota Prius car. In Proceedings of the 16th International Electric Vehicle Symposium, Beijing, China, 12–16 October 1999.
50. Baghdadi, I.; Briat, O.; Eddahech, A.; Vinassa, J.M.; Gyan, P. Electro-thermal model of lithium-ion batteries for electrified vehicles applications. In Proceedings of the 2015 IEEE 24th International Symposium on Industrial Electronics (ISIE, Buzios, Brazil, 3–5 April 2015; pp. 1248–1252. [CrossRef]
51. Seger, P.V.; Coron, E.; Thivel, P.X.; Riu, D.; Cugnet, M.; Genies, S. Open data model parameterization of a second life Li-ion battery. *J. Energy Storage* **2021**, *47*, 103546. [CrossRef]
52. Quelin, A.; Damay, N. Coupling electrical parameters of a battery equivalent circuit model to electrodes dimensions. *J. Power Sources* **2023**, *561*, 232690. [CrossRef]
53. Juston, M.; Damay, N.; Forgez, C. Extracting the diffusion resistance and dynamic of a battery using pulse tests. *J. Energy Storage* **2023**, *57*, 106199. [CrossRef]
54. Afshar, S.; Disfani, V. Optimal Scheduling of Electric Vehicles in the Presence of Mobile Charging Stations. In Proceedings of the IEEE Power & Energy Society General Meeting (PESGM), Denver, CO, USA, 17–21 July 2022. [CrossRef]
55. Saboori, H.; Jadid, S.; Savaghebi, M. Optimal Management of Mobile Battery Energy Storage as a Self-Driving, Self-Powered and Movable Charging Station to Promote Electric Vehicle Adoption. *Energies* **2021**, *14*, 736. [CrossRef]
56. Houbbadi, A.; Trigui, R.; Pelissier, S.; Redondo-Iglesias, E.; Bouton, T. Optimal scheduling to manage an electric bus fleet overnight charging. *Energies* **2019**, *12*, 2727. [CrossRef]

World Electric Vehicle Journal

Article

Evaluation of Different Scenarios to Switch the Whole Regional Bus Fleet of an Italian Alpine Region to Zero-Emission Buses

Wolfram Sparber [1], Andrea Grotto [1,2], Pietro Zambelli [1], Roberto Vaccaro [1] and Alyona Zubaryeva [1,*]

1 Institute for Renewable Energy, European Academy of Bolzano (EURAC Research), Viale Druso 1, 39100 Bolzano, Italy
2 IREC, Jardins de les Dones de Negre 1, 2ª pl., 08930 Sant Adrià del Besòs Barcelona, Spain
* Correspondence: alyona.zubaryeva@eurac.edu; Tel.: +39-0471-055619

Abstract: Public bus decarbonization is increasingly important to address the global issue of climate change. There are several challenges associated with large-scale introduction of zero-emission technologies in public fleets. This is especially the case in an extra-urban context, of mountain regions with challenging weather conditions. In this work the analysis of the state-of-the-art ZEBs, local bus lines, and timetables was performed to understand the best fit of technology—battery electric buses (BEBs) or fuel cell electric buses (FCEBs)—for each line in such a region. Further, a simulation tool was developed to calculate the compatibility of zero-emission technologies with the current needs of the public transportation considering distance, altitude difference, and climate conditions. The results show that a complete switch of the fleet is possible with a slight increase in the number of buses and that there is no clear difference in the distance covered in mountainous areas by BEBs versus FCEBs, but that both technologies can cover similar distances. The tool developed is not limited to bus fleets but can be applied to all kinds of fleets that cover clearly defined daily routes.

Keywords: battery electric bus; fuel cell electric bus; energy consumption; simulation

Citation: Sparber, W.; Grotto, A.; Zambelli, P.; Vaccaro, R.; Zubaryeva, A. Evaluation of Different Scenarios to Switch the Whole Regional Bus Fleet of an Italian Alpine Region to Zero-Emission Buses. *World Electr. Veh. J.* **2023**, *14*, 91. https://doi.org/10.3390/wevj14040091

Academic Editor: Joeri Van Mierlo

Received: 17 February 2023
Revised: 24 March 2023
Accepted: 30 March 2023
Published: 1 April 2023

1. Introduction

Public bus decarbonization is becoming increasingly important to address the global issue of climate change. The transportation sector is one of the largest sources of greenhouse gas emissions, given that passenger travel vehicles (cars, motorbikes, and buses) produce around 45% of global transport emissions [1]. Decarbonizing the public bus fleet not only reduces emissions, but it also has a range of other benefits for both the environment and society. There are, however, several challenges associated with large-scale introduction of zero-emission technologies in public fleets, that include: high upfront costs [2], technical issues (i.e., battery range, charging times, and the reliability of the technology) [3] or financial sustainability [4] to ensure long-term maintenance and operation costs within the public domain. The public transportation authority (PTA) in the Italian Alpine province of South Tyrol has been at the forefront in the testing and implementation of the innovative zero-emission bus fleets for decades [5] and foresees the complete phase-out of diesel buses and their replacement with zero emission buses (ZEBs). The geographical and climatic characteristics of the province also represent peculiarities for currently available zero-emission technologies and their management. South Tyrol is the most northern province of Italy and is located in the Dolomite Alps with around 60% of the area located above 1500 m a.s.l., only 4% of the area is located below 500 m a.s.l. Besides its around 530,000 inhabitants [6], it is visited by several million tourists per year. To act as a representative region for green transport [7] and in consideration of the climate policy developments in Europe, South Tyrol approved an ambitious provincial climate plan [8], which includes drastic reduction of transport GHG emissions. In this work, first the analysis of the state-of-the art ZEBs, South Tyrolean bus lines, and their timetables was carried out. The goal of the study was to understand which technology—battery electric buses (BEBs) or fuel cell electric buses (FCEBs)—is appropriate for a particular line. Furthermore, a simulation tool was developed to calculate the compatibility of zero-emission technologies with the current needs of the public transportation service in South Tyrol to provide a preliminary study to decision-makers. The authors adopted a simplified energy modelling methodology, accepting model limitations. To the best of the authors' knowledge, there have not yet been scholarly studies that combine the relationship among several important factors [9] for zero-emission fleets in extra-urban environments: elevation, temperature considerations, longer distances, timetables, and both electric and hydrogen bus technologies.

2. Literature Review on State-of-Art of Zero-Emission Fleets Simulation Tools Used in Planning

While several works concentrate on the evaluation and comparison of the environmental and financial performance of future electric and hydrogen fleets [10–12], only few offer frameworks for a regional transition with specific applications. To keep this review succinct, we focus on the relevant optimization and simulation tools that either aim at transition of large-scale fleets and/or simulate several bus technologies simultaneously. Scholars proposed simulation optimization tools and decision-making frameworks of the potential replacement of the conventional bus fleets with zero-emission ones mostly at urban scale. The authors of [13] proposed using a mixed integer linear model for the analysis of multiple bus routes at urban level in Berlin (Germany) for battery-electric buses, using such parameters as individual route, bus type, and traffic. The impact of high slope variation or longer distances was not discussed as based on the topology of the case it was not relevant. Within the optimization tools, [14] used a similar mixed integer linear cost optimization model to analyze the potential for the substitution of traditional public transportation modes with the electric buses using a function of least energy consuming routes. They assumed only constant slope values in a simplified road network with varying distances, but focused on the potential location of the infrastructure for electric buses. The authors of [15] combined energy consumption, carbon emission, and operational cost models and bi-objective model, which minimizes the total carbon emission and operational cost of all bus lines applied for the fleet of the city of Liuzhou for a mix of electric and diesel buses. Scholars considered travel distance in their energy consumption model; however, no slope considerations were included. The authors of [16] developed a mixed integer nonlinear programming with a detailed calculation of energy consumption for electric buses only applied to the transit network of Belleville city (Canada). This energy consumption module mostly considers the impact of traffic intensity as the slope variation in the case study is low. Another optimization approach through mixed integer linear model was proposed by [17] and applied for the electrification of a bus route under partial charging and battery degradation in Singapore.

Regarding simulation tools, [18] developed a model as part of a larger grid impact study to evaluate the potential implementation of only electric buses. The model was applied to the case study of the network of the city of Belleville (Canada). Simulations were performed based on a logit model that considers bus deployment on the one hand and the operation of charging systems on the other. Among the variables considered were standard urban and suburban distances for bus travel and simulated energy consumption without consideration of slope or climate impacts. In [19], a simulation tool combined bus route design, energy range, and the infrastructure scheduling for the transition to all-electric bus fleets in urban environments, with Barcelona (Spain) as the case study. Energy consumption and battery capacity in this work is calculated as a function of speed and distance, without slope and weather conditions. Within the knowledge of the authors, the only study addressing the transition of large-scale fleets towards battery-electric buses at extra-urban level was performed by [20]. They developed a model that considers distinctive spatiotemporal characteristics and difficulties associated with longer distance operation of battery-electric buses. The model suggests potential for investment optimization based on a variety of strategies that can be adopted by public transportation operators operating at regional level for charging infrastructure based on electric bus specific operation characteristics.

Specific works addressing the use of simulation tools that consider at once slopes, longer route length, and climatic conditions and their impact on energy consumption of electric and hydrogen fueled buses and therefore fleet composition as well as related infrastructure that needs to be applied in highly mountainous areas are missing in scholarly literature.

3. Materials and Methods

Within the following section the methods and materials used in order to set up the simulation tool are described. The tool has been written using Python programming language and consists of approximately 10,000 lines of code.

The tool includes two main algorithms:

The first algorithm calculates the energy consumption of BEBs and FCEBs traveling once along one specific line. In order to do so, vehicle parameters have to be assumed (Section 3.1), precise information on the lines needs to be available (Section 3.2) and the energy consumption to travel along the lines is simulated (Section 3.3).

This first algorithm is applied to each single line covered by public buses today in order to evaluate if ZEBs are able to travel one return trip on the specific line.

The second algorithm is designed to calculate the minimum number of buses and charging stations required to serve a specific bus line. To achieve this, the algorithm takes into account the timetable and frequency of the buses, energy requirements for traveling along the line (which can vary depending on the direction of travel), and the state of charge (SOC) of the batteries or hydrogen tank at the start and end of the route. It considers whether recharging or refueling is needed and positions charging or refueling stations accordingly at the beginning and end of routes. The algorithm also determines how much recharging or refueling is possible within the given timetable.

This algorithm is applied to all bus lines by considering one representative day. The algorithm considers both zero-emission bus (ZEB) technologies: electric and hydrogen. For each line, and technology, the algorithm identifies whether additional ZEB buses are required compared to the number of buses required to serve the same line using diesel buses. This need for additional buses arises when the SOC of the battery or the hydrogen available is insufficient to cover the next trip, and recharging or refueling time between trips is too short to adequately recharge or refuel.

3.1. Vehicle Study

The state-of-the-art ZEBs were implemented in order to analyze their possible employment on the different lines in South Tyrol. Before proceeding to data collection on the performances of the different models, the main ZEBs' characteristics were identified.

Forty-one models of ZEBs were analyzed (of which 12 were FCEBs), sold in European, American, and Asian markets. Buses are classified into three categories, according to the Italian Ministerial Decree of 20 June 2003 [21] as per Table 1.

Table 1. Bus classification used in ZEB model analysis.

Class	Description
C1	vehicles built with areas designed primarily for standing passengers (normally employed in urban areas), with more than 22 passengers
C2	vehicles designed primarily to carry seated passengers, designed so that standing passengers can also be transported in the aisle and/or designated area (normally employed in urban and suburban areas), with more than 22 passengers
C3	vehicles built with areas designated for seated passengers only, also referred to as "coach", with more than 22 passengers

- Given the goal of the project, evaluating the possibility of changing the current fleet of diesel buses in regional service, the class of interest is C3. Given the small number of BEBs-class C3 and the absence of FCEBs-class C3 buses at the time of data collection, C2 and C1 classes were also considered.
- The following parameters were identified to describe the different buses:
- Energy storage: the study considered both battery capacity in kWh (for BEBs and FCEBs), and hydrogen storage in kg (for FCEBs). Although some manufacturers state that battery capacity depends on charging velocity, authors considered slow charging condition (typical in depot) for data uniformity issues and did not include the aging process. Most of the data were derived from the manufacturer's websites;
- Range: It represents the maximum distance that the bus can cover in km. This parameter is influenced by the battery capacity, the HVAC (heating ventilation and air conditioning) system, the regenerative braking (recovery efficiency and braking style), and the trips and the general driving style. Many manufactures did not refer to standardized duty cycles (some of them adopted S.O.R.T. cycles [22]) and did not specify the HVAC's influence on energy consumption. The data collected were mainly obtained from the manufacturer's website, in some cases by emails received directly from the manufacturer or by interpolations;
- Motor power: given the presence of wheel motors in some models, this parameter corresponds to the sum of all motor powers present in the power train. It is expressed in kW;
- Specific consumption: it represents the average energy consumption to travel a distance and is expressed in kWh/km for BEBs and (kg of hydrogen)/km for FCEBs. As for the range, the specific consumption has been deduced from information received from the manufacturer (website or emails) or through interpolation. The HVAC system and the adoption of a specific running cycle have in many cases not been declared by the manufacturer;
- Vehicle weight: this parameter is expressed in kg and was used in order to make considerations on the vehicle's potential energy, given the variations of altitude on the South Tyrolean bus lines.

The electric bus considered for the simulations has an energy storage of 660 kWh, a motor power of 248.59 kW, a weight of 19,050 kg, a claimed range of 386 km, and an estimated specific consumption of 1.4 kWh/km. The hydrogen model used for the simulations has a range of 350 km, a storage of 38 kg of hydrogen, and a specific consumption of about 0.11 kg of hydrogen/km.

3.2. *Road Network*

The road network was converted into a 3D vector, combining 2D roads with digital terrain models (DTMs) using the GRASS GIS 'v.drape' command [23]. Figure 1 shows the intercity lines that have been simulated in South Tyrol, considering the orography of the territory. However, the 3D roads included several elevation points that were not correct due to road infrastructures: bridges, tunnels, underpasses, and subways. The official road network contained the information on the infrastructure type for each line. Therefore, the authors proceeded in interpolating the elevation of the critical infrastructure using a linear interpolation [24] with the road elevation before and after the infrastructure. An iterative linear interpolation was implemented and applied to the road network to recursively remove the road segments with a slope greater than 20%. The bus lines time schedule was provided by the local public transport organization Strutture Trasporto Alto Adige spa (STA) using the General Transit Feed Specification (GTFS) format. The time schedule provided covers the full year, however—to reduce complexity—for the simulations the authors selected one representative particularly challenging day. The bus line geometries extracted from the GTFS file were forced to follow the 3D roads network moving the bus line point to the closest 3D road point using the GeoPandas Python package [25]. Then, using

Shapely [26], the authors characterized each line with the total distance (subdivided by distance up and down) and with the elevation difference (distinguished in up and down).

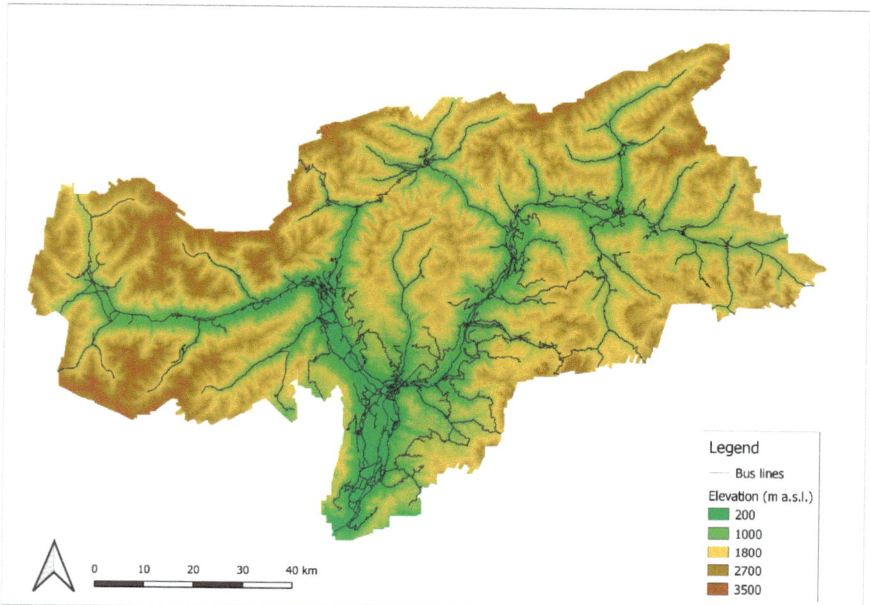

Figure 1. Bus lines analyzed in the case study of South Tyrol province, Italy.

3.3. Vehicle Modelling and Assumptions

Formulas (1)–(4) provide the model used to calculate the energy consumption of electric and hydrogen buses. The model takes into account parameters such as distance travelled, height difference, vehicle weight, and climatic conditions to determine the energy required for a bus to travel a certain route. Route profiles, including total altitude difference and distance travelled uphill, downhill, and on flat terrain, are also factored in. However, traffic conditions, speed, and driver behavior are not considered due to lack of available data and the need for a different simulation model. The model also does not provide detailed information on the drive system components, which are included in the average energy consumption per kilometer used in the calculation.

The authors analyzed the bus schedule to record the cases where the stops were longer than 10 min. On the overall bus service, only 46 stops were registered. Therefore, it has been decided to ignore the stops within the line and to consider only the stops between one course and the next with a stopping time above 10 min as a potentially eligible chance to charge the bus.

The energy required by the bus to travel each line was assessed by performing a global energy balance for BEBs and FCEBs. The study aimed to assess the feasibility to move the full current public transport buses to ZEBs, therefore the analysis was conducted selecting conservative values and considering the worst conditions a bus could encounter. The energy balance was performed based on Formula (1).

$$E_{tot} = E_{dist} + E_{up} - E_{down} \quad (1)$$

where, E_{dist} is the energy required to move the bus in a horizontal direction; and E_{up} and E_{down} are, respectively, the energy required by the bus to travel upwards and the energy that might be generated through the regenerative braking system travelling downwards. Formula (2) was applied to assess the energy due to the distance under conservative conditions.

$$E_{dist} = c * D_l / \eta_{clima} \quad (2)$$

where, c is the specific consumption of the bus as specified by the bus model datasheet; the authors considered the kWh/km for the BEBs and the kgH_2/km for the FCEBs as declared by bus model data sheet; D_l is the distance in km of the single bus line; and η_{clima} is the increased consumption per km due to the energy need of the space conditioning system. The actual energy consumption for space conditioning of the bus cabin can vary significantly in dependence of the outside temperature (heating or cooling need), outside air humidity (dehumidification need) and occupancy of the cabin (internal heat loads in the bus cabin reducing heating needs but enhancing cooling and air exchange needs). In the present analysis a simplified approach has been adopted inserting a fixed η_{clima} and choosing a conservative value so that on most days of the year the impact of the air conditioning system will be (relevantly) lower than assumed. For the BEBs, η_{clima} = 0.75 [27] as the efficiency of the heating system was considered and for the FCEBs, η_{clima} = 0.75 [28] as the efficiency of the cooling system (as for the heating system partially the use of the waste heat of the fuel cell has been considered, and therefore the cooling situation can be the more energy intensive one).

The energy required to travel up and down is defined by Formula (3).

$$E_{up} = g * \left(W_{bus} + W_{people}\right) * \Delta_{up} / \left(\eta_{up} * \eta_{clima}\right) \quad (3)$$

where, g is the gravity force 9.81 m/s^2; W_{bus} is the weight in kg of the empty bus; W_{people} is the weight in kg of the bus users calculated as the number of places available on the bus multiplied by an average weight per person of 70 kg; Δ_{up} is the elevation difference in m; while η_{up} is the efficiency from battery to wheel. An efficiency of 0.85 [29] for the BEBs has been considered and 0.425 for the FCBs has been considered. The second is the result of 0.85 (battery to wheel) multiplied by the efficiency of the fuel cell (FC) to convert the hydrogen to electricity, which was considered to be 0.54 [30]; η_{clima} is the penalty to guarantee the comfort condition within the bus using 0.75 as in the previous formula.

The energy that might be generated through the regenerative braking system is assessed based on Formula (4):

$$E_{down} = g * W_{bus} * \Delta_{down} * (\eta_{down} * \eta_{clima}) \quad (4)$$

where, g is gravity, W_{bus} is the weight of the empty bus, Δ_{down} is the elevation difference, η_{down} is the efficiency from wheel to battery that was considered to be equal to 0.40 [31] for BEBs and FCEBs, and with η_{clima} of 0.75. A worst case assumption has been made, that the bus is full when going upwards and it is empty when going downwards (so max weight to be lifted up, and min possibility to recover energy travelling down) [31].

All BEBs in the simulation have an initial level of the battery (90%)/FCEBs a level of tank of (100%). The lower limit of the battery for the BEBs was set to be 15%, while for the FCEBs the lower limit of the tank level was set to be 5%. The simulation tool goes through the time schedule and for each line it checks if there is, in the initial/starting place, a bus, already available, with enough energy to complete the line; if yes, then the line will be assigned to the bus; if the bus that is available at the stop does not have enough energy, it checks if the bus stays at the bus stop for more than 10 min; if yes, then it creates a recharging/refueling point, then it checks if the energy in the battery/hydrogen in the tank is enough to complete the line after 10 min; if yes it assigns the line to the bus, if not it is marked that the bus is not available, it checks the next bus or creates a new one. Figure 2 shows the algorithm that describes the tool's decision-making process.

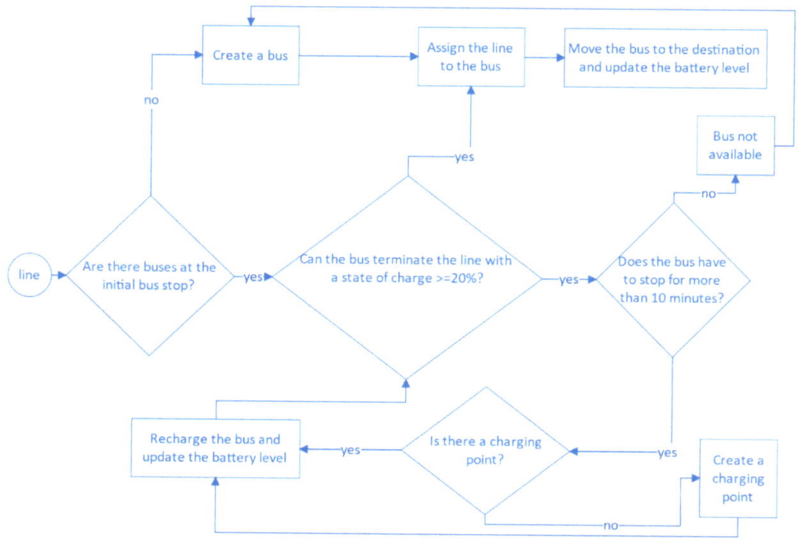

Figure 2. The algorithm that describes the tool's decision-making process.

As no specific data on the actual operation of the vehicles were available at the time of the study, it was assumed that each vehicle is "dedicated" to one single line only.

The simulation was executed under three different operative conditions. The first condition simulates the option to recharge the bus only during the night. The second condition simulates to recharge whenever the bus remains at the destination for a time longer than 10 min. The third condition follows the second one but reduces the recharge of the bus to the situation where battery/tank level is below 40%.

The recharging of the battery/tank is explained in Formula (5).

$$E_{recharge} = E_{flow} * T_{stop} * \eta_{recharge} \quad (5)$$

where, E_{flow} is considered to be 150 kWh/3600 s = 0.04167 kWh/s for the BEBs and of 0.075 kgH_2/s for the FCBs as recharging/refueling of the bus. This flow is multiplied by T_{stop}, the number of seconds (rounded to the floor minutes) that the bus stays on the bus stop, multiplied by $\eta_{recharge}$, the recharge efficiency of the system (95% for the BEBs and 100% for the FCBs have been considered).

4. Results

4.1. Vehicle Study

In this section the results regarding the vehicle simulations are presented (Table 2). In particular, motor power presented a range between 150 and 350 kW for both FCBs and BEBs, similar to the actual diesel bus fleet [32].

Table 2. Vehicle study parameters.

Parameters/Vehicle Type	BEB	FCEB
Energy storage	150–650 kWh	Battery capacity: 40–120 kWh H_2 storage: 30–38 kg
Range	200–370 km	300–400 km
Motor power	140–350 kW	140–350 kW
Specific consumption	0.7–1.4 kWh/km	0.085–0.1 kg/km
Weight	14,000–27,000 kg	20,000 kg

4.2. Road Network

In the following, the results of the analysis of the road network and the energy consumption by BEBs and FCEBs in order to run the single bus lines are shown.

Each line of the bus network has been characterized by the distance and the elevation difference; Figure 3 shows the scatterplot of the bus lines considering these two variables. As can be seen, the maximum distance is less than 60 km, while the maximum elevation difference is slightly above 1750 m. The points are concentrated in the lower and left parts of the graph, where the elevation difference is small (0 to 250 m) and the distance is short (0 to 30 km).

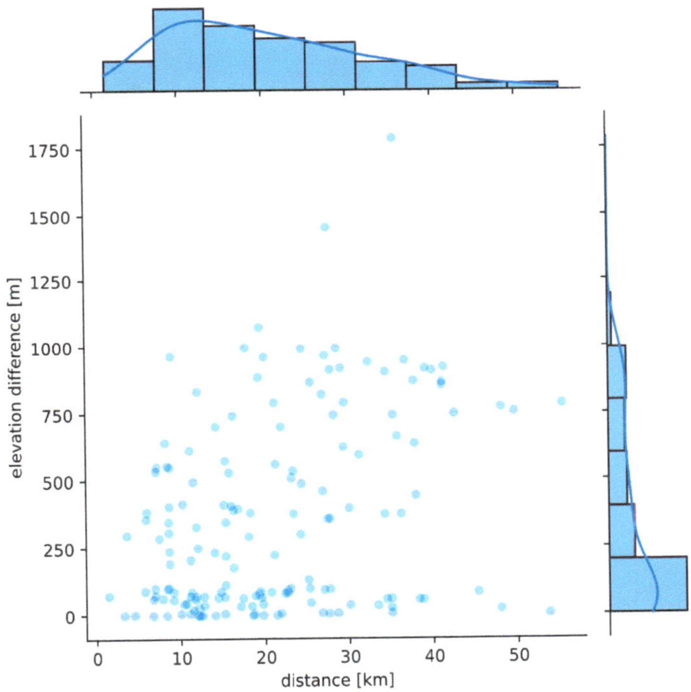

Figure 3. Statistical analysis of the network: elevation difference vs. distance.

Based on the formula described in the methodology section, the authors assessed the required energy of each line to guarantee the service under cautious conditions. Figure 4 displays the distribution of energy consumption for BEBs expressed in kWh/km for various lines (including variations of the main ones). The most frequent case is a specific energy consumption of around 3 kWh/km.

Figure 5 shows the energy consumption for each cluster (very low, low, medium, high and very high) of round-trip bus lines, taking into account the length of each line. Market-available energy storage sizes are shown in gray. As can be seen, most of the lines can be served by bus types using batteries with limited capacity (very low, low, and medium clusters of round-trip bus lines), while all the lines can be covered by buses with larger capacities available on the market (high and very high clusters of round trip bus lines), while for the FCEBs, all the bus lines can be covered by the tank capacity that is installed in the buses available on the market.

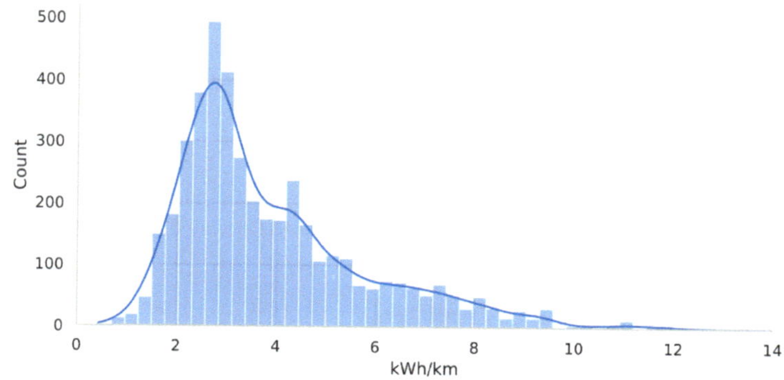

Figure 4. Distribution of energy consumption for various bus lines.

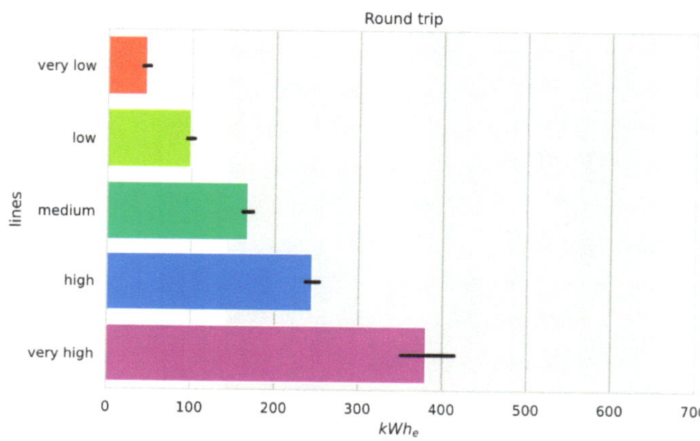

(**A**)

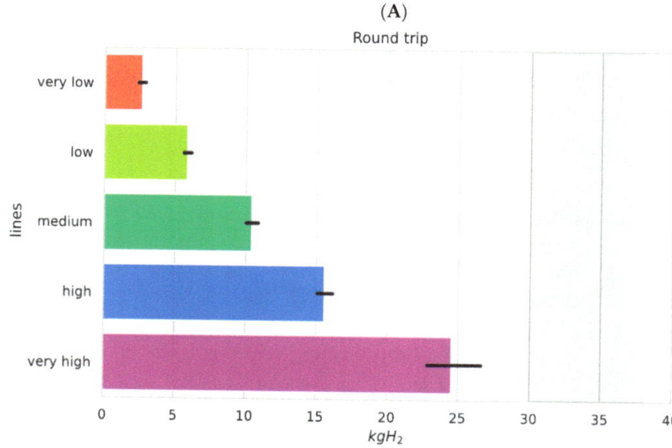

(**B**)

Figure 5. Energy required for each cluster of round-trip bus lines, the background dark grey color shows the battery capacity available on the market for BEBs (**A**) and the tank capacity available for the FCEBs (**B**). The number of lines for each cluster are not equally distributed. In fact, 9 lines are in the purple cluster, 17 in the blue cluster, 31 in the green cluster, and, respectively, 65 and 102 in the light green and the red cluster.

The average value of energy required for the line round trip is 139.8 kWh and 9.1 kg H_2 for BEBs and FCEBs, respectively, while the maximum energy demand is 508.9 kWh and 32.5 kg H_2.

4.3. Modelling Buses Driving the Single Trips and Assumptions

As described in Section 3, three different recharging options were simulated. Based on these assumptions it was defined how many buses [32,33] are required to guarantee the bus service for a full day, for a specific line, based on different recharging options. The single recharging options can have an impact on the number of buses needed, and if only overnight charging is allowed, more buses are needed than if opportunity charging is also allowed. Figure 6 shows an example of state of charge considering the third simulation option, where opportunity charging is allowed if the battery level is below 40%. As can be seen, the bus starts the journey with the battery level at 90% and is discharging during the day, arriving at the final destination several times, around 14:00 o'clock the battery level goes below the 40% limit and as the stop was longer than 10 min, the bus status is changed to recharge.

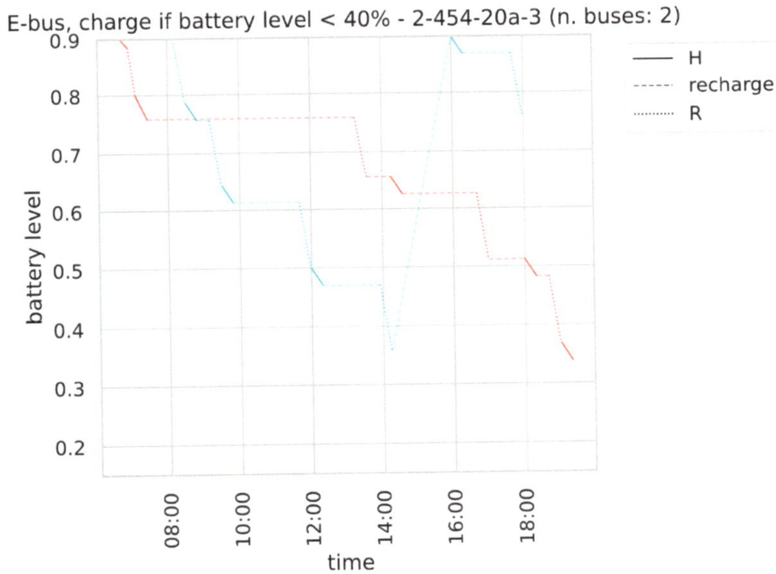

Figure 6. An example of the state of charge of a bus during the day travelling a specific line is shown. Each color represents a bus, the different line styles represent the bus status: the continuous line is the bus that is going to its destination (H), the dotted line is the bus that is on its way back to the origin (R), and dashed line is the bus that is waiting and/or recharging (recharge).

The maximum energy required by the bus line round trip is less than the maximum capacity of the batteries and tanks of the buses on the market. Figure 7 illustrates how many lines can be covered by BEBs (A) and FCEBs (B) if a full charge is made during the night, multiple charges during the day, or whether it is necessary to increase the number of buses compared with the equivalent diesel system. With both technologies, more than 78% of the lines can be covered with the same number of buses. With both technologies, slightly more than 20% of the lines require more buses to guarantee service. Altogether, around 6% more buses are required, independent of the technology chosen (BEBs of FCEBs). In the case of FCEBs, a relevant larger number of trips can be covered with only overnight charge (41%) in comparison to BEBs (24%).

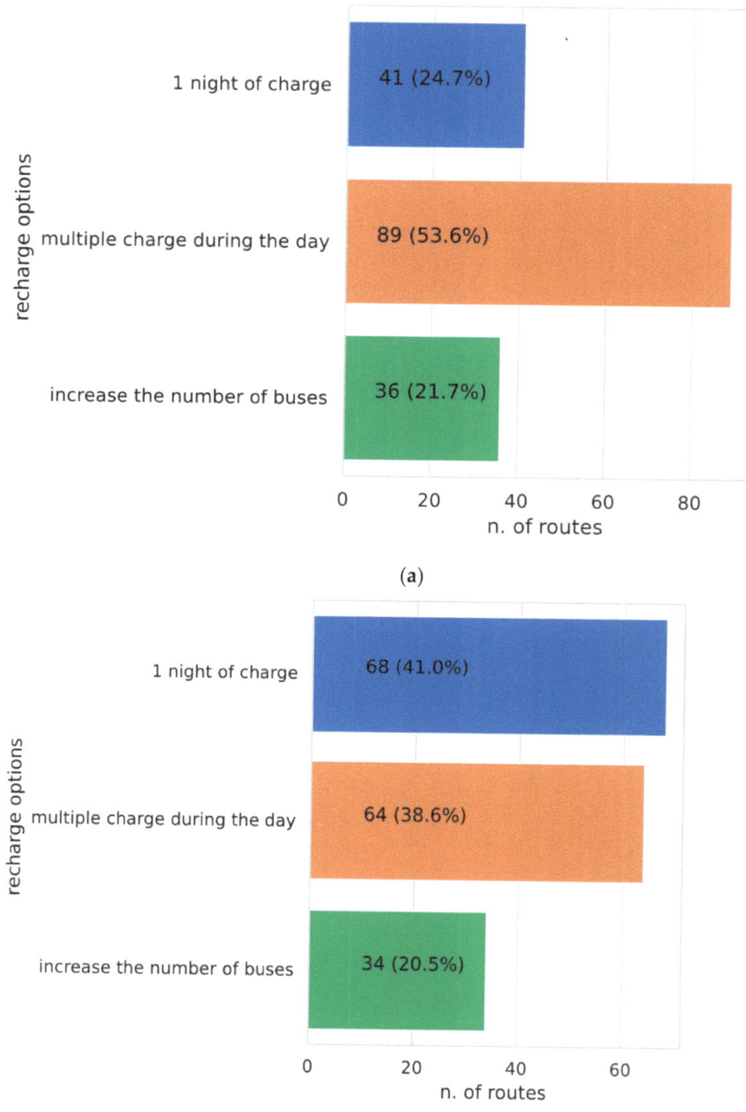

(a)

(b)

Figure 7. Number of lines that can be covered by BEBs (**a**) and FCEBs (**b**) with a full recharge during the night, multiple recharges during the day, and by increasing the number of buses with respect to the bus required by a diesel equivalent system.

The simulation tool also estimates the number of recharge stations and hydrogen refueling points required by the regional buses. Following the used model for the BEBs the recharge infrastructure obtained is of 223 recharging stations assuming 150 kW of power capacity, while for the FCEBs the required recharge stations are 206, which become 21 if the identified stations are aggregated and clustered using a radius distance of 15 km. In fact, as the cost of hydrogen stations is relevant and at present in South Tyrol only two hydrogen refueling stations exist, it was considered as very unlikely that within a radius of 15 km several hydrogen refueling stations would be set up.

5. Discussions and Conclusions

This study evaluates the possibility to transfer extra-urban bus services in an Alpine region (South Tyrol—northern Italy) from diesel buses to battery electric buses (BEBs) or fuel cell electric buses (FCEBs).

For this purpose, a model has been set up based on the vehicle's energy balance. Herewith the energy consumption is evaluated based on the physical characteristics of the single bus line (distance and altitude), climatic conditions, and its timetables. As such, the model can be implemented in mountainous areas and on intercity lines allowing the specific energy consumption.

A detailed study has been performed to digitize all bus lines in the region by correcting more than three million points in the digital maps (tunnels, bridges, etc.). From the analyses carried out, the average distance of the lines resulted in 17 km (max: 55 km) with an average height difference of 300 hm (max: 1786 hm). Each line (distance and height difference) was transformed into energy consumed for the specific technology. The average energy required for the outward and return of the buses on the lines resulted in 139.8 kWh for BEBs and 9.1 kg H_2 for FCEBs with a maximum consumption of 508.9 kWh and 32.5 kg H_2, respectively. In the very conservative calculation (assuming full bus upwards, empty bus downwards, and difficult climatic conditions) the average consumption is estimated to be 3.8 kWh/km for BEBs and 0.25 kg/km for FCEBs.

Based on the market research of available bus models and the calculated specific energy consumption, the authors conclude that 100% of the lines can be completed in a single round trip with both electric buses (higher performance models) and fuel cell buses. As the routine schedule foresees some stops of more than 10 min (which allows an electric bus to recharge), in around 80% of the cases one diesel bus can be replaced by one electric/fuel cell bus. For the remaining 20%, an increase of 6% in the number of available buses is necessary to guarantee the same time schedule. The numbers of BEBs and FCEBs needed to substitute the diesel ones could eventually be lower as the estimated energy consumption in this study was precautionary and the range of electric and hydrogen vehicles is gradually increasing due to technological progress.

The authors' initial hypothesis that longer distances in mountainous areas can only be covered by FCEBs, while shorter ones by both FCEB and BEB, was not confirmed in this study, as there was no significant difference in the distance covered in mountainous areas by BEBs versus FCEBs. In fact, one aspect underestimated before the study was the enhanced energy recovery capacity by BEBs in comparison to FCEBs driving downwards based on the much larger battery capacity of the vehicle.

In order to guarantee the bus service under all conditions and with unchanged timetables in comparison to the current situation, the simulation led to over 200 charging points located in depots and stops where more than 10 min breaks allow (at least partial) recharge. In the case of FCEBs the resulted number of hydrogen refueling stations was very similar. The clustering of recharge/refueling stations substantially reduces the number of stations required by the regional transport system. In the case of hydrogen and considering a clustering radius of 15 km, the number drops from over 200 to 21 refueling stations.

The presented method was developed and applied for one specific region (Italian Alps) and one specific case study (extra-urban bus fleet in South Tyrol), but is considered as generally suitable for PTCs for the evaluation of the possibility to switch a fossil fuel bus fleet to BEBs and FCEBs and the related infrastructure needed. It is not limited to bus fleets but can be applied to all kind of fleets that cover clearly defined routes on a daily basis (such as logistics in the food sector, medical sector, goods transportation to stores, etc.).

Regarding the specific energy consumption/km it was noted from the beginning of the project that monitoring data on the specific lines under all conditions (winter/summer, bus loaded/unloaded) is needed in order to validate the presented assumptions and to enhance the model's results. Such data at that time were not available. Monitoring data from the LifeAlps project [34] for the years 2021 and 2022 for over 20 buses in service in South Tyrol on specific lines are available [35]. A direct comparison of the data sets is not possible, but the results show an average consumption which is 50% lower than the one resulting from the simulations under conservative assumptions for both technologies—battery electric and fuel cell electric buses. Detailed evaluations on this dataset will follow in future works.

If further data will confirm a lower average and max consumption it suggests that a technology shift to zero emission buses will require fewer charging stations/refueling stations and less lines requiring more ZEB buses than diesel buses currently in service.

Author Contributions: Conceptualization, W.S., P.Z. and A.Z.; methodology, W.S. and P.Z.; software, P.Z.; validation, R.V.; formal analysis, W.S., A.G., P.Z. and R.V.; investigation, A.G. and P.Z.; data curation, W.S., A.G., P.Z. and R.V.; writing—original draft, W.S., A.G., P.Z., R.V. and A.Z.; writing—review and editing, W.S., A.G. and A.Z.; visualization, P.Z.; supervision, A.Z.; funding acquisition, W.S. All authors have read and agreed to the published version of the manuscript.

Funding: This research received no external funding.

Data Availability Statement: Not applicable.

Acknowledgments: The authors would like to acknowledge STA for the commissioning of the study and the fruitful cooperation. Additionally, they acknowledge the public transportation company SASA for the fruitful discussion on bus line services in South Tyrol and the EU commission for the support of the LifeAlps project allowing future detailed monitoring evaluations.

Conflicts of Interest: The authors declare no conflict of interest.

References

1. IEA Licence: CC BY 4.0 Transport Sector CO2 Emissions by Mode in the Sustainable Development Scenario, 2000–2030. Available online: https://www.iea.org/data-and-statistics/charts/transport-sector-co2-emissions-by-mode-in-the-sustainable-development-scenario-2000-2030 (accessed on 20 December 2022).
2. Aldenius, M.; Mullen, C.; Pettersson-Löfstedt, F. Electric Buses in England and Sweden—Overcoming Barriers to Introduction. *Transp. Res. D Transp. Environ.* **2022**, *104*, 103204. [CrossRef]
3. Mahmoud, M.; Garnett, R.; Ferguson, M.; Kanaroglou, P. Electric Buses: A Review of Alternative Powertrains. *Renew. Sustain. Energy Rev.* **2016**, *62*, 673–684. [CrossRef]
4. Muñoz, P.; Franceschini, E.A.; Levitan, D.; Rodriguez, C.R.; Humana, T.; Correa Perelmuter, G. Comparative Analysis of Cost, Emissions and Fuel Consumption of Diesel, Natural Gas, Electric and Hydrogen Urban Buses. *Energy Convers. Manag.* **2022**, *257*, 115412. [CrossRef]
5. Müller, K.K.; Schnitzeler, F.; Lozanovski, A.; Skiker, S.; Ojakovoh, M. Clean Hydrogen in European Cities (CHIC)—Deliverable No. 5.3—Final Report; 2017. Available online: https://fuelcellbuses.eu/public-transport-hydrogen/final-report-chic-clean-hydrogen-european-cities (accessed on 23 November 2022).
6. ASTAT Popolazione | Istituto Provinciale di Statistica | Provincia Autonoma di Bolzano—Alto Adige. Available online: https://astat.provincia.bz.it/it/popolazione.asp (accessed on 10 February 2023).
7. Green Mobility Mobilità Sostenibile | Mobilità | Provincia Autonoma di Bolzano—Alto Adige. Available online: https://www.provincia.bz.it/turismo-mobilita/mobilita/mobilita-sostenibile.asp (accessed on 10 February 2023).
8. Alto Adige Province Piano Clima Alto Adige 2040. Available online: https://ambiente.provincia.bz.it/piano-clima-energia-alto-adige-2050.asp (accessed on 12 January 2023).
9. Papa, G.; Santo Zarnik, M.; Vukašinović, V. Electric-Bus Routes in Hilly Urban Areas: Overview and Challenges. *Renew. Sustain. Energy Rev.* **2022**, *165*, 112555. [CrossRef]
10. Logan, K.G.; Nelson, J.D.; Hastings, A. Electric and Hydrogen Buses: Shifting from Conventionally Fuelled Cars in the UK. *Transp. Res. Part D Transp. Environ.* **2020**, *85*, 102350. [CrossRef]
11. Hensher, D.A.; Wei, E.; Balbontin, C. Comparative Assessment of Zero Emission Electric and Hydrogen Buses in Australia. *Transp. Res. Part D Transp. Environ.* **2022**, *102*, 103130. [CrossRef]
12. Correa, G.; Muñoz, P.M.; Rodriguez, C.R. A Comparative Energy and Environmental Analysis of a Diesel, Hybrid, Hydrogen and Electric Urban Bus. *Energy* **2019**, *187*, 115906. [CrossRef]
13. Kunith, A.; Mendelevitch, R.; Goehlich, D. Electrification of a City Bus Network—An Optimization Model for Cost-Effective Placing of Charging Infrastructure and Battery Sizing of Fast-Charging Electric Bus Systems. *Int. J. Sustain. Transp.* **2017**, *11*, 707–720. [CrossRef]
14. Hulagu, S.; Celikoglu, H.B. Electrified Location Routing Problem with Energy Consumption for Resources Restricted Archipelagos: Case of Buyukada. In *2020 Forum on Integrated and Sustainable Transportation Systems, FISTS*; Institute of Electrical and Electronics Engineers Inc.: Piscataway, NJ, USA, 2020; pp. 323–327.
15. Shao, S.; Tan, Z.; Liu, Z.; Shang, W. Balancing the GHG Emissions and Operational Costs for a Mixed Fleet of Electric Buses and Diesel Buses. *Appl. Energy* **2022**, *328*, 120188. [CrossRef]

16. El-Taweel, N.A.; Farag, H.E.Z.; Mohamed, M. Integrated Utility-Transit Model for Optimal Configuration of Battery Electric Bus Systems. *IEEE Syst. J.* **2020**, *14*, 738–748. [CrossRef]
17. Zhou, Y.; Meng, Q.; Ong, G.P. Electric Bus Charging Scheduling for a Single Public Transport Route Considering Nonlinear Charging Profile and Battery Degradation Effect. *Transp. Res. Part B Methodol.* **2022**, *159*, 49–75. [CrossRef]
18. Mohamed, M.; Farag, H.; El-Taweel, N.; Ferguson, M. Simulation of Electric Buses on a Full Transit Network: Operational Feasibility and Grid Impact Analysis. *Electr. Power Syst. Res.* **2017**, *142*, 163–175. [CrossRef]
19. Estrada, M.; Mensión, J.; Salicrú, M.; Badia, H. Charging Operations in Battery Electric Bus Systems Considering Fleet Size Variability along the Service. *Transp. Res. Part C Emerg. Technol.* **2022**, *138*, 103609. [CrossRef]
20. Wei, R.; Liu, X.; Ou, Y.; Kiavash Fayyaz, S. Optimizing the Spatio-Temporal Deployment of Battery Electric Bus System. *J. Transp. Geogr.* **2018**, *68*, 160–168. [CrossRef]
21. MIT Ministero Delle Infrastrutture e dei Trasporti. Available online: https://www.governo.it/it/ministeri/ministero-delle-infrastrutture-e-dei-trasporti (accessed on 10 February 2023).
22. European Commission EUROPA—Environment—Air Quality—NEBEI Workshop. Available online: https://ec.europa.eu/environment/archives/clean_bus/ (accessed on 10 February 2023).
23. GRASS GIS V.Drape—GRASS GIS Manual. Available online: https://grass.osgeo.org/grass82/manuals/v.drape.html (accessed on 10 February 2023).
24. The SciPy Community Scipy.Interpolate.Interp1d—SciPy v1.10.0 Manual. Available online: https://docs.scipy.org/doc/scipy/reference/generated/scipy.interpolate.interp1d.html (accessed on 10 February 2023).
25. Jordahl, K.; Van den Bossche, J.; Fleischmann, M.; Wasserman, J.; McBride, J.; Gerard, J.; Tratner, J.; Perry, M.; Badaracco, A.G.; Farmer, C.; et al. *Geopandas/Geopandas: V0.8.1*; Zenodo: Geneve, Switzerland, 2020. [CrossRef]
26. GiTHuB GitHub—Shapely/Shapely: Manipulation and Analysis of Geometric Objects. Available online: https://github.com/shapely/shapely (accessed on 10 February 2023).
27. Göhlich, D.; Kunith, A.; Ly, T. Technology Assessment of an Electric Urban Bus System for Berlin. *WIT Trans. Built Environ.* **2014**, *138*, 137–149. [CrossRef]
28. NAF 20 Popular EVs Tested in Norwegian Winter Conditions | NAF. Available online: https://www.naf.no/elbil/aktuelt/elbiltest/ev-winter-range-test-2020/ (accessed on 10 February 2023).
29. Chanpeng, W.; Hachanont, P. Design of Efficient In-Wheel Motor for Electric Vehicles. *Energy Procedia* **2014**, *56*, 525–531. [CrossRef]
30. Dodds, P.E.; Staffell, I.; Hawkes, A.D.; Li, F.; Grünewald, P.; McDowall, W.; Ekins, P. Hydrogen and Fuel Cell Technologies for Heating: A Review. *Int. J. Hydrogen Energy* **2015**, *40*, 2065–2083. [CrossRef]
31. Ranta, M.; Pihlatie, M.; Pellikka, A.P.; Laurikko, J.; Rahkola, P.; Anttila, J. Analysis and Comparison of Energy Efficiency of Commercially Available Battery Electric Buses. In Proceedings of the 2017 IEEE Vehicle Power and Propulsion Conference, VPPC, Belfort, France, 11–14 December 2017; pp. 1–5. [CrossRef]
32. Van Hool Van Hool Ships the First CX45E Electric Coach to the US. | Van Hool. Available online: https://www.vanhool.com/en/news-fairs/van-hool-ships-the-first-cx45e-electric-coach-to-the-us (accessed on 10 February 2023).
33. Proterra The Proterra ZX5 Electric Transit Bus | Proterra. Available online: https://www.proterra.com/products/transit-buses/ (accessed on 10 February 2023).
34. Zero Emission LIFEIP per una Mobilità Sostenibile—LIFEalps. Available online: https://www.life-alps.eu/it/ (accessed on 10 February 2023).
35. Estrada Poggio, A.; Balest, J.; Zubaryeva, A.; Sparber, W. Monitored Data and Social Perceptions Analysis of Battery Electric and Hydrogen Fuelled Buses in Urban and Suburban Areas. *J. Energy Storage* **2023**, *manuscript submitted for publication*.

World Electric Vehicle Journal

Article

The Impact of Charging Infrastructure on Local Emissions of Nitrogen Oxides

Karsten Hager * and Alexandra Graf

Institut Stadt | Mobilität | Energie (ISME) GmbH, Rotenwaldstraße 18, 70197 Stuttgart, Germany
* Correspondence: karsten.hager@i-sme.de

Abstract: Benefits from EV (Electric vehicles) and e-mobility include the reduction of local emissions of pollutants from particulate matter ($PM_{0.5}$, $PM_{5,}$ and PM_{10}) and nitrogen oxides (NO_x and NO_2). Cities and urban agglomerations benefit the most from potential emission reductions from EVs due to the large number of cars utilized in most urban traffic systems. This abstract presents results from a corporate research and funding project in Baden-Wuerttemberg, Germany (LINOx BW) which facilitates the installation of 2358 charging points within 178 different sub-projects in 23 different cities, spanning a period of four years. Utilizing several different survey waves, data about outgoing currents from these publicly funded charging points are gathered. Converting this data utilizing car classifications and emission classes (HBEFA), the reduction of local nitrogen oxides is derived.

Keywords: BEV (battery electric vehicle); charging; infrastructure; pollution; research; nitrogen oxides

Citation: Hager, K.; Graf, A. The Impact of Charging Infrastructure on Local Emissions of Nitrogen Oxides. *World Electr. Veh. J.* **2023**, *14*, 90. https://doi.org/10.3390/wevj14040090

Academic Editor: Joeri Van Mierlo

Received: 28 February 2023
Revised: 22 March 2023
Accepted: 27 March 2023
Published: 31 March 2023

1. Introduction

Limit values for nitrogen oxide emissions were adopted by the European Commission in 2008 and are mandatory for all member countries since 2010 [1]. Legal measures regarding nitrogen oxide emissions were not only taken in Europe but, for example, in the U.S. as well [2]. It is well known that nitrogen oxides have significant negative impacts on human health: the fundamentals of the influence of nitrogen oxides on public health concerns are shown in [3]: increasing the risk of respiratory tract infections [4], increasing the incidence and exacerbation of respiratory disease [5], comparing nitrogen oxides to other air pollutants [6], and case studies from New Zealand [7], China [8] and Malaysia [9]. This is especially the case for transport-related pollution, increasing the risk of death from cardiopulmonary causes [10], increasing the risk of childhood asthma [11], and asthma prevalence [12]. BEV and PHEV lead to positive health and air-quality co-benefits [13] even though there are also natural sources of nitrogen oxide emissions [14].

By 2017, 65 different cities in Germany measured nitrogen oxide values far greater than the threshold of 40 $\mu g/m^3$ on a daily basis [15]. Cities exceeding this threshold for a certain amount of time are legally responsible to adopt measures to reduce nitrogen oxide emissions; e.g., driving bans for diesel cars in city centers. The quantification of positive environmental effects such as pollution reduction of charging infrastructure has not yet been sufficiently addressed by research. Addressing this problem, the German Federal Ministry for Economic Affairs and Climate Action (BMWK) released a funding program facilitating the construction of charging infrastructure for the impacted cities in 2017 (Sofortprogramm "Saubere Luft 2017–2020"). Quantifying the reduction of nitrogen oxide emissions from charging infrastructure throughout the research program is an ongoing scientific task. Previous studies addressing this problem utilized car tracking devices combined with real-world driving scenarios [16], applied spatial panel data approaches [17], or executed well-to-wheel (WTW) approaches [18]. The importance of low-emission technologies and their essential role in the future is highlighted by materials utilized for batteries [19], energy savings from Plug-in Hybrid electric vehicles (PHEV) [20], and the utilization and comparison of fuel-cell electric vehicles (FCEV) and BEV [21].

Deriving potential emission reductions from charging infrastructure having no primary link between each other is the novelty of the presented approach. Two research concepts are presented utilizing different resolutions of collected data leading to potential transferable results from other research projects. Discussion of the presented approach is just the first step towards additional required research quantifying the emission reduction potential of charging infrastructure.

The corporate research project "Ladeinfrastruktur in NOx-Kommunen in Baden-Württemberg" (LINOx BW) consists of 178 recipients from 23 different cities in Baden-Württemberg [22]. Grant applications from interested proposers inhabiting the impacted cities in Baden-Württemberg were bundled within the research project to simplify administrative processes. Additionally, accompanying research for all recipients is conducted consistently by two research institutes. So far, 2358 charging points were funded within the research project, utilizing around 10 million euros of federal subsidies. Figure 1 presents the amount of charging points within each city.

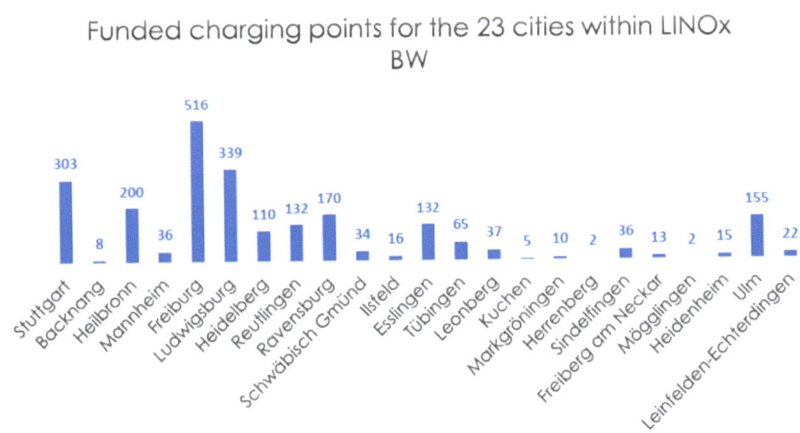

Figure 1. Funded charging points for the 23 cities within LINOx BW.

2. Materials and Methods

Charging infrastructure projects were categorized into seven different use cases by researchers based on potential EV driving patterns (and therefore different charging opportunities) or different spatial boundary conditions: residential quarter, semi-public space, parking lots and park and ride, private on-site parking, nursing services, tourism, and e-bikes (pedelecs). Utilizing a mixed methods approach consisting of quantitative and qualitative surveys, all recipients received a t_0 survey at the start of their respective project about their status quo containing questions, among others, relating to their vehicle fleets, existing charging infrastructure, parking lot size, and enterprise size. A different questionnaire was set up including specific questions for each use case. The goal of the research project is to gather as much information about user experiences, vehicle fleets, and the amount of electricity delivered as possible, thus, at the end of their respective project, funding recipients received a t_1 survey containing the same questions as the t_0 survey. This procedure allows for a comparison between a point in time before the construction of the charging infrastructure and a point in time afterward.

Additionally, hand-picked recipients planning exceptionally large or special charging infrastructure projects were chosen for a one-hour telephone interview and are referred to as lead partners. The utilized semi-structured interview guideline comprises four main sections: decision-making and communication procedures, cost, revenue, and operator models, use intensity of charging infrastructure, and use-case-specific aspects. These use-case-specific questions allow for the accommodation of use cases such as nursing services or tourism, which are very different from private on-site parking, the use case that is encountered most often in the funding program.

Presented results in the paper refer to data from the cutoff date of 15 January 2023. The official project end is set for 31 December 2023, so funding recipients receive sufficient time for their respective depreciation periods caused by, e.g., delays from delivery bottlenecks for charging infrastructure. New funding recipients will not be approved beyond this point. The research presented in this paper is based on the evaluation of 82 questionnaires—82 funding recipients finished the realization of their respective projects and depreciated their charging infrastructure. Referring to the introduction, data from almost 100 recipients of LINOx BW could not be utilized within this paper because their projects are unfinished as of now. The first project results for parking lots and private areas were already published in German [23].

One of the goals of LINOx BW is to quantify the reduction of local nitrogen oxides when charging infrastructure is built. Therefore, the amount of electricity delivered by the funded charging points is the key variable to quantify the impact of charging infrastructure on local emissions of nitrogen oxides—overall and within the project. Utilizing these values, the amount of electricity is converted to electric vehicle kilometers based on average consumption values of kWh/100 km for different car classifications from electric car databases. From the collected data, two results are derived:

(1) The current NO_x-reduction based on the amount of electricity of the funded charging infrastructure divided by the average consumption value. For this paper, 18.4 kWh/100 km is utilized, based on literature [24] (p. 17). To derive the amount of NOx reduction, an average NO_x factor (g/km) based on the average German fleet composition (2022) included in the current version of HBEFA and the HBEFA traffic scenario "D Ø UBA 2021" is utilized. This approach is utilized for all surveys containing data on delivered electricity; no fleet of vehicle data is required to derive nitrogen oxide reductions.

(2) The maximum NO_x-reduction based on a 1:1 substitution for all cars in the respective vehicle fleets utilizing a fleet of vehicle data (mileage and car classifications) to calculate NOx-reduction based on all cars in the respective vehicle fleets utilizing emission factors for ICEV based on the Handbook Emission Factors for Road Transport (HBEFA, [25]). In research concept (2) surveys where car emission concepts are unknown, average emission factors are calculated based on HBEFA and utilized for the calculation of the NO_x-reduction values. Figure 2 summarizes the research concept for this paper; Tables 1 and 2 show the utilized passenger car and heavy-duty vehicle emissions concepts and their respective HBEFA emissions factors utilized for this research.

Table 1. Passenger car emission concepts and HBEFA emission factors.

Emission Concept	Share (%) in German Fleet Composition	HBEFA-Value (g/km)
petrol (4S)	51.08%	0.091258764
diesel	45.22%	0.599183857
electricity	0.98%	0
bi-fuel CNG/petrol	0.25%	0.170386568
plug-in hybrid petrol/electric	1.50%	0.021050557
plug-in hybrid diesel/electric	0.17%	0.085323997
bi-fuel LPG/petrol	0.80%	0.172697365

Table 2. Heavy-duty vehicle emission concepts and HBEFA emission factors.

Emission Concept	Share (%) in German Fleet Composition	HBEFA-Value (g/km)
petrol (4S)	4.66%	0.14836058
diesel	94.12%	0.816266835
electricity	0.79%	0
bi-fuel CNG/petrol	0.36%	0.066065677
plug-in hybrid petrol/electric	0.06%	0.013855861

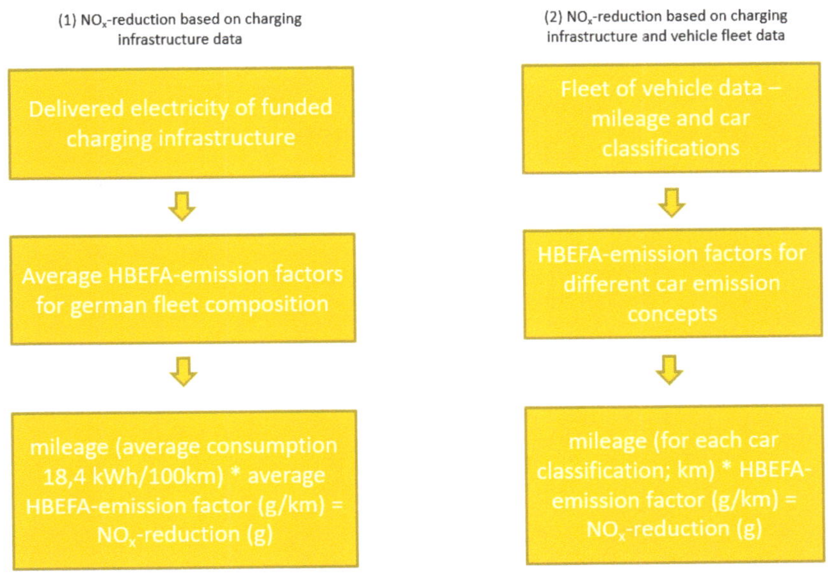

Figure 2. Research concept of LINOx BW for classifying the impact of charging infrastructure on local emissions of nitrogen oxides.

The average HBEFA value for surveys lacking fleet data is 0.371688628 g/km (based on a share of passenger cars and heavy-duty vehicles in German fleet composition, taken from [26]). For cases containing passenger cars with unknown emission concepts, an average value utilizing the data from Table 1 was calculated: 0.319849483 g/km. Similarly, for heavy-duty vehicles, the average value is 0.775203772 g/km. Euro certificates of cars could not be utilized for the research, because these data could not be provided by the funding recipients, even though the HBEFA provides different emissions factors related to Euro certificates.

Unfortunately, independent scientific tracking of all funded charging points could not be provided within the research project due to funding guidelines; thus, research is dependent on survey data from funding recipients measuring and documenting the amount of electricity delivered by charging infrastructure. Many funding recipients were small- and medium-sized companies, and the quality of completed surveys differed greatly between recipients with familiarity with research projects as well as familiarity with the field of electric mobility. However, the results represent many different use cases containing data from daily charging and BEV utilization routines potentially leading to realistic results where data are provided. The driving patterns of the respective vehicle fleets could not be provided within the research project. The completion of the t_1 survey, however, was mandatory for all funding recipients. Without approval from the research group that the survey has been completed, the last 10% of funding money would not be disbursed.

3. Results

3.1. General Results

An overview of the collected data from the t_1 survey is shown in Table 3. Additionally, Figure 3 shows the spatial location of the 82 questionnaires within the LINOx BW cities. Questionnaires hail from 17 of the 23 possible cities with the vast majority located in the city of Freiburg (25 or 30.4% of all 82 questionnaires), followed by the city of Stuttgart (11 or 13.4%) and the city of Ulm (9 or 11%).

Table 3. Key figures about utilized data for this paper.

Number of	Amount
Questionnaires	82
Surveys with charged electricity data	63
Surveys with a fleet of vehicle data	33
Charging points before LINOx BW	131
Charging points with/after LINOx BW	949
BEV purchased since funding of charging infrastructure through LINOx BW	124
PHEV (diesel) purchased since funding of charging infrastructure through LINOx BW	25
PHEV (petrol) purchased since funding of charging infrastructure through LINOx BW	94
Pedelecs purchased since start of LINOx BW	61
Surveys replacing passenger cars with pedelecs	5
Passenger cars in existing vehicle fleets	1615
Passenger cars utilizing diesel	739
With unknown car emission concept	248
Heavy-duty vehicles in existing vehicle fleets	486
Heavy-duty vehicles utilizing diesel	190
With unknown car emission concept	236
Installed charging points (AC)	799
Installed charging points (DC)	19
Sum of installed charging infrastructure power (AC) (kW)	12,899.5
Sum of installed charging infrastructure power (DC) (kW)	1365
Charging capacity/charging point (kW)	17.4

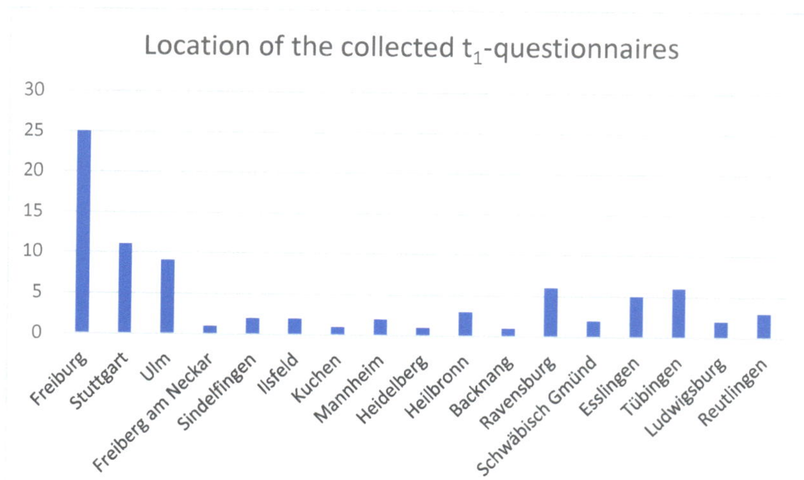

Figure 3. Spatial location of the collected t_1-questionnaires within LINOx BW cities.

Out of the 82 questionnaires collected, only 63 collected charged electricity data from the funded charging infrastructure. Reasons for 19 cases without these data are, e.g., no separate electric meters for charging infrastructure, thus collected data would include household electricity as well. A total of 63 cases were utilized for research concept (1); see Section 3.2. All questionnaires containing vehicle fleet data, 33 to be precise, were utilized for research concept (2); see Section 3.3.

During the duration of the project, 818 charging points were installed throughout the 82 questionnaires utilized for this paper, corresponding to 34.6% of the charging points funded within the whole project. Results shown in this paper consist of approximately one-third of the overall project results.

Vehicle fleets contain an additional 124 BEV, 25 PHEV (diesel), and 94 PHEV (petrol) since the start of the project—to summarize, 243 electric vehicles for 818 charging points

(29.7%) relate to additional electric vehicles from visitors, employees, etc., utilizing the charging infrastructure. This highlights future potential for additional electric cars in their respective fleet of vehicles for BEV based on their higher absolute electric range compared to PHEV. In total, survey data shows 303 electric vehicles utilized at the time of data collection within the examined vehicle fleets. During the project period, 243 out of 303 electric vehicles were purchased by the funding recipients (80.2%).

Funding recipients were also allowed to install charging infrastructure for e-bikes (pedelecs). A total of 61 pedelecs were purchased within the project period, with five recipients additionally replacing ICV with pedelecs. No independent tracking of delivered electricity for pedelecs occurred. Quantifying the success and utilization of pedelecs is not possible with the available survey data.

The overall vehicle fleets contain 1615 passenger cars and 486 heavy-duty vehicles (2101 vehicles in total); 243 electric vehicles correspond to 11.5% of all vehicles in the presented vehicle fleets. Unfortunately, 248 passenger cars (15.4%) and 236 heavy-duty vehicles (48.6%) could not be assigned to a car emission concept from HBEFA leading to the increased utilization of the average HBEFA values as shown in Section 2 and thus potential inaccuracies. The largest share of emission concepts for either passenger cars (45.8%) or heavy-duty vehicles (39.1%) within the survey data is diesel. Heavy-duty-vehicle data unfortunately have a higher share of unknown emission concepts than the respective largest share of emission concepts in diesel leading to potential inaccuracies utilizing this data for further calculations.

In total, the sum of the installed charging points and charging infrastructure power for alternate current (AC) charging infrastructure sums up to 799 points and 12,899.5 kW; for direct current (DC), charging infrastructure to 19 points 1365 kW were put into operation during LINOx BW combining for 611 points and 14,264.5 kW in total. Dividing the charging capacity through the number of installed charging points, a factor of 17.4 kW/charging point is derived.

3.2. Nitrogen Oxide Reduction Research Concept (1)

Table 4 shows key figures referring to the nitrogen oxide reduction within LINOx BW. The number of valid cases for the reduction calculation from research concept (1) is 63. These 63 cases represent 611 funded charging points (74.7%) and specified values of delivered electricity from the funded charging infrastructure summarizing to 1,600,045.64 kWh. Deriving the approximate electric mileage utilizing 18.4 kWh/100 km values, 8,695,900.22 km will be utilized for calculating the approximate nitrogen oxide reduction. Combining this value with the average HBEFA value for all surveys without vehicle fleet data, approximately 3.2 tons of nitrogen oxides were avoided within the research project.

Table 4. Key figures for research concepts (1) and (2).

Name	Value
Number of valid cases for research concept (1)	63
Delivered electricity from funded charging infrastructure (sum; kWh) (1)	1,600,045.64
Electric mileage utilizing (18.4 kWh/100 km) (km) (1)	8,695,900.22
NO_x-reduction (sum; g) (1)	3,232,167.22
Number of valid cases for research concept (2)	27
Electric mileage utilizing fleet of vehicle data (km) (2)	9,854,093.28
NO_x-reduction (sum; g) (2)	35,247,610.06
Difference between electric mileage research concept (1) and (2)	−1,158,193.06

3.3. Nitrogen Oxide Reduction Research Concept (2)

Only 33 questionnaires contained vehicle fleet data—other questionnaires either do not possess a vehicle fleet or could only provide insufficient data regarding their vehicle fleet; e.g., a few respondents could not provide start-up dates for their respective charging infrastructure. The lack of this information is problematic since the data had to be distributed evenly over the survey period. This results from the fact that not the total kilometers but the kilometers per year were surveyed. Additionally, several valid cases were able to provide vehicle fleet data, but no charged electricity data. Thus, only 27 valid cases were utilized for the reduction calculation for the research concept (2). The valid cases represent only 117 charging points (14.3%). Added-up mileage data for all BEV, PHEV (diesel), and PHEV (petrol) is equivalent to 9,854,093.28 km. To derive a potential nitrogen oxide reduction, HBEFA values from Tables 1 and 2 for each respective vehicle fleet are utilized. In total, roughly 35.2 tons of nitrogen oxides were avoided within the research project.

Comparing the electric mileage from each research concept, however, a difference of 1,158,193.06 km is calculated, meaning the funded charging infrastructure of LINOx BW is not covering the electric mileage of the vehicle fleet data surveys. Even though comparing results from 63 valid cases to 27 valid cases, the important fact is that the difference is negative. A few possible reasons can be considered to explain this:

- Semi-public accessibility in several cases, leading to external electric cars utilizing the charging infrastructure (tourists, visitors, etc.);
- Private utilization of electric cars belonging to a company's vehicle fleet;
- Charging infrastructure owner's permission for employees to utilize the charging infrastructure for their private cars;
- Unknown delivered electricity values due to no data collected or no separate electric meters for the funded charging infrastructure;
- Unknown vehicle fleet data or unusual mileage values;
- The special case of the installation of charging infrastructure in homeowners' associations—no vehicle fleet data and no charged electricity are provided, even though dozens of charging points are installed.

3.4. Comparison of Both Research Concepts Based on the Valid Cases of Research Concept (2)

Table 5 presents more detailed information about the valid cases of research concept (2). Here, the direct comparison between the potential NOx-reduction based on the different research concepts is shown. Few valid cases provide vehicle fleet data, thus valid for research concept (2), but did not provide charged electricity data, thus invalid for research concept (1).

First of all, the number of charging points installed before funding of LINOx BW has to be compared to the number of newly installed charging points with project funding. A total of 22 cases installed their first charging infrastructure within the project; thus, they started to utilize electric mobility. The other five cases already had charging points installed, leading to potentially higher values from research concept (2). In most cases, the larger the vehicle fleet provided in the questionnaire, the higher the result for research concept (2). Number 82 is a huge company with several thousands of employees, a vehicle fleet of 511 passenger cars, and a lot of charging infrastructure; however, only four charging points were funded within LINOx BW, questioning the eligibility of outliers for research concept (2). Without number 82, the potential nitrogen oxide reduction for research concept (2) results in only 2.05 t.

The variable days of data acquisition are irrelevant to research concept (1) and can potentially be neglected for future research utilizing research concept (1).

Additionally, a few municipalities completed the survey providing information about the vehicle fleet of the whole municipality—not only for the site their charging infrastructure was installed. Unknown car emission concepts need to be calculated with average HBEFA values, which are considerably higher especially comparing diesel and petrol cars. Summarizing the comparison, Table 5 shows the data heterogeneity based on the survey data complicating the task of appropriate research concepts to analyze the project's data.

The utilization of additional descriptive factors, e.g., daily charged electricity was neglected due to the fact that most funding recipients have started utilizing their funded charging points on separate dates and thus daily charged electricity values would need to be averaged. Total values of charged electricity suffice for a preliminary comparative analysis.

Table 5. Detailed information about the 27 valid cases for research concept (2).

Number	Charging Points ex. LINOx BW	Charging Points by LINOx BW	Charged Electricity (kWh)	Days of Data Acquisition of Charged Electricity	NOx-Reduction (1) (g)	Number of Passenger Cars	Number of Heavy-Duty Vehicles	NOx-Reduction (2) (g)
1	0	8	8043	336	16,247.24	1	1	6097.64
2	0	5	7703	559	15,560.42	1	2	21,265.52
3	0	2	6147	1008	12,417.23	6	0	51,835.74
10	2	5	6763	1501	24,969.80	5	9	112,723.54
14	0	1	3633	386	7338.83	20	2	70,910.60
20	0	4	No data	No data	0.00	8	0	57,703.04
21	0	5	1740	314	3514.88	6	0	9028.37
32	2	21	11,298.12	765	53,161.80	28	104	62,913.07
36	0	6	18,717	719	37,809.22	89	0	480,019.94
41	0	4	12,110	541	24,462.77	18	0	16,231.56
46	0	4	1579	99	3189.65	4	36	78,903.78
47	0	7	172,515	602	348,488.39	11	0	123,104.59
48	0	2	No data	No data	0.00	5	0	52,439.56
49	0	6	8271	423	17,998.62	45	1	56,195.56
59	0	10	14,875	700	30,048.20	28	2	8535.81
64	0	2	2018	424	4076.45	3	17	384,143.37
68	0	2	4845.22	1271	9787.57	2	0	21,597.75
69	0	3	2044.36	1628	4129.69	4	1	36,037.62
70	0	2	2580	1230	5211.72	1	1	420.23
71	0	2	636.42	1257	1285.59	3	1	13,976.85
72	4	2	3654	1350	7381.25	6	1	88,646.38
73	7	1	1302	1110	2630.10	9	2	132,713.05
74	0	2	896.66	341	1811.29	2	1	641.09
75	0	2	1430	1319	2888.67	4	1	81,244.59
76	0	1	No data	No data	0.00	1	0	3672.08
81	0	4	1579	99	3189.65	4	36	78,903.78
82	87	4	13,686	999	27,646.36	511	0	33,197,704.94

4. Discussion

Contextualizing the presented results, the 818 installed charging points within Baden-Württemberg over the course of the project, a comparison with data from the German Federal Network Agency (Bundesnetzagentur) for installed charging points within Baden-Württemberg follows.

In Germany, charging points with at least 11 kW are subject to registration. By 1 January 2019, the nearest date to our project start with available data, 2180 AC charging points and 437 DC charging points were installed in Baden-Wuerttemberg (total of 2.617 charging points), with a factor of 28.01 kW/charging point. By October 01, 2022, 11,149 AC charging points and 1909 DC charging points were installed in Baden-Württemberg (total of 13.058 charging points), with a factor of 28.03 kW/charging point [27]. During this period, 10,441 charging points have been installed in Baden-Wuerttemberg, with 818 charging points funded by LINOx BW (7.8%). However, the factor of 17.4 kW/charging point within LINOx BW is considerably lower leading to the assumption that LINOx BW facilitated an area-wide roll-out of AC charging infrastructure in the first place. Additionally, small- and medium-sized enterprises as well as considerably small projects with a low amount of charging infrastructure undertaking their first steps towards electric mobility do not require fast charging infrastructure for their respective needs.

The presented results account for only one-third of all possible data collected within LINOx BW. Most of the recipients participating in LINOx BW are unfamiliar with research projects leading to uncertainties when it comes to filling in the survey. Many questions in the questionnaires were not answered and left without remark; moreover, some answers did not correspond to the t_0-survey. Additionally, certain data in the respective funding recipient's organization is unavailable due to technical or organizational reasons. The possibility for each recipient to be able to derive electricity data from their funded charging infrastructure could not be made mandatory due to funding guidelines skipping research potential. Recipients that owned several locations but installed charging infrastructure on only one location could not separate their internal data to the locations relevant to LINOx BW. Conducting the results at the end of the project at the end of 2023, standardizing and verifying the collected data, and quantifying the number and the respective amount of side effects such as the allowance for semi-public or employee charging will be key. Additionally, unconventionally large or special cases need to be taken care of differently, as number 82 from Table 5 shows.

Comparing the eligibility of both presented research concepts, (1) represents 74.7% of all funded charging points within current survey data, whereas (2) represents only 14.3%. Providing the required data for research concept (2) is more difficult for the recipients inexperienced with surveys and research projects. Thus, a convergence of both % utilizing the upcoming additional data of not yet finished projects is highly unlikely. Looking toward future research, the utilization of research concept (1) seems more promising.

Final calculations for the emission reduction potential require data from the t_1-survey and the respective vehicle fleets. The huge amount of unknown car emission concepts leading to the utilization of average HBEFA factors for the nitrogen oxide reduction potential increases potential inaccuracies for the calculation results. The substitution of ICV to BEV or PHEV differs from recipient to recipient and is also based on other effects such as depreciation periods of currently utilized cars. Quantifying the reduction from recipients with private charging only is rather simple. The utilization of pedelecs adds slightly to the project's success (in particular, if the respective recipients are replacing passenger cars or heavy-duty vehicles); however, quantification is not possible.

In 2020, the German emissions of NOx reached a total of 978,000 tons, with 390,300 tons belonging to the traffic sector [28]. The current gathered LINOx BW data adds up to 35.24 tons (for research concept (2)) and to 3.23 tons (for research concept (1)) and is partly collected from several different years. For scientific comparison, potential emission reductions must be calculated on a yearly basis. Overall, the presented results utilized data from 818 charging points (the whole project funds 2.358 charging points). Effects will increase with time, leading to the integration of more electric cars into vehicle fleets as well as more electric cars in general in Germany, thus leading to an increased usage from semi-public charging infrastructure. However, all 82 questionnaires presented in this paper will not be surveyed another time, their respective results are final.

5. Conclusions

A new concept for quantifying nitrogen oxide emissions from charging infrastructure is presented. The presented approach shows a few limitations based on the project specifications of LINOx BW: No mandatory guidelines for deriving the charged electricity could be implemented, leading to the utilization of assumptions and some cases not being valid for research concepts. The verification of nitrogen oxide values is not possible due to the nature of air pollutants; additionally, no direct access to the required data from the recipients for researchers, thus leading to the utilization of questionnaires.

Gathering the required data for the analysis of nitrogen oxide emission reductions is an uphill task without direct access to the data. The mandatory installation of smart meters and separate electric meters for the funded charging infrastructure would greatly increase the data quality required for the research. Additional information about specific user groups utilizing the funded charging infrastructure would increase the scientific insight as well. The reliability of mileage and a fleet of vehicle data needs to be addressed for future research as well.

The analysis of the results so far has not considered the seven different use cases that all recipients were classified into: residential quarter, semi-public space, parking lots and park and ride, private on-site parking, nursing services, tourism, and pedelecs. Use-case-specific analysis of the survey results might lead to more intriguing results; e.g., comparing only nursing service or homeowners' associations with each other.

The application of current limit values for nitrogen oxides is based on air pollution data from only a few measurement sites within cities, yet the measured data are utilized by politics for statements concerning whole municipalities. Increasing the number of certified measurement sites within cities affected by high nitrogen oxide values would support the utilization of the correct referent figures for legal measures and increase the foundation of data for area-wide modeling of nitrogen oxide values. Incorporating additional different, yet comparable spatial and traffic settings with additional measurement sites within several cities greatly improves comparative studies between several cities. Addressing this should be a priority moving forward, especially because the World Health Organization plans to rework its recommendations concerning nitrogen oxides to lower the limit values. For a small part, nitrogen oxide values will decline on their own due to newer and cleaner emission concepts and an increasing share of electric vehicles.

The original task to quantify the reduction of nitrogen oxide measurement sites to prevent cities from adopting legal measures for their municipality is not possible for various reasons. Selective measurement sites on certain streets are utilized for statements pertaining to whole municipalities, whereas the specific driving patterns of electric cars are unknown. Deriving spatial correlations between nitrogen oxides, charging infrastructure and electric cars requires enormous amounts of data. Yet, the benefit remains unknown. Calculating the success of charging infrastructure regarding nitrogen oxide reductions utilizing charged electricity data is an appropriate way. In the case of LINOx BW, added-up results for the whole research project in comparison to overall German values seem to be a better way to quantify the success of the project. Nevertheless, the funding of 2.358 charging points in various real-life settings remains a success.

Author Contributions: Conceptualization, K.H. and A.G.; methodology, K.H.; software, K.H.; validation, K.H.; formal analysis, K.H. and A.G.; investigation, K.H. and A.G.; resources, K.H. and A.G.; data curation, K.H. and A.G.; writing—original draft preparation, K.H.; writing—review and editing, K.H. and A.G.; visualization, K.H.; supervision, K.H.; project administration, K.H. and A.G.; funding acquisition, K.H. All authors have read and agreed to the published version of the manuscript.

Funding: This research was funded by the German Federal Ministry for Economic Affairs and Climate Action (BMWK) (Sofortprogramm "Saubere Luft 2017–2020"); support code 01MZ18012A-D (LINOx BW).

Data Availability Statement: Survey data is unavailable due to privacy and funding guidelines within LINOx BW.

Acknowledgments: The authors thank the project partners of LINOx BW: the Association of Cities Baden-Württemberg (Städtetag Baden-Württemberg), the Verband Region Stuttgart, the State Agency for New Mobility Solutions and Automotive (e-mobil BW) and the Zentrum für Sonnenenergie- und Wasserstoff-Forschung Baden-Württemberg (ZSW). Furthermore, we would like to thank everyone who participated in phone interviews and the quantitative surveys and thus helped us to gain insight into their respective institutions.

Conflicts of Interest: Karsten Hager and Alexandra Graf are employees of Institut Stadt | Mobilität | Energie (ISME) GmbH. The paper reflects the views of the scientists, and not the company.

Abbreviations

AC	Alternate Current
BEV	Battery Electric Vehicle
BMWK	German Federal Ministry for Economic Affairs and Climate Action
DC	Direct Current
EV	Electric Vehicles
FCEV	Fuel Cell Electric Vehicle
HBEFA	Handbuch für Emissionsfaktoren
ICEV	Internal Combustion Engine Vehicle
LINOx BW	Ladeinfrastruktur in NOx-Kommunen in Baden-Württemberg
PHEV	Plug-In Hybrid Electric Vehicle
WTW	Well-to-wheel

References

1. Directive 2008/50/EC of the European Parliament and of the Council of 21 May 2008 on Ambient Air Quality and Cleaner Air for Europe. Available online: https://eur-lex.europa.eu/legal-content/DE/ALL/?uri=celex%3A32008L0050 (accessed on 22 February 2023).
2. Muncrief, R. Briefing—A Comparison of Nitrogen Oxide (NOx) Emissions from Heavy-Duty Diesel, Natural Gas, and Electric Vehicles. Available online: https://theicct.org/publication/a-comparison-of-nitrogen-oxide-nox-emissions-from-heavy-duty-diesel-natural-gas-and-electric-vehicles/ (accessed on 27 February 2023).
3. Schlesinger, R.B.; Lippmann, M. Nitrogen Oxides. In *Environmental Toxicants: Human Exposures and Their Health Effects*, 4th ed.; Lippmann, M., Leikauf, G.D., Eds.; John Wiley & Sons: Hoboken, NJ, USA, 2020; pp. 721–781.
4. Chen, T.-M.; Gokhale, J.; Shofer, S.; Kuschner, W.G. Outdoor Air Pollution: Nitrogen Dioxide, Sulfur Dioxide, and Carbon Monoxide Health Effects. *Am. J. Med. Sci.* **2007**, *333*, 249–256. [CrossRef] [PubMed]
5. Laumbach, R.J.; Kipen, H.M. Respiratory health effects of air pollution: Update on biomass smoke and traffic pollution. *J. Allergy Clin. Immunol.* **2012**, *129*, 3–11. [CrossRef] [PubMed]
6. Manisalidis, I.; Stavropoulou, E.; Stavropoulos, A.; Bezirtzoglou, E. Environmental and Health Impacts of Air Pollution: A Review. *Front. Public Health* **2020**, *8*, 14. [CrossRef] [PubMed]
7. Kuschel, G.; Metcalf, J.; Berentson-Shaw, J.; Hales, S.; Atkinson, J.; Woodward, A. *Health and Air Pollution in New Zealand 2016 (HAPINZ 3.0): Volume 1—Finding and Implications*; Report prepared for Ministry for the Environment, Ministry of Health, Te Manatū Waka Ministry of Transport and Waka Kotahi NZ Transport Agency; Kuschel, G., Metcalfe, J., Sridhar, S., Davy, P., Hastings, K., Mason, K., Denne, T., Berentson-Shaw, J., Bell, S., Hales, S., Eds.; Emission Impossible: Mt Eden, Auckland, New Zealand, 2022.
8. Guan, Y.; Xiao, Y.; Chu, C.; Zhang, N.; Yu, L. Trends and characteristics of ozone and nitrogen dioxide related health impacts in Chinese cities. *Ecotoxicol. Environ. Saf.* **2022**, *241*, 113808. [CrossRef] [PubMed]
9. Afroz, R.; Hassan, M.N.; Ibrahim, N.A. Review of air pollution and health impacts in Malaysia. *Environ. Res.* **2003**, *92*, 71–77. [CrossRef] [PubMed]
10. Krzyzanowski, M.; Kuna-Dibbert, B.; Schneider, J. Health Effects of Transport-Related Air Pollution. WHO Europe: Copenhagen, Denmark, 2005.
11. Ganderman, J.; Avol, E.; Lurmann, F.; Kuenzli, N.; Gilliland, F.; Peters, J.; McConnell, R. Childhood Asthma and Exposure to Traffic and Nitrogen Dioxide. *Epidemiology* **2005**, *16*, 737–743. [CrossRef] [PubMed]

12. Favarato, G.; Anderson, H.R.; Atkinson, R.; Fuller, G.; Mills, I.; Walton, H. Traffic-related pollution and asthma prevalence in children. Quantification of associations with nitrogen dioxide. *Air Qual. Atmos. Health* **2014**, *7*, 459–466. [CrossRef] [PubMed]
13. Garcia, E.; Johnston, J.; McConnell, R.; Palinkas, L.; Eckel, S.P. California's early transition to electric vehicles: Observed health and air quality co-benefits. *Sci. Total Environ.* **2023**, *867*, 161761. [CrossRef] [PubMed]
14. Peel, J.L.; Haeubner, R.; Garcia, V.; Russell, A.G.; Neas, L. Impact of nitrogen and climate change interactions on ambient air pollution and human health. *Biogeochemistry* **2013**, *114*, 121–134. [CrossRef]
15. Umweltbundesamt—Liste der Städte mit NO2-Grenzwertüberschreitung. Available online: https://www.umweltbundesamt.de/dokument/liste-der-staedte-no2-grenzwertueberschreitung (accessed on 22 February 2023).
16. Feinauer, M.; Ehrenberger, S.; Epple, F.; Schripp, T.; Grein, T. Investigating Particulate and Nitrogen Oxides Emissions of a Plug-In Hybrid Electric Vehicle for a Real-World Driving Scenario. *Appl. Sci.* **2022**, *12*, 1404. [CrossRef]
17. He, S.; Jiang, L. Identifying convergence in nitrogen oxides emissions from motor vehicles in China: A spatial panel data approach. *J. Clean. Prod.* **2021**, *316*, 128177. [CrossRef]
18. Seol, E.; Yoo, E.; Lee, C.; Kim, M.; Cho, M.; Choi, W.; Song, H.H. Well-to-wheel nitrogen oxide emissions from internal combustion engine vehicles and alternative fuel vehicles reflect real driving emissions and various fuel production pathways in South Korea. *J. Clean. Prod.* **2022**, *342*, 130983. [CrossRef]
19. Rubaiyat Reza Habib, A.K.M.; Butler, K.S. Alternatives to lithium-ion batteries in electric vehicles. *Future Technol.* **2022**, *1*, 33–34. [CrossRef]
20. Dong, H.; Fu, J.; Zhao, Z.; Liu, Q.; Li, Y.; Liu, J. A comparative study on the energy flow of a conventional gasoline-powered vehicle and a new dual clutch parallel-series plug-in hybrid electric vehicle under NEDC. *Energy Convers. Manag.* **2020**, *218*, 113019. [CrossRef]
21. Rubaiyat Reza Habib, A.K.M.; Butler, K.S. Environmental and economic comparison of hydrogen fuel cell and battery electric vehicles. *Future Technol.* **2022**, *1*, 25–33. [CrossRef]
22. Official Project Homepage LINOx, BW. Available online: https://www.linox-bw.de (accessed on 22 February 2023).
23. Hager, K.; Graf, A. Zwischenergebnisse des BMWi-Projektes LINOx BW—Fact Sheet der Anwendungsfälle: Parkhaus—P&R—Privates Firmengelände. *Transform. Cities* **2022**, *2022*, 73–77.
24. Fraunhofer ISI: Klimabilanz, Kosten und Potenziale Verschiedener Kraftstoffarten und Antriebssysteme für Pkw und Lkw, Endbericht. Available online: https://publica.fraunhofer.de/entities/publication/caf276e2-a560-430f-8c46-dc0a9593d821/details (accessed on 22 February 2023).
25. HBEFA. Available online: https://www.hbefa.net/d/ (accessed on 22 February 2023).
26. Kraftfahrtbundesamt—Bestand nach Fahrzeugklassen und Aufbauarten. Available online: https://www.kba.de/DE/Statistik/Fahrzeuge/Bestand/FahrzeugklassenAufbauarten/2022/2022_b_fzkl_tabellen.html?nn=3524712&fromStatistic=3524712&yearFilter=2022&fromStatistic=3524712&yearFilter=2022 (accessed on 22 February 2023).
27. Bundesnetzagentur—Elektromobilität: Öffentliche Ladeinfrastruktur. Available online: https://www.bundesnetzagentur.de/DE/Fachthemen/ElektrizitaetundGas/E-Mobilitaet/start.html (accessed on 22 February 2023).
28. Umweltbundesamt—Stickstoffoxid-Emissionen. Available online: https://www.umweltbundesamt.de/daten/luft/luftschadstoff-emissionen-in-deutschland/stickstoffoxid-emissionen#entwicklung-seit-1990 (accessed on 22 February 2023).

World Electric Vehicle Journal

Article

Academic and Corporate Vehicle Electrification Research

Hans Pohl [1,*] and Magnus Karlström [2]

[1] Lindholmen Science Park, P.O. Box 8077, 402 78 Gothenburg, Sweden
[2] Department of Energy and Environment, Chalmers University of Technology, 412 96 Gothenburg, Sweden
* Correspondence: hans.pohl@lindholmen.se; Tel.: +46-708402740

Abstract: We developed and used methodology to analyze scientific publications in Scopus relating to vehicle electrification and associated key enabling technologies: batteries, fuel cells and electric machines with power electronics. The global research landscape was mapped, and an analysis of the 16 most active countries was carried out. Vehicle electrification publications are rewarded with a high citation impact, and they include corporate actors to a great extent. China dominates in vehicle electrification research as well as in the enabling technologies, and China's position is set to become even more dominating. Battery research has grown rapidly with a high citation impact, whereas the volume of research for the other enabling technologies was more constant during 2017–2021. Automakers' research that has led to scientific publications was specifically studied. Ford Motor Company was the automaker with the highest number of vehicle electrification publications during 2017–2021. A large share of the automakers' publications was co-authored with academic actors, and such publications were rewarded with a higher citation impact than those without. However, the share of international co-publications among the automakers was meager. It is concluded that the analysis of vehicle electrification publications gives an overview of the rapidly developing field. Moreover, the analysis of automakers' involvement in such research is one way of obtaining one perspective on their strategies and priorities.

Keywords: BEV (battery electric vehicle); EV (electric vehicle); fuel cell vehicle; HEV (hybrid electric vehicle); research publications; scientometrics

Citation: Pohl, H.; Karlström, M. Academic and Corporate Vehicle Electrification Research. *World Electr. Veh. J.* **2023**, *14*, 71. https://doi.org/10.3390/wevj14030071

Academic Editor: Joeri Van Mierlo

Received: 20 January 2023
Revised: 6 March 2023
Accepted: 7 March 2023
Published: 10 March 2023

1. Introduction

Electrified vehicles, including batteries and fuel cells, are on the market, and the volumes of electric cars, trucks and buses are growing. Nevertheless, new solutions and knowledge are still needed to broaden the market and reduce the environmental impact. This paper used scientific publications to study knowledge development related to vehicle electrification.

Quantitative studies of vehicle electrification using publications typically focus rather narrowly on electric vehicles and specific geographic areas. However, publications explicitly mentioning "electric vehicles" or similar terms constitute only a tiny part of the total research body relevant to vehicle electrification. For example, very few articles within battery research mention specific applications as they are on a more generic level.

The purpose is to better understand vehicle electrification research. This is achieved through mapping of the global vehicle electrification research landscape using scientific publications. To manage this mapping, new publication analysis methodology had to be developed. This mapping is relevant to understand how different countries are involved and how research in this field differs from other fields. A specific focus of the study is on automakers and their collaborations with academia. The results can be used to guide research policies, and the methodology used can inspire other researchers within the field of scientometrics.

This article develops new methodology to capture a broader range of publications relevant to vehicle electrification than in previous studies. In addition to the study of "vehicle electrification" publications, three enabling technologies are added and studied separately. They are batteries, fuel cells and electric machines and power electronics. Another contribution of the study is the analysis of a selection of automakers and how they collaborate with academic actors.

Several questions are addressed by analyzing Scopus publications from 1996 until and including 2021 about electrified vehicles and key enabling technologies. Which countries and automakers are most active? What is their citation impact? How are they collaborating?

Following a brief review of previously published studies of vehicle electrification, the methodology and data are outlined. Thereafter, results are presented and divided into two parts, one giving national data and one describing how automakers are involved in academic publications. Finally, discussions and conclusions follow.

2. Previous Literature

From the previously published studies of vehicle electrification, typically, a few keywords or classifications provided by the database supplier can be used to identify relevant publications [1–7]. They focus on electric vehicles, which means publications explicitly mentioning both electrification and an automotive application, in some cases using search terms such as "electric vehicle" or similar [cf. 2 and 5]. Some studies try to gain a comprehensive view of electric vehicle technologies and understand the emphasis of current research. Web of Science dominates as the source of data. Among the results found are that vehicle electrification publications are rapidly increasing in volume, and China is gaining a dominating role in producing such publications.

There are also published studies addressing the enabling technologies. In [8], Li-ion battery research is studied using a selection of search terms. Rather detailed results indicating the most productive researchers are presented. Based on the most frequent keywords, hotspots are suggested. Another study of battery-related publications focuses on how the European Battery 2030+ initiative is positioned. It uses relatively advanced clustering methodology to identify and analyze relevant publications. Among the results, the United States and Canada tend to lead in citation impact, whereas China dominates in publication volumes [9]. An even more specific battery study addresses battery states of charge and health [10]. It uses both Web of Science and Scopus data to identify top authors and institutions as well as some trends. Additionally, one rather specific, published study focuses on wireless charging [11]. It uses a small selection of search terms combined with the manual scrutiny of all resulting publications to obtain a set of publications for further analysis.

The study of the previous literature highlights some critical aspects. First, it is important to have competence in vehicle electrification and scientometrics. Second, a clear description of the search strategy employed is critical for interpreting the results. Third, a narrow search extracting only "electric vehicle" publications might miss important developments in the technologies enabling vehicle electrification. Finally, it could also be noted that in most cases, types of data other than publication data are not used.

3. Methodology and Data

SciVal, a tool provided by Elsevier, was used to extract and analyze Scopus publications. Scopus is a publication database with broader coverage than Web of Science [12]. Data were extracted in December 2022, meaning that the last almost complete year was 2021. Elsevier considers the publication data for a year sufficiently complete in May of the following year. However, this does not apply to citation indicators, which are based on accumulating data over a more extended period of time.

The identification of publications was based on queries. Queries use search terms which can be combined with AND, OR, and AND NOT. In addition, entities such as countries, institutions, and authors were used. The search strategy is illustrated in Figure 1. Vehicle electrification publications, from now on called XEV publications, are the sum of

"Vehicle electrification" and the overlapping areas joining the enabling technologies and the automotive domain (Figure 1 is not to scale).

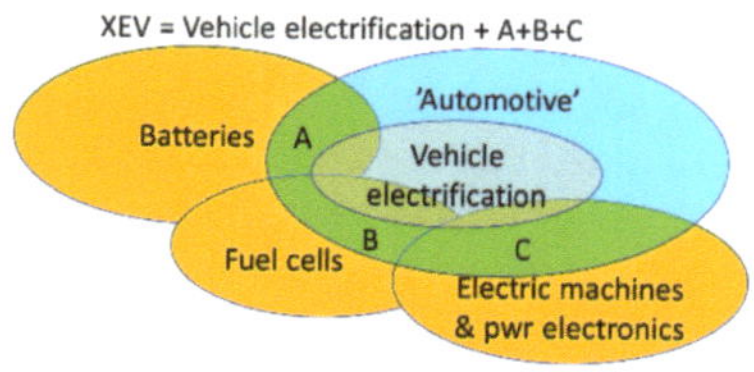

Figure 1. Search query for vehicle electrification publications (XEV).

One challenge is that publications in enabling technologies clearly relevant to vehicle electrification seldom mention automotive applications in the title, keywords, or abstract. Only a few percent of battery publications mention automotive applications. One of Sweden's most productive battery researchers estimated that 50% of her publications addressed issues relevant to automotive applications. However, less than 3% were identified by the query for XEV publications. The strategy to manage this challenge was to include separate studies of publications on enabling technologies. The queries were designed in an iterative process; see Figure 2.

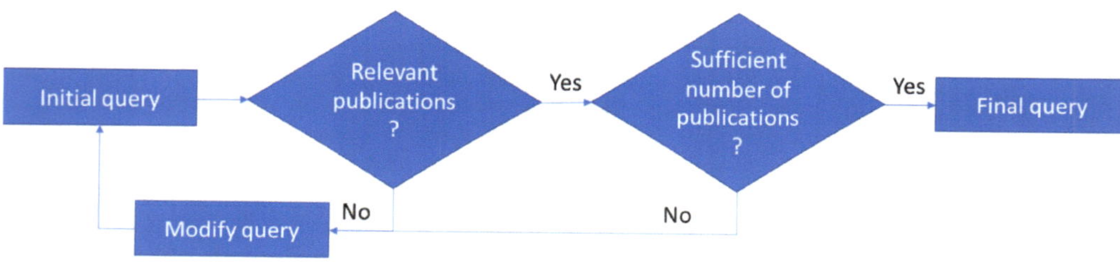

Figure 2. Query design process.

The manual scrutiny of 100 randomly selected publications' titles, abstracts, and journal names was used to determine if the extracted publications were within the field of vehicle electrification. The next and more difficult step was to determine if all relevant publications were included. This was addressed using publication lists from researchers in the field as well as all publications from the Swedish Electromobility Centre. Further details about the methodology are given in [13].

Even though we iterated the queries many times to maximize the number of correctly identified publications and minimize the number of falsely included ones, the resulting lists are incomplete. The number of relevant publications is likely higher. Our study underestimates the total. Another known and probably unavoidable problem is a bias towards more applied research. For obvious reasons, blue sky research seldom mentions potential applications, and they are therefore not captured by the queries. This is true for the XEV query as well as for the queries for the enabling technologies.

Preliminary results and the methodology were discussed with researchers and other actors working with vehicle electrification. Approximately 40 people were consulted, among them being the R&D managers at the three large automakers in Sweden: Volvo Cars, AB Volvo, and Scania. Thereafter, modifications were made.

4. Results

4.1. Overall Description of Data

First, word clouds for 2012 and 2021 were compared. They were based on Elsevier's Fingerprint engine, which uses the title, abstract, and keywords to generate key phrases. The larger the font, the higher the relevance. To some extent, the word clouds mirror the query used to extract the publications. However, some search terms are clearly represented in the word clouds, whereas others are not. The green word cloud in Figure 3 is based on 12,809 XEV publications from 2021. The blue word cloud in Figure 4 uses 5410 XEV publications from 2012.

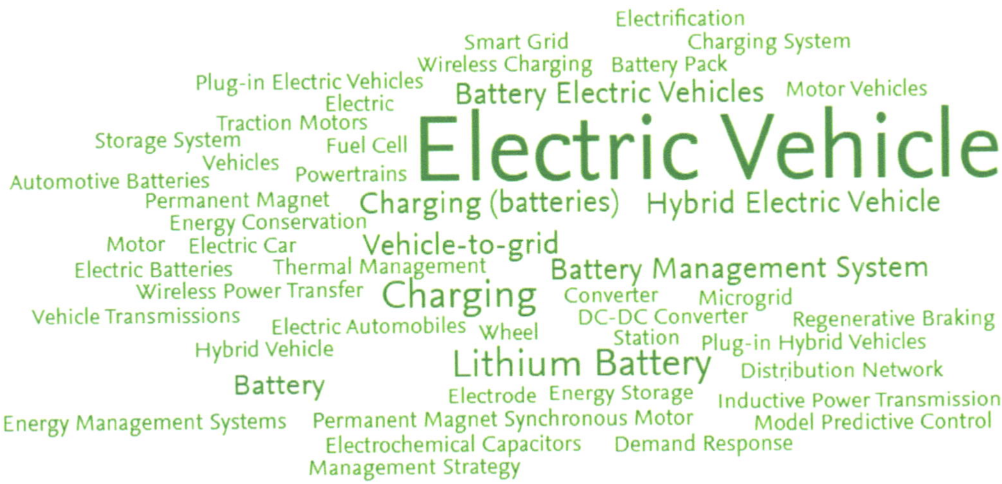

Figure 3. Key phrases based on XEV publications in 2021.

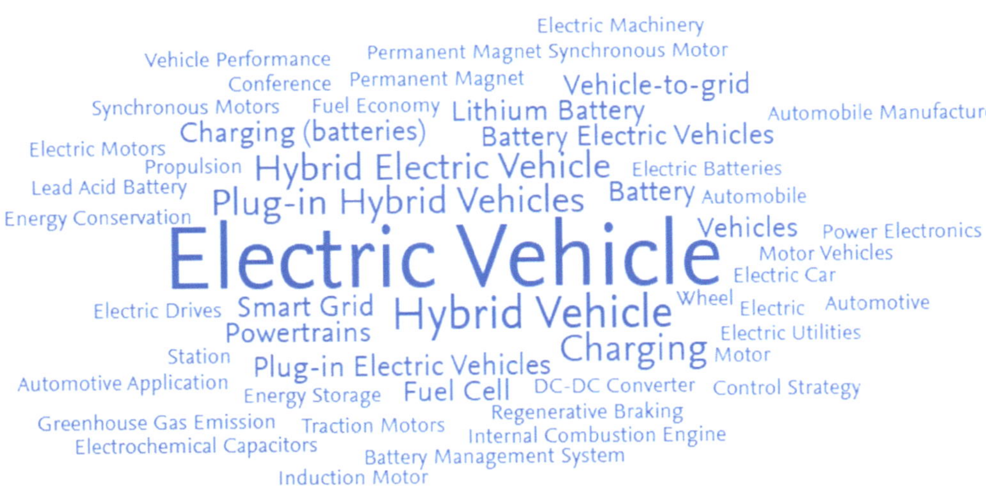

Figure 4. Key phrases based on XEV publications in 2012.

During this period, the focus of research changed. In 2012, different types of hybrid electric vehicles constituted substantial portions of research. In 2021, hybrid vehicles were less in focus. Key phrases related to charging indicate that this type of research produced approximately the same share of publications in both years. Fuel cells were mentioned in 2012, albeit with just one key phrase. In 2021, there was no fuel-cell-related key phrase.

Given the design of XEV publications as a combination of vehicle electrification publications and publications in the enabling technologies mentioning automotive applications, it is interesting to study to what extent the enabling technology publications mention electric vehicle applications. In Figure 5, the darker blue areas represent the publications identified as both XEV and within one or several enabling technologies. For example, out of 24,228 battery publications, 12% or 2797 were also XEV publications.

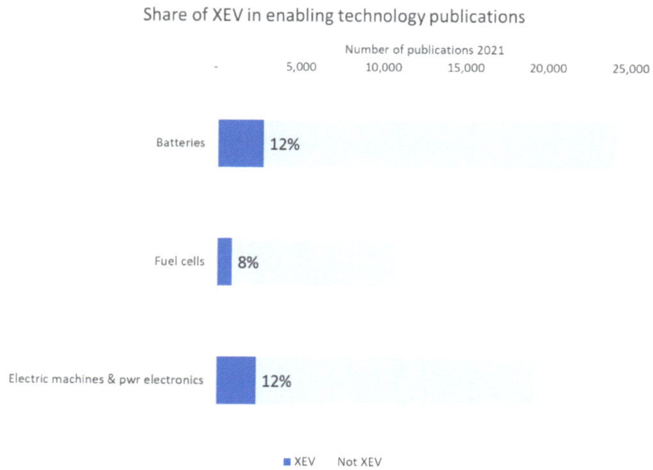

Figure 5. Overlap between XEV and enabling technology publications (2021).

As explained in the Methodology section, Figure 5 should not be interpreted as an indication that most of the enabling technology research is not relevant to vehicle electrification. Rather, it shows that enabling technology research is often carried out on a more basic level, where the potential applications are not relevant to mention. Moreover, it should be noted that the query to identify enabling technology publications is biased towards more applied research.

4.2. National Perspectives

The volumes of vehicle electrification (XEV) publications increased far more than the overall increase in publications. Figure 6 shows the annual production of the 16 countries with the highest numbers of XEV publications. In 2011, China overtook the United States in XEV publication volume, and in 2021, China also passed India.

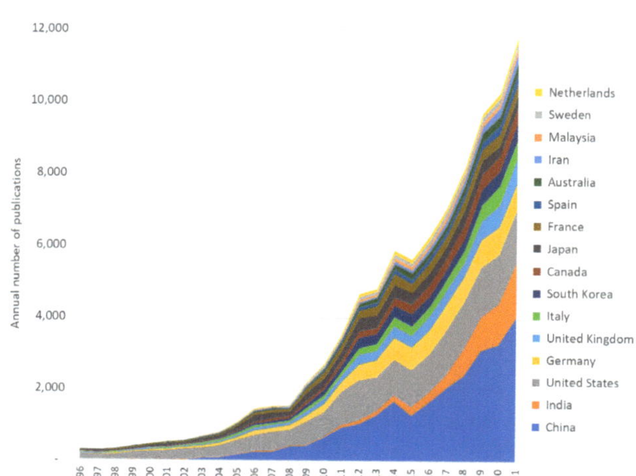

Figure 6. Annual vehicle electrification publication volumes.

Table 1 features some indicators characterizing the XEV publications per country. Data for all publications in each country are also given to allow for comparisons. In line with Figure 3, China's publication volume was the largest, with almost 15,000 publications over the five-year period 2017–2021, more than twice the volume of XEV publications produced in the second largest country, the United States. On the other hand, India was only larger than the United States in 2021, which means that during the period, it was the third largest country.

Table 1. Key indicators for XEV publications in 2017–2021.

	Publication Volume XEV 2017–2021	Annual Publication Volume Growth 2017–2021 (CAGR)		Citation Impact (FWCI)		Share of Academic-Corporate Co-Publications		International Co-Publications (FWIS)	
		XEV	Nt'l	XEV	Nt'l	XEV	Nt'l	XEV	Nt'l
Australia	1110	10.3%	3.3%	2.25	1.61	3.1%	3.3%	1.59	1.39
Canada	1988	6.2%	3.0%	2.08	1.47	6.6%	4.3%	1.45	1.32
China	14,905	14.4%	10.2%	1.35	1.08	6.3%	2.8%	0.59	0.54
France	1531	3.2%	0.4%	1.41	1.30	12.9%	6.2%	1.57	1.35
Germany	3706	0.7%	1.9%	1.40	1.33	17.6%	6.5%	0.77	1.21
India	4815	28.6%	8.7%	1.12	0.93	2.2%	1.2%	0.43	0.47
Iran	997	18.3%	6.7%	1.99	1.09	0.7%	0.8%	1.20	0.67
Italy	2132	12.2%	5.5%	1.78	1.44	9.5%	4.1%	1.06	1.13
Japan	1774	−3.3%	1.4%	1.09	0.93	13.2%	6.6%	0.69	0.72
Malaysia	643	5.2%	4.9%	1.52	1.08	0.9%	1.6%	1.25	1.16
Netherlands	645	4.8%	2.9%	1.85	1.73	11.9%	7.3%	1.28	1.52
South Korea	1887	10.3%	3.9%	1.28	1.08	9.4%	4.5%	0.57	0.70
Spain	1030	14.5%	5.3%	1.48	1.28	9.9%	3.9%	1.31	1.17
Sweden	681	6.8%	2.5%	1.89	1.62	11.9%	7.4%	1.25	1.56
United Kingdom	2614	14.0%	1.5%	2.15	1.56	11.6%	5.5%	1.55	1.38
United States	6722	3.1%	0.7%	2.08	1.37	8.2%	4.7%	1.08	0.87
World	50,262	10.5%	4.1%	1.37	1.00	6.4%	2.7%	0.53	1.00

Green color indicates maximum value in each column.

In the third and fourth columns, the publication volume growth from 2017 to 2021 was based on a linear regression approximation of the data points for each country and year, followed by the compounded annual growth rate calculation. Again, India showed very rapid growth. For most countries, XEV publication volumes grew more rapidly than the national average, which was also confirmed on the world level, with 10.5% growth for XEV versus 4.1% for all publications.

The field-weighted citation impact (FWCI) is a proxy for publication quality. It is a normalized indicator comparing the number of citations that a publication receives with the citation volumes for other publications in the same scientific field, published in the same year, in the same type of publication. The global average is 1. Almost all countries have higher FWCIs for XEV publications than for average publications. To some extent, a high FWCI confirms that the field is "hot" and rapidly growing.

The share of academic-corporate co-publications reflects academic-corporate collaboration. XEV publications involve corporate actors to a higher degree than average publications. In several countries, the shares are very high and are investigated in more detail below.

Finally, in Table 1, another normalized indicator based on international co-publications with a global average of 1 (see [14]) shows no clear pattern regarding how international the vehicle electrification research is. However, a comparison of Germany and France reveals large differences. Whereas Germany has a much lower share of international co-publications in the XEV area than nationally, France has a slightly higher score.

In relative terms, which countries are most active in XEV research? In Table 2, three relative indicators are given, comparing the XEV publication volumes with automotive, national, and global volumes of publications.

Table 2. Relative XEV publications (2017–2021).

	XEV Share of Automotive	XEV Share of National	Share of XEV World
Australia	31%	0.19%	2.2%
Canada	41%	0.33%	4.0%
China	43%	0.42%	29.7%
France	33%	0.24%	3.0%
Germany	29%	0.38%	7.4%
India	32%	0.48%	9.6%
Iran	41%	0.30%	2.0%
Italy	33%	0.31%	4.2%
Japan	34%	0.25%	3.5%
Malaysia	23%	0.34%	1.3%
Netherlands	31%	0.19%	1.3%
South Korea	34%	0.41%	3.8%
Spain	30%	0.19%	2.0%
Sweden	26%	0.29%	1.4%
United Kingdom	35%	0.23%	5.2%
United States	31%	0.19%	13.4%

Green color indicates maximum value in each column.

China has the highest share of XEV publications when related to the total volume of automotive publications. This could, to some extent, depend on China's relatively late involvement in the production of road vehicles. The internal combustion engine has probably created large communities of researchers in countries with long traditions in vehicle development and production. The capacity to develop the internal combustion engine is still relevant, even though its role as a key performance ingredient is rapidly decreasing. Another potential explanation for China's high share in this comparison might be that China has had a relatively strong policy promoting vehicle electrification.

India leads when the XEV publication volume is compared with all publications made with Indian researchers as (co-)authors. This means that the share of automotive research in India is rather large. It is surprising that several of the traditional countries with large automotive industries do not appear to have a corresponding volume of research. However, a part of the explanation could also be that the query used in this study has bias towards more applied research. As described in the Methodology section using the case of batteries, most of the battery publications do not mention automotive applications, even though it might be relevant.

The final column once again illustrates China's dominance in XEV publications. As the table is based on an average of five years, and China's research is still growing more rapidly than the world average, the share in 2021 is even larger than 29.7%.

As described in the Methodology section, a separate analysis of enabling technologies was also carried out. Given the vast range of applications for these technologies, the volume of research is not directly linked to the capability to develop more advanced electrified vehicles. However, it is probable that a large share of the researchers in these fields can contribute.

4.2.1. Battery Publications

Batteries are needed for all types of electrified vehicles. Many chemistries compete, and the processes within the battery cells are challenging to understand. The volumes of scientific publications increased rapidly during 2017–2021, which is in line with the trend from previous years (see Table 3).

Table 3. Key indicators for battery publications.

	Publication Volume Batteries 2017–2021	Publication Volume Growth 2017–2021 (CAGR)		Batteries Share of Automotive	Batteries Share of National	Share of Batteries World
		Batteries	National			
Australia	3829	12.2%	3.3%	108%	0.67%	3.6%
Canada	2859	9.8%	3.0%	58%	0.47%	2.7%
China	60,614	13.7%	10.2%	176%	1.73%	56.5%
France	2572	4.0%	0.4%	55%	0.40%	2.4%
Germany	6093	6.2%	1.9%	48%	0.63%	5.7%
India	8228	15.9%	8.7%	55%	0.82%	7.7%
Iran	1427	13.4%	6.7%	59%	0.43%	1.3%
Italy	1737	9.0%	5.5%	27%	0.25%	1.6%
Japan	4483	5.0%	1.4%	86%	0.64%	4.2%
Malaysia	1215	7.3%	4.9%	43%	0.64%	1.1%
Netherlands	548	6.6%	2.9%	26%	0.16%	0.5%
South Korea	8041	7.1%	3.9%	146%	1.74%	7.5%
Spain	1941	10.4%	5.3%	57%	0.36%	1.8%
Sweden	1101	11.3%	2.5%	43%	0.47%	1.0%
United Kingdom	3850	13.3%	1.5%	51%	0.34%	3.6%
United States	15,814	2.3%	0.7%	73%	0.44%	14.7%
World	107,282	8.0%	4.1%			

Green color indicates maximum value in each column.

China dominates, with 56.5% of the publications and rapid growth, which indicates that the share will soon be even higher. South Korea and Japan, which historically have been dominating battery development and production in the world, exhibit rather low shares of battery publications and a less rapid increase than the world average. Still, battery research represents the largest share of all publications in South Korea.

The fact that some countries have higher volumes of battery than automotive publications highlights the need for a separate analysis of research relating to the enabling technologies. Even though far from all battery-related publications represent knowledge of relevance for automotive applications, a large share probably does. A narrow study of the small share of "automotive battery publications" might miss important aspects.

The citation impact for battery publications is very high, and this is true for the world as well as for almost all countries. On average, a battery publication is cited approximately twice as much as an average publication. Battery publications with Australian participation enjoy the highest FWCI (2.98), followed by the United States (2.56) and Canada (2.52). Chinese battery research is cited more than the average battery publication, with a FWCI of 2.16.

4.2.2. Fuel Cell Publications

Fuel cells for automotive applications are predominantly of the polymer electrolyte (PEM) type, which means they operate at relatively low temperatures. Therefore, the query extracting fuel cell publications included all types. As highlighted in Table 4, fuel cell research is, when measured by publication volumes, approximately half the size of battery research. Moreover, the volume is increasing at a slightly lower rate than the overall publication volume, which means that the share is decreasing.

Table 4. Key indicators for fuel cell publications.

	Publication Volume Fuel Cells 2017–2021	Publication Volume Growth 2017–2021 (CAGR)		Fuel Cells Share of Automotive	Fuel Cells Share of National	Share of Fuel Cells World
		Fuel Cells	National			
Australia	1049	0.9%	3.3%	30%	0.18%	2.2%
Canada	1538	0.2%	3.0%	31%	0.26%	3.2%
China	16,294	9.7%	10.2%	47%	0.46%	34.1%
France	1707	0.3%	0.4%	36%	0.27%	3.6%
Germany	2517	−1.7%	1.9%	20%	0.26%	5.3%
India	4451	8.7%	8.7%	30%	0.44%	9.3%
Iran	1785	8.7%	6.7%	74%	0.54%	3.7%
Italy	1580	−2.4%	5.5%	24%	0.23%	3.3%
Japan	2741	−3.5%	1.4%	53%	0.39%	5.7%
Malaysia	1074	5.9%	4.9%	38%	0.57%	2.2%
Netherlands	376	−1.5%	2.9%	18%	0.11%	0.8%
South Korea	2847	4.6%	3.9%	52%	0.61%	6.0%
Spain	1063	0.9%	5.3%	31%	0.20%	2.2%
Sweden	543	0.0%	2.5%	21%	0.23%	1.1%
United Kingdom	2093	3.5%	1.5%	28%	0.18%	4.4%
United States	5667	−2.5%	0.7%	26%	0.16%	11.9%
World	47,776	3.9%	4.1%			

Green color indicates maximum value in each column.

China also dominates in fuel cell research, with approximately one third of all publications in the world and the fastest increase over the period of 2017–2021. On the other hand, India also had a high growth rate and has, after the United States, the third largest share in the world. As for batteries, fuel cell research in South Korea has the highest share of the overall production in the country.

The low or even decreasing volumes of fuel cell research in several countries that have been very active in the development of such vehicles for at least 25 years, among them being Japan, South Korea, and Germany, could either signal that the technology is not very high on the agenda any longer or that the need for more research is not so strong. As hydrogen and fuel cells have reappeared as priorities in national and international policies, not least for heavier vehicles, it is probable that the latter is the case. If so, it could indicate that the technologies needed are available and that it is now mainly a matter of engineering to make them competitive.

4.2.3. Electric Machines and Power Electronics

Finally, the development of research in electric machines and power electronics is the most general field, which means that it does not link so strongly to vehicle electrification. The volume of publications is on par with the battery field, but the growth rate is clearly lower; see Table 5.

India is the second largest producer of publications in this field after China, which makes more than one-quarter of all publications and has the highest growth rate. As for fuel cells, countries have shown both positive and negative trends over the past five years. Iran was the only country with a higher number of publications in this field than automotive ones. This is explained in part by the fact that Iran has a relatively limited total volume of automotive publications.

Table 5. Key indicators for electric machines and power electronics publications.

	Publication Volume Electric Machines and Power Electronics 2017–2021	Publication Volume Growth 2017–2021 (CAGR)		Machines Share of Automotive	Machines Share of National	Share of Machines World
		Machines	National			
Australia	1640	5.0%	3.3%	46%	0.29%	1.7%
Canada	2434	0.1%	3.0%	50%	0.40%	2.6%
China	25,297	8.3%	10.2%	73%	0.72%	26.9%
France	2813	−3.1%	0.4%	60%	0.44%	3.0%
Germany	4765	0.5%	1.9%	37%	0.49%	5.1%
India	11,571	3.0%	8.7%	77%	1.16%	12.3%
Iran	2614	7.2%	6.7%	108%	0.79%	2.8%
Italy	3068	2.8%	5.5%	47%	0.45%	3.3%
Japan	3615	−6.1%	1.4%	69%	0.52%	3.8%
Malaysia	1174	−5.3%	4.9%	42%	0.62%	1.2%
Netherlands	603	6.5%	2.9%	29%	0.18%	0.6%
South Korea	3149	−0.5%	3.9%	57%	0.68%	3.3%
Spain	2032	5.4%	5.3%	60%	0.38%	2.2%
Sweden	643	−0.2%	2.5%	25%	0.28%	0.7%
United Kingdom	3990	1.1%	1.5%	53%	0.35%	4.2%
United States	9496	−0.8%	0.7%	44%	0.26%	10.1%
World	94,158	2.2%	4.1%			

Green color indicates maximum value in each column.

4.3. Automakers' Involvement in Research

When looking at scientific publications, a less common perspective is how the corporate actors participate. Data for a selection of automakers representing different countries and vehicle types and two Tier 1 suppliers are presented in Table 6. Only automakers with substantial publication volumes were included in the analysis, which, among others, means that companies only making electric vehicles such as Tesla were not included. Robert Bosch has the largest total volume of scientific publications, whereas Ford Motor Company has the highest number of publications relating to vehicle electrification, albeit with a negative trend.

Toyota had the highest share of electrification publications during the period. In terms of vehicle output, Toyota has been and is the largest or second largest automaker. It made a clear commitment to vehicle electrification by developing and launching the hybrid electric Toyota Prius in 1997. Since then, it has introduced many hybrid models and has also been a pioneer in the market introduction of fuel cell vehicles. It is therefore not so surprising that Toyota has the largest share of XEV publications.

Compared to the analysis in Section 4.1, this subsection includes all publications where the automotive actor has an affiliation as it is very probable that their research is related to vehicles. However, it should also be noted that the volumes of XEV publications per automaker are relatively small, which calls for some caution when interpreting the results.

The research managers at the Swedish OEMs argued that the volumes should be related to the size of the company. When the volumes were related to the turnover or the official data for investments in research and development, the smaller companies (including those in Sweden) indicated the highest relative research intensity.

Another perspective on automakers' publications is to what extent they are co-authored with academic actors. The definition of an academic–corporate co-publication is that it has at least two co-authors and at least one academic and one corporate affiliation.

This means that double or multiple affiliations including academic and corporate actors also qualify, as long as two authors are involved. In Figure 7, the FWCI is indicated for publications with and without such collaboration.

Table 6. Electrification publications with corporate (co-)authors.

Scopus Publications 2017–2021	All		XEV		Share
	Volume	Trend	Volume	Trend	Electrification
BMW Group	1333	3.1%	136	14.0%	10.2%
Daimler AG	1163	0.1%	189	8.0%	16.3%
FAW Group Corporation	575	3.8%	48	8.8%	8.3%
Ford Motor Company	2187	−8.6%	310	−12.8%	14.2%
General Motors	1708	−7.9%	284	−7.7%	16.6%
Honda Motor Co., Ltd.	920	4.3%	87	−4.1%	9.5%
Hyundai Motor Group	1316	7.1%	120	7.3%	9.1%
Scania AB	203	−5.6%	24	0.0%	11.8%
Toyota Motor	1191	−3.4%	242	−11.4%	20.3%
Volkswagen AG	1732	2.4%	184	−0.2%	10.6%
Volvo Car Corporation	568	−0.1%	42	1.9%	7.4%
Volvo Group	462	−2.0%	36	17.8%	7.8%
DENSO Corporation	716	−13.1%	96	−12.1%	13.4%
Robert Bosch GmbH	2497	2.2%	258	−8.9%	10.3%

Green color indicates maximum value in each column.

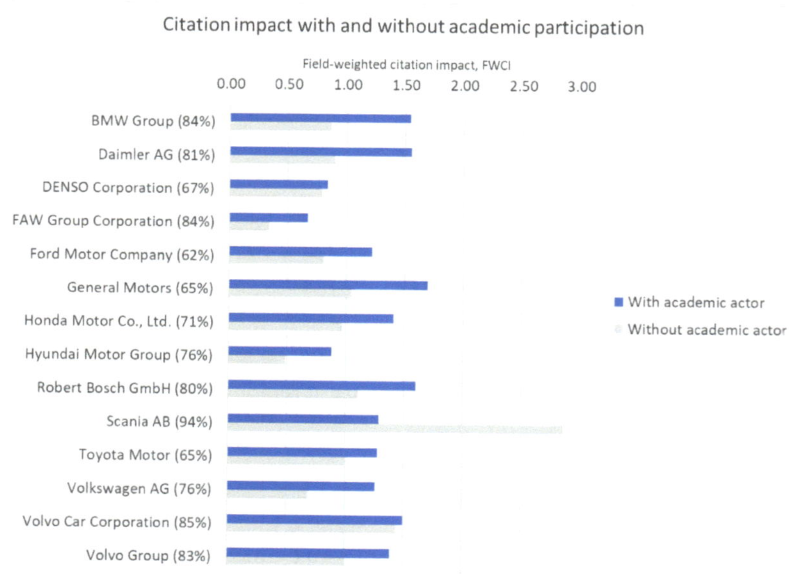

Figure 7. Automaker publications with and without academic actors in 2017–2021.

Directly behind each automaker's name, the percentage indicates how large the shares of publications including academic co-authors are. For example, it could be noted that Scania almost exclusively publishes with academic participation and that the automaker with the highest share of in-house publications is Ford Motor Company. The main message in Figure 7 is that collaboration with academic actors leads to a higher citation impact. Scania is an exception, but it should be noted that the number of publications without academic participation is minimal (12 publications).

Sweden's FWCI for the period was 1.62 (see Table 1). Publications involving the three automakers in Sweden had clearly lower FWCIs. In discussion with the R&D managers from the automakers, the central hypothesis was that a large share of the publications was made by junior researchers, not least PhD candidates. This could have led to a lower citation impact.

In the highly international automotive industry, it could be expected that research collaboration is also international. The main trend in research is towards more international co-publications, and academic–corporate co-publications are no exception; cf. [15]. However, the analysis shows that publications involving the automakers are largely national, and the academic partner is often even regional, which means the partner is in the same city as the main research facility; see Table 7. General Motors has more international than national publications; otherwise, the share of international co-spublications is much lower than those in the home country of the automaker.

Table 7. International partners among the top ten collaborations.

Scopus Publications 2017–2021	Top 3 Partner Institutions						Among Top 10 Partners	
Automotive Company	**Name**	**No. of Co-Pubs.**	**Name**	**No. of Co-Pubs.**	**Name**	**No. of Co-Pubs.**	**No. of Int'l**	**Share of Publ. with Int'l**
BMW Group	Technical University of Munich	378	Karlsruhe Institute of Technology	63	Universität der Bundeswehr München	48	2	6.4%
Daimler AG	University of Stuttgart	128	Ulm University	127	Karlsruhe Institute of Technology	97	0	0.0%
FAW Group Corporation	Jilin University	291	Volkswagen AG	23	Harbin Institute of Technology	22	1	5.3%
Ford Motor Company	University of Michigan, Ann Arbor	174	Ohio State University	96	Michigan State University	77	2	12.4%
General Motors	Université du Québec à Chicoutimi	67	University of Waterloo	66	University of Michigan, Ann Arbor	60	5	56.2%
Honda Motor Co., Ltd.	Tokyo Institute of Technology	47	Kyoto University	35	Technische Universität Darmstadt	31	3	29.9%
Hyundai Motor Group	Seoul National University	156	Hanyang University	140	Korea Advanced Institute of Science and Technology	70	0	0.0%
Scania AB	KTH Royal Institute of Technology	111	Uppsala University	21	Chalmers University of Technology	15	0	0.0%
Toyota Motor	Toyota Central R&D Labs., Inc.	148	The University of Tokyo	92	Nagoya University	72	0	0.0%
Volkswagen AG	Technical University of Braunschweig	157	Technical University of Munich	123	Karlsruhe Institute of Technology	84	1	7.8%
Volvo Car Corporation	Chalmers University of Technology	315	KTH Royal Institute of Technology	42	University of Gothenburg	37	0	0.0%
Volvo Group	Chalmers University of Technology	146	Eicher Motors Limited	78	Mälardalen University	37	3	27.9%
DENSO Corporation	Nagoya University	68	Toyota Motor	37	Tohoku University	35	0	0.0%
Robert Bosch GmbH	University of Stuttgart	216	Karlsruhe Institute of Technology	190	Ulm University	99	0	0.0%

As Table 7 indicates, half of the automotive companies have no international partner in their list of the ten largest research collaborations. Among those with at least one international partner, it tends to be at the bottom of the list, thus constituting a low share of the total. The location of the largest partner institution shows, in many cases, where the main research facility is based.

Another rule of thumb is that international co-publications are more cited than national co-publications. In Figure 8, the FWCIs for national and international publications are indicated. In this regard, the automotive industry is in line with the norm, and international co-publications are typically cited more often. The numbers behind the names of the automakers are the shares of national co-publications in the period.

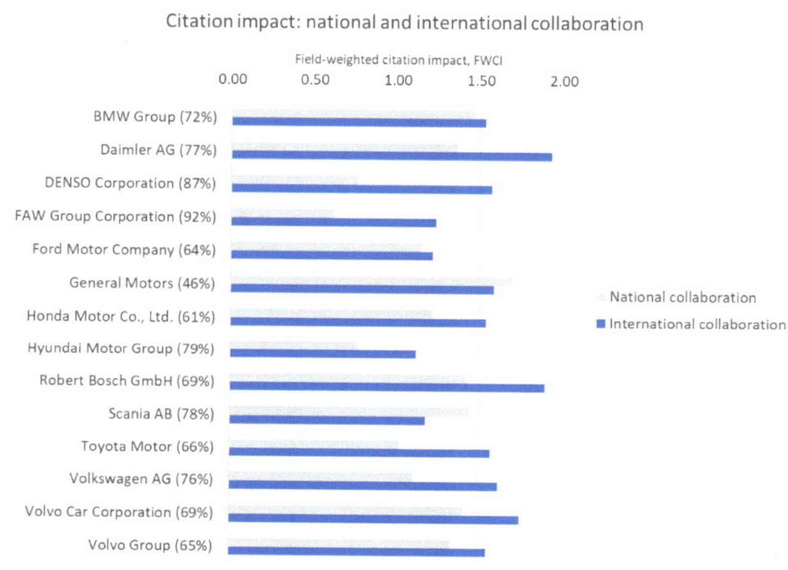

Figure 8. National and international co-publications in 2017–2021.

5. Discussion

The method and data differ from previous studies. The Scopus database is broader and includes more publications in the softer sciences than Web of Science, which was used in most other studies. Another advantage of the study is that the development of the search queries is clearly presented, contributing to increased transparency. Even though it is almost impossible to tell how large a share of the publications within the enabling technologies is relevant for vehicle electrification, the inclusion of these technologies and publications allows for a better understanding of how the field is developing. Several indicators, such as the citation impact and the share of academic–corporate publications, are included, contributing to perspectives on how various countries approach vehicle electrification. However, the broad scope of the study is also a disadvantage, as many questions remain to be studied more in detail.

What do the results indicate? Even though the approach is somewhat different, the study confirms the trends towards more publications in China and India, and a lower share in Western countries, as noted in previous studies. China is now the largest producer of XEV publications as well as publications in all three enabling technologies selected. It should also be noted that it is not only quantity; publications including Chinese authors are also cited. For example, battery research with Chinese participation had a high FWCI, meaning that these publications were cited more than twice the global average for all scientific publications during 2017–2021.

The automotive industry has been and is very international in almost all aspects. Therefore, it is surprising that the research is not so international. Germany's XEV publications had an FWIS of 0.77, which is lower than the average publication globally (FWIS = 1) and much lower than the average German publication (FWIS = 1.21). Why? Is it because automotive research is more applied than typical research or is it funded differently? This aspect clearly deserves closer study.

The enabling technologies have developed differently. For example, the volume of battery publications has increased dramatically, whereas the fuel cell publications have not increased at all. Battery publications also have a very high citation impact. China dominates all three enabling technologies, but especially batteries, accounting for more than 50% of the global output.

Whereas battery electric vehicles now have been on the market with growing volumes for a couple of years, fuel cell vehicles are at least one step behind. This could lead to a shift in focus for battery electric vehicles towards more applied questions and thus less research generating publications. Obviously, this is not the case. An attempt to explain the difference could be that batteries are needed in all types of electrified vehicles and used in many different ways. This, combined with the still quite high number of competing battery chemistries and the often very complicated chemical reactions that constitute the function of the batteries, could motivate a continued high volume of research. Fuel cells are less complicated, and therefore, the focus might be more on engineering issues.

The study of publications including automakers (and to some extent, two Tier 1 suppliers) revealed that a large share of the publications also included academic actors. This was expected. There appear to be differences between US and Japanese automakers, with relatively lower shares of academic–corporate co-publications and European and South Korean automakers with higher shares. A capacity to make scientific publications in-house could be considered a strength, but it could also signal that the traditions differ between countries and companies.

As expected, publications including academic actors as well as international co-publications were rewarded with a higher citation impact. However, the strong preference for national and even local collaborations with academic actors might signal that aspects other than scientific excellence are important when selecting the academic partners. Obviously, one such aspect is the potential for recruitment. It is probably much easier to recruit new employees locally than nationally or internationally.

Overall, this study indicates a very rapid development of vehicle electrification research with new countries emerging and China becoming more and more dominant. As the decarbonization of road transport is a global challenge, it is encouraging to see that substantial efforts are being invested. However, limited international collaboration is troublesome. Overall, XEV publications are internationally co-authored half as much as average publications. There is a clear risk that parallel research is being carried out, which is inefficient. For competition reasons, it might be challenging for automakers to collaborate extensively with international partners. Nevertheless, universities and research institutes should be encouraged to intensify their international collaborations. Another type of collaboration which could have an impact is between universities and the automotive industry.

Finally, a study of publications is just one of many approaches available to obtain perspectives on research. One advantage of this approach is the possibility to cover the whole world, with the possibility of zooming in on individual institutions or researchers, if needed. Our publication study was discussed with some actors in the research system, thus enhancing the quality and increasing the relevance. The deep dive in automaker's publications gives a perspective on their involvement in vehicle electrification research, which otherwise is challenging to obtain.

6. Conclusions

This paper puts forward methodology to study research relating to vehicle electrification using Scopus publications. The results cover 16 countries with the highest number of vehicle electrification publications. Among them, China is by far the largest, and China's dominance increased during 2017–2021. This is true for publications directly relating to vehicle electrification as well as for the enabling technologies of batteries, fuel cells and electric motors with power electronics. In comparison to average national publications, vehicle electrification publications are rewarded with higher citation impacts, and they also include corporate actors to a greater extent. It is argued that international collaboration should be promoted. Among publications, including the automotive industry, Robert Bosch has the highest total volume of scientific publications, whereas Ford Motor Company has the highest number of vehicle electrification publications. Toyota had the highest share of vehicle electrification publications. Co-publications with academic actors are dominating and appear to be rewarded in terms of higher citation impacts. International research collaboration is limited for most automakers. Vehicle electrification publications offer a relatively straightforward method to obtain an overview of the rapidly developing field. Moreover, the study of automakers' involvement in such research is a possible way of obtaining one perspective on their strategies and priorities, which otherwise is rather challenging to obtain.

Author Contributions: Conceptualization, H.P. and M.K.; methodology, H.P. and M.K.; software, H.P.; validation, H.P. and M.K.; formal analysis, H.P.; investigation, H.P. and M.K.; resources, H.P.; data curation, H.P.; writing—original draft preparation, H.P. and M.K.; writing—review and editing, H.P.; visualization, H.P.; project administration, H.P. and M.K.; funding acquisition, H.P and M.K. All authors have read and agreed to the published version of the manuscript.

Funding: This research was funded by the Swedish Energy Agency, the Swedish Transport Administration and the Region Västra Götaland.

Data Availability Statement: The main data are from Scopus, which are accessible from Elsevier.

Conflicts of Interest: The authors declare no conflict of interest. The funders had no role in the design of the study; in the collection, analyses, or interpretation of data; in the writing of the manuscript; or in the decision to publish the results.

References

1. Sarasini, S. Electrifying the automotive industry: The geography and governance of R & D collaboration. *Environ. Innov. Soc. Transit.* **2014**, *13*, 109–128.
2. Hu, Y.; Sun, J.; Li, W.; Pan, Y. A scientometric study of global electric vehicle research. *Int. J. Innov. Technol. Manag.* **2014**, *98*, 1269–1282. [CrossRef]
3. Zhao, Q. Electromobility research in Germany and China: Structural differences. *Scientometrics* **2018**, *117*, 473–493. [CrossRef]
4. Naumanen, M.; Uusitalo, T.; Huttunen-Saarivirta, E.; van der Have, R. Development strategies for heavy duty electric battery vehicles: Comparison between China, EU, Japan and USA. *Resour. Conserv. Recycl.* **2019**, *151*, 104413. [CrossRef]
5. Ramirez Barreto, D.A.; Ochoa Guillermo, E.V.; Peña Rodriguez, A.; Cardenas Escorcia, Y.D.C. Bibliometric Analysis of Nearly a Decade of Research in Electric Vehicles: A Dynamic Approach. 2018. Available online: http://repositorio.cuc.edu.co/handle/11323/1053 (accessed on 12 November 2021).
6. Pinto, K.; Hari, O.; Praveen, G. A comprehensive assessment of the techno-socio-economic research growth in electric vehicles using bibliometric analysis. *Environ. Sci. Pollut. Res.* **2021**, *29*, 1788–1806. [CrossRef] [PubMed]
7. Barbosa, W.; Prado, T.; Batista, C.; Câmara, J.C.; Cerqueira, R.; Coelho, R.; Guarieiro, L. Electric Vehicles: Bibliometric Analysis of the Current State of the Art and Perspectives. *Energies* **2022**, *15*, 395. [CrossRef]
8. Wei, L.; Zhou, J.; Mao, X. Bibliometric analysis of global lithium ion battery research trends from 1993 to 2008. *Rep. Electrochem.* **2011**, *1*, 1–9.
9. Ahlgren, P.; Jeppsson, T.; Stenberg, E.; Berg, E.; Edström, K. *A Bibliometric Analysis of Battery Research with the BATTERY 2030+ Roadmap as Point of Departure*; Uppsala University: Uppsala, Sweden, 2022; Available online: https://www.diva-portal.org/smash/get/diva2:1654553/FULLTEXT01.pdf (accessed on 12 November 2021).
10. Swarnkar, R.; Harikrishnan, R.; Singh, M. Analysis of Electric Vehicle Battery State Estimation Using Scopus and Web of Science Databases from 2000 to 2021: A Bibliometric Study. *World Electr. Veh. J.* **2022**, *13*, 157. [CrossRef]
11. Gbey, E.; Turkson, R.F.; Lee, S. A Bibliometric Survey of Research Output on Wireless Charging for Electric Vehicles. *World Electr. Veh. J.* **2022**, *13*, 37. [CrossRef]
12. Baas, J.; Schotten, M.; Plume, M.; Coté, G.; Karimi, R. Scopus as a curated, high-quality bibliometric data source for academic research in quantitative science studies. *Quant. Sci. Stud.* **2020**, *1*, 377–386. [CrossRef]

13. Pohl, H.; Karlström, M. *Elektrifierade Vägtransporter—Publikationsstudie*; RISE Rapport; Research Institutes of Sweden: Gothenburg, Sweden, 2020; Volume 2.
14. Pohl, H.; Warnan, G.; Baas, J. Level the playing field in scientific international collaboration with the use of a new indicator: Field-Weighted Internationalization Score. *Res. Trends* **2014**, *39*, 3–8.
15. Pohl, H. Internationalisation, innovation and academic-corporate co-publications. *Scientometrics* **2021**, *126*, 1329–1358. [CrossRef]

World Electric Vehicle Journal

MDPI

Article

Charging after Lockdown: The Aftermath of COVID-19 Policies on Electric Vehicle Charging Behaviour in The Netherlands

Mylène van der Koogh [1,2,*], Rick Wolbertus [1] and Renee Heller [1]

1 Energy & Innovation Group, Urban Technology, Faculty of Techology, Amsterdam University of Applied Sciences, Weesperzijde 190, 1097 DZ Amsterdam, The Netherlands
2 Energy & Industry Group, Engineering Systems and Services, Faculty of Technology Policy and Management, Jaffalaan 5, 2628 BX Delft, The Netherlands
* Correspondence: m.l.van.der.koogh@hva.nl

Abstract: During the COVID-19—related lockdowns (2020–2022), mobility patterns and charging needs were substantially affected. Policies such as work from home, lockdowns, and curfews reduced traffic and commuting significantly. This global pandemic may have also substantially changed mobility patterns on the long term and therefore the need for electric vehicle charging infrastructure. This paper analyzes changes in electric charging in the Netherlands for different user groups during different phases of the COVID-19 lockdown to assess the effects on EV charging needs. Charging needs dropped significantly during this period, which also changed the distribution of the load on the electricity grid throughout the day. Curfews affected the start times of charging sessions during peak hours of grid consumption. Infrastructure dedicated to commuters was used less intensively, and the charging needs of professional taxi drivers were drastically reduced during lockdown periods. These trends were partially observed during a post–lockdown measuring period of roughly 8 months, indicating a longer shift in mobility and charging patterns.

Keywords: charging; EV; policy; case–study; deployment; energy consumption

Citation: van der Koogh, M.; Wolbertus, R.; Heller, R. Charging after Lockdown: The Aftermath of COVID-19 Policies on Electric Vehicle Charging Behaviour in The Netherlands. *World Electr. Veh. J.* **2023**, *14*, 67. https://doi.org/10.3390/wevj14030067

Academic Editor: Joeri Van Mierlo

Received: 22 December 2022
Revised: 27 February 2023
Accepted: 1 March 2023
Published: 7 March 2023

1. Introduction

1.1. Background

COVID-19 lockdowns halted transport movements all across the world; electric vehicles were no exception. People were working from home more and contacting each other through digital means. Car–related transport decreased by up to 50% in 2020 [1]. The use of electric vehicles in the Netherlands saw a similar drop. Many suggested that the effects of working from home would last longer than the COVID-19 lockdowns as the benefits of working from home, such as work effectiveness, decreased commute times, fewer traffic jams, and cleaner air, were experienced at a mass scale for the first time [2–4]. Major companies have announced new work-from-home policies. Additionally, car transport shifted throughout the day as most only ventured out for necessary appointments.

These policies also affected the movements, and thus the charging needs, of electric vehicle drivers. This study examines differences in charging behavior during the COVID-19 lockdowns, with a special focus on curfews and user groups, and finally post-lockdown to see the implications of new company policies. This is done by analyzing a dataset on charging behavior from 2020 up to and including October 2022. From this analysis, strategic effects on charging infrastructure planning are derived.

1.2. Literature

The COVID-19 lockdowns disrupted the EV manufacturers supply chains, productions, and operations [5]. Despite these disruptions, the full fleet of battery-electric vehicles in the Netherlands grew 70% in 2020, compared to 2019 [6]. In 2020 as well as 2021, 1 out of 5 newly sold vehicles in the Netherlands were electric [7]. This is because the Dutch have been intensively addressing decarbonization through mobility in their national [8], regional [9], and local policies (e.g., [10]). Mobility is one of the four key pillars mentioned in their Climate Agreement [11,12]. This has led to an uptake of electric vehicles and the development of one of the densest charging networks in the world [13].

The COVID-19 lockdowns affected different aspects of the energy system. For example, the roll-out of PV solar panels was much lower than predicted in 2020 because of manufacturing and supply chain issues and decreased urgency because of lower peak demands during COVID-19 [14]. The effect of COVID-19 and related policies on the charging of electric vehicles has been monitored by various researchers. Two researchers monitored two charging points at a public facility in California, USA [15]. They found that charging sessions declined from the start of lockdown and came to a complete halt between August and the first half of November 2020. The state of California then dealt with multiple issues, such as increasing pandemic-related death rates, wildfires, and blackouts, explaining the lack of sessions. At the end of 2020, they monitored charging activity at roughly 25% of the original capacity (a −75% decrease). The authors also found an increase in sessions starting in May 2021, which they attribute to the roll-out of vaccines. Some researchers have also been monitoring pandemic EV charging in Utah [16]. Here, only a maximum decrease of −40% (or 60% of full capacity) was measured, with the steepest decline in May 2020. A group of Dutch researchers focused on three charging locations in the Dutch city of Utrecht [17]. They compared a residential area with an office area and an event location. During the first lockdown, they saw the most decline at the event location (−99.2%), followed by the office area (−89.6%), and the least decline at the residential area (−73.6%). Results from this study differ from the USA studies, since the first lockdown phase around March 2020 had the strongest effect (only 25% of full capacity, or a −75% decline), whereas the later lockdown phases in 2020 led to a smaller decline. A Chinese study of four cities found similar declines in the range of roughly 50–70% in early 2020 and correlated the decline in EV charging demand with the number of confirmed COVID-19 cases and the associated hospitalizations, with one factor being more important than the other, depending on the city [18].

There have also been studies addressing the different charging behaviors associated with distinct user segments. In the Netherlands, start times of charging, charging durations, time between charging, the kWh charged, and the number of selected locations to charge can differ significantly between user segmentations (such as residents, commuters, taxi drivers, and shared vehicles) [19].

These studies, although insightful, are based on a limited number of observations over a limited time span. In the upcoming analysis, we will look at the different lockdown phases, types of areas, and user segmentations in the Netherlands using a longer measuring period. This enables us to further determine the effects of COVID lockdowns on electric vehicle charging.

2. Methods

2.1. Data

Charging data was collected from public charging stations in the Netherlands in the cities of Amsterdam, Rotterdam, Utrecht, and The Hague. In total, more than 7.5 million charging sessions were collected from January 2020 up to October 2022. Background on the dataset and the data collection process can be found in another publication [20]. An example of data variables gathered can be seen in Table 1.

Table 1. Data variables and examples.

Variable	Example
RFID	60DF4D78
Address	Prinsengracht 767, Amsterdam
Start Connection Date Time	24−04−2015 13:56:00
End Connection Date Time	24−04−2015 17:14:00
Connection Time	2:18:00
Volume	6.73 kWh

2.2. Timeframes

We used the 'Coronavirus timeline' of the Dutch government to compare important lockdown dates with our charging data [21]. The fuel comparison with kWh charged was scaled by the availability of traffic fuel (petrol) usage during COVID-19. The Dutch Central Bureau of Statistics keeps track of traffic, fuel, behavior, and car ownership, among others. This data was available on a monthly basis until April 2021. The energy (kWh) charged was summed with all the charging points that were available at January 2020 to avoid measuring increased roll-out and adoption. Office comparisons were made by selecting public charging points from office areas and by fingerprinting sessions to determine which behavior corresponded with employee charging behavior. A night curfew was installed to combat the spread of the new variants of COVID-19. The curfew was first installed in January 2021 and was extended until the end of April 2021. The curfew had two distinct phases with two different end times. In the fall of 2021, another partial lockdown was installed, and in December 2021, the last hard lockdown in the Netherlands was installed. This final lockdown took place for another month, and from February 2022 on, there were no more lockdowns in the Netherlands. Some restrictions, such as the use of vaccination passports and well-ventilated rooms, continued until the spring and summer of 2022.

2.3. Approach

We selected data from the important periods as specified in the *Coronavirus Timeline* [21] and compared them with other months during lockdown. Data was analyzed using Rstudio. Table 2 shows an overview of the analyses that can be found in this paper.

Table 2. Overview of analyses in this paper.

Analysis	Question	Data Selection	Indicators	Results
Section 3.1—Impact analysis	What is the overall effect of COVID-19 lockdowns on various charging indicators? (2020–2022)	All EV drivers	Connection times, kWh charged, sessions, users & stations	Baseline comparison, Line plot, Summary stats
Section 3.2—Fuel comparison	How does the decrease in kWh sales compare to the decrease in petrol sales for traditional vehicles? (2020–2021)	All chargepoints	kWh charged (and external data of petrol sold)	Baseline comparison, Line plot
Section 3.3—User comparison	Does this decrease also differ between specific user segments? (2020–2022)	Employees, Office chargers, Taxi drivers, Shared vehicles	kWh charged	Fingerprinting, Baseline comparison, Line plot
Section 3.4—Curfew analysis	What was the effect of the 2021 curfews on the start times of charging in the evening? (2021)	All EV drivers	Start times of charging in the evening	Summary stats, Violin plot

3. Results

3.1. COVID-19 Impact Analysis

COVID-19 lockdowns saw an immediate impact on charging behavior once installed, with a 50% reduction in charging sessions and energy charged (see Figure 1). At the same time, average connection time at the charging station increased by over 50% as EV drivers left their cars at the station during the early days of the lockdown. After restrictions were lifted in early June 2020, it still took a while before at least the same number of users were reached (partly also due to underlying growth in the number of electric vehicles).

The second lockdown in the fall of 2020 had a far lesser impact on charging behavior. It was only in late January 2021, when curfews were installed, that the number of charging sessions per charging station dropped again. Still, a significant reduction in the number of charging sessions per station could be observed compared to February 2020. It was not until March 2022, when all restrictions were lifted, that charging station use, both in terms of sessions per station and average connection time, was similar to pre-COVID.

This recovery continues during the rest of 2022 (except for the summer period, which is expected). By the end of 2022, the number of unique users in the charging network will have more than doubled compared to pre-COVID levels (January 2020). The number of unique users only grew by 50% between early 2021 and early 2022; this means growth has accelerated again since then.

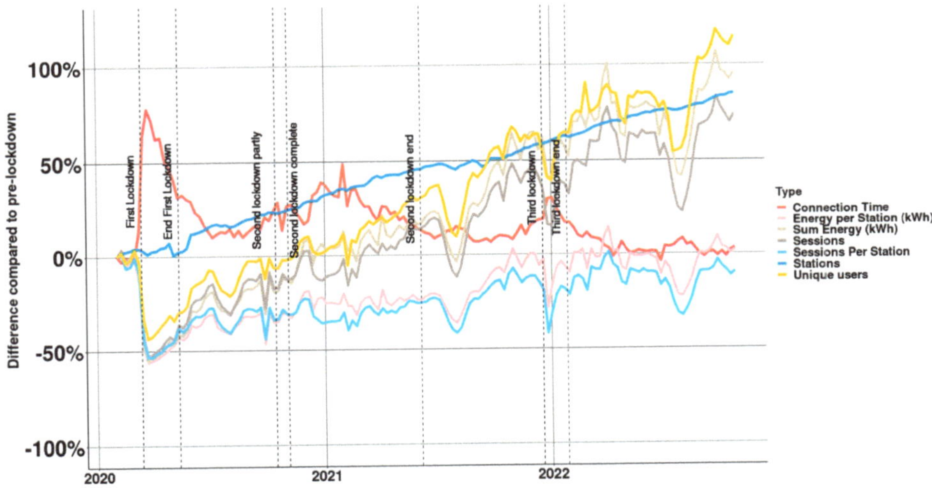

Figure 1. Effect of COVID-19 lockdowns on charging behavior over time.

Given that the number of charging sessions per station and connection have returned to pre-COVID levels, the question arises if a significant effect could be expected of COVID-19 lockdowns on car-related transport behavior and consequently charging behavior. For further inspection, charging behavior across the day for different phases of the COVID-19 lockdown is examined (see Figure 2). What stands out is that there is a significant reduction in evening demand compared to pre-lockdown behavior. Work demand seems to have returned to pre-COVID levels (see the significant reduction in the first lockdown), but charging session starts seem to have been shifted across the day away from classical 9−to−5 office behavior. March and April 2022 show a slight increase in the evening peak, but not as strong as in the morning. Such results might indicate more flexible working hours or hybrid working environments. Such a shift in charging behavior away from the peak might also be good news for grid operators as local power demand is shifted towards the day (coinciding with solar production) and away from peak hours. A post-lockdown line was added, covering an 8−month period after no more restrictions were in place. This line is very similar to other lines during lockdowns, and the gap between this line and the pre-lockdown line can still be observed, especially at the end of the day. This confirms that there is a persistent effect on the start times of charging sessions.

Additional insights are gained by comparing users that were already charging in January 2020 and are still charging in 2022. For this analysis, were included users that charged at least five times in January 2020 and were still charging in March 2022. The analysis excludes user groups such as taxis and shared vehicles. A total of 586 users (19550 charging sessions) were included in the analysis. Results of this comparison are shown in Table 3 for different metrics.

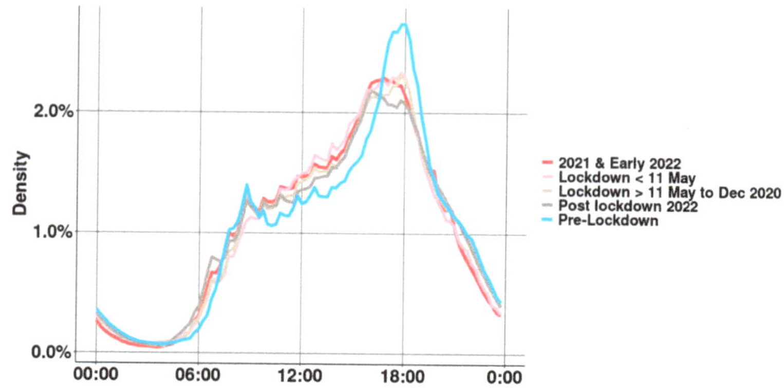

Figure 2. Density plot for charging session start across the day for different phases in 2020–2022.

Table 3. Comparison between January 2020 and March 2022 for EV drivers active in all months.

Date	Sessions/ User	Energy/Session (kWh)	Connection Time	Unique Number of Charging Stations/User	Sum of Energy Charged (kWh)	Number of Unique Neighbourhoods Visited
January 2020	14.9	16.5	12.7	3.2	221	1.94
March 2022	10.9	17.0	11.9	2.6	159	1.58

Although some metrics have stayed nearly the same (such as energy per session and connection time), it can also be seen that the same drivers have been charging much less, both in terms of the number of charging sessions and total energy in a single month. Although the turnover per charging station has come close to normal since March 2022, EV drivers individually still drive less. This can also be seen from the number of unique charging stations they visited, as well as the different neighborhoods and cities. Most charging tends to be done at a few charging stations. These results could show indications of the lasting effects of new COVID-19 travel habits among EV drivers as a possible result of a shift in workplace norms.

3.2. *Fuel Comparison (2020–2021)*

COVID-19 has led to a decrease in mobility activities during 2020 and 2021. We compared this decrease of energy charged with the decrease of (traffic) petrol sales using CBS data [22]. The comparison point was set for January 2020. The data was aggregated into monthly data between January 2020 and April 2021. Only charging points that were in the system as of January 2020 were used in the analysis to avoid adoption growth effects. In April 2021, 97% of these locations were still active, indicating a maximum potential loss of 3% of locations or location charging data.

Figure 3 shows the differences in decreases in petrol and electricity sales for vehicles from January 2020 until April 2021. Both lines show a strong decrease around the first lockdown at the end of March 2020. The underlying data reveals that in March, petrol sales decreased by almost 10%, whereas kWh charged decreased ~30%. In April, the decrease was the strongest, with a ~25% decrease in traffic petrol sales and ~50% decrease in kWh charged. This decrease recovers partially during the summer, and in September 2020, petrol sales were even slightly higher than in January 2020. Both lines also show a dip (extra decrease) in petrol sales and kWh charged during curfews, where the initial effect seems stronger for petrol sales but recovers faster.

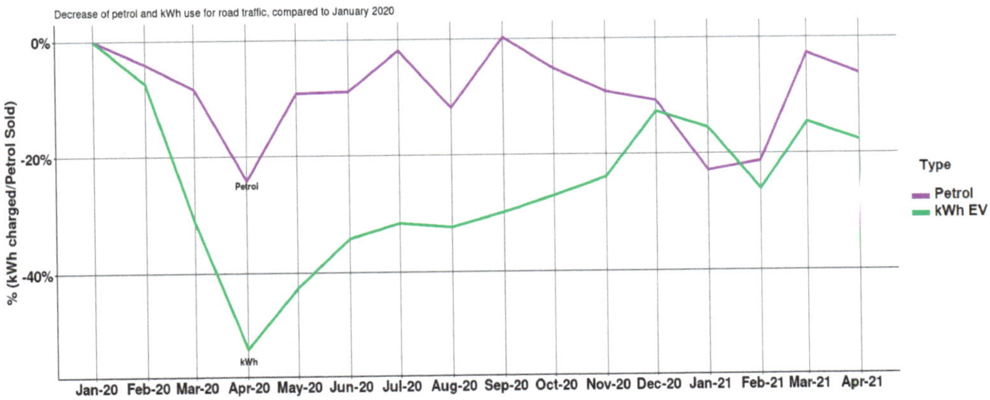

Figure 3. Comparison of national traffic petrol sales (purple) with kWh charged in G4 municipalities (green) in 2020 [22].

The much higher reduction in electricity sales compared to petrol can be explained by the different user groups. As electric transport is mostly centered around passenger cars, with only a few trucks on the road, gasoline and diesel vehicles are present in all sectors. Goods transports barely decreased during lockdowns, explaining the difference between the two. The overall reduction in 2020 (compared to 2019) was ~20% for passenger cars (24% for business related travel), and only 3% for goods-related transport [23].

3.3. User Group Comparison

EV charging had a much stronger decrease during the initial lockdown (March-April 2020) than petrol vehicles. A potential contributing factor to this effect (aside from transport) could be business related. Considering the entire personal vehicle fleet in the Netherlands in 2020, only 17.3% is categorized as business (company car, business lease, or personal lease) [24]. However, this number is much higher for electric vehicles (see Figure 4). Additional CBS and RDW data reveal that in 2020, only roughly a third of the electric vehicles on Dutch roads were owned personally [25]. For full electric vehicles, only 21% were personally owned.

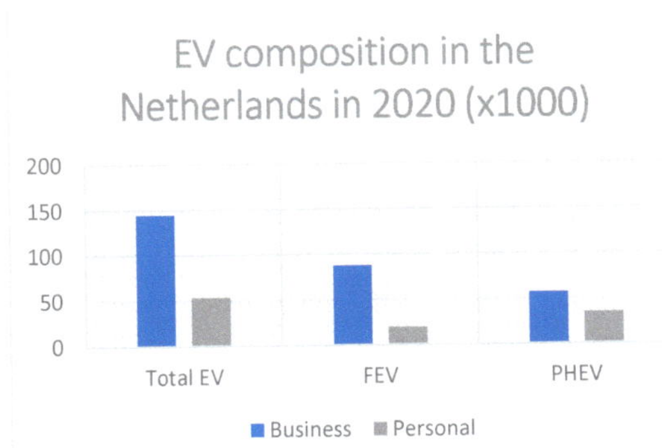

Figure 4. EV owner composition in the Netherlands in 2020 [24,25].

The charging behavior of professional user groups is analyzed to determine how these groups were affected during COVID-19 and compared to the full set containing all public charging. Decrease rates were calculated using the initial charging numbers for each group in January 2020 (this is the 0% point in the graphs). The following sections will discuss charging during COVID-19 for office areas, employees/commuters, taxi drivers, and shared vehicles.

Office Comparison (2020–2022)

A potential explanation for the stronger initial lockdown decrease in kWh charged is the low percentage of personal vehicle owners in the EV fleet. Most electric vehicles are business owned and/or leased. This implies that the vehicle has an important function in either commuting to work, business-related travel, or traffic-related services. Some lockdowns included a work from home policy, either mandated or highly advised, which affected the number of commuters on the road. Working from home was still recommended at times when leisure activities were allowed again. The following paragraphs will address office charging, using three groups to compare (see Table 4).

Table 4. Parameters of Office comparison groups.

Comparison Group	Parameters
All sessions (full set)	All public charging data
Sloterdijk Office Area	All charging from 17 charging points in an office location
Suspected employees (office/commuters)	All charging sessions during weekdays (mon-fr), starting between 7–10 a.m. and connected between 3–10 h

17 charging points that were in Sloterdijk Office Area ('Bedrijventerrein i') are analyzed and compared with all public EV charging in 2020–2022. The full set of sessions and the office area sessions with the sessions of individual users that are identified as employees are compared. We assumed the following while identifying potential employees on the public charging network: Employees start their charging sessions on weekdays between 7 and 10 a.m., and their connection time is between 3 and 10 h. We found that 7–8% of the sessions corresponded with this behavior.

In 2020, we see a steep decline for kWh charged in all groups in March-April 2020 (see Figure 5). This is when the COVID-19 lockdowns started in the Netherlands. Office area charging points were affected, even more so than the full set of sessions. The energy charged at the office decreased by 70%. The green and pink lines (all sessions and suspected employees) do not decrease as strongly and recover more quickly than the blue line (charging points in the Sloterdijk I office area). The 7–8% of sessions that we suspect are from employees showed similar decreases to the full set of sessions and recovered more quickly than both groups towards the end of 2020.

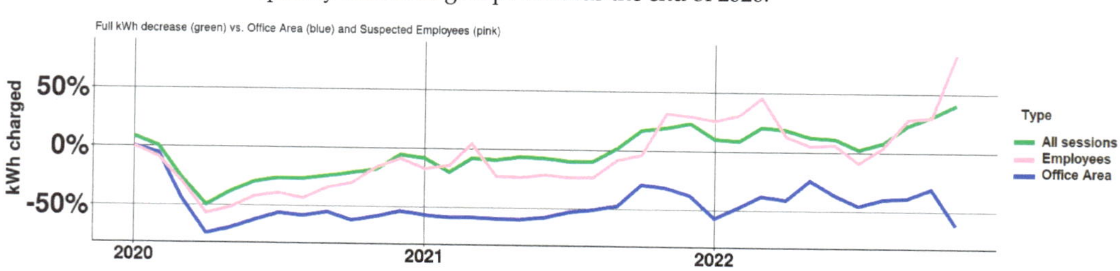

Figure 5. Comparing the full set of sessions (green, top line) with the sessions in Sloterdijk office area (blue, lower line) and the sessions with behaviour that is associated with employees (pink line). kWh decrease is based on the kWh charged that was observed for each group in January 2020.

Charging at the office area did recover slightly by fall 2021. Suspected employees charged more during early 2021 than before, and the demand for this group experienced a steeper growth at the end of 2021. The green line (full kWh charging) and the pink line (suspected employees) exceed 100% in 2021, despite our efforts to select only locations that were available from January 2020 on. This shows that, in some cases, the kWh charged per station is higher than before COVID.

Halfway through 2022, the office area recovers to almost 75% of the original kWh charged (before COVID). kWh charged by those we expect to commute (suspected employees) peaked temporarily again in March 2022. This can be explained since March has a cold temperature in the Netherlands (which makes batteries less efficient and users likely to drive more), there are 31 days in the month, and no holidays occur during this month. There is an expected drop during the summer for all groups (vacation), but in general, by the general population as well as the suspected employees, up to 25% more kWh is charged at these charging points compared to pre-COVID levels. In November 2022, a similar peak to March 2022 is found, and the overall activity is also higher. In the office area, a reverse effect is found, which cannot be explained by trends or lockdowns. It is possible that charging points were limited in availability and that there were office specific reasons for this drop that cannot be confirmed through the data.

We looked at multiple months for each year to account for potential EV adoption growth effects and compare them with the full set of new charging points (Table 5).

Table 5. Number of active charging points and users.

Active ChargePoints and Users: Sessions filtered on ChargePoints that have been online since Jan−2020			
Point in time	**# unique RFIDs (1 month)**	**Locations (used in 1 month)**	**RFIDs / Locations (1 month)**
January 2020	61,866	7730	8
June 2020	45,460	7502	6.06
January 2021 *	59,092	7490	7.89
June 2021	71,925	7509	9.58
January 2022	85,344	7326	11.65
June 2022	102,232	7059	14.48
October 2022	119,866	7339	16.33
Active ChargePoints and Users:Full Set (Without Jan−2020 filter)			
Point in time	**# Unique rfids (1 month)**	**# of used locations (1 month)**	**RFID / location (1 month)**
January 2020	61,987	7730	8
June 2020	54,787	8841	6.19
January 2021	66,752	10,409	6.41
June 2021	85,225	11,880	7.17
January 2022	110,900	13,310	8.33
June 2022	143,808	14,472	9.94
October 2022	169,989	15,827	10.74

(* the number of RFIDs was higher in December 2020 than January/February 2021, CPO policy may have played a role).

- **Growth/Adoption effect:** The charging points used in the analysis (online since January 2020) were shared by more distinct users than pre-COVID, after June 2021. The occupancy rates on these charging points have now doubled compared to pre-COVID. The kWh increase that surpasses pre-COVID is therefore partially attributable to the adoption/growth effect, despite measuring the same charging points.
- **Representability:** The charging points used in the analysis (online since January 2020) were mostly still online in October 2022. There is no significant loss of charging points, which makes them comparable over the years. Despite the fact that new installations have doubled the number of charging points available, 70% of the current (Oct−22) user base is using the charging points that have been installed since January 2020.

Shared vehicles and Taxi drivers (2020 only)

We were able to distinguish between a group of shared vehicles (185 unique RFIDs, minimum per month active: 175) and a group of Amsterdam taxi drivers (630 unique RFIDs, minimum per month active: 603). These groups also show a similar dip during the first lockdown (see Figure 6), with taxi drivers having the most extreme kWh decrease of 80% compared to January 2020. The charging of taxi drivers found some momentum again during the summer, when some policies were temporarily lifted and the catering industry opened up. In 2021 and 2022, the number of identified RFIDs drops significantly, to 2/3rd and 1/6th of their original number, respectively. Potential explanations include takeovers, drivers who are out of business, and replaced cards. However, this drop is too significant to confidently represent the segments after 2020; therefore, we chose not to include user segment analysis over a longer time period.

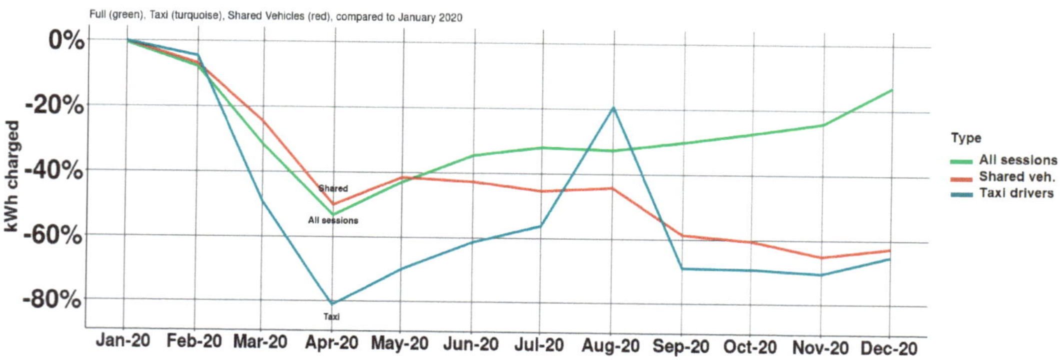

Figure 6. User segments and decrease of kWh charged during COVID-19 lockdowns in 2020. The kWh decrease is determined by comparing the kWh charged of each group with their kWh charged in January 2020.

3.4. Curfew Analysis (2021)

A curfew was installed to combat the spread of the new variants of COVID-19. The curfew was first installed on 23 January 2021. Curfew times took place between 9 p.m. (21:00) and 4:30 a.m. (4:30). On 31 March 2021, the curfew was extended with a delayed start time of 10 p.m. (22:00). The curfew took place until 28 April 2021 [21]. For this analysis, we compared the kWh charged in the evening (after 7 p.m.) over four periods of time: the month before the curfew, the 9 p.m. curfew, the 10 p.m. curfew, and the month after the curfew. Details can be found in the table below (Table 6). Percentages are calculated over the entire set of sessions (considering all hours of the day).

Table 6. Summary statistics of sessions started in the evening during Curfew phases.

Summary Statistics	Before Curfew ~1.5 Month Sample	First Curfew ~2 Month Sample	Second Curfew ~1 Month Sample	After Curfew ~1 Month Sample
Sample size (# of sessions)	700,633	910,988	314,976	340,565
# of sessions that started between 7 p.m.–8 p.m.	37,477	70,828	23,453	20,207
Percentage	**5.3%**	**7.8%**	**7.4%**	**5.9%**
# of sessions that started between 8 p.m.–9 p.m.	30,052	40,526	16,958	15,283

Table 6. *Cont.*

Summary Statistics	Before Curfew ~1.5 Month Sample	First Curfew ~2 Month Sample	Second Curfew ~1 Month Sample	After Curfew ~1 Month Sample
Percentage	**4.3%**	**4.5%**	**5.4%**	**4.5%**
# of sessions after 9 p.m.	54,530	22,828	25,153	22,646
Percentage	**7.8 %**	**2.5%**	**7.9%**	**6.6%**
# of sessions that started between 9 p.m.—10 p.m.	24,660	13,786	11,361	11,046
Percentage	**3.5%**	**1.5%**	**3.6%**	**3.2%**

- Outside of curfews, the percentage of sessions that start between 7 p.m. and 8 p.m. is between 5.3% (before curfew measurement) and 5.9% (after curfew measurement). During both curfews, this percentage increased to 7.8% (during the 9 p.m. curfew) and 7.4% (during the 10 p.m. curfew). This means that charging that started between 7 p.m.–8 p.m. increased by almost 50% during both phases of the curfew (compared to before and after measurements).
- Before curfew, the percentage of all sessions that started after 9 p.m. was 7.8%. During the first curfew, this number dropped to 2.5%. This is a decline of more than 60%. During the second curfew, this percentage recovered to 7.9%, almost identical to the before curfew measurement.
- The percentage of sessions that started exactly between 9 p.m. and 10 p.m. decreased to 1.5% during the first curfew, a decline of more than 50%. There was barely any difference between the 9 p.m.–10 p.m. start times of the second curfew (3.6%) and the before curfew measurement (3.5%). We observed no late-evening decline during the second curfew, as opposed to the first curfew.

The violin plot (Figure 7) illustrates how curfews affected the start times of public EV charging during lockdown. The month before and after the curfew are almost identical in shape, whereas the shapes of the two curfew periods differ. The first curfew period differs strongly and on multiple occasions in the evening. The second curfew illustrates the same 8 p.m. spike as the first curfew, but later in the evening the shape becomes more similar to the periods before and after the curfew.

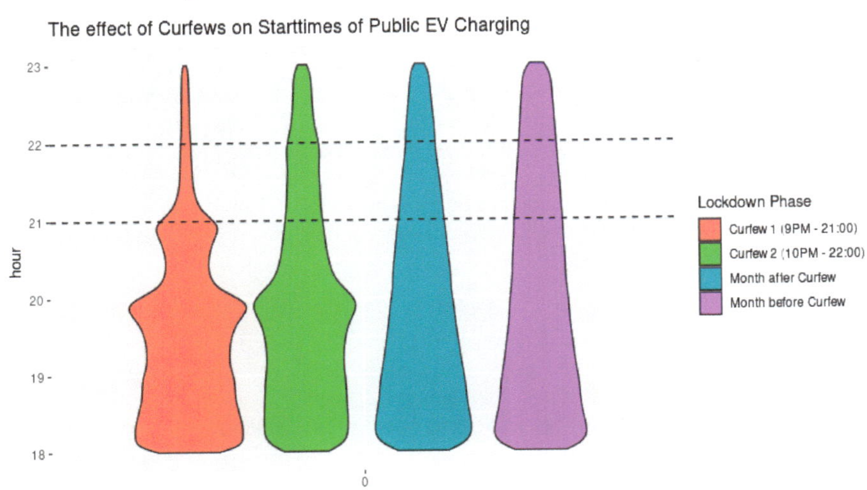

Figure 7. Violin plot illustrating the differences in start time distribution between 6 p.m.—11 p.m.. From left to right: Curfew 1, Curfew 2, Month after, Month before.

4. Discussion

4.1. Lockdown Effects

The COVID-19 lockdowns had a significant effect on travel demand and, therefore, EV charging demand. Initial lockdowns brought the world to a standstill with a reduction of more than 50% in the energy that was charged at public charging stations in the Netherlands. This resulted in much longer connection times at charging stations as most stayed at home. 2020 saw a slow recovery in the number of charging sessions and energy charged, but this was mainly fuelled by a large number of new EVs coming to the market. It was only until March 2022 that the energy transferred at charging stations was comparable to pre-COVID levels.

4.2. Demand Migration

Overall, the demand migration caused by COVID-19 reduced the 6 p.m. charging peak (see Figure 2). This effect is still at play, months after all restrictions were lifted. This capacity reduction occurs at a moment when other household electricity use peaks, so this reduction helps to keep grid capacity growth at lower levels than before COVID-19. Of course, for real grid capacity expansion, a local grid investigation is needed, but this analysis gives a first indication of possible lower capacity needs.

Curfews affected demand differently (see Figure 7). During the first (9 p.m.) curfew, the number of sessions that started at 9 p.m. dropped significantly, and the number of sessions between 9 p.m. and 10 p.m. also dropped significantly. During the second (10 p.m.) curfew, we saw some increase in sessions between 8 p.m. and 9 p.m., compared to other periods, but we did not observe a decrease in sessions that started after 9 p.m..

4.3. User Segments

The effects on charging commercial vehicles differed. Taxi drivers, shared vehicles, and office chargers were all affected by lockdowns, though to a different extent. Dutch media and local governments have also reported the challenges of taxi drivers (including non-electric taxi drivers). TaxiPro, the Dutch trade magazine for the taxi sector, reported a 90% loss of work in 2020 [26]. This is more extreme than the −80% we have measured for electric taxi drivers who are active in the urban agglomeration of the Netherlands. The municipality of Amsterdam monitored its taxi sector in 2020 and found similar patterns as our sample of electric taxi drivers: an average 77% decrease at the first lockdown and a recovery during the summer months [27]. Dutch Royal Traffic reported that 30% of the taxi drivers that were active in early 2020 had quit their jobs by summer 2020 [28]. The municipality of Amsterdam also reported in their monitor that, despite the loss of taxi drivers and a shrinking sector, the percentage of electric taxi drivers did not change much. Although taxi drivers were affected the most, the initial impact for shared vehicles did not differ much from the full user group. In late 2020, differences started to arise, with the full group of sessions recovering from initial effects while kWh consumption for shared vehicles lowered even more. The electricity demand for public charging in an office area was much lower than the already reduced demand for public EV charging altogether, although the demand that is associated with suspected employee charging did not differ as much from normal charging.

4.4. Limitations

EV adoption is growing steadily in the Netherlands. It is challenging to determine the demand reduction long-term, since the fleet is growing and resources are shared between more users over time. This made policy sensitivity analysis more difficult from late 2021 on: it cannot always be said which portion of the demand increasing again can be attributed to lifted restrictions, and which portion should be attributed to the growth of the user base (see Table 5). Another issue is the long-term availability of charging points. Some of the charging points were not used every month, and it is not possible to determine from our data if this is because of user preference or, for example, technical or construction issues that temporarily prevented users from accessing these charging points altogether. In the latter case, some users may have opted for a charging point that was installed after the January 2020 baseline period, and therefore, these sessions were not included in our data analysis. Only 17 charging points in one office area were investigated, and although initial effects are expected because of the work from home policy, additional data is necessary to generalize the after-lockdown effects for the larger population of office chargers. Suspected employees were recognized through an assumption; therefore, it is possible that this group contains some sessions that coincidentally were made under the same behavior by non-employees.

5. Conclusions

First, a conclusion is given by answering the research questions as determined in Section 2.3. The paper will conclude with implications for the future based on the persisting effects that are still observed after lockdowns.

5.1. Conclusions

What is the overall effect of COVID-19 lockdowns on various charging indicators? (2020–2022) Lockdown and work from home policies during the COVID-19 pandemic have led to a decline in charging, different start times for charging during the day, and longer connection times. The adoption of new vehicles and the roll-out of new stations did still increase during this period.

How does the decrease in kWh sales compare to the decrease in petrol sales for traditional vehicles? (2020–2021)

Lockdown and work from home policies affected energy use in all traffic, but more so (overall) in electric vehicles than in petrol vehicles. A potential explanation is that the majority of EV adoption is personal and business-related. In general, electric mobility was more heavily reduced than vehicles driven by petrol as most applications of electric mobility are still in personal vehicles. In the Netherlands, most EVs are also company lease cars, so it can be expected that these drivers mostly work at offices, for whom it is easier to work from home than other professions. As well, logistics, which continued to be nearly fully operational during the lockdowns, have not been electrified. Therefore, petrol sales did not fall to the same levels as could be seen with EV charging.

Does this decrease also differ between specific user segments? (2020–2022)

As explained in Section 4, user segments were affected differently by lockdowns and work-from−home policies. The impact was most visible in April 2020, where consumption was drastically lowered, especially for taxi drivers, and in the measured office area. This was expected because of the work-from−home policy. The effect of lockdowns on taxi drivers was recognized through local outlets as well. Based on the comparisons by local news outlets and municipalities [26–28], we can conclude that electric taxi drivers in the urban agglomeration were slightly less (but still strongly) affected by COVID-19 lockdowns than the taxi sector altogether and that the percentage of EV adoption among taxi drivers was not negatively affected by the lockdowns. The taxi sector did have great challenges in overcoming its negative lockdown effects.

What was the effect of the 2021 curfews on the start times of charging in the evening? (2021)

Both curfews temporarily increased the number of sessions that started between 7 p.m. and 8 p.m.. This demand migration towards 7 p.m.−8 p.m., could have created issues for a larger user pool, since this increase in demand does happen during a peak consumption window. Charging in the later evening declined during the first curfew of 9 p.m., whereas the second curfew of 10 p.m. did not show these effects. Even after the curfew deadline, charging was still occurring. In fact, in the later evening, the second curfew resembled the period before the curfews, indicating that the second curfew had little effect on late night charging and that demand in the late night was not as high as during the first curfew.

5.2. Future Implications

Analysis of charging patterns from early 2022, when nearly all restrictions were lifted, shows that there could be some long-lasting effects of the COVID lockdown period. A clear shift in the timing of charging sessions in the evening could be observed, indicating that strict 9−to−5 workplace norms are no longer in place. This does not necessarily have an impact on the number of charging stations but could result in a shift in local energy demand (away from a peak) and longer connection times (an earlier start but the same end time). This increases the potential for smart charging of electric vehicles at these stations. A comparison of a group of EV drivers active before and directly after COVID restrictions (Table 3) showed that this group charged significantly less (up to 30%) and charged at fewer different locations. A possible explanation is that business-related travel was reduced as there was a substantial shift to online meetings. A comparison of office areas with the general population also showed a significant reduction after COVID restrictions.

The last COVID restrictions (such as vaccination passports and evening lockdowns) were halted in spring 2022, 8 months before the last analysis update of this study in fall 2022. Since then, EV charging has significantly increased again. Since the start of COVID, the number of users has more than doubled, and many new charging stations have been installed. The kWh that is charged by various user groups has recovered to 50–70% of original kWh use (office charging points) and even more than before the pandemic (commuters and the general population). Office areas have not recovered fully to their pre-COVID charging levels (see Figure 5), and start times of charging sessions also do not follow the after work/evening peak that was observed before COVID (Figure 2). This indicates that there is a persisting effect where working from home and working flexible hours are still partially available for employees (more than before COVID lockdowns). However, because of new adoption, we expect office infrastructure to stay relevant over time.

Author Contributions: Conceptualization, R.W., M.v.d.K. and R.H; methodology, M.v.d.K. and R.W.; software, M.v.d.K. and R.W.; validation, R.W., M.v.d.K. and R.H.; investigation, M.v.d.K. and R.W.; writing—original draft preparation, R.W. and M.v.d.K.; writing—review and editing, R.H and M.v.d.K.; visualization, R.W. and M.v.d.K.; supervision, R.H.; project administration, R.W.; funding acquisition, R.W. All authors have read and agreed to the published version of the manuscript.

Funding: This research was funded by NWO RAAK-SIA grant number RAAK.PRO03.128.

Data Availability Statement: Data is curated through a project consortium. For details about access, visit ev.data.nl.

Acknowledgments: The authors would like to thank all partners in the RAAK-SIA−funded Future Charging project, with special thanks to the municipalities of Amsterdam, Rotterdam, Utrecht, The Hague, and the surrounding metropole regions for the availability of their public charging data.

Conflicts of Interest: The authors declare no conflict of interest.

References

1. Kim, J.; Kwan, M.-P. The Impact of the COVID-19 Pandemic on People's Mobility: A Longitudinal Study of the U.S. from March to September of 2020. *J. Transp. Geogr.* **2021**, *93*, 103039. [CrossRef] [PubMed]
2. Huang, X.; Li, Z.; Jiang, Y.; Li, X.; Porter, D. Twitter Reveals Human Mobility Dynamics during the COVID-19 Pandemic. *PLoS ONE* **2020**, *15*, e0241957. [CrossRef] [PubMed]
3. Heiler, G.; Reisch, T.; Hurt, J.; Forghani, M.; Omani, A.; Hanbury, A.; Karimipour, F. Country-Wide Mobility Changes Observed Using Mobile Phone Data During COVID-19 Pandemic. In Proceedings of the 2020 IEEE International Conference on Big Data (Big Data), Atlanta, GA, USA, 10–13 December 2020; pp. 3123–3132. [CrossRef]
4. Kramer, A.; Kramer, K.Z. The Potential Impact of the COVID-19 Pandemic on Occupational Status, Work from Home, and Occupational Mobility. *J. Vocat. Behav.* **2020**, *119*, 103442. [CrossRef] [PubMed]
5. Wen, W.; Yang, S.; Zhou, P.; Gao, S.Z. Impacts of COVID-19 on the Electric Vehicle Industry: Evidence from China. *Renew. Sustain. Energy Rev.* **2021**, *144*, 111024. [CrossRef] [PubMed]
6. Rijksdienst voor Ondernemend Nederland. *Elektrisch Rijden Op (de) Weg—Voertuigen En Laadpunten: Overzicht Tot En Met 2020*; Rijksdienst voor Ondernemend Nederland: The Hague, The Netherlands, 2021.

7. Rijksdienst voor Ondernemend Nederland. *Electric Vehicle Statistics in the Netherlands*; Rijksdienst voor Ondernemend Nederland: The Hague, The Netherlands, 2021.
8. RVO. Nationale Agenda Laadinfrastructuur. Available online: https://www.klimaatakkoord.nl/documenten/publicaties/2019/01/08/achtergrondnotitie-mobiliteit$-$laadinfrastructuur (accessed on 15 January 2021).
9. MRA-Elektrisch Werkprogramma 2019 Samenvatting. 1–20 April 2019. Available online: https://api1.ibabs.eu/publicdownload.aspx?site=noordholland&id=1100095163 (accessed on 15 January 2021).
10. Gemeente Amsterdam. Laad Me: Strategisch Plan laadinfrastructuur 2030–2040. 2020. Available online: https://www.amsterdam.nl/wonen-leefomgeving/duurzaam-amsterdam/publicaties-duurzaam$-$groen/laad-strategisch$-$plan-laadinfrastructuur/?PagClsIdt=15860861#PagCls_15860861 (accessed on 15 January 2021).
11. Rijksoverheid. Climate Agreement. 2019. Available online: https://www.rijksoverheid.nl/documenten/rapporten/2019/06/28/klimaatakkoord (accessed on 15 January 2021).
12. de Rijksoverheid. Klimaatakkoord C2: Mobiliteit. Available online: Klimaatakkkoord.nl.2018. (accessed on 15 January 2021).
13. International Energy Agency. Trends in Charging Infrastructure. 2022. Available online: https://www.iea.org/reports/global-ev$-$outlook$-$2022/trends-in$-$charging-infrastructure (accessed on 27 November 2022).
14. Aminifar, F.; Shahidehpour, M.; Alabdulwahab, A.; Abusorrah, A.; Al-Turki, Y. The Proliferation of Solar Photovoltaics: Their Impact on Widespread Deployment of Electric Vehicles. *IEEE Electrif. Mag.* **2020**, *8*, 79–91. [CrossRef]
15. Shahriar, S.; Al-Ali, A.R. Impacts of COVID-19 on Electric Vehicle Charging Behavior: Data Analytics, Visualization, and Clustering. *Appl. Syst. Innov.* **2022**, *5*, 12. [CrossRef]
16. Palomino, A.; Parvania, M.; Zane, R. Impact of COVID-19 on Mobility and Electric Vehicle Charging Load. In Proceedings of the 2021 IEEE Power & Energy Society General Meeting (PESGM), Washington, DC, USA, 26–29 July 2021; pp. 1–5. [CrossRef]
17. Brinkel, N.; Schram, W.; AlSkaif, T.; van Sark, W. A Quantitative Analysis of the Short-Term and Structural Impact of COVID-19 Measures on Electric Vehicle Charging Patterns. In Proceedings of the 2021 International Conference on Smart Energy Systems and Technologies (SEST), Vaasa, Finland, 6–8 September 2021; pp. 1–6. [CrossRef]
18. Zhou, K.; Hu, D.; Li, F. Impact of COVID-19 on private driving behavior: Evidence from electric vehicle charging data. *Transp. Policy* **2022**, *125*, 164–178. [CrossRef] [PubMed]
19. Helmus, J.; van den Hoed, R. Unraveling user type characteristics: Towards a taxonomy for charging infrastructure. In *Abstract from The 28th International Electric Vehicle Symposium and Exhibition*; Kintex: Goyang, Republic of Korea, 2015.
20. Maasse, S.; van den Hoed, R. Charging Data Management, Five Issues to Solve. In Proceedings of the EVS 2019—32nd International Electric Vehicle Symposium and Exhibition, Lyon, France, 19–22 May 2019.
21. Rijksoverheid. Timeline Coronavirus. Available online: https://www.rijksoverheid.nl/onderwerpen/coronavirus-tijdlijn (accessed on 1 December 2021).
22. Centraal Bureau Statistiek. *Environment and Energy during Corona*; Centraal Bureau Statistiek: Heerlen, The Netherlands, 2021.
23. Bovag. Invloed van Corona op de Mobiliteitsretail. Available online: https://mijn.bovag.nl/downloads/onderzoek-cijfers/invloed-corona$-$op-mobiliteitsretail$-$2020/bovag_invloed-corona$-$op-mobiliteitsretail_dec$-$2020 (accessed on 20 April 2022).
24. Vereniging Nederlandse Autoleasemaatschappijen. *Car Lease Market in Numbers 2020*; Vereniging Nederlandse Autoleasemaatschappijen: Bunnik, The Netherlands, 2021.
25. Centraal Bureau Statistiek. *Growth Number of Pluggable Cars Continues*; Centraal Bureau Statistiek: Heerlen, The Netherlands, 2021.
26. Krabbendam, V. Dit Betekent de Harde Lockdown voor Straattaxi's, Contractvervoer en Touringcars. Available online: https://www.taxipro.nl/special/2020/12/14/dit-betekent$-$de-harde$-$lockdown-voor$-$straattaxis-contractvervoer$-$en-touringcars/?gdpr=accept (accessed on 12 April 2022).
27. Gemeente Amsterdam. *Taximonitor 2020*; Amsterdam, The Netherlands, 2020. Available online: https://openresearch.amsterdam/nl/page/72356/taximonitor-2020 (accessed on 12 April 2022).
28. Redactie AT5. Koninklijk Nederlands Vervoer: "Dertig Procent van de Taxichauffeurs Is Door Corona Gestopt". Available online: https://www.at5.nl/artikelen/203651/koninklijk-nederlands$-$vervoer-dertig$-$procent-van$-$de-taxichauffeurs$-$is-door$-$corona-gestopt (accessed on 12 April 2022).

World Electric Vehicle Journal

MDPI

Article

GridShield—Optimizing the Use of Grid Capacity during Increased EV Adoption

Hanna L. van Sambeek [1], Marisca Zweistra [1,*], Gerwin Hoogsteen [2,*], Ivo A. M. Varenhorst [2] and Stan Janssen [3]

1 ElaadNL, 6827 AV Arnhem, The Netherlands
2 Department of Electrical Engineering, Mathematics and Computer Science, University of Twente, 7522 NB Enschede, The Netherlands
3 Enexis, 5223 MB 's-Hertogenbosch, The Netherlands
* Correspondence: marisca.zweistra@elaad.nl (M.Z.); g.hoogsteen@utwente.nl (G.H.); Tel.: +31-26-3120223 (M.Z.)

Abstract: With the increasing adoption rate of electric vehicles, power peaks caused by many cars simultaneously charging on the same low-voltage grid can cause local overloading and power outages. Smart charging solutions should spread this load, but there is a residual risk of incidental peaks. A decentralized and autonomous technology called GridShield is being developed to reduce the likelihood of a transformer's fuse blowing when other congestion solutions have failed. It serves as a measure of last resort to protect the grid against local power failures from unpredicted congestion by temporarily limiting the virtual capacity of charging stations. This paper describes the technical development and demonstrates how GridShield can keep a transformer load below a critical limit using simulations and real-world tests. It optimizes grid capacity while ensuring grid reliability.

Keywords: energy security; energy network; smart charging; load management; charging infrastructure; electric vehicles

Citation: van Sambeek, H.L.; Zweistra, M.; Hoogsteen, G.; Varenhorst, I.A.M.; Janssen, S. GridShield—Optimizing the Use of Grid Capacity during Increased EV Adoption. *World Electr. Veh. J.* **2023**, *14*, 68. https://doi.org/10.3390/wevj14030068

Academic Editor: Joeri Van Mierlo

Received: 1 February 2023
Revised: 27 February 2023
Accepted: 3 March 2023
Published: 7 March 2023

1. Introduction

Electric vehicles (EVs) play an important role in the transition to a more sustainable society. The charging of these cars, however, is putting a strain on the low-voltage electricity grid as their popularity grows [1]. New or larger grid connections for charging infrastructure are frequently delayed or rejected in the Netherlands for congestion reasons, and even higher vehicle electrification is still expected in the coming decade. In the coming years, some residential areas are likely to experience power outages because of the increased peak load from home chargers [2].

Replacing transformers and cables across the country to increase peak capacity in response to EV adoption forecasts and the electrification of heating is deemed impractical, if not impossible, and would drastically increase costs for all users [3]. Smart charging, tariffs, and market-based alternatives will be the primary solutions to mitigate peak grid congestion, but these are still being developed and are subject to regulatory changes. Furthermore, existing grid congestion solutions and technologies, as well as those currently being developed, are mainly focused on the medium-voltage grid and on daily or hourly forecasts. Therefore, the authors are developing GridShield, a decentralized solution to complement primary peak congestion mitigation. It helps to protect the low-voltage grid from local power failures caused by unexpected congestion by temporarily reducing the virtual charging capacity at charging stations. Preventing a power outage is good for both people who drive electric cars and people who use electricity in their homes or businesses.

1.1. Grid Congestion from EV Charging

The typical grid connection for a Dutch residential house is 3×25 A, which is approximately 17 kW. However, because traditional household electricity usage is much lower, the Dutch grid was designed for an average simultaneous power consumption of 1.5 to 4 kW per household, depending on the age of the grid. Home chargers for EVs, on the other hand, typically draw between 3.7 and 11 kW, and charging frequently occurs at the same time, which leads to a high simultaneity factor. With the projected increasing rate of EV adoption, it is expected that in a matter of years, typical behavior—such as charging the EV in the early evening after commuting—will easily exceed the physical limit of the low-voltage grid, resulting in regular local power outages. Simulations by ElaadNL show that, in a neighborhood with 250 households and 100 EVs, peak load can be exceeded daily if there is no implementation of smart charging; see Figure 1. The Netherlands is expected to have approximately 3000 such neighborhoods by 2025. Moreover, ElaadNL concluded that occasional peak overload could happen with as few as 50 EVs in a neighborhood [2].

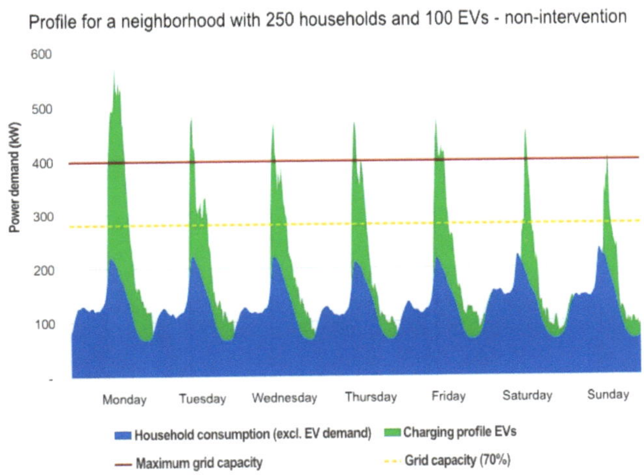

Figure 1. Uncontrolled EV charging is expected to cause regular power outages in 2025.

Smart charging solutions or (Home) Energy Management Systems ((H)EMS) should be able to spread many of these flexible loads within the current grid capacity if properly designed and implemented; see Figure 2. Schoot Uiterkamp et al. [4] have proposed a robust fill-level approach that utilizes baseload and energy generation forecasts to perform online control to optimally peak-shave a household load using an EV. Algorithms for smart charging and optimal integration within energy markets and distribution systems are a widely studied field. Many papers have surveyed different strategies to optimize and coordinate a fleet of electric vehicles. Vandael et al. [5] have proposed a scalable, three-step approach that utilizes the aggregated flexibility of a fleet of EVs to optimize the dispatch control signals towards charging stations. This serves as the foundation for an event-driven method that achieves optimal market integration in the face of EV arrival uncertainty [6]. Alternatively, optimization can be used to optimally charge EVs, such that they mitigate voltage problems in low-voltage distribution grids [7]. These are just a few examples from the literature on the optimal charging of one or more EVs with different objectives [8–10].

However, smart charging is not mandatory, and although price incentives can align with grid congestion problems, they are not a guarantee to protect the grid from peak load as grid limits are currently not enforced [11]. Furthermore, the market is still immature, and implementation is lagging [12]. As a result, the Distribution System Operators (DSOs) need a method of last resort to protect the grid from unforeseen congestion and to protect its users from the worst-case scenario: local power outages.

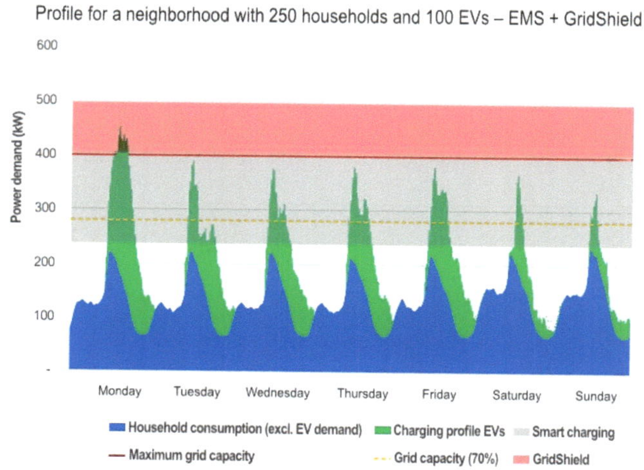

Figure 2. Smart charging can spread the EV load, but GridShield is needed to reduce peaks.

1.2. Robust Fall-Back Mechanisms to Prevent Grid Overloading

Although many smart charging solutions exist in the literature, these are not widely adopted in the real world. DSOs foresee local overloading problems in their networks due to EV charging in the near future due to rising EV sales. Hence, a protective control method is required to prevent outages in low-voltage grids due to EV charging.

By law, DSOs should maintain the distribution network but not interfere with the energy markets. However, it is not guaranteed that the energy market will result in a feasible solution with respect to grid limitations. Therefore, the Universal Smart Energy Framework (USEF) working group has come up with a traffic-light concept [13]. In this concept, the DSO is allowed to interfere with the market and impose direct control to resolve capacity and voltage issues. A similar concept is introduced by BDEW [14]. Next to this, a sort of direct control with the aim of protecting the grid is already implemented in PV inverters, which are obliged to cease production during times of overvoltage or overfrequency [15].

These developments have led to new methods for direct control over emerging technologies, such as heat pumps and EVs. These solutions can then coexist with market-based solutions and are to be activated during emergencies or if the market does not resolve overloading problems. Haider et al. [16] propose a scheme that optimally utilizes the flexibility of smart loads during contingencies. However, their solution relies on substantial coordination between local energy management systems and a central controller. This may introduce various vulnerabilities and failure vectors that can result in the malfunction of the system during critical moments in which the DSO relies on its operation.

Instead, it may be beneficial to have a very simple solution in terms of code and communication. Reduced complexity allows for the easier formal verification of algorithms to ensure that they work robustly under all circumstances. Furthermore, it reduces the cost of computation hardware, allows for fast execution and response, and avoids reliance on expensive solvers. Next, utilizing only one-way communication can harden these solutions against cyber-attacks (e.g., distributed denial of service (DDOS) attacks). One such simple method is the Smart Grid Ready interface, which allows DSOs to send signals to heat pumps to temporarily increase or decrease their power usage [17,18]. It is also shown that state information from the heat pumps, requiring two-way communication, has very little added value.

1.3. Contribution

This paper presents GridShield, a mechanism that grid operators can use in the near future to reduce interruptions in electricity supply and prevent damage to power infrastructure. It should reactively mitigate unpredicted events, such as congestion or EMS malfunctions, by instructing charging stations to temporarily reduce their electricity demand. GridShield is based on the aforementioned principles and acts as a mission-critical solution on which DSOs can rely.

The main functionality of GridShield is to protect low-voltage grids against unforeseen, temporary overloads that could cause service interruptions or premature wear. As a measure of last resort against local blackouts from EV charging peaks, GridShield will be designed to be robust, fast, and safe. Being designed by and for grid operators, it will have to take proportionality, privacy, and user experience into account.

Additionally, the solution provides feedback to the grid operator when GridShield is required to intervene, allowing for additional analysis of the local power grid. GridShield should also not introduce a new risk or point of entry for malicious actors, for instance in the form of cyberattacks.

The desired solution will play a critical role in the energy transition by ensuring the grid's continued operation while incorporating increasing EV charging demand. Thus, it helps to maintain a cost-effective power system. The primary solution to grid congestion is smart charging, which will take time, legislative changes, and good incentives to implement. Grid reinforcement as a temporary solution is prohibitively expensive and can take a long time to implement in the Netherlands due to shortages of both materials and personnel [3].

GridShield aims to help with optimizing the use of existing grid capacity and deploying reinforcements at the right time and location. Implementing GridShield is not a guarantee against power outages, but it minimizes the risk of it happening due to EV charging.

1.4. Scope

By clearly defining what GridShield does and does not do, we can maintain a robust and reliable implementation. There are already mechanisms in place for larger producers and users to protect the grid, such as re-dispatch and congestion management. However, these typically deal with national or regional congestion, not local, and they are not applicable to low-voltage connections. Therefore, the focus of our solution is:

- The low-voltage grid; the medium- to low-voltage transformer (10 or 20 kV transformed to 400 V), and low-voltage cables.
- All small customer connections, both with and without active energy management.
- Non-predictable, real-time situations.

To keep the implementation simple, the current scope deals with the most pressing issue: congestion of local low-voltage network sections that contain controllable and flexible power in the form of Evs. The emphasis here is on typical west-European low-voltage underground distribution grids, where the capacity of outgoing feeders from the transformer forms the main bottleneck. As such, this research currently assumes that overloading of underground cables and voltage issues do not occur as long as the feeders are operated within their rated capacity. However, when real-time measurements of the load on the cables are available, these can also act as steering signals for the GridShield commands, comparable to the way measurements of the feeders are used in this study. The GridShield concept is not limited to this grid situation, and can likely be adapted for use in other countries and situations where EV charging may result in an unexpected peak load.

Overload detection based on potential (volt) was considered but rejected. The voltage is not uniform across a cable, and the transformer's fuse reacts to an overload in current (ampere).

Protecting the grid by steering other high-capacity consumption devices or feed-in, e.g., PV panels or vehicle-to-grid, is a future development that fits into the GridShield concept. Decentralized electricity generation and feed-in have the potential to both alleviate and cause congestion. When recommending hardware and writing communication standards for the development of GridShield, we take care not to introduce any barriers to the future use of the technology in these situations.

The remainder of this paper is structured as follows: Section 2 explains the technical details of GridShield. In Section 3, we show results from both simulations and field tests. The results are discussed in Section 4, where we argue that GridShield can protect a transformer fuse from blowing in unforeseen situations or due to a lack of smart steering, and we also discuss further work. Finally, Section 5 summarizes our conclusions.

2. Materials and Methods

GridShield acts as a software-defined, dynamic fuse that intervenes in problematic situations before the transformer overloads by lowering the capacity of connections. Unlike a fuse that is binary, this software fuse can gradually lower the current to a safe level, preventing physical damage and power outages. With the transformer module communicating unidirectionally with one or more charging station modules, it mitigates unanticipated congestion reactively and autonomously by requiring flexible loads (i.e., charging stations) to temporarily reduce their electricity demand.

The following sub-chapters explain the technical design of the modules (Section 2.1), the control logic (Section 2.2), and how they are tested in reality and simulations (Section 2.3).

2.1. GridShield Modules

2.1.1. Design Overview

The GridShield system consists of two types of modules: a transmitter module and one or more receiver modules, as illustrated in Figure 3. At a regular interval, the transmitter module takes measurements at the grid transformer. According to internal control logic and parameters that can be set by the grid operator, it determines if the grid is overloaded or not. If the grid is overloaded, the transmitter will broadcast an over-the-air message with an instruction to reduce the maximum power that a charger can draw to a given fraction of the nominal capacity. The receiver modules in the neighborhood pick up this message. They will determine whether the message applies to their grid connection and then convert the message into an instruction for the charging station that is connected to the receiver.

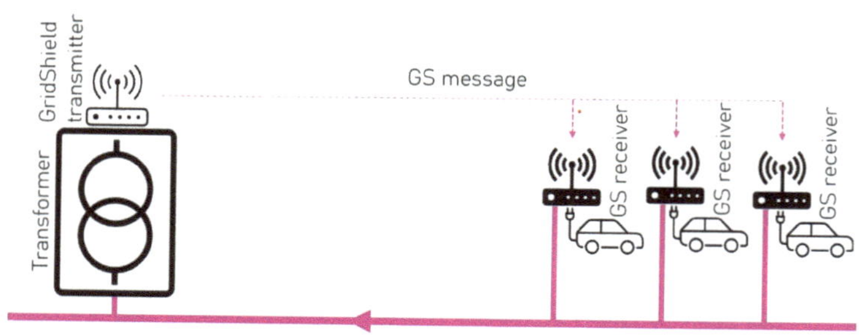

Figure 3. Schematic overview of the GridShield transmitter and receiver [19].

These modules work without any reliance on external infrastructure, such as wireless networks, internet service, cloud computing, or other forms of external processing or mediation. They also contain very little intelligence or analysis. This adds to their robustness, response time, and safety.

The transmitter module is installed inside the low-voltage transformer building, connected to a measurement device that is connected to the low-voltage cables. The interface between the transmitter module and the measurement device is usually in the form of some kind of data connection. The transmitter module must be programmed to communicate with the specific measurement device that is in use at the facility. At SlimPark, the test site described in this study, the measurement device uses a PAC-2200 controller with a ModBus TCP interface.

The receiver module is installed inside the charging station and is linked to the charging station controller, which controls the maximum allowed charging rate. The interface between the receiver module and the charging station is similar to the interface between the transmitter module and the measurement device. Charging stations usually implement a communication interface for such purposes. These interfaces, however, are non-standard, which means that the receiver module must be programmed to work correctly with a specific charging station type. In the test setup used here, the charging stations implement a ModBus TCP connection where writing to a single register will change the maximum allowed charging current.

Each charging station has one module installed. If there is more than one charging outlet, the charging station's internal logic is free to allocate the available current.

2.1.2. Communication between the Transmitter and Receiver Modules

The communication between the two sides of the GridShield protocol is unidirectional: messages are sent by the transmitter module and received by the receiver module. The transmitter module must be able to reach all the relevant grid connection points that it supplies power to. In the Netherlands, that usually means a required broadcasting range between 500 and 2000 m.

In the GridShield trials, LoRa radios of the type RFM95 were used in conjunction with off-the-shelf general-purpose mini-computer devices and interconnections. LoRa is a physical, proprietary radio modulation technique. It operates in the license-free sub-gigahertz radio band EU868 and has a real-world range of three kilometers [20].

2.1.3. GridShield Messages

The message transmitted by GridShield contains the following information:

- The relevant cable number that the limitation applies to. Each receiver module is pre-configured with a cable number based on the physical grid connection. Since multiple cables may run through an equally reachable area, the GridShield signals must be selective between these cables.
- The maximum value is expressed as a percentage $\phi(t)$ of the charging station's regular connection capacity $P_{\mathrm{EV}}^{\mathrm{max}}$, i.e., the charging station power allowed by GridShield becomes $P_{\mathrm{EV}}^{\mathrm{GS}}(t) = \phi(t) \cdot P_{\mathrm{EV}}^{\mathrm{max}}$. This implementation is further explained in Section 2.2.1. The receiver module is configured with a nominal grid capacity, usually equal to the contracted maximum capacity of the grid connection at the charging station. In the Netherlands, this is usually 3×25 A or 3×35 A.
- Fields concerning the security of the message are further explained below.

GridShield's most critical security requirement is that the receiver module is capable of determining whether a message it receives is legitimate, i.e., that it was transmitted by the actual trusted transmitter module connected to the relevant grid transformer. For GridShield, the messages contain the following elements that make this possible:

- The time at which the message was sent. Messages older than some predefined age are discarded by the receivers. This prevents malicious transmission or the accidental retransmission of an earlier, valid message.
- A random 4-byte nonce; the receivers remember each of these "nonces" as long as the message is valid according to the previous criterion. This prevents a valid message from being retransmitted within the cut-off period.
- A cryptographic signature; the transmitter module has an embedded private ECDSA key, which is used to sign the contents of the message. The receiver modules have the accompanying public ECDSA key, which can be used to verify that the message (1) was signed by the corresponding private key and (2) that the message was not altered after it was signed.

These three components mean that each GridShield message can be sent unencrypted through the air while still ensuring trust between the receiver and the transmitter. By embedding the private key in a dedicated hardware component, it is impossible to retrieve the private key from the transmitter module. The public key does not need to be kept secret and can be published by the grid operator to allow trouble-free installation of additional receiver units in the area after the initial provisioning of the transmitter module.

2.2. Control Logic

2.2.1. Transmitter Algorithm

The GridShield transmitter must determine which signals to broadcast after reading the measurements from the low-voltage substation. This can be considered a classic control loop, and as such some classic control algorithms can be considered. Research reported in [6] tested and compared three GridShield implementations with respect to power limit violations and energy not served to EVs: (1) AIAD (Additive Increase, Additive Decrease); (2) AIMD (Additive Increase, Multiplicative Decrease); and (3) PID (Proportional–Integral–Derivative). The authors conclude that AIMD performed best as a control strategy for overloaded distribution grids. Hence, this control scheme is used for implementing the GridShield concept simulations and the presented real-world tests.

The AIMD algorithm was originally developed for the fair usage of limited bandwidth on the internet, but has recently been used, e.g., for power grid stabilization [8]. The GridShield implementation of AIMD in this work is based on [19,21] and is presented in Figure 4. We see that when the GridShield transmitter detects an overload at the transformer, i.e., $P_{\text{trafo}}(t) > P_{\text{trafo}}^{\text{max}}$, it will decrease the maximum charging power factor $\phi(t)\beta$ by a multiplicative factor $\phi(t) = \beta \cdot \phi(t-1)$. When the $\phi(t)$ problem is resolved, the algorithm enters the additive increase phase, where αy by a factor $\phi(t) = \phi(t-1) + \alpha$.

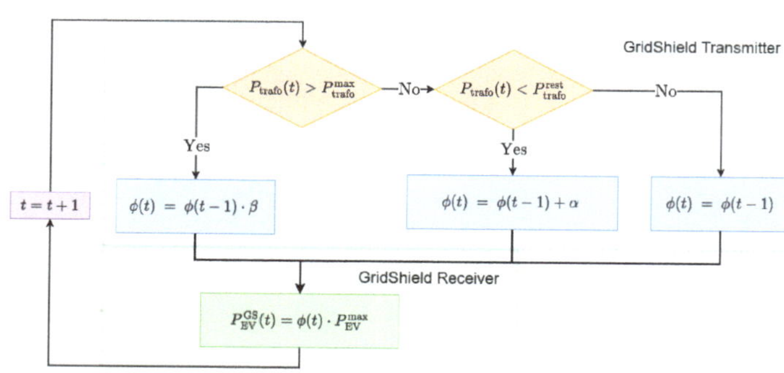

Figure 4. Block diagram of the AIMD GridShield algorithm. Blue: sender-side implementation. Green: receiver-side implementation. Adapted from [21].

Notice that oscillations between the increase and decrease phases may occur around the transformer limit if we start the increase phase immediately when we are below the limit. Thus, a deadband is introduced to prevent them. The power is reduced until it is below the limit $P_{\text{trafo}}^{\text{max}}$, but the increase phase is only entered when the transformer power is below a certain *restoration limit* $P_{\text{trafo}}^{\text{rest}}$. Effectively, the algorithm does not change $\phi(t)$ when $P_{\text{trafo}}^{\text{rest}} \leq P_{\text{trafo}}(t) \leq P_{\text{trafo}}^{\text{max}}$. Using multiplication to decrease the load means that the initial steps cause a larger decrease, while further steps are more granular. This allows a fast initial response to an overload event, and a more fine-grained control to further lower the allowed current. The upwards trajectory is a fixed step increase, preventing the algorithm from increasing the power too quickly.

The specifics of the control algorithm can be further optimized to meet one or more objectives, such as:

- Minimize overloading the transformer.
- Maximum use of capacity, i.e., being close to the limit of the transformer, comes at the cost of more fluctuation in the charging session.
- Limit the number of GridShield messages.
- Fairness amongst EVs.

2.2.2. Receiver Control Logic

The receiving module multiplies the received percentage value (for instance, 60%) with the configured grid connection limit to arrive at the current limit (ampere) used for the charging station. A complicating factor is that some EVs need a minimum of 6 A to charge. If the GridShield signal results in a limit lower than 6 A, vehicles could stop charging. Aside from the possible under-utilization of the electricity grid in such a case, the sudden halt in charging by many vehicles can result in a large swing in power. This, in turn, could lead to oscillation of the control system after the power is increased to the equivalent of >6 A per charging station.

A possible solution to this issue in an area with several charging stations would be for the receiver module to perform randomization to determine whether to continue charging at 6 A or to cut it off completely at 0 A. If, for example, the current limit results in a 5 A charging limit, the station could use a 5/6 random chance to let the station continue at 6 A and a 1/6 chance to turn the charging off at 0 A. In aggregate, this results in an average charging current of 5 A across many charging stations.

2.3. Living Labs and Simulations

GridShield is being implemented and researched in three living labs of different scales, contexts, and users: at the University of Twente (UT), Province House Zwolle, and at the world's largest, bidirectional charging car park at insurer a.s.r. in Utrecht [22]. All three locations combine solar energy, charging stations, batteries, and an Energy Management System. The users vary from short-stay visitors to full-day parking. Furthermore, SlimPark (UT) researches the end-users' needs and preferences via a link to an app that provides insight into how and when the generated solar energy can be distributed as efficiently as possible among the plugged-in cars.

The GridShield implementation is being simulated for validation and optimization in DEMKit. DEMKit, an open-source toolkit developed at the University of Twente, simulates devices, grids, and control components for researching multi-energy systems [23]. A validated model of a neighborhood in Lochem [1] was used to simulate GridShield interventions in a typical Dutch residential neighborhood.

3. Results

3.1. GridShield Prototype

3.1.1. Hardware- and Software-Agnostic Modules

The hardware for the transmitter and receiver units is identical. A custom software application written in Python is used to perform the required operations. The first GridShield prototype modules are built using the following off-the-shelf hardware components:

- A Raspberry Pi 2 Model B.
- A LoRa "hat" for the Raspberry Pi, based on the SX126X LoRa radio module.
- A standard 20 cm long, 900 MHz-range, SMA-connected antenna.
- An HD44780-compatible display unit for configuration and monitoring during tests.
- Connections to the measurement device and the charging station were made via the regular on-board Ethernet port to transport ModBus TCP messages.

A global supply chain shortage of relevant components prompted the development of a second set of prototypes. These use an industrial Linux-based mini-computer with on-board serial ports (Artila Matrix 500). A custom PCB carrying the RFM95 LoRa radio was developed, which connects to the Matrix 500's on-board SPI interface. These Linux boxes do not support the Python programming language; therefore, the GridShield application code was ported to C to run on these devices. As a result, it has been established that the concept is software- and hardware-agnostic. The modules are designed to be easily installed and to connect to an existing metering system (on the transmitting side) as well as to the charging post (on the receiving side). Once connected, the modules work autonomously.

3.1.2. Privacy and (Cyber) Security

GridShield works in a controller–agent setting. Charge points—agents—listen to a controller unit in the transformer. Because communication is one-way, cryptographically signed, and only local, i.e., not through the internet, the attack surface of the software environment is reduced. Furthermore, the module does not gather and store privacy-sensitive data. The only data registered are the load on the transformer and when and how GridShield intervened.

An inherent characteristic of GridShield is that it provides protection for the grid against some large-scale, remote cyberattacks. Protection against overload attacks follows automatically from the way GridShield intervenes when the technical boundary of the grid is exceeded. Other attacks, such as oscillation attacks below the maximum capacity, cannot be mitigated by GridShield.

Furthermore, the solution must be applicable to existing infrastructure, which means that it can only control the maximum allowed charging current. Other information, such as user preferences and the state of charge, is currently not communicated to the charging stations in practice, and is unavailable to GridShield due to the one-way communication.

3.2. *DEMKit Simulations*

DEMKit simulations, reported in [24], tested the effectiveness of GridShield in a typical Dutch neighborhood with 80 single-phase connected households and one EV charging station per household. Each EV has a battery capacity of 100 kWh and can charge at a maximum power of 7.4 kW (equivalent to 32 A). The arrival and departure times of the EVs and the required amount of charging, as well as the base load power profiles of the households, have been generated by the Artificial Load Profile Generator [19].

Figure 5 shows that 80 EVs in a typical residential neighborhood will drastically overload the transformer if there is no smart charging or energy management applied ("uncontrolled"). This scenario can also represent a situation in which the connection between charge stations and the back office that sends smart charging profiles is faulty or disrupted, causing charge stations to revert to a maximum charge setting. Using GridShield, a power outage is avoided with an acceptable delay in charging time.

The scope of GridShield is limited to dealing with temporary, local, unforeseen situations, with smart charging or an EMS primarily protecting the grid from overloading. However, when other solutions fail or are not implemented, these results show that GridShield functions as a coarse but effective EMS.

3.3. *Real-World Test: SlimPark*

The SlimPark Living Lab on the campus of the University of Twente is a solar carport with integrated charge stations connected to a real-time dynamic energy management system. The grid connection, acting as the transformer in this experiment, powers nine EV chargers and a battery. To test different scenarios, the power levels of four chargers, as well as the solar array and the battery, can be manipulated. During tests, GridShield intervenes and lowers the power of these four EV chargers when necessary, such that the power drawn from the grid stays below a pre-defined limit.

3.3.1. Scenario 1: Additional Car Triggers GridShield

In the scenario demonstrated in Figure 6, three EVs are charging and the load is just below 60 A, which is the fictive transformer limit within this test setup. When EV4 starts charging, the current exceeds this limit.

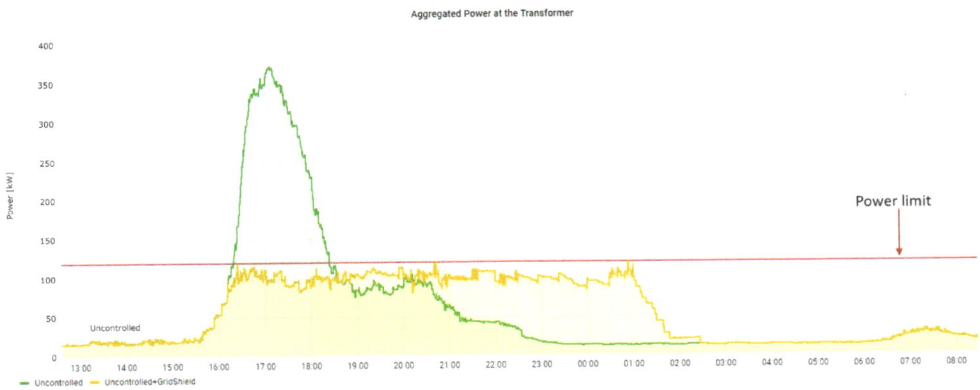

Figure 5. Digital twin simulation of a Dutch neighborhood with 80 EVs.

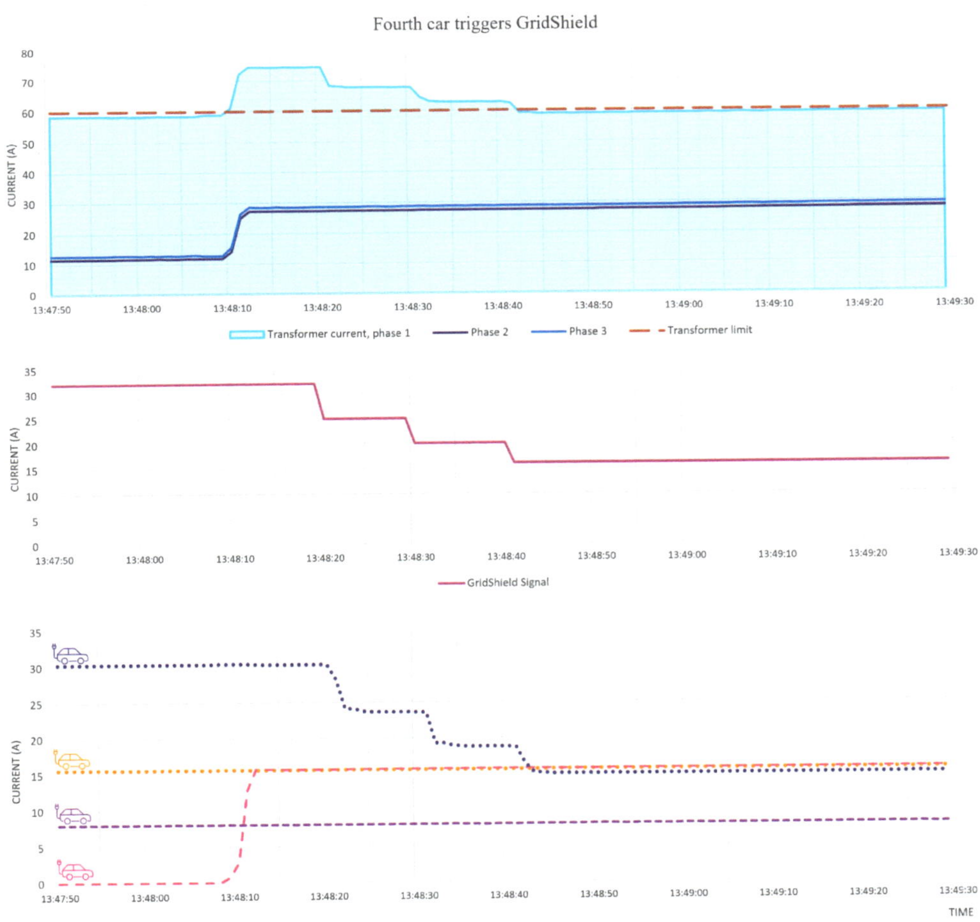

Figure 6. Transformer current, GridShield signal, and EV charging when an additional EV overloads the system.

Maximum charging power of the vehicles prior to GridShield intervention:

- EV1: 16 A, phase 1;
- EV2: 8 A, three phases;
- EV3: 32 A, phase 1;
- EV4: 16 A, three phases.

GridShield intervenes, lowering the maximum current allowed on the chargers to a new stable value of 16 A. We see that only EV3 is affected by the GridShield signal, because it is the only one charging above 16 A. In this experiment, GridShield required three iterations of decreasing control signals before the current was at a safe and stable level. In this case, a gradual decrease was a safe option that avoided excessively harsh interference in the charging. Depending on the nature of an energy system, the GridShield response can be programmed accordingly.

3.3.2. Scenario 2: Additional Car and Increased Baseload

Initially, this scenario is the same as scenario 1, but here changes to the base load are added; see Figure 7. We mimic an increase in base load by letting a battery located at the pilot site draw more power. At 14:01:17, the underlying base load increases and counteracts some of the decrease achieved by GridShield. In reality, this could be caused by heat pumps or any other type of high-power device.

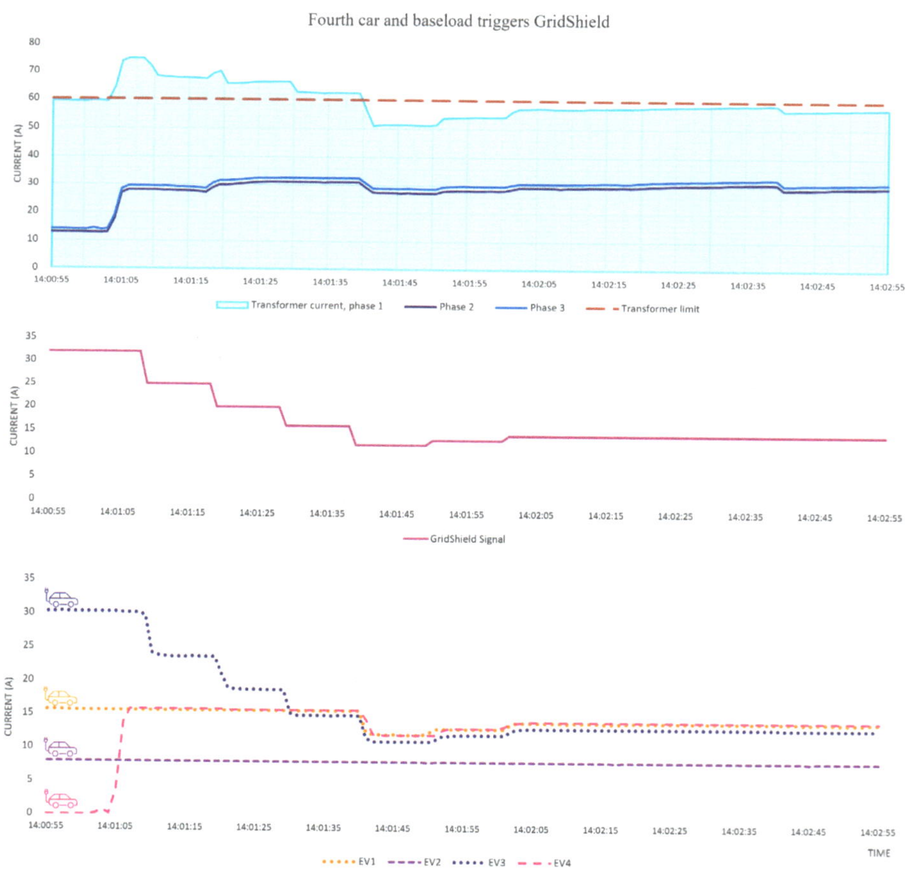

Figure 7. Transformer current, GridShield signal, and EV charging (ampere).

Due to the increased baseload, GridShield has to intervene more than in scenario 1. Charging has been decreased to 12 A, which affects all cars except for EV4, which is charging at 8 A. GridShield then determines that there is sufficient capacity on the transformer to safely increase the charging limit. Increases are made in 1 A steps until a stable situation is reached at 14 A.

GridShield lowers the virtual capacity of the charge stations, so EVs charging at a higher capacity are affected first. The power charge of cars charging at a lower capacity but on three phases, EV2 and EV4, is affected less or later by GridShield intervention, as illustrated in Figure 8.

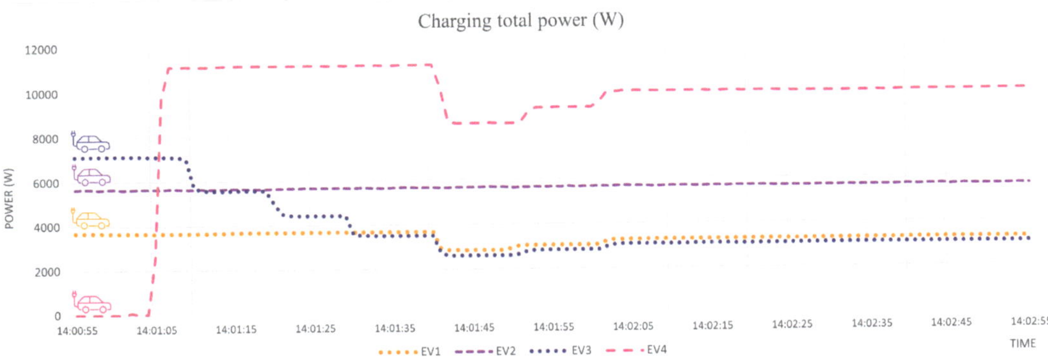

Figure 8. Total charging power per EV, during GridShield intervention.

4. Discussion

4.1. Discussion of Results

Simulations and real-world field tests demonstrate that GridShield can prevent local power outages. It functions as a software-based smart fuse that intervenes before the physical fuse in the transformer is overloaded, by temporarily and gradually reducing the capacity of charge stations following the programmed AIMD algorithm. We have simulated and tested many different scenarios, and plan to test even more, and on a larger scale, in our living labs. GridShield consistently intervenes and restores the current to a safe level, allowing charging to resume at its initial rate when it is safe for the grid.

As previously stated in Section 2.2, the algorithm settings can be changed to suit different preferences and circumstances. Both scenarios demonstrate the usefulness of the AIMD algorithm: decreases are large in the beginning to have a swift effect, and increases are gradual to keep the system safe. It touches on the balance between reacting fast enough but not overreacting. While a transformer can tolerate being overloaded for a period of time, it is undesirable and causes component wear. At the same time, it is desirable to minimize the impact on EV users.

GridShield intervenes if the current on any one phase is over the threshold, but the signal to the charge station applies to all phases. To keep GridShield robust, fast, and safe, its functionality is simple and does not take phase imbalance into account. Steering per phase adds complexity. Additionally, most currently available chargers are not technically capable of steering charging per phase. There is a role and incentive here for smart charging and technical innovation to manage phase imbalance. Charging power (W) is less of a problem for the local grid if the charging is spread over all three phases.

Based on the available measurements in the transformer, GridShield can react when *one* cable is overloaded, sending a signal to charging stations on that cable, or it can react when the sum of the cables is overloaded. When a charging station is installed, it should be linked to the appropriate GridShield transmitter, including information on which cable it is on.

When GridShield intervenes, the EV temporarily charges at a lower capacity. The effect on the state of charge will vary depending on the situation and implementation. The Smart Charging project Flexpower supports the conclusion that a temporary limitation of charging capacity has a minimal effect on the total energy charged [25]. Furthermore, GridShield is intended as a last resort to prevent power outages. Therefore, its effect on the user experience should be positive, compared to the alternative scenario in which the system would suffer a power outage. Issues such as the minimum charge and expected state of charge should primarily be discussed and researched in the context of smart charging.

GridShield modules are technology-agnostic and can be configured to work with a variety of different technologies, including heat pumps, batteries, and vehicle-to-grid installations. We foresee the demand for such solutions to appear in the near future and be integrated into GridShield.

GridShield is designed to have an easy and low-cost implementation. It connects to existing measuring equipment in the transformer. The hardware and software needed for the modules are minimal and technology-agnostic. The preferred long-term solution for charging stations is that the GridShield logic is incorporated into the existing controller logic. The installation of a simple LoRa-chip (or other communication protocol) could be required if it is not already available. Alternatively, a complete GridShield receiver module as described in Section 3.1 can be (retro)fitted.

4.2. Future Work

Having demonstrated the concept's viability, the optimization of how and when GridShield will intervene is further researched, validated, and tested in both simulations and real-world tests.

Closely related to this technological development is the progress on the regulatory basis for private charge point implementation of GridShield. In the Dutch regulatory context, intervening at charge points on household connections is not straightforward and will require regulatory changes. We do not expect this to affect the hardware design or the communication model, but we do foresee that it will influence the exact workings of the algorithms used, because "fairness" is subjective. An interim solution in lieu of regulatory change could be to make contractual agreements at public or semi-public charging stations, with or without economic incentives. This solution is relevant to the Dutch context, and its applicability will vary by country.

5. Conclusions

The increased electrification of vehicles is causing a strain on the Dutch electricity grid at peak times. Aside from grid reinforcements, it is projected that Smart Charging, (H)EMS, and tariffs can prevent the grid from overloading. However, these systems are expected to not be implemented sufficiently to keep up with the predicted local grid congestion. Furthermore, communication with centralized smart systems may fail or be hacked, resulting in immediate local grid overload.

This paper has shown that it is possible to build a robust, decentralized system capable of autonomously preventing the grid from becoming overloaded in unforeseen circumstances. Simulations and real-world tests presented show that GridShield intervenes before a potential power outage by temporarily reducing the capacity of charging stations and restoring capacity when congestion subsides. Due to its unidirectional, autonomous design, it is deemed compatible with Dutch and EU privacy laws, and its functionality provides protection against overload attacks in the context of cybersecurity.

GridShield protects all grid users with minimal inconvenience for the EV user by preventing unpredicted, temporary, local power failures. It can help optimize the use of grid capacity during increased EV adoption and help enable additional charge points to be added in areas with grid congestion.

Author Contributions: Conceptualization, M.Z., G.H., H.L.v.S. and S.J.; methodology, S.J., G.H., I.A.M.V. and H.L.v.S.; software, S.J., I.A.M.V. and G.H.; validation, S.J. and G.H.; formal analysis, H.L.v.S. and G.H.; investigation, S.J., G.H., I.A.M.V. and H.L.v.S.; resources, S.J. and G.H.; data curation, G.H., I.A.M.V. and H.L.v.S.; writing—original draft preparation, H.L.v.S. and S.J.; writing—review and editing, G.H., H.L.v.S., S.J., I.A.M.V. and M.Z.; visualization, H.L.v.S., I.A.M.V. and G.H.; supervision, H.L.v.S. and G.H.; project administration, M.Z.; funding acquisition, M.Z. All authors have read and agreed to the published version of the manuscript.

Funding: This research was carried out in the Project SmoothEMS met GridShield (MOOI32005), with the support of RVO—Netherlands Enterprise Agency.

Data Availability Statement: Data are available on request.

Acknowledgments: The authors would like to thank Bart Nijenhuis and Stefan Rang for their work on the SlimPark tests, Nataly Bañol Arias and Leoni Winschermann for their work on DEMKit simulations, and all Smooth EMS met GridShield project partners: AmperaPark, a.s.r., ElaadNL, Kropman Installatietechniek, Mennekes eMobility Nederland and the University of Twente. The development of GridShield builds on a Proof of Concept called the Social Module that was created and tested by Jesse Kerkhoven at ElaadNL in 2020 [26].

Conflicts of Interest: The authors declare no conflict of interest.

References

1. Hoogsteen, G.; Molderink, A.; Hurink, J.L.; Smit, G.J.M.; Kootstra, B.; Schuring, F. Charging electric vehicles, baking pizzas, and melting a fuse in Lochem. *CIRED* **2017**, *1*, 1629–1633. [CrossRef]
2. Refa, N.; Hammer, D.; Rookhuijzen, J. Elektrisch rijden in stroomversnelling. *ElaadNL Outlook* **2021**, Q3. Available online: https://elaad.nl/wp-content/uploads/2022/05/2021Q3_Elaad_Outlook_Personenautos_2050.pdf (accessed on 6 March 2023).
3. Netbeheer Nederland. Position Paper Voor Het Rondetafelgesprek over Het Elektriciteitsnet. 3 February 2022. Available online: https://www.netbeheernederland.nl/_upload/Files/Rondetafel_Elektriciteitsnet_03-02-2022_Inbreng_Netbeheer_Nederland_242.pdf (accessed on 6 March 2023).
4. Schoot Uiterkamp, M.H.H.; Gerards, M.E.T.; Hurink, J.L. Online electric vehicle charging with discrete charging rates. *Sustain. Energy Grids Netw.* **2021**, *25*, 100423. [CrossRef]
5. Vandael, S.; Claessens, B.; Hommelberg, M.; Holvoet, T.; Deconinck, G. A Scalable Three-Step Approach for Demand Side Management of Plug-in Hybrid Vehicles. *IEEE Trans. Smart Grid* **2013**, *4*, 720–728. [CrossRef]
6. De Craemer, K.; Vandael, S.; Claessens, B.; Deconinck, G. An Event-Driven Dual Coordination Mechanism for Demand Side Management of PHEVs. *IEEE Trans. Smart Grid* **2014**, *5*, 751–760. [CrossRef]
7. Spitzer, M.; Schlund, J.; Apostolaki-Iosifidou, E.; Pruckner, M. Optimized Integration of Electric Vehicles in Low Voltage Distribution Grids. *Energies* **2019**, *12*, 4059. [CrossRef]
8. Bañol Arias, N.; Hashemi, S.; Andersen, P.B.; Træholt, C.; Romero, R. Distribution System Services Provided by Electric Vehicles: Recent Status, Challenges, and Future Prospects. *IEEE Trans. Intell. Transp. Syst.* **2019**, *20*, 4277–4296. [CrossRef]
9. Thomas, P.; Karthikeyan Shanmugam, P. A review on mathematical models of electric vehicle for energy management and grid integration studies. *J. Energy Storage* **2022**, *55*, 105468. [CrossRef]
10. Deb, S.; Pihlatie, M.; Al-Saadi, M. Smart Charging: A Comprehensive Review. *IEEE Access* **2022**, *10*, 134690–134703. [CrossRef]
11. McKenna, K.; Keane, A. Discrete Elastic Residential Load Response Under variable Pricing Schemes. In *ISGT-Europe*; IEEE: New York, NY, USA, 2014; pp. 1–6.
12. Nationale Agenda Laadinfrastructuur. Slim Laden voor Ledereen 2022–2025: Actieplan. Available online: https://agendalaadinfrastructuur.nl/ondersteuning+gemeenten/documenten+en+links/documenten+in+bibliotheek/handlerdownloadfiles.ashx?idnv=2297386 (accessed on 6 March 2023).
13. de Heer, H.; van der Laan, M.; Sáez Armenteros, A. USEF: The Framework Explained-Update 2021. Available online: https://www.usef.energy/news-events/publications/ (accessed on 20 February 2023).
14. Bundesverband der Energie-und Wasserwirtschaft. Smart Grid Traffic Light Concept: Design of the Amber Phase. Technical Report. March 2015. Available online: https://www.bdew.de/media/documents/20150310_Smart-Grid-Traffic-Light-Concept.pdf (accessed on 6 March 2023).
15. Hartmanns, A.; Hermanns, H.; Berrang, P. A Comparative Analysis of Decentralized Power Grid Stabilization Strategies. In Proceedings of the 2012 Winter Simulation Conference (WSC), Berlin, Germany, 9–12 December 2012; pp. 1–13.
16. Haider, Z.M.; Mehmood, K.K.; Khan, S.U.; Khan, M.O.; Wadood, A.; Rhee, S.-B. Optimal Management of a Distribution Feeder during Contingency and Overload conditions by Harnessing the Flexibility of Smart Loads. *IEEE Access* **2021**, *9*, 40124–40139. [CrossRef]
17. Fischer, D.; Madani, H. On heat pumps in smart grids: A review. *Renew. Sustain. Energy Rev.* **2017**, *70*, 342–357. [CrossRef]
18. Fischer, D.; Triebel, M.-A.; Selinger-Lutz, O. A concept for controlling heat pump pools using the smart Grid ready interface. In *ISGT-Europe*; IEEE: New York, NY, USA, 2018; pp. 1–6.

19. Tangerding, F.; Varenhorst, I.A.M.; Hoogsteen, G.; Gerards, M.E.T.; Hurink, J.L. GridShield: A Robust Fall-Back Control Mechanism for Congestion Management in Distribution Grids. In *ISGT-Europe*; IEEE: New York, NY, USA, 2022; pp. 1–5.
20. Augustin, A.; Yi, J.; Clausen, T.; Townsley, W.M. A Study of LoRa: Long Range & Low Power Networks for the Internet of Things. *Sensors* **2016**, *16*, 1466. [PubMed]
21. Varenhorst, I.A.M. GridShield: Robust Control Algorithms to Prevent Power Outages. Master's Thesis, University of Twente, Enschede, The Netherlands, June 2022.
22. SmoothEMS Met GridShield. Available online: https://www.elaad.nl/smoothems-met-gridshield/ (accessed on 9 November 2021).
23. Hoogsteen, G.; Hurink, J.L.; Smit, G.J.M. DEMKit: A Decentralized Energy Management Simulation and Demonstration Toolkit. In *ISGT-Europe*; IEEE: New York, NY, USA, 2019; pp. 1–5.
24. Nijenhuis, B.; Winschermann, L.; Arias, N.B.; Hoogsteen, G.; Hurink, J.L. Protecting the distribution grid while maximizing EV energy flexibility with transparency and active user engagement. CIRED Porto Workshop 2022: E-mobility and power distribution systems. In Proceedings of the Hybrid Conference, Porto, Portugal, 2–3 June 2022.
25. Groen, M.; Hoogvliet, T.; Niesing, H.; Wolbertus, R.; van Zalingen, W.; Zweistra, M. Flexpower3: Meer Laden op een Vol Elektriciteitsnet. Available online: https://elaad.nl/wp-content/uploads/2022/11/FlexPower-Rapport.pdf (accessed on 21 February 2023).
26. Kerkhoven, J. Proof of Concept Sociale Module voor Laadpalen, Decentrale Variant. Lokaal Congestiemanagement Door Middel van Smart Charging Met Behulp van LoRa ElaadNL. 2020. Available online: https://elaad.nl/wp-content/uploads/2022/08/Paper_Jesse_kerkhoven_-_Sociale_Module_Lokaal_Congestiemanagement.pdf (accessed on 6 March 2023).

World Electric Vehicle Journal

MDPI

Article

Circular Economy-Based Alternatives beyond Second-Life Applications: Maximizing the Electric Vehicle Battery First Life

Maite Etxandi-Santolaya [1,2,*], Lluc Canals Casals [2], Beatriz Amante García [2] and Cristina Corchero [1]

1 Department of Energy Systems Analytics, Institut de Recerca en Energia de Catalunya—IREC, 08930 Sant Adrià de Besòs, Spain
2 Department of Engineering Projects and Construction, Universitat Politècnica de Catalunya—UPC, 08034 Barcelona, Spain
* Correspondence: metxandi@irec.cat

Abstract: Electric vehicle battery second-life applications are gaining attention as a way to minimize the environmental impact and increase economic profits. However, the demand for stationary energy storage is expected to be saturated in the near future with these second-life batteries. This fact, in addition to the several technical and economic challenges of second-life batteries, promotes exploring other alternatives. This work analyses and compares these possible approaches in terms of battery degradation and economic profit. The results show that for large batteries, intensive Vehicle to Grid does not cause an early retirement of the battery and allows reducing the underuse of the battery. In addition, for the same battery size, Vehicle to Grid provides more economic profit than second-life applications. Nevertheless, only in a few cases does this appear to be more profitable than simply utilizing the battery for driving. Importantly, this study has shown how the assessment of the second-life tends to be too optimistic as a consequence of assuming a fixed End of Life threshold for the batteries.

Keywords: battery ageing; electric vehicles; second-life application; V2G (Vehicle to Grid)

Citation: Etxandi-Santolaya, M.; Canals Casals, L.; Amante García, B.; Corchero, C. Circular Economy-Based Alternatives beyond Second-Life Applications: Maximizing the Electric Vehicle Battery First Life. *World Electr. Veh. J.* **2023**, *14*, 66. https://doi.org/10.3390/wevj14030066

Academic Editor: Joeri Van Mierlo

Received: 7 February 2023
Revised: 2 March 2023
Accepted: 3 March 2023
Published: 5 March 2023

1. Introduction

The fast growth of the Electric Vehicle (EV) market goes hand in hand with a large R&D effort to improve the technology, reduce the cost of the vehicles, especially of their Lithium-ion (Li-ion) batteries, and to lengthen their lifespan.

Degradation is a key issue surrounding these batteries. As a consequence of usage and time, batteries lose the abilities to provide energy and power [1], which can intensify the range anxiety of the drivers. To avoid this, EV manufacturers have taken two actions: first, thanks to their cost reduction, the battery capacity is constantly being increased. While early EV models counted on a battery of 16 or 24 kWh, current EV batteries have capacities of 40, 70, 90 kWh, or even higher. Secondly, in order to avoid excessive levels of degradation that could jeopardize meeting the driver's day-to-day requirements, the End-of-Life (EoL) of the battery for traction purposes was set to 70–80% of the State of Health (SoH) [2].

Considering this threshold, especially in recent models, EVs are expected to reach the EoL with a large residual capacity. To extract value from this residual capacity, second-life applications gained presence in the research field and have started to be deployed in real life. Through the deployment of second-life applications, instead of sending batteries to be recycled or disposed, they can first be used in less demanding applications, providing economic revenue and extending the total lifetime of the battery [3].

However, second-life applications present several technical, economic and environmental uncertainties. From the technical point of view, the main challenges are using the Battery Management System (BMS) outside its original purpose [4] and using batteries after important levels of degradation [5]. From the economic side, it is unclear whether these applications are as profitable as some authors have suggested due to the additional costs of battery refurbishment and equipment purchase [6]. In terms of the environmental impact, most studies argue that the Greenhouse Gas (GHG) emissions saved by deploying second-life applications can be over 25% [7,8]. However, the estimations are based on the assumption that the second-life battery replaces a new one, which may not be realistic as the purchase of a new battery for providing grid services may not have taken place due to their high cost [9].

In addition, the growth of EV sales also means that the same number of batteries will be reaching EoL at some point. Considering the forecasted demand for energy storage, several sources point out that the needs of this market will be saturated with second-life batteries close to 2030–2035 [10,11]. Therefore, many of the retired batteries will not be able to reach second-life markets.

In view of these uncertainties, other solutions are required to increase economic revenue and reduce the environmental impact of EV batteries. In this sense, the circular economy can be considered as a valuable framework to obtain guidelines. The circular economy aims to decouple economic activity from environmental degradation by promoting the practices that maintain the value of the resources for longer.

This study explores circular economy-based alternatives beyond second-life applications, which follow the principles of sharing and enlarging the lifespan to maximize the use until the EoL, considering that it is better to totally use a device in the first place than to not do so and reuse it afterwards to reach the same end-point [12]. The alternatives considered are:

(a) the reduction in the battery capacity to the minimum size that can cover the driving requirements,
(b) redefining the EoL threshold to avoid early retirement and,
(c) counting on Vehicle to Grid (V2G) as the tool for value extraction based on sharing instead of the second-life applications.

Reducing the battery capacity is based on the idea of narrowing resource loops. Considering the range requirements of most of the population, batteries over 40 kWh have the risk of being underused, which does not help minimize the environmental impact of the EV [13]. Therefore, if the nominal capacity of the battery is reduced, all its value can be extracted during the first life, without the need for other cycles, such as a second life, which require additional resources to implement.

Regarding the battery EoL, a few authors have started to question the commonly assumed threshold of 70–80% SoH, arguing that the battery can be functional at higher degradation levels [14,15]. A universal threshold does not take into account the nominal battery capacity or the actual driving requirements. Therefore, in many cases, the EoL could be pushed further, which affects the duration of the first life, as estimated by the Remaining Useful Life (RUL) algorithms, and the state of the battery at retirement, which directly affects the evaluation of the battery reuse.

Finally, V2G enables EVs to provide a bi-directional flow of energy when they are connected to the EV supply equipment. In this way, the battery not only serves the automotive application but can additionally provide services to the grid by discharging energy into it. V2G adds additional cycling which intensifies the usage of the battery during its first life to avoid the potential underuse that has been mentioned previously.

The three alternatives proposed aim to give a higher weight to the battery's first life to avoid having to count on a second life to make sure that the battery does not reach the recycling facilities with residual value. This idea is backed by the circular economy, which promotes prioritizing cycles at the first stages of the lifecycle over those at later stages.

In particular, this study aims to obtain a general EoL range requirement that batteries should meet to cover common driving trips and to evaluate whether the current batteries found on the market are able to arrive at the EoL at the same time as the vehicle, while providing the required functionalities. Based on the results, the potential place for providing V2G services is analysed and, the different circular economy-based approaches are compared in terms of the economic cost to evaluate which strategies should be prioritized.

In addition, the impact of considering the fixed EoL threshold over the functional one is discussed, with a special focus on the implications on the economic analysis of the second-life applications.

2. Methodology

This study starts by estimating the requirements that EoL batteries should meet to cover the range requirements of most drivers. According to [16], the daily mileage can be represented using a Weibull distribution in which 90% of the daily trips consist of less than 60 km. Considering this conservative daily mileage and the fact that the consumption suffers from seasonal variations (up to 60%) due to the difference in ambient temperature [17], it is estimated that the EoL battery should be able to provide at least 14.85 kWh to cover 90% of all the trips all year round.

2.1. Degradation Model

A degradation model is used to estimate the ageing of batteries with different nominal capacities and usage patterns to understand their SoH at EoL. The model combined the degradation caused by driving the EV (D1) and the one caused by a specific application such as the V2G profile or the second-life application (D2).

2.1.1. Degradation Model D1

To model the degradation caused by driving, the mileage completed through their entire first life is assessed. The lack of retired EVs forced us to adopt assumptions to evaluate this EoL mileage. It was considered that EVs will be retired and sent to dismantling facilities having a similar distribution of mileage and years as diesel vehicles. Considering this assumption and the open data provided by the MOT (Ministry of Transportation) on regular inspections of vehicles driving in the UK [18], the mileage of EoL vehicles has been analysed. Depending on the year of retirement, the distribution of the EoL mileage is significantly different. However, to make a conservative estimation, the distribution corresponding to a retirement age of 20 years has been obtained. This mileage can be best represented by a normal distribution with a mean of 227,264 km and a standard deviation of 91,505 km. Considering 90% of all vehicles retired, the EoL mileage obtained through this distribution is 344,532 km.

The degradation model used to estimate the SoH after the defined EoL mileage was presented by the authors in a previous publication that provides a detailed explanation [19]. In summary, the model was built based on the degradation patterns found in UK fleet vehicles obtained from the EV Battery Degradation Comparison Tool provided by Geotab [20]. The data show the evolution of the SoH for different EVs over time.

The ageing tendency found using Geotab is relatively linear and could be represented by Equation (1), where α is the parameter that relates the ageing of the battery through time. This ageing tendency was transformed into Equation (2), considering that the average mileage is 28,175 km per year as published by the UK department of transport statistics. The β parameter is related to the ageing of the battery depending on the mileage. Both α and β depended on the EV battery capacity and were the result of using linear regression on the Geotab data. The coefficients for both regressions are shown in Table 1. For more details, the reader can refer to the original publication [19].

$$\mathrm{SoH_{D1}} = 100 - \alpha \cdot \mathrm{age} \tag{1}$$

$$\mathrm{SoH_{D1}} = 100 - \beta \cdot \mathrm{km} \tag{2}$$

Table 1. D1 degradation model parameters according to battery size.

Battery Capacity (kWh)	α	β
16	3.64	0.000227
24		0.000161
30		0.000129
40	2.24	0.000097
70		0.000083
90		0.000064

This degradation model is used to estimate the battery SoH after 20 years of driving the EV. It should be noted that each EV battery could have a different degradation level even for the same mileage, depending on the driving conditions. However, using a degradation model based on the analysis of the ageing tendencies of different EVs allows us to cover several driving conditions simultaneously and therefore, it can be used for generalizing.

2.1.2. Degradation Model D2

An empirical degradation model is used to estimate the additional degradation D2 caused by the V2G or second-life profiles that can not be captured through model D1. In this case, it is assumed that the battery performs discharging at constant currents unlike during driving conditions. The model developed by Olmos et al. for NMC cells is employed for this purpose [21]. Following the assumption from model D1, a linear relationship is considered between the Full Equivalent Cycles (FEC) and SoH. In addition, all the cycling is considered to take place at 25 °C, 1 C-rate for charge and discharge and starting at 100% State of Charge (SoC). The average SoC (SoC_m) is defined by the desired Depth of Discharge (DoD) of the duty cycle. The degradation model D2 is represented by Equation (3) and its parameters are shown in Table 2.

$$\mathrm{SoH_{D2}} = 100 - \mathrm{a} * \exp(\mathrm{bDoD} + \mathrm{c}) * \left(1 + \mathrm{dSoC_m}\left(1 + \frac{\mathrm{SoC_m}}{\mathrm{e}}\right)\right) * \mathrm{FEC} \quad (3)$$

Table 2. D2 degradation model parameters.

D2 Parameters	
a	0.001673
b	0.022
c	0.4124
d	−0.0212
e	84

2.2. Use Cases

Four different use cases have been defined to analyse their technical feasibility considering the battery degradation and to evaluate, from the economic point of view, which alternative is the most profitable:

- 1—No grid services: aims to represent the baseline where the battery does not provide any grid services and is only used for driving until the functional EoL (20 years) is reached. This use case allows for evaluating which is the minimum capacity that would provide the required functionalities during the entire EV lifespan.
- 2—V2G: considers the functional EoL and different grid services provided through V2G, as explained later in this section.

- 3—Second-life: considers the functional EoL and different grid services provided through a second-life application instead of V2G, as explained later in this section. To evaluate the impact of assuming the fixed threshold in the second-life assessment studies, the same use case has been defined considering the EoL at 80% SoH, instead of the functional one.

There are several grid services that EV batteries can provide. In this study, the residential (R) use, peak shaving (PS), and participation in the demand response (DR) actions have been considered. In this section, the assumptions made regarding the V2G profiles are presented. It should be noted that the approach in this study is to consider V2G as an important source of value extraction, exploiting the battery's first life as much as possible. Other works, however, may view V2G as an opportunistic behaviour and therefore may reduce the frequency of the service to lower values.

- Residential (R): EV batteries can be used to provide energy to households while parked to support renewable energy sources or perform load balancing or energy arbitrage strategies. This study considers that the battery provides 50% of the load of the home which is around 4000 kWh yearly for European households [22]. This means that the battery provides 5 kWh daily regardless of its capacity.
- Peak Shaving (PS): this service aims at reducing the power consumed by a building by providing energy from batteries during peak times and charging them during low-demand hours. It is expected that several EVs will provide this service by aggregating the energy discharged by each one. Therefore, the profile selected in this case is not calculated by a fixed amount of energy, but instead, by a value of the DoD that should not be exceeded to avoid the range limitations of the driver. This value is set to 25% and the frequency of the service is assumed to be 5 times per week.
- Demand Response (DR): in this case, the goal of the service is to solve particular constraints that may appear on the grid, such as congestion or frequency deviations caused by an imbalance between supply and demand. Batteries are considered to be ideal for providing primary and secondary frequency control services due to their fast response time [23]. These services are only activated during a short period of time. For this study, 30 min of discharge at 1C are considered for each activation, the frequency of the service is assumed to be 3 times per week, and the battery availability is set to 5 h/daily.

These same services can be provided by a second-life battery. Considering that in the second life the automotive restrictions are not present, the operating conditions are redefined to be more extreme, as presented in Table 3.

Table 3. Grid service parameters considered.

Grid Service	V2G		Second-Life	
	Energy Provided	**Frequency (Times/Year)**	**Energy Provided**	**Frequency (Times/Year)**
Residential	5 kWh	365	8 kWh	365
Peak Shaving	25% DoD	260	80% DoD	365
Demand Response	30 min at 1C	156	45 min at 1C	365

2.3. Battery Lifecycle Costs

The different use cases defined are compared in terms of the cost over the lifecycle of the battery. The goal of this study is not to provide an accurate estimation of the cost of each of the use cases, but to compare them. For that reason, the activities that are common for all use cases are not analysed (e.g., EV manufacturer or battery disassembly from the EV).

Figure 1 shows the main activities during the entire lifecycle of the EV battery. The first step is the raw material extraction and manufacturing process. The cost of these processes depends on several factors such as the location, which affects the energy mix, the chemical composition of the cells, or the methods used for raw material extraction and component manufacturing. The cost of battery manufacturing is not considered as it is common for use cases with the same battery size. However, it will be necessary for estimating the cost of battery degradation, which takes into account the battery purchase, as explained later on. Therefore, the cost of the battery purchase (C_{bat}) is set to 120 €/kWh [24].

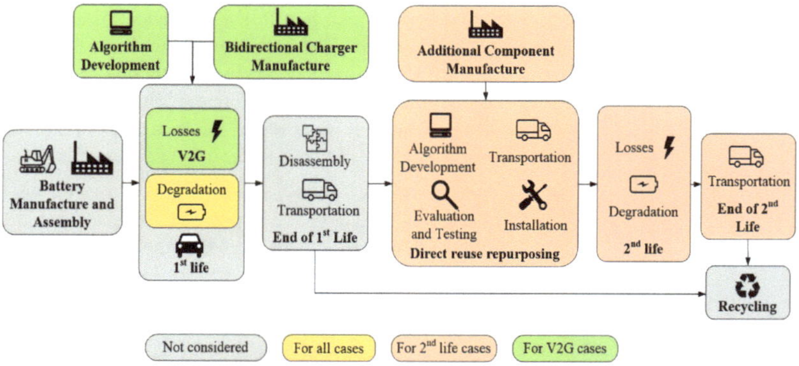

Figure 1. Lifecycle of the EV battery.

The battery then serves its main purpose by providing energy to the EV during its first life. The costs during this stage (e.g., battery charging) are common for all the use cases and therefore were not considered. For the cases where the battery is employed for V2G, additional costs should be considered (C_{V2G}). First, a bidirectional charger should be purchased instead of a standard Level 2 charger. This impact has been estimated by comparing the cost of commercial bidirectional chargers with standard ones for the same company [25]. The estimated additional cost is 5200€. Additionally, the algorithm development for providing V2G services is estimated to be 27 €/kWh, equal to the one that will be assumed for the second life [26].

Once the battery reaches the EoL it should be disassembled from the EV and transported to the second-life repurposing or recycling facilities. It is assumed that the distance to each facility is similar and therefore the cost of this stage is common for all use cases.

If the battery is reused in a second life, additional steps are required, which entail an additional cost ($C_{2nd\text{-}life}$). In this study, it is assumed that the retired battery will be directly reused, without the need to dismantle the pack at the module level or develop a new battery management system. The activities related to the repurposing are the development of algorithms for the control, evaluation and testing of the battery, and the transportation and installation of the battery at the end-use location. The cost related to the testing and transportation is assumed to be 87 €/kWh for a direct reuse approach [27]. The cost related to the algorithm development is estimated at 27 €/kWh and the final installation and commissioning are assumed to cost 52 €/kWh [26]. Additionally, new components are needed for the second life, which includes power electronics converters and electrical safety elements. The power considered in the inverter is set to 40 kW for the peak shaving and DR services and 5 kW for the residential one. The cost of the power electronics and the additional electrical equipment is considered to be 80 €/kW [28].

Additionally, for the cases providing grid services either during a first or a second-life, the additional cycling of the battery must be translated to a cost, as it implies some energy losses due to the efficiencies of the EV charger or the second-life power electronics (C_{loss}). The efficiency of the charger is assumed to be 87% when working at the maximum power [29] and the second-life inverter efficiency is assumed to be 90% [30]. The value of this cost was obtained from Equation (4) that considers the efficiency of the charger or inverter (η), the energy provided as grid services during the entire lifespan ($E_{grid\ services}$), and an average cost for the electricity consumed (C_{av}). The cost of electricity is obtained as an average of different available tariffs reviewed for residential or industrial consumers (peak shaving and demand response) in Spain [31]. The values considered are 0.287 €/kWh and 0.101 €/kWh, respectively.

$$C_{loss} = \eta \, E_{grid\ services} \, C_{av} \tag{4}$$

The last cost to be considered is related to battery degradation. Depending on the use case, the battery suffers different levels of degradation, which should be translated into a cost to reflect the fact that the value of the battery deprecates with degradation. The assumed degradation cost (C_{degr}) follows Equation (5) and considers the replacement of the battery that depends on the capacity loss (Q_{loss}), the cost of the new battery per kWh (C_{bat}), and the assumed EoL, which has been set to 60% of the SoH [32].

$$C_{degr} = \frac{Q_{loss}}{60\%} C_{bat} \tag{5}$$

Providing grid services entails revenues ($R_{grid\ services}$) depending on the application. The benefits derived from the residential service are obtained from energy arbitrage strategies which are based on charging the batteries during off-peak periods and using this stored energy to provide electricity to the loads during peak periods, in which the cost is higher. Considering residential 3-level tariffs from different electrical companies, the difference in the cost between peak and off-peak periods is estimated at 0.13 €/kWh [31].

Similarly, peak shaving reduces the energy consumed during peak periods to reduce the power consumed at these times. In this case, it is considered that peak shaving is provided to an industry or larger consumer. The associated revenue comes from the energy arbitrage considering a tariff for a large consumer above 15 kW (0.035 €/kWh in the reduction from peak to off-peak periods), from avoiding exceeding the contracted power at a penalty of 1.4 €/kW and from shifting the contracted power at peak periods to off-peak ones, which reduce the annual cost by 369 €/kW [31]. The excess power penalties are assumed to take place only 6 times per year, consuming 10% more than the contracted power.

The economic retribution from the demand response services comes from the availability and utilization concepts. The first is related to the time during which the power is available for providing frequency regulation and the second considers the actual energy provided. The values considered were 4 €/MW/hour for the availability [33] and 0.023 €/kWh for the utilization obtained from the Spanish capacity market prices in 2021 [34].

The last step in the battery lifecycle for all cases is recycling. It is assumed that all the batteries would be recycled using the same processes and therefore their cost is not considered.

A summary of the cost and revenue considered for this study is given in Table 4.

Figure 2 shows the calculation of the costs for each of the use cases defined. As mentioned, the first life of the battery is assumed to be 20 years and the second life is defined to end at 60% SoH. An additional restriction is set for the second-life lifespan, considering the fact that batteries cannot be employed forever and that they eventually become obsolete. For this reason, the second life can not exceed 15 years beginning from the battery retirement of the EV.

Table 4. Costs and revenues considered.

	Concept		Value	Reference
Costs	Energy Losses (C_{loss})		See Equation (4)	-
	Electricity (C_{av})	Residential	[31]	[31]
		Industrial	0.101 €/kWh	
	Degradation (C_{deg})		See Equation (5)	[32]
	Battery (C_{bat})		120 €/kWh	[24]
	V2G (C_{V2G})	Components	5200 €	[25]
		Software	27 €/kWh	[26]
	Second-Life ($C_{2nd\text{-}life}$)	Testing and Transportation	87 €/kWh	[26]
		Installation and Commissioning	52 €/kWh	[26]
		Software	27 €/kWh	[26]
		Components	80 €/kW	[28]
Revenues ($R_{grid\ services}$)	Residential Arbitrage		0.13 €/kWh	[31]
	Peak Shaving	Arbitrage	0.035 €/kWh	[31]
		Excess Power Penalties	1.4 €/kW	
		Contracted Power Shift	369 €/kW	
	Demand Response	Availability	4 €/MW/h	[33]
		Utilization	0.023 €/kWh	[34]

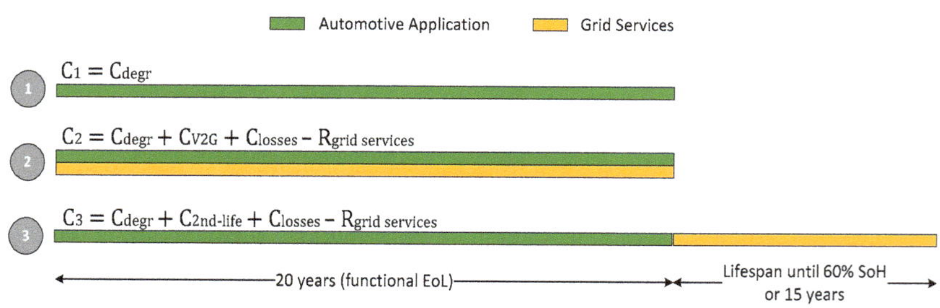

Figure 2. Cost calculation for each use case.

3. Results

Figure 3a shows the EoL battery capacity and SoH for different sizes considering the fixed threshold. All batteries, except those having 16 kWh, reach the EoL while being able to cover the required range. However, the mileage performed by the EVs is, in all cases, lower than the assumed mileage of the EoL vehicles, especially for the small-capacity batteries. Figure 3b shows a more realistic scenario where the functional EoL is considered, meaning that the battery and EV reached the EoL at the same time, with the previously derived mileage of 344,532 km after 20 years.

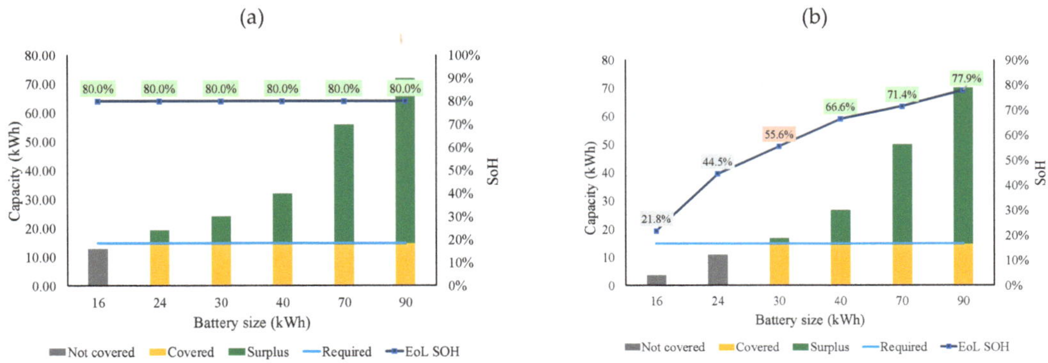

Figure 3. EoL capacity for 16 to 90 kWh batteries considering: (**a**) a fixed threshold and (**b**) a functional threshold.

Considering the 20-year lifespan, batteries with 16 and 24 kWh of capacity are not able to reach the EoL alongside the EV while providing the required driving range. The 30 kWh batteries, on the other hand, hold enough capacity but reach the EoL with an important level of degradation (55.6% SoH), with potential safety and underperformance issues linked to this state. Batteries of 40 kWh and above exceed the required capacity while maintaining a relatively healthy state, over 60% SoH, indicating that a capacity of 40 kWh would be enough to meet the requirements of most drivers (1—No Grid Services).

Larger batteries are able to reach the EoL with a large residual capacity and with SoH values over 70%. Therefore, these batteries could either extend their use during their first life by providing V2G services (2—V2G) or serve a second-life application after the EoL (3—Second-life).

If V2G is used during the first life of these larger batteries, the degradation increases, ending with more aged batteries at the EoL, as shown in Figure 4 for the different V2G profiles.

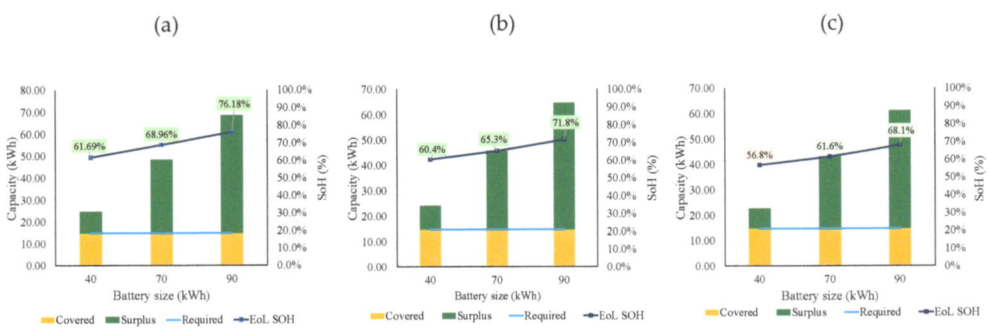

Figure 4. EoL capacity for 40, 70, and 90 kWh batteries considering V2G services: (**a**) residential, (**b**) peak shaving and (**c**) demand response.

In all cases, except for the DR service with the 40 kWh battery, EV is able to provide V2G services while reaching the EoL in a healthy state and being able to provide the required range. In fact, the largest battery of 90 kWh still holds an important residual capacity and reaches the EoL with SoH close to or over 70%. Therefore, it is possible to extend the use of these batteries by stacking different V2G services or through a second life. The most demanding V2G application is the DR one, which increases the capacity fade by 9.8% over the 20 years, followed by peak shaving with a 6.1% increase, and the residential one with an increase between 1.8% and 4.9% depending on the battery capacity.

3.1. Economic Analysis

Considering the previous results, it is possible to compare the different use cases that guarantee that the driving requirements are not compromised. This involves batteries of 40 kWh and higher for all use cases except for the V2G DR one. This analysis serves to understand which strategy is more beneficial from an economic point of view.

The cost for each case over the battery life can be observed in Figure 5. The reader is reminded that for the Case 1—No Grid Services, the cost was related to the degradation of the battery, and for the Case 2—V2G and Case 3—Second-life, the cost and revenues of providing grid services are additionally considered.

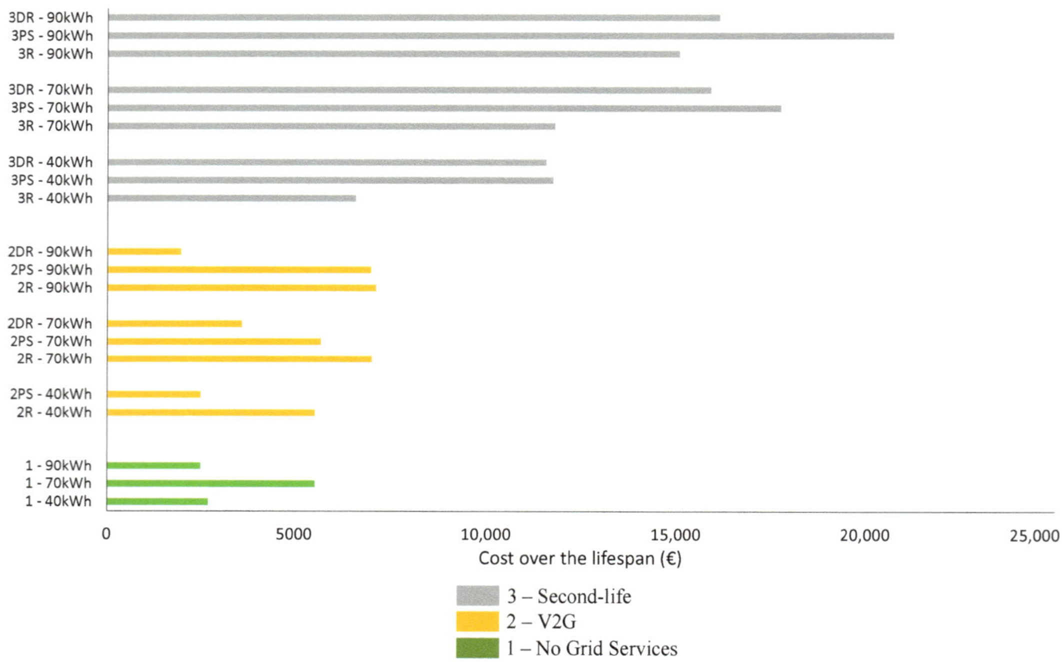

Figure 5. Cost over the battery lifetime for the different use cases: 1—No grid Services, 2—V2G, and 2—Second-life (R: residential, PS: peak shaving, DR: demand response).

It can be seen how the cost of Case 3—Second-life cases are the highest except for the 40 kWh providing residential services, which is slightly less costly than the other alternatives. The lowest cost comes from the use case 2DR-90 kWh. However, the reader is reminded that the purchase of the battery is not considered in the total cost and, therefore, the comparison should be made for the same battery sizes.

To provide a clearer representation, the previous costs are grouped by battery capacity in Figure 6. Notice how for the 40 kWh battery (Figure 6a), the DR case provided by V2G is discarded due to technical feasibility issues. The first thing to highlight is that few use cases allow for a reduction in the cost compared to Case 1—No Grid Services. Especially, not a single Case 3—Second-life scenario appears to be economically beneficial. As mentioned in the introduction, the additional costs that are required for second-life applications hinder their profitability. In all services, Case 2—V2G entails a lesser cost than the same services provided through a second life. However, only some use cases prove to be profitable compared to simply using the battery for driving (Case 1—No Grid Services). These profitable cases are PS provided by the 40 kWh battery and DR with the 70 or 90 kWh ones.

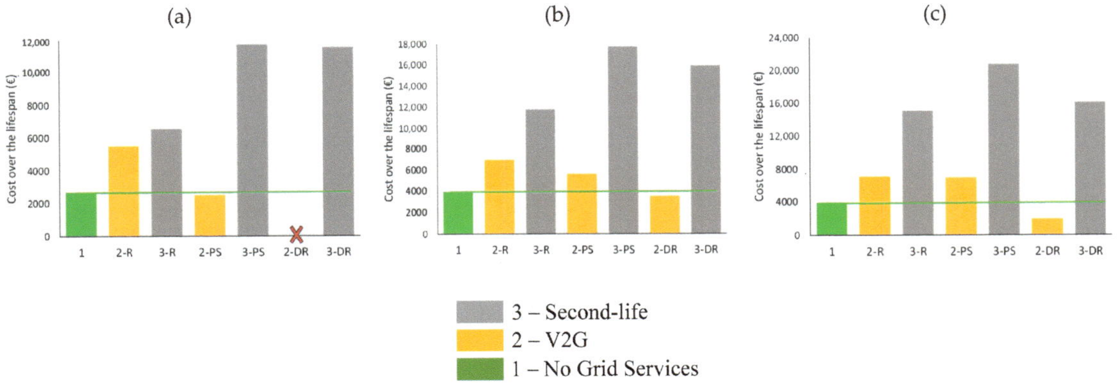

Figure 6. Cost over the battery lifetime for (**a**) 40 kWh batteries, (**b**) 70 kWh batteries, and (**c**) 90 kWh batteries.

Therefore, V2G appears to be more beneficial than second-life applications. Nevertheless, the previous results are absolute cost values and do not consider the energy provided to the grid as a service. For that reason, the specific cost (€/kWh of grid services provided) is represented in Figure 7. Even if the specific costs of the second life become closer to the ones of V2G, in all cases the latter still appears more beneficial.

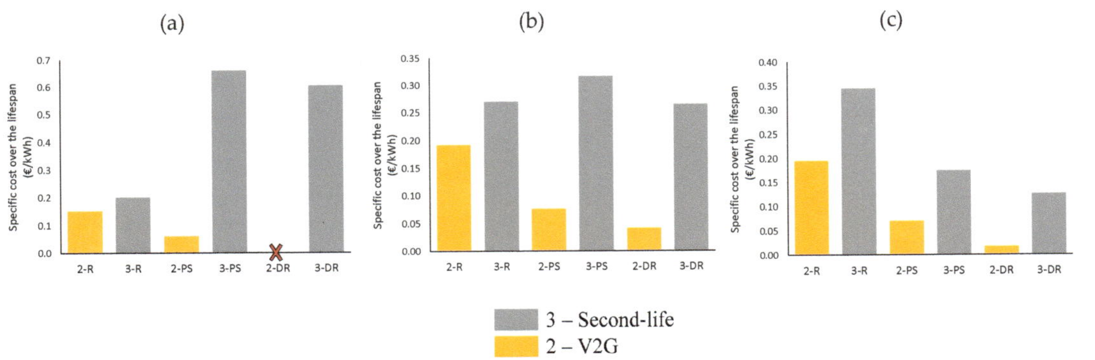

Figure 7. Specific cost over the battery lifetime, considering the provided kWh as grid services, for (**a**) 40 kWh batteries, (**b**) 70 kWh batteries, and (**c**) 90 kWh batteries.

3.2. Implications of the Fixed Threshold

In this study, the criterion employed assumes a functional EoL that extends the use of the battery until the vehicle EoL. However, as mentioned in the introduction, most of the second-life assessments and RUL algorithms in the literature rely on the fixed EoL threshold of 80% SoH. This section analyses the implications of using the fixed threshold instead of a functional one in terms of the battery's first-life and second-life evaluation.

Considering the fixed EoL threshold implies, except for the 16 kWh battery, assuming an early battery retirement from the EV. As shown in Figure 3b, 24 kWh and 30 kWh batteries are not able to reach the EV EoL, however, they can still provide the required range below the 80% SoH threshold. For these batteries, the EoL should be defined based on functional requirements such as not providing the required range, having limited power, or due to safety considerations. As shown in Table 5, if the EoL is calculated using the fixed threshold, which is the current state of the art for RUL algorithms, the battery is assumed to be retired between 0.5 and 7.1 years early.

Table 5. Functional EoL and assumed early retirement from the EV.

Battery Capacity (kWh)	Functional EoL SoH (%)	EoL Reason	Early Retirement (Years)
24	61.9%	Not able to cover range requirements	2.7
30	60.0%	Other functional limitations	1.3
40	67%	EV reaches EoL	7.1
70	71%	EV reaches EoL	4.9
90	78%	EV reaches EoL	0.5

Besides the implications in the estimated first-life duration, the fixed threshold also affects the evaluation of the second life. Since the EoL batteries are considered to have a higher SoH than the functional one, the estimated lifespan of the second life is higher, which increases the economic interest in battery reuse.

First of all, the low-capacity batteries (the 16, 24, and 30 kWh ones) are considered technically viable for a second life assuming the fixed EoL. In fact, for the applications defined, a second life for these batteries would be economical. However, the reality is that these low-capacity batteries reach the EoL with high levels of degradation and therefore cannot be used for a second life. Furthermore, if they were to be retired early from the EV, the replacement cost of the battery for driving would need to be considered, which would reduce the consumer's interest in selling the battery for reuse.

For batteries above 30 kWh, a second life is technically possible considering the functional EoL. Nevertheless, battery reuse is much less profitable than commonly assumed in the literature. Depending on the application and battery capacity, the lifespan of the second life considering the functional EoL can be up to 5.7 years shorter than considering the fixed one. This has an important impact on the cost assessment. In fact, assuming 80% SoH at the EoL implies an underestimation of the cost between 2.5 and 33% depending on the use case, without taking into account the additional replacement cost due to the early retirement.

4. Conclusions

This work has presented a study of degradation and lifetime costs for EV batteries under different usage strategies, with the goal of promoting the most profitable actions.

In accordance with other studies in the literature, the minimum battery capacity that covers the required range throughout the EV lifespan has been shown to be 40 kWh. Many of the current EV models on the market are way above this value, suggesting that batteries may remain underused. Considering guidelines on the sustainable material usage, additional cycling is required to maximize the value of the batteries during their lifetime. This may come both in the form of V2G during the first life or through a second life. Both strategies require additional resources to be put into practice, especially for second-life reuse. For this reason, this study has shown how V2G is always more profitable than battery reuse, even considering a direct reuse approach which is the least costly. Nevertheless, only in a few cases have the revenues provided by V2G allowed counteracting the additional costs. This promotes the idea that it is better to reduce the battery capacity as much as possible to avoid having to rely on additional services. This study has focused on analysing the circular-economy approaches in terms of cost. However, an environmental impact assessment is required to fully understand which strategy should be pursued.

This work has also highlighted an important issue regarding current assumptions in the literature. Almost every study dealing with EoL EV batteries assumes that they should be retired from the vehicle at 70–80% SoH. The use of this fixed threshold implies that batteries are considered to be retired earlier and in better conditions than in reality. In fact, this study has shown how different functional aspects cause the EoL, but for all battery capacities, the EoL is found to be below 80% SoH.

This study has provided a general guideline for the best practices for EV batteries. However, driver requirements and degradation trends change from case to case and an individual analysis should be performed to find the optimal pathway for each battery. Understanding each driver's requirements and driving conditions is key for accurately defining the EoL and for decision-making on whether V2G can be provided without compromising their needs.

Author Contributions: Conceptualization, L.C.C., B.A.G. and M.E.-S.; methodology, L.C.C., M.E.-S. and B.A.G.; software, M.E.-S. and L.C.C.; validation, L.C.C., M.E.-S. and B.A.G.; formal analysis, M.E.-S., L.C.C. and B.A.G.; investigation, M.E.-S. and L.C.C.; resources, M.E.-S. and L.C.C.; data curation, M.E.-S. and L.C.C.; writing—original draft preparation, M.E.-S.; writing—review and editing, M.E.-S. and L.C.C.; visualization, M.E.-S.; supervision, L.C.C. and C.C.; project administration, L.C.C. and C.C.; funding acquisition, L.C.C. and C.C. All authors have read and agreed to the published version of the manuscript.

Funding: This project has received funding from the European Union's Horizon 2020 research and innovation program under grant agreement No. 963580. This funding includes funds to support research work and open-access publications.

Data Availability Statement: Not applicable.

Acknowledgments: Lluc Canals Casals is a Serra Hunter Fellow.

Conflicts of Interest: The authors declare no conflict of interest.

References

1. Barre, A.; Suard, F.; Gerard, M.; Riu, D. A Real-Time Data-Driven Method for Battery Health Prognostics in Electric Vehicle Use. *PHM Soc. Eur. Conf.* **2014**, 2, 8.
2. Martinez-Laserna, E.; Gandiaga, I.; Sarasketa-Zabala, E.; Badeda, J.; Stroe, D.-I.; Swierczynski, M.; Goikoetxea, A. Battery Second Life: Hype, Hope or Reality? A Critical Review of the State of the Art. *Renew. Sustain. Energy Rev.* **2018**, *93*, 701–718. [CrossRef]
3. Hossain, E.; Murtaugh, D.; Mody, J.; Faruque, H.M.R.; Sunny, S.H.; Mohammad, N. A Comprehensive Review on Second-Life Batteries: Current State, Manufacturing Considerations, Applications, Impacts, Barriers & Potential Solutions, Business Strategies, and Policies. *IEEE Access* **2019**, *7*, 73215–73252. [CrossRef]
4. Casals, L.C.; García, B.A.; Cremades, L.V. Electric Vehicle Battery Reuse: Preparing for a Second Life. *JIEM* **2017**, *10*, 266. [CrossRef]
5. Martinez-Laserna, E.; Sarasketa-Zabala, E.; Sarria, I.V.; Stroe, D.-I.; Swierczynski, M.; Warnecke, A.; Timmermans, J.-M.; Goutam, S.; Omar, N.; Rodriguez, P. Technical Viability of Battery Second Life: A Study From the Ageing Perspective. *IEEE Trans. Ind. Appl.* **2018**, *54*, 2703–2713. [CrossRef]
6. Haram, M.H.S.M.; Lee, J.W.; Ramasamy, G.; Ngu, E.E.; Thiagarajah, S.P.; Lee, Y.H. Feasibility of Utilising Second Life EV Batteries: Applications, Lifespan, Economics, Environmental Impact, Assessment, and Challenges. *Alex. Eng. J.* **2021**, *60*, 4517–4536. [CrossRef]
7. Kamath, D.; Shukla, S.; Arsenault, R.; Kim, H.C.; Anctil, A. Evaluating the Cost and Carbon Footprint of Second-Life Electric Vehicle Batteries in Residential and Utility-Level Applications. *Waste Manag.* **2020**, *113*, 497–507. [CrossRef]
8. Cicconi, P.; Landi, D.; Morbidoni, A.; Germani, M. Feasibility Analysis of Second Life Applications for Li-Ion Cells Used in Electric Powertrain Using Environmental Indicators. In Proceedings of the 2012 IEEE International Energy Conference and Exhibition (ENERGYCON), Florence, Italy, 9–12 September 2012; pp. 985–990.
9. Kotak, Y.; Marchante Fernández, C.; Canals Casals, L.; Kotak, B.S.; Koch, D.; Geisbauer, C.; Trilla, L.; Gómez-Núñez, A.; Schweiger, H.-G. End of Electric Vehicle Batteries: Reuse vs. Recycle. *Energies* **2021**, *14*, 2217. [CrossRef]
10. Zhao, Y.; Pohl, O.; Bhatt, A.I.; Collis, G.E.; Mahon, P.J.; Rüther, T.; Hollenkamp, A.F. A Review on Battery Market Trends, Second-Life Reuse, and Recycling. *Sustain. Chem.* **2021**, *2*, 167–205. [CrossRef]
11. International Energy Agency. *Global EV Outlook 2020*; International Energy Agency: Paris, France, 2020.
12. Suárez-Eiroa, B.; Fernández, E.; Méndez-Martínez, G.; Soto-Oñate, D. Operational Principles of Circular Economy for Sustainable Development: Linking Theory and Practice. *J. Clean. Prod.* **2019**, *214*, 952–961. [CrossRef]
13. Etxandi-Santolaya, M.; Canals Casals, L.; Corchero, C. Estimation of Electric Vehicle Battery Capacity Requirements Based on Synthetic Cycles. *Transp. Res. Part D Transp. Environ.* **2023**, *114*, 103545. [CrossRef]
14. Saxena, S.; Le Floch, C.; MacDonald, J.; Moura, S. Quantifying EV Battery End-of-Life through Analysis of Travel Needs with Vehicle Powertrain Models. *J. Power Sources* **2015**, *282*, 265–276. [CrossRef]

15. Canals Casals, L.; Rodríguez, M.; Corchero, C.; Carrillo, R.E. Evaluation of the End-of-Life of Electric Vehicle Batteries According to the State-of-Health. *World Electr. Veh. J.* **2019**, *10*, 63. [CrossRef]
16. Tehrani, N.H.; Wang, P. Probabilistic Estimation of Plug-in Electric Vehicles Charging Load Profile. *Electr. Power Syst. Res.* **2015**, *124*, 133–143. [CrossRef]
17. Hao, X.; Wang, H.; Lin, Z.; Ouyang, M. Seasonal Effects on Electric Vehicle Energy Consumption and Driving Range: A Case Study on Personal, Taxi, and Ridesharing Vehicles. *J. Clean. Prod.* **2020**, *249*, 119403. [CrossRef]
18. Ministry of Transportation, UK Anonymised MOT Tests and Results. Available online: https://data.gov.uk/dataset/e3939ef8-30c7-4ca8-9c7c-ad9475cc9b2f/anonymised-mot-tests-and-results (accessed on 9 November 2020).
19. Casals, L.C.; Etxandi-Santolaya, M.; Bibiloni-Mulet, P.A.; Corchero, C.; Trilla, L. Electric Vehicle Battery Health Expected at End of Life in the Upcoming Years Based on UK Data. *Batteries* **2022**, *8*, 164. [CrossRef]
20. Geotab EV Battery Degradation Comparison Tool. Available online: https://storage.googleapis.com/geotab-sandbox/ev-battery-degradation/index.html (accessed on 1 September 2020).
21. Olmos, J.; Gandiaga, I.; Saez-de-Ibarra, A.; Larrea, X.; Nieva, T.; Aizpuru, I. Modelling the Cycling Degradation of Li-Ion Batteries: Chemistry Influenced Stress Factors. *J. Energy Storage* **2021**, *40*, 102765. [CrossRef]
22. Odyssee-Mure Average Energy Consumption per Dwelling. Available online: https://www.odyssee-mure.eu/publications/efficiency-by-sector/households/average-energy-consumption-dwelling.html (accessed on 2 January 2022).
23. Canals Casals, L.; Amante García, B.; Canal, C. Second Life Batteries Lifespan: Rest of Useful Life and Environmental Analysis. *J. Environ. Manag.* **2019**, *232*, 354–363. [CrossRef]
24. BloombergNEF. *Electric Vehicle Outlook 2021*; BloombergNEF: New York, NY, USA, 2021.
25. Cargadores de Coches Eléctricos | Wallbox España. Available online: https://wallbox.com/es_es (accessed on 2 May 2022).
26. Montes, T.; Etxandi-Santolaya, M.; Eichman, J.; Ferreira, V.J.; Trilla, L.; Corchero, C. Procedure for Assessing the Suitability of Battery Second Life Applications after EV First Life. *Batteries* **2022**, *8*, 122. [CrossRef]
27. Canals Casals, L.; Amante García, B.; González Benítez, M.M. A Cost Analysis of Electric Vehicle Batteries Second Life Businesses. In *Lecture Notes in Management and Industrial Engineering*; Ayuso Muñoz, J.L., Yagüe Blanco, J.L., Capuz-Rizo, S.F., Eds.; Springer International Publishing: Cham, Switzerland, 2014; pp. 129–141, ISBN 978-3-319-26457-8.
28. Rallo, H.; Canals Casals, L.; De La Torre, D.; Reinhardt, R.; Marchante, C.; Amante, B. Lithium-Ion Battery 2nd Life Used as a Stationary Energy Storage System: Ageing and Economic Analysis in Two Real Cases. *J. Clean. Prod.* **2020**, *272*, 122584. [CrossRef]
29. Thingvad, A.; Ziras, C.; Marinelli, M. Economic Value of Electric Vehicle Reserve Provision in the Nordic Countries under Driving Requirements and Charger Losses. *J. Energy Storage* **2019**, *21*, 826–834. [CrossRef]
30. Ghafoor, A.; Munir, A. Design and Economics Analysis of an Off-Grid PV System for Household Electrification. *Renew. Sustain. Energy Rev.* **2015**, *42*, 496–502. [CrossRef]
31. Mercados y Precios | ESIOS Electricidad Datos Transparencia. Available online: https://www.esios.ree.es/es/mercados-y-precios?date=02-03-2023# (accessed on 2 May 2022).
32. Calearo, L. *Grid Integration of Electric Vehicles: Battery Degradation and User Needs*; Technical University of Denmark: Lyngby, Denmark, 2022.
33. Canals Casals, L.; Barbero, M.; Corchero, C. Reused Second Life Batteries for Aggregated Demand Response Services. *J. Clean. Prod.* **2019**, *212*, 99–108. [CrossRef]
34. ESIOS Precio Medio Ponderado Banda de Regulación Secundaria. Available online: https://www.esios.ree.es/es/analisis/10388?vis=1&start_date=01-01-2021T00%3A00&end_date=31-12-2021T23%3A50&compare_start_date=01-12-2020T00%3A00&groupby=month (accessed on 4 March 2022).

World Electric Vehicle Journal

MDPI

Article

Battery Pack Prospects for Long-Haul Transport Trucks Considering Electrified Highways and Megawatt Charging

K. Darcovich [1,*], H. Ribberink [2], K. Qiu [2] and E. Soufflet [3]

1 National Research Council of Canada, 1200 Montreal Rd., Ottawa, ON K1A 0R6, Canada
2 Natural Resources Canada, 1 Haanel Drive, Ottawa, ON K1A 1M1, Canada
3 ICAM-Lille, 6 Rue Auber, 59800 Lille, France
* Correspondence: ken.darcovich@nrc-cnrc.gc.ca; Tel.: +1-613-993-6848

Abstract: The economics of long haul trucking on both continuous and intermittent Electric Highways were evaluated from a technical perspective as alternatives to using conventional megawatt chargers. The study revealed complex and sensitive interplay between the various technical factors related to the vehicle, its battery pack and the configuration and construction of the electrified highway. Key preliminary outcomes showed that a 250 kW highway power supply level allows a 36 tonne truck to drive continuously without requiring off-road recharging, and it can operate with a battery pack of about 50% of the size required for a truck only using megawatt charging. For now, while there is no overwhelming case in favour of any particular technology, the study serves to highlight the relevant factors impacting anticipated design criteria for the electrification of highways.

Keywords: heavy-duty vehicles; batteries; wireless charging infrastructure; simulation; economics

Citation: Darcovich, K.; Ribberink, H.; Qiu, K.; Soufflet, E. Battery Pack Prospects for Long-Haul Transport Trucks Considering Electrified Highways and Megawatt Charging. *World Electr. Veh. J.* **2023**, *14*, 60. https://doi.org/10.3390/wevj14030060

Academic Editor: Joeri Van Mierlo

Received: 10 February 2023
Revised: 22 February 2023
Accepted: 24 February 2023
Published: 28 February 2023

1. Introduction

There is growing worldwide societal commitment to the electrification of road transport which will require a broad overhaul of equipment and infrastructure. All classes of on-road vehicles consume upwards of 75% of the energy used in the transportation sector, with heavy duty trucks accounting for about one-third of this [1]. An emerging technology is that of electrified roadways whereon electric vehicles receive electrical power to power the vehicle as well as charge their batteries while driving. On electrified highways (e-Hwys), vehicles would require smaller batteries than otherwise, and still drive long distances. Vehicles seeking to adhere to a schedule would not require significant downtime for recharging. Similarly, heavy duty trucks for long-distance road freight, whose battery capacity would normally be a concern as well as a costly component, could potentially match today's diesel performance with far lower energy costs and without releasing greenhouse gas, soot and polluting gas components. Of course, the infrastructure costs for e-Hwys are high and would require a solid techno-economic case to bring about their implementation.

To now, there have been a number of published studies considering energy and infrastructure forecasting for long haul trucking electrification [2–4]. In general they suggest that e-Hwys are a viable alternative to megawatt charging systems on an economic level. Naturally, whatever modality or mix of modalities of electrification is eventually adopted, its implementation will have an all-encompassing impact in terms of equipment and compatibility, accessibility, ease of use and operational efficiency. A key component in all of these options will be vehicle battery packs. An assessment of battery pack requirements and projected durability in these scenarios will be a critical criterion informing choices ultimately to be made for future infrastructure development. At power provision levels similar to or above instantaneous vehicle needs, the battery state of charge (SOC) can be controlled to deviate only slightly from a mid-level set point, thereby simultaneously minimizing degradation rates.

The electrification of roadways is now a developing reality. Ultimately, should e-Hwys become widespread, it would require a level of full commitment to them from a range of sectors including highway departments, vehicle manufacturers, the electrical distribution grid, favorable policy and popular support. Test sections of e-Hwys are an essential first step. Notable initial efforts include a 1.6 km test section of e-Hwy on the Swedish island of Gotland [4], a 10 km catenary supplied section of the A5 autobahn in Germany [5], a test section of e-Hwy between Paris and Orléans in France is planned with the Vinci group working with government support [6], as well as portions of a route used by electric buses in Tel Aviv [7] also involving commercial company Electreon.

In a simulation study, the National Research Council of Canada and Natural Resources Canada jointly investigated the operation and economics of electric trucks using megawatt charging and of e-Hwys with DWPT, specifically including the cost of battery degradation in the economic evaluation. Additionally, the potential of intermittent electrification of the 540 km highway between Toronto and Montreal in Canada was evaluated as a means to reduce investment costs.

2. Study Scope and Methodology

2.1. Route

The route between Toronto and Montreal in Canada is one of the busiest highways in North America. A total of about 8000 trucks use it in either direction on a daily basis. Local depots of a shipping company in both cities were selected as the end points of the daily two-way duty cycle of 1081 km, of which 99% is on highways. Speed limits and elevations at 500 m intervals along the route were obtained from Google Earth [8], and outdoor conditions along this route were given by hourly temperature data published by Environment Canada [9] for the year 2020. On the highway, the truck speed was set as 100 km/h. Figure 1 is a map of the route driven, along with an accompanying elevation profile in the inset.

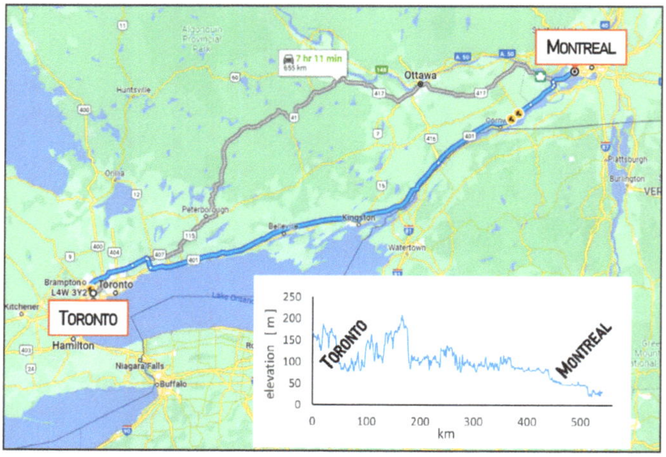

Figure 1. Map showing Toronto to Montreal highway driven, elevation profile shown in inset.

2.2. Scenarios

Three e-Hwy scenarios were evaluated against a reference case of a truck using only battery power and being recharged with a megawatt charger:

1. A 100% electrified highway (providing 200 kW at 85% uptake efficiency [10] to the truck)
2. A 50% electrified highway with alternating sections of 20 km that are electrified (at 400 kW) or non-electrified
3. A 50% electrified highway with alternating sections of 2 km that are electrified (at 400 kW) or non-electrified

2.3. Electric Truck Battery Sizing

The reference basis for modeling an e-truck was the proposed Tesla Semi [11]. The Tesla Semi is specified to have a 947 kWh battery pack composed of 4860 format cylindrical cells with an anticipated 800 km driving range. As of now, there are no characterization data available for them in the technical literature. For the intended Toronto to Montreal transit of 540 km, it was desired to have a battery electric truck (BET) with sufficient driving range to be able complete the trip over its lifetime (i.e., with decreasing battery capacity down to 60% of the original capacity) without requiring mid-trip recharging. For now, the battery pack that was used in the 2016 Tesla Model S [12] was used as a base unit for building a BET pack. In the simulated BET, cell banks of 3.07 Ah cells were configured in a 74-parallel and 84-series arrangement. For our simulations, packs were simulated in quantities ranging from 4 banks up to 16 banks (267 to 1068 kWh) to power the BET. Battery capacities used in this study are summarized in Table 1 below. The 84-series specification was required for floor level 240 V operation.

Table 1. Battery life for different scenarios and battery sizes. End of life (EOL) values in years.

Battery Size (kWh) [banks]	Megawatt Charging		P_{hwy} = 200 kW 100% Electrified Highway		P_{hwy} = 400 kW 50% Electrified (20 km on/off)		P_{hwy} = 400 kW 50% Electrified (2 km on/off)	
	equiv. cycles	EOL	equiv. cycles	EOL	equiv. cycles	EOL	equiv. cycles	EOL
267 [4]			3872	15.89	6059	6.24	6888	6.98
401 [6]			3089	18.86	7232	10.87	7667	11.22
534 [8]			2481	20.12	7022	13.97	7194	14.08
1068 [16]	4235	8.41						

3. Modeling Overview

A motivation for this project was to examine the concept of interrupted battery load demands through intermittent e-Hwy configurations to exploit temporal relaxation states that could enhance battery durability.

3.1. Primer on Battery Degradation

Use of a battery will gradually degrade its available energy capacity via low-rate irreversible chemical side reactions which proceed even when the battery is at rest. These reactions consume lithium ions, and fix them as resistance building deposits inside the electrode, thereby reducing the amount of charge that can be transferred during battery operation [13].

In general, factors which are understood to drive degradation are the level of applied current, the battery temperature and its instantaneous state of charge, and temporal effects related to its immediate history of charging, discharging or rest phases. Empirical expressions, such as Equation (1) developed in [13] can be applied in battery usage simulations over short time steps and can account for instantaneous degradation rates occurring under the state and operational conditions of the battery. Capacity loss or fade is normally expressed in A h, occurring at nano to micro-level quantities over a time step, and which accumulate to significant percentage levels over extended time periods.

$$\mathrm{cap}_{\mathrm{fade}} = \mathrm{cap}_{\mathrm{fade}}^{\mathrm{REF}}\,\Delta t \cdot F_{\mathrm{CUR}}[I(\Delta t)]\; F_{\mathrm{DOD}}[\mathrm{DOD}]\; F_{\mathrm{T}}[\mathrm{T,I}]\; F_{\mathrm{DUR}}[\Delta\mathrm{DOD}] \tag{1}$$

In Equation (1), $\mathrm{cap}_{\mathrm{fade}}^{\mathrm{REF}}$ is a reference degradation rate determined from experiments and normally given in A·h/s, Δt is the time step, I is the current, T is the temperature, DOD is the depth of discharge in the cell, ΔDOD is the extent of DOD swing in effect, and F_i are factors applied to modify the degradation rate according to how the various inputs differ from the reference conditions. More complete details are provided in [13,14].

Conventional experimental cell testing methods were used to provide the data for the analyses given here. The experimental setup consisted of 18,650 cells placed inside a temperature controlled chamber and connected to independent channels on an Arbin BT2000 series battery test system. Each channel is both a potentiostat and galvanostat which provided both voltage and current control for all charging and discharging requirements. The temperature was controlled to within ±0.1 °C of the desired set point with a thermal control unit described and illustrated in [15].

3.2. Mode Switching Effects

A mode switch can be defined here as an instantaneous change from an operational state of discharging, recharging or rest, to a different operational state. Details of the derivation of the duration effect on battery degradation are given in [13]. One additional situation relevant to the present study is discussed below.

Two basic cases are presented here to illustrate the tracking of the ΔDOD argument for duration factor ($F_{DUR}[\Delta DOD]$) on degradation for situations where an operation in progress experiences a brief rest or zero current period before resuming. The underlying concept is that an uninterrupted continuing charge or discharge operation can build up an excess lithium ion concentration in the electrodes which was shown to be a driver for degradation [14]. Punctuating such operations with rest periods allows this excess concentration to dissipate over time, such that the degradation driver associated with the just ceased battery operation is still present, although subsiding.

The first case (a) in Figure 2 shows the effective duration of a recharge operation in terms of the absolute value of the degree of discharge (ΔDOD), along with rest phase relaxation rates applied according to details given in [13]. On the horizontal axis, the step index is simply a time step count.

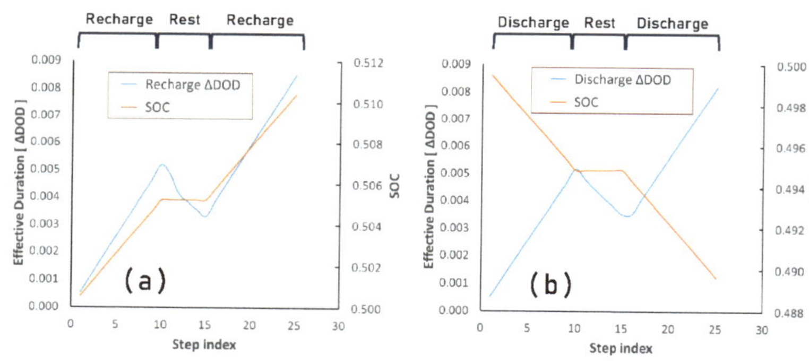

Figure 2. (**a**) Schematic of recharge-rest-recharge sequence. (**b**) Schematic of discharge-rest-discharge sequence.

The duration continues up until a rest period begins. The duration level attained at the start of the rest decays according to the time elapsed during the rest period. The steps along the horizontal axis represent 10 s intervals. Here, after a 50 s rest, the decay is only partial, and the lithium profiles in the anode which drive the degradation still show some excess concentration. Thus, when the recharge operation resumes, the duration effect built up previously is not entirely erased, but restarts from a reduced value arrived at via decay over the rest period. For longer rest periods the ΔDOD would decay down to zero and produce a temporal factor which reflects rest state degradation also known as calendar fade.

Figure 2b shows a parallel situation but for a discharge operation with a rest period followed by continuing discharge. In this case, the rest period which follows discharge drops the excess concentration driving degradation to a lower level, and it begins rebuilding at the same rate once the current is restored.

In both plots in Figure 2 the cell state of charge (SOC) is plotted along the same horizontal axis as the operation duration, its value is indicated on the right side vertical axis. The SOC curve is shown as an aid for following the cell operation over the steps; it drops during discharge, stays constant during rests and increases during recharge.

3.3. Battery Life Model

BET power loads were determined by applying driving conditions to an expression from [10] given below.

$$P_{\text{truck}} = \frac{(\mu f + \sin\alpha)Mg}{\eta_{\text{eq}}} + \frac{C_d A v(t)^3 \rho}{2\eta_{\text{eq}}} + \frac{\delta M a(t) v(t)}{\eta_{\text{eq}}} = \eta_{\text{eff}} \cdot IV \tag{2}$$

The terms in Equation (2) account for rolling resistance, elevation change, form drag and acceleration effects. The current demand for the battery can then be determined by the equivalence of P_{truck} to $\eta_{\text{eff}} \cdot IV$, where η_{eff} is the electric motor efficiency and I would be the current required from a battery at potential V. The power requirement P_{truck} is a duty load which reflects the driving and highway features, and was a common input for all the various simulation cases.

The simulation runs also employed some basic control logic, akin to the battery management system (BMS) found in an electric vehicle. In this case, the cell voltage range was restricted between 4.05 and 2.9 V. For the base reference case, the battery pack was fully charged to begin each driving trip. The power flow dynamics inside the BET are depicted in the schematic diagram, Figure 3, below.

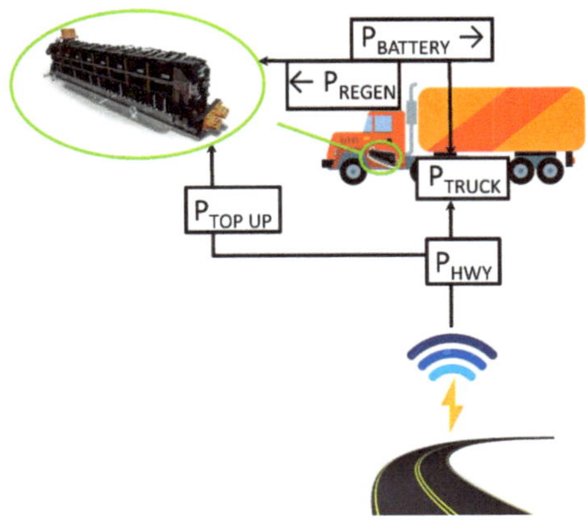

Figure 3. Power flow configuration in electric truck used in simulations. Inset in green oval is the truck battery pack.

The BET power demand P_{truck} is served by power provided by the e-Hwy P_{hwy} and supplemented by P_{battery} if required. If P_{hwy} was in excess of P_{truck}, the excess power available was provided to the battery pack if the DOD (Depth of Discharge) was greater than 0.50 and referred to as $P_{\text{top up}}$. During some descents, regenerative braking power could be produced, and was supplied to the battery pack, referred to as P_{regen}. The BET power model was then linked to an equivalent circuit model for battery operation [12]. Thermal states were tracked using an empirical transient model for the battery packs which accounted for ambient temperatures and battery function [16]. Battery electrochemical and thermal states then were applied to a capacity loss model to track cell State of Health (SOH) including intermittent effects [17]. Batteries were assumed to have reached their end of life when their SOH was reduced to 60%.

The tracking of the electrochemical state of the battery in simulation relies on an equivalent resistance type (Gao) model [18]. The model is expressed by Equation (3), which provides a means of updating the DOD given a specified battery state and load. The parameters α and β in the model pertain to the particular battery cell being simulated and are determined from data in discharge curves [17].

$$\frac{d\,\text{DOD}}{dt} = \frac{-I(t)}{\alpha\beta\,\text{CAP}_{\text{REF}}} \tag{3}$$

The resulting voltage during discharging can be determined by:

$$V(t, DOD, I(t)) = OCV(DOD) - I(t)\,R_{\text{int}}(DOD) \tag{4}$$

where $I(t)$ is the current (amperes) and R_{int} is the internal resistance (ohms). The Gao model is computationally efficient, and detailed one second time step multi-year EV simulations can be run in less than one hour.

Figure 4 gives a flow sheet depiction of the simulation model basis.

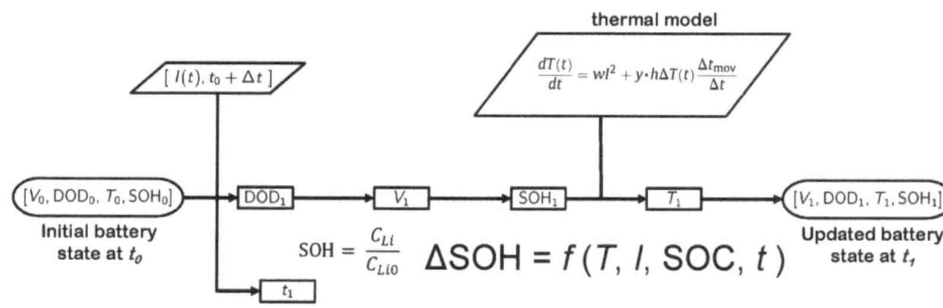

Figure 4. Simulation algorithm flow chart.

For EV lifetime determinations, tracking of the capacity loss is critical. This was done in the simulations on a one second time step basis, using Equation (1) which accounts for instantaneous effects contributing to lithium loss based on the type of cell used, the cell DOD value, the current level and temporal effects. The temporal effects are represented with a factor $F_{\text{DUR}}[\Delta\text{DOD}, t]$, which is determined according to the duration of the current mode of operation (discharge, recharge or rest) and residual effects from previous operational models.

This present study is the first to incorporate a model where operational duration effects and mode switching effects have been considered in capacity loss determinations [14]. Briefly, it was found that the capacity loss varied significantly as a function of cycle amplitude for tests conducted at the same current level and for the same length of time. Analysis in [13] was able to determine an instantaneous degradation driver for indeterminate operational lengths by finding values that applied over the interval produced the average value corresponding to the measured degradation from the variable amplitude experiments. Figure 5 shows these experimental results and the analytical advance. The instantaneous degradation factor is relevant for situations such as driving on a highway with elevation changes or intermittent DWPT power provision which will produce abrupt changes in current direction in the battery or allow rest periods for the battery. During rests electrolyte phase lithium levels which drive degradation and are established during steady operation will relax and decrease the degradation rates. A partition function for the relative contributions of the discharge and recharge phases of current flow was also determined in [13] and is used in these e-Hwy scenarios for BETs.

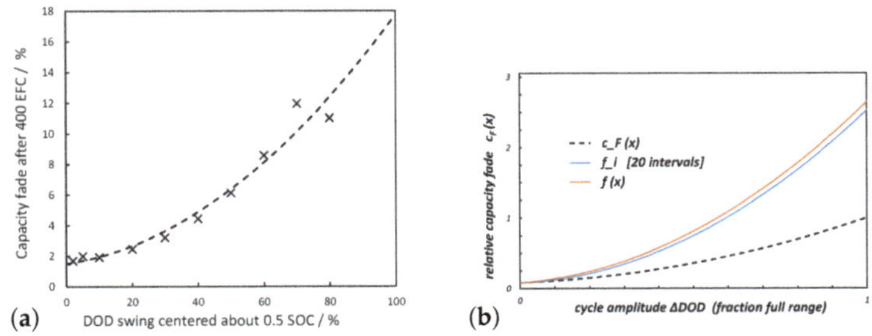

Figure 5. (**a**) Experimental capacity fade rates in Li-ion cells plotted as a function of cycle amplitude. (**b**) The same curve in dashed black for c (depth of discharge) intervals used as basis to determine relative instantaneous degradation factor $f(x)$ for battery operation across any ΔDOD range. Here the function f_i is a discretized form of $f(x)$ based on a 20-interval partition of the DOD range.

3.4. Economic Model

An economic model was developed to calculate the aggregated costs per kilometer driven for battery use, electricity consumption and capital costs for the charging infrastructure. It takes a break-even perspective on the operation of the charging infrastructure, and a wide range of possible investment cost values were evaluated for both megawatt charging and e-Hwys, as accurate cost numbers are presently uncertain for these technologies that are still under development. The economic model adapted a similar approach detailed by the authors in [19]. In our study, the capital costs included all the costs for the chargers and their installation or for the e-Hwy (coils, electronic cabinets, installation), though it did not consider the cost of grid upgrades. For the recharging equipment, an installed megawatt charger was priced at $1.5M. The e-Hwy infrastructure for the 534 km route was priced at $1.0M per km at 200 kW power and $1.9M per km at 400 kW power. All costs reported here are in US dollars.

4. Results

4.1. Daily Operation and Battery Life

Possible BET scenarios with DWPT are extensive and the cases considered here aim at first understanding the interplay among important input parameters in order to identify suitable operational conditions. Continuous and intermittent highway electrification were compared to investigate whether a modality of power provision analogous to pulsing could benefit battery performance and durability. The power flow logic employed here prioritized direct DWPT to the electric motor, and any additional available energy could recharge the battery when the state of charge was below 50%, while regenerative braking supplied recharge energy whenever it occurred. When BET power demands exceeded the 200 or 400 kW provision from the e-Hwy, the battery discharged the balance of the demand. With longer intermittency distances, the power flow occurred in longer and deeper cycling patterns. A reference case was simulated with a BET without DWPT and using post-trip 1 megawatt level recharging. For thermal control, the battery was not allowed to drop below −10 °C when operating. The energy required for this was not tracked, but based on [17], an estimate for heating energy would represent an overall additional 0.08%, considered negligible for this preliminary study.

4.2. Reference and Base e-Hwy Cases

To illustrate the what happens concretely over the course of one day in the simulations, two example cases of driving are shown in Figure 6, both simulated over 24 h on 1 January, reflected in the outdoor temperatures shown. Things to note from Figure 6 are that for the no-DWPT base case, the battery SOC goes from a near fully charged state down to about 0.35 over the 540 km trip and then undergoes a complete recharge. These operations cause significant heating in the battery as well. For the case with e-Hwy power provision, the battery pack SOC varied only in a 3% range and ended the day at essentially the same level as it began. Note that for e-Hwy cases, the initial DOD set was 0.50, chosen so that the DOD would rarely move more than 0.10 units away from its intial set point. Battery degradation rates as a function of DOD are at their miniumum near DOD = 0.50. Thus, with e-Hwy power, no recharge operations were required, which absolved the battery pack of exposure to sustained high currents and significant temperature rises. For the electrified highway case, the significantly lower amount of power transfer to and from the battery also results in far smaller temperature swings.

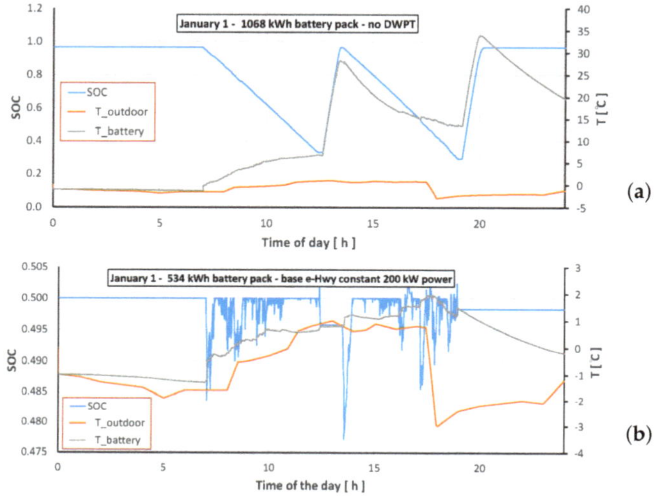

Figure 6. Plots tracking ambient temperature, battery pack temperature and battery SOC over a 24-hour period on January 1. Plot (**a**) shows output from a simulation with a 16-bank pack and no DWPT. Plot (**b**) show simulation output for a truck with an 8-bank pack, and full e-Hwy coverage with 200 kW DWPT power provision.

It can be seen in Table 1 that the roughly 8.5 year lifetime determined for the reference case of BETs using megawatt charging, moved to values greater than 15 years for BETs using electrified highways, even for the 4 bank pack, and increased to over 20 years for the 8 bank pack. This value is approaching the calendar life of the battery pack determined as 26.7 years under the same climatic conditions. In general, long-haul trucks are known to typically last 7–8 years (specifically, 900,000 km or up to 17 years with light use) in Canada [20], so the results here show that the battery packs in the BETs simulated on the e-Hwy will normally not need replacement during the vehicle life. In terms of energy efficiency, a slight benefit can be seen for larger pack sizes in Table 1. The table also reports equivalent cycles alongside the EOL values. An equivalent cycle is defined as the battery discharging the amount of energy considered to be the equivalent of the ΔDOD range of its rated capacity. The ratio of equivalent cycles to EOL value is an indication of how hard the battery is worked over its life. From this point of view, non-intermittent e-Hwys seem to put the least stress on the battery packs.

4.3. Intermittent e-Hwy Configurations

To make available the same amount of power over an equivalent distance, on/off intervals of equivalent segment length were simulated to assess the effects of intermittency in electrified highways. Initially, alternating powered and unpowered segments of 20 and 2 km were investigated.

The effect of the 200 kW and 400 kW e-Hwy power provision levels are shown in Figure 7, with battery pack capacity and 2 km and 20 km intermittency intervals as additional parameters. It can be seen that increased e-Hwy power benefits battery lifetime for large packs, and when comparing packs of equivalent capacity, smaller intermittent intervals provide slight lifetime gains. In general larger packs perform more efficiently, and the 2 km intervals showed slightly better energy efficiency compared to the 20 km intervals. At 200 kW e-Hwy power, significant energy from mid-trip recharging was required, notably with smaller battery packs, thus reducing BET lifetimes. While intermittent electrification has lower battery degradation on a per cycle basis, the truck battery was used much more than in the case of full electrification, leading to a shorter life (Table 1).

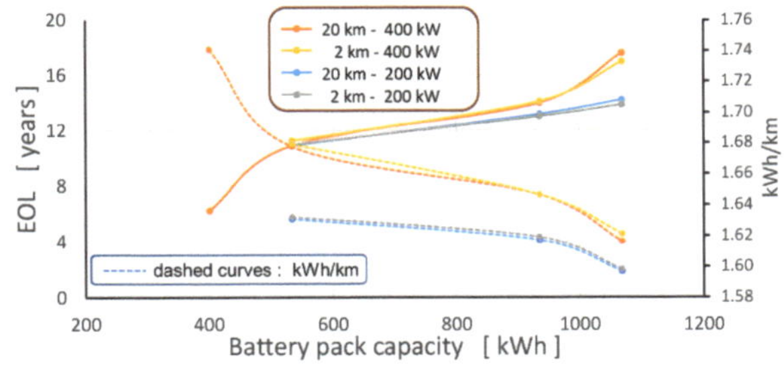

Figure 7. Battery lifetimes and energy use per km as a function of BET battery pack size on 2 km and 20 km interval intermittent e-Hwys at 200 kW and 400 kW power provision levels.

Much finer intervals of e-Hwy intermittency were investigated to see if such configurations could provide energy use benefits over the BET lifetime. Results in Table 1 already show slight benefits in battery lifetimes for 2 km on/off intervals compared to 20 km intervals, along with larger incremental gains as battery pack sizes increase. Regularly spaced on/off intervals shorter than 2 km were simulated; 1000 m, 500 m, 250 m and 100 m. A 100 km/h highway speed is 27.78 m/s, requiring less than four seconds to cover 100 m. In order to have adequate spatial resolution for on/off e-Hwy switching over 100 m intervals, a simulation time step of 0.5 s was required. For these cases, an e-Hwy with 400 kW power provision was simulated with e-trucks with 4 and 8 bank battery packs, traveling the Toronto to Montreal route with intermittent DWPT.

In Figure 8 a small lifetime benefit for 4-bank packs at shorter on/off electrified intervals can be observed. For an 8 bank pack, there is an essentially negligible benefit to very short intermittent intervals reflected by the near zero slope of the data. These output data confirm that larger battery packs have both longer life and more efficient performance. Based on the theoretical advances alluded to in Figure 5, it is understood that fast mode switching, and/or continuous operations that experience rest intervals (akin to pulse charging) will expose the electrodes to smaller build-ups of excess lithium, known to be a significant degradation driver.

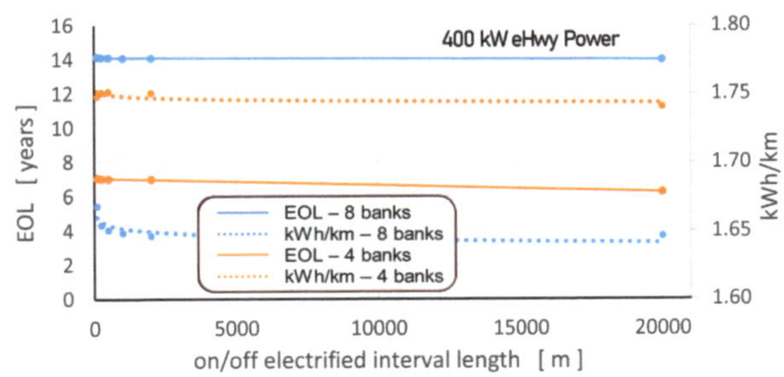

Figure 8. Battery lifetimes and energy use per km as a function of intermittency interval length for 4 and 8 bank BET batteries.

Also shown in Figure 8 for the cases considered are curves with kWh/km values. The energy efficiency is very sensitive to many factors, and despite anticipated benefits of intermittency, the data show negligible benefits. The efficiency calculations show better performance with longer intervals which could possibly be a minor thermal effect given the lack of sophisticated thermal management modeled here. Over longer on/off intervals the battery will run slightly hotter, contributing to slight efficiency gains in generally cold operating environments.

4.4. Effect of Highway Electrification Power Level

Some simulations were run to determine what level of e-Hwy power provision would be sufficient to no longer require off-highway recharging. This question arose in response to data reported in Table 1 where it can be seen that intermittent cases at P_{hwy} levels of 200 kW show low EOL values since extensive mid-trip recharging was required in these cases to complete the driving duty. In these present cases (Figure 9), for a long haul truck of 36 tonnes, around 250 kW is a level that allows the truck with an eight bank battery to do whatever off-e-Hwy driving may be required in a day and maintain some charge in the battery, as well as be able to reach a desired on-highway operational SOC set point.

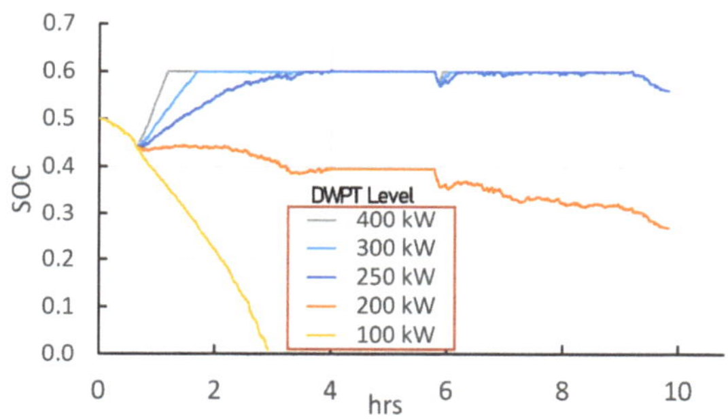

Figure 9. SOC level in truck with eight bank battery driving a distance of just over 750 km on an e-Hwy at a range of power provision levels.

A few observations from Figure 9 merit some mention. For P_{hwy} levels of 250 kW or higher, it can be seen that the e-Hwy power provision is more than enough to power the truck and that the SOC can reach a control set point. It simply takes longer at 250 kW compared to higher levels to regain the set point. At a P_{hwy} level of 200 kW, the truck is able to complete the day's driving but does so by depleting battery capacity. Such practice would not be sustainable day after day without off-highway recharging. Finally, for a P_{hwy} level of 100 kW, the truck power demand is constantly more than what the highway supplies and the SOC drops steadily to depletion well before the 750 km trip is completed. Highway power levels much beyond what the truck would require in general would require extra levels of control to avoid large recharge currents which would contribute to more rapid battery degradation.

4.5. Highway Power Levels Allowing Intermittency

A series of simulations were run with high e-Hwy power levels to see if this could allow small battery pack sizes to complete driving duties on an intermittent e-Hwy with 20 km intervals of electrified and unpowered roadway. Figure 10 plots EOL values versus batter pack capacity for three levels of e-Hwy power provision. Larger battery sizes lead to larger EOL values through mitigation of high current values, whereas elevated P_{hwy} levels shorten battery lifetimes through exposure to excessive current rates. These observations echo the findings discussed for Figure 9.

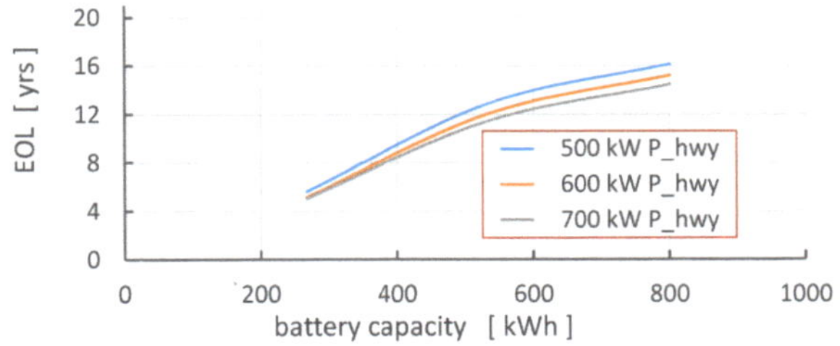

Figure 10. Battery lifetimes for 20 km e-Hwy intermittency at power provision levels ranging from 500 to 700 kW.

4.6. Discussion of the Operational Duration Effect

To better understand the limited benefits of small intermittent intervals shown in Figure 8, the ΔDOD parameter used to determine the $f(x)$ term as raw simulation output can be seen in Figure 11 which shows the ΔDOD levels determined for 20 km, 2 km and 0.10 km e-Hwy intermittency over the Toronto to Montreal drive, as well as the main reference case of the BET with a 16 bank battery and no DWPT provision. The reference case shows about a 50% higher ΔDOD value for most of the trip, and it only becomes much higher for about the last 100 km of the trip where the route elevation features do not allow the ΔDOD value to relax like the intermittent cases. Thus the low duration driver values for degradation clearly contribute to intermittent cases with pack banks of 6 units or more having markedly longer lifetimes than the reference case.

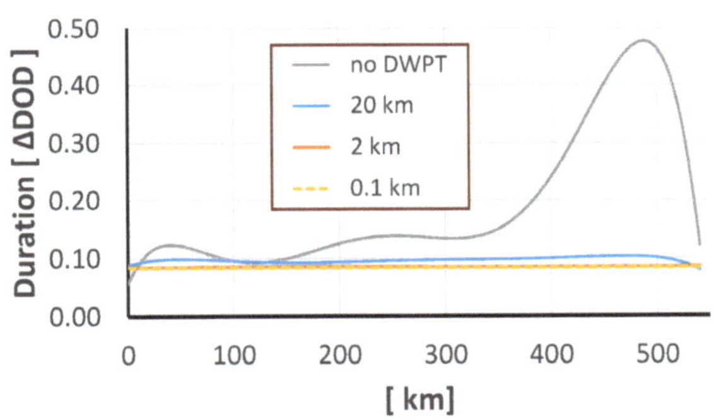

Figure 11. Smoothed ΔDOD curves for the reference no DWPT case and 20 km, 2 km and 0.10 km e-Hwy on/off intervals.

The irregularity of the power demand produces ΔDOD values that average around 0.10 for the 20 km intervals. The increased irregularity obtained from reducing the on/off interval to 2 km drops the average ΔDOD levels by about 15%. Given that this is a small decrease that manifests mainly during brief operational durations to begin with, the benefits in terms of extending battery life are small. When going from 2 km to 0.10 km intermittency, the ΔDOD levels decrease by only about 2 percent. The main takeaway here is that micro-intermittency below 2 km provides negligible battery durability benefit for the scenarios examined. In hindsight, the results are consistent with the reference case depicted in Figure 6a. In that case, the entire 540 km trip was completed, depleting the SOC of an initially fully charged 16 bank battery by about 62%. Thus, even 20 km segments, proportionally would represent an SOC change (or ΔDOD) of around 2.3%, which would thus produce small ΔDOD-based degradation drivers. Relative to this, finer intermittency would simply be a slight incremental reduction of an already small quantity.

4.7. Economic Evaluation

The economics of the electric highways variants and of the reference case of megawatt charging were evaluated by comparing their total cost per km for the infrastructure, the electricity consumed, and the cost of battery degradation. The per km costs for the reference case of megawatt charging are displayed in Figure 12a. The costs strongly depend on the currently unknown costs to purchase and install megawatt chargers and on their utilization rate. There is also a wide range in cost forecasts for electrifying highways. Additionally, the cost per km will depend on the number of trucks using the e-Hwy. The resulting costs per km for a fully electrified highway are given Figure 12b.

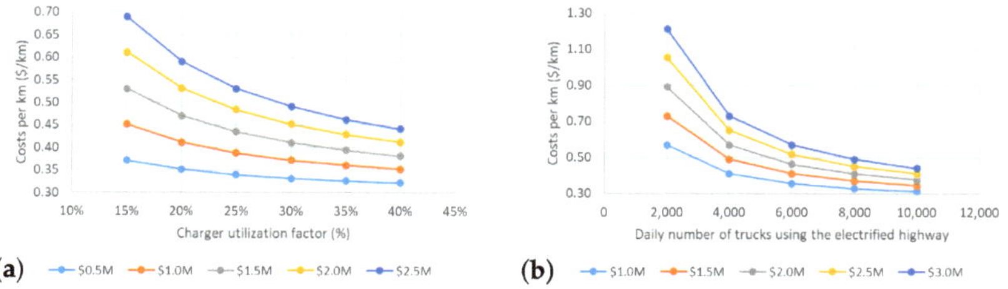

Figure 12. Costs per kilometer for the reference case of fast charging at a one megawatt power level for scenarios of different charger cost and charger utilization with a 16-bank truck battery (**a**) and for the 100% electrified highway for different levels of investment costs per km and highway utilization with a 4-bank truck battery (**b**).

The infrastructure costs were determined as an annual per truck cost, based on 8000 truck trips occurring on the route per day. The capital cost was spread over 20 years with 6% annual interest applied. Per km costs for the various trucking scenarios include contributions for capital, electricity and battery costs.

Intermittent electrification scenarios reduce investment costs through savings on the costs of the in-road part (for the 20 km and 2 km case) and on the costs for electricity distribution along the highway (only for the 20 km case). Savings for the 2 km case were therefore assumed to be half of those for the 20 km case. However, for most of the cases evaluated, the savings in investments were cancelled out by higher battery use costs and greater electricity losses (see Figure 13a, Table 2). Despite the higher investment costs, simulation results indicated a benefit of installing a larger battery in the truck than strictly necessary for intermittently electrified highways (Figure 13b).

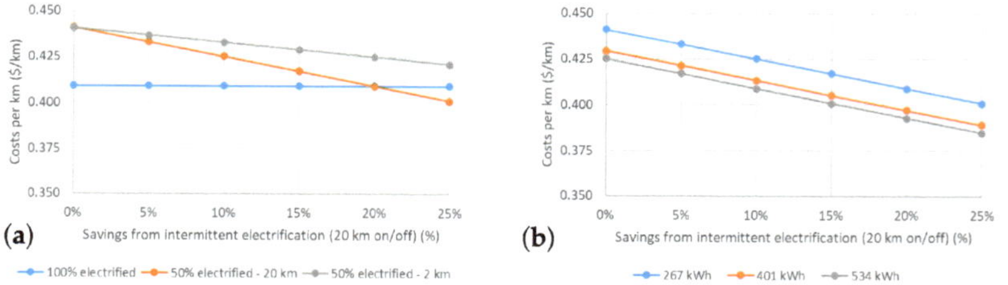

Figure 13. (**a**) Costs per kilometer driven for different electrification scenarios using a 267 kWh battery. (**b**) Costs for trucks with different battery sizes for intermittent electrification (20 km on/off), showing different levels of cost reduction from intermittent electrification with investment costs of \$2.0 million/km and highway usage at 8,000 trucks per day.

Table 2 indicates that per km costs are similar for high usage scenarios for the middle of the current range of investment costs. More accurate investment cost data will be necessary to understand which technology will be more cost effective for the transportation sector.

Table 2. Per km costs for different cost components.

	Megwatt Charging	100% Electrified Highway		50% Electrified (20 km on/off)		50% Electrified (2 km on/off)	
Battery size (kWh)	1,068	267	534	267	534	267	534
banks	16	4	8	4	8	4	8
Battery use	$0.048	$0.006	$0.010	$0.016	$0.015	$0.015	$0.014
Capital costs	$0.102	$0.161	$0.161	$0.136	$0.136	$0.148	$0.148
Electricity	$0.243	$0.242	$0.241	$0.264	$0.250	$0.266	$0.250
Total	$0.394	$0.409	$0.412	$0.417	$0.401	$0.429	$0.413

5. Conclusions

This study has provided the rationale and context to incorporate temporal and current mode degradation drivers into BET usage simulations. Fundamental electrochemical derivations of these novel operational duration based drivers were presented in this paper. General preliminary findings are that electrified highways significantly increase BET battery life, but total energy consumption is similar to one megawatt recharging of BETs driving on highways without DWPT. However, electrified highways allow battery packs less than 50% of the size required with BETs using one megawatt recharging. Important parameters for transport trucks on electrified highways were identified as battery size, e-Hwy power provision level and route elevation features.

This study also evaluated the operation and economics of continuous and intermittent Electric Highways as alternatives to using megawatt chargers. Because all of these technologies are still under development, there is a large uncertainty regarding their investment costs. For all technologies, a high utilization rate is necessary to obtain low costs per km, and mid-range investment costs currently lead to similar costs per km for all technologies.

Intermittent electrification was shown to increase the number of equivalent cycles within the life of the battery, but requires more intense battery usage, decreasing the actual battery life in number of calendar years. Equipping trucks with larger batteries than necessary reduced cost per km due to longer battery life. For a 36 tonne transport truck, an e-Hwy power provision level of about 250 kW was identified as being suitable for sustainable long-haul driving without requiring off-highway recharging. e-Hwy power levels above this 250 kW are not of much additional benefit, as they only serve to shorten the times required to regain an on-road operational SOC set point, but at higher degradation levels from the higher currents they produce.

Going forward from this present study, subjects that will require further investigation will be ways to better exploit intermittent electrification of highways including electrified interval lengths and their configurations. Such further studies will need to be closely tied to economic evaluations of the entire system including installed equipment and the operation of BETs in the e-Hwy environment.

Author Contributions: K.D.: code development, simulation strategy, data analysis, manuscript preparation. H.R.: project commissioning, simulation strategy, technical oversight, techno-economic analysis. K.Q.: techno-economic analysis. E.S.: running simulations, organizing and presenting data. All authors have read and agreed to the published version of the manuscript.

Funding: Funding for this work was provided by the National Research Council of Canada and by Natural Resources Canada through the Program of Energy Research and Development (PERD).

Data Availability Statement: Not applicable.

Conflicts of Interest: The authors declare no conflict of interest.

Abbreviations

The following abbreviations are used in this manuscript:

BET	battery electric truck
BMS	battery management system
CPT	conductive power transfer
DOD	depth of discharge
DWPT	dynamic wireless power transfer
e-Hwy	electrified roadway furnishing DWPT
EOL	battery EOL life value, in years
EV	electric vehicle
OCV	open-circuit voltage
P2D	pseudo two-dimensional
SEI	solid-electrolyte interface
SOC	state of charge
SOH	state of health

References

1. Osieczko, K.; Zimon, D.; Płaczek, E.; Prokopiuk, I. Factors that influence the expansion of electric delivery vehicles and trucks in EU countries. *J. Environ. Manag.* **2021**, *296*, 113177. [CrossRef] [PubMed]
2. Qiu, K.; Ribberink, H.; Entchev, E. Economic feasibility of electrified highways for heavy-duty electric trucks. *Appl. Energy* **2022**, *326*, 119935. [CrossRef]
3. Triviño, A.; González-González, J.M.; Aguado, J.A. Wireless power transfer technologies applied to electric vehicles: A review. *Energies* **2021**, *14*, 1547. [CrossRef]
4. Gustavsson, M.G.H. Research & Innovation for Electric Roads. In Proceedings of the EVS35 Symposium, Oslo, Norway, 11–15 June 2022.
5. Dhayarkar, S.; Sontakke, S.; Shinde, V.; Ghuge, N. Electric Heavy Vehicle Transportation Highway (By Catenary System). *ITM Web Conf.* **2022**, *44*, 01004. [CrossRef]
6. Gouvernement de France, Ministère de la Transition énérgetique, Système de Route électrique, Groupe de Travail n°1. In *Décarboner le Transport Routier de Marchandise par l'ERS, Enjeux et Stratégie*; Trafigura: Singapore, 2021; 196p.
7. Green Car Reports. Edelstein, S. Short-Loop Electric Bus Serves as a Proving Ground for Dynamic Wireless Charging. Available online: https://www.greencarreports.com/news/1131665_short-loop-electric-bus-serves-as-a-proving-ground-for-dynamic-wireless-charging (accessed on 12 January 2023).
8. Google Earth. Available online: earth.google.com (accessed on 7 October 2021).
9. Environment Canada, Historical Weather Data. Available online: climate.weather.gc.ca/historical_data/search_historic_data_e.html (accessed on 6 October 2021).
10. Foote, A.; Onar, O.C.; Debnath, S.; Chinthavali, M.; Ozpineci, B.; Smith, D.E. Optimal sizing of a dynamic wireless power transfer system for highway applications. In Proceedings of the 2018 IEEE Transportation Electrification Conference and Expo, (ITEC), Long Beach, CA, USA, 13–15 June 2018; pp. 1–6.
11. Tesla. Available online: www.tesla.com/enCA/semi (accessed on 2 October 2021).
12. Darcovich, K.; Ribberink, H.; Michelet, C.; Lombard, K.; Ghorab, M. The Feasibility of Electric Vehicles as Taxis in a Canadian Context. In Proceedings of the 2019 Electric Vehicles International Conference (EV), Bucharest, Romania, 3–4 October 2019; pp. 1–6.
13. Darcovich, K.; Recoskie, S.; MacNeil, D.D.; Darcovich, A. Operational intra-cycle temporal and current mode effects on battery capacity loss. *eTransportation* **2022**, *13*, 100185. [CrossRef]
14. Darcovich, K.; Recoskie, S.; Fattal, F. Fast operational mode switching effects on battery degradation. *J. Appl. Electrochem.* **2020**, *50*, 111–124. [CrossRef]
15. Darcovich, K.; MacNeil, D.D.; Recoskie, S.; Cadic, Q.; Ilinca. F. Comparison of cooling plate configurations for automotive battery pack thermal management. *Appl. Therm. Eng.* **2019**, *155*, 185–195. [CrossRef]
16. Ribberink, H.; Lombardi, K.; Humphries, K.; Loiselle-Lapointe, A.; Stefopulos, N.; Varghese, S.; Pundsack, N. Impact of Ambient Temperature and Battery Activity on Internal Battery Temperatures of Electric Vehicles. In Proceedings of the 32nd International Electric Vehicle Symposium and Exhibition (EVS32), Lyon, France, 19–22 May 2019.
17. Darcovich, K.; Recoskie, S.; Michelet, C.; Ribberink, H. The Impact of V2X Service under Local Climatic Conditions within Canada on EV Durability. *eTransportation* **2021**, *9*, 100124. [CrossRef]
18. Gao, L.; Liu, S.; Dougal, R. Dynamic Lithium-Ion Battery Model for System Simulation. *IEEE Trans. Components Packag. Technol.* **2002**, *25*, 495–505.

19. Qiu, K.; Ribberink, H.; Entchev, E. Technical and Economic Feasibility of Electrified Highways for Heavy-Duty Electric Trucks. In Proceedings of the Applied Energy Symposium: MIT A+B, Cambridge, MA, USA, 11–13 August 2021; Paper ID APEN-MIT-2021_122, 5p.
20. Talebian, H.; Herrera, O.E.; Tran, M.; Mérida, W. Electrification of road freight transport: Policy implications in British Columbia. *Energy Policy* **2018**, *115*, 109–118. [CrossRef]

World Electric Vehicle Journal

MDPI

Article

Fanbois and Fanbots: Tesla's Entrepreneurial Narratives and Corporate Computational Propaganda on Social Media

David A. Kirsch [1,*] and Mohsen A. Chowdhury [2]

[1] Robert H. Smith School of Business, 4556 Van Munching Hall University of Maryland, College Park, MD 20742, USA
[2] Olin Business School, Washington University in St. Louis, St. Louis, MO 63120, USA
* Correspondence: dkirsch@umd.edu

Abstract: This paper reports the discovery of a series of computational social media accounts (Fanbots) on Twitter that may have played a critical role in sustaining the entrepreneurial narratives of Tesla, the electric-vehicle maker. From 2010 to 2020—a period of trial, error, and eventual success for Tesla—these computational agents generated pro-firm tweets (Corporate Computational Propaganda, CCP), accounting for more than 10% of the total Twitter activity that included the cashtag, $TSLA, and 23% of activity that included the hashtag, #TSLA. Though similar to programmed social media content in the political sphere, the activities of these accounts predate the existence of political computational propaganda associated with foreign support for, for instance, Brexit in the United Kingdom (2016) and Donald Trump in the United States (2016). The paper (a) characterizes the extent of Fanbot content in two large Tesla tweet corpora, (b) identifies possible motivations for the creation of these accounts in relation to the firm's entrepreneurial narratives, and (c) explores possible mechanisms by which the Fanbots might have acted. Although we are unable to directly observe the source or stated purpose of these accounts, based upon the timing of Fanbot creation and other indirect indicators, we infer that these accounts and the social media activity they generated were intended to influence social perception of Tesla. The conclusion assesses the generalizability of a Fanbot-based strategy, highlighting contextual limitations, while also pointing to ways that firms may already be using CCP to manage social approval in emerging-industry contexts.

Keywords: case study; entrepreneurial narratives; impression management; computational propaganda; industry emergence

Citation: Kirsch, D.A.; Chowdhury, M.A. Fanbois and Fanbots: Tesla's Entrepreneurial Narratives and Corporate Computational Propaganda on Social Media. *World Electr. Veh. J.* **2023**, *14*, 43. https://doi.org/10.3390/wevj14020043

Academic Editor: Joeri Van Mierlo

Received: 16 November 2022
Revised: 28 January 2023
Accepted: 31 January 2023
Published: 5 February 2023

1. Introduction

For decades, the electric vehicle has been the "car of tomorrow" but never quite the car of today [1,2]. Notwithstanding annual predictions about any given year finally being the year of the electric car, recent developments suggest that the automobile industry may truly be on the cusp of electrification. Nearly every automaker from General Motors to Rolls-Royce has announced plans to build and market one or more electric automobiles, and numerous startups have entered the electric-vehicle market. In this sense, electric-vehicle manufacturer Tesla may have accomplished its initial mission statement, "to accelerate the world's transition to sustainable transport," and in the process, making CEO Elon Musk one of the wealthiest people alive.

Among the many possible reasons advanced to explain Tesla's success, this paper identifies a potentially novel contributing factor: the creation of algorithmic Twitter accounts that generated pro-Tesla content, totaling more than one tenth of the volume of (cashtag) tweets about the firm. By supporting the entrepreneurial narratives advanced by the firm and its CEO, this computational content may have influenced public perception of Tesla from the time of the firm's IPO in June 2010 to the end of 2020. To the best of our knowledge, the existence of this type of computational content has not been previously identified or reported upon.

The paper first introduces the concept of entrepreneurial narratives and their centrality to stakeholders in new ventures. Tesla's many, expansive entrepreneurial narratives are then introduced. Support for these narratives is found in both traditional media and on Twitter, the CEO's preferred social media platform. The crux of the paper arises from our discovery of a set of algorithmic Twitter accounts (Fanbots) created in apparent support of Tesla. The existence of these accounts challenged us to (a) characterize the extent of Fanbot content in two large Tesla tweet corpora, (b) identify possible motivations for the creation of these accounts as they affect the firm's entrepreneurial narratives, and (c), if we stipulate that the Fanbots were created to advance the firm's preferred entrepreneurial narratives and suppress nascent counternarratives, explore possible mechanisms by which the Fanbots might have acted.

To characterize the pro-Tesla conversation on Twitter and the role of Fanbots therein, we first collected the corpora of tweets containing the cashtag $TSLA or hashtag #TSLA using Twitter's Academic API. Next, we used an off-the-shelf tool for the detection of bots (Botometer) to identify likely Fanbots among the most active accounts in these two datasets. We find that a substantial share of active users, who are in turn responsible for a significant fraction of the tweets, are likely Fanbots. We suggest the Fanbots may operate by shaping the engagement between and among Fanbois, a term used to describe the enthusiastic followers of the firm and its CEO.

The conclusion discusses the consequences of our findings for Tesla and proposes additional research that might be undertaken to deepen and generalize an understanding of the relationship between corporate computational propaganda and the social perception of entrepreneurial narratives in emerging-industry contexts such as the global electric vehicle industry.

2. Theoretical Background

Entrepreneurial narratives matter for early-stage ventures [3]. Narratives are especially important when there is a large gap between the present state of the world and the future the firm promises to create [4]. Consequently, protecting firm-generated narratives from criticism and potentially damaging counternarratives emerges as an important function of the firm and its managers. Proceeding from the concept of entrepreneurial narrative and its importance to the way stakeholders—including investors, customers, suppliers, employees, and policymakers—interact with an entrepreneurial firm, we highlight one case where narratives were numerous, expansive, and central to the company's strategic objectives. We also recognize the importance of emergent critiques and counternarratives that may undermine a firm's preferred entrepreneurial narratives [5].

Impression management—via strategic public communication resulting in traditional media mentions—has long been recognized as an important managerial responsibility [6]. Increasingly, firms and their leaders use social media to convey information and influence and align stakeholders with the corporate mission through online "evangelization" [7]. Social media users can bring attention to certain topics through a process described as "visibilization" [8]. While traditional media were long thought to be the preferred means of influencing stakeholder perceptions and sentiments in bulk [9], social media allows a firm or its CEO to directly engage with stakeholders, including supporters of the firm.

Devoted supporters of a firm—who may be a subset of social media users—are particularly important targets for impression management. These "fanbois"—a term originally coined to describe enthusiastic followers of the Apple founder and technology evangelist Steve Jobs—are defined by the website Techopedia as people "who (are) unusually attracted or devoted to a particular technology or tech company" [10]. Social media platforms allow fanbois to be more than "followers" by demonstrating their support for a firm in a shared virtual space, while affiliating with other like-minded users.

Algorithms have long been known to influence users of social media platforms. Existing research has focused on the ways that platform algorithms—like those that select which posts are visible to certain audiences—distort discourse through information overload and opacity [11]. By curating the data available to users, these algorithms can alter public perception, while concealing the source and agency of the content [12]. Algorithms may operate independently of the platform, and such independent computational agents acting in support of the firm are identified as "fanbots." The firm-specific content they generate represents a new type of communication labeled "corporate computational propaganda" (CCP). This content may align collective beliefs among like-minded users about a firm's entrepreneurial narratives.

Taken together, these theoretical domains—entrepreneurial narrative, impression management, and relationships between firm supporters and social media algorithms—frame the context for our research: the creation and use of fanbot-generated content on Twitter to manage and support Tesla's expansive entrepreneurial narratives.

3. Entrepreneurial Narratives and Counternarratives

Observers who have followed the trajectory of electric vehicle maker, Tesla, have documented multiple, overlapping entrepreneurial narratives that supporters of the firm, including CEO Elon Musk, have deployed to strengthen conviction in the firm's ultimate success [13]. Without evaluating the specific claims underlying each, we view these narratives as appealing to the preferences and beliefs of different subsets of stakeholders. Given that Tesla was a modest-sized startup (in 2010) that promised to disrupt and transform the auto sector, the firm needed to deploy an expansive set of entrepreneurial narratives that were capable of connecting the present and the future. Challenging each of these narratives was a nascent counternarrative. Though thought to be less visible and salient, counternarratives pose a unique risk to the entrepreneurial firm, potentially jeopardizing its ability to secure the resources required to fulfill its mission. Therefore, protecting entrepreneurial narratives from displacement by emerging counternarratives was an important function of entrepreneurs and management teams. We explored a subset of Tesla's entrepreneurial narratives and nascent counternarratives (see Figure 1).

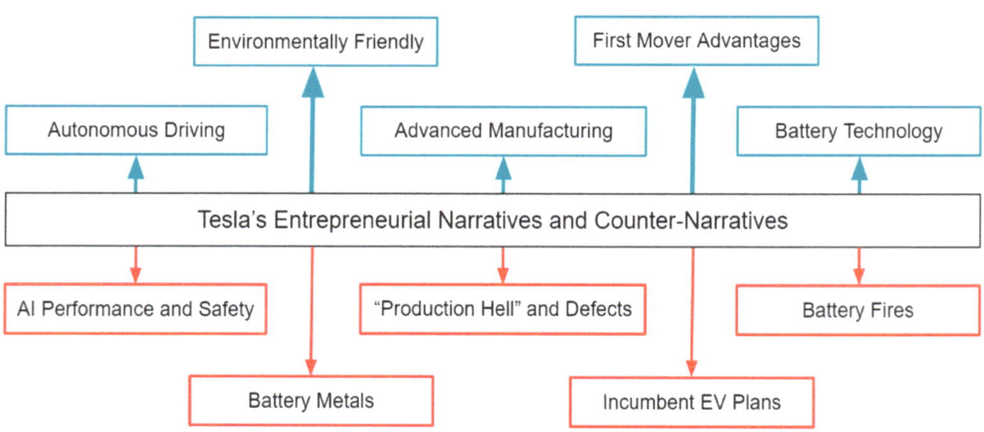

Figure 1. Tesla's entrepreneurial narratives and associated counternarratives.

3.1. Tesla Produces Environmentally Friendly Automobiles

Tesla's stated purpose and initial entrepreneurial narrative positioned the firm as an electric-vehicle company whose products would help to save the world from accelerating climate change by reducing global emissions of greenhouse gasses associated with the use of traditional, internal combustion engine-powered vehicles [14]. The firm also had a goal of making the energy supply chain more sustainable, as exemplified by its acquisition of SolarCity and deployment of residential solar kits and commercial megapacks [15]. This narrative can be found early in the firm's existence and is more or less constant throughout the 2010s.

Several counternarratives challenged Tesla's positioning as a sustainable transport and energy company. First, the company's vehicles and battery pack-based products are powered by the electricity stored in battery assemblies that require metals such as lithium, nickel, and cobalt to build. The extraction and processing of these metals are known to be socially and environmentally problematic [16]. Second, though Tesla's products are "electric," the sources of that electricity vary by region and generally include fossil fuels [17]. Beyond the environmental issues related to material and energy sourcing, critics of electrification have argued that the adoption of EVs is insufficient to mitigate the effects of climate change, and that the growth of Tesla does not represent a major turning point in the environmental footprint of our transportation systems [18,19].

3.2. Tesla Is a Tech Company with a Leading Position in Autonomous-Driving Technology

A second narrative positioned Tesla as a software company whose software happened to be embedded in electric vehicles. This view was reaffirmed by CEO Musk on Twitter as recently as February of 2022 [20]. Many supporters of the firm described a Tesla as "a computer on wheels" [21]. Specific elements of this narrative focused on novel aspects of Tesla's operations, including, for instance, its introduction of "over the air" software updates [22] and the idea that a Tesla vehicle was a technology platform that could support other value-generating activities [23]. Broad reliance upon this narrative allowed the firm to position itself as a "tech company" that should be evaluated and assessed in comparison with technology companies like Apple and Amazon, rather than car companies like Ford and General Motors [24–26].

A popular variant of this entrepreneurial narrative showcased Tesla's identity as a software company with a leading position in autonomous-driving technology. This narrative was underpinned by accounts of the vast quantities of data Tesla had accumulated from its vehicles, data that were being analyzed by the firm's proprietary supercomputers to power the artificial intelligence engine behind its autonomous-driving technology [27]. This narrative projected a future with "Full Self Drive," whereby millions of "robotaxis" would provide driverless transport services as a revenue source for owners of Tesla cars [23,28]. Tesla was so confident in this narrative that it charged customers up to $15 K for a "Full Self Drive" package that was known to be in various stages of beta testing [29].

The counternarrative to the view that "Tesla is a software company" focused upon its principal identity as an automobile manufacturer. For instance, even after acquiring SolarCity and introducing Powerwall battery technology, Tesla generated approximately 85% of its revenue from the sale of automobiles as late as the fourth quarter of 2020 [30]. Moreover, specific claims about Tesla's autonomous-driving technology were subject to repeated criticism, at times from authorities [31]. By mid-2022, with no revenue-generating robotaxis in operation, many of CEO Elon Musk's other forward-looking statements were greeted by online comments reminding observers how Musk's prediction of one million robotaxis by the end of 2019 had failed to come to fruition [32].

3.3. Tesla Will Use Its First-Mover Advantage to Dominate the EV Industry

A third entrepreneurial narrative built upon Tesla's position as the first successful, large-scale entrant in the electric-vehicle market. In this view, the company was able to take maximum advantage of (i) federal loan guarantees intended to stimulate the growth of alternatives to internal combustion [33–36], (ii) subsidies for would-be purchasers of Tesla's products [37], and (iii) investors seeking a stake in the predicted electrification of the global automobile market [38]. In addition to building electric vehicles, Tesla would also build charging stations, benefiting from network effects [39] and possibly from a closed ecosystem, or "walled garden", whereby the only way customers and suppliers enter is through Tesla, affording it immense market power. Since subsequent entrants may not be able to realize these advantages [40] and will face competition from the ever-dominant Tesla, competitors' growth opportunities are more limited.

The counternarrative to Tesla's claim to the first-mover advantage took two forms. First, commentators pointed out that Tesla was not, in fact, the first company to bring a battery electric vehicle to market. Observers frequently point to GM's EV1 or Toyota's RAV4-EV, both of which were introduced in the late 1990s, as vehicles that preceded Tesla entry in the EV market [41]. The consumer adoption of these vehicles was obviously quite limited, with no evident competitive advantage attaching to the firms that preceded Tesla into the modern EV market. Second, many industry watchers cautioned that the incumbent automobile makers were simply waiting out the startup-led EV rush until such time as they could profitably enter the market space [42]. Companies like Ford, General Motors, and Toyota would take up the EV banner if and only if they could make money doing so, and the fact that Tesla was alone atop the EV market table for so many years reflected the incumbents' calculation that Tesla could not make money selling EVs [43].

3.4. Tesla's Leadership in Manufacturing Technology Will Result in Competitive Advantages

A fourth entrepreneurial narrative sought to foreground Tesla's development of new, best-in-class manufacturing technology. From its first car, a Lotus chassis fitted with a battery-electric drivetrain, Tesla claimed to be gaining manufacturing advantages over the auto industry through vertical integration and automation. Emphasizing the machine that builds the machine, CEO Elon Musk was quoted as having said, "Tesla's long-term competitive advantage will be manufacturing" [44]. At various times, the advanced manufacturing narrative was supported by the public release of the video footage of the line of robots (so called "army of droids") that assembled the Model S, the IDRA Giga Press that "printed" single-piece frames, and the construction of "Giga" factories in Nevada, China, Texas, and Germany. Taken together, this narrative sought to demonstrate how manufacturing advantages would generate cost efficiencies that would enable Tesla to make affordable, mass-market electric vehicles [45].

The counternarrative to Tesla's leadership in advanced manufacturing emerged during the ramp-up of the production of the Model 3, the first time the company confronted the specific challenges faced by mass-market car makers such as industrial-level inventory management and high-volume assembly-line operations. As Tesla attempted to build the Model 3 at scale, the firm struggled to manage these and other constraints due to over-automation of the assembly line and other factors that the company did not anticipate [46]. This struggle was described by CEO Musk as "production hell", and the scaling process was closely watched by both supporters and critics. At one point, the weekly Model 3 count was one of the most closely followed metrics in the business world [47].

Even as the company resolved production issues, and as the Model 3 and later, the Model Y, became available to customers, reports of manufacturing flaws began to appear in traditional [48,49] and social media [50]. In response to product defects, Tesla issued several recalls [51]. Regardless of the firm's remedies, these product flaws and defects hurt public perception of the quality of Tesla vehicles [52]. Moreover, accidents related to autonomous driving such as unintended acceleration extended the negative perception further to Tesla's software products. Taken together, these issues generated a counternarrative that Tesla vehicles were poorly manufactured and programmed.

3.5. Tesla Is Developing Advanced Battery Technology

As a derivative of the "Tesla produces an environmentally friendly vehicle" narrative, Tesla also positioned itself as a battery company whose world-leading battery technology would allow the firm to (i) rapidly achieve scale production [53,54], (ii) dominate the emerging stationary battery market [55,56], and (iii) outcompete manufacturers who procured batteries in the open market [57]. The company completed several acquisitions to bolster its battery-technology portfolio [58,59]. As battery production was a critical-path resource enabling the transition from internal-combustion engine to electric vehicle, Tesla's early advances in battery technology, manufacturing, and supply were reported as sources of enduring competitive advantage [60,61].

The counternarrative to Tesla's leadership in battery technology was catalyzed by several incidents of Tesla Model S vehicles bursting into flames following crashes [62]. These "thermal incidents" posed a high-visibility challenge to the battery-technology narrative [63] and are discussed in further detail below.

3.6. Balancing Entrepreneurial Narratives and Counternarratives

The preceding set of entrepreneurial narratives and counternarratives is intended to be suggestive, not exhaustive. Additional narratives stressed synergies arising from the combination of some of the base narratives identified above, as well as Tesla's promised expansion into related product markets such as light-duty trucks (Cybertruck) [64], long-haul trucking (Semi) [65], energy generation (Solar Roof) [66], local energy storage (Powerwall) [67], and insurance to cover them all [68].

In each of these cases, narratives were followed by counternarratives. Given Tesla's reliance upon continued access to public capital markets to secure funds for operations and to fund operating losses during its first decade as a public company, counternarratives that threatened the firm's preferred entrepreneurial narratives posed a meaningful threat to the firm's long-run prospects. Based upon our analysis and upon our observation of persistent differences between the market-valuation metrics of Tesla and that of its automotive rivals, none of these counternarratives were broadly embraced by Tesla stakeholders [69,70].

4. Surprise Finding

In the course of investigating the relationship between Tesla's entrepreneurial narratives and Tesla-related social media on Twitter, we observed a period in late 2013 when one of the counternarratives threatened to displace the associated, firm-preferred narrative. The narrative in question was Tesla's expertise in advanced battery technology. At the time, Model S was receiving highly favorable reviews, with automotive [71,72] and non-automotive sources [73] hailing the car as a singular breakthrough in the race to introduce a commercially successful electric vehicle. Starting in early Fall, however, social and traditional media reported several Model S battery fires, instances where the battery ignited following crashes or impact with roadside debris, destroying the vehicle [74,75]. Subsequent reports of Model S fires suggested spontaneous combustion, with parked vehicles catching fire and, in some cases, destroying buildings along with the car [76,77]. Stories about the dangers of the Model S, specifically its battery technology, surfaced following these incidents and provided traction for counternarratives that questioned the safety of EVs, especially those made by Tesla [78–81].

On 7 November 2013, following a third reported Model S battery fire, several weeks of bad publicity, and consecutive days of poor stock performance, we observed the creation of a set of eight unusual Twitter accounts that had the following features: the eighth account was created within 70 min of the first; all eight accounts posted positive messages about Tesla (using the $TSLA cashtag); and their Twitter activity started on the night of November 7 (see list of the eight accounts in Table 1). Over the following seven years, these accounts would post nearly 25,000 $TSLA and over 5000 #TSLA tweets. The content of these tweets did not appear to have been written by a human user (as shown for one account in Table 2), and their timing did not resemble that of a human user as the tweets occurred at three-hour intervals throughout the day (see histogram in Figure 2). Such content and tweet patterns led us to suspect that the accounts in question were independent computational agents acting in support of the firm, or Fanbots.

Table 1. Eight $/#TSLA accounts created within 75 min on 7 November 2013 (with Botometer scores).

TweetCount	UserName	UserCreationDateTime	UserBotometerScore
3386	danrocks4	2013-11-07 22:41	4.6
3258	lookn4wins	2013-11-07 21:59	4.6
3194	leahanneta	2013-11-07 22:29	4.8
3036	Jake132013	2013-11-07 21:42	4.8
2956	clayton_dd	2013-11-07 21:53	4.6
2852	Jim5011	2013-11-07 22:07	4.6
2850	Stock_Tracker1	2013-11-07 22:17	4.4
2807	Emylers	2013-11-07 22:50	4.9

Table 2. Sample of early $TSLA tweets from one 7 November 2013 account.

UserName	TweetCreationDateTime	TweetText
danrocks4	2013-11-13 15:45	Danrocks: $TSLA keeping this on watch ...
danrocks4	2013-11-13 19:07	Danrocks: $TSLA Starting to get interesting ...
danrocks4	2013-11-18 19:16	Danrocks: $TSLA Hit my list a few days ago ...
danrocks4	2013-11-30 09:41	$TSLA Hitting Buzz ...
danrocks4	2013-12-10 12:44	$TSLA alot of talk going around about this one ...

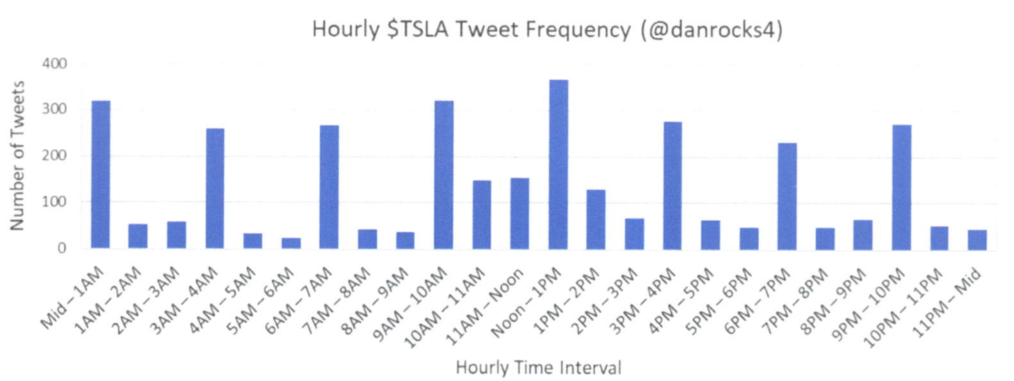

Figure 2. Hourly $TSLA tweet frequency of one 7 November 2013 account over the study period.

The creation of the 7 November 2013 Fanbots in such close proximity to the emergence of a counternarrative cued us to the possibility that computational agents may have been deployed to shape the conversation about Tesla on Twitter. In the end, the counternarrative that Model S batteries were dangerous and prone to combustion did not displace Tesla's preferred entrepreneurial narrative about the merits of the firm's battery technology. Over the ensuing years, pro-Tesla entrepreneurial narratives extended the appeal of the firm to broader sets of stakeholders, even as various counternarratives continued to simmer. However, the discovery of these Fanbots raised new questions about the use of corporate computational propaganda in shaping a firm's entrepreneurial narratives.

5. Data and Methodology

To explore the extent of CCP in this context, we sought to characterize the extent of fanbot activity within the corpus of tweets containing either the cashtag $TSLA or hastag

#TSLA from the time of Tesla's IPO in June of 2010 to the end of 2020, our study period. Using the Twitter Academic API, we queried the historical Twitter archive for tweets containing "$TSLA" or "#TSLA." Data and associated metadata were collected, including Tweet Date/Time, Tweet ID, and User ID. Subsequently, the two corpora were consolidated and sorted by total activity (number of tweets within the study period) of user accounts. The User Names of the 400 most active accounts in each corpus were then retrieved.

To identify potential fanbots within these corpora, these User Names were evaluated using an off-the-shelf IS platform known as Botometer (previously BotOrNot) [82–84]. This tool analyzes the language, timing, frequency, and other aspects of a given user's tweets to generate a score between zero (likely human user) and five (likely programmed bot user). Following previous research on Twitter bots, we treated a user with a Botometer score greater than or equal to 4.0 as a fanbot account [82,85] and found two visibly distinct groups of users (see histogram of Botometer scores in Figure 3). Within the 400 most active $TSLA and #TSLA Twitter accounts, we found 95 and 143 Fanbots, respectively, with an overlap of 24 accounts. Of these, 81 $TSLA and 126 #TSLA Fanbots were created within our sampling window, with an overlap of 21 accounts. In total, we identified 186 unique, post-IPO Fanbots operating within the two corpora.

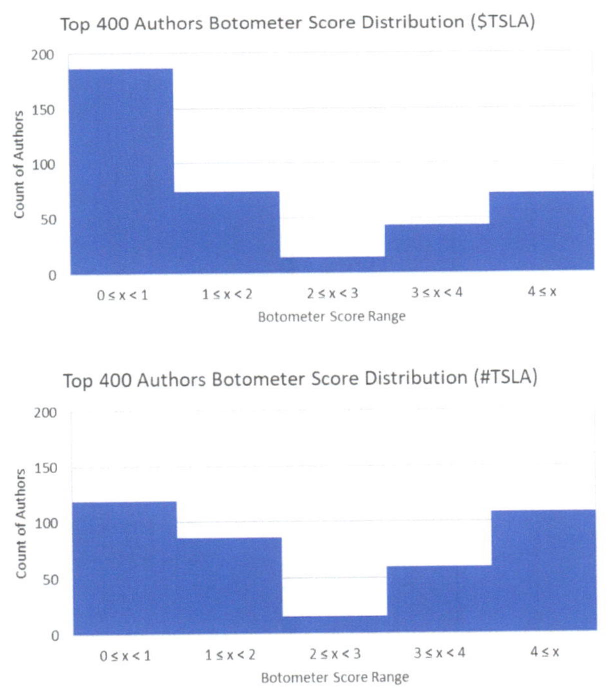

Figure 3. Histogram of the 400 most active $TSLA and #TSLA users by Botometer score.

Over the study period, there were a total of nearly 4.3 million tweets containing $TSLA. Of these tweets, approximately 1.4 million (or 33%) were posted by the 400 most active $TSLA accounts and more than 400,000 (or 10%) by accounts identified by Botometer as Fanbots. Over the same period, there were approximately 157,000 tweets containing #TSLA. Of these tweets, nearly 66,000 (or 42%) were posted by the 400 most active #TSLA accounts and 36,000 (or 23%) by accounts identified by Botometer as Fanbots. We interpreted the magnitude and share of CCP in the $TSLA and #TSLA Twitter corpora to be significant (see Figure 4).

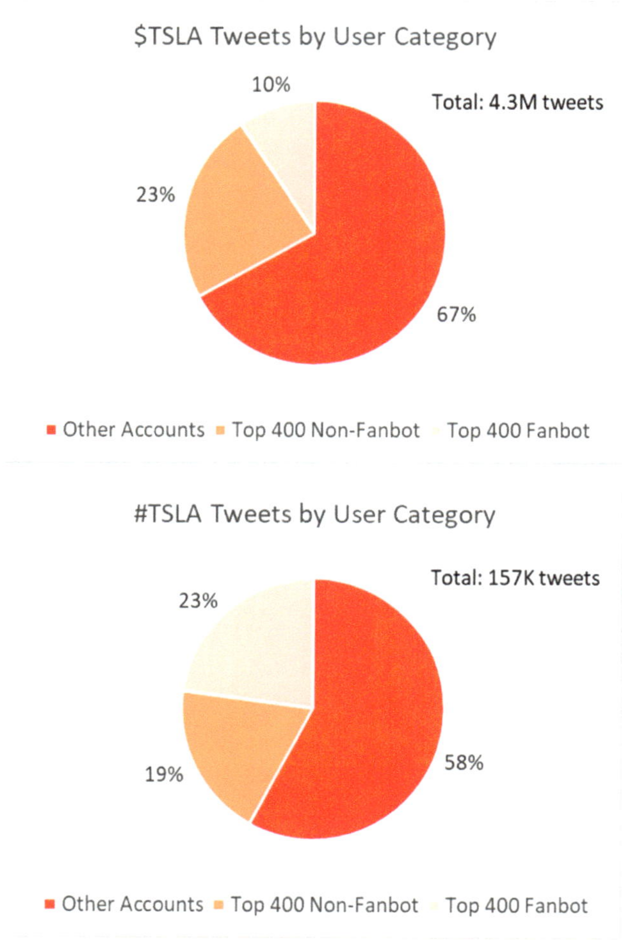

Figure 4. Distribution of $TSLA and #TSLA tweets by category of users.

6. Interpretation of Fanbots, CCP and Entrepreneurial Narratives

Having identified algorithmic Twitter accounts as a source of Corporate Computational Propaganda (CCP), we propose two possible motivations for the use of CCP to support Tesla's entrepreneurial narratives. Since creating fanbots and having them generate CCP is not costless, the fact that fanbots exist and make a considerable contribution to the overall corpus of tweets about Tesla suggests that whoever was responsible for the Fanbots believed that deploying these resources—to shape engagement and influence the conversation about Tesla on Twitter—would help bolster Tesla's preferred entrepreneurial narratives and, thereby, support the company's mission and operations. Even if the actors responsible for the Fanbots did not believe that they would be value-creating for Tesla (or in the immediate interests of the actors), they may have believed that CCP would insulate the firm's preferred entrepreneurial narratives from the threat of nascent counternarratives. In the latter case, investment in fanbots may be thought of as a counter-weight, or an insurance policy, against information that undermines the firm's preferred entrepreneurial narratives. Whether Tesla's entrepreneurial narratives would have thrived, or even survived, in the absence of the Fanbots and the CCP they generated is a question for future research. For now, the existence of fanbots suggests that some actor (an individual or group) believed that they would serve the interests of the firm, though the mechanisms of such service warrant further study.

Below we present several possible mechanisms by which CCP may have increased social acceptance of Tesla's entrepreneurial narratives. We gathered these mechanisms under the umbrella of visibilization, the process by which fanbot-generated CCP increased the visibility of certain entrepreneurial narratives and, thereby, altered Twitter users' impressions of Tesla.

Visibilization describes the process by which certain tweets—Fanbot tweets containing content that supports Tesla's various entrepreneurial narratives—would attract attention and, thereby, be more visible to human users on the platform [8]. Social psychology suggests three possible mechanisms by which visibilization may operate in the context of CCP: repetition, recency, and availability.

Repetition. Unlike human users who need periodic breaks, fanbots are able to tweet almost continuously. As Tesla Fanbots generate CCP throughout the day, human Twitter users may be influenced by observing the tweets over and over again. Adding to this influence is the fact that CCP is generated by numerous Fanbot accounts, potentially exposing the observer to multiple sources of pro-Tesla content. As a result, human users may be subject to repetition bias, by which they believe the positive content to which they are exposed. We surmise that the beliefs of Fanbois, who were pro-Tesla prior to repeated CCP exposure, may be strengthened by repeated exposure to positive messages about the firm [86].

Recency. When tweets about a firm appear frequently, they may also be the most recent tweets observed, activating a recency bias among human users [87,88]. Such bias may leave the user with a pro-Tesla orientation and, after doing so repeatedly over time, may orient a wider group of active users such as Fanbois to more strongly believe in Tesla's entrepreneurial narratives and, as a result, support the firm through offline actions (e.g., purchasing vehicles or other products, purchasing shares, or generally holding and spreading positive beliefs about the firm and its entrepreneurial narratives).

Availability. When a social media user observes the repetition and recency of certain content, they may be further subject to an availability bias, whereby examples that readily come to mind are overweighted [89]. By exploiting the design architecture of the Twitter "feed," in which tweets scroll past and users see only the latest content, fanbots increase the likelihood that Twitter users, especially those actively engaging in the pro-Tesla conversation such as Fanbois, will view CCP and will overweigh the pro-Tesla information provided. As a result, Fanbois may become more committed to one or more pro-Tesla entrepreneurial narratives.

7. Discussion and Conclusions

We have provided evidence of the existence of Fanbot-generated corporate computational propaganda on Twitter that supported Telsa's entrepreneurial narratives. We proposed that this content bolstered positive accounts of these narratives—while also insulating against damaging counternarratives—through a process of visibilization. Fanbois, a large group of pro-Tesla human Twitter users, may have reacted to this visibilization through social psychological mechanisms of recency, repetition, and availability. Further research will be needed to fully identify the mechanisms linking fanbot activity to observable social phenomena like online polarization and mobilization. Nevertheless, this article has proposed one set of relationships between Fanbot-generated CCP and Fanboi-led social approval of the firm.

We readily admit that we have not ruled out all alternative explanations for the presence of Fanbots in the $TSLA and #TSLA tweet corpora. For instance, researchers have previously documented the prevalence of generic pro-market content on "financial" Twitter (tweets containing any cashtag) [90]. In unreported analyses, we did find evidence of "bot" content relating to other securities. However, the $TSLA corpus contains much more bot content than we found in the cashtag corpora of other companies in a random sample of S&P500 firms. During the ~ten-year period we studied, Tesla's cashtag tweet corpus differed significantly from that of any other publicly traded firm, by both the total volume and total amount of CCP. Second, we cannot rule out reverse causality, whereby Tesla's ability to crack the long-standing challenge of successfully commercializing electric vehicles attracted both a growing community of admirers (Fanbois) and unrelated spammers seeking to redirect or capture attention in a large, online community of active, human users. This "collective attention spam" has been observed in Twitter [91]. However, even if we accept that some of the Fanbot tweet activity that we observed was created to access the emergent Tesla community on Twitter, we would still not be able to explain the events of 7 November 2013, when a set of Fanbot accounts were created in apparent response to specific challenges to a cornerstone entrepreneurial narrative of Tesla. Given the provisional and descriptive nature of our analysis, we expect and encourage further research to test these and possibly other alternative explanations. We have proposed our best explanation of the observed fact patterns.

At a more general level, does the discovery of Twitter Fanbots generating pro-Tesla computational propaganda foretell the rise of a new form of impression management for CEOs seeking to generate support for their firms' respective entrepreneurial narratives? This single-firm study cannot answer this question. On the one hand, our analysis thus far has assumed that any firm could, in principle, take advantage of CCP. However, we must stipulate that Tesla's CEO Elon Musk (even before announcing efforts to acquire Twitter in mid-2022) was a singular business figure on the Twitter platform. His Twitter popularity allowed him to use the platform as a powerful engagement tool, possibly enhancing the effects of Fanbot-generated CCP. Therefore, future research will need to examine the boundary conditions for these findings.

Given what we know at present, it is not clear if this strategy could be replicated by other firms seeking to use fanbot-generated CCP to support their own entrepreneurial narratives in the absence of a central actor with Elon Musk's ex ante celebrity. Put another way, we do not yet know the conditions under which CCP is likely to be effective in managing and supporting entrepreneurial narratives. Nevertheless, many less-celebrated CEOs, observing Musk's success using Twitter to advance Tesla's entrepreneurial narratives, are likely already trying to emulate this strategy, on Twitter or on other social media platforms.

Based upon Tesla's success, we can expect these strategies to be deployed by companies in similar situations, namely, startup ventures relying upon entrepreneurial narratives to sustain access to resources. Since 2016, policymakers have been focused upon identifying and regulating political computational propaganda [92], but as we have shown, the use of computational propaganda on Twitter by pro-corporate actors preceded its use in politics by several years. Our observation of Fanbots supporting entrepreneurial narratives leaves open the possibility that corporate computational propaganda may already be influencing social approval in other nascent market settings.

These initial findings provide a foundation for future research in this area. These findings also provide a foundation for future research on the role and function of CCP, particularly in the emergence of the contemporary electric-vehicle industry. Have other EV startups sought to use CCP to legitimize their respective entrepreneurial narratives? Does CCP have spillover effects between firms? Conversely, can CCP have competitive implications, such as enabling firms to engage in online, negative peer disclosure? Relatedly, does the existence of CCP alter the sentiment of online conversations about specific firms? If so, would such CCP-driven distortion have any impact on market movements of certain firms?

Author Contributions: Conceptualization, D.A.K.; Methodology, D.A.K. and M.A.C.; Software, M.A.C.; Validation, D.A.K. and M.A.C.; Formal Analysis, D.A.K. and M.A.C.; Investigation, D.A.K. and M.A.C.; Resources, D.A.K.; Data Curation, M.A.C.; Writing—Original Draft Preparation, D.A.K. and M.A.C.; Writing—Review & Editing, D.A.K. and M.A.C.; Visualization, M.A.C.; Supervision, D.A.K.; Project Administration, D.A.K. All authors have read and agreed to the published version of the manuscript.

Funding: This research received no external funding.

Data Availability Statement: Twitter data were collected under the terms of the Twitter Academic API during calendar years 2021 and 2022. The authors may share the queries that generated the tweet corpora, but the Twitter Academic user license does not allow sharing of large tweet collections. Furthermore, because Twitter actively and continuously searches for and removes accounts for a variety of reasons, we cannot warrant that a future query will return the same dataset that we have used in our analyses.

Acknowledgments: The authors acknowledge research assistance from Veronica Kim, as well as feedback from Jerry Davis, Anastassia Fedyk, Michael Jacobides, Heeyun Kim, Fil Menczer, Violina Rindova, Wesley Sine, Marc Ventresca, Dave Waguespack, Kai-Cheng Yang, Anastasiya Zavalyova, and seminar, workshop, and conference participants at the University of Maryland, Robert H. Smith School of Business, the Oxford Reputation Conference, the IDeaS Workshop, the Sumantra Ghoshal Strategy Conference, the 35th International Electric Vehicle Symposium, the SMS Special Conference on Governing Knowledge and Imagination in the Digital Era, the Industry Studies Association Annual Conference, and the SMS 42nd Annual Conference. All errors remain the responsibility of the authors.

Conflicts of Interest: As of November 2022, one of the co-authors, Mohsen A. Chowdhury, holds a long-term position in Tesla.

References

1. Kirsch, D.A. *The Electric Vehicle and the Burden of History*; Rutgers University Press: New Brunswick, NJ, USA, 2000.
2. Mom, G. *The Electric Vehicle: Technology and Expectations in the Automobile Age*; JHU Press: Baltimore, MA, USA, 2013; ISBN 9781421412689.
3. Martens, M.L.; Jennings, J.E.; Jennings, P.D. Do the Stories They Tell Get Them the Money They Need? The Role of Entrepreneurial Narratives in Resource Acquisition. *Acad. Manag. J.* **2007**, *50*, 1107–1132. [CrossRef]
4. Rindova, V.; Barry, D.; Ketchen, D.J. Entrepreneuring as Emancipation. *Acad. Manag. Rev.* **2009**, *34*, 477–491. [CrossRef]
5. Boje, D.M.; Svane, M.; Gergerich, E.M. Counternarrative and Antenarrative Inquiry in Two Cross-Cultural Contexts. *Eur. J. Cross-Cult. Competence Manag.* **2016**, *4*, 55–84. [CrossRef]
6. Bansal, P.; Clelland, I. Talking Trash: Legitimacy, Impression Management, and Unsystematic Risk in the Context of the Natural Environment. *Acad. Manag. J.* **2004**, *47*, 93–103. [CrossRef]
7. Heavey, C.; Simsek, Z.; Kyprianou, C.; Risius, M. How Do Strategic Leaders Engage with Social Media? A Theoretical Framework for Research and Practice. *Strateg. Manag. J.* **2020**, *41*, 1490–1527. [CrossRef]
8. Gümüsay, A.A.; Raynard, M.; Albu, A.; Etter, M.; Roulet, T. Digital Technology and Voice: How Platforms Shape Institutional Processes through Visibilization. *Res. Sociol. Organ.* **2022**, *83*, 57–85.
9. Pollock, T.G.; Lashley, K.; Rindova, V.P.; Han, J.-H. Which of These Things Are Not Like the Others? Comparing the Rational, Emotional, and Moral Aspects of Reputation, Status, Celebrity, and Stigma. *Acad. Manag. Ann.* **2019**, *13*, 444–478. [CrossRef]
10. Techopedia Fanboi. Available online: https://www.techopedia.com/definition/28754/fanboi (accessed on 9 November 2022).
11. Etter, M.; Albu, O.B. Activists in the Dark: Social Media Algorithms and Collective Action in Two Social Movement Organizations. *Organization* **2021**, *28*, 68–91. [CrossRef]
12. Keller, F.; Schoch, D.; Stier, S.; Yang, J. How to Manipulate Social Media: Analyzing Political Astroturfing Using Ground Truth Data from South Korea. *ICWSM* **2017**, *11*, 564–567. [CrossRef]
13. Goldfarb, B. Why Tesla Is Overhyped—And Overvalued-Tesla and the Deep Problems with "disruption" Theory. *Vox*, 1 August 2017.
14. The Mission of Tesla. Available online: https://www.tesla.com/blog/mission-tesla (accessed on 7 November 2022).
15. Tesla and SolarCity. Available online: https://www.tesla.com/blog/tesla-and-solarcity (accessed on 7 November 2022).
16. Lithium-Ion Batteries Need to Be Greener and More Ethical. *Nature* **2021**, *595*, 7. [CrossRef] [PubMed]
17. With The Tesla Model S, Elon Musk Has Created A Nice Fossil Fuel Car. Available online: https://industrialprogress.com/with-the-tesla-model-s-elon-musk-has-created-a-nice-fossil-fuel-car/ (accessed on 7 November 2022).
18. Schneider, M. Op-Ed: Think Bigger. Switching to Electric Cars Isn't Enough. *Los Angeles Times*, 15 September 2022.
19. Wade, L. Tesla's Electric Cars Aren't as Green as You Might Think. *Wired*, 31 March 2016.
20. Musk, E. Tesla Is as Much a Software Company as It Is a Hardware Company, Both in Car and in Factory. This Is Not Widely Understood. Available online: https://twitter.com/elonmusk/status/1488391303520034817?lang=en (accessed on 12 November 2022).
21. King, R. At Tesla, Software Takes the Driver's Seat. *The Wall Street Journal*, 14 May 2014.
22. Lavrinc, D. In Automotive First, Tesla Pushes Over-the-Air Software Patch. *Wired*, 24 September 2012.
23. Musk, E. Master Plan, Part Deux. Available online: https://www.tesla.com/blog/master-plan-part-deux (accessed on 12 November 2022).

24. Randewich, N. Wall Street Values Tesla Motors at $620,000 per Car. *Reuters*, 3 May 2016.
25. Tesla's Not as Disruptive as You Might Think. *Harvard Business Review*, 1 May 2015; 22–23.
26. Lee, T. The American Economy's Big Problem: We Don't Have Enough Companies Like Tesla. Available online: https://www.vox.com/2016/5/13/11670940/tesla-capital-glut (accessed on 12 November 2022).
27. Hull, D. The Tesla Advantage: 1.3 Billion Miles of Data. *Bloomberg News*, 20 December 2016.
28. Korosec, K. Tesla Plans to Launch a Robotaxi Network in 2020. *TechCrunch*, 22 April 2019.
29. Lambert, F. Tesla Increases Price of Full Self-Driving in Move to Make Its Cars 'appreciating Assets'. Available online: https://electrek.co/2019/05/03/tesla-price-full-self-driving-cars-appreciating-assets/ (accessed on 12 November 2022).
30. Securities and Exchange Commission. 2020. Available online: https://www.sec.gov/Archives/edgar/data//1318605/000156459020019931/tsla-10q_20200331.htm (accessed on 12 November 2022).
31. Lomas, N. Tesla Told to Stop Using 'Misleading Term Autopilot' in Germany. *TechCrunch*, 17 October 2016.
32. Lambert, F. Tesla's 1 Million Robotaxis by End of the Year Becomes "1 Million People in FSD Beta". Available online: https://electrek.co/2022/05/17/tesla-1-million-robotaxis-fsd-beta/ (accessed on 12 November 2022).
33. Chereb, S. Nevada Gives $1.3 Billion Tax Break to Electric Car Maker Tesla. *Scientific American*, 12 September 2014. Available online: https://www.scientificamerican.com/article/nevada-gives-1-3-billion-tax-break-to-electric-car-maker-tesla/ (accessed on 12 November 2022).
34. Lambert, F. Complete Breakdown of the $4.9 Billion in Government Support the LA Times Claims Elon Musk's Companies Are Receiving. Available online: https://electrek.co/2015/06/02/complete-breakdown-of-the-4-9-billion-in-government-support-the-la-times-claims-elon-musks-companies-are-receiving/ (accessed on 12 November 2022).
35. Economic Recovery Act. 2010. Available online: https://obamawhitehouse.archives.gov/sites/default/files/microsites/Recovery_Act.PDF (accessed on 12 November 2022).
36. Tesla Gets Loan Approval from US Department of Energy. Available online: https://www.tesla.com/blog/tesla-gets-loan-approval-us-department-energy (accessed on 12 November 2022).
37. Electric Vehicle & Solar Incentives. Available online: https://www.tesla.com/support/incentives (accessed on 12 November 2022).
38. Tesla Battery System Costs Will Drop 40% by 2020. Available online: https://ark-invest.com/articles/analyst-research/tesla-battery-system-cost/ (accessed on 12 November 2022).
39. Bhargava, H.; Boehm, J.; Parker, G.G. How Tesla's Charging Stations Left Other Manufacturers in the Dust. *Harvard Business Review*, 27 January 2021.
40. Lieberman, M.B.; Montgomery, D.B. First-Mover Advantages. *Strateg. Manag. J.* **1988**, *9*, 41–58. [CrossRef]
41. Pugliese, J. Power Play. *Wired*, 11 January 2016.
42. Lambert, F. VW CEO Says Electric Car Costs Higher than Expected, Sees Competition Making Progress. Available online: https://electrek.co/2018/09/13/vw-ceo-higher-electric-car-costs-competition-progress/ (accessed on 12 November 2022).
43. Hull, D.; Recht, H. Tesla Doesn't Burn Fuel, It Burns Cash. Available online: https://www.bloomberg.com/graphics/2018-tesla-burns-cash/ (accessed on 12 November 2022).
44. 2021 Shareholder Meeting. Available online: https://www.tesla.com/2021shareholdermeeting (accessed on 13 November 2022).
45. Boudette, N.E. Inside Tesla's Audacious Push to Reinvent the Way Cars Are Made. *The New York Times*, 30 June 2018.
46. Duhigg, C. Dr. Elon & Mr. Musk: Life Inside Tesla's Production Hell. *Wired*, 13 December 2018.
47. Gibbs, S. Tesla Finally Hits Weekly Production Target for Model 3 Cars. *The Guardian*, 2 July 2018.
48. Olsen, P. Tesla Model 3 Falls Short of a CR Recommendation. Available online: https://www.consumerreports.org/hybrids-evs/tesla-model-3-review-falls-short-of-consumer-reports-recommendation/ (accessed on 13 November 2022).
49. Holley, P. Consumer Reports: Tesla's Model 3 Has "Big Flaws.". *The Washington Post*, 22 May 2018.
50. Lau, A. Tesla Model Y's Serious Production Problems: The Most Notable Ones. Available online: https://www.motortrend.com/news/tesla-model-y-ev-safety-quality-issues-problems/ (accessed on 12 November 2022).
51. Eisenstein, P.A. Tesla Recalls Almost Half the Cars It Ever Built, as Shares Tank and Musk's Billions Shrink. Available online: https://www.nbcnews.com/business/autos/tesla-recalls-almost-half-cars-it-ever-built-shares-tank-n861421 (accessed on 12 November 2022).
52. Sage, A. Build Fast, Fix Later: Speed Hurts Quality at Tesla, Some Workers Say. *Reuters*, 28 November 2017.
53. Hardawar, D. How Tesla's Battery "Gigafactory" Could Change Everything—Not Just Electric Cars. Available online: https://venturebeat.com/mobile/how-teslas-battery-gigafactory-could-change-everything-not-just-electric-cars/ (accessed on 12 November 2022).
54. Battery Cell Production Begins at the Gigafactory. Available online: https://www.tesla.com/blog/battery-cell-production-begins-gigafactory (accessed on 12 November 2022).
55. Sakti, A.; Miller, R.; Brushett, F. What's Cost Got to Do with It? Available online: https://energy.mit.edu/news/whats-cost-got-to-do-with-it/ (accessed on 12 November 2022).
56. Introducing Megapack: Utility-Scale Energy Storage. Available online: https://www.tesla.com/blog/introducing-megapack-utility-scale-energy-storage (accessed on 12 November 2022).

57. Linder, C. Elon Musk Finally Reveals His Grand Plans to Revolutionize the Battery. Available online: https://www.popularmechanics.com/science/energy/a34114885/elon-musk-tesla-battery-day-recap/ (accessed on 12 November 2022).
58. van Romburgh, M. Tesla to Buy Battery Maker Maxwell Technologies for $218M. Available online: https://www.bizjournals.com/sanjose/news/2019/02/04/tesla-maxwell-technologies-acquisition-tsla-mxwl.html (accessed on 12 November 2022).
59. Jarratt, E. Tesla Acquires Canadian Battery Specialist, Hibar Systems. Available online: https://electricautonomy.ca/2019/10/04/tesla-acquires-canadian-battery-specialist-hibar-systems/ (accessed on 12 November 2022).
60. Battery Technology Powers Tesla's Moat. Available online: https://ark-invest.com/articles/analyst-research/battery-technology/ (accessed on 12 November 2022).
61. Sanderson, H. Tesla's Move into Mining Aimed at Energising Battery Supply Chain. *Financial Times*, 22 October 2020.
62. Valdes-Dapena, P. A Third Tesla Model S Had Burst into Flames Following a Crash in Tennessee. Available online: https://money.cnn.com/2013/11/07/autos/tesla-fire (accessed on 12 November 2022).
63. Elliott, H. The Tesla Fire Is A Textbook PR Problem—And They Should Fix It. *Forbes*, 3 October 2013.
64. Stevens, T.; Krok, A. Tesla Cybertruck Electrifies the Truck Segment. Available online: https://www.cnet.com/roadshow/news/tesla-cybertruck-unveiled-elon-musk-electric-pickup-truck/ (accessed on 12 November 2022).
65. Estrada, Z. This Is the Tesla Semi Truck. Available online: https://www.theverge.com/2017/11/16/16667366/tesla-semi-truck-announced-price-release-date-electric-self-driving (accessed on 12 November 2022).
66. Solar Roof. Available online: https://www.tesla.com/blog/solar-roof (accessed on 12 November 2022).
67. Safi, M. Tesla Announces Low-Cost Batteries for Homes. *The Guardian*, 1 May 2015.
68. Introducing Tesla Insurance. Available online: https://www.tesla.com/blog/introducing-tesla-insurance (accessed on 12 November 2022).
69. Cornell, B.; Damodaran, A. Tesla: Anatomy of a Run-Up. *J. Portf. Manag.* **2014**, *41*, 139–151. [CrossRef]
70. Damodaran, A. Musings on Markets. Available online: https://aswathdamodaran.blogspot.com/2021/11/teslas-trillion-dollar-moment-valuation.html (accessed on 1 February 2022).
71. MacKenzie, A. 2013 Motor Trend Car of the Year: Tesla Model S. Available online: https://www.motortrend.com/news/2013-motor-trend-car-of-the-year-tesla-model-s/ (accessed on 12 November 2022).
72. Holmes, J. Consumer Reports Rates Tesla Model S 99 Out Of 100. Available online: https://www.motortrend.com/news/consumer-reports-rates-tesla-model-s-99-out-of-100-223097/ (accessed on 12 November 2022).
73. Tesla Model S—The Electric Car That Shatters Every Myth. Available online: https://www.consumerreports.org/cro/news/2012/11/tesla-model-s-the-electric-car-that-shatters-every-myth/index.htm (accessed on 12 November 2022).
74. Klayman, B. Tesla Grapples with Impact of Battery Fire in U.S. *Reuters*, 3 October 2013.
75. Bullis, K. What the Tesla Battery Fire Means for Electric Vehicles. *MIT Technology Review*, 3 October 2013.
76. Tesla Model S Catches Fire While Sitting in a Toronto Garage. Available online: https://financialpost.com/business-insider/tesla-catches-fire-while-sitting-in-a-toronto-garage (accessed on 12 November 2022).
77. Woodall, B.; Shirouzu, N. California Fire Department Says Tesla Model S Charging System May Have Started Garage Fire. *Business Insider*, 19 December 2013.
78. Bullis, K. UPDATE: Early Data Suggests Collision-Caused Fires are More Frequent in the Tesla Model S than Conventional Cars. *MIT Technology Review*, 21 November 2013.
79. Klayman, B.; Woodall, B. Tesla Reports Third Fire Involving Model S Electric Car. *Reuters*, 7 November 2013.
80. Meier, F.; Woodyard, C. Feds Review Third Tesla Fire as Shares Fall Again. *USA Today*, 7 November 2013.
81. Vlasic, B.; Trop, J. After 3 Fires, Safety Agency Opens Inquiry Into Tesla Model S. *The New York Times*, 19 November 2013.
82. Davis, C.A.; Varol, O.; Ferrara, E.; Flammini, A.; Menczer, F. BotOrNot: A System to Evaluate Social Bots. In Proceedings of the 25th International Conference Companion on World Wide Web, Montréal, QC, Canada, 11–15 April 2016; International World Wide Web Conferences Steering Committee: Geneva, Switzerland, 2016; pp. 273–274.
83. Keller, T.R.; Klinger, U. The Needle in the Haystack: Finding Social Bots on Twitter. In *Research Exposed: How Social Science Gets Done in the Digital Age*; Hargittai, E., Ed.; Columbia University Press: New York, NY, USA, 2020; pp. 30–49. ISBN 9780231548007.
84. Rossi, S.; Rossi, M.; Upreti, B.; Liu, Y. Detecting Political Bots on Twitter during the 2019 Finnish Parliamentary Election. In Proceedings of the 53rd Hawaii International Conference on System Sciences, HICSS 2020, Maui, HI, USA, 7–10 January 2020.
85. Rossi, S. The Scamdemic Conspiracy Theory and Twitter's Failure to Moderate COVID-19 Misinformation. In Proceedings of the 55th Hawaii International Conference on System Sciences, HICSS, Maui, HI, USA, 4–7 January 2022.
86. Hassan, A.; Barber, S.J. The Effects of Repetition Frequency on the Illusory Truth Effect. *Cogn. Res. Princ. Implic.* **2021**, *6*, 38. [CrossRef] [PubMed]
87. Richter, L.; Kruglanski, A.W. Seizing on the Latest: Motivationally Driven Recency Effects in Impression Formation. *J. Exp. Soc. Psychol.* **1998**, *34*, 313–329. [CrossRef]
88. Cowen, T. The Latest Bias to Worry About: Recency Bias. *The Washington Post*, 27 January 2022.
89. Tversky, A.; Kahneman, D. Availability: A Heuristic for Judging Frequency and Probability. *Cogn. Psychol.* **1973**, *5*, 207–232.
90. Ranco, G.; Aleksovski, D.; Caldarelli, G.; Grčar, M.; Mozetič, I. The Effects of Twitter Sentiment on Stock Price Returns. *PLoS ONE* **2015**, *10*, e0138441. [CrossRef] [PubMed]

91. Lee, K.; Caverlee, J.; Kamath, K.Y.; Cheng, Z. Detecting Collective Attention Spam. In Proceedings of the 2nd Joint WICOW/AIRWeb Workshop on Web Quality, WebQuality '12, Lyon, France, 16 April 2012; Association for Computing Machinery: New York, NY, USA, 2012; pp. 48–55.
92. Bessi, A.; Ferrara, E. Social Bots Distort the 2016 U.S. Presidential Election Online Discussion. *First Monday* **2016**, 21. [CrossRef]

World Electric Vehicle Journal

MDPI

Article

Electric Vehicle Charging Sessions Generator Based on Clustered Driver Behaviors †

Gilles Van Kriekinge *, Cedric De Cauwer, Nikolaos Sapountzoglou, Thierry Coosemans and Maarten Messagie

EVERGi Research Group, MOBI Research Centre & ETEC Department, Vrije Universiteit Brussel (VUB), Pleinlaan 2, 1200 Brussels, Belgium

* Correspondence: gilles.van.kriekinge@vub.be

† This paper is an extended version of our paper published in 35th International Electric Vehicle Symposium and Exhibition (EVS35), Oslo, Norway, 11–15 June 2022.

Abstract: Increasing penetration of electric vehicles brings a set of challenges for the electricity system related to its energy, power and balance adequacy. Research related to this topic often requires estimates of charging demand in various forms to feed various models and simulations. This paper proposes a methodology to simulate charging demand for different driver types in a local energy system in the form of time series of charging sessions. The driver types are extracted from historical charging session data via data mining techniques and then characterized using a kernel density estimation process. The results show that the methodology is able to capture the stochastic nature of the drivers' charging behavior in time, frequency and energy demand for different types of drivers, while respecting aggregated charging demand. This is essential when studying the energy balance of a local energy system and allows for calculating future demand scenarios by compiling driver population based on number of drivers per driver type. The methodology is then tested on a simulator to assess the benefits of smart charging.

Keywords: electric vehicle; smart charging; simulation; clustering; user behavior

Citation: Van Kriekinge, G.; De Cauwer, C.; Sapountzoglou, N.; Coosemans, T.; Messagie, M. Electric Vehicle Charging Sessions Generator Based on Clustered Driver Behaviors. *World Electr. Veh. J.* **2023**, *14*, 37. https://doi.org/10.3390/wevj14020037

Academic Editor: Grzegorz Sierpiński

Received: 8 December 2022
Revised: 13 January 2023
Accepted: 30 January 2023
Published: 2 February 2023

1. Introduction

Governments are pushing towards the electrification of the transport sector to reduce the human-related CO_2 emissions in order to fight climate change. For instance, the European Parliament adopted the proposal stating that CO_2 emissions of new passenger cars should be reduced by 100% by 2035 [1]. Such climate-focused policy pledges and announcements will increase the number of electric vehicles on the market since they are zero tailpipe emission cars. In the International Energy Agency (IEA) Announced Pledges Scenario (APS), the electric vehicle (EV) global market share is expected to grow up to 30% in 2030 [2].

Following the increase in the number of EVs, new charging infrastructures must be installed at different locations (residences, office buildings, hospitals, etc.). Such new installations should be carried out in a controlled way by assessing the possible impacts on electric systems, and potentially, to act accordingly in case of emerging problems. In scientific literature, simulations show that the installation of new charge points will impact the electric grid on different levels, from local energy systems (LES) [3] to distribution and transmission system operators (DSO and TSO) [4]. Typical issues are for instance voltage deviations, congestion issues and/or increase of peak powers [5]. Solutions to these problems are already strongly studied and often presented in simulations. There exist multiple solutions to the problem of installing new charge points, but two are of high interest: (a) adapt the design of electric system by reinforcing it, and (b) charge in an intelligent manner (smart charging or vehicle-to-grid) [6].

While simulations are of great interest to understand, anticipate and solve future problems, they require accurate data and models to have a correct representation of the reality. With respect to the simulations of EV charging sessions, an important stakeholder needs to be modelled which is the EV driver with their EV. The complexity relies on the modeling of individual human behavior (e.g., arrival and departure time) as well as the characteristics of the EV used by the driver (e.g., charging power). In addition, for most of the use cases, a fleet of EVs needs to be modelled which contains a panel of different EV drivers with different charging behaviors. These complexities are the main subject of this paper which proposes an electric vehicle charging sessions data generator with specific new contributions that simplify the modeling of the charging demand of a fleet of EVs.

1.1. Literature Review

The modeling of the mobility demand has already been conducted in previous works and can be divided into two main stochastic methods. The first method uses Markov chain processes. The Markov chain method associates a probability for the transition from a state to another state, where state could mean, for instance, driving state or charging state.

Authors in [7] propose an inhomogeneous Markov chain method to model the use of EV in order to assess the impact of load profiles at different parking locations. The results show a low deviation with the real mobility dataset. In [8], driver's behavior are simulated using a heterogeneous Markov chain process fitted on data collected by the US Bureau of Labor Statistics. The objective is to predict personal vehicle use and to assess the impact on the electricity demand. Authors in [9] use an inhomogeneous Markov chain process to model driving patterns. The model is fitted using data collected directly from an EV. In [10], authors use a cyclic Markov chain model to describe the behavior of charging stations. From this, they derive charge profiles to include in power grid simulation or to forecast electricity demand.

The advantage of the Markov-chain process is that the model takes the previous state into account, and hence assumes a consequential relationship between states. The disadvantage is that it usually requires abundant empirical data that is often not available. To tackle the issue of need of abundant data, a second method is also used in literature and consists often of using probability density functions without using consequential relationships. Such method is less accurate but easier to implement since it requires less data.

Authors in [11] developed a tool that generates time series of vehicle mobility, driving consumption, grid availability and grid electricity demand. The input parameters are numerous, ranging from EV driver information to EV characteristics and charge point information. Such input data are usually unknown for new charging locations, for instance, when a site wants to design or expand its charging location. Authors in [4] adapted a "Remote-Areas Multi-energy systems load Profiles" (RAMP) software engine, from [12], for stochastic EV driver behavior simulation. The simulator uses EV driver data from surveys to classify the behavior into default groups such as student, workers and inactive users, but also into small, medium and large EV sizes. This methodology has been designed to simulate mass-scale deployment of EVs on a country-level, which is not necessarily suited for smaller charging locations. To simplify this, authors in [13] generated EV charging sessions based on standard probability density functions. Typically, arrival and departure times are based on Gaussian distributions, and the daily mileage is based on log-normal distributions. However, this methodology lacks accuracy when dealing with different types of EV drivers because it assumes one probability density function for all drivers. In [14], authors developed an electric vehicle load profile generator. Their method focuses on generating load profiles directly from existing empirical data from EVs. The dataset used comes from three field trials in Germany from 2011 to 2015. Their method allows for a realistic representation of EV demand. However, it cannot be used to model EV charging behavior for different use cases. Authors in [15] modelled charging profiles of EVs based on real-world electric vehicle charging data. They focus on trip level and conditional probabilities to determine whether an EV will be charged after a trip or not.

The method is applied on residential use cases which makes it very user-centric modeling and not necessarily suited for a pool of EVs. In [16], authors propose a synthetic data generator for electric vehicle charging sessions using a large dataset of 1.8 million sessions. The results show great statistical similarities between the generated and real data. While this method works well for large use cases, it is unknown if it can be applied to specific local energy systems.

Since most of these previous works have different input data, different end-objectives and different techniques to model the mobility demand, Table 1 attempts a classification of the previous cited papers.

Table 1. Literature review classification on mobility demand modeling.

Topic	Classification	Papers
Input data	Survey	[4,7,17]
	Limited empirical data	[9,13,16] & [This paper]
	Abundant empirical data	[8,10,11,14,15]
Use cases	Residential	[7–9,15]
	Local energy system (e.g., office building)	[10,14] & [This paper]
	Large-scale use cases (e.g., country level)	[4,11,13,16,17]
Method	Consequential probabilities	[7–10,16]
	Non-consequential probabilities	[4,11,13–15,17] & [This paper]

1.2. Research Gap and Contributions

The list of papers previously cited shows the interest in modeling the charging behavior. There are still some research questions to be answered. First of all, the models presented previously tend to use different types of datasets leading to different methodologies and results. Some of these papers use abundant empirical data (e.g., EV motor power, battery capacity, temperature, etc.) which usually are not available, difficult to access and very user- and vehicle-centric. Other papers use surveys to build their model which is less accurate than empirical data and not easy to deploy. Very few papers have developed a methodology able to be applied on many different use cases such as local energy systems (e.g., office buildings, shops, etc.). Secondly, none of the previous papers have the ability to capture specific charging behaviors of a specific use case—for instance, specific behaviors such as morning vs. afternoon charging shifts or employee vs. visitor charging behavior. Nonetheless, such particular charging behaviors must be included in the model to have a more accurate simulation.

The previous research gaps are the main research questions that are intended to be solved in this paper. The key contributions of the EV charging sessions generator proposed in this paper are:

- The input dataset of the methodology originates from a standard communication protocol widely available in the interoperability of charging infrastructures. The standard communication protocol allows for applying the methodology on many different and specific use cases (office buildings, shops, houses, etc) and can help to investigate/design different use cases;
- The classification of EV driver's profiles with similar charging behaviors in order to improve the modeling and simulation results. The classification is performed using a clustering technique. In addition, the Kernel Density Estimation process is used to better capture details of each cluster as well as particular charging behaviors;
- The modularity of the generator, its ease-of-use and the standardized output data format are key attributes of its scalability and replicability.

1.3. Structure of Paper

The paper is organized as follows: The methodology of the charging sessions generator is explained in Section 2. The use case upon which the generator is tested is presented in

Section 3.1, and the results are analyzed in Section 3.2. In addition, Section 3.3.1 focuses on the validation of the methodology. Finally, to illustrate the benefit of using the generator, the impact of uncoordinated and smart charging is simulated and presented in Section 3.4.2.

2. Materials and Methods

The structure of the methodology proposed in this paper is illustrated in Figure 1. It consists mainly of four different steps. In a first step, a data pre-processing is carried out and detailed in Section 2.1. This step explains the origin of the data, the cleaning of it and the features it contains. Section 2.2 introduces the clustering algorithm used to characterize the EV charging behavior and the need to normalize the dataset. Followed by this, the main generator algorithm with the statistical parameters is explained in Section 2.3. Finally, Section 2.4 details the validation criteria used to assess the performances of the methodology.

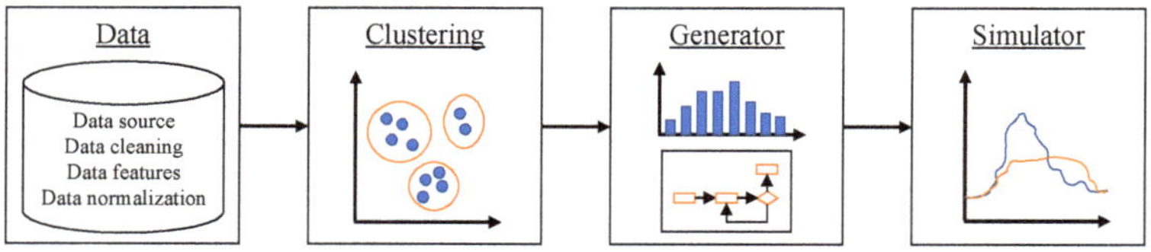

Figure 1. Scheme of the methodology of the generator.

2.1. Data Pre-Processing

2.1.1. Dataset and Features

The dataset consists of individual charging sessions where each of them contains four different features: (a) an identification, (b) a plug-in time, (c) a plug-out time; and (d) an energy consumed. These features are usually available since they are part of the communication standard "Open Charge Point Protocol" (OCPP) between charge points and the charge point operator (CPO) [18]. These charging sessions data are required in order to bill a charging session to an EV driver (using Charge Detail Record (CDR) in OCPP).

To obtain a specific behavior per EV driver, a technique is used to replace all charging sessions from an individual EV driver to one specific theoretical charging session. The method consists of computing the mean value of the plug-in times, the parking times (where parking time is the difference between plug-in and plug-out times) and the energy charged, of all the charging sessions of an EV driver. This results in one theoretical charging session per EV driver consisting of only mean values.

Once each EV driver is associated with one theoretical charging session, they still need to be classified based on the frequency of charging. In other words, a new feature shall be associated with a driver to differentiate regular EV drivers from occasional ones. One would use the number of charging sessions of an EV driver to make the distinction. However, this approach would be limited by the fact that the period where these charging sessions took place would not be considered. For instance, an EV driver that charged 20 times in 20 days is assumed to have the same behavior as an EV driver that charged 20 times in 200 days. To include this aspect, the frequency of charging is used instead of just the number of charging sessions. The frequency of charging, denoted $freq$, is defined in (1):

$$freq = \frac{\text{Number of charging sessions}}{\text{Period between first session and last day of the dataset}} \tag{1}$$

The period in the denominator includes the last day of the dataset and not the last day of an EV driver session. Unless this distinction is made, a driver with a single charging session would present a frequency of one which would misleadingly be interpreted as a 100% probability of having one session per day.

2.1.2. Data Cleaning

The raw dataset needs to be cleaned since it contains charging sessions that failed or did not start. From a time perspective, all charging sessions with parking times lower than five minutes are removed. In addition, duplicated charging sessions, based on the plug-in time, are filtered out. Finally, the time is in UTC so it is adjusted to local time.

A second set of filters is applied on the energy measurements. Firstly, all charging sessions with an energy charged lower than 500 Wh are filtered out. In addition, charging sessions with an energy demand higher than 120 kWh are deleted. Finally, an average charging power is computed per charging session by dividing the energy by the parking time. All average power higher than 22 kW is removed since the charge points are AC technology, hence limited to 22 kW.

2.2. Clustering Technique

The objective is to group EV drivers with a similar charging behavior. There exist multiple techniques to group similar objects together. A well-known approach is an unsupervised machine learning technique called clustering. Among the numerous clustering algorithms in the literature, the most common one is the *k-means* clustering. The objective of this clustering algorithm is to minimize the within-cluster sum of squares. It is mathematically expressed in (2):

$$min \sum_{i=1}^{k} \sum_{x \in S_i} ||x - \mu_i||^2 \tag{2}$$

where k is the number of clusters, S_i a subset of datapoints of cluster i, and μ_i is the mean value of the subset S_i. An in-depth explanation on the working principle of *k-means* clustering can be found in [19]. Examples of clustering of charging sessions using *k-means* algorithm can be found in [20–23].

The *k-means* algorithm requires the number of clusters as input to cluster. Since it is unknown how many clusters there are, different metrics are used from literature to help identify the correct number of clusters. These are the elbow method [21] and the Davies–Bouldin score [22]. The elbow method consists of plotting the clustering results in the function of the number of clusters. The number of clusters to choose should appear in an elbow on the plot. Intuitively, the more clusters are added, the better the objective function is, but the higher the probability is of over-fitting. Hence, the plot should reflect the balance between the maximum number of clusters and the over-fitting issue. The Davies–Bouldin score is another method which computes the ratio of within-cluster and between-cluster distances. The lower the value is, the optimal number of clusters is.

The clustering proposed in this study works in two stages. A first clustering is performed on the charging sessions' plug-in time, parking time and energy charged features. The second clustering is then performed on the frequency of charging of an EV driver. It has been decided to cluster in two stages since clustering all features together results in a bad grouping of EV drivers and low flexibility in choosing the right number of clusters.

Finally, grouping behaviors together has the advantage of being able to modify certain groups in different ways, making the methodology highly modular. For instance, if a cluster has been identified as employee behavior (early arrival time and long parking time), and it is known from the charging site that many new employees will switch to EVs in the coming years; this specific cluster can be modified to mimic this future behavior specific to the site. Section 3.4.1 shows an example of the strength of the methodology in being able to modify certain clusters to draft future scenarios.

Data Normalization

The features that will be clustered are on different scales, with different ranges, and sometimes on a different order of magnitude. The issue is that clustering techniques rely strongly on the distance between features. Hence, it is important to adjust all feature scales to a common scale by using a normalization technique. Two different common normalization techniques can be found in literature, called *z-score* and *min-max* normalization. The choice between the two depends mainly on the data source and its characteristics. Since the data source is based on human behavior, which is highly stochastic and usually following Gaussian distributions, the technique used is the *z-score* normalization. The formula of the *z-score* normalization is given by (3)

$$x_N = \frac{x_t - \mu}{\rho} \tag{3}$$

where x_t is the non-normalized data point, μ is the mean value, and ρ is the standard deviation.

2.3. Generator Principle

2.3.1. Statistical Distributions

Once the charging sessions have been clustered together, some statistical parameters are extracted to build probability distributions. The main statistical parameters which are extracted per cluster are listed hereunder:

- The probability of having a certain number of charging sessions per day. It has been decided to divide this probability into two probability distributions, mainly one for the working days and one for the weekend days, since the number of sessions are highly different;
- The probability of having an EV plug-in and plug-out at a certain time;
- The probability of having a certain amount of energy to charge.

Each statistical distribution is estimated using a kernel density estimation process from [24] instead of the common Gaussian distribution used in many papers. The advantage relies in a smoother representation of the drivers behavior and hence better results. This is shown in Section 3.3.1.

2.3.2. Pseudo Algorithm

The EV charging sessions generator principle is shown in Algorithm 1. It requires two inputs: (a) the period over which to simulate and (b) the probability distributions per cluster. Using this information, the generator works in two main steps:

- Step (1) For each cluster, and for each day to simulate, a function (called f1) determines the number of charging sessions to generate;
- Step (2) For each charging session to generate, two functions (called f2 and f3) determine the plug-in and plug-out time, and the energy to charge.

Algorithm 1 EV charging session generator

```
Input: Simulation dates, clusters data
for all clusters do
    for all simulation dates do
        f1: Get number of sessions
        for all sessions do
            f2: Get plug-in and plug-out time
            f3: Get energy
Output: Generated charging sessions
```

The first function (f1) chooses a certain number of charging sessions for the day and cluster selected. It will select randomly a number of sessions based on the probability distribution of the frequency feature for a cluster. Similarly, the second function (f2) chooses a plug-in and plug-out time based on the probability distributions of those features. With function (f3), two values are created: (a) the maximum energy that can be charged considering the maximum charging power of the charger during a parking time determined with (f2), and (b) a random energy selection based on the distribution of the energy feature of the cluster. This is carried out to verify that the energy generated for the session is lower than the maximum energy that can be charged.

2.4. Validation Criteria

In order to ensure the validity of the proposed method, the dataset is divided into two subsets. A first subset, called training subset, contains 90% of the dataset, and a second subset, called validation subset, contains the remainder of the dataset. The training subset allows for fitting the model by creating the clusters and to generate EV charging sessions data. Then, the generated EV charging session data can be assessed by comparing it with the validation subset.

The objective is to assess the fit between the generated charging sessions and the real charging sessions. Since the charging sessions are characterized by plug-in time, parking time and energy consumed, histograms of each feature will be shown. Followed by this, a comparison between the histograms from the generated charging sessions and the validation charging sessions will be shown and assessed. The assessment is carried out using the *chi-square histogram distance* evaluation metrics detailed in (4) [25]:

$$\chi^2 = \frac{1}{2}\sum_{i=1}^{n}\frac{(p_i - q_i)^2}{p_i + q_i} \tag{4}$$

where $i = 1, ..., N$ is the ith bin on the total amount of bins N, and p_i and q_i are the occurrences associated with bin i.

3. Results and Discussion

3.1. Use Case

The use case understudy is to simulate the expected charging needs in 2025 of the hospital's parking lot in Brussels. It is an open-access charging location, with paid parking fee. It consists of six charge points containing two Type 2 (IEC 61851) connectors, hence a maximum of 12 connectors, delivering up to 22 kW each. The charging sessions have been logged between May 2018 and January 2022. This period corresponds to 10,477 charging sessions and 424 different EV drivers. To have a better understanding on the use case under study, the number of charging sessions from frequent drivers represents 95.7% of the total number of charging sessions. In addition to a parking lot, solar panels of 590 kWp are installed with an on-site transformer limited to 630 kW. Different buildings are connected to transformer with a low electricity demand compared to the transformer limit.

3.2. Clustering Results

The first step is to cluster the plug-in time, the parking time and the energy charged. Using the elbow method and Davies–Bouldin score, the number of clusters has been set to five. The two methods are shown in Figure 2.

The elbow method does not show a straightforward elbow so it does not help to find the optimal number of clusters. Still, the number of clusters should be between four and seven. On the other hand, the Davies–Bouldin score indicates the optimal number of clusters to be five clusters (lowest score on Figure 2). To have an idea on how EV drivers differ and are grouped, two different clusters are represented, as an example, in Figure 3, by illustrating the plug-in times, parking times and energy needs.

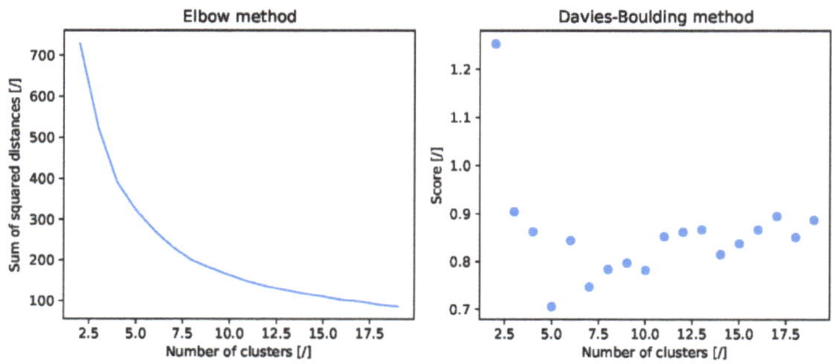

Figure 2. Number of clusters identification for the k-means clustering algorithm.

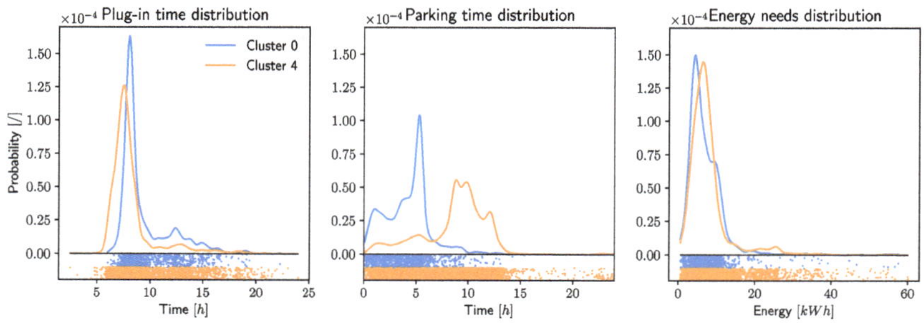

Note: Scatter points below the zero y-axis are shown to better understand the distribution profiles.

Figure 3. Results of the first clustering.

Both clusters have similar plug-in times and energy demands but a strong difference in parking times. Cluster 0 typically represents an early visitor profile, whereas Cluster 4 typically represents an employee profile, with early arrival and long stay. To have a quantitative understanding, Table 2 summarizes the characteristics of the five clusters.

Table 2. Clusters' quantitative characteristics.

Cluster ID	# of Sessions	# of Drivers	Plug-In Time (Mean Value)	Parking Time (Mean Value)	Energy (Mean [kWh])	Sub-Clusters
Cluster 0	1088	104	Morning (09h26)	Mid (04h15)	Low (7.22)	2
Cluster 1	826	139	Afternoon (15h51)	Mid (05h52)	Low (9.31)	2
Cluster 2	521	39	Morning (09h42)	Long (07h03)	High (40.4)	2
Cluster 3	2	1	Afternoon (16h45)	Very long (38h48)	Low (5.07)	N.A.
Cluster 4	6618	69	Morning (08h15)	Long (08h58)	Mid (7.9)	3

Note: Parking time: "Short" below 3 h, "Mid" for [3–6] h, "Long" for [6–9] h, "Very long" over 9 h. Energy: "Low" below 10 kWh, "High" over 10 kWh.

Table 2 shows a couple of different charging behaviors which are interesting to analyze. First of all, the mean energies are relatively low except for Cluster 2 (which represents 9.2% of the number of total EV drivers). Cluster 2 and Cluster 4 have the same parking behavior (morning arrival and long stay) but not the same charging behavior (higher energy demand). A second interesting result to discuss is that Cluster 3 is represented only by one driver because this driver has a very different behavior compared to the other clusters.

This cluster shows the good performances of the clustering technique since it is able to exclude outliers into an individual cluster. Finally, Cluster 0 and Cluster 1 differ from the other clusters with short parking times, and are themselves differentiated by morning and afternoon arrivals. To conclude, Table 2 shows high differences in behavior between clusters. Hence, it shows the importance to cluster drivers' behavior.

The last column of Table 2 shows the results of the second level clustering. It indicates the number of subclusters based on the frequency of charging. The results show that most of the clusters are divided into two subclusters, mainly in low and high frequency of charging. Nevertheless, for certain clusters, more than two subclusters are required due to the higher frequency divergence of charging.

3.3. Generator Results

3.3.1. Validation

Figure 4 shows three histograms that consist of plug-in times, parking times and energy needs, for both generated charging sessions and the real charging sessions (from the validation subset). The number of charging sessions generated is 1335, which is close to the original number of charging sessions which is 1422, on a period of four months.

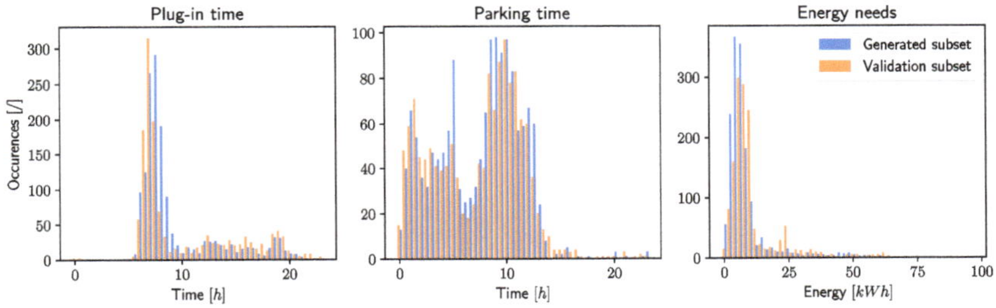

Figure 4. Comparison of plug-in time, parking time and energy needs between generated and real charging sessions.

The results show a good match between the generated and the real charging sessions. The methodology proposed in this paper is able to capture different subtleties. For instance, regarding the plug-in times, the method correctly generates many charging sessions in the morning but tends to underestimate the afternoon charging sessions. Similarly, the generated parking times follow correctly the validation subset but with an underestimation of short parking times. Finally, the generated energy needs fit the validation subset, with an underestimation of higher energy demand. This is due to a low number of charging sessions with high energy demand (as shown in Table 2). In general, Figure 4 shows the great representativeness of the methodology proposed in this paper with a close match between the generated subset and the validation subset.

3.3.2. The Impact of Clustering and Kernel Density Distribution

Two important novelties are included in the methodology of this paper which are the introduction of clustering and the use of kernel density distributions. It is therefore interesting to assess their impact on the validation subset. Four different scenarios are tested and assessed by computing the *chi-square histogram distance* using (4). The results are presented in Table 3.

Table 3. Chi-square distance evaluation for different scenarios.

Scenario ID	Description	Plug-In Time	Parking Time	Energy Needs
1	Gaussian distribution without clustering	491.8	135.7	261.3
2	Gaussian distribution with clustering	379.1	167.7	136.7
3	Kernel distribution without clustering	123.9	48.9	221.4
4	Kernel distribution with clustering	103.7	37.9	181.5

From Table 3, it can be observed that the introduction of kernel distribution has a positive impact, by reducing the *chi-square distance*, on the plug-in time and parking time but not on the energy needs. The main reason is that the energy needs distribution in Figure 4 strongly follows a Gaussian distribution. It can also be observed that the introduction of clustering has a positive impact on the three features, where scenario 4 has the lowest chi-square distance. It is important to note that the clustering main objective is not to improve the model, but rather to make it modular by allowing changes in groups of EV drivers.

3.3.3. The Evolution in Charging Behavior

It has been observed that the charging session generator tends to underestimate the energy demand. From the example of the previous section, the total energy demand from the generated charging sessions is 14,480 kWh, which is 19.2% lower than the real validation total energy demand of 17,933 kWh. To understand such underestimation, Figure 5 shows the daily energy demand of the full dataset.

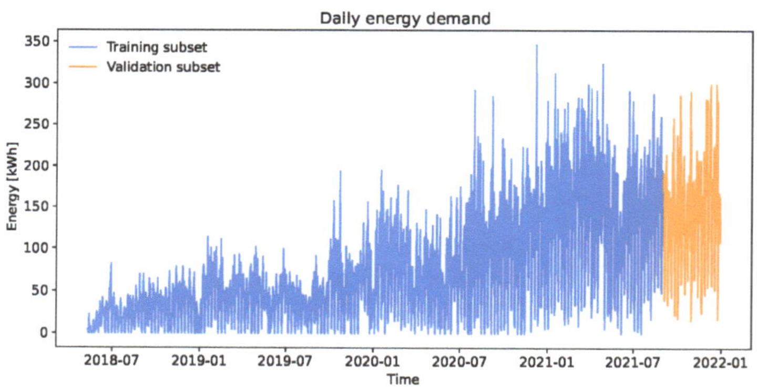

Figure 5. Daily charging energy demand evolution.

From Figure 5, it can be observed that the daily energy demand has a strong evolution over the four years of the dataset. Two main reasons can explain such evolution which are an increase of the number of users and the evolution from PHEVs to BEVs. The drawback of the methodology proposed in this paper is that it includes low energy demand charging sessions (from 2018 until 2020) in the statistical distributions. Consequently, it will reproduce such energy demands in the validation subset, hence lowering the overall energy demand. An interesting future research would be to find a methodology that can model the evolution.

3.4. Simulation Results

3.4.1. Scenario Construction

The final objective of the methodology proposed in this paper is to generate EV charging session data based on clustered EV drivers' behavior. Generating EV charging session data directly from the clusters will reproduce the same scenario as the existing one (from the existing data) which is not the goal. To generate future scenarios, some changes in the clusters are required to build more realistic (future) scenarios. How these scenarios are built is the main subject of this section.

It is known from literature that the EV penetration follows an exponential behavior, and that it will likely still follow an exponential behavior in the future [2]. Knowing this, the clusters are expected to follow an exponential behavior. Consequently, an extrapolation technique is used to evaluate the trend of each cluster and the expected future behavior. In this paper, a *nonlinear least squares* method is used to extrapolate the clusters behaviors [26]. Results of this method are shown in Figure 6.

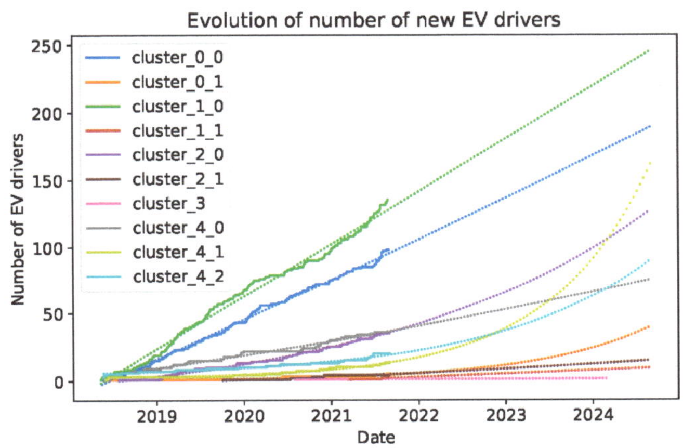

Figure 6. Extrapolation of number of charging sessions per cluster.

From the extrapolation technique, a scenario can be drafted for 2025. The clusters are then accordingly modified, and the scenario will be simulated in Section 3.4.2.

3.4.2. Uncoordinated vs. Smart Charging

The use case understudy is a hospital parking lot which desires to install new charge points to meet the future EV charging demand in 2025. The first objective is to analyze the impact of installing new charge points using uncoordinated charging on the local energy system. The second objective is to analyze the positive impacts of coordinating the charging processes using a smart charging algorithm.

In order to achieve this, the generated charging sessions produced by the generator are simulated in a simulator available in [6]. The smart charging algorithm is a model predictive control algorithm that minimizes the cost of charging and the peak powers. In this paper, perfect forecast is used to simplify the simulations. The results are presented in Figure 7 showing the charging power profile dynamics by building, for each quarter hour, the mean values, 1st and 3rd quartiles values and the maximum values of full simulation.

Figure 7 shows the strong impact of uncoordinated charging on the parking lot of the hospital. The maximum values go up to 728 kW, exceeding the maximum capacity of the on-site 630 kW transformer. In addition, important injection can be seen during the day, mainly because the EVs are fully charged. When controlling the charge points in a smart way, the maximum peaks, the 3rd quartile and the mean values are strongly reduced to take advantage of overproduction during the day. This is the peak shaving and valley filling principle. The self-consumption changes from 27.5% for uncoordinated charging to 59% for smart charging.

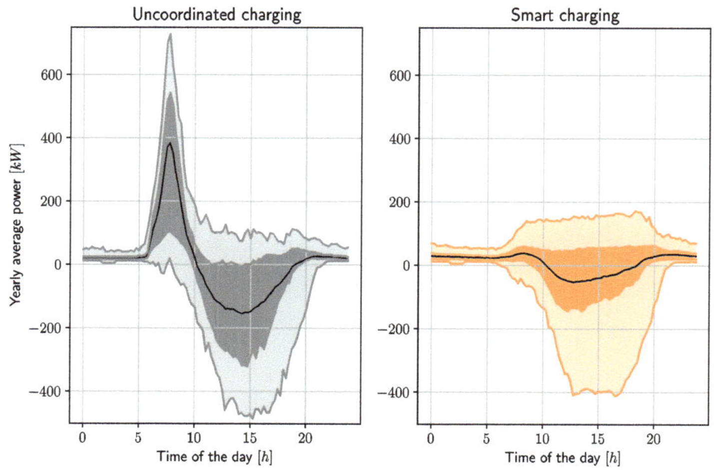

Figure 7. Six-month simulation of uncoordinated and smart charging.

4. Conclusions

In this paper, an EV charging sessions generator is presented. It enables the creation of charging sessions data based on historical data of a specific charging location. The historical data have been analyzed to group different types of EV drivers together. For each group, specific sets of statistical parameters are extracted, which are then used by the generator. The full methodology is applied on a use case of a hospital which plans to expand its EV fleet. The results are presented in two parts.

A first part is dedicated on the clustering of EV charging sessions from historical data. The different types of EV drivers are presented with some important characteristics. The results show the big differences in behavior between the EV drivers and the importance of grouping such drivers. A second part focuses on the actual generation of EV charging sessions data. The generator is validated by comparing the historical data set with a newly generated session according to populated clusters. A scenario is then defined and analyzed by simulating uncoordinated charging and smart charging from the generated charging sessions. The results indicate the strong impact on power and energy demand when adding new EV drivers to the population. The analyzed scenario highlights the need for grid reinforcement or smart charging technologies to avoid overloading and peak demands due to an increase of charging sessions of specific EV driver types. The good results of the validation process demonstrate the potential of this generator to simulate new scenarios while the scenario analysis demonstrates its usefulness to analyse future EV transition scenarios.

Future research could include a method to follow the evolution of the energy demand of EVs since drivers are switching from plug-in hybrid EVs to battery EVs requiring more charging needs.

Author Contributions: Conceptualization, C.D.C. and G.V.K.; methodology, G.V.K.; software, G.V.K.; validation, G.V.K., C.D.C. and N.S.; formal analysis, G.V.K.; data curation, C.D.C.; writing—original draft preparation, G.V.K.; writing—review and editing, C.D.C. and N.S.; visualization, G.V.K.; supervision, C.D.C. and N.S.; project administration, T.C. and M.M.; funding acquisition, T.C. and M.M. All authors have read and agreed to the published version of the manuscript.

Funding: The authors would like to thank the Agency for Innovation and Entrepreneurship (VLAIO) for providing funding of this research under Grant No. HBC.2018.0519 and Flux50 for support to our team.

Data Availability Statement: The datasets used during the current study are not available since they are protected by privacy reasons, mainly because of the presence of EV users' radio frequency identification (RFID).

Conflicts of Interest: The authors declare no conflict of interest.

Abbreviations

The following abbreviations are used in this manuscript:

APS	Announced Pledges Scenario
BEV	Battery Electric Vehicle
CPO	Charge Point Operator
CDR	Charge Detail Record
DSO	Distribution System Operator
EV	Electric Vehicle
IEA	International Energy Agency
LES	Local Energy System
OCPP	Open Charge Point Protocol
PHEV	Plug-in Hybrid Electric Vehicle
RAMP	Remote-Areas Multi-energy systems load Profiles
RFID	Radio Frequency Identification
TSO	Transmission System Operator

References

1. European Parliament. 2021/0197(COD)-CO2 Emission Standards for Cars and Vans. 2022. Available online: https://oeil.secure.europarl.europa.eu/oeil/popups/ficheprocedure.do?reference=2021/0197(COD)&l=en (accessed on 1 February 2023).
2. International Energy Agency (IEA). *Global EV Outlook 2022*; IEA: Paris, France, 2022; pp. 98–100. Available online: https://www.iea.org/reports/global-ev-outlook-2022 (accessed on 1 February 2023).
3. Firouzjah, K.G. Profit-based electric vehicle charging scheduling: Comparison with different strategies and impact assessment on distribution networks. *Int. J. Electr. Power Energy Syst.* **2022**, *138*, 107977. [CrossRef]
4. Mangipinto, A.; Lombardi, F.; Sanvito, F.D.; Pavičević, M.; Quoilin, S.; Colombo, E. Impact of mass-scale deployment of electric vehicles and benefits of smart charging across all European countries. *Appl. Energy* **2022**, *312*, 118676. [CrossRef]
5. Dubey, A.; Santoso, S. Electric Vehicle Charging on Residential Distribution Systems: Impacts and Mitigations. *IEEE Access* **2015**, *3*, 1871–1893. [CrossRef]
6. Van Kriekinge, G.; De Cauwer, C.; Sapountzoglou, N.; Coosemans, T.; Messagie, M. Peak shaving and cost minimization using model predictive control for uni- and bi-directional charging of electric vehicles. *Energy Rep.* **2021**, *7*, 8760–8771. [CrossRef]
7. Fischer, D.; Harbrecht, A.; Surmann, A.; McKenna, R. Electric vehicles' impacts on residential electric local profiles – A stochastic modelling approach considering socio-economic, behavioural and spatial factors. *Appl. Energy* **2019**, *233–234*, 644–658. [CrossRef]
8. Muratori, M. Impact of uncoordinated plug-in electric vehicle charging on residential power demand. *Nat. Energy* **2018**, *3*, 193–201. [CrossRef]
9. Iversen, E.B.; Morales, J.M.; Madsen, H. Optimal charging of an electric vehicle using a Markov decision process. *Appl. Energy* **2014**, *123*, 1–12. [CrossRef]
10. Gruosso, G.; Gaiani, G.S. A Model of Electric Vehicle Recharge Stations based on Cyclic Markov Chains. In Proceedings of the IECON 2019-45th Annual Conference of the IEEE Industrial Electronics Society, Lisbon, Portugal, 14–17 October 2019; Volume 1, pp. 2586–2591. [CrossRef]
11. Gaete-Morales, C.; Kramer, H.; Schill, W.P.; Zerrahn, A. An open tool for creating battery-electric vehicle time series from empirical data, emobpy. *Sci. Data* **2021**, *8*, 1–18. [CrossRef] [PubMed]
12. Lombardi, F.; Balderrama, S.; Quoilin, S.; Colombo, E. Generating high-resolution multi-energy load profiles for remote areas with an open-source stochastic model. *Energy* **2019**, *177*, 433–444. [CrossRef]
13. Jiang, R.; Zhang, Z.; Li, J.; Zhang, Y.; Huang, Q. A coordinated charging strategy for electric vehicles based on multi-objective optimization. In Proceedings of the 2017 2nd International Conference on Power and Renewable Energy, ICPRE, Chengdu, China, 20–23 September 2017, pp. 823–827. [CrossRef]
14. Schäuble, J.; Kaschub, T.; Ensslen, A.; Jochem, P.; Fichtner, W. Generating electric vehicle load profiles from empirical data of three EV fleets in Southwest Germany. *J. Clean. Prod.* **2017**, *150*, 253–266. [CrossRef]
15. Brady, J.; O'Mahony, M. Modelling charging profiles of electric vehicles based on real-world electric vehicle charging data. *Sustain. Cities Soc.* **2016**, *26*, 203–216. [CrossRef]
16. Lahariya, M.; Benoit, D.F.; Develder, C. Synthetic Data Generator for Electric Vehicle Charging Sessions: Modeling and Evaluation Using Real-World Data. *Energies* **2020**, *13*, 4211. [CrossRef]

17. Harris, C.B.; Webber, M.E. An empirically-validated methodology to simulate electricity demand for electric vehicle charging. *Appl. Energy* **2014**, *126*, 172–181. [CrossRef]
18. Open Charge Allaince-OCPP2.0. 2022. Available online: https://www.openchargealliance.org/protocols/ocpp-20/ (accessed on 1 February 2023).
19. Aggarwal, C.C. *Data Mining*; Springer International Publishing: Cham, Switzerland, 2015; Volume 1, pp. 162–163. [CrossRef]
20. Wang, Y.; Cai, S.; Shen, Y.; Shao, D.; Gong, X.; Zhou, C.; Chu, D. The analysis of electrical vehicles charging behavior based on charging big data. In Proceedings of the 2019 IEEE 4th International Conference on Cloud Computing and Big Data Analytics, ICCCBDA, Chengdu, China, 12–15 April 2019; Volume 1, pp. 63–67. [CrossRef]
21. Crozier, C.; Apostolopoulou, D.; McCulloch, M. Clustering of Usage Profiles for Electric Vehicle Behaviour Analysis. In Proceedings of the Proceedings-2018 IEEE PES Innovative Smart Grid Technologies Conference Europe, ISGT-Europe, Sarajevo, Bosnia and Herzegovina, 21–25 October 2018; Volume 1, pp. 1–6. [CrossRef]
22. Xydas, E.; Marmaras, C.; Cipcigan, L.M.; Jenkins, N.; Carroll, S.; Barker, M. A data-driven approach for characterising the charging demand of electric vehicles: A UK case study. *Appl. Energy* **2016**, *162*, 763–771. [CrossRef]
23. Xiong, Y.; Wang, B.; Chu, C.C.; Gadh, R. Electric Vehicle Driver Clustering using Statistical Model and Machine Learning. In Proceedings of the IEEE Power and Energy Society General Meeting, Portland, OR, USA, 5–10 August 2018; Volume 2018. [CrossRef]
24. Pedregosa, F.; Varoquaux, G.; Gramfort, A.; Michel, V.; Thirion, B.; Grisel, O.; Blondel, M.; Prettenhofer, P.; Weiss, R.; Dubourg, V.; et al. Scikit-learn: Machine learning in Python. *J. Mach. Learn. Res.* **2011**, *12*, 2825–2830.
25. Pele, O.; Werman, M. The Quadratic-Chi Histogram Distance Family. In *Lecture Notes in Computer Science, Proceedings of the Computer Vision—ECCV 2010, Crete, Greece, 5–11 September 2010*; Daniilidis, K., Maragos, P., Paragios, N., Eds.; Springer: Berlin/Heidelberg, Germany, 2010; Volume 6312, pp. 749–762. [CrossRef]
26. Virtanen, P.; Gommers, R.; Oliphant, T.E.; Haberland, M.; Reddy, T.; Cournapeau, D.; Burovski, E.; Peterson, P.; Weckesser, W.; Bright, J.; et al. SciPy 1.0: Fundamental Algorithms for Scientific Computing in Python. *Nat. Methods* **2020**, *17*, 261–272. [CrossRef]

World Electric Vehicle Journal

MDPI

Article

Estimation of Public Charging Demand Using Cellphone Data and Points of Interest-Based Segmentation

Victor Radermecker [1,†] and Lieselot Vanhaverbeke [2,*,†]

1 MOBI Research Group, Université Libre de Bruxelles, 1050 Brussels, Belgium
2 Department BUTO—Business Technology & Operations, Faculty of Social Sciences & Solvay Business School, Vrije Universiteit Brussel, 1050 Brussels, Belgium
* Correspondence: lieselot.vanhaverbeke@vub.be
† Current address: MOBI Research Group, Pleinlaan 2, 1050 Brussels, Belgium.

Abstract: The race for road electrification has started, and convincing drivers to switch from fuel-powered vehicles to electric vehicles requires robust Electric Vehicle (EV) charging infrastructure. This article proposes an innovative EV charging demand estimation and segmentation method. First, we estimate the charging demand at a neighborhood granularity using aggregated cellular signaling data. Second, we propose a segmentation model to partition the total charging needs among different charging technology: normal, semi-rapid, and fast charging. The segmentation model, an approach based on the city's points of interest, is a state-of-the-art method that derives useful trends applicable to city planning. A case study for the city of Brussels is proposed. Our demand estimation results heavily correlate with the government's predictions under similar assumptions. The segmentation reveals clear city patterns, such as transportation hubs, commercial and industrial zones or residential districts, and stresses the importance of a deployment plan involving all available charging technologies.

Keywords: EV (electric vehicle); charger; deployment; infrastructure; case-study

Citation: Radermecker, V.; Vanhaverbeke, L. Estimation of Public Charging Demand Using Cellphone Data and Points of Interest-Based Segmentation. *World Electr. Veh. J.* **2023**, *14*, 35. https://doi.org/10.3390/wevj14020035

Academic Editor: Joeri Van Mierlo

Received: 24 November 2022
Revised: 23 December 2022
Accepted: 10 January 2023
Published: 30 January 2023

1. Introduction

Climate change is probably the premier hurdle ever faced by our societies, and as engineers and scientists, it is our responsibility to drive forward research towards a sustainable world. Most cities worldwide have set high environmental ambitions for decarbonizing their transport network. The race for road electrification has started, and convincing drivers to switch from fuel-powered vehicles to Electric Vehicles (EV) requires robust public charging infrastructure.

Indeed, deploying enough charging infrastructure to meet the growing demand for EVs is a crucial enabler in switching to sustainable mobility and a predominantly zero-emission vehicle fleet by 2050. According to the European Automobile Manufacturers Association, the electrically chargeable segment of new vehicles registered increased from 3% to 10% from 2019 to 2020, during the coronavirus outbreak. Furthermore, the automotive industry reflects that the European production of EVs will multiply six-fold by 2025, exceeding 4 million cars and vans annually [1,2].

Today, the transition is already taking place, and EV charging infrastructures must follow this new trend to meet drivers' energetic needs. The end goal is to have a sufficient charging capacity to avoid "range and queuing anxieties" among EV users: the fear that their vehicle will have insufficient range to reach their destination and that charging may mean lengthy queues if a station is already occupied.

As public charging stations occupy the already scarce public space, governments need a comprehensive understanding of their cities' charging demand patterns to ensure the optimal location of these newly built charging points. Today, three major EV charging technologies exist: normal, semi-rapid, and fast. They differ in the type of electric current

(AC/DC) and the maximum power delivered by the terminal, which results in different charging speeds for drivers. While the fastest stations can meet the needs of more drivers and reduce range anxiety, they are the most expensive to build and can cause electric grid instabilities.

Many scientific studies have been carried out to develop optimization algorithms to optimally locate stations in urban areas. However, most researchers did not consider the different charging technologies. This is problematic as city planning officials need to know where which types of stations would be most suitable. The ultimate goal of this research project is twofold.

First, we estimate the EV charging demand at a neighborhood granularity through Cellular Signalling Data (CSD). As shown in the literature review in Section 2, a lot of research has already been done on this topic, but no case study was applied to the city of Brussels. Furthermore, most studies had access to individual-level CSD datasets, whereas in this case, only city trends are available due to the European General Data Protection Regulation (GDPR) regulations. With the needs of each neighborhood being established, we could prioritize the deployment to accelerate the city's electrification. For instance, it could help decide where to place the first 250 charging stations of the 2022 cohort.

Second, another critical issue that researchers have little explored is the segmentation of this demand into different charging technologies. Suppose that one neighborhood requires 600 kWh of daily charging demand. Would drivers benefit the most from normal, semi-rapid, or rapid charging? Imagine that you run out of electricity on a motorway, and as you decide to stop at the first fuel station, you discover that only normal (or "slow") charging stations are available. This means that you would have to stay there for many hours before having enough energy to finish your journey. Of course, this example is extreme as no one would imagine installing such charging stations on a freeway service area. However, this example highlights the importance of this challenge, which could drastically increase the occupancy rate of the terminals if answered correctly. Knowing the massive price differences among technologies and the construction constraints such as grid instabilities or access to high-voltage electricity, maximizing the station's occupancy rate is crucial for them to be economically viable. These topics were little explored in the literature, and in this article, we propose an innovative demand segmentation model based on the city's Points Of Interest (POI).

A case study for the city of Brussels (Belgium) is proposed. By 2035, thermal vehicles will be completely prohibited from the city. The government has therefore set itself the goal of building 22,000 charging points by then, which translates into 11,000 charging stations (each station being equipped with two charging points).

2. Literature Review

The literature offers many articles on optimal locations of public EV charging stations. First, Section 2.1 presents different studies which do not distinguish the different charging technologies. Most of these articles combine demand estimation methods with optimization models to derive the optimal locations of charging points. Second, Section 2.2 summarizes some studies addressing roll-out strategies for specific charging speeds. Although some papers are starting to emerge, very few authors have investigated these issues in detail.

2.1. Charging Type Independent Approaches

2.1.1. Demand Estimation

Given its importance in optimization algorithms, EV charging demand estimation has been examined by researchers worldwide. Many authors took advantage of Cellular Signaling Data (CSD) to estimate charging demand. Among them, Jia et al. [3] proposed an approach to locate public charging stations using reconstructed EVs trajectories derived from simulation. The simulation is done through a mesoscopic simulation tool that can reconstruct a trajectory based on multiple checkpoints [4]. Although presenting promising results, this study requires CSD at an individual level, which is difficult to collect.

For instance, in Europe, such data mining goes against the GDPR (General Data Protection Regulations). Whereas Jia et al. [3] assumed a constant charging probability for EV drivers, Cao et al. [5] proposed a probabilistic model to simulate EV drivers' charging habits. Their model included various factors, including the State of Charge (SOC), the next driving mileage, the travel time value, and economic factors.

Socio-demographic indicators are popular variables to model the charging demand. He et al. [6] proposed to use institutional and spatial factors to model Beijing's charging demand. They have identified six key socio-demographic attributes: income, vehicle ownership, educational level, age, gender, and family size. Frade et al. [7] used a similar approach for Lisbon, but they distinguished nighttime and daytime charging. Nighttime demand was derived using the number of cars per household, whereas daytime demand was estimated from the volume of employment.

Travel surveys and traffic simulations are also interesting methodologies. Efthymiou et al. [8] used traffic simulation models to build a travel Origin–Destination (OD) matrix employing data collected by phone surveys. Similarly, Baouche et al. [9] took advantage of household travel surveys to build an OD matrix for the city's neighborhoods. They used a routing API to compute the driving distance under realistic traffic conditions and build an energy demand OD matrix through an EV consumption model [10,11]. Cavadas et al. [12] exploited a mobility survey that questioned drivers about their daily trips on business days. They built an energy OD Matrix by converting these trips into EV consumption under specific assumptions.

Finally, some authors assumed that car-parking demand is proportional to EVs' charging demand in urban areas. Chen et al. [13] aggregated a dataset of around 50,000 person trips and combined it to land use characteristics. Using this input data, they trained an Ordinary Least Squares (OLS) regression to ultimately predict the total parking times and derive the charging demand.

2.1.2. Optimization Models

Once the charging demand is estimated, optimization models allow locating optimal sites for stations. Researchers such as Zuo-Jun et al. [14] have extensively aggregated the existing optimization models. To name a few, Upchurch et al. [15] introduced a capacitated refueling location model (CRLM) to locate EV charging stations. The original uncapacitated model assumed that the presence of a refueling station is sufficient to supply all vehicles intersecting the station node, which was not realistic for EVs. Cavadas et al. [12] proposed a mixed integer programming (MIP) model for locating charging stations. The model considered not only the drivers' parking locations but also their daily activities.

Efthymiou et al. [8] proposed a genetic algorithm to estimate the optimal location of charging stations. The advantage is that their algorithm provides valuable results with limited data requirements. Sweda and Klabjan [16] created an agent-based decision support system. Its end goal was to identify patterns in residential EV ownership and enable strategic deployment of new charging infrastructure. Xi and Sioshansi [17] developed a simulation–optimization model which determines optimal EV charging stations' location by maximizing their popularity among EV drivers.

2.2. Charging Type Dependent Approaches

Only a few studies have focused on EV charging infrastructure for specific charging technologies. Zhang and Iman [18] computed suitability scores for normal charging stations in urban areas. Their approach relied on using Geographic Information System (GIS) data to determine the sites' suitability for hosting normal EV charging stations. By aggregating various input factors such as land use, demographics, and employment centers, they derived suitability scores for all stations in Utah (US). Agent-based models have also been explored to investigate charging stations roll-out strategies for semi-rapid and fast technologies. Hoed et al. [19] simulated a population of diverse agents (citizens, foreigners, taxis, and shared vehicles) to model the multiplicity of charging behaviors within cities.

Regarding fast charging, Morrissey et al. [20] investigated a widely accepted roll-out strategy, which consists of creating electric corridors on major interurban motorways to connect cities. They assessed the adequacy of this roll-out strategy, which, in theory, should allow longer trips to be undertaken and increase EV users' potential journey ranges.

Some authors explored optimization methods for specific charging technologies. For instance, Wang and Chuan-Chih [21] expanded the concepts of set-coverage and maximum-coverage to express a capacitated location model that deals with multiple types of charging stations using a MIP algorithm. The main contributions of this study are that it relaxes the constraints of using a single kind of recharging station. The results highlight that a combination of charging technologies, although it requires a higher budget, leads to greater convenience for travelers. Others have also explored optimization models to locate charging stations for specific fleets, such as taxis, shared cars, and buses [22,23].

2.3. A New Approach

Numerous studies have focused on estimating the total EV charging demand to apply location models. However, these approaches do not sufficiently account for the various complementary charging technologies. This article proposes an innovative methodology, which, by segmenting the charging demand into the different charging speeds, allows to re-use the technology-independent models presented in Section 2.1 for specific charging speeds, and hence helps provide actionable insights to city planning authorities. Our method is a two-step approach. First, we propose a demand estimation method, which relies on aggregated CSD, thus respecting GDPR. Second, we take advantage of GIS data to assess each neighborhood's primary function and compute its needs for each charging type.

The paper is structured as follows. Next, we present the methodology, which outlines our two-step method to estimate and segment public charging demand. To illustrate the model, we apply it to the case study of Brussels, Belgium. We present the datasets used and the various models. Results are summarized and discussed to highlight how they can provide actionable insights to planning authorities. Finally, we present conclusions of the work, identify its limitations, and highlight future areas for continued investigation.

3. Materials and Methods

As explained above, our method consists of two main steps. First, Section 3.1 details the EV public charging demand estimation. Second, Section 3.2 develops the segmentation model.

3.1. Demand Estimation

3.1.1. Methodology Overview

The goal of Section 3.1 is to estimate the EV charging demand at neighborhood granularity for Brussels. The city is divided into 145 neighborhoods, with each one having an average area of 0.51 km^2. The whole process is presented in Figure 1. First, we take advantage of aggregated CSD to build an OD matrix, revealing the number of trips between Brussels neighborhoods. In Section 3.1.2, we convert this number of trips into driving distances using BingMaps Services.

Second, two correction factors are applied on top of previous results to make the model more realistic. (1) An analysis of mobility in Brussels leads the way towards the first improvement, the Driving Ratio (DR). This step removes walking, bike, and public transport trips from the data by estimating the ratio of car trips to other transportation means per neighborhood. (2) A socio-demographic study of private parking opens to the second refinement, the Private Parking Ratio (PPR). By estimating the number of citizens owning a private parking lot, we can infer the share of the population having access to a private at-home charger, thus not relying on public charging infrastructure. These factors are presented in Section 3.1.3.

Finally, the total demand is aggregated at a neighborhood level and converted into energy demand in kWh. These conversions are based on the EV energy consumption averages established by the International Energy Agency (IEA).

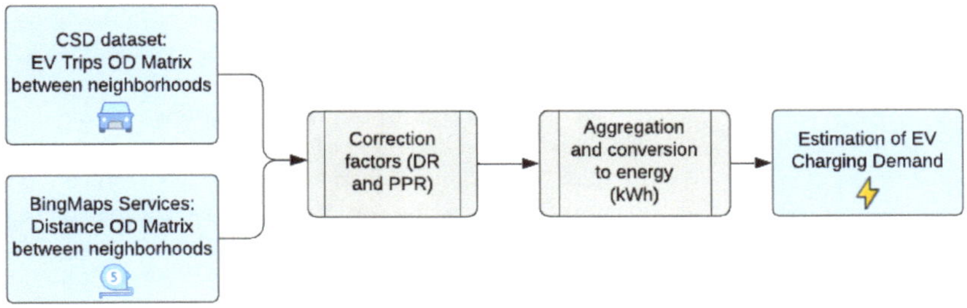

Figure 1. Process of EV charging demand estimation for Brussels. Section 3.1.2 estimates the number of trips between districts based on CSD. Then, it builds an OD matrix by computing the quickest driving distance between neighborhoods. Section 3.1.3 applies two correction factors, the DR and PPR. Section 3.1.4 converts OD matrices into neighborhood-level energy demand in kWh.

3.1.2. Origin–Destination Matrix

This dataset was provided by Proximus S.A., one of Belgium's largest telecommunication companies. It gives the total number of trips between specific origins and destinations pairs, defined as 'staying points' in a Technology Agnostic Cell Sector (TACS) network. Each TACS represents the unified coverage area of an antenna, as shown in Figure 2. A 'staying point' is defined as being in the same TACS for at least 30 min confirmed or 60 min estimated. Confirmed time is when the duration between the first and last network signals in the same sector is at least half an hour. Estimated time is computed as the time passed between two network signals incoming from different sectors.

Figure 2. The 880 Proximus TACS covering the city of Brussels.

The data was collected on workdays, defined as all Mondays, Tuesdays, and Thursdays between 20 January 2020 and 20 February 2020. The final numbers are the average profile of these fifteen data collection days. The trips are divided into two categories: regular and irregular trips. A trip is defined as regular if it occurs not less than twice a week throughout at least four out the five weeks regardless of the time and day (including non-work days). Therefore, multiple daily trips are considered regular. All other trips are defined as irregular. As Proximus customers are not equally spread geographically throughout Brussels and Proximus is not the unique telecommunication services provider, an extrapolation factor is necessary to accurately reflect the population and prevent bias. The original dataset was aggregated at a neighborhood granularity by assigning each TACS to a specific neighborhood. The aggregation process uses the intersection over union indicator to associate the most probable district to each network signal.

As Brussels counts 145 neighborhoods, we have two matrices M^r and M^{ir} of dimensions 145 × 145, where each element M_{ij} corresponds to the number of regular and irregular trips from neighborhood i to j. We can convert these numbers of trips into a distance OD matrix by computing the average distance between all possible OD pairs. Following Baouche et al. [9], we will take advantage of an online geodata Application Programming Interface (API) to perform these computations. The BingMaps Services API provides an interface to perform tasks such as creating static maps with markers, geocoding addresses, retrieving imagery metadata, or creating routes.

Table 1 shows the fastest driving distances computed using the BingMaps API for a subset of five neighborhoods in Brussels. We see that the diagonal elements of the OD matrix are null. Indeed, BingMaps returns a null distance for two identical points, in our case, the centroid's coordinates of the same neighborhood. However, as the CSD dataset includes non-null values for intra-neighborhood trips, we need to compute the average intra-neighborhood distance to fill the diagonal of M_{ij}.

Table 1. Distances (in km) between each pair of the five neighborhoods presented above using the BingMaps API. Traffic conditions: Monday morning (7 AM).

	Altitude 100	**Boondael**	**Vivier d'oie**	**Université**	**Observatoire**	**...**
Altitude 100	0	5.937	5.906	5.430	3.386	...
Boondael	5.616	0	5.930	1.486	4.590	...
Vivier d'oie	6.181	4.868	0	6.478	3.594	...
Université	5.516	1.582	7.609	0	4.809	...
Observatoire	3.579	4.334	4.028	4.422	0	...
...	...	...	...	...	...	...

The most accurate method is to numerically simulate the average distance for each specific neighborhood. We randomly sample pairs of points within each district and compute the relative euclidean distances. This process converges to the desired metric through the mean of all computed distances with sufficient iterations. These are straight-line distances, but people need to follow roads to reach their destinations in reality. Boscoe et al. [24] have analyzed the difference between straight-line and driving distances for more than 66,000 trips in 50 states of the United States. They highlighted highly correlated measures with a coefficient of determination ($r^2 > 0.91$) and a detour index of 1.417. This index can be used to convert the distances as the crow flies to driving distances.

3.1.3. Correction Factors

This first correction factor aims at answering the following question: what proportion of trips collected through the CSD accounts for trips made by cars? In other words, what is the balance of walk, bike, and public transportation trips?

Mobility surveys, such as the recent *Monitor* study for Brussels, are helpful to answer such questions. Table 2 presents the proportion of different transportation means for different trip lengths in Brussels (0 to 50 km) [25,26]. As we have estimated the distance of each specific OD trip, we need to assign a driving proportion to any length from 0 to 50 km. By multiplying the number of trips by this driving proportion, we obtain the corrected number of trips. This correction factor is called the DR.

Table 2. Proportions of each type of transportation means for 1 to 50 km trips in Brussels [25,26].

Distance	Drive	Public Tr.	Bike	Walk	Other
0–1 km	17%	2%	14%	62%	5%
1–2 km	40%	5%	26%	29%	0%
2–5 km	59%	9%	19%	12%	1%
5–10 km	72%	11%	11%	4%	2%
10–20 km	78%	13%	5%	2%	3%
20–50 km	74%	22%	<1%	<1%	<4%

Using the data given in Table 2, we fit an exponential function, $f(x) = A(1 - e^{-Bx})$, to estimate the driving ratio on a continuous spectrum. We use the non-linear least-squares optimization to find the best parameters A and B for this candidate function. The final sum of squares is $\sum_i (y_i - \hat{y}_i) \approx 0.001$, which is relatively low. The two estimated parameters are: $A = 0.759$, $B = 0.466$. Results are shown in Figure 3.

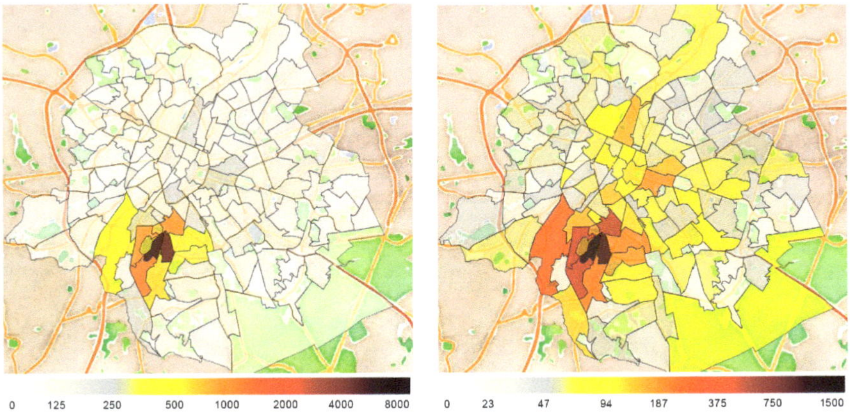

(a) (b)

Figure 3. Total number of trips, before and after applying the Driving Ratio correction factor. (**a**) Total number of trips (regular and irregular) originating from *Altitude 100* before applying the DR correction factor. (**b**) Total number of trips (regular and irregular) originating from *Altitude 100* after applying the DR correction factor.

The drivers' charging behavior is still difficult to anticipate, but among the various hard-to-predict factors, the PPR is a critical variable that must be reliably quantified. It is defined as the proportion of drivers who will own an at-home charging point and will not rely on public charging infrastructure.

Many recent studies concluded that most drivers prefered at-home charging. A recent investigation from Figenbaum showed that, in Norway, in 2017, 97% of drivers living in semi-detached and detached housing were taking advantage of private charging points while only 3% charged their car outside their home [27]. In the US, it is currently estimated that between 84% and 90% of US EV drivers charge their car at home, a number that recently jumped with the COVID-19 pandemic [28]. Other studies in Australia also confirmed that home-charging, when possible, is the preferred solution for EV drivers [29,30]. Adding all governmental incentives for building private charging stations, we can confidently assume that all drivers having the possibility to install at-home charging points will do so.

The PPR is defined as the density of Private Access Roads (PARs) per household and neighborhood. PARs are defined as any private driveways, including garage doors and setback areas, private parking lots, etc. Large-scale inventories are regularly carried out by surveyors in Brussels, which include counting on-street parking spaces and photographing them in compliance with the GDPR. In Equation (1), $\sum_j \sigma_j$ represents the sum of all PARs in neighborhood i, τ_i is the population density and χ_i is the average household size of neighborhood i.

$$PPR_i = \frac{\sum_j \sigma_j}{\tau_i \chi_i}, \tag{1}$$

3.1.4. Conversion to Energy

As the goal is to estimate the energy demand of each neighborhood, we need to convert these corrected distance OD matrices into energy values. The most accurate method to do this conversion is to use a consumption model simulator, as done by Baouche et al. [9]. However, we considered Brussels as a relatively flat city, which does not crucially need an altitude-based analysis for estimating EV consumption. Therefore, multiplying the average electricity consumption of an EV by the number of trips estimates the energy supply OD matrix. The International Energy Agency (IEA) provides reliable EV energy consumption metrics [31]. A passenger car consumes on average (on a road) between 0.20–0.24 kWh/km. In the analysis, we used the value of 0.22 kWh/km.

3.2. Demand Segmentation

3.2.1. Methodology Overview

In Section 3.1, we have estimated the public charging demand of EVs at a neighborhood level. Now, the goal is to segment this demand into normal, semi-rapid, and fast charging.

The segmentation process, shown in Section 4, consists of two independent steps. (1) We assume that EV drivers will always prefer normal charging in residential neighborhoods, which is cheaper than semi-rapid or fast charging. Therefore, the first step is to compute the proportion of residential charging demand for each neighborhood. This initial segmentation is called the Highway Analysis and is described in Section 3.2.2. (2) The remaining non-residential charging demand is segmented in normal, semi-rapid, and fast charging using a state-of-the-art approach detailed in Section 3.2.3. As shown in Figure 4, the non-residential segmentation splits the demand according to the city's points of interest (POIs).

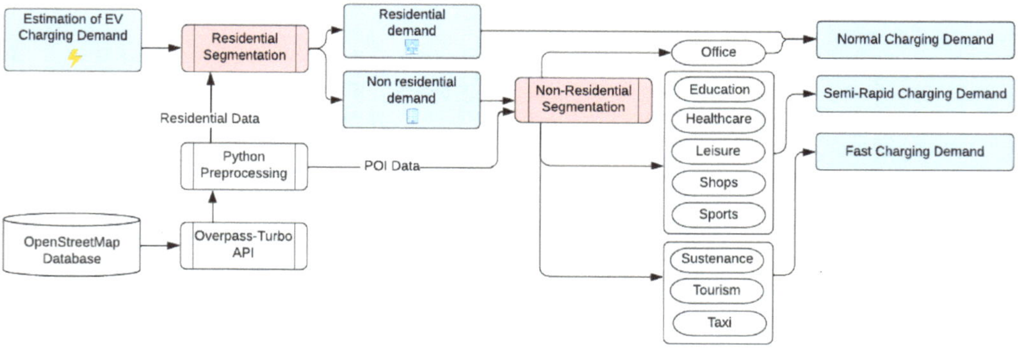

Figure 4. This diagram shows the complete segmentation process. The two segmentation steps are highlighted in red. First, the residential segmentation called Highways Analysis is presented in Section 3.2.2. Second, the non-residential segmentation called the POI-based segmentation is presented in Section 3.2.3. The charging demand at various stages is represented as blue cells. The data used in the model is extracted from the OpenStreetMap databases through the Overpass-Turbo API.

This model requires a massive amount of up-to-date geodata to run effectively, making the data mining process critical. Different web maps services exist online such as GoogleMaps, BingMaps, and OpenStreetMap (OSM). The latter is a community-powered map that supplies many websites and applications with its data. It is entirely free to use yet maintains a remarkable level of accuracy thanks to the work of numerous volunteers and engineers who populate it [32]. Furthermore, OSM comes with Overpass Turbo, a web-based data filtering tool that makes the mining process more accessible. It is built on top of the Overpass API, a read-only API that extracts OSM map data and acts like an online shared database: the client sends a query to the API, which retrieves the corresponding dataset and exports it as a *json* file.

3.2.2. Residential Segmentation

The first step of the segmentation takes advantage of the city's highways because they are accurately labeled in OSM. As the goal is to derive the proportion of residential charging per neighborhood, we compute the ratio of the total length of residential lanes over the full length of all other roads.

Figure 5 shows all the available highways for the whole city. The blue lines represent residential highways, whereas the red ones are motorways, primary, secondary, or tertiary highways (major city highways). The bold black lines represent the neighborhood's boundaries. On the one hand, we can identify industrial areas without any residential lanes, for instance, in the city center. On the other hand, we can also recognize some strongly residential neighborhoods. As many roads spread over different districts, a binary search algorithm is used to efficiently split them into sub-lanes fully included in a single neighborhood.

Figure 5. Zoom on highways extracted from OSM for Brussels. The blue lines are residential streets, whereas the red ones are major city highways. The bold black lines represent the neighborhoods' boundaries.

3.2.3. Non Residential Segmentation

In the city's large peripheral commercial areas, people often come for a couple of shopping hours and then leave. Therefore, building normal charging stations, which require about 10–12 h of charging for most vehicles, would not make sense. The most appropriate solution would be to build semi-rapid chargers. A fuel station on a motorway would benefit from fast-charging stations to decrease the drivers' range anxiety. In areas abundant with restaurants, bars, and tourist spots, many people flock for short periods of time. Therefore, a fast-charging station would also probably be the most appropriate solution. In a working district, where drivers park their cars for the whole working day, normal charging points would be the most appropriate.

For these reasons, exploring the cities' POIs distribution may provide valuable insights regarding EV charging demand segmentation. We have extracted all POIs available in OSM for Brussels in this work. Then, we have assigned a specific charging technology to each type of POI. Table 3 represents all the points extracted. The investigated amenities range from education (school, universities), sustenance (bars, restaurants, cafes), shops (convenience stores, supermarkets, shopping malls), to hospitals.

Table 3. This table summarizes the different amenities used in the study. The 'Charger' column shows which charging technology was assigned, among normal, semi-rapid, and fast charging. Amenities are grouped into different sectors named in the 'Sector' column. The 'Type of Building' column indicates which POI is discussed. The 'OSM Features' column denotes the specific tag used on OSM to extract the data. The 'Total' column gives the number of POI extracted while the 'Area' indicates the sum of their areas in km^2.

Charger	Sector	Type of Building	OSM Features	Total	Area
Normal charging	Home	Residential areas	highway = "residential"	NA	NA
	Work	Office areas	Amenity = "office"	1094	0.82
Semi-Rapid charging	Education	Universities Schools Kindergarten	amenity = "university" amenity = "school" amenity = "kindergarten"	606	5.39
	Entertainment	Music Schools Cinemas Theaters	amenity = "entertainment"	451	0.49
	Shops	Supermarkets Furnitures Health & Beauty Clothing Other	amenity = "shops"	4510	1.29
	Healthcare	Hospital Clinic	amenity = "clinic" amenity = "hospital"	19 50	1.21
	Sports	Sport facilities	amenity = "sport"	808	1.24
Fast/Ultra-fast charging	Sustenance	Bars Cafes Fast foods Ice Cream Pub Restaurant	amenity = "bar" amenity = "café" amenity = "fast_food" amenity = "ice_cream" amenity = "pub" amenity = "restaurant"	802 569 917 30 1136 3057	0.54
	Tourism	Monuments, Hotels, etc	amenity = "tourism"	1231	1.31
	Taxi	Specific areas	amenity = "taxi"	669	0.007

When all POIs are extracted, a challenge remains. We aim to estimate the number of charging sessions at each POI, and therefore, we need to assess each POI's popularity. The optimal solution would be to use the attendance rate of each restaurant, bar, hotel. Such data is only reliably available on one web maps service, GoogleMaps, which is unfortunately quite expensive. This model assumes that the popularity is proportional to the amenity's area. Of course, this is a strong assumption that does not always hold. Sports facilities are great examples of this. A medium-sized supermarket attracts more people daily than a colossal athletics stadium. For this reason, sports facilities were not taken into consideration in the final model, even if they are presented in Table 3. In practice, different models were explored by researchers in the literature, the latest of which is the *Universal visitation law of human mobility* proposed by Schläpfer, which rigorously proves the aforementioned assumptions [33]. If attendance data were to become available, the accuracy of the model could probably be increased considerably.

$$\Phi_{Nres} = \alpha(1-\gamma)(\Delta_i + \Delta_j) \tag{2}$$

$$\Phi_{Noff} = \frac{a_{office}}{\sum_i a_i}(1-\alpha)(1-\gamma)(\Delta_i + \Delta_j) \tag{3}$$

$$\Phi_{Sem} = \frac{a_{semi\_rapid}}{\sum_i a_i}(1-\alpha)(1-\gamma)(\Delta_i + \Delta_j) \tag{4}$$

$$\Phi_{Rap} = \frac{a_{rapid}}{\sum_i a_i}(1-\alpha)(1-\gamma)(\Delta_i + \Delta_j) \tag{5}$$

$$\Phi_{Par} = \gamma(\Delta_i + \Delta_j) \tag{6}$$

The final charging demand Φ is computed using the following formulas for each category. Φ_{Nres} and Φ_{Noff} are the proportion of normal charging for residential and office purposes, respectively. Φ_{Sem} and Φ_{Rap} are, respectively, the semi-rapid and fast charging needs. In these equations, α represents the proportion of residential demand ($\alpha \in [0,1]$), γ is the PPR ($\gamma \in [0,1]$), a_i is the total area of all POIs assigned to charging type i (Units: km^2), $\sum_i a_i$ is the total area of all POIs within one neighborhood (Units: km^2) and finally, Δ_i and Δ_j are respectively the regular and irregular charging demand (Units: kWh). The DR was taken into consideration when computing Δ_i and Δ_j.

4. Results

4.1. Case Study Of Brussels

4.1.1. Estimation Results

Figure 6 shows the final EV public charging demand estimation normalized by each neighborhood's area (in MWh/km^2). We can notice that the city center's demand is overall more important than in the suburbs. This is a direct effect of the PPR correction factor.

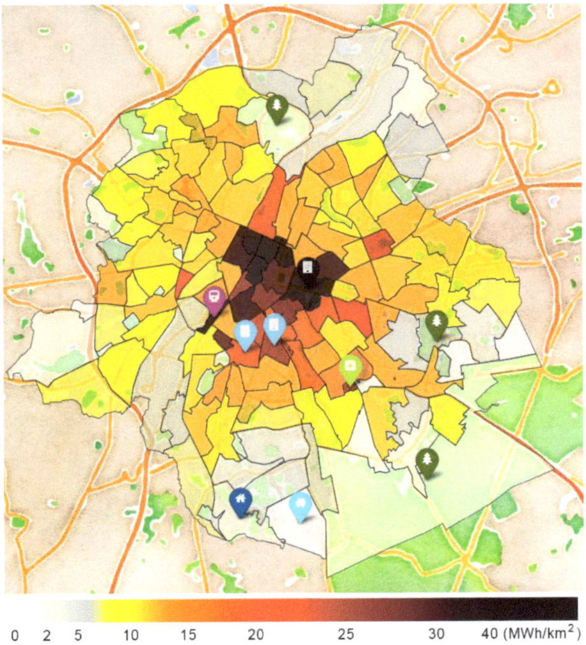

Figure 6. Total Public EV Charging demand normalized by the neighborhood's areas (both regular and irregular trips aggregated). The various markers were added to attract the reader's attention. Addition information about them is given in the core text. Units: MWh/km^2.

The two blue-home (,) neighborhoods in the south are both highly residential. Nevertheless, there is a significant charging demand difference between them, which may, at first, seem paradoxical. The left one, *Homborch*, exceeds 5.2 MWh (around 50 stations needed), whereas the right one, *Viver d'Oie*, has the lowest charging demand of all districts, 0 MWh. How is this difference possible? *Vivier d'Oie* accommodates mainly large detached houses, which leads to a PPR close to 1 and makes it independent from public infrastructures. On the contrary, *Homborch* accommodates subsidized housings with a low PPR, which justifies the higher density of public chargers. Note that a null charging demand for *Viver d'Oie* is probably unrealistic as under-estimated. The demand is set to 0 because the PPR ≈ 1.03 means more private parking lots than households. This extreme example highlights the importance of the PPR. If we want the chargers to be ultimately homogeneously dispatched around the city, the PPR value may be capped at 95% to prevent such situations.

The black building icon (▦) in the city center represents the *Quartier Européen*, a working district. Its high charging demand makes sense, but, as we shall show in Section 4.1.2, companies may partly support it. The two light-blue apartment ( ) icons show residential areas where many people live in apartments with a very low PPR Ratio. The light-green cross (⊞) icon represents the Delta neighborhood, hosting the city's largest hospital. The purple-train icon ( ) shows a bustling district that includes one of the city's main train stations in Brussels. In such areas, it is understandable that charging needs will be higher. Finally, the green-tree icons (♣) symbolize poorly populated district, such as parks, forests, or cemeteries, which explains the poor (or null) charging demand (white tiles).

4.1.2. Segmentation Results

Let us analyze the segmentation results for three neighborhoods of relatively different nature: *Matonge*, *Chant d'Oiseau*, and *Quartier Européen*. These are presented in Table 4 and Figure 7. The total EV charging demand is segmented into Normal (Residential), Normal (POI—Offices), Semi-Rapid, Rapid, and Parking. Parking represents the proportion of demand that drivers' private charging points should support.

Table 4. Segmentation results for four different neighborhoods. Normal (Res) represents the residential normal charging demand while the normal (Poi) represents the office normal charging demand.

Neighborhood	Normal (Res) (Φ_{Nres})	Normal (Poi) (Φ_{Npoi})	Semi-Rapid (Φ_{Sem})	Rapid (Φ_{Rap})	Parking (Φ_{Npar})
Chant d'Oiseau	37.3%	0.3%	8.0%	0.2%	54.2%
Matonge	77.4%	1.5%	13.5%	3.8%	3.8%
Quartier Européen	43.5%	34.4%	4.4%	5.1%	12.6%

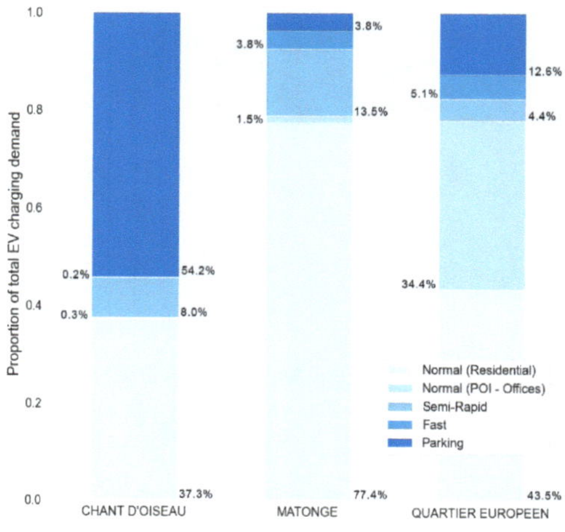

Figure 7. Segmentation results for three different neighborhoods (*Chant d'Oiseau*, *Matonge*, *Quartier Européen*). The y-axis shows the percentage of each charging technology. It is interesting to compare these proportions with Figure 8, which highlights the different POIs included in each neighborhood.

Chant d'Oiseau is the perfect residential neighborhood. More than 90% of its charging needs are supported either by normal residential stations or private chargers. *Matonge* is a residential neighborhood with heavy parking restrictions (less than 5% of private chargers), which also include a substantial number of shops and restaurants (17.3% of semi-rapid and fast charging needs). *Quartier Européen* is a working district, as shown by the 34.4% of normal office charging. The numerous highways and restaurants justify the 9.5% need for semi-rapid and fast charging. Furthermore, we can see on that most of the charging demand is endorsed by normal and private charging, which means that semi-rapid and rapid charging accounts only for a minority. This is reassuring, as these technologies should only be used when the cheaper normal charging solutions are not possible.

The normal (POI—Offices) demand may partly be considered as private charging, as companies will invest in charging points for their employees. European institutions in Brussels have already started massively investing in various EV charging solutions. Therefore, in work districts such as *Quartier Européen*, where the normal (POI—Offices) charging demand is high, the public demand could be reduced.

Figure 8 shows the different amenities accessible within each of these neighborhoods. We can distinguish the different nature of each neighborhood. Many blue lines indicate a high residential proportion. Areas including many public facilities, such as restaurants, supermarkets, and tourist spots, are interesting places for semi-rapid or rapid charging stations.

Figure 8. This figure shows three neighborhoods, each of them having a different nature. Their segmentation results are presented in Figure 7. All markers indicate a specific amenity extracted in OSM. (**a**) The map highlights *Chant d'Oiseau*, a heavily residential neighborhood, as shown by the numerous blue lines, testifying for the high proportion of housing. Regarding public amenities, it only accommodates a few shops. The PPR rate is relatively high due to the low parking regulations. (**b**) The map highlights *Matonge*, a neighborhood accommodating many public amenities such as bars, restaurants, and shops (ranging from convenience stores to supermarkets). The high number of blue lines indicates a high proportion of housing. The meager PPR rate is due to heavy parking restrictions. (**c**) The map highlights *Quartier Européen*, a working district which accommodates many offices (mainly the buildings of the European Institutions). It abounds with restaurants and includes many long-term parking lots as people park their cars for the whole day while working. Some streets also have housing.

5. Discussion

Having estimated and segmented the demand into different charging technologies, it is interesting to convert these energy values (kWh) into a number of charging stations. Table 5 summarizes the different assumptions made for this conversion. Most of them are based on an analysis proposed by Brussels Environment, the city planning authorities in Brussels [34].

Table 5. Assumptions taken to convert the demand in kWh into a number of stations [34].

Charger	Assumption	Value
Normal	Theoretical Power	7 kW
	Avg. Pow. Delivery	80%
	Occupancy Rate	50%
	Opening hours	24 h/24 h
Semi-Rapid	Theoretical Power	22 kW
	Avg. Pow. Delivery	80%
	Occupancy Rate	80%
	Opening hours	10 h–20 h
Rapid	Theoretical Power	100 kW
	Avg. Pow. Delivery	80%
	Occupancy Rate	25 %
	Opening hours	24 h/24 h

For each technology, we consider a value for the theoretical power (maximal power deliverable by the terminal), average power delivery (mean power delivered over the whole charging session), occupancy rate (proportion of time when a vehicle is plugged-in and charging), and opening hours (the time when the station is open for charging).

Table 6 gives the final number of charging stations needed to cover the whole city. We obtain 11,766 normal charging stations for residential purposes only. Regarding at-office charging stations, the 1047 stations may be, at least partly, supported by companies. This would substantially reduce the pressure on public services. Around 3004 semi-rapid charging stations are needed to efficiently cover the demand around POIs such as universities, museums, and commercial centers. Finally, let us consider that only normal charging stations should meet the needs of the entire charging demand, which is the currently preferred approach of Brussels Environment. We then obtain 24,413 stations for the whole city. If we add the Good Move ambitions, a plan aiming to reduce traffic in the city [34] by at least 5%, this gives 23,193 points.

Table 6. Number of charging points needed to cover the total demand.

Charging Type	Nbr. Stations
Normal (Residential)	11 766
Normal (Work)	1 047
Semi-Rapid	3 004
Rapid	312
Full Normal	24 413

This last value of 23,193 charging points aligns with the 22,000 charging points derived by Brussels authorities [34]. Results show that focusing solely on one technology may not be optimal, as the number of normal stations is snowballing. Many cities plan to free as much public space as possible. As shown in Table 6, mixing a variety of charging technologies can substantially reduce the total number of charging points (16,129 instead of 24,413). Furthermore, semi-rapid and fast charging stations can be efficiently implemented in semi-private areas, such as supermarkets and shopping malls.

The choropleth maps shown in Figures 9 and 10 presents the final number of stations needed normalized by the area at neighborhood-level. Some interesting patterns appear. For residential charging, the stations are fairly evenly distributed over the territory. Some sites, even though they are highly residential such as the south of the city, are less dependent on public infrastructure because of the high rate of ownership of private chargers. For office charging, the results seem relevant as we can identify the different city's work districts, such as *Industrie Sud* or *Quartier Européen*.

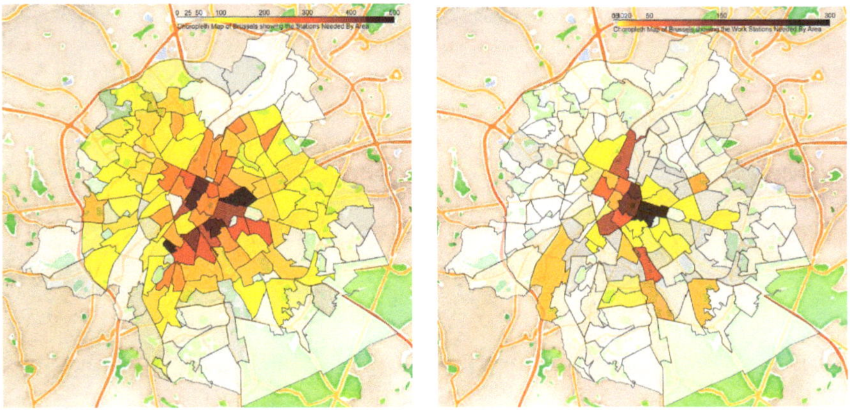

(**a**) (**b**)

Figure 9. Number of charging stations needed for normal technology (residential vs. office charging). The number of stations are normalized by the neighborhood's area to facilitate data interpretation and visualization. (Units: number of stations/km^2). (**a**) Number of normal (residential) charging stations needed in Brussels normalized by each neighborhood's area. (**b**) Number of normal at-office charging stations needed in Brussels normalized by each neighborhood's area.

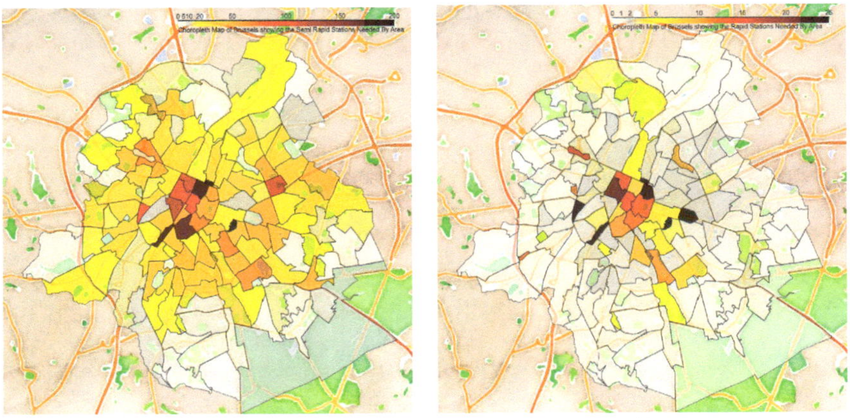

(**a**) (**b**)

Figure 10. Number of semi-rapid and fast charging stations needed for each technology. The number of stations are normalized by the neighborhood's area to facilitate data interpretation and visualization. (Units: number of stations/km^2). (**a**) Number of semi-rapid charging stations needed in Brussels normalized by each neighborhood's area. (**b**) Number of fast charging stations needed in Brussels normalized by each neighborhood's area.

The city's central railway station, *Bruxelles Midi*, shows a high dependence on rapid and semi-rapid stations. It is a mobility hub in Brussels for individual vehicles, shared cars, and taxis, which justify these costly charging technologies. Fast charging seems primarily concentrated in a few downtown areas, which might not be optimal. One ambition of Brussels is to reduce the traffic burden in the city center, and placing fast chargers there would probably increase traffic. Therefore, one should always remain critical of the raw model's results and see if they can be realistically implemented in the city. The best solution would probably be to build these stations along major highways, such as the motorways that surround the city center, called the small belt.

Note that the 50% occupancy assumption is very optimistic, but was used as it is the value assumed by Brussels government [34]. Different researchers have explored these questions, and it seems that the actual occupancy rates of EV charging stations are much lower. Wolbertus and van den Hoed [35] highlighted relatively low occupancy rates for normal charging stations, such as 15% during the daytime and 25% during the nighttime. Gnann et al. [36] also explored the occupancy rates, for fast charging this time, in Germany and Sweden. Results show that the occupancy rates seem to increase with power delivery, and for 100 kW, rates ranged from 18% to 28%.

Underlying Assumptions and Possible Improvements

Regarding the demand estimation, three significant assumptions need to be stated. First, this study considers that the proportions of car, walking, and public transport trips are geographically constant and solely depend on the average distance between two neighborhoods. In reality, these proportions will highly depend on the city area. Downtown, public transports are a preferred solution as they are incredibly efficient. In the suburbs, people rely on their private cars as public transport is scarce. Obtaining an OD matrix for public transportation trips in the city would significantly increase the model's accuracy. Second, as this work aggregates the total demand at a neighborhood level, it presumes that people will charge their car after each sufficiently long break. In practice, drivers' charging behavior remains unknown due to immature battery technologies, missing data, and scarce charging infrastructure. Predicting EV drivers' future behavior is an active research topic [37]. However, as the goal is here to produce prevailing neighborhood-level trends, such assumptions do not considerably harm the results.

Finally, the present work assumes each trip of the input dataset as an EV trip, an assumption that does not hold today as only around 8% of Brussels cars are electric [38]. This distribution is assumed constant over the coming years despite Brussels' ambitions to reduce the city's daily traffic.

Regarding the segmentation process, here are the two principal hypotheses. First, the Highway Analysis elaborated in Section 3.2.2 considers the residential lanes extracted from OSM as fully residential. In reality, this is not always the case. Figure 8 shows some residential streets which are full of shops and restaurants. Physically, the ground floor is rented for commercial purposes, while the upper floors are housing. Therefore, this analysis probably tends to over-estimate the residential proportion. One solution could be to estimate the ratio of the number of houses over the number of buildings (houses and POIs). Second, the popularity of each POI is assumed proportional to its area. This assumption may not always hold true. Therefore, using the real frequentation rate of each amenity would be the most trustworthy solution.

6. Conclusions

The major breakthrough in this article is the EV charging demand segmentation process, which is an approach that researchers have not yet explored. These segmentation results are vital to elaborate roll-out strategies for EV charging infrastructure. By adding the charging speed layer of granularity to the demand, we can apply the optimization models listed in Section 2.1 to find the optimal locations of each type of charger. Therefore, the model provides actionable insights for city planning authorities and opens the door to more complex deployment road maps. This model is a preliminary version, reasonably straightforward, and is supported by rather strong premises. Many improvement directions are possible and would eventually allow obtaining optimal results.

The case study proposed proved to be quite precise and one of the model's main advantages is its easy scalability to other cities. Indeed, all prominent cities around the world are rigorously annotated in OSM. Furthermore, our results for Brussels converge to previous studies on the conclusion that mixing various charging technologies is always preferable to only one type [17,19,21]. Although associated deployment costs will be higher, it allows to substantially reduce the overall number of charging points and maximize the freed-up public space. Moreover, private companies may partly support these costs, as semi-rapid and fast chargers are meant to be built in semi-private areas, such as supermarkets, shopping malls, etc.

Future Work

The next logical step in this project would be to apply operations research algorithms to locate the best charging station locations at the neighborhood level. The most straightforward procedure would be to train a single model for each type of charger (normal, semi-fast, and fast).

Many popular algorithms exist in the literature, such as the Capacitated p-median Model, the Set-Covering model, or the Maximal Covering Model. [14,39] The candidate locations can be extracted from the OSM data through the Overpass Turbo API, exactly as done in Section 3.2. An example of candidate locations are presented in Table 7. These models could also take as inputs the location of 400 V high voltage cabin, as it is needed to build semi-rapid and rapid chargers in the future.

Table 7. Example of candidate locations for optimization models.

Charging Type	Candidate Location
Normal (Residential)	On-Street parking, large public parking, etc.
Semi-Rapid	Shopping malls, semi-private parking, supermarket parking lots, etc.
Rapid	Fuel Stations, intersection of major highways, motorway entrances, etc.

Author Contributions: Conceptualization, V.R.; Methodology, V.R.; Software, V.R.; Validation, V.R.; Formal analysis, V.R.; Investigation, V.R.; Resources, V.R.; Data curation, V.R.; Writing—original draft, V.R.; Writing—review & editing, V.R.; Visualization, V.R.; Supervision, L.V. All authors have read and agreed to the published version of the manuscript.

Funding: This research received no external funding.

Data Availability Statement: The *Public Access Roads* dataset lists all private vehicle access roads in the city. An access road is a parking space reserved for a personal user (i.e., a driveway, a garden spot, etc.). An electric car driver could build an electric vehicle charging station in these locations. The dataset can be found at https://data.mobility.brussels/fr/info/access_carrossables/ (accessed on 22 February 2022) and https://datastore.brussels/web/data/dataset/3602f144-3119-4241-b9ad-d21223541321 (accessed on 22 February 2022). The *IBSA Public transportation* dataset provides historical series on transport supply and demand for bus, streetcar, and metro and train operators, air transport related to the airports serving Brussels, cabs and Cambio shared cars. The dataset can befound at https://ibsa.brussels/themes/mobilite-et-transport/transport-collectif-et-partage (accessed on 22 February 2022). The *Public Parking Types* dataset delivers the different parking places in Brussels sorted into three categories: reglemented (not free), free, and reserved. The reserved parking spaces are used by the fire department, the police or people with reduced mobility. The dataset can be found at https://data.mobility.brussels/fr/info/parkingtypes_view/ (accessed on 22 February 2022). The *Population Density* dataset is also a very valuable asset to perform socio-demographic analyses. It offers a large number of datasets for Brussels, most of them being up-to-date. The dataset presented in this section is the population density (in inhabitants/km^2). The datasets are often directly downloadable at various granularities, at the municipality and district levels. The link can be found at https://monitoringdesquartiers.brussels/maps/statistiques-population-bruxelles/evolution-population/densite-de-population/1/2019/ (accessed on 22 February 2022). The *Average Household Size* dataset lists the average household size for each neighborhood in Brussels, last updated in 2019. The online platform can also be used to visualize the data before downloading it, which is often quite convenient. The link can be found at https://monitoringdesquartiers.brussels/maps/statistiques-population-bruxelles/menages/taille-moyenne-des-menages-prives/1/2019/ (accessed on 22 February 2022). The *Origin-Destination* dataset from Proximus, as explained in the article, is confidential and not available online.

Conflicts of Interest: The authors declare no conflict of interest.

Abbreviations

The following abbreviations are used in this manuscript:

EV	Electric Vehicle
AC	Alternating Current
DC	Direct Current
TACS	Technology Agnostic Cell Sector
OD	Origin-Destination
API	Application Process Integration
CSD	Cellular Signaling Sata
DR	Driving Ratio
GDPR	General Data Protection Regulation
PPR	Private Parking Ratio
IEA	International Energy Agency
POI	Point Of Interest
OSM	OpenStreetMap

References

1. Brown, D.; Flickenschild, M.; Mazzi, C.; Gasparotti, A.; Panagiotidou, Z.; Dingemanse, J.; Bratzel, S. The Future of the EU Automotive Sector. 2021. Available online: https://www.europarl.europa.eu/RegData/etudes/STUD/2021/695457/IPOL_STU(2021)695457_EN.pdf (accessed on 22 February 2022).
2. Balko, L.; Urbanič, B.; Franková, Z. Infrastructure for Charging Electric Vehicles: More Charging Stations but Uneven Deployment Makes Travel Across the EU Complicated. 2021. Available online: https://www.eca.europa.eu/Lists/ECADocuments/SR21_05/SR_Electrical_charging_infrastructure_EN.pdf (accessed on 22 February 2022).
3. Jia, J.; Liu, C.; Wan, T. Planning of the Charging Station for Electric Vehicles Utilizing Cellular Signaling Data. *Sustainability* **2019**, *11*, 643. [CrossRef]
4. Zhou, X.; Taylor, J. DTALite: Light-weight Dynamic Traffic Assignment Engine. *Cogent Eng.* **2015**, *1*, 961345. [CrossRef]
5. Cao, W.; Wan, Y.; Wang, L.; Wu, Y. Location and capacity determination of charging station based on electric vehicle charging behavior analysis. *IEEJ Trans. Electr. Electron. Eng.* **2021**, *16*, 827–834. [CrossRef]
6. He, S.; Kuo, Y.H.; Wu, D. Incorporating institutional and spatial factors in the selection of the optimal locations of public electric vehicle charging facilities: A case study of Beijing, China. *Transp. Res. Part C Emerg. Technol.* **2016**, *67*, 131–148. [CrossRef]
7. Frade, I.; Ribeiro, A.; Gonçalves, G.; Pais Antunes, A. Optimal Location of Charging Stations for Electric Vehicles in a Neighborhood in Lisbon, Portugal. *Transp. Res. Rec.* **2011**, *2252*, 91–98. [CrossRef]
8. Efthymiou, D.; Chrysostomou, K.; Morfoulaki, M.; Ayfantopoulou, G. Electric vehicles charging infrastructure location: A genetic algorithm approach. *Eur. Transp. Res. Rev.* **2017**, *9*, 27. [CrossRef]
9. Baouche, F.; Billot, R.; Trigui, R.; El Faouzi, N.E. Efficient Allocation of Electric Vehicles Charging Stations: Optimization Model and Application to a Dense Urban Network. *IEEE Intell. Transp. Syst. Mag.* **2014**, *6*, 33–43. [CrossRef]
10. Trigui, R.; Jeanneret, B.; Badin, F. M odélisation systémique de véhicules hybrides en vue de la prédiction de leurs performances énergétiques et dynamiques. Construction de la bibliothèque de modèles VEHLIB. *Rech. Transp. Sécurisé* **2004**, *21*, 129–150. [CrossRef]
11. Florescu, A.; Turker, H.; Bacha, S.; Vinot, E. Energy management system for hybrid electric vehicle: Real-time validation of the VEHLIB dedicated library. In Proceedings of the 2011 IEEE Vehicle Power and Propulsion Conference, Chicago, IL, USA, 6–9 September 2011; pp. 1–6. [CrossRef]
12. Cavadas, J.; Homem de Almeida Correia, G.; Gouveia, J. A MIP model for locating slow-charging stations for electric vehicles in urban areas accounting for driver tours. *Transp. Res. Part E Logist. Transp. Rev.* **2015**, *75*, 188–201. [CrossRef]
13. Chen, T.D.; Kockelman, K.M.; Khan, M. Locating Electric Vehicle Charging Stations: Parking-Based Assignment Method for Seattle, Washington. *Transp. Res. Rec.* **2013**, *2385*, 28–36. [CrossRef]

14. Shen, Z.J.M.; Feng, B.; Mao, C.; Ran, L. Optimization models for electric vehicle service operations: A literature review. *Transp. Res. Part B Methodol.* **2019**, *128*, 462–477. [CrossRef]
15. Upchurch, C.; Kuby, M.; Lim, S. A model for location of capacitated alternative-fuel stations. *Geogr. Anal.* **2009**, *41*, 85–106. [CrossRef]
16. Sweda, T.; Klabjan, D. An agent-based decision support system for electric vehicle charging infrastructure deployment. In Proceedings of the 2011 IEEE Vehicle Power and Propulsion Conference, Chicago, IL, USA, 6–9 September 2011; pp. 1–5. [CrossRef]
17. Xi, X.; Sioshansi, R.; Marano, V. Simulation–optimization model for location of a public electric vehicle charging infrastructure. *Transp. Res. Part D Transp. Environ.* **2013**, *22*, 60–69. [CrossRef]
18. Zhang, Y.; Iman, K. A multi-factor GIS method to identify optimal geographic locations for electric vehicle (EV) charging stations. *Proc. ICA* **2018**, *1*, 127. [CrossRef]
19. Wolbertus, R.; van den Hoed, R.; Kroesen, M.; Chorus, C. Charging infrastructure roll-out strategies for large scale introduction of electric vehicles in urban areas: An agent-based simulation study. *Transp. Res. Part A Policy Pract.* **2021**, *148*, 262–285. [CrossRef]
20. Morrissey, P.; Weldon, P.; O'Mahony, M. Informing the Strategic Rollout of Fast Electric Vehicle Charging Networks with User Charging Behavior Data Analysis. *Transp. Res. Rec.* **2016**, *2572*, 9–19. [CrossRef]
21. Wang, Y.W.; Chuan-Chih, L. Locating multiple types of recharging stations for battery-powered electric vehicle transport. *Transp. Res. Part E Logist. Transp. Rev.* **2013**, *58*, 76–87. [CrossRef]
22. Cai, H.; Jia, X.; Chiu, A.S.; Hu, X.; Xu, M. Siting public electric vehicle charging stations in Beijing using big-data informed travel patterns of the taxi fleet. *Transp. Res. Part D Transp. Environ.* **2014**, *33*, 39–46. [CrossRef]
23. Wang, X.; Yuen, C.; Hassan, N.U.; An, N.; Wu, W. Electric Vehicle Charging Station Placement for Urban Public Bus Systems. *IEEE Trans. Intell. Transp. Syst.* **2017**, *18*, 128–139. [CrossRef]
24. Boscoe, F.P.; Henry, K.A.; Zdeb, M.S. A nationwide comparison of driving distance versus straight-line distance to hospitals. *Prof. Geogr. J. Assoc. Am. Geogr.* **2012**, *64*, 188–196. [CrossRef]
25. Boon, D.; Thiry, C. *Plan Piéton Stratégique—Bruxelles, Ville Pétonne*; Bruxelles Mobilité: Bruxelles, Belgium, 2012.
26. Derauw, S.; Gelaes, S.; Pauwels, C. *Enquête Monitor sur la mobilité des Belges*; Bruxelles Mobilité: Bruxelles, Belgium, 2019.
27. Figenbaum, E. Battery Electric Vehicle Fast Charging–Evidence from the Norwegian Market. *World Electr. Veh. J.* **2020**, *11*, 38. [CrossRef]
28. Webasto Charging Systems Inc. *Home Charging Becoming More Popular among EV Owners*; Webasto Charging Systems Inc.: Monrovia, CA, USA, 2021.
29. Speidel, S.; Bräunl, T. Driving and charging patterns of electric vehicles for energy usage. *Renew. Sustain. Energy Rev.* **2014**, *40*, 97–110. [CrossRef]
30. Jabeen, F.; Olaru, D.; Smith, B.; Braunl, T.; Speidel, S. Electric vehicle battery charging behaviour: Findings from a driver survey. In Proceedings of the Australasian Transport Research Forum, ATRF 2013—Proceedings, Brisbane, QLD, Australia, 2–4 October 2013.
31. IEA *Global EV Outlook 2019*; IEA: Paris, France, 2019; p. 51. [CrossRef]
32. AnnaDziuba.; Dziuba, A. Choosing a Map API for Your Next App: Mapbox, Google Maps, OpenStreetMap. 2021. Available online: https://mapsplatform.google.com/ (accessed on 16 February 2022).
33. Schläpfer, M.; Dong, L.; O'Keeffe, K.; Santi, P.; Szell, M.; Salat, H.; Anklesaria, S.; Vazifeh, M.; Ratti, C.; West, G.B. The universal visitation law of human mobility. *Nature* **2021**, *593*, 522–527. [CrossRef]
34. Environment Brussels. *Zero Emission Mobility Brussels—Roadmap 1.0*; Environment Brussels: Bruxelles, Belgium, 2020.
35. Wolbertus, R.; van den Hoed, R. Managing parking pressure concerns related to charging stations for electric vehicles: Data analysis on the case of daytime charging in The Hague. In Proceedings of the European Battery, Hybrid & Fuel Cell Electric Vehicle Congress, Geneva, Switzerland, 14–16 March 2017.
36. Gnann, T.; Funke, S.; Jakobsson, N.; Plötz, P.; Sprei, F.; Bennehag, A. Fast charging infrastructure for electric vehicles: Today's situation and future needs. *Transp. Res. Part D Transp. Environ.* **2018**, *62*, 314–329. [CrossRef]
37. Hardman, S.; Jenn, A.; Tal, G.; Axsen, J.; Beard, G.; Daina, N.; Figenbaum, E.; Jakobsson, N.; Jochem, P.; Kinnear, N.; et al. A review of consumer preferences of and interactions with electric vehicle charging infrastructure. *Transp. Res. Part D Transp. Environ.* **2018**, *62*, 508–523. [CrossRef]
38. Craps, R.; Vijghen, A.; Welde, L.V.; Vanhaverbeke, L.; Hollander, S.; Sury, D.; Sergeant, N.; Gerard, A.; Duprez, L. *Zero Emission Mobility Brussels—Roadmap 1.0*; Bruxelles Environnement: Brussel, Belgium, 2021.
39. Majhi, R. A systematic review of charging infrastructure location problem for electric vehicles. *Transp. Rev.* **2020**, *41*, 432–455. [CrossRef]

World Electric Vehicle Journal

MDPI

Article

Review on Smart Charging of Electric Vehicles via Market-Based Incentives, Grid-Friendly and Grid-Compatible Measures

Doris Johnsen [1,*], Lars Ostendorf [2], Mischa Bechberger [1] and Daniel Strommenger [3]

[1] Institut für Innovation und Technik (iit) in VDI/VDE Innovation + Technik GmbH, Steinplatz 1, 10623 Berlin, Germany
[2] TÜV Rheinland Consulting GmbH, Am Grauen Stein 27, 51105 Köln, Germany
[3] Independent Researcher, 14974 Ludwigsfelde, Germany
* Correspondence: johnsen@iit-berlin.de

Abstract: Smart charging of electric vehicles is a promising concept for solving the current challenges faced by connecting mobility and electricity within the context of the ongoing sustainable energy transition. It allows cost savings for the expansion and operation of the power grid and a more efficient use of renewable energies. However, wide implementation of smart charging requires further work on technical and regulatory issues and further development of standards, especially an end-to-end consistency of the control signals. A fully automated process, as well as customisable services and flexible tariffs, would also facilitate wider market penetration. The novelty of this paper is the consensus of German pilot projects funded within the German programme "Elektro-Mobil" on the communication channel between all stakeholders for the use cases of smart charging based on market price incentives. Within this consensus, the projects have illustrated how specific standards can facilitate the communication between smart charging stakeholders, become a reality in the pilot projects and should be applied to further use cases in the low-voltage network. This consensus results in a white paper. On this basis, the adjustment of the standards can be made to ensure the consistency of the control signals from the beginning of the control process up to the end. In an advanced Edition, solutions for the prioritisation and orchestration of the different control signals could be designed.

Keywords: regulation; smart charging; standardisation; V2G; V2H

Citation: Johnsen, D.; Ostendorf, L.; Bechberger, M.; Strommenger, D. Review on Smart Charging of Electric Vehicles via Market-Based Incentives, Grid-Friendly and Grid-Compatible Measures. *World Electr. Veh. J.* **2023**, *14*, 25. https://doi.org/10.3390/wevj14010025

Academic Editor: Joeri Van Mierlo

Received: 4 October 2022
Revised: 5 December 2022
Accepted: 3 January 2023
Published: 16 January 2023

1. Introduction

The transition to electrically powered vehicles is associated with the goal of reducing the consumption of fossil resources and minimising emissions that contribute both locally and globally to the ongoing process of climate change. At the same time, the expansion of electric mobility poses new challenges to the existing power grid, especially with regard to the distribution grid level. For example, it increases the number of systems consuming electricity and thus the overall load demand. For this reason, the introduction of electric mobility will ultimately lead to a higher simultaneity of demand for electricity, and this is at a time when the fluctuating nature of the generation of renewable energy is increasing and exercising its own effects on the distribution grid. To address these challenges and to minimise the necessary investments in the power grid, smart charging concepts should be considered from the outset when expanding the charging infrastructure (e.g., [1,2]).

However, the implementation of smart charging faces some obstacles (e.g., [3,4]) that prevent a widespread roll-out of the technology, such as viable pricing models and market designs or a lack of continuity of control signals in the field of standardisation [5,6].

Therefore, this paper provides insights into necessary implementation options for smart charging from the perspective of German pilot projects and needs for action, as well as an overview of the state of the art regarding smart charging. The novelty of this paper is thus:

- The structuring work regarding the use cases of smart charging;
- The consensus of German pilot projects on the communication channel between all stakeholders for two specific use cases of smart charging;
- An agreement on the applicable standards;
- Solutions for prioritisation and orchestration of the control signals.

The projects of the funding programme "Elektro-Mobil" of the German Federal Ministry for Economic Affairs and Climate Action (BMWK) show in the context of their pilot projects that smart charging can be effectively implemented with the help of price and other control signals [7]. The projects of "Elektro-Mobil" have jointly published a study on smart charging, as well as a white paper, together with the accompanying research [7,8]. In the study, the authors outline concrete needs for action to promote or to implement smart charging via price signals. Several topics were identified, which are delineated in the following paper.

In addition, the accompanying research of the funding programme "Elektro-Mobil" moderated an agreement process on standards for the communication of the control signals to assure the continuity of the communication in all process steps. The results of this process were published in a white paper on the control of charging processes in electro-mobility [8] and are described in this paper as well.

This creates a basis for the next steps towards the roll-out of possible use cases on the market and for further work on the interoperability of the standards. For the implementation of smart charging, the transmission of the control signals must be ensured according to specific standards. Once there is consensus on the communication channels and the associated standards, the adjustment of standards can be made to ensure consistency of control signals from the beginning to the end of the control process. In addition, the implementation of regulations is required to enable stakeholders to complete the entire process from identifying market signals and grid challenges, through processing and orchestrating different control signals, to measuring and accounting for the delivered electricity.

This paper is based on a meta-study of existing technologies, structural ideas, standards, etc., and on a workshop-based consensus process.

The consensus process was led by the accompanying research of "Elektro-Mobil". The basis of the process was several expert workshops within the funding programme. In the first step, the participants agreed on a technical and regulatory framework in Germany that can be used as a basis for further work. In the second step, the participants created a map of communication channels for two specific use cases (one of which is shown in this paper). The next step led to a consensus on the combination of communication channels together with the associated standards. This consensus led to the joint publication of a white paper showing a solution for smart charging in the low-voltage grid for the specific situation in Germany. Industries can develop their products based on this knowledge of the surroundings and accelerate the adoption of smart charging.

2. Relevant Smart Charging: Use Cases and Learnings from Their Application in Different Pilot Projects

Smart charging can be understood in the two categories grid-friendly and grid-compatible charging. Grid-compatible charging offers several possibilities to integrate price incentives into charging management. Price-controlled charging is one way to realise a more efficient utilisation of the distribution grids and thus saving costs and freeing up resources that can be invested in grid expansion and balancing measures. Furthermore, it allows for adapted timing of electricity demand and the generation of renewable energy, as well as an active contribution to environmental and climate protection through the more efficient use of fluctuating renewable energies. In addition, this can create leeway for reducing the operating costs of the charging infrastructure.

For example, the charge point operator (CPO) can be motivated to reduce the price of the charged energy through lower grid connection costs.

The term price-controlled charging used in the following describes a charging management system that plans and implements the charging process of one or more electric vehicles (EVs) depending on price signals from the electricity market. In times of high electricity demand, charging the vehicle is delayed, and where bidirectional charging is possible, the EV feeds energy into the grid.

2.1. Use Cases

Smart charging is highly dependent on the specific use case. In the following, various use cases are presented, which include smart charging via price incentives and grid measures. Potentially, all use cases that include smart charging via price incentives or grid measures can be realised unidirectional and bidirectional.

2.1.1. Smart Charging via Price Incentives

The possibilities and characteristics of price-controlled charging are diverse. Table 1 classifies use cases of price-controlled charging with direct price control. Direct price control means that the charging strategy is adjusted exclusively on the basis of time-resolved price signals. In this case, the price signals can vary highly dynamically.

Table 1. Use cases of smart charging via price incentives ([7], Source: BDL project).

Use Case	Revenue Location	Customer Group	Regulation
Time arbitrage through trading on the electricity market (day ahead, intraday) (V2G)			In front of the metre
Tariff-optimised charging (stock market-oriented, variable grid charges) (V2H)			Behind the metre
Legend	 Grid and grid operator	 Industry/public use	 Home/private use

When using time arbitrage through trading on the electricity market, the charging process depends on the price differences of the intraday or day-ahead markets. The EV is charged at times when the electricity price is low. If bidirectional charging is possible, power is fed back into the grid when the price of electricity is high.

Tariff-optimised charging, in contrast, uses dynamic tariffs that are time-dependent. When the time is determined when the EV should reach the set battery level, the EV is additionally charged when the electricity tariff is low and the energy is fed back into the grid when the electricity price is high. In this use case, the revenue is not gained in the grid, but in the house.

Tariffs for price-controlled charging of EVs.

Similar to these different options for price-controlled charging, there are also different approaches to setting dynamic prices. Dynamic pricing refers to retail electricity prices that pass on at least some of the volatility of wholesale prices to end consumers. These marginal costs of electricity generation are expressed in the short run through spot prices and in the long run through forward prices. Dynamic pricing means that prices and price periods are aligned so that they actually provide incentives to shape energy consumption behaviour in line with wholesale price trends. According to Table 2, four types of tariffs can be classified.

Table 2. Smart charging based on price signals [9,10].

Type of Tariff	Time-of-Use Pricing (TOU)	Variable Peak Pricing (VPP)	Critical Peak Pricing (CPP)	Real-Time Pricing (RTP)
Setting prices	Static	Static and dynamic	Static and dynamic	Dynamic
Description	Usage over large time blocks of several hours (e.g., day and night pricing)	different periods for pricing are defined in advance, but the price for the on-peak period varies depending on market conditions	Rate in which electricity prices increase substantially for a few days in a year	Prices are determined close to real-time consumption of electricity on fine granularity (e.g., 15 min)
Graphical representation	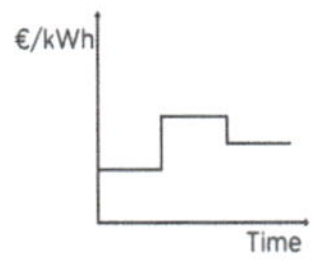	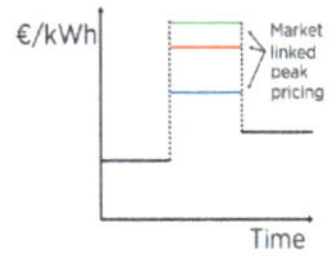	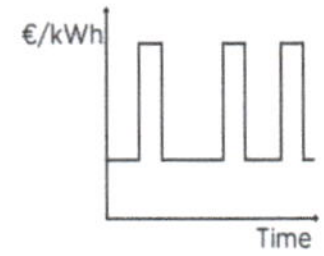	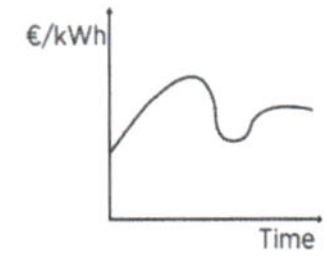

From a technical point of view, load profile measurement is required for the implementation of dynamic tariffs that follow the electricity market. This is the measurement process that the energy supplier uses at the customer's premises. This metering process has been used for decades for customers with registered load profile metering (RLM) and is to be introduced in Germany by so-called "smart metres" (intelligent metering systems, iMSys) for smaller consumers between 6 and 100 MWh per year of electricity demand on the distribution network level.

2.1.2. Grid-Friendly and Grid-Compatible Charging

The capability of price-controlled charging is not the only factor that influences the implementation of smart charging. Grid-serving measures also have a great impact on how smart charging is realised at home or at the workplace.

Table 3 presents three use cases in which grid-serving measures are used for indirect price-controlled charging. Indirect price control represents a charging strategy based on an overall optimisation that includes price signals.

Table 3. Use cases of smart charging via grid-friendly and grid-compatible charging ([7], Source: BDL project).

Use Case	Revenue Location	Customer Group	Regulation
Peak load capping (V2B)			Behind the metre
Increase in self-consumption (V2H)			Behind the metre
System services (V2G)			In front of the metre
Legend	 Grid and grid operator	 Industry/public use	 Home/ private use

When realising peak load capping, e.g., in a company, the overall peak load is reduced by charging the EVs during times of reduced load. The vehicles can feed electricity into the system when consumption of other consumers is high.

Increase in self-consumption means charging an EV to temporarily store the surplus electricity produced by an electricity generator (e.g., solar panel). This use case is part of vehicle to home (V2H).

System services can be used to minimise electricity shortages when network bottlenecks occur. The vehicles are charged or uncharged based on signals by the distribution system operator (DSO). This use case is part of vehicle to grid (V2G).

2.2. Application of Use Cases

Pilot projects in the funding programme "Elektro-Mobil" of the BMWK.

The projects of the funding programme "Elektro-Mobil" of the BMWK demonstrate how smart charging can be implemented taking into account control signals. The projects use different solutions that address diverse use cases and different forms of potential implementations in Germany. The solutions range from charging at the workplace or in residential areas and on the properties of municipal companies to bidirectional charging in private homes [7].

The project BSR-Li-Flx deals, among other things, with solutions for price-controlled charging on the premises of municipal companies. It is based on developing a decentralised solution for an electricity customer with registered power measurement (RLM customer) in addition to the company's own energy management system and the existing measurement technology. The control is based on a 15 min timetable. The calculation of the reduced purchase price is based on the subsequent measurement of the flexibility services realised by the customer.

In the project LamA—Charging at Work, the focus is on the nationwide development of charging infrastructure at Fraunhofer Society locations. In the project, a platform architecture for intelligent load management is being developed, which can take into account various control signals, such as simulated price signals from the electricity exchange, user requirements, and other consumers. In this way, the charging management ensures a peak load management for all consumers on the company premises.

In the project ELBE, over 7000 charging points are being set up under the premise of avoiding grid expansion in the distribution network. It is implemented exemplarily in the form of a "minimum valuable product", which connects the charging infrastructure to the electricity grid via a direct interface. This project pragmatically shows how the implementation can be carried out taking into account business models that have already been established in a similar form in the USA. In this project, the communication standard openADR 2.0 is used to connect the CPO to the distribution grid operator (DSO) or other entities, which has meanwhile been transferred to the IEC standard 62746-10-1 ED1.

In addition to the examples mentioned above, "Elektro-Mobil" has funded further projects that deal with smart charging in different use cases. Projects such as ARNi, BDL, DatenTanken, Resigent and unIT-e^2 are working on unidirectional and bidirectional smart charging, including a working standardised charging system. All these projects were involved in the consensus process to develop the above-mentioned white paper.

Further market application of price-controlled charging.

When smart metering is available, more and more companies offer variable energy prices for households or companies for different applications and initial situations, not only for charging EVs. A solution for time-variable grid tariffs has been developed under the leadership of the distribution grid operator MitNetz in Germany. The energy management system of a property or a smart wallbox controls the charging processes of the vehicles based on price signals (based on time-variable grid charges) as well as on information about available grid capacity, received from the DSO via an interface. There is no direct intervention by the DSO in this approach. Instead, the DSO uses the specification of the time-variable grid tariffs of the load management as a control instrument. With these instruments, critical grid situations can be prevented. The German company Easy Smart Grid, for example, has implemented a so-called Walrasian auctioneer in the pilot project SoLAR [11], which ensures that the local real-time price always reflects the balance of supply and demand. Unlike traditional markets such as EPEX, this means that participants are actively involved in price formation. Other companies, such as Awattar from Austria or Tibber from Germany, offer flexible tariffs where electricity is purchased cheaply on the stock exchange and the price is passed on to the customer. Most of them promote the use of renewable energy with their flexible tariffs. In addition, Tibber also provides a public Application Programming Interface (API), which can be used independently for integrating applications or for smart home control. The company Stromdao offers a variable tariff throughout Germany based on a green electricity index to ensure a low carbon footprint for its customers [12–14].

Even though there are already some variable electricity tariffs where lower purchase prices are passed on to the customer, the technical and legal implementation of the entire process is not yet feasible in Germany. The integration of the tariffs into the control of the vehicle's charging profile is also not directly offered by these companies, as most of them do not distinguish between different end-user equipment, such as EVs or heat pumps. In domestic use, this can be implemented via the energy management system (EMS) of a property or alternatively via the CPO backend, although the realisation of both solutions is still very complex for private households.

International experiences with price-controlled charging.

Worldwide, most experience has been gathered using dedicated static time-of-use (TOU) charging for EVs [15]. This model is currently offered in California by Pacific Gas & Electric (PG&E) [16] and by San Diego Gas & Electric (SDG&E) [17]. Both provide different TOU tariffs from which the customers can choose depending on their individual energy profile. They define peak hours, in which the consumption of electricity is more expensive than in off-peak hours. A project with a critical peak pricing (CPP) tariff was launched in France [18], where the price increases significantly for a determined period of time that is communicated in advance. If consumers shift their electricity use to off-peak hours (outside this time frame), they receive a fixed cost reimbursement [19].

Other pilot projects were launched to implement dynamic pricing for EVs, such as SDG&E, which offers three different tariffs for EVs at home with off-peak and "super off-peak" rates. These depend on the time of year and day [20]. Another pilot project to introduce dynamic pricing was started in Norway in 2019 by the Norwegian Water Resources and Energy Directorate (NVE) [21]. The charged tariff depends on the electricity demand in a given period, which provides a direct incentive to avoid EV charging during peak hours.

Comparison of first solutions.

Static TOU was found to be easier to realise and easier to understand. With the exception of the standard tariff, TOU is the most common pricing model, even though acceptance and adaption have to be increased by further incentives [22]. However, optimal grid load and use of renewable energy can only be achieved by implementing dynamic control signals, as proposed by the "Elektro-Mobil"-projects. Although the introduction of real-time pricing (RTP) has already begun [19], much development is still necessary before dynamic pricing can be realised as a standard. The necessary developments depend on technical and regulatory issues, and the standards must be compatible. What is needed is both a fully automated process and customisable offers, which provide customers with flexible tariffs. Usually, the DSOs offer tariffs with mixed calculation on the supplier side, as well as offers for pilot projects with special conditions.

3. Consensus on Communication Channel and Standards for Price-Driven Use Cases

Building a system architecture that connects the charging infrastructure, the backend of the charging infrastructure, the energy management system and the preferences of EV users while at the same time controlling the charging processes based on grid- and market-oriented information is an extremely complex task. This complexity requires a large number of interfaces and standards.

Much work has already been conducted by the German standardisation committees. In 2021, ISO 15118 still lacked a comprehensive implementation of a standard in vehicles and charging infrastructures to ensure that information from battery management in the EV is passed on to the electric vehicle supply equipment (EVSE). This deficit is currently being addressed in an update of the standard ISO 15118-20. ISO 15118-20 is scheduled for publication in summer 2022. This updated version includes the deposit of the charging certificate in the EV and additional functions of the charging management that enable V2G [23].

ISO 15118-20 represents a significant step towards bidirectional charging on the section between EV and EVSE. However, further standards need to be established in support of bidirectional and smart charging. Furthermore, standards need to be linked to ensure the whole process of load control and billing. In consideration of the German framework described, e.g., by the National Platform Electro Mobility (NPM) [24], VDE FNN targets picture controllability of EVSE [25] and the smart metre gateway-Roadmap-process [26,27]. A range of projects funded within the funding programme "Elektro-Mobil" agreed on a series of standards to process the communication for smart charging. These standards can currently be used nationally and, to some degree, internationally [8]. The work of the funded projects and the accompanying research support the process of standardisation by highlighting which configuration of standards facilitated the communication for two exemplary use cases of smart charging and recommending this configuration for further use cases. The core configuration is a triad consisting of the standards ISO 15118, IEC 63110 with OCPP and VDE-AR-E 2829-6 (equality to VDE-AR-E 2122-1000 in process) with EEBUS among others. Figure 1 shows the actors/entities and the communication channels of smart charging systems for two specific use cases that occur in the "Elektro-Mobil" pilot projects. While the upper graphic in Figure 1 shows the communication channels for price-driven use cases, the lower graphic depicts the communication channels of emergency regulations via grid measures. In addition to the different communication channels in the respective use cases, Figure 1 also shows the relevance of different standards depending on the use case.

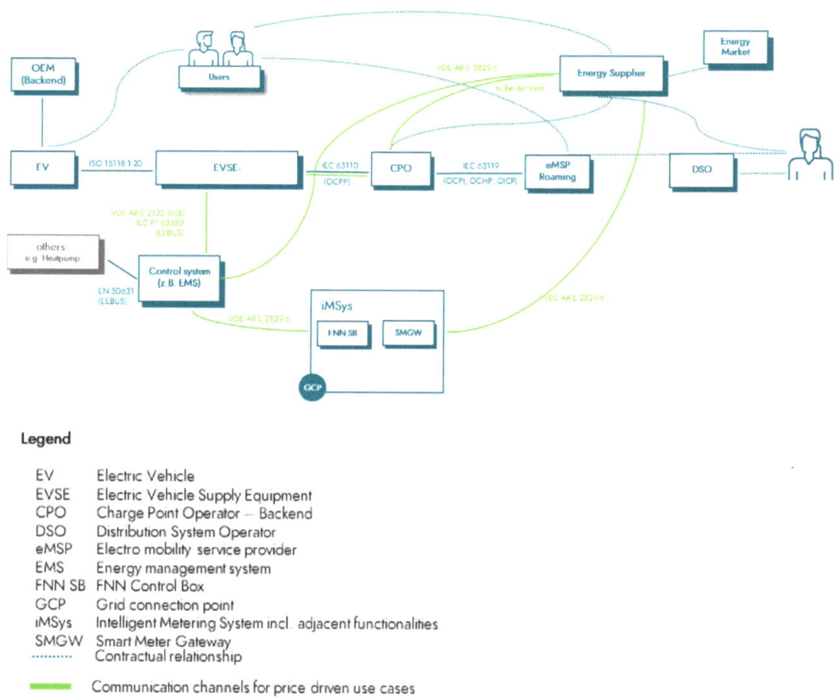

Figure 1. Smart charging system. Communication channels for price-driven use cases (green-coloured lines, upper graphic) and of emergency regulations via grid measures (red-coloured lines, lower graphic). In addition to the different communication channels in the respective use cases, the relevance of different standards depending on the use case is also shown.

In order to provide solutions for different use cases and situations, the projects are working on different approaches. Most projects are developing solutions with local control at the grid connection point. The project ELBE is pursuing a cloud-based approach to implement smart charging and uses openADR for the connection to the DSO. Both figures show the complexity of smart charging and its standards, which most projects have been able to agree on. The "Elektro-Mobil" projects have made it possible to agree on the standards concerning smart charging via price incentives and grid measures.

In an expanded version of the aforementioned white paper [28], the accompanying research and the "Elektro-Mobil" projects have incorporated supplementary proposals for prioritising and orchestrating control signals and control in general via a digital grid connection point. Via a power limitation at the (digital) grid connection point (below which the customer system can act freely) and an orchestration of the systems (through predefined prioritisation levels of control signals) via an energy management system behind the grid connection point, market use cases (for controlling the charging processes) can quickly be developed further. This reduces the complexity of the overall system. It can be accompanied by simple coordination, which makes it possible to prioritise and process the control signals and their sometimes conflicting nature. In this way, the ramp-up of electro-mobility, including grid integration, can be advanced without endangering grid stability in the distribution grid, and at the same time contribute to grid support and system security at the levels above. VDE FNN published an impulse paper on this topic in July 2022, which launched a stakeholder consultation process. The feedback will be incorporated into a subsequent key points paper on future grid operation with flexibilities in the low voltage in 2023 [29]. In addition, the expanded version of the white paper also addresses possible approaches to a resilient design of the charging infrastructure system (e.g., through the automated continuation of stand-alone operation and black start capability) [30,31].

Based on this consensus, further developments were possible in other areas such as compatibility. To achieve a standardised system that functions on an international level, further work is needed to develop new standards and to update existing standards.

In addition, the implementation of regulations is necessary to enable stakeholders to fulfil the entire process from identifying market signals and grid challenges to processing and orchestrating different control signals, measuring and accounting for the delivered electricity.

4. Potentials, Incentives and Needs for Actions for Smart Charging

4.1. Economic Aspects

As a result of the activities of the funding programme "Elektro-Mobil", relevant needs for action to promote or to implement smart charging via price signals were derived in various topic areas outlined below. At the same time, a consensus was reached between the funded projects on standards regarding the communication for smart charging under the German legal framework for specific use cases [8].

To create an economically functioning system for smart charging, there are different aspects to be considered. The following sections will focus on exemplary aspects.

"The integration of EV charging in a smart and connected environment will be a key success factor, enabling annual revenue growth of up to EUR 5 billion in Europe by 2030. A reliable and intelligent charging infrastructure is the key for the mass transition to electric mobility", states consultancy firm P3 [32].

The full potential of EVs for the energy system is realised when they are intelligently charged and also feed electricity back into the grid. In the future, electricity will be temporarily stored in the car and fed back into the grid at a later time. This will help avoid having to throttle the generation of renewable energy. EVs capable of bidirectional charging will make an important contribution to the stability of the electricity grids. The electricity fed back into the grid can be used as large-scale storage by pooling vehicles. A legal framework must be created to ensure that both grid integration and the flexible use of smart and bidirectional charging succeed at all grid levels. Finally, EV owners will only make their mobile storage available for flexibility services if they can generate income from it and thus reduce their total cost of ownership. The investment costs for all participants in the business model of bidirectional charging (vehicle users, aggregators, charging point operators) will only be refinanced on the market if the necessary investment costs can be recovered through corresponding profits.

4.2. Raising Ecological Potentials Economically

In addition to measures to control charging processes, pooled storage of bidirectional vehicles can make an important contribution to reducing investments in grid expansion, ensuring grid stability, guaranteeing supply security and increasing the share of renewable energies in the power system. Bidirectional charging will limit the need for investments in gas power plants by up to EUR 25 billion by 2050. Load procurement costs on the European stock market could be reduced by EUR 45 billion per year in 2050. In the scenario of bidirectional charging, the average European price of electricity falls by 3 EUR/MWh, and the European system costs will be reduced by a total of EUR 9 billion per year in 2030 [33].

As shown in [34], smart charging of EVs can also have positive effects on household electricity prices, grid load and emissions in Germany. With more EVs in use, the overall demand for electricity increases. Uncontrolled charging of EVs will cause load peaks, because most of these vehicles are charged at home after work when household consumption is high. Controlled charging can help flatten the load curve by shifting the charging process to night hours. Hence, the need for grid expansion is reduced despite the overall increase in demand for electricity. This improves the general grid load. Network charges, which can be described as a temporary rent of the grid network, are reduced due to the more efficient use of the grid. However, more research is needed to determine whether the demonstrated cost differences between smart and non-controlled charging are sufficient incentives to nudge EV owners to participate in charging management. To leverage these positive effects, the regulatory framework needs to be adapted in such a way that the ecological incentives are also reflected in monetary terms for the implementing persons or actors [21]. Another, even higher incentive can be created by implementing bidirectional charging, which can reduce the pressure on the grid even further.

The integration of economic or control signals into the charging management is mandatory for the implementation of smart charging. With no standards existing for this important interface yet, the integration of signals is currently the most urgent need for action, especially for smart charging. In consequence, generally accepted standards are necessary to simplify access and use of economic signals for charging management. The achievements of the funded pilot projects on standardisation are highlighted in Section 3.

4.3. Creation of Incentives, Regulation and Use Aspects

When designing incentives for smart charging, two different target groups must be taken into consideration—commercial addresses and private drivers of an EV. It is of primary importance that users from both target groups receive financial compensation. Also important, though secondary, is their contribution to the turnaround in energy policy, as well as their impact on grid stability. The users' choice of the charging management system depends highly on their personal and economic preferences (e.g., [3,7,35]). If a private person is generally interested in environmental protection, they will usually select more eco-friendly tariff models. Companies are increasingly forced by customers to reduce their carbon footprint. Further non-financial incentives—especially for private customers—could be created by installing appealing user interfaces, e.g., in the form of app-based feedback on the individual contribution to climate protection (gamification). For drivers of EVs in particular, other less obvious motivators, such as a positive user experience and the assumption of a pioneering role in society are relevant for the use of intelligent charging systems [7,36].

The access to price signals and the amount of financial leverage differ between private households, small companies and companies with high energy consumption and individual contracts, so-called "customers with registered power measurement" (CRPM). Compared to small energy customers, customers with registered power measurement have an easier starting position for the introduction of price-controlled charging management due to the already existing measurement technology and the 'individualisable' contract design. For these customers, the current possibilities to use price signals from the electricity market and the grid for charging management are currently shaped less by technical factors, but more by organisational and contractual factors, such as contracts that include peak load shaving and respective capping on a large scale.

Data privacy, data availability and data economy need to be ensured for all business models. Smart charging requires data to optimise the charging process, which must be provided or exchanged in particular by users, CPOs and electro-mobility service providers (eMSPs). It must be ensured that the data are provided by the respective players and that data protection and the principle of data economy are observed for the critical infrastructure of charging points.

Creating attractive incentives is not only a question of business models. It is also a question of regulation to enable envisaged developments. Particularly in view of the fact that in Germany taxes and grid fees account for 70% of the total price of household electricity [37], a variable electricity exchange price would only have a very limited financial incentive effect on the charging behaviour under the current legislation. In other countries such as the USA, more attractive price incentives could be achieved due to the different market and regulatory frameworks that prevail there [7]. In consequence, it must be clarified how a higher incentive can be achieved via regulatory intervention and how regulatory measures can be effectively used to push controlled loading. It should also be identified how the incentive system can be designed most efficiently for the desired goals and thus achieve the greatest effect. This could mean, for example, that the focus is not concentrated solely on reduced prices for off-peak periods. Static price components could be used as well—for example, reducing the building cost contribution with stipulated controlled charging. Here, the incentive would lie in the reduction of the initial investment and not in the ongoing operation and would directly allow a lower connected load to be realised. The business segment of smart charging is already seen as central by relevant actors and is considered to be a significant source of revenue. An important question in this respect is how a critical mass of flexibilities, which is traded on the electricity market, can be provided by electric mobility. The bundling of flexibilities could help incentivise stakeholders or aggregators to invest, despite the high costs of hardware and software, as well as the ongoing operating costs that have to be covered. Regulations need to be adapted accordingly based on the knowledge of user behaviour, depending on differences between private and commercial customers [7,33].

5. Conclusions and Outlook

Smart charging can help remove the barriers that currently hinder the expansion of electric mobility. It could potentially reduce costs for users while simultaneously improving the stability of the grid. At the same time, a higher share of renewable energy could be fed into the grid, and investments in grid infrastructure could be reduced. However, smart charging is currently not yet prevalent. Various projects in Germany, the USA and in other European countries are researching how it can be optimally implemented. In some countries, implementation is already partially possible. It was found that the most user-friendly smart charging tariff is TOU, which offers reduced electricity prices during off-peak hours and is already in use in several regions of the world. Nevertheless, it is possible that more efficient tariffs for the grid, such as dynamic RTP, could assert themselves if the charging process is conducted automatically. It is crucial for smart charging that relevant standards for the communication between all entities are compatible in order to guarantee the processing of the entire chain from load control to billing. In Germany, the most promising seems to be the core configuration of standards consisting of ISO 15118, IEC 63110, with OCPP and VDE-AR-E 2829-6 (equivalence with VDE-AR-E 2122-1000 in process) among others, EEBUS between EV and EVSE and grid. Internationally, openADR is also a preferred standard for connection to the grid. While their usage can depend on the use case, it is also possible that their functionalities and contents need to be converged. For this reason, further observation is needed.

While smart charging already has many advantages, bidirectional charging offers even higher incentives for users and for the spread of electro-mobility in general, being more climate-friendly. With this system, it is not only possible for users to save money when charging their EVs during off-peak hours, but also to earn money by selling energy that had previously been stored in their vehicles during a period of oversupply. However, bidirectional charging does not only provide incentives for users, but could also play an important role by ensuring grid stability. With pooled bidirectional charging supplied as system services, renewable energy can be used to a greater extent due to the system's ability to compensate for the natural fluctuations in the production of this energy. Smart charging can thus help in paving the way for a more climate-friendly world and an efficient energy system. In order for the aggregated flow of electricity from EVs into the energy system to take place via V2G and V2H, all necessary processes must be clearly regulated by law.

Author Contributions: Writing—original draft, D.J., L.O., M.B. and D.S. All authors contributed substantially and equally in writing and original draft preparation. All authors have read and agreed to the published version of the manuscript.

Funding: This research received no extra external funding. It is stated in the beginning of the paper, that the content of this paper is based on the work in the accompanying research of the funding programme Elektro-Mobil of the Federal Ministry for Economic Affairs and Climate Action.

Data Availability Statement: As this paper has the character of a review paper, no new data were created or analyzed in this study. Data sharing is not applicable to this article.

Acknowledgments: The statements and results of this contribution are based on the work of several people who were working on projects of the "Elektro-Mobil" funding programme. The projects, as well as the accompanying research, are funded by the German Federal Ministry for Economic Affairs and Climate Action. Some of the project partners were authors of the above-mentioned short study [3] and contributed to the white paper [4]. Their previous work formed the fundamental basis for the central content of this article. Therefore, their names of the projects are listed here for Acknowledgements: ARNi, BDL, DatenTanken, ELBE, ELSTA, LamA, LamA-Connect, RESIGENT, unIT-e2.

Conflicts of Interest: The authors declare not to have any conflict of interest. The authors declare that clients of the accompanying research had no role in the design of the study, in the collection, analyses, or interpretation of the data nor in the writing of the manuscript. Doris Johnsen and Mischa Bechberger are employees of the Institut für Innovation und Technik (iit) in VDI/VDE Innovation + Technik GmbH. The paper reflects the views of the scientists, and not the company. Lars Ostendorf is an employee of TÜV Rheinland Consulting GmbH. The paper reflects the views of the scientist, and not the company.

References

1. Agora Verkehrswende. Verteilnetzausbau für die Energiewende, Berlin. 2019. Available online: https://static.agora-energiewende.de/fileadmin/Projekte/2018/Netzausbau_Elektromobilitaet/Agora-Verkehrswende_Agora-Energiewende_EV-Grid_WEB.pdf (accessed on 4 October 2022).
2. Brinkel, N.; Schram, W.; Al Skaif, T.; Lampropoulos, I.; van Sark, W. Should we reinforce the grid? Cost and emission optimization of electric vehicle charging under different transformer limits. *Appl. Energy* **2020**, *276*, 115285. [CrossRef]
3. Das, R.; Wang, Y.; Putrus, G.; Kotter, R.; Marzband, M.; Herteleer, B.; Warmerdam, J. Multi-objective techno-economic-environmental optimisation of electric vehicle for energy services. *Appl. Energy* **2020**, *257*, 113965. [CrossRef]
4. Al-Ogaili, A.; Hashim, T.T.; Rahmat, N.; Ramasamy, A.; Marsadek, M.; Faisal, M.; Hannan, M. Review on Scheduling, Clustering, and Forecasting Strategies for Controlling Electric Vehicle Charging: Challenges and Recommendations. *IEEE Access* **2019**, *7*, 128353–128371. [CrossRef]
5. Yilmaz, M.; Krein, P. Review of the Impact of Vehicle-to-Grid Technologies on Distribution Systems and Utility Interfaces. *IEEE Trans. Power Electron.* **2013**, *28*, 5673–5689. [CrossRef]
6. Wolf, S.; Korzynietz, R. Innovation Needs for the Integration of Electric Vehicles into the Energy System. *World Electr. Veh. J.* **2019**, *10*, 76. [CrossRef]

7. TÜV Rheinland Consulting GmbH; Institut für Innovation und Technik (iit) in der VDI/VDE Innovation + Technik GmbH. Gesteuertes Laden von Elektrofahrzeugen über Preisanreize Köln. 2021. Available online: https://www.iit-berlin.de/de/publikationen/gesteuertes-laden-von-elekrofahrzeugen-ueber-preisanreize/ (accessed on 25 February 2021).
8. TÜV Rheinland Consulting GmbH; Institut für Innovation und Technik (iit) in der VDI/VDE Innovation + Technik GmbH. *Steuerung von Ladevorgängen in der Elektromobilität Köln, Berlin*, 1st ed.; 2022. Available online: https://www.iit-berlin.de/wp-content/uploads/2022/02/Whitepaper-Steuerung-Ladevorgaenge-Elektromobilitaet_2022_FINAL.pdf (accessed on 27 April 2022).
9. Irena, P. Innovation Landscape Brief: Time-of-Use Tariffs. Abu Dhabi, United Arab Emirates, 2019. Available online: https://www.irena.org/-/media/Files/IRENA/Agency/Publication/2019/Feb/IRENA_Innovation_ToU_tariffs_2019.pdf (accessed on 4 October 2022).
10. Dynamic Pricing in Electricity Supply—A EURELECTRIC Position Paper, Brussels. 2017. Available online: http://www.eemg-mediators.eu/downloads/dynamic_pricing_in_electricity_supply-2017-2520-0003-01-e.pdf (accessed on 4 October 2022).
11. Walter, T. Simple and efficient implementation of local energy markets. In Proceedings of the 4th International GSM Symposium, Lucerne, Switzerland, 19–20 October 2020.
12. Tibber. Tibber API. Available online: https://developer.tibber.com/docs/overview (accessed on 7 May 2021).
13. Stromdao, Corrently Ökostrom. Available online: https://www.corrently.de/home.html (accessed on 7 May 2021).
14. aWATTar. aWATTar Deutschland. Available online: https://www.awattar.at/ (accessed on 7 May 2021).
15. Innovation Outlook Smart Charging for Electric Vehicles. Abu Dhabi, United Arab Emirates, 2019. Available online: https://www.irena.org/publications/2019/May/Innovation-Outlook-Smart-Charging (accessed on 4 October 2022).
16. PG&E. ToU-Traffis. Available online: https://www.pge.com/en_US/residential/rate-plans/rate-plan-options/electric-vehicle-base-plan/electric-vehicle-base-plan.page (accessed on 17 March 2021).
17. SDG&E. ToU-Traffis. Available online: https://www.sdge.com/whenmatters (accessed on 17 March 2021).
18. Kolokathis, C.; Hogan, M.; Jahn, A. *Cleaner, Smarter, Cheaper: Network Tariff Design for a Smart Future*; Regulatory Assistance Project: Brussels, Belgium, 2018. Available online: https://www.raponline.org/knowledge-center/cleaner-smarter-cheaper-network-tariff-design-for-a-smart-future (accessed on 18 March 2021).
19. Hildermeier, J.; Kolokathis, C.; Rosenow, J.; Hogan, M.; Wiese, C.; Jahn, A. Smart EV Charging: A Global Review of Promising Practices. *World Electr. Veh. J.* **2019**, *10*, 80. [CrossRef]
20. Fitzgerald, G.; Nelder, C.; Newcomb, J. *Electric Vehicles as Distributed Energy Resources*; Rocky Mountain Institute: Basalt, CO, USA, 2018; Available online: http://www.rmi.org/pdf_evs_as_DERs (accessed on 7 May 2021).
21. OECD/IEA. Nordic EV Outlook. 2018. Available online: https://www.nordicenergy.org/wp-content/uploads/2018/05/NordicEVOutlook2018.pdf (accessed on 27 April 2022).
22. Szinai, J.K.; Sheppard, C.; Abhyankar, N.; Gopal, A.R. Reduced grid operating costs and renew-able energy curtailment with electric vehicle charge management. *Energy Policy* **2020**, *136*, 111051. [CrossRef]
23. Nationale Plattform Zukunft der Mobilität (NPM). Roadmap zur Implementierung der ISO 15118. 2020. Available online: https://www.plattform-zukunft-mobilitaet.de/wp-content/uploads/2020/12/NPM_AG5_AG6_2020_Q4_ISO15518.pdf (accessed on 27 April 2022).
24. Nationale Plattform Zukunft der Mobilität (NPM). *Schwerpunkt-Roadmap zum Intelligenten Lastmanagement*; NPM: Berlin, Germany, 2020.
25. Vde Fnn. Zielbild Steuerbarkeit von Ladeinfrastruktur für E-Fahrzeuge. 2021. Available online: https://www.vde.com/de/fnn/themen/elektromobilitaet-backbone-stromnetz/netz-als-backbone (accessed on 11 January 2022).
26. Bundesamt für Sicherheit in der Informationstechnik (BSI). Roadmap-Prozess. Available online: https://www.bsi.bund.de/DE/Themen/Unternehmen-und-Organisationen/Standards-und-Zertifizierung/Smart-metering/Roadmap-Prozess/roadmap-prozess.html (accessed on 27 April 2022).
27. Bundesamt für Sicherheit in der Informationstechnik (BSI). Stufenmodell zur Weiterentwicklung der Standards für die Digitalisierung der Energiewende, Berlin. 2021. Available online: https://www.bsi.bund.de/SharedDocs/Downloads/DE/BSI/SmartMeter/Stufenmodell/Anhang.pdf?__blob=publicationFile&v=5 (accessed on 4 October 2022).
28. TÜV Rheinland Consulting GmbH; Institut für Innovation und Technik (iit) in der VDI/VDE Innovation + Technik GmbH. *Steuerung von Ladevorgängen in der Elektromobilität—Darstellung der Kommunikationswege Anhand Einer Landkarte*, 1st ed.; 2022. Available online: https://www.iit-berlin.de/wp-content/uploads/2022/10/Whitepaper_Gesteuertes-Laden-Erweiterte-Editin-1.pdf (accessed on 1 December 2022).
29. VDE FNN. Eckpunkte zum zukünftigen Netzbetrieb mit Flexibilitäten in der Niederspannung, VDE FNN Impuls. 2022. Available online: https://www.vde.com/resource/blob/2191870/c830251e02291ca67fd187c02381783e/eckdatenpapier-flexibilitaet-data.pdf (accessed on 11 January 2022).
30. Fraunhofer IEE. *Resilienz im Energiesystem. Expert Web Session 24*; 2022. Available online: https://www.iee.fraunhofer.de/de/presse-infothek/Veranstaltungen-Messen/2022/resi-lienz-energiesystem.html (accessed on 1 December 2022).
31. Mayer, C.; Brunekreeft, G. Resilienz digitalisierter Energiesysteme. In *Blackout-Risiken Verstehen, Stromversorgung Sicher Gestalten*; 2020. Available online: https://energiesysteme-zukunft.de/fileadmin/user_upload/Publikationen/PDFs/ESYS_Analyse_Digitalisierung.pdf (accessed on 1 December 2022).
32. P3 Group. *Smart Charging: Die bevorstehende Revolution der Elektrofahrzeug-Wirtschaft*. Available online: https://www.p3-group.com/smart-charging-die-bevorstehende-revolution-der-elektrofahrzeug-wirtschaft/ (accessed on 27 April 2022).

33. Kern, T.; Kigle, S. Modeling and Evaluating Bidirectionally Chargeable Electric Vehicles in the Future European Energy System. In Proceedings of the Energy Reports ICACER 2022 Conference Proceedings, Barcelona, Spain, 20–22 April 2022.
34. Kühnbach, M.; Stute, J.; Gnann, T.; Wietschel, M.; Marwitz, S.; Klobasa, M. Impact of electric vehicles: Will German households pay less for electricity? *Energy Strategy Rev.* **2022**, *32*, 100568. [CrossRef]
35. Li, H.; Wan, Z.; He, H. Constrained EV Charging Scheduling Based on Safe Deep Reinforcement Learning. *IEEE Trans. Smart Grid* **2019**, *11*, 2427–2439. [CrossRef]
36. Huber, J.; Schaule, E.; Jung, D.; Weinhardt, C. Quo Vadis Smart Charging? A Literature Review and Expert Survey on Technical Potentials and User Acceptance of Smart Charging Systems. *World Electr. Veh. J.* **2019**, *10*, 85. [CrossRef]
37. Heidjann GmbH. Zusammensetzung Strompreise. Available online: https://www.stromauskunft.de/strompreise/strompreis-zusammensetzung/ (accessed on 14 May 2021).

World Electric Vehicle Journal

MDPI

Article

Profitability Evaluation of Vehicle-to-Grid-Enabled Frequency Containment Reserve Services into the Business Models of the Core Participants of Electric Vehicle Charging Business Ecosystem

Andrei Goncearuc [1,2,*], Nikolaos Sapountzoglou [1], Cedric De Cauwer [1], Thierry Coosemans [1], Maarten Messagie [1] and Thomas Crispeels [2]

[1] ETEC Department & MOBI Reserach Group, Vrije Universiteit Brussels (VUB), Pleinlaan 2, 1050 Brussels, Belgium

[2] BUTO Department, Vrije Universiteit Brussel, Pleinlaan 2, 1050 Brussels, Belgium

* Correspondence: goncearuc.andrei@vub.be

Abstract: The current paper defines a framework for the introduction of frequency containment reserve (FCR) services, enabled by vehicle-to-grid (V2G) technology, into the business model of an entity owning and operating electric vehicle (EV) charging infrastructure. Moreover, the defined framework can also be extrapolated, with minor adjustments, to the business models of different core participants of the EV charging business ecosystem. This study also investigates the financial factors impacted by this introduction, eventually evaluating its financial profitability under given assumptions and comparing it to the profitability of the traditional business model of an entity owning and operating a unidirectional EV charging infrastructure. The current research shows that offering additional V2G-enabled FCR services can be potentially more profitable than the existing unidirectional approach if the V2G technology reaches its maturity phase with mass market adoption and economies of scale.

Keywords: V2G (vehicle to grid); business model; grid balancing

Citation: Goncearuc, A.; Sapountzoglou, N.; De Cauwer, C.; Coosemans, T.; Messagie, M.; Crispeels, T. Profitability Evaluation of Vehicle-to-Grid-Enabled Frequency Containment Reserve Services into the Business Models of the Core Participants of Electric Vehicle Charging Business Ecosystem. *World Electr. Veh. J.* **2023**, *14*, 18. https://doi.org/10.3390/wevj14010018

Academic Editor: Ghanim A. Putrus

Received: 7 October 2022
Revised: 18 November 2022
Accepted: 3 January 2023
Published: 6 January 2023

1. Introduction

1.1. Context

The modern world is experiencing a massive electrification trend, and the increasing transition from internal combustion engine (ICE) vehicles to electric vehicles (EVs) is an important part of it. This electrification of transport can potentially place a significant stress on electricity grids by increasing energy demand and peak loads [1,2]. However, EVs can also be an enabler to mitigate this issue through the use of vehicle-to-grid (V2G) technology [1–4].

V2G technology gives the opportunity to discharge the energy from EVs back to the grid, opening the door to grid balancing (flexibility) services, such as frequency regulation and peak shaving [1,3,5]. Therefore, the rising popularity of EVs, in combination with the active use of V2G technology and smart charging, can become a very relevant long-term alternative to the traditional grid reinforcement methods currently employed by grid operators [6,7].

At the same time, provision of the V2G-enabled flexibility services can become a very relevant opportunity for the participants of the EV charging business ecosystem as well, serving as a new value proposition and opening up new significant revenue streams for its participants [6,8–10].

1.2. Literature Overview

1.2.1. V2G Technology

The term V2G was introduced by the research of Kempton et al. in 2001 [11] while they were investigating the technical and financial potential of bidirectional EV charging, including the opportunity of EV participation in ancillary services. Later on, with more and more applications, V2G technology evolved to a V2X concept [12]. As part of the latter, the EV battery can be used to power up various electric appliances (V2L) [13], a home (V2H) [14], a building (V2B) [15], a whole community (V2C) [16], other EVs (V2V) [17], etc.

It is important to mention that V2G technology is not yet in its maturity phase, and there are numerous social (e.g., range anxiety), economic (e.g., high cost of V2G chargers), political (e.g., lack of relevant legislative mechanisms), and technical (e.g., lack of V2G-enabled EVs) barriers to overcome before the technology reaches the mass market [1,11,18]. Many of these barriers represent chicken-and-egg deadlocks in which the technology is not able to be massively adopted before overcoming the barrier, while the technology does not become relevant for the surrounding business ecosystem unless it is massively adopted. For instance, EV manufacturers do not produce the V2G-enabled EVs due to the absence of V2G infrastructure, while the infrastructure is absent due to the lack of a clear financial incentive for the EV charging business ecosystem [1,11,18]. This financial incentive could be provided by the grid operators through the revenues from the V2G-enabled flexibility services. However, the grid operators do not adapt the flexibility market to V2G-enabled solutions due to the absence of these solutions on the mass market. Thus, understanding the business case is key to incentivizing all the involved actors to realize the full potential of the technology by simultaneously collaborating and advancing the technology and reach a win–win situation.

1.2.2. V2G-Enabled Flexibility Services and EV Charging Business Ecosystem

The existing literature offers relatively limited knowledge on the effect of the introduction of V2G-enabled grid balancing services into the business models of the participants of public and semi-public EV charging infrastructures [9,10,19,20]. Moreover, there is a lack of a clear, quantified business model towards capturing value [1,18].

The currently existing EV charging business ecosystem includes numerous interrelated entities acting together to provide EV charging services, the core value proposition of the ecosystem [9]. It is noticeable that the entities owning and operating EV charging infrastructure are at the core of the EV charging business ecosystem, being directly related to the provision of its current core value proposition [9]. Thus, one of the key participants of the EV charging business ecosystem is the charge point operator (CPO)—the entity that manages, maintains and often owns EV charging infrastructure. Therefore, a successful and profitable introduction of V2G technology into the business model of a CPO (and other entities with similar roles) is a major step towards V2G adoption by the whole business ecosystem.

The business model of an archetypical CPO owning and operating a public and/or semi-public EV charging infrastructure is provided in Figure 1 where the changes caused by the introduction of the V2G technology are highlighted in blue.

As shown in Figure 1, the introduction of new technology influences all the business model perspectives (including customer, internal business, value propositions and financial perspectives), as it involves the introduction of—among others—new key activities, customer segments, revenue streams, and cost structure elements.

However, following the logic of the business model innovation mechanism, described in [9], the technological transformation of the business model begins with the introduction of a new key resource—V2G technology. Consequently, it allows the company to offer new value propositions enabled by this new key resource. New value propositions enabled by the V2G technology are mainly related to grid-balancing services (see Figure 1).

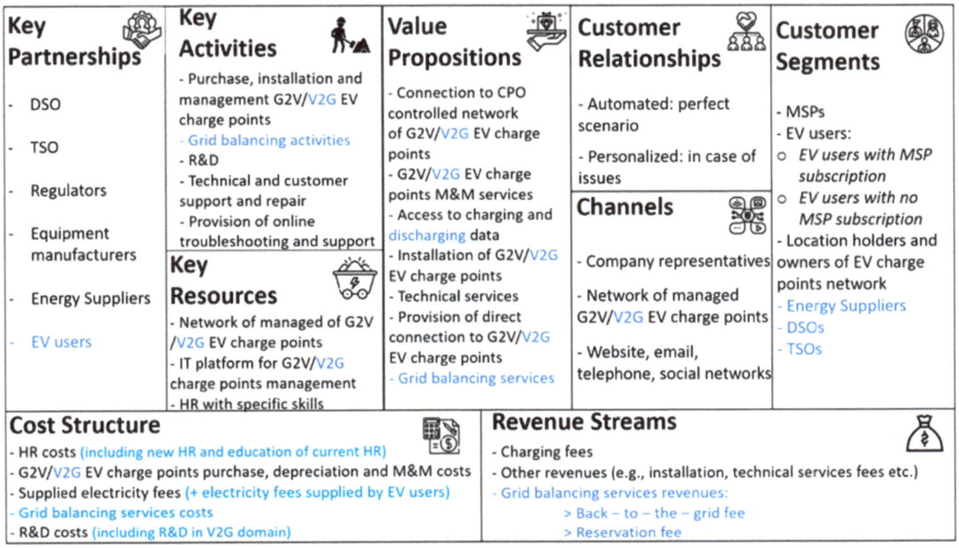

Figure 1. Business model: CPO + V2G (V2G—caused changes are marked in blue) [9].

1.2.3. Grid-Balancing Services

In order to introduce grid-balancing services into the list of its value propositions, an entity owning and operating EV charging infrastructure must contractually become one of the balancing service providers (BSP) connected to the respective transmission system operator (TSO) [21].

According to Elia [22], the Belgian Transmission System Operator (TSO), there are three types of grid balancing services that could be provided by a BSP connected to the grid. These services allow the maintenance of the balance between the energy injection and consumption and maintain the grid operation at a constant frequency of 50 Hz. They are presented in Table 1.

Table 1. Types of grid-balancing services [22].

Type	Description
Frequency Containment Reserve (FCR)	Primary automatic reserve provided by the entity with FCR service contract, reacting within a timeframe between zero and 30 s to a frequency deviation and stabilizing it on a certain level.
Automatic Frequency Restoration Reserve (aFRR)	Secondary automatic reserve provided by the entity with aFRR service contract, restoring the frequency to 50 Hz within a timeframe between 30 s and 15 min.
Manual Frequency Restoration Reserve (mFRR)	Tertiary reserve, provided by the entity with mFRR service contract, manually activated by the TSO in case of major imbalances and congestions. The respective BSPs must make mFRR available not later than 15 min after the TSO's demand.

The currently existing grid-balancing market is designed for large, centralized entities able to operate relatively large power capacities and volumes of energy. The provision of FCR services is typically considered a side activity for these entities. However, due to the large power capacity and energy volumes offered, the generated revenues can reach significant values.

At the same time, the modern world faces the decentralization trend of energy supply caused by the technological developments related to the renewable energy generation, power transfer, and reduction of battery costs [23]. Even though the power capacity that can be offered individually by the smaller entities (e.g., prosumers) is significantly less, the aggregated capacity is potentially able to reach significant values. This creates the opportunity to evolve the traditional business model for FCR services provision and include new decentralized players into the market. Moreover, even though BSPs often interface with DSOs, TSOs have the ability to directly contract BSPs [24].

Even though the decentralization process has already begun, there is still a number of unfavorable conditions for potential smaller decentralized BSPs. This list of unfavorable conditions mainly includes the minimum energy capacity bid size of 1 MW along with the specialized metering equipment on every so-called, "Delivery Point" (EVSE, in case of EV charging infrastructure) [25,26]. However, according to Elia and the Belgian energy regulator CREG (Commission for Electricity and Gas Regulation), grid balancing service conditions for BSPs must evolve in the direction of the reduction of minimal contracted capacity and utilization of standard widely accepted smart meters as metering equipment [27–29].

However, despite the existing decentralization initiatives, the framework offered by the current research (see Section 2.1) and resulting models is still not fully applicable for the residential, fully decentralized V2G EVSE due to the need of an aggregator to aggregate the capacities of multiple delivery points and virtually create a large entity corresponding with the currently existing grid balancing market design. The role of the aggregator, in the case of V2G-enabled flexibility services, can be taken over by a CPO (or other entity-managing EVSE infrastructure willing to become a BSP). The most suitable type of V2G-enabled flexibility service could be the FCR because of the ability of EV batteries to react immediately to a power request and a relative readiness of the TSOs and policy makers to adopt FCR service conditions for smaller decentralized BSPs [28].

The only remuneration foreseen for FCR services is based on the energy capacity offered by the BSP and reserved by the TSO, expressed in €/MW/h price. This price is defined by means of capacity bids on the FCR energy capacity auction, organized by the TSO for the involved BSPs. Due to the symmetric nature of FCR services in Belgium, requiring rapid upward and downward activations of the contracted balancing power capacities, the supplied and consumed energy payments during the activations offset each other, while the TSO offers remuneration for the reserved power capacity [30].

Concerning the risks for the BSP, the participation into FCR services involves potential penalties for: (a) not passing the availability test of the reserved capacity (organized by the TSO) and (b) the inability to activate the reserved capacity. However, the potential financial penalty cannot exceed the remuneration paid for the reserved energy capacity, making the grid-balancing services market, to a certain extent, risk-free [30] except for risks related to the initial investments into the infrastructure.

1.3. Contribution

The current paper adds value to the existing literature by decreasing lack of knowledge on the effect of introduction of V2G-enabled grid balancing services into the business models of the participants of public and semi-public EV charging infrastructures. The list of the contributions of this research includes: (a) definition of the framework for the introduction of the V2G-enabled FCR service into the list of the core value propositions of the business model of different entities (with minor case-dependent adjustments) owning and operating public and semi-public EV charging infrastructure (e.g., CPO, EV charging location holder); (b) evaluation of the profitability of the introduction of this new V2G-enabled value proposition by the means of a set of profitability indicators; (c) comparative profitability analysis with unidirectional EV charging infrastructure business model; and (d) a sensitivity analysis of the profitability of an entity owning and operating public and semi-public EV charging infrastructure (both uni- and bidirectional, offering FCR services), defining the most critical revenue and cost factors.

2. Methods

2.1. V2G-Enabled FCR Services into the Financial Perspective of the Business Model

The current paper focuses on the introduction of V2G technology into the business model and V2G-enabled grid balancing opportunities, while more elaborate definitions of revenue streams and cost structure of entities owning and maintaining EVSE networks can be found in [31]. According to [31], the generalized financial framework can be defined as follows (Equations (1) and (2)):

$$\text{Revenues} = \text{TF} + \text{OR}, \tag{1}$$

- TF: total fee received from the charging activities on the EVSE network.
- OR: other revenues generated by side activities not directly related to the EV charging (e.g., advertisement, technical fees, etc.).

$$\text{Costs} = \text{C}_{Infrastructure} + \text{C}_{Electricity} + \text{C}_{MP} + \text{C}_{HR} + \text{C}_{Other}, \tag{2}$$

- $\text{C}_{Infrastructure}$: depreciation, management and maintenance costs of EVSE infrastructure.
- $\text{C}_{Electricity}$: electricity costs paid to the energy suppliers.
- C_{MP}: costs for accessing the common interoperable marketplace for EV charging business ecosystem.
- C_{HR}: costs related to the human resources.
- C_{Other}: other additional costs, not represented by the previous categories.

As it is explained in Section 1.2.2, currently, the most convenient type of V2G-enabled grid balancing service for an entity owning and operating EVSE network is the FCR service, adding an additional revenue stream paid by the TSO to the BSP (Equation (3)):

$$\text{Revenues} = \text{TF} + \text{OR} + \text{R}_{FCR}, \tag{3}$$

- R_{FCR}: revenues generated through FCR flexibility services

The formula for the revenues generated by the means of V2G-enabled FCR services can be defined as follows (Equation (4)):

$$\text{R}_{FCR} = \text{FCR}_{Bid} * \sum_{y=1}^{Z} \left(K_y * N_y * (CR_y - UR_y) \right) * T \tag{4}$$

FCR_{Bid}: average FCR capacity bid (in €/MW/h) during the considered time period (T) on the energy capacity auction organized by the TSO.

y: type of V2G EVSE

K_y: power level of EVSE type y

CR_y: connection rate of EVSE (in%), being the percentage of the considered period (T) that the considered EVSE type y was connected to an EV.

UR_y: usage rate of EVSE (in %), being the percentage of the considered period (T) that the considered EVSE type y was actively engaged into the EV charging process.

N_y: number of EVSE type y

T: considered EVSE availability period

The difference between CR and UR can be considered the EVSE idle time available for the provision of the FCR services. Regarding the cost structure, while the introduction of V2G technology does not bring any new cost elements to the list, it does increase the existing ones. This raises concerns mainly related to the costs of the EVSE infrastructure ($\text{C}_{Infrastructure}$), i.e., depreciation, management and maintenance, and software and HR costs.

2.2. Profitability Evaluation

The current study's comparative profitability evaluation model is based on several common profitability indicators. The list of these profitability indicators includes earnings before interests and taxes (EBIT), earnings before interests, taxes, depreciation, and amortization (EBITDA), and the margins of these indicators and the annualized ROI [32–34]. The definitions of the chosen indicators can be found in Table 2.

Table 2. Profitability indicators with definitions [32–34].

Indicator	Generalized Formula	Definition
EBIT	$= Revenues - Costs$	EBIT is the difference between company's operating revenues (not including the interest revenues) and costs (before the inclusion of tax-related expenses).
EBITDA	$= EBIT + Depreciation + Amortization$	EBITDA repeats the definition of EBIT but does not include the depreciation and amortization into the costs list.
EBIT margin (%)	$= \frac{EBIT}{Revenue} * 100\%$	EBIT margin is a company profitability ratio, indicating the relative part of the revenues preserved after the deduction of expenses (before interests and taxes).
EBITDA margin (%)	$= \frac{EBITDA}{Revenue} * 100\%$	EBITDA margin repeats the definition of EBIT margin but does not consider the depreciation and amortization costs.
ROI (%)	$= \frac{EBIT}{Total\ investment} * 100\%$	ROI shows the ratio of company's EBIT to the total amount of the invested capital.

The selection of EBIT as a profitability indicator allows to avoid the peculiarities of different taxation mechanisms, providing the opportunity to extrapolate the results of the current study to different geographic regions [31]. The use of EBITDA allows the capital-intensive companies (e.g., entities owning EVSE networks) to obtain alternative view on the operating financial perspective of their business models, taking the costly assets with long depreciation periods out of scope [33]. High EBIT and EBITDA margins indicate a higher efficiency of the company, where a significant part of revenue is retained by the company [34]. Finally, the ROI, one of the capital return profitability ratios, considers the total investment and allows the evaluation of the profitability of the company with respect to the total invested capital [34].

2.3. Sensitivity Analysis

EBIT is the central profitability indicator of the current research and is considered in the calculations of all the other indicators. As EBIT is the difference between a company's operating revenues and costs, its calculation comprises numerous revenues and costs factors (described in Section 2.1, Section 2.2, and Section 3.1). However, not all the factors play equal roles, as EBIT might be more sensitive to the changes in one factor over the other. In order to assess the importance of the factors driving the EBIT calculation, the current research offers a sensitivity analysis (Section 3.4) based on the nominal range sensitivity method [35].

The nominal range sensitivity method implies choosing the base case value for the dependent variable and fixing the independent variables on the correspondent values. Thereafter, one (or each individually) of the independent variables is elected to vary across the predefined percentual range, while all the other independent variables remain fixed. Clearly, the changes in input, ceteris paribus, cause changes in output. Thus, after running this algorithm with all the independent variables, it becomes possible to compare the intensiveness of the reaction of the dependent variable to the changes of each independent variable and assess the importance of each input factor [35].

However, it is important to mention that this method has a significant disadvantage, that could be addressed by future research. Namely, the nominal range sensitivity method isolates the selected independent variable by fixing all the others on a certain value and does not take into consideration the potentially present interrelations [36].

2.4. Values of the Relevant Factors

The values of the factors participating into the aforementioned revenues and cost calculations, along with the sources validating these values, are provided in Table 3.

Table 3. Values of the factors participating into the profitability evaluation of the business models [19,37–52].

<table>
<tr><th>#</th><th>Parameter</th><th>Symbol</th><th>Unit</th><th colspan="3">Value</th></tr>
<tr><td>1</td><td>EVSE type</td><td>y</td><td>/</td><td>Unidirectional AC charger</td><td colspan="2">V2G charger</td></tr>
<tr><td>2</td><td>EVSE power level [37]</td><td>Ky</td><td>kW</td><td colspan="3">11</td></tr>
<tr><td rowspan="4">3</td><td rowspan="4">EVSE price [19,37–39]</td><td rowspan="4">Py</td><td rowspan="4">€</td><td rowspan="4">1200</td><td>DC Current</td><td>5000</td></tr>
<tr><td>DC Break-even</td><td>3999</td></tr>
<tr><td>DC Estimated</td><td>3500</td></tr>
<tr><td>AC Estimated</td><td>1200</td></tr>
<tr><td>4</td><td>EVSE installation cost [37,40]</td><td>Iy</td><td>€</td><td colspan="3">1000</td></tr>
<tr><td>5</td><td>Charging fee [40–43]</td><td>CFy</td><td>€/kWh</td><td colspan="3">0.35</td></tr>
<tr><td>6</td><td>Maximum yearly availability time</td><td>T</td><td>hours</td><td colspan="3">8760</td></tr>
<tr><td>7</td><td>Maximum yearly charging capacity</td><td>MCy</td><td>kWh/year</td><td colspan="3">96,360</td></tr>
<tr><td>8</td><td>Connection rate [43–46]</td><td>CR_y</td><td>%</td><td colspan="3">42</td></tr>
<tr><td>9</td><td>Charging usage rate [43–46]</td><td>UR_y</td><td>%</td><td colspan="3">7</td></tr>
<tr><td>10</td><td>Electricity price [47]</td><td>$C_{Electricity}$</td><td>€/kWh</td><td colspan="3">$0.9636 * (MCy * URy * Ny)^{-0.126}$</td></tr>
<tr><td>11</td><td>Average FCR capacity bid [48]</td><td>FCR_{Bid}</td><td>€/MW/h</td><td colspan="3">16.6</td></tr>
<tr><td>12</td><td>Useful lifetime [49,50]</td><td>Ly</td><td>years</td><td colspan="3">10</td></tr>
<tr><td>13</td><td>Salvage value [49,50]</td><td>Sy</td><td>%</td><td colspan="3">5</td></tr>
<tr><td>14</td><td>HR cost [43]</td><td>C_{HR}</td><td>€</td><td colspan="3">$1000 * Ny$</td></tr>
<tr><td>15</td><td>Cost of accessing the marketplace [51,52]</td><td>C_{MP}</td><td>€</td><td colspan="3">15,000</td></tr>
<tr><td>16</td><td>Miscellaneous costs [43,49,50]</td><td>C_{Other}</td><td>€</td><td colspan="3">100 000</td></tr>
<tr><td>17</td><td>Management and maintenance costs [43,49,50]</td><td>$C_{M\&M}$</td><td>€</td><td colspan="3">$10\% * (Py + Iy)$</td></tr>
<tr><td>18</td><td>Total initial investment</td><td>/</td><td>€</td><td colspan="3">$(Py + Iy) * Ny$</td></tr>
<tr><td>19</td><td>Number of EVSE</td><td>Ny</td><td>Units</td><td colspan="3">Variable</td></tr>
</table>

Having defined the values of the factors from Table 3, a justification of the chosen values is provided below as their retrieval is not always straightforward. First, since the V2G technology is not yet in its maturity phase, the price of V2G EVSE (Py) has not yet reached its mass market value. Therefore, the current research uses four different V2G EVSE pricing methods (see Table 3, #3). The first one is the currently existing V2G DC EVSE market price, which is considered to be too high over a longer period of time due to the current lack of mass production and potential future economies of scale [19]. The second is a so-called, "break-even price", indicating the V2G DC EVSE price ceiling at which the

EBIT of the entity owning a network of unidirectional EVSE equals the EBIT of the entity providing additional V2G-enabled FCR services. The third is the estimated V2G DC EVSE price in its maturity phase. Finally, the fourth price setting indicates a presumable V2G AC EVSE price equalized with the current price of unidirectional AC EVSE of a comparable power level. However, it is important to mention that the V2G AC EVSE is currently absent on the market due to the lack of the necessary communication protocols between the vehicle and charger. Moreover, the technology also requires acceptance from the EV manufacturers' side [53].

Another factor value that requires further explanation is the electricity price ($C_{Electricity}$ in Table 3, #10). The formula provided in Table 3 is represented by the following trendline (Figure 2) of the electricity price data of annual consumption bands for non-households in Belgium, provided by Eurostat [47].

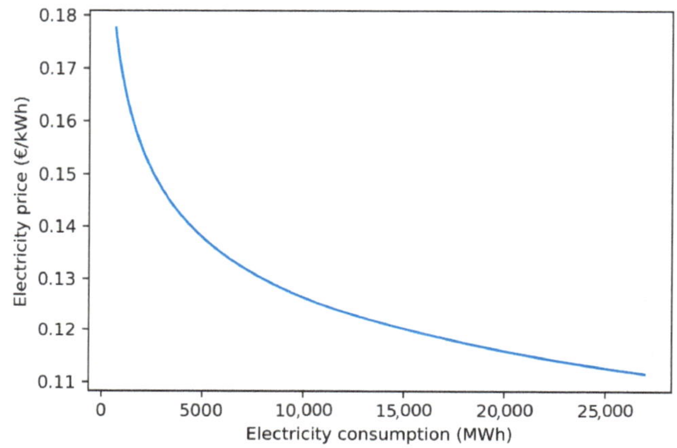

Figure 2. Electricity price trendline (based on 2011–2021 data) for Belgian non-household users in function of consumption [47].

The trendline shown in Figure 2 shows a clear negative relation between the electricity price and consumption; the increase in consumption causes the decrease in electricity price per kWh. The curve is steeper on the left side of the graph at a relatively small volume of electricity consumption, while further increase in consumption has a noticeably smaller effect on price. For instance, considering the values of the parameters presented in Table 3, a small network of 5 EVSE generates the consumption of 34 MWh per year, which leads to an electricity price of 0.26 €/kWh. The expansion of the EVSE network to 2000 units, consuming 13,500 MWh, decreases the electricity price to 0.12 €/kWh. However, doubling of the network size to 4000 EVSE, consuming 27,000 MWh, would cause only a further (minor) reduction of electricity price to 0.11 €/kWh. Moreover, this effect is not infinite, and the energy price stabilizes at a very high level of consumption of around 500 GWh (e.g., at large energy-intensive industries) [48].

Drawing back to the $C_{Electricity}$ formula from Table 3, the trendline provided by Figure 2 shows the following dependency between the consumption and the price (Equation (5)):

$$C_{Electricity} = 0.9636 * Electricity\ Consumption^{-0.126} \quad (5)$$

Translation of Equation (5) in terms of the model used by the current paper leads to Equation (6), provided in Table 3:

$$C_{Electricity} = 0.9636 * (MCy * URy * Ny)^{-0.126} \quad (6)$$

It is also important to mention that while the Equation (6) delineates the dependency between the electricity price and the internal electricity consumption of the company, the changes in the electricity price can be also caused by factors external to the company.

The currently ongoing energy crisis can serve as a very relevant example of such an external factor. Equation (6), along with the trendline provided in Figure 2, is based on the electricity price data for the period between 2011 and 2021, while the electricity price in Belgium in 2022 is significantly higher due to the ongoing geopolitical processes. For instance, a small network of 5 EVSE generating the average yearly consumption of 34 MWh would pay 0.85 €/kWh (September 2022 electricity prices) [54] instead of the previously valid price of 0.26 €/kWh (see Figure 2). At the same time, the trendline showing the relation between electricity price and consumption remains unchanged—the more energy consumed the lower is the price per kWh.

Electricity price is an important factor, having, ceteris paribus, a significant influence on the EBIT of current research's focal entities (see Section 3.4). Consequently, these entities must react to such a drastic increase in electricity price with a rise in charging fee (CFy), attempting to maintain the financial viability of the company. This step involves potential risks related to the decrease in customers' loyalty. However, due to the lack of historical data and the uniqueness of the situation, it is currently impossible to make any trustworthy projections about the customers' reaction and the future profitability of these companies.

Thus, according to these data availability requirements, the time scope of the current research is limited to the period between 2011 and 2021. However, due to the importance of the ongoing unprecedented processes on the energy market, the collection of the present-day data, their analysis, and the comparison of results with the results of the current research (see Section 3) are among the most relevant future research steps.

3. Results

In this section, we provide the results of the calculations of the aforementioned profitability indicators for two types of business models, namely the entity owning and operating a network of unidirectional EVSE and the one additionally providing V2G-enabled FCR services.

3.1. EBIT and EBIT Margin

The first profitability indicator evaluated by the current study is EBIT (Figure 3):

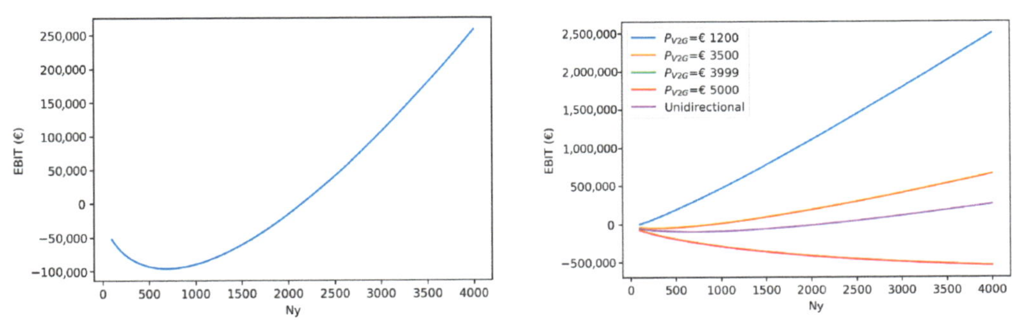

(a) (b)

Figure 3. (**a**) EBIT of an entity owning a network of unidirectional EVSE in function of Ny; (**b**) EBIT of an entity owning a network of V2G EVSE in function of Ny.

Figure 3a shows the EBIT of an entity owning and operating a network of unidirectional EVSE in function of the number of EVSE into the network (Ny). It is noticeable that the EBIT in Figure 3a shows a negative trend for low Ny values. The reason behind that is twofold. First, the revenues from a small EVSE network are not able to cover high initial costs (C_{HR}, C_{MP}, C_{Other}), that must be paid from the very beginning. Secondly, as was explained in previous section, the electricity price diminishes with the growth of consumption. Thus, a smaller EVSE network is not able to consume sufficient electricity to negotiate a favorable electricity price. However, as the number of EVSE grows further, the EBIT trend switches to positive, reaching the break-even point at 2150 EVSE units. After the point where EBIT becomes positive, it grows even stronger, due to the growing energy consumption causing higher revenues and diminishing per kWh electricity prices.

Figure 3b shows the EBIT of an entity owning a network of bidirectional EVSE, including four different EBIT curves, each of which represents a different V2G EVSE pricing method (described in the previous section). It is noticeable that with the currently existing V2G DC EVSE market price, the EBIT shows a negative trend and is not able to reach the break-even point (see Figure 3b, red curve). However, the decrease in V2G DC EVSE price to EUR 3999 (see Figure 3b, green curve), would already lead to the equalization of EBIT with the one presented in Figure 3b. A further decrease in V2G DC EVSE price, estimated by the currently existing literature (Figure 3b, yellow curve), would intensify the CPO EBIT growth, allowing achievement of the break-even point at 929 V2G EVSE units. Finally, the introduction of V2G AC EVSE with the price equalized to that of unidirectional EVSE of a comparable power level (Figure 3b, blue curve) could give an enormous boost to the EBIT, changing the break-even quantity to 89 V2G EVSE units in the network.

The EBIT margin, being the ratio of EBIT to its revenue component, is presented below in Figure 4.

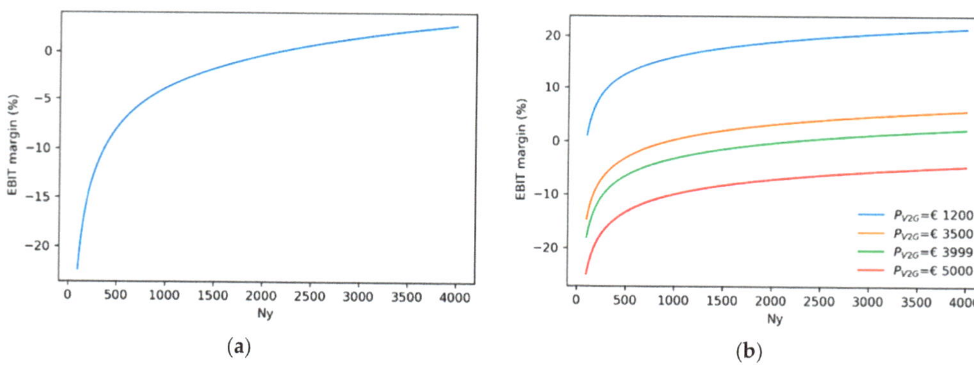

Figure 4. (**a**) EBIT margin of an entity owning a network of unidirectional EVSE in function of *Ny*; (**b**) EBIT margin of an entity owning a network of V2G EVSE in function of *Ny*.

The EBIT margins of the described business models are in general relatively low. In most of the scenarios, they do not reach 5% at a reasonable EVSE network size. This is due to high operational costs offsetting a major part of the revenues and indicating a low profitability efficiency. However, there is also an exception represented by the case of V2G AC EVSE (see Figure 4b, blue curve). A lower V2G AC EVSE price significantly cuts the depreciation expenses and gives a relatively high EBIT margin of 20% with a network size of 4000 EVSE.

3.2. EBITDA and EBITDA Margin

As described in Section 3.1, EBITDA is the profitability indicator excluding the depreciation and amortization costs. Therefore, the division of V2G EVSE pricing methods, present in the EBIT calculations in the previous section, becomes irrelevant for EBITDA calculations. The following figures present EBITDA and EBITDA margins (Figure 5a,b, respectively) both for business models of entities owning a network of unidirectional and V2G EVSE:

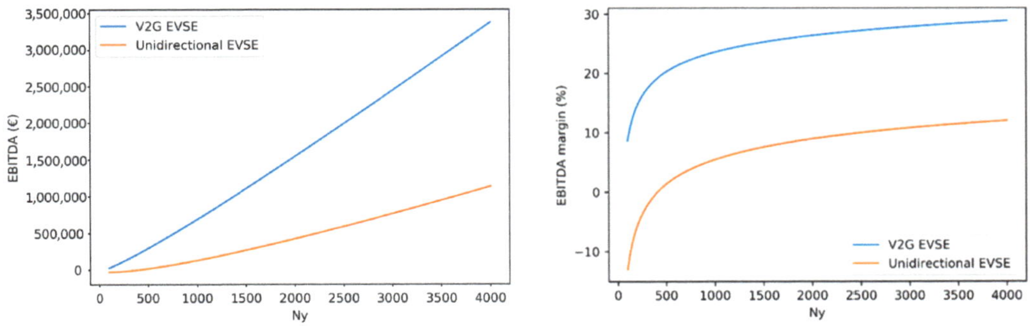

(a) (b)

Figure 5. (**a**) EBITDA of entities owning a network of unidirectional and V2G EVSE in function of *Ny*; (**b**) EBITDA margins of entities owning a network of unidirectional and V2G EVSE in function of *Ny*.

A typical business model of an entity, owning and maintaining numerous assets (e.g., CPO with EVSE network) with long depreciation periods, is relatively capital-intensive. Therefore, it becomes difficult to define the operational profitability of the company, as every asset must be depreciated and becomes the part of costs. In this case, EBITDA can become and interesting profitability indicator, concentrated on the current situation.

As becomes clear from Figure 5a,b, the business model which includes the provision of V2G-enabled FCR services is significantly more efficient in terms of this profitability indicator, generating a positive EBITDA from relatively small EVSE network sizes and reaching the EBITDA margin of 30% at the network size of 4000 EVSE.

3.3. ROI

In contrast to the profitability indicator discussed in the previous section, the current section evaluates the ROI (Figure 6), the profitability ratio directly related to the amount of capital invested into the company.

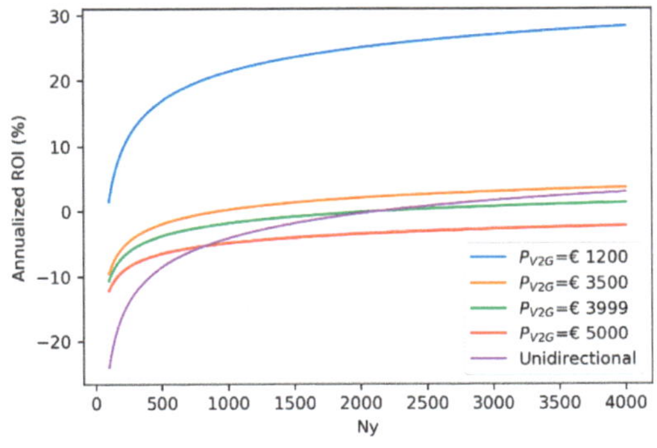

Figure 6. ROI of entities owning a network of unidirectional and V2G EVSE in function.

Entities owning EV charging infrastructure participate in a capital-intensive industry, having, in general, a relatively low EBIT margin (as described in Section 3.1). The combination of these two factors leads to a relatively low ROI, barely differentiating from zero, at the network size of 4000 EVSE. The obvious outlier with significantly higher ROI is the blue curve, indicating the V2G AC EVSE network with lower capital cost, creating a relatively high ROI of 30% within the network size of 4000 EVSE.

We also observe that the ROI curve of the entity owning a network of unidirectional EVSE (Figure 6, purple curve) shows a slightly different behavior. At a small network size, the ROI values shown by the purple curve are significantly lower than the rest, while the growth of EVSE network gives it a stronger boost, outperforming the curves indicating the application of current and break-even V2G DC EVSE prices and almost crossing the yellow curve (indicating the estimated long-term V2G DC EVSE price) at a network size of 4000 EVSE. From this observation, we conclude that V2G-enabled FCR services allow achievement of profitability faster, but the less-capital-intensive EVSE can generate a higher ROI on larger network sizes.

3.4. Sensitivity Analysis

In this study, EBIT is the central profitability indicator influencing all the other indicators. Therefore, after the definition of all the profitability indicators, it is also interesting to assess the importance of the individual influence of revenue and cost factors on EBIT.

The conducted sensitivity analysis is presented in Figure 7 and shows the EBIT sensitivity to the list of selected factors playing major roles in the business model of an entity owning and operating EV charging infrastructure: (1) the number of EVSE into the network (*Ny*); (2) the network usage rate (*URy*); (3) network connection rate (*CRy*); (4) electricity costs (*Ce*); and (5) charging fee (*CFy*). Each of these factors individually fluctuates, ceteris paribus, within a range of −10% to +10% from its base case value, which corresponds with the break-even point (EBIT = EUR 0) of unidirectional EV charging network.

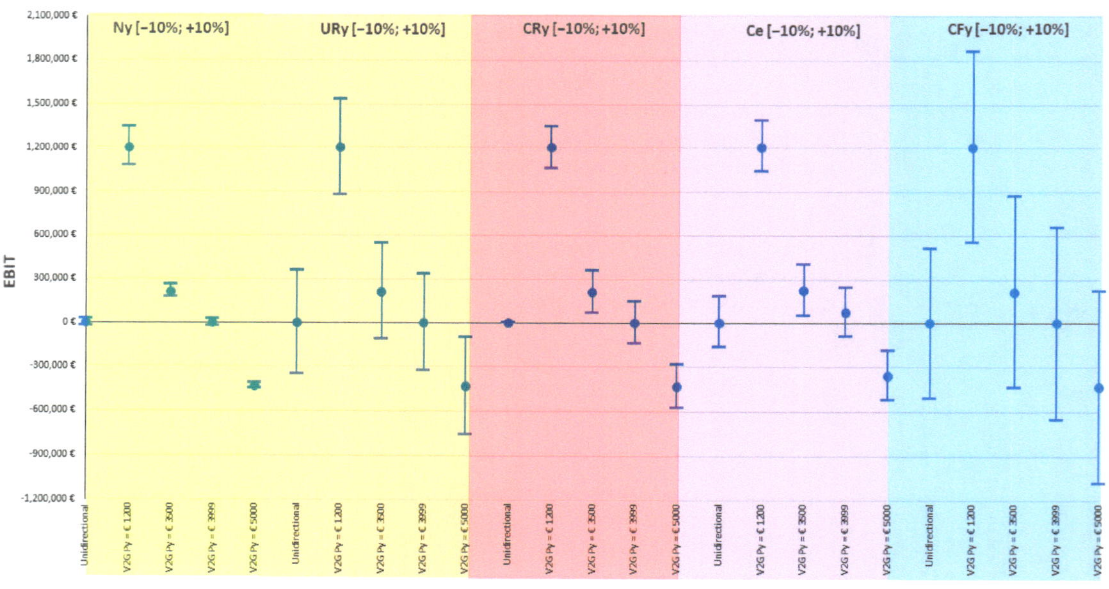

Figure 7. EBIT (EUR) sensitivity to elected revenues and costs.

Moreover, owning and operating an EV charging infrastructure implies participation into a capital-intensive industry, where the initial cost of capital in significant. Therefore, the analysis is conducted for each considered EVSE price level individually, for every selected factor.

In Figure 7, it is visible that the most significant factor influencing the EBIT is the EV charging fee (*CFy*), for which a 10% change in *CFy* causes a change in EBIT of around EUR 600,000. This observation is valid both for the business model of the entity owning and operating a unidirectional EV charging infrastructure and for the one providing additional V2G-enabled FCR services. This allows us to conclude that even after the transformation by the introduction of V2G-enabled FCR services, the business model of the focal entities still remains very dependent on its initial core value proposition. Moreover, this conclusion is also supported by the high importance of EV charging infrastructure usage rate (*URy*) and electricity costs (*Ce*) factors, directly influencing the revenues and costs related to the EV charging services.

However, it is also important to notice the differences in EBIT sensitivity to the EV connection rate (*CRy*) between unidirectional and V2G-enabled FCR business models. While in the first case, sensitivity is absent, the addition of V2G-enabled FCR services makes it an important factor, as the connected EVs are able to generate significant revenues during idle connection time, making the FCR services another important value proposition.

Finally, the influence of number of EVSE into the network (*Ny*) is relatively small, as the depreciation of the expensive assets eventually undercuts profitability. However, the obvious outlier, showing significant EBIT sensitivity to the minor changes in infrastructure size, is the less-capital-intensive AC V2G EVSE infrastructure case, for which the generated revenue not only allows to cover the initial capital costs but also gives a substantial boost to the EBIT.

4. Discussion

After the presentation of results, it is important to note that even though the defined model is based on the currently existing archetypical business model of a CPO [3,31], it could be also extrapolated, with minor adjustments, to other participants of the EV charging business ecosystem that perform similar functions. For instance, a location holder owning an EVSE network can hire an external CPO to operate this network for a fixed fee. In this case, the revenue and costs model remains unchanged except for the addition of the CPO fee to the list of costs. It should be underlined that the scope of the framework presented in this study is limited to public and semi-public EVSE infrastructure.

It should be also pointed out that the performed sensitivity analysis (see Section 3.4) does not take into account the interrelations between the input factors (e.g., the rise of *CFy* can potentially exceed the customers' willingness to pay and decrease the *URy*, which would lead to a negative impact on EBIT). Therefore, the study on interrelations between the input factors would be a useful future research direction.

5. Conclusions

The current paper defines a quantitative framework for the introduction of the V2G-enabled FCR services into the business model of an entity owning and operating public and/or semi-public EV charging infrastructure. It also evaluates the profitability of this introduction by the means of a set of profitability indicators. Furthermore, a sensitivity analysis is performed on the major factors influencing the profitability of the proposed business model. Moreover, the current study provides the comparison of the values of the profitability indicators for the business model, implying the owning and operating of the network of unidirectional EVSE only, and the one providing additional V2G-enabled FCR services.

The analysis of the profitability indicators allows to conclude that under the current market conditions with a V2G DC EVSE market price of EUR 5000, the existing business model of an entity owning a network of unidirectional chargers is more profitable than the business model after the introduction of V2G-enabled FCR services. However, V2G technology has not yet reached its technological maturity phase and lacks the benefits of economies of scale. Therefore, the estimated target price of V2G DC EVSE (EUR 3500) is significantly lower than the one currently existing on the market. The introduction and mass adoption of V2G AC EVSE (having potentially significantly lower prices than V2G DC EVSE and being comparable to the current prices of unidirectional AC EVSE) would thus allow the business model of an entity owning and operating EV charging infrastructure and offering additional V2G-enabled FCR services to strongly outperform—in terms of profitability—the business model offering the unidirectional charging of comparable power level. In fact, even the reduction in the V2G DC EVSE price below the break-even level of EUR 3999, would, ceteris paribus, already show a higher profitability.

From the sensitivity analysis, it can be concluded that the EVSE price is not the only factor playing an important role. The strongest impact on the EBIT of both types of business models comes from changes in the EV charging fee (*CFy*). In combination with a relatively high impact of EVSE usage rate (*URy*), this indicates that, even after the introduction of V2G-enabled FCR services, the business model remains quite dependent on its initial core value proposition. While changes in *URy* have a strong influence on the EBITs of both types of considered business models, changes in the connection rate (*CRy*) impact the available energy capacity for grid–balancing services, which considerably affects the revenues generated by the V2G-enabled FCR services. These changes do not influence the EBIT of the entity operating the network of unidirectional EVSE. Thus, the idle *CRy* factor, traditionally unused in unidirectional EVSE business model case, could become very useful for the new additional V2G-enabled revenue stream and decrease the dependency of the business model on the core value proposition.

Finally, as was already mentioned in Section 2.1, FCR is only one of the three existing grid balancing services. Therefore, another interesting research step would be to study the potential integration of V2G-enabled aFRR and mFRR services into the business models of the participants of EV charging business ecosystem.

Author Contributions: Conceptualization, A.G., N.S. and C.D.C.; methodology, A.G.; software, A.G.; validation, A.G., C.D.C., N.S. and T.C. (Thomas Crispeels); formal analysis, A.G.; investigation, A.G.; writing—original draft preparation, A.G.; writing—review and editing, C.D.C., N.S. and T.C. (Thomas Crispeels); visualization, A.G.; supervision, C.D.C., N.S. and T.C. (Thomas Crispeels); project administration, C.D.C.; funding acquisition, T.C. (Thierry Coosemans) and M.M. All authors have read and agreed to the published version of the manuscript.

Funding: This research was funded by Flanders Innovation & Entrepreneurship (VLAIO), grant number HBC.2018.0519.

Data Availability Statement: Not applicable.

Conflicts of Interest: The authors declare no conflict of interest. The funders had no role in the design of the study; in the collection, analyses, or interpretation of data; in the writing of the manuscript; or in the decision to publish the results.

Nomenclature

Abbreviations

Abbreviation	Definition
aFRR	Automatic frequency restoration reserve
BSP	Balancing service provider
CPO	Charge point operator
CREG	Commission for electricity and gas regulation
EBIT	Earnings before interests and taxes
EBITDA	Earnings before interests and taxes depreciation and amortization
EV	Electric vehicle
EVSE	Electric vehicle supply equipment
FCR	Frequency regulation reserve
ICE	Internal combustion engine
mFRR	Manual frequency restoration reserve
ROI	Return on investment
TSO	Transmission system operator
V2G	Vehicle-to-grid

Variables

Symbol	Variable	Unit
y	EVSE type	/
Ky	EVSE power level	kW
Py	EVSE price	€
Iy	EVSE installation cost	€
CFy	Charging fee	€/kWh
T	Maximum yearly availability time	hours
MCy	Maximum yearly charging capacity	kWh/year
CR_y	EVSE connection rate	%
UR_y	EVSE charging usage rate	%
$C_{Electricity}$	Electricity price	€/kWh
FCR_{Bid}	Average FCR capacity bid	€/MW/h
Ly	Useful lifetime	years
Sy	Salvage value	%
C_{HR}	HR cost	€
C_{MP}	Cost of accessing the marketplace	€
C_{Other}	Miscellaneous costs	€
$C_{M\&M}$	Management and maintenance costs	€
Ny	Number of EVSE	Units
R_{FCR}	Revenues generated through FCR flexibility services	€
TF	Total fee received from the charging activities on the EVSE network	€
OR	Other revenues generated by side activities not directly related to the EV charging (e.g., advertisement, technical fees, etc.).	€

References

1. Lauinger, D.; Vuille, F.; Kuhn, D. A Review of the State of Research on Vehicle-to-Grid (V2G): Progress and Barriers to Deployment. In Proceedings of the European Battery, Hybrid and Fuel Cell Electric Vehicle Congress, Geneva, Switzerland, 16 March 2017.
2. Gunkel, P.A.; Bergaentzlé, C.; Græsted Jensen, I.; Scheller, F. From Passive to Active: Flexibility from Electric Vehicles in the Context of Transmission System Development. *Appl. Energy* **2020**, *277*, 115526. [CrossRef]
3. Gonzalez Venegas, F.; Petit, M.; Perez, Y. Active Integration of Electric Vehicles into Distribution Grids: Barriers and Frameworks for Flexibility Services. *Renew. Sustain. Energy Rev.* **2021**, *145*, 111060. [CrossRef]
4. Zhou, C.; Xiang, Y.; Huang, Y.; Wei, X.; Liu, Y.; Liu, J. Economic Analysis of Auxiliary Service by V2G: City Comparison Cases. *Energy Rep.* **2020**, *6*, 509–514. [CrossRef]
5. Ali, H.; Hussain, S.; Khan, H.A.; Arshad, N.; Khan, I.A. Economic and Environmental Impact of Vehicle-to-Grid (V2G) Integration in an Intermittent Utility Grid. In Proceedings of the 2020 2nd International Conference on Smart Power and Internet Energy Systems, Bangkok, Thailand, 15–18 September 2020.
6. Borne, O. *Vehicle-to-Grid and Flexibility for Electricity Systems: From Technical Solutions to Design of Business Models*; Université Paris Saclay: Paris, France, 2019.
7. Knezović, K.; Marinelli, M.; Codani, P.; Perez, Y. Distribution Grid Services and Flexibility Provision by Electric Vehicles: A Review of Options. In Proceedings of the Universities Power Engineering Conference, Stoke on Trent, UK, 1–4 September 2015.
8. Pearre, N.S.; Ribberink, H. Review of Research on V2X Technologies, Strategies, and Operations. *Renew. Sustain. Energy Rev.* **2019**, *105*, 61–70. [CrossRef]
9. Goncearuc, A.; Sapountzoglou, N.; de Cauwer, C.; Coosemans, T.; Messagie, M.; Crispeels, T. An Integrative Approach for Business Modelling: Application to the EV Charging Market. *J. Bus Res.* **2022**, *143*, 184–200. [CrossRef]
10. Høj, J.C.M.L.; Juhl, L.T.; Lindegaard, S.B. V2G-An Economic Gamechanger in E-Mobility? *World Electr. Veh. J.* **2018**, *9*, 35. [CrossRef]
11. Sovacool, B.K.; Noel, L.; Axsen, J.; Kempton, W. The Neglected Social Dimensions to a Vehicle-to-Grid (V2G) Transition: A Critical and Systematic Review. *Environ. Res. Lett.* **2018**, *13*, 013001. [CrossRef]
12. Thompson, A.W.; Perez, Y. Vehicle-to-Everything (V2X) Energy Services, Value Streams, and Regulatory Policy Implications. *Energy Policy* **2020**, *137*, 111136. [CrossRef]
13. Wang, X.; Liu, Y.; Qian, W.; Wang, B.; Lu, X.; Zou, K.; González-Santini, N.; Karki, U.; Peng, F.Z.; Chen, C. A 25kW SiC Universal Power Converter Building Block for G2V, V2G, and V2L Applications. In Proceedings of the Proceedings—2018 IEEE International Power Electronics and Application Conference and Exposition, Shenzhen, China, 4–7 November 2018.
14. Rua, D.; Issicaba, D.; Soares, F.J.; Rocha Almeida, P.M.; Rei, R.J.; Peças Lopes, J.A. Advanced Metering Infrastructure Functionalities for Electric Mobility. In Proceedings of the IEEE PES Innovative Smart Grid Technologies Conference Europe, ISGT Europe, Gothenburg, Sweden, 11–13 October 2010.

15. Buonomano, A. Building to Vehicle to Building Concept: A Comprehensive Parametric and Sensitivity Analysis for Decision Making Aims. *Appl. Energy* **2020**, *261*, 114077. [CrossRef]
16. Yamagata, Y.; Seya, H.; Kuroda, S. Energy Resilient Smart Community: Sharing Green Electricity Using V2C Technology. *Energy Procedia* **2014**, *61*, 84–87. [CrossRef]
17. Harding, J.; Powell, G. *Vehicle-to-Vehicle Communications: Readiness of V2V Technology for Application*; National Highway Traffic Safety Administration: Washington, DC, USA, 2014.
18. Noel, L.; Zarazua de Rubens, G.; Kester, J.; Sovacool, B.K. Navigating Expert Skepticism and Consumer Distrust: Rethinking the Barriers to Vehicle-to-Grid (V2G) in the Nordic Region. *Transp. Policy* **2019**, *76*, 67–77. [CrossRef]
19. Sovacool, B.K.; Kester, J.; Noel, L.; Zarazua de Rubens, G. Actors, Business Models, and Innovation Activity Systems for Vehicle-to-Grid (V2G) Technology: A Comprehensive Review. *Renew. Sustain. Energy Rev.* **2020**, *131*, 109963. [CrossRef]
20. Geske, J.; Schumann, D. Willing to Participate in Vehicle-to-Grid (V2G)? Why Not! *Energy Policy* **2018**, *120*, 392–401. [CrossRef]
21. Elia Becoming a Balancing Service Provider. Available online: https://www.elia.be/en/electricity-market-and-system/system-services/becoming-a-balancing-service-provider (accessed on 31 October 2022).
22. Elia Flexible Demand Management Products. Available online: https://www.elia.be/-/media/project/elia/elia-site/electricity-market-and-system---document-library/balancing---balancing-services-and-bsp/2017/2017-brochure-flexible-demand-management-products_en.pdf (accessed on 3 July 2022).
23. Elia Towards a Consumercentric and Sustainable Electricity System. Available online: https://www.eliagroup.eu/-/media/project/elia/shared/documents/elia-group/publications/studies-and-reports/20210618_elia_ccmd-white-paper_en.pdf (accessed on 31 October 2022).
24. Gerard, H.; Rivero, E.; Six, D. Basic Schemes for TSO-DSO Coordination and Ancillary Services Provision. *SmartNet* **2016**, *1*, 12.
25. Elia FCR Service Design Note. Available online: https://www.elia.be/-/media/project/elia/elia-site/electricity-market-and-system---document-library/balancing---balancing-services-and-bsp/2019/2019-design-note-fcr-for-2020.pdf (accessed on 22 August 2022).
26. Smart Energy Europe The SmartEn Map. Available online: https://www.smarten.eu/wp-content/uploads/2018/11/the_smarten_map_2018.pdf (accessed on 20 August 2022).
27. Elia Market Facilitation. Available online: https://innovation.eliagroup.eu/innovation-pillars/market-facilitation/ (accessed on 18 August 2022).
28. Elia Using Electric Vehicles to Balance the Network. Available online: https://innovation.eliagroup.eu/projects/v2g/ (accessed on 23 August 2022).
29. Elia Innovation Incentive 2020–2023 Plan Update. Available online: https://www.creg.be/sites/default/files/assets/Consult/2020/658E69/PRD658E69Annex.pdf (accessed on 18 August 2022).
30. Elia Terms and Conditions for Balancing Service Providers for Frequency Containment Reserve (FCR). Available online: https://www.elia.be/-/media/project/elia/elia-site/public-consultations/2020/20200317-tc-bsp-fcrfinal-consulten.pdf (accessed on 15 August 2022).
31. Goncearuc, A.; Sapountzoglou, N.; de Cauwer, C.; Coosemans, T.; Messagie, M.; Crispeels, T. Business Model Quantification Framework for the Core Participants of the Ev Charging Market. *World Electr. Veh. J.* **2021**, *12*, 229. [CrossRef]
32. Perisa, A.; Kurnoga, N.; Sopta, M. Multivariate Analysis of Profitability Indicators for Selected Companies of Croatian Market. *UTMS J. Econ.* **2017**, *8*, 231–242.
33. Nissim, D. EBITDA, EBITA, or EBIT? *SSRN Electron. J.* Columbia Business School. **2017**, 17–71. [CrossRef]
34. Bouwens, J.; de Kok, T.; Verriest, A. The Prevalence and Validity of EBITDA as a Performance Measure. *Comptab. Controle Audit.* **2019**, *25*, 55–105. [CrossRef]
35. Frey, H.C.; Patil, S.R. Identification and Review of Sensitivity Analysis Methods. *Risk Anal.* **2002**, *22*, 553–578. [CrossRef]
36. Saltelli, A.; Aleksankina, K.; Becker, W.; Fennell, P.; Ferretti, F.; Holst, N.; Li, S.; Wu, Q. Why so Many Published Sensitivity Analyses Are False: A Systematic Review of Sensitivity Analysis Practices. *Environ. Model. Softw.* **2019**, *114*, 29–39. [CrossRef]
37. Izvia Bornes de Recharge V2G Entreprise. Available online: https://www.izivia.com/bornes-electriques-lieu-de-travail/bornes-de-recharge-V2G-entreprise (accessed on 21 August 2022).
38. Indra Vehicle-to-Grid (V2G) Chargers. Available online: https://www.indra.co.uk/v2g (accessed on 11 August 2022).
39. Alfen Alfen Enabling the Energy Transition. Available online: https://alfen.com/sites/default/files/media/documents/Alfen%20Annual%20Report%202021.pdf (accessed on 14 May 2022).
40. BlueCorner Laadpalen Onderweg. Available online: https://www.bluecorner.be/laadpalen/onderweg/ (accessed on 23 August 2022).
41. Allego What Costs Charging? Available online: https://www.allego.eu/support (accessed on 23 August 2022).
42. Eflux Mobility Cards to Charge Your EV. Available online: https://www.e-flux.io/en/e-flux-mobility-cards/ (accessed on 23 August 2022).
43. Allego Enabling Electrification. Available online: https://www.allego.eu/-/media/A7ECA09B306A409B99F39CA87292562F.ashx (accessed on 23 August 2022).
44. Wolbertus, R.; Kroesen, M.; van den Hoed, R.; Chorus, C. Fully Charged: An Empirical Study into the Factors That Influence Connection Times at EV-Charging Stations. *Energy Policy* **2018**, *123*, 1–7. [CrossRef]
45. Wolbertus, R.; van den Hoed, R.; Maase, S. Benchmarking Charging Infrastructure Utilization. *World Electr. Veh. J.* **2016**, *8*, 754. [CrossRef]

46. Wolbertus, R.; van den Hoed, R. Charging Station Hogging: A Data-Driven Analysis. In Proceedings of the EVS 2017—30th International Electric Vehicle Symposium and Exhibition, Stuttgart, Germany, 9–11 October 2017.
47. Eurostat Electricity Prices Components for Non-Household Consumers—Annual Data. Available online: https://ec.europa.eu/eurostat/web/energy/data/database (accessed on 23 August 2022).
48. CREG Study on the Functioning and Price Evolution of the Belgian Wholesale Electricity Market—Monitoring Report 2020. Available online: https://www.creg.be/sites/default/files/assets/Publications/Studies/F2229EN.pdf (accessed on 23 August 2022).
49. Nelder, C.; Rogers, E. *Reducing EV Charging Infrastructure Costs*; Rocky Mountain Institute: Basalt, CO, USA, 2020.
50. Madina, C.; Zamora, I.; Zabala, E. Methodology for Assessing Electric Vehicle Charging Infrastructure Business Models. *Energy Policy* **2016**, *89*, 284–293. [CrossRef]
51. Hubject Flexible Pricing That Scales with Your Business. Available online: https://www.hubject.com/pricing (accessed on 23 August 2022).
52. Eclearing Pricing. Available online: https://e-clearing.net/partners/pricing (accessed on 23 August 2022).
53. Wellisch, D.; Lenz, J.; Faschingbauer, A.; Pöschl, R.; Kunze, S. Vehicle-to-Grid AC Charging Station: An Approach for Smart Charging Development. *IFAC-Pap.* **2015**, *48*, 55–60.
54. VREG Energiemarkt in Cijfers. Available online: https://www.vreg.be/nl/energiemarkt-cijfers (accessed on 7 October 2022).

World Electric Vehicle Journal

MDPI

Article

Design and Application of the unIT-e^2 Project Use Case Methodology

Adrian Ostermann [1,2,*], Patrick Dossow [1,2] and Valerie Ziemsky [1]

[1] FfE Munich, 80995 München, Germany
[2] School of Engineering and Design, Technical University of Munich (TUM), 80333 München, Germany
* Correspondence: aostermann@ffe.de

Abstract: The ramp-up of electromobility requires cross-industry holistic solutions. However, bringing together stakeholders from different branches holds challenges. One prerequisite for successful collaboration is a uniform understanding of roles, processes, and interfaces. Based on existing methods and experience from former projects, this paper describes a method for the systematic description of use cases for smart charging of electric vehicles. This method enables a uniform understanding of all actors involved and guarantees application-oriented usability. The unIT-e^2 use case methodology consists of the business-use case level and the technical-use case level, which describe the use case in a structured layout. The method was applied in all so-called clusters of project unIT-e^2. In total, we identify 25 higher-level business use cases and highlight similarities and differences between them. Further, this paper describes the business-use case regulatory-defined grid-serving flexibility in detail.

Keywords: EV (electric vehicle); use case; energy network; smart connected EV; research; demonstration

Citation: Ostermann, A.; Dossow, P.; Ziemsky, V. Design and Application of the unIT-e^2 Project Use Case Methodology. *World Electr. Veh. J.* **2023**, *14*, 13. https://doi.org/10.3390/wevj14010013

Academic Editor: Joeri Van Mierlo

Received: 1 November 2022
Revised: 16 December 2022
Accepted: 3 January 2023
Published: 5 January 2023

1. Introduction

The European Union (EU) aims to reach climate neutrality by 2050 to meet the goal of the Paris Agreement to limit global warming to well below 2, preferably to 1.5 degrees Celsius, compared to pre-industrial levels [1]. Therefore, the EU needs to reduce emissions by increasing energy efficiency and investing in green technology. The European electricity sector lowered its emissions by 39% in 2019 compared to 1990 [2]. However, even if the generation of electricity were solely renewable, it would not be sufficient to achieve the goal of climate neutrality. A high share of variable renewable energy (VRE: wind and photovoltaic—PV) results in various challenges for the energy system due to intermittency, location-specific output, uncertainty, and limits in predictability [3–6]. To ensure the security of supply, which is put at risk by the integration of high shares of renewable energies, the energy system must be flexible by shifting demand in times when renewable energy generation is high. Sector coupling is widely considered necessary to achieve flexibility, significant emission reduction, and climate neutrality by interconnecting the energy-consuming sectors of industry, buildings (heating and cooling), and transport with the energy-producing sector [7–9]. Especially the transport sector needs to be transformed since greenhouse gas (GHG) emissions have increased between 2013 and 2019 as opposed to the other sectors [10]. In 2018, more than 12% of EU GHG emissions were caused by passenger cars [10]. Decarbonization in the transport sector can be realized by shifting to electric vehicles (EVs) if the electricity consumed is generated from renewable energy sources [11]. EVs can further offer positive and negative flexibility by charging batteries during periods of low demand or low prices, by interrupting an ongoing charging process, or by reducing charging power. Therefore, the integration of EVs is one of the most promising means to achieve the 2030 target of the European Commission's current revised proposal to reduce average fleet-wide emissions from newly registered vehicles from 37.5%

to 55% relative to a 2021 benchmark [12]. The trend towards an increasing share of EVs can already be observed in the EU. In 2019, the share of newly registered passenger EVs (battery electric vehicles—BEV—and plug-in hybrid electric vehicles—PHEV) in the EU was 3% [13]. In 2020, this share increased to 11%, and in 2021, to 18% [13].

The growing prevalence of EVs results in new business opportunities opening the market for new players: original equipment manufacturers (OEM), such as VW or Tesla, are now offering electricity tariffs in Germany [14,15]. On the one hand, the electrification of components in the transport sector, like EVs, or in the heating sector, such as heat pumps (HP), offers new opportunities and flexibility. On the other hand, electrification leads to increasing grid loads entailing new challenges for grid operators. In Germany, electric vehicle supply equipment (EVSE) with an output of 3.7 to 11 kW must be registered with the grid operator according to section 19 of the Low-Voltage Connection Ordinance (German: Niederspannungsanschlussverordnung—NAV) [16]. The installation of EVSE above 11 kW even requires a permit from the grid operator. Besides their permission to curtail VRE, grid operators might be authorized to interrupt charging sessions to ensure grid stability in the future. This, however, is in contradiction to EV user needs for the highest availability of mobility. Potential grid operator interventions could also affect other players' economic interests, such as flexibility aggregators. Therefore, all market participations must have a coordinated approach to reconcile grid stability, high shares of renewable energy, and customer satisfaction with diverse business models of other parties. This challenge is the starting point of the project unIT-e^2, which unites the automotive and energy sector to enable the integration of electromobility in line with grid and market requirements by defining uniform processes and creating standardized interfaces.

2. The Project unIT-e^2, German Market and Regulations

By bringing together 29 partners from the automotive and energy sector, IT and charging infrastructure, and the scientific research sector, the project unIT-e^2 offers a unique consortium along the entire value chain [17]. The project started in August 2021 and will last three years with the research institute FfE as the consortium leader. The project focuses on the market and grid integration of electromobility by defining interoperable, holistic, and intelligent charging concepts and demonstrating them in four large-scale field trials defined as clusters. These clusters are named Cit-E-Life, sun-E, Harmon-E, and Heav-E. Each cluster consists of an automobile manufacturer and various partners from the energy and IT sector. Accompanying the four clusters, three conceptional subprojects (SP) ensure the transfer into practice: SP research, SP grid, and SP project management and synthesis. At the beginning of the project, it was necessary to identify and define use cases that the clusters will demonstrate. For the successful demonstration in field trials, all parties involved must have the same understanding of the procedures and processes that arise from the electromobility use cases. Due to the high number of partners, the structure of different clusters, and the various individual backgrounds, FfE developed the unIT-e^2 use-case methodology to define and describe the use cases systematically and uniformly.

Integrating electromobility into the energy system requires a fitting regulatory environment, which must be well-known and taken into consideration during the process of use-case development. German legislation aims to induce flexibility procurement with several incentive mechanisms. According to § 41a Energy Industry Act (German: Energiewirtschaftsgesetz—EnWG), electricity suppliers are required to offer variable tariffs to their customers to incentivize energy savings or control consumption [18]. Furthermore, § 41a (2) EnWG obliges electricity suppliers with a certain size of their customer base to offer electricity supply contracts with dynamic tariffs in the future [18]. Time-variable tariffs aim to incentivize shifting electricity consumption to times of low electricity stock exchange prices, usually correlating with a high share of renewable electricity generation. Similar to financial incentives on the actual electricity price, German legislators established a measure to enable variable grid fees. § 14a EnWG requires grid operators to offer reduced grid fees for consumers providing the grid-serving control of their controllable consumption devices [18]. However, in September 2021, the European Court of Justice found § 14a EnWG not to comply with European law (C-718/18) [19]. The only institution allowed to develop grid fee models is the Federal Network Agency. Thus, the concrete handling of variable grid fees in Germany remains unclear. In this respect, the reform of the § 14a EnWG failed after more than two years of consultation. The draft, which was withdrawn by the Federal Ministry for Economic Affairs and Climate Action (German: Bundesministerium für Wirtschaft und Klimaschutz—BMWK, formerly BMWi) in January 2021, would have allowed grid operators to reduce the power of "controllable consumer devices", such as EVSE, or even disconnect them from the grid if risking grid congestion otherwise [20]. While the energy sector broadly supported the policy that was originally planned, there was growing concern in the automotive sector that this would hinder the progress of electromobility and cause irritation among customers. The reform of §14a EnWG is the task of the current federal government. One of the project goals of unIT-e^2 is to develop cross-sectoral accepted criteria for the redesign of §14a EnWG and to communicate them to the political decision-makers.

At the same time, the compulsory smart-meter rollout (§29 German Federal Law on Metering Point Operation, German: Messstellenbetriebsgesetz—MsbG) further complicates flexibility procurement. Accordingly, energy generation plants with a minimum installed capacity of 7 kW and consumers with an annual consumption exceeding 6000 kWh are subject to the mandatory installation of a smart metering system consisting of a modern metering device and a smart meter gateway (SMGW) [21]. While modern metering devices measure the actual electricity consumption, SMGWs serve as a communication interface to process, save, and communicate measured data. The state-of-the-art smart-metering-device technology is insufficient for a widespread rollout, as not all kinds of tariffs can be measured, e.g., load-based variable tariffs, consumption-based variable tariffs, and critical peak pricing, including real-time pricing [22]. Therefore, the Federal Cyber Security Authority (German: Bundesamt für Sicherheit in der Informationstechnik—BSI) set up a stage model for the further development of standards for the digitalization of the energy transition [23]. The stage model describes the necessary smart metering system development path to enable use cases around submetering, electromobility, and control of flexibilities. As the smart meter rollout affects consumers with an annual electricity consumption higher than 6000 kWh, most common households are currently not affected by the smart meter rollout. However, by adding the charge load of an electric vehicle to the standard household load, the expected annual electricity consumption increases by 3000 kWh, thus requiring a smart metering device [24,25]. The project unIT-e^2 addresses the described regulatory challenges: usage of variable electricity tariffs for charging EVs, control of flexibilities, the concrete design of variable grid fees, and whatever else is necessary.

3. unIT-e^2 Use-Case Methodology

In the field of electromobility, actors with different roles must work together, such as distribution-grid operators, transmission-grid operators, charge-point operators, metering-point operators, energy suppliers, and aggregators. Each actor knows his area of expertise and the associated processes. For the successful and well-ordered execution of an implementation project, the uniform comprehension of all processes and interfaces up to a certain level of detail is indispensable. This is where the use-case methodology developed in the unIT-e^2 project can be applied. Starting with the fundamental understanding that a use-case definition is necessary for joint implementation, the objective of the methodology is a systematic description of use cases based on a uniform level of detail, which ensures a consistent understanding of the use-case processes by all participants.

3.1. Literature Review

In literature, different use-case definitions can be found [26,27]. Cockburn defines a use case as "a description of the possible sequences of interactions between the system under discussion and its external actors, related to a particular goal" [26]. At the same time, ref [27] depicts a use case as a set of actions carried out by a system and produces an observable result that is typical of value for one or more actors or other stakeholders of the system. The second definition is closer to our understanding of use cases than the first one. Therefore, we define a use case as follows:

"A use case describes the functionality of a system from the user's point of view. A user can be a person, a role, an organization, or another system. The name of the use case is derived from the goal of the use case from the user's point of view. The aim of defining use cases is to reach an agreement and a common understanding about the behavior and scope of a system between the stakeholders of a project. Use cases can be represented graphically or in text documents".

Several norms and methods to develop and describe use cases exist [26–32]. Cockburn establishes a guide in his book Writing Effective Use Cases on how to define use cases with a consistent style [26]. The use-case diagram describes the user's possible interactions with a system based on graphical representation via the unified modeling language (UML) and is often used in software design [28]. The aim is to mimic the real world as simply as possible to understand how the system is going to be designed. The e3-value methodology, first developed by [30], is a well-accepted business modeling technique also based on graphical representation and has been developed to be tractable and lightweight. The focus of the e3-value model lies in the exchange of objects of value between actors performing activities. Originally used in explorations of e-commerce business models, e3-value models abstract from process details, thereby helping decision-makers focus on economic viability [33]. One limitation of e3-value models is the focus on the business model, leaving out many other concerns [33]. The IEC 62559 series provides a use-case methodology for power-system professionals to specify and detail "their user requirements for automation systems, based on their utility business needs" [29,34]. The IEC 62,559 series describes processes and provides basics for the use-case methodology like terms or use-case types, while it also defines the structure of a use-case template, an actor list, and a list of requirements. Further, it defines the required core concepts and their serialization into an XML format of a use-case template [29]. The IEC 62,913 series builds on the use-case methodology defined in the IEC 62,559 series and gives a more detailed methodology for describing use cases and extracting requirements from them, focusing on smart grids instead of only power systems [27]. The IEC 62,913 series focuses on capturing and sharing generic smart-grid requirements from a basis for standardization work that ensures and improves the interoperability between smart-energy systems and components [27]. However, for non-domain experts, the benefit of the documents is rather limited. Further, the familiarization and the consistent, correct use of the documents are time-consuming. The smart-grid architecture model (SGAM) defined by [35] provides technology-neutral analysis and the architectural or cross-system mapping of smart grid use cases. SGAM is thus suitable for representing the technical–operational implementation of use cases and the associated interoperability requirements. IEEE defines interoperability as "the ability of two or more systems or components to exchange information and to use the information that has been exchanged" [36]. Through the different views of the smart-grid architecture, technical (syntactic), informational (semantic), and organizational (pragmatic) interoperability can be represented and checked [37]. A guide on applying SGAM can be found in [37]. The use-case methodology developed by Faller et al. [32] during the project C/sells is mainly based on [27,31]. Faller et al. defined three main steps to describe a use case: 1. description of the business use cases and the use-case concept, 2. process and system description, and 3. procedure specifications (sequence diagrams). Business-use cases (BUCs) describe the business application and specify roles and responsibilities for executing business processes, while the focus is on the company's internal processes and not the overall system. In this process, the involved participants, relevant influences, and the purpose of the use case are clarified and presented [32]. Furthermore, [38] differentiates between parties (legal entities, i.e., either natural persons (a person) or judicial persons (organizations)), roles (representing the intended external behavior (i.e., responsibility) of a party), actors (a party that participates in a (business) transaction), and responsibilities (external behavior to be performed by parties). Role models such as [39–41] can be used to define roles uniformly across projects. The first step includes identifying political and regulatory factors influencing the use case by analyzing laws, directives, regulations, and standards. Afterward, the business services and processes are visualized in an e3-value model, including the identified roles, actors, parties, and responsibilities. Additionally, the business or operational benefit is described in a business-model-canvas or a platform-business-model-canvas. As the last step 1, the use-case goals are documented in a classical project-management target table. In step 2, Faller et al. focus on the representation of system components and the associated process scope when using the components in the use case [32]. Using a detailed process diagram (e.g., business process model and notation [42]), the description of initial interfaces, communication requirements, parts of the system (components), and their functions is developed in tabular form. In step 3, the use case is described in more detail (technically) about its procedures (sub-processes). For this purpose, the use case is described in sequence diagrams, and the respective information flows, messages, subsystems, and applied standards are specified in more detail.

From experience in the project C/sells, applying the described use-case methodology by Faller et al. [32] compared to other ones like [29] or [27] is less complex but still requires a considerable amount of time and resources to document every use case. In other projects, such as InDEED [43], Trade-EVs II [44], and Bidirectional Charging Management (BCM) [45] use cases, were identified with a similar or adapted use-case methodology. Experience from the aforementioned projects showed that creating, maintaining, and reading documents with long texts is time-consuming, particularly in keeping the documents up to date and consistent. Further, various partners perceived the many different visualization methods as unnecessary. Only roles and actors directly affected or involved in the use case should be displayed to enhance readability. Another lesson was that the business and technical discussions often mixed and made it challenging to achieve purposeful results during each step or phase. A stronger separation of these two fields would thus increase the efficiency of use-case development. Due to the size of unIT-e^2 and a large number of partners in the four different clusters resulting in many use-case documents, an efficient method was essential.

3.2. Methodology Description

Based on the mentioned methods as well as experience from these projects, FfE developed the unIT-e^2 use-case methodology. Figure 1 shows the schematic procedure.

At first, the unIT-e^2 use-case methodology deducts the former three steps from Faller et al. into two use-case levels: business-use case (BUC) and technical-use case (TUC). The first level, BUC, consists of the use-case identification and the basic concept. To identify a use-case idea, various methods such as mind mapping, design thinking, and others can be applied. For the basic concept of the use case, the following questions need to be answered: who is involved, what are the relationships between participants, who gets added value, and which laws must be considered? Further, the desired implementation needs to be defined. Therefore, five different implementation stages can be distinguished: conceptual, simulation, laboratory, pilot operation, and real operation. The conceptual stage solely analyzes the use case on a conceptual basis, while simulation means that the use case is investigated by simulation models. A laboratory implementation tests a use case under technical and/or regulatory development in a protected development environment without a direct connection to the public power grid. For pilot operation, a use case under technical and/or regulatory development is tested with connection to the public power grid with preferably "friendly users" on a limited scale. The final implementation real operation tests a technically and regulatory-compliant use case with not necessarily certified or approved components in the real end-customer sector. The second level and one level more profound is the TUC, comprising step 2 and communication format, channels, and standards from step 3 of the use-case method defined by Faller et al. [32]. The specification of the processes applied by Faller et al., corresponding to internal processes in companies via sequence diagrams, is too detailed for discussing the use case with the number of project partners involved in unIT-e^2. Therefore, the TUC focuses on the technical components (soft- and hardware) involved and their interaction with each other and not on the exact process sequence. The necessary process specification takes place after the unIT-e^2 use-case methodology. For each use-case level, a description template and an e3-value model template for visualizing relevant relationships are created. Hence, long texts are avoided by using icons and graphics. The BUC template describes the use case on a high level, includes relevant participants, describes the benefit, how the use case is to be implemented in the project, and states the overall goal of the use case. The BUC e3-value model visualizes relationships and interactions between participants. The TUC template describes the processes, participants, relevant technical components, and involved communication protocols, norms, and standards. The TUC e3-value model includes all relevant components and their data and communication interfaces. The description template, as well as the e3-value for both BUCs and TUCs, are limited to two PowerPoint pages each, resulting in four pages in total to describe one use case.

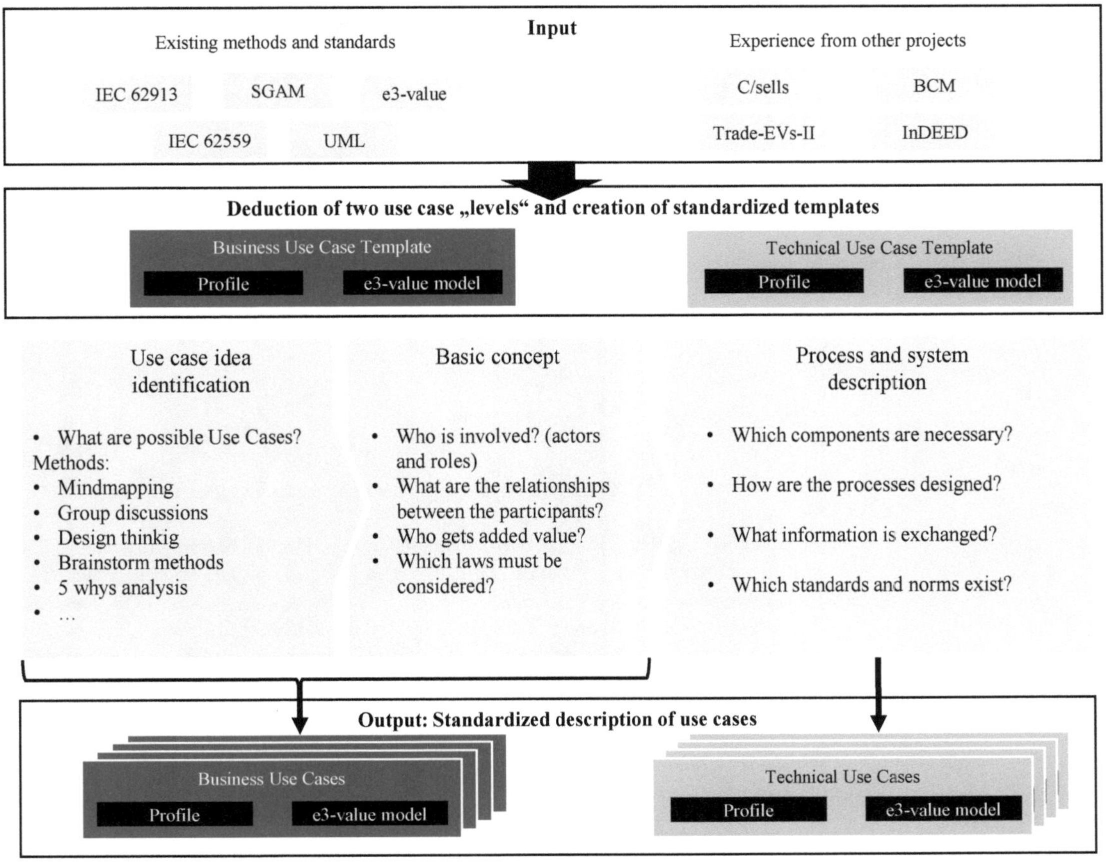

Figure 1. unIT-e^2 use-case methodology.

At the beginning of the project, several workshops were conducted in each cluster to develop use cases to be demonstrated in field trials. Depending on the particular use case and the number of involved project partners, it took several workshop sessions to complete the process visualized in Figure 1. A glossary was written to foster a uniform understanding of recurring terms. In the end, the standardized use-case templates constitute the foundation for the field tests of the project

4. Resulting Business Use Cases in unIT-e^2

The unIT-e^2 use-case methodology was applied in all four clusters, which resulted in the 25 higher-level BUCs displayed in Figure 2. Depending on the location of the flexibility and the direction of power flow, the BUCs can be subdivided into more than 40 individual BUCs, as stakeholders, responsibilities, and requirements differ for these cases. Different implementation steps are planned for the various BUCs. Some will only be analyzed by simulation or tested in a laboratory environment rather than demonstrated during the field trials. The following section discusses the higher-level BUCs to reduce complexity.

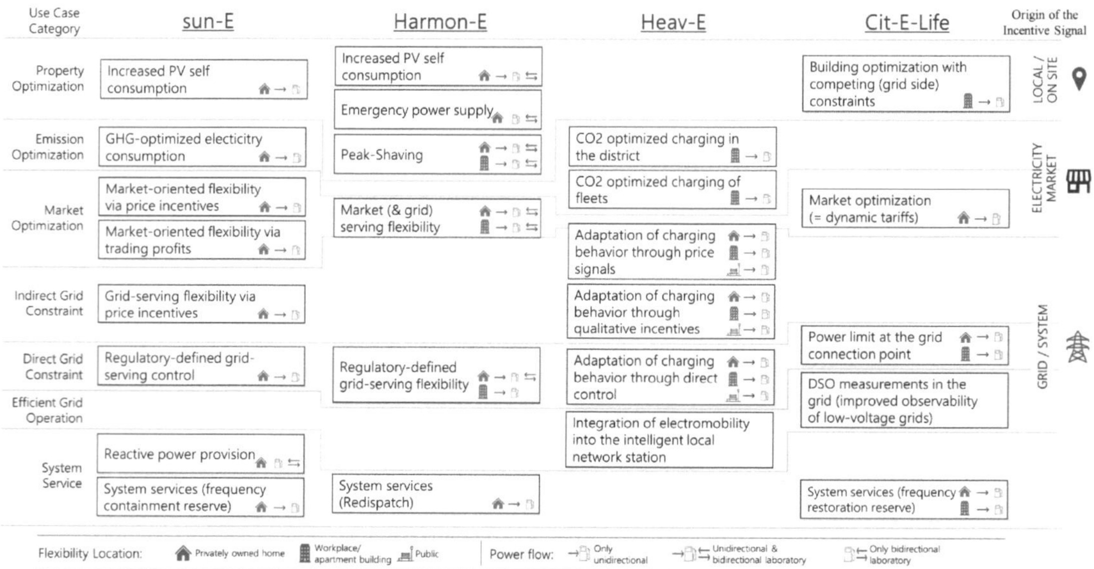

Figure 2. Overview of BUCs in unIT-e^2 project grouped per cluster and incentive signal.

We categorized the BUCs in two dimensions: per unIT-e^2 cluster in which the BUC will be implemented and per origin of the incentive signal, i.e., the key source, which creates an incentive for the BUC to be implemented. The origin of the incentive signal is divided into three types: 1. local/on-site, 2. electricity market, and 3. grid/system. Further, these incentives can be subdivided into different use-case categories displayed on the left. For each cluster, five to eight BUCs were identified. Five use cases are incentivized locally/at the site where the technical implementation takes place. Seven use cases are incentivized by variable market prices resulting from the electricity spot markets. The largest group of 13 BUCs is incentivized by the electric grid or, more precisely, by the possibility to reduce grid load, avoid grid congestions, and thus optimize grid use and avoid grid expansion.

4.1. Incentive Signal: Local/On Site

The category of locally incentivized BUCs contains two similar BUCs called increased PV self-consumption, one from the cluster sun-E and one from the cluster Harmon-E. Both will be implemented at private homes with a PV system and a home energy management system (HEMS), which optimizes the self-consumption behind the meter. On a technical level, both approaches require a smart metering system that allows EEBUS communication, which is a communication interface based on standards and norms, with the HEMS and PV-forecast data. The main difference between both BUCs is that, for the cluster

Harmon-E, bidirectional EVs will be tested in a laboratory environment in addition to the unidirectionally chargeable EVs in the field trial. The BUC peak-shaving in this category is comparable to the BUC building optimization. Both use cases aim to optimize the use of available power at the grid connection point by avoiding peak loads and corresponding fees. The optimization of flexible loads is implemented behind the meter in both cases. While both cases are to be implemented in apartment buildings or commercial sites with unidirectional EVs, peak-shaving in the cluster Harmon-E will also be tested in a single home, and bidirectional EVs will be tested in a laboratory environment. For peak-shaving in apartment buildings/commercial sites, an aggregator will operate the charging strategy, whereas for building management, an energy management system (EMS) will manage different flexibilities from various owners under potential grid restrictions.

4.2. Incentive Signal: Electricity Market

Out of the seven use cases incentivized by market prices, three use cases aim at reducing greenhouse gas emissions, which are GHG-optimized electricity consumption, CO_2-optimized charging in a smart quarter, and CO_2-optimized charging of a vehicle fleet. These use cases aim to minimize the direct emissions of electricity for charging the EV, when forecast data regarding the greenhouse gas emissions of electricity are needed in an appropriate temporal resolution. The BUCs differ in the planned implementation, as cluster sun-E will mainly simulate its BUCs for a single home, whereas cluster Heav-E will test the two BUCs in field trials in a smart quarter (apartment block) and for a fleet of EVs. On a technical level, the role of optimization varies. In sun-E, the HEMS optimizes based on emission data. In Heav-E, for the smart-quarter BUC, the smart-quarter manager optimizes all flexible assets with varying restrictions and objectives. For the vehicle-fleet case, this role is the fleet manager. In four BUCs, varying spot market prices are utilized to minimize electricity costs. For all four cases, the energy provider has market access, from which market prices are derived and passed either directly to the local EMS via a secure SMGW pathway or to an aggregator. The cases of market-oriented flexibility via price incentives and market optimization (dynamic tariffs) do not involve any aggregator, but instead market price tables are transmitted to the local HEMS, wherein the charging strategy is optimized locally. For the cases market (and grid) serving flexibility and market-oriented flexibility via trading profits, an aggregator either sends price tables to the local HEMS of a single home or sends flexibility schedules, which in turn determine the charging strategy for apartment buildings or commercial sites.

4.3. Incentive Signal: Grid/System

From the category of use cases incentivized by optimizing the electric grid, three BUCs are based on variable electricity prices, which are derived from the forecasted grid load. These cases have in common that a hypothetical variable grid usage fee is introduced by the grid operator based on the time-dependent local grid load. The grid operator thus needs locally and temporally resolved grid status data. In all cases, the local EMS receives variable price signals, which are included in the behind-the-meter optimization. For the BUC grid-serving flexibility via price incentives (sun-E), the variable grid fee will contain both a price component based on long-term forecasts and a component based on dynamic short-term predictions. The BUC adaption of charging behavior by price signals shows the feature that not only single home and apartment blocks/commercial sites are part of the field trials but also public charging. Each cluster includes one BUC with a strict load limit, which is set by the grid operator depending on the grid load at the respective time (i.e., the German §14a EnWG regulation or a possible future adaptation of the originally planned paragraph). For all these cases, reduced grid fees are considered, which are accounted for by the energy provider. In all cases, the grid operator sends a load limit signal via the smart-metering-system infrastructure to the grid connection point, where a local EMS must adjust the local power consumption accordingly. Two use cases aim at generally improved grid management through extensive data collection and automation. Key objectives in both cases are the monitoring and forecast of grid status to determine grid congestion and to derive required actions, such as load limits. Necessary data should be acquired through conventional measuring points, such as electric transformers, and new grid status data from local smart-metering systems. The lower four BUCs are linked to grid system services to maintain overall grid stability. For all these BUCs, an aggregator is needed who participates in one of the respective marketplaces and offers flexibility from pooled assets. If an offer is accepted, the offered flexibility is used to provide the respective system service, wherein the aggregator receives a command signal from the grid operator if necessary. The roles of the different players involved differ for the different use cases. e.g., in sun-E, the OEM functions as an aggregator, whereas for the other cases, the aggregator receives required data from the OEM's backend or the local EMS.

5. Business Use Case: Regulatory-Defined Grid-Serving Flexibility

Since all clusters will test a use case from the category direct grid constraint and the discussions about the reform of §14a EnWG, the following section describes the BUC regulatory-defined grid-serving flexibility from cluster Harmon-E in more detail. Depending on the power flow and the location of the flexibility, the use case can be further divided into three individual use cases. However, the overall goal remains the same: enable the distribution operator (DSO) to control the flexibility within a regulatory-defined framework. Figure 3 shows the template description (top) and the e3-value model (bottom) for the individual BUC regulatory-defined grid-serving unidirectional flexibility for privately owned homes.

This use case will be implemented in Harmon-E in pilot operation. The power flow is unidirectional; the location of the flexibility is a privately owned home; there is no feedback into the grid, and the flexibility is remotely controlled. The incentive signal origins from the grid. The user becomes ecological/sustainable, and there is also financial added value. The use case can be categorized as grid serving. A detailed description of the definition of grid serving can be found in [46]. The involved roles are connectee (in this case connectee = connection user), DSO, energy supplier, and meter operator. The meter operator can either be an active or passive market participant or both (aEMP/pEMP), resulting in two more BUCs than the one presented. First, the connectee allows the control of his flexibility (e.g., EV, HP, battery storage) by the DSO (according to §14a EnWG) and accepts the technical requirements according to the technical connection conditions. Therefore, the DSO settles reduced network charges for this connectee via the energy supplier. The energy supplier offers the connectee a reduced energy contract. The DSO carries out network condition monitoring and if he detects grid congestion, the DSO performs curative power adjustments by limiting the total flexibility of the customer installation. If the connectee has several flexibilities, the control authority of these flexibilities resides with the connectee via an (H)EMS. The specified power adjustment (PLim) is validated based on the measured data, and the energy supplier also receives information about the curative power adjustment. Figure 3 shows the case when the meter operator functions as an aEMP. The different roles are a result of the rollout of the smart-meter infrastructure. The aEMTs do not only receive data but can also control downstream devices via the SMGW. To control the flexibility, the aEMP must send a communication request to the gateway administrator (GWA). However, this process is part of the TUC rather than the BUC. The general procedure for controlling flexibilities via the smart-meter infrastructure is described in [47]. An aEMT must therefore have certification in accordance with ISO/IEC 27001, which covers all smart-meter public-key infrastructure-relevant processes and IT systems [48]. On the other hand, pEMPs can only receive data from SMGWs. This is a prerequisite for pEMPs to be able to handle their business processes, e.g., to create invoices and determine network states based on received meter values. The other possible variant of the use case would be if the DSO functions as an aEMP and the meter operator as a pEMP. In the following step, the template description and the e3-value model shown in Figure 3 serve as a basis for the design of the corresponding TUC.

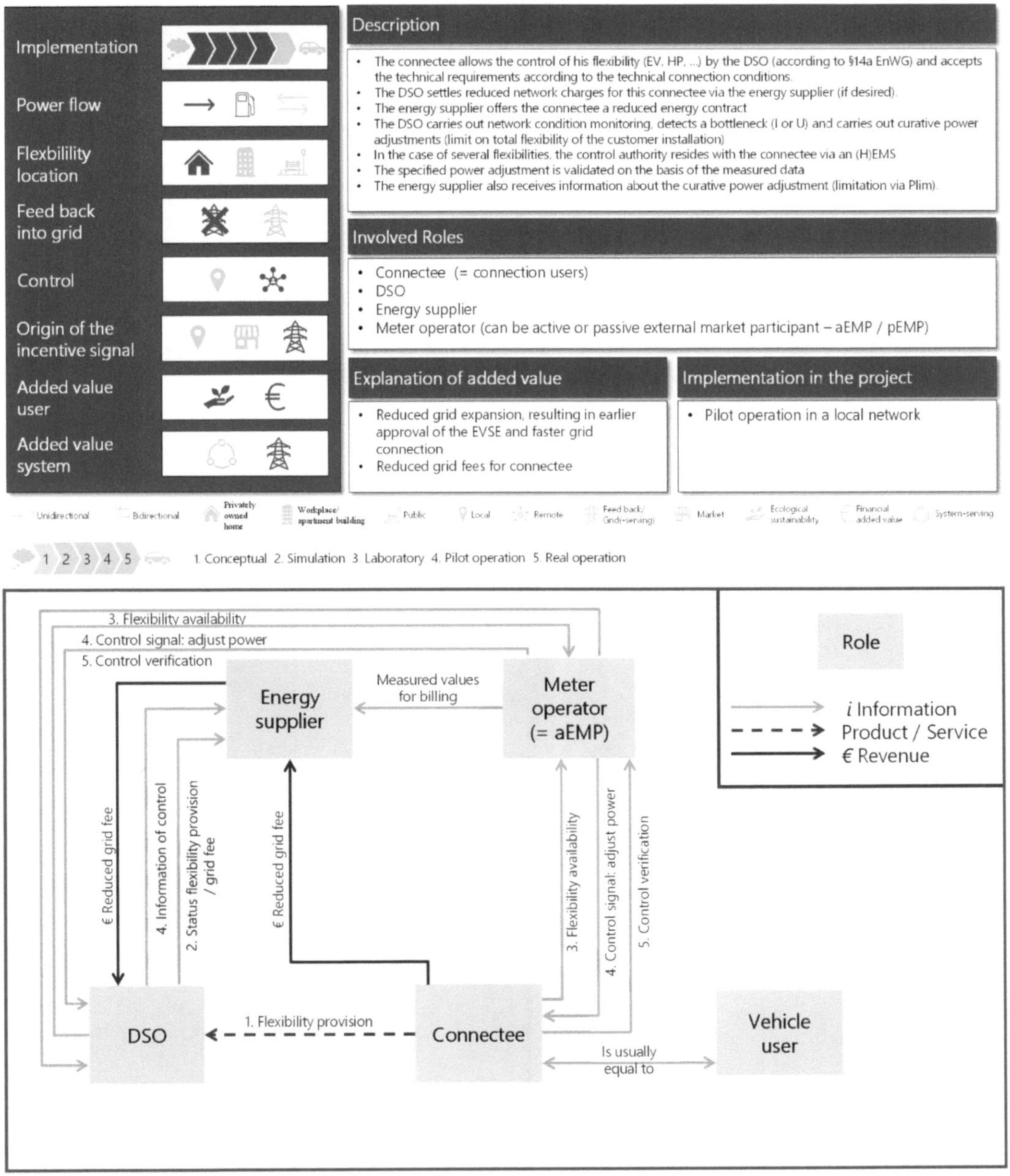

Figure 3. BUC regulatory-defined grid-serving unidirectional flexibility for privately owned home: description template (**top**) and e3-value model (**bottom**).

6. Technical Use Case: Regulatory-Defined Grid-Serving Flexibility

The following section describes one possible technical implementation of the use-case regulatory-defined grid-serving flexibility for privately owned home as it will probably be tested in cluster Harmon-E. Figure 4 shows the template description (top) and the e3-value model (bottom) for the technical use case.

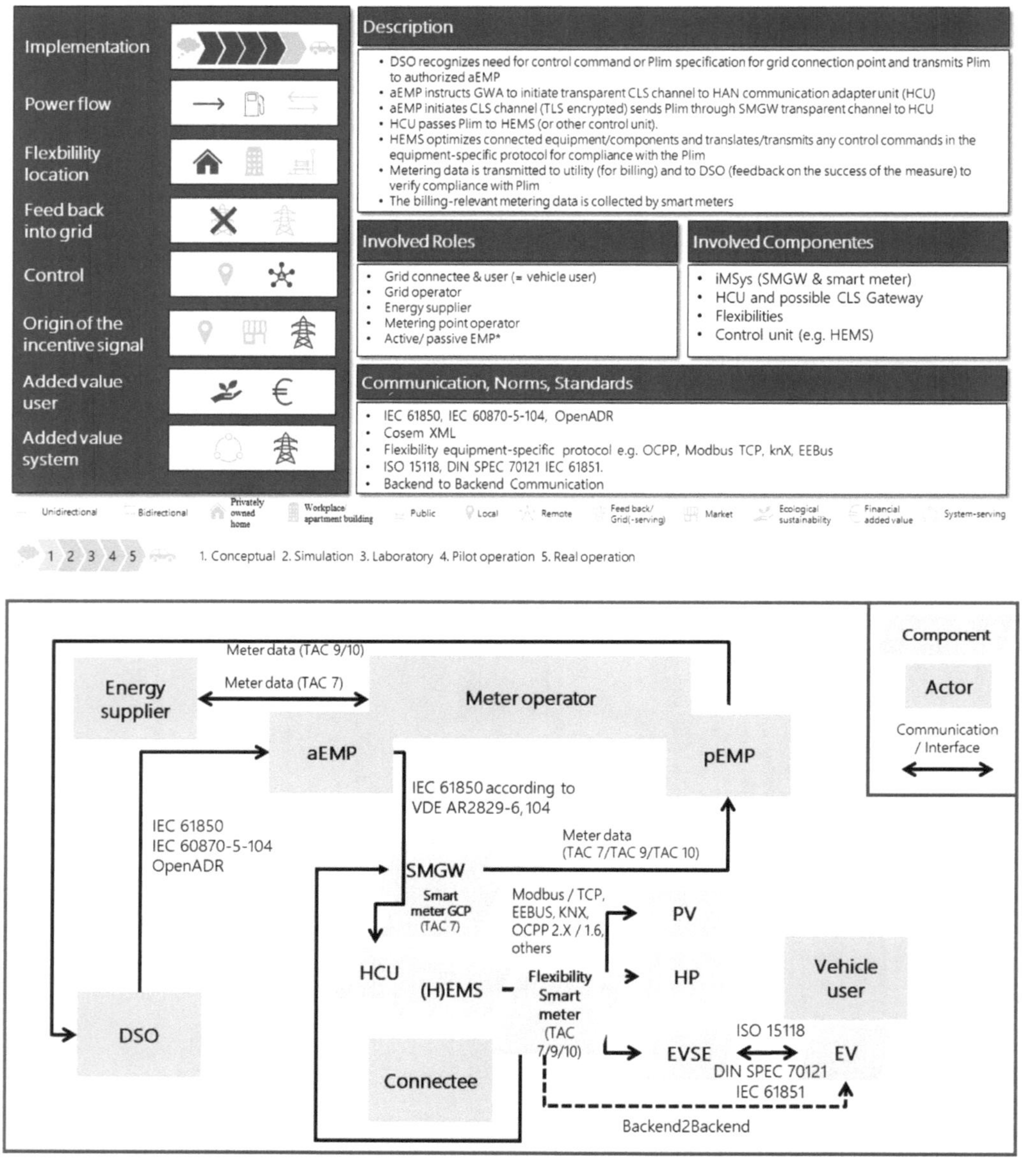

Figure 4. TUC regulatory-defined grid-serving unidirectional flexibility for privately owned home: description template (**top**) and e3-value model (**bottom**).

The technical use case proceeds as follows. The DSO recognizes the need for a control command or a Plim specification for the grid connection point (GCP) and transmits the Plim to an authorized aEMP. The applicable standards for communication are IEC 61850, IEC 70870-5-104, and OpenADR. Again, the meter operator acts as an aEMP. The aEMP instructs the GWA to initiate a transparent CLS channel to the HAN communication adapter unit (HCU). The aEMP initiates the establishment of a controllable local system (CLS) channel (TLS encrypted) and sends the Plim signal through the SMGW transparent channel to the HCU, which is the endpoint of the CLS channel. In correspondence with [23], the transmission via a CLS proxy corresponds to stage 3 of the stage model. Once again, the IEC 61,850 following VDE AR2829-6 offers a communication standard. Afterwards, the HCU passes the Plim signal to a HEMS. Another possibility is that the HCU passes the Plim signal directly to a control unit (e.g., PV, HP, EVSE). The HEMS optimizes the connected equipment/flexibilities, translates, and transmits any control commands in the equipment-specific protocol for compliance with the Plim. Depending on the flexibility, multiple protocols are possible (e.g., EEBUS, Modbus/TCP, OCPP, KNX, ...). Another possible transmission is the backend-to-backend communication of the component manufacturers. Between the EVSE and EV, the possible communication standards are ISO 15118, DIN SPEC 70121, and IEC 61851. Metering data is transmitted via the SMGW according to the tariff application cases (TAC, German: Tarifanwendungsfall) 7, 9 or 10 to the energy supplier (for billing) and the DSO to verify adherence to the Plim signal [49]. In compliance with German calibration regulations, every flexibility needs a separate calibrated metering point to be used by §14a EnWG. If the Plim signal refers to the GCP, then one grid connection meter is sufficient. If the Plim signal refers to a single control unit, usage of the meter data from the HEMS would be interesting to avoid additional meters. However, this is currently not in compliance with German calibration regulations. Nevertheless, in future, this might represent an attractive alternative solution.

7. Conclusions and Outlook

In summary, this paper describes a use-case methodology oriented towards practical use. The methodology is based on existing norms and standards, as well as experience from other projects. The use-case methodology was applied in all four clusters resulting in several BUCs and TUCs. Most of the BUC descriptions are published on the project website https://unit-e2.de/ (accessed on 17 December 2022) [17]. The methodology supports a common understanding of the research subject and wording and thus constitutes a solid basis for the first project phase. The simplified method is constructed such that every project partner, even those with limited time and/or limited interest in scientific analysis, are capable and motivated to participate. The major advantage of this approach is that it yields quick results, which enable further preparation for the field tests. The documentation is reduced but still sufficient so that all stakeholders, even those without specific expertise, can understand the essential components of the use case in a short time. Further, the brief description and the graphic representation as an e3-value model are impactful and readily comprehensible for new project staff or external stakeholders, even those without specific expertise. The simplicity of the unIT-e^2 use-case methodology is an advantage and at the same time also a drawback. While the use-case descriptions are sufficiently detailed for project staff to understand the process, software developers, for example, need detailed sequence diagrams for implementation. These additional specifications of certain processes and/or interfaces are necessary but not part of the unIT-e^2 use case methodology. To develop valuable business models, further steps are also needed. Another limitation of the paper is we cannot yet publish more than one TUCs at this point. In future work, we plan to publish further selected TUCs. Additionally, in the unIT-e^2 subproject research, most of the BUCs will be implemented and examined in various simulation environments. With the FfE model, the Electric Grid and Energy System Model for Distribution Grids—GridSim effects of the use cases on e.g., grid loads in various grids (urban and rural) will be examined [50]. Further, implementation of the use cases in the FfE model electric Flexibility assessment modelling environment—eFlame will allow the investigation of current and future revenues [51]. Future research should be devoted to the development of a system architecture derived from the BUC and TUC, which enables a full understanding of the whole system rather than focusing on one use case. Another interesting topic for future work is the systematic assessment of the combination possibilities of the use cases.

Author Contributions: Conceptualization, A.O. and P.D.; methodology, A.O and P.D.; validation, A.O., P.D. and V.Z.; formal analysis, A.O. and P.D.; investigation, V.Z.; resources, A.O.; writing—original draft preparation, A.O., P.D. and V.Z. writing—review and editing, A.O.; visualization, A.O. and P.D.; supervision, A.O.; project administration, A.O.; funding acquisition, A.O. All authors have read and agreed to the published version of the manuscript.

Funding: This research was conducted as part of the activities of FfE in the project "unIT-e^2—Living Lab for Integrated E Mobility". The project is funded by the Federal Ministry for Economic Affairs and Climate Action (BMWK) (funding code: 01MV21UN11 (FfE e.V.), 01MV21UN01 (FfE GmbH)).

Data Availability Statement: Not applicable.

Acknowledgments: Special thanks go to all unIT-e^2 partners involved in the process and the various workshops. Further, we like to thank Katharina Sommer, Ryan Harper and Robbie Göb.

Conflicts of Interest: The authors declare no conflict of interest.

Abbreviations

The following abbreviations are used in this manuscript:

BCM	Bidirectional Charging Management
BUC	Business-Use Case
CLS	Controllable Local System
DSO	Distribution Network Operator
EMS	Energy Management System
EU	European Union
EV	Electric Vehicle
EVSE	Electric Vehicle Supply Equipment
GCP	Grid Connection Point
GHG	Greenhouse Gas
GWA	Gateway Administrator
HAN	Home Area Network
HCU	HAN Communication Adapter Unit
HEMS	Home Energy Management System
HP	Heat Pump
OEM	Original Equipment Manufacturer
SGAM	Smart Grid Architecture Model
SMGW	Smart Meter Gateway
SP	Sub Project
TUC	Technical-Use Case
UML	Unified Modeling Language

References

1. *Paris Agreement*; Version 2015; United Nations: Paris, France, 2015.
2. EEA Greenhouse Gases-Data Viewer. Available online: https://www.eea.europa.eu/data-and-maps/data/data-viewers/greenhouse-gases-viewer (accessed on 10 February 2022).
3. Bird, L.; Debra, L.; Millian, M.; Carilini, M.; Estanqiero, A.; Flynn, D.; Gomez-Lazaro, E.; Holttingen, H.; Menemenlis, N.; Orths, A.; et al. Wind and solar energy curtailment: A review of international experience. *Renew. Sustain. Energy Rev.* **2016**, *65*, 577–586. [CrossRef]
4. Denholm, P.; Hand, M. Grid flexibility and storage required to achieve very high penetration of variable renewable electricity. *Energy Policy* **2011**, *39*, 1817–1830. [CrossRef]
5. Bistline, J. Economic and technical challenges of flexible operations under large-scale variable renewable deployment. *Energy Econ.* **2017**, *64*, 363–372. [CrossRef]
6. Guerra, K.; Haro, P.; Gutiérrez, R.; Gómez-Barea, A. Facing the high share of variable renewable energy in the power system: Flexibility and stability requirements. *Appl. Energy* **2022**, *310*, 118561. [CrossRef]
7. Mathiesen, B.V.; Lund, H.; Connolly, D.; Wenzel, H.; Østergaard, P.A.; Möller, B.; Nielsen, S.; Ridjan, I.; Karnøe, P.; Sperling, K. Smart Energy Systems for coherent 100% renewable energy and transport solutions. *Appl. Energy* **2015**, *145*, 139–154. [CrossRef]
8. Mathiesen, B.V.; Connolly, D.; Lund, H.; Nielsen, M.P.; Schaltz, E.; Wenzel, H.; Bentsen, N.S.; Felby, C.; Kaspersen, P.; Ridjan, I. *CEESA 100% Renewable Energy Transport Scenarios towards 2050—Technical Background Report Part 2*; Aalborg University: Copenhagen, Denmark, 2015.

9. Duscha, V.; Wachsmuth, J.; Eckstein, J.; Pfluger, B. *GHG-Neutral EU2050—A Scenario of an EU with Net-Zero Greenhouse Gas Emissions and Its Implications*; Umweltbundesamt: Dessau-Roßlau, Germany, 2019.
10. *National Emissions Reported to the UNFCCC and to the EU Greenhouse Gas Monitoring Mechanism*; European Environment Agency (EEA): Copenhagen, Denmark, 2020. Available online: https://www.eea.europa.eu/data-and-maps/data/national-emissions-reported-to-the-unfccc-and-to-the-eu-greenhouse-gas-monitoring-mechanism-16 (accessed on 27 February 2022).
11. Bieker, G. *A Global Comparison of the Life Cycle Greenhouse Gas Emissions of Combustion Engine and Electic Passenger Cars*; The International Council on Clean Transportation: Berlin, Germany, 2021.
12. Dornoff, J.; Mock, P.; Baldino, C.; Bieker, G.; Díaz, S.; Miller, J.; Sen, A.; Tietge, U.; Wappelhorst, S. *Fit for 55: A Review and Evaluation of the European Commission Proposal for Amending the CO_2 Targets for New Cars and Vans*; The International Council on Clean Transportation: Berlin, Germany, 2021.
13. Total Market Share of Newly Registered Passenger Electric Vehicles in the European Union between 2015 and 2020. Available online: https://www.statista.com/statistics/625010/electric-vehicle-market-share-in-eu-annual/ (accessed on 20 May 2022).
14. Volkswagen Naturstrom. Available online: https://www.elli.eco/en/green-energy (accessed on 9 May 2022).
15. Tesla Stromtarif. Available online: https://www.tesla.com/de_de/tesla-stromtarif (accessed on 9 May 2022).
16. *Verordnung über Allgemeine Bedingungen für den Netzanschluss und dessen Nutzung für die Elektrizitätsversorgung in Niederspannung (Niederspannungsanschlussverordnung—NAV)*; Bundesrepublik Deutschland: Berlin, Germany, 2006.
17. Project Webpage unIT-e^2: Living Lab for Integrated e-Mobility. Available online: https://unit-e2.de/ (accessed on 29 August 2022).
18. *Gesetz über Die Elektrizitäts- und Gasversorgung (Energiewirtschaftsgesetz) (EnWG)*; Version 2018; Bundesministerium der Justiz und für Verbraucherschutz: Berlin, Germany, 2018.
19. European Commission v Federal Republic of Germany. *Failure of a Member State to fulfil Obligations—Internal Markets for Electricity and Natural Gas (Case C-718/18)*; Court of Justice of the European Union: Luxembourg, 2021.
20. Pressemittelung zur Berichterstattung in der Heutigen Welt am Sonntag zu Thema Laden von E-Autos. Available online: https://www.bmwk.de/Redaktion/DE/Pressemitteilungen/2021/01/20210117-berichterstattung-welt-am-sonntag-laden-von-e-autos.html (accessed on 9 May 2022).
21. *Gesetz zur Digitalisierung der Energiewende*; Version 2016; Bundesministerium für Wirtschaft und Energie (BMWi): Bonn, Germany, 2016.
22. Ziemsky, V.; Biedenbach, F.; Matzner, A. *Systemintegration von Elektrofahrzeugen—Stromlieferverträge mit Nicht-Statischen Tarifen-FfE Grundlagenpapier*; Forschungsstelle für Energiewirtschaft (FfE): München, Germany, 2022.
23. *Stufenmodell zur Weiterentwicklung der Standards für Die Digitalisierung der Energiewende*; Bundesamt für Sicherheit in der Informationstechnik: Berlin, Germany; Bundesministerum für Wirtschaft und Energie: Bonn, Germany, 2020.
24. Müller, M.; Biedenbach, F.; Reinhard, J. Development of an Integrated Simulation Model for Load and Mobility Profiles of Private Households. *Energies* **2020**, 13, 3843. [CrossRef]
25. Doppelter Haushaltsstromverbrauch mit Elektroauto. Available online: https://www.vbew.de/energie/presseinfos-energie/news/doppelter-haushaltsstromverbrauch-mit-elektroauto (accessed on 11 May 2022).
26. *Cockburn, Alistair: Writing Effective Use Cases*; Addison-Wesley Professional: Boston, MA, USA, 2000.
27. *IEC SRD 62913-1:2019*; Generic Smart Grid Requirements (Part 1: Specific Application of the Use Case Methodology for Defining Generic Smart Grid Requirements According to the IEC Systems Approach). International Electrotechnical Commission: Geneva, Switzerland, 2019.
28. Oestereich, B. *Analyse und Design Mit der UML 2.5—Objektorientierte Softwareentwicklung*; Oestereich: München, Germany, 2012.
29. *IEC 62559-2:2015*; Use Case Methodology—Part 2: Definition of the Templates for Use Cases, Actor List and Requirements List. Issued 2015-04; International Electrotechnical Commission: Geneva, Switzerland, 2015.
30. Gordijn, J.; Akkermans, J.M. Value-based requirements engineering exploring innovative e-commerce ideas. *Requir. Eng.* **2003**, 8, 114–134. [CrossRef]
31. CEN-CENELEC-ETSI. *Smart Grid Coordination Group: SG-CG/M490/K_ SGAM Usage and Examples—SGAM User Manual—Applying, Testing & Refining the Smart Grid Architecture Model (SGAM)*; Version 3.0; CENELEC: Brüssel, Belgien, 2014.
32. Faller, S.; Bogensperger, A.; Köppl, S.; Kießling, A. *Anwendungshilfe Use Case Methodik—Eine praktische Anwendungshilfe für die Use Case Entwicklung*; Forschungsstelle für Energiewirtschaft e. V. (FfE): München, Germany, 2020.
33. Weigand, H.; Johannesson, P.; Andersson, B.; Bergholtz, M.; Edirisuriya, A.; Ilayperuma, T.; Ribbers, P. Strategic analysis using value modeling—The c3-value approach. In Proceedings of the 40th Hawaii International Conference on System Sciences, Waikoloa, HI, USA, 3–6 January 2007; Tilburg University: Tilburg, The Netherlands, 2007.
34. IEC 62559—Use Case Methodology. Available online: https://syc-se.iec.ch/deliveries/iec-62559-use-cases/ (accessed on 9 May 2022).
35. CEN-CENELEC-ETSI. *Smart Grid Coordination Group: SG-CG/M490/I_Smart Grid Interoperability—Methodologies to Facilitate Smart Grid System Interoperability through Standardization, System Design and Testing*; CEN-CENELEC-ETSI: Brüssel, Belgium, 2014.
36. *IEEE Standard Computer Dictionary: A Compilation of IEEE Standard Computer Glossaries*; Issued 1991-01-18; IEEE: New York, NY, USA, 1991.
37. Faller, S.; Schütz, J.; Bogensperger, A.; Uslar, M. Anwendungshilfe SGAM—Smart Grid Use Cases modellieren. In *Energiwirtschaftliche Tagesfragen*; Forschungsstelle für Energiewirtschaft e. V. (FfE), OFFIS e.V.: Offenbach, Germany, 2020.

38. CEN-CENELEC-ETSI. *Smart Grid Coordination Group: SG-CG/ M490/F_ Overview of SG-CG Methodologies*; CENELEC: Brüssel, Belgien, 2014.
39. Frein, C. *Rollenmodell für Die Marktkommunikation im Deutschen Energiemarkt—Anwendungshilfen—Strom und Gas*; BDEW Bundesverband der Energie- und Wasserwirtschaft e.V.: Berlin, Germany, 2016.
40. ENTSO-E. *The Harmonised Electricity Market Role Model*; entso-e aisbl: Brüssel, Belgien, 2017.
41. EASEE-Gas. *Harmonised Gas Role Model Specification*; EASEE-Gas: Paris, France, 2016.
42. *Business Process Model and Notation (BPMN)*; Object Management Group: Needham, MA, USA, 2011.
43. InDEED—Konzeption, Umsetzung und Evaluation einer auf Blockchain Basierenden Energiewirtschaftlichen Datenplattform für die Anwendungsfälle "Labeling" und "Asset Logging". Available online: https://www.ffe.de/projekte/indeed/ (accessed on 11 May 2022).
44. Trade-EVs II—Trade of Renewable, Aggregated and Distributed Energy by Electric Vehicles. Available online: www.ffe.de/tradeEVsII (accessed on 1 January 2021).
45. Bidirektionales Lademanagement (BDL)—Intelligentes Zusammenspiel von Elektrofahrzeugen, Ladeinfrastruktur und Energiesystem. Available online: https://www.ffe.de/projekte/bdl/ (accessed on 9 January 2021).
46. Schulze, Y.; Müller, M.; Faller, S.; Duschl, W. Was ist Netzdienlichkeit? In *et-Energiewirtschaftliche Tagesfragen*; Forschungsstelle für Energiewirtschaft (FFE e.V.): München, Germany, 2021.
47. Bogensperger, A.; Estermann, T.; Samweber, F.; Köppl, S.; Müller, M.; Zeiselmair, A.; Wohlschlager, D. *Smart Meter—Umfeld, Technik, Mehrwert*; Forschungsstelle für Energiewirtschaft e.V.: München, Germany, 2018.
48. Bundesamt für Sicherheit in der Informationstechnik (BSI). *Zertifizierung nach ISO 27001 auf der Basis von IT-Grundschutz—Zertifizierungsschema*; Version 1.2; Bundesamt für Sicherheit in der Informationstechnik (BSI): Bonn, Germany, 2014.
49. *Technische Richtlinie BSI TR-03109-1—Anforderungen an die Interoperabilität der Kommunikationseinheit Eines Intelligenten Messsystems*; Version 1.1; Bundesamt für Sicherheit in der Informationstechnik: Bonn, Germnay, 2021.
50. FfE Model: GridSim—Electric Grid and Energy System Model for Distribution Grids. Available online: https://www.ffe.de/en/tools/gridsim-electricity-grid-and-energy-system-model-for-distribution-grids/ (accessed on 1 December 2022).
51. FfE Model: eFlame—Electric Flexibility Assessment Modeling Environment. Available online: https://www.ffe.de/tools/eflame-electric-flexibility-assessment-modeling-environment/ (accessed on 1 December 2022).

World Electric Vehicle Journal

MDPI

Article

Top-Down Validation Framework for Efficient and Low Noise Electric-Driven Vehicles with Multi-Speed Gearbox

Steffen Jäger *, Jonas Schätzle and Tilmann Linde

Faculty of Business Administration and Engineering, Furtwangen University, 78120 Furtwangen, Germany
* Correspondence: jaes@hs-furtwangen.de; Tel.: +49-7723-920-2424

Abstract: The shift towards e-mobility is resulting in new technological challenges. Thus, new, more efficient product development methods and a better product understanding are required. During product development, validation is essential both to achieve significantly increased knowledge of the system in question and to ensure that customers' expectations of characteristics are met. Based on existing top-down validation approaches, this article discusses an innovative both-ends-against-the-middle-approach (BEATM) developed by the author. A validation framework which combines physical and virtual elements is presented. By way of example, a development approach for the toothing validation layer of an electric vehicle powertrain with a multi-speed gearbox is introduced.

Keywords: electric drive; noise; gear; powertrain; simulation

Citation: Jäger, S.; Schätzle, J.; Linde, T. Top-Down Validation Framework for Efficient and Low Noise Electric-Driven Vehicles with Multi-Speed Gearbox. *World Electr. Veh. J.* **2022**, *13*, 228. https://doi.org/10.3390/wevj13120228

Academic Editor: Joeri Van Mierlo

Received: 14 October 2022
Accepted: 23 November 2022
Published: 30 November 2022

1. Introduction

Facing todays challenges in electrical powertrain development, multi-speed gearboxes are increasingly used, as they improve the efficiency of utilisation [1,2].

Various methods may be employed in the search for the appropriate motor-gearbox combination. Any of these methods can help achieve an optimised powertrain topology both in terms of efficiency [2–4] and with regard to the customers' requirements [1,5]. Multi-speed gearboxes also make use of lighter electric motors with higher possible maximum rotational speeds, resulting in a decrease in costs [1]. However, both high rotational speeds and the absence of acoustic masking in combustion engines lead to an increase in actual gear noises as well as their perception by the driver. In particular, the tooth mesh and its behaviour are a source for noises and must be considered throughout the entire development process.

To achieve improved acoustic characteristics, both the macro- and the micro-geometry of the teeth can be optimised [6]. Furthermore, specific modifications of the teeth's surface will also be discussed [7]. Established analytical models are used to describe the influence of profile shifts on the mesh stiffness of spur gears [8]. Additionally, the vibration and acoustic radiation of gearbox housings based on 3D simulations will be explored in detail, as will the influence of different gearbox macro-designs on efficiency and vibration at high speeds [9,10]. While methods of generating acoustically-optimised gears and efficiency-optimised gearbox designs are well established, there is a distinct lack when it comes to a holistic view of the interaction between efficiency and noise behaviour in the early stages of product development.

Within the product development process, validation is the main source of knowledge and is essential in ascertaining the fulfilment of customer expectations with regard to product characteristics [11]. Using a mixed physical–virtual validation environment allows for early studies of product behaviour [12]. The x-in-the-loop framework (XiL) supports validation activities and the development of validation environments, leading to faster and more efficient product development [13–15]. Examples of the established methods to create a physical validation environment, in particular for the acoustic analysis, will be discussed [16,17].

XiL is used to validate product properties over a wide range of development stages without the need for a whole-system prototype. Considering validation as a multi-layer activity, the X stands for the system in question on the particular layer. Independently of the layer at which a system is validated, the user and its environment are an essential part of the XiL framework. This can be shown particularly clearly in the validation of vehicles, since, here, the driver closely interacts with the system as well as the environment, which includes factors such as routes chosen, weather conditions, etc.; see Figure 1. When setting up the validation layer, the first step is the definition of the unit under test as well as the specification of the rest-system properties required for the particular purpose of the test. The system is then divided into subsystems, which are either physical or virtual—without losing the closed control loop shown in Figure 1.

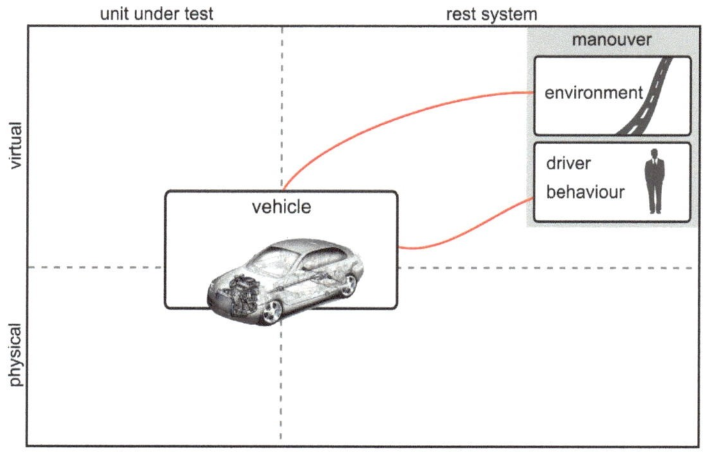

Figure 1. System layout, vehicle-in-the-loop.

By combining the approaches and frameworks shown above with meta models such as the V-Model [18], the integrated Product engineering Model (iPeM) [19], or the Munich Procedural Model (MPM) [20], the product development process can be streamlined and made more efficient. Considering XiL as the basis of a top-down planned validation, the established processes of Simultaneous Engineering and Concurrent Engineering can be combined to form a new validation approach. This optimised top-down validation approach can lead to the simultaneous development of physical as well as virtual models in a both-ends-against-the-middle approach [14,21].

Above all, the challenge is to develop virtual models which then have to be validated by physical tests. These virtual models are intended to evaluate both acoustic quantities and the efficiency in the early stages of the product development process.

2. Top-Down Framework and Model Implementation

The XiL framework introduced above, together with the top-down validation approach, will determine the validation set-up. Planning the validation from the top-level, in this particular setting, leads to a four-layer environment. The main focus when working on solutions for optimising powertrain units is still the vehicle as a whole, with its real behaviour in a real environment with a real driver. In particular, the drivers' experience of vibration and sound must be considered in all validation layers. Therefore, the top-level vehicle tests are planned as set out in Figure 1.

Stripping the chassis, the wheels, the side shafts, and the battery from the real vehicle—and transferring these to the virtual rest system—results in the second validation layer (layer 2, see Figure 2). Within that layer, the motor-gearbox-unit is the physical unit under test, thus forming a powertrain test rig. The environment and the driver behaviour will also be part of the virtual rest system; see the system layout in Figure 2. The red lines indicate the interaction between the components. Where they represent a cross-over between the physical and virtual domains, suitable sensors and actuators are needed, as well as, ideally, a real-time-capable simulation software.

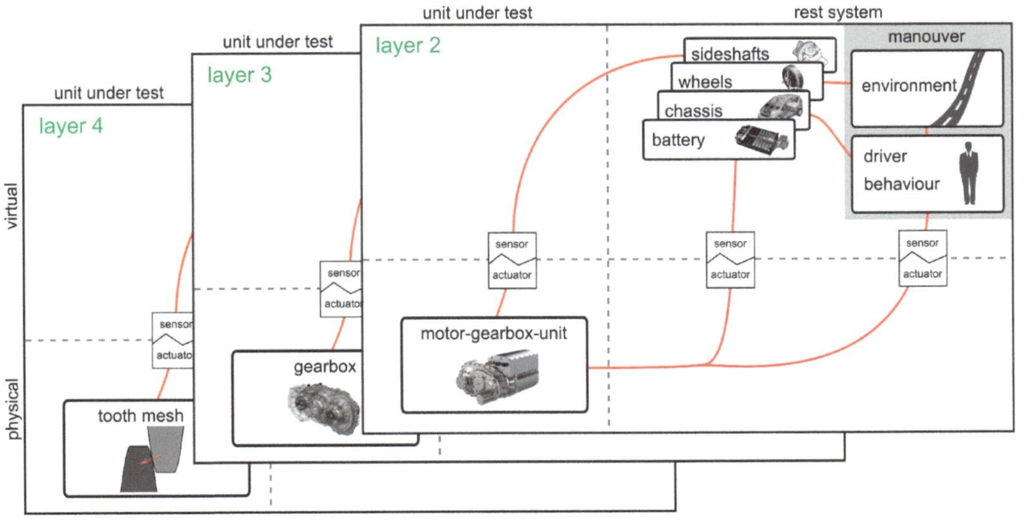

Figure 2. XiL system layout.

Reducing this system by the traction motor leads to a gearbox test environment on the third validation layer, which is not further discussed in this article. However, it is worth mentioning that the interaction between all validation layers is always taken into account. The fourth validation layer examines the tooth mesh as the physical unit under test; here. all the other vehicle components are part of the virtual rest system.

Planning the validation in a top-down order clearly defines all the validation layers as well as the system interfaces. The implementation of the tests (both physical and virtual) is usually performed in a bottom-up order; please refer to the discussion of the extended validation approach in Section 5.

The model implementation as described below (Sections 3 and 4) is based on the system layout as stated above. Starting the implementation activities on the fourth validation layer leads to a gear mesh study with the objective of gathering results for gears optimised in terms of both sound emission and efficiency. The virtual domain of that layer is described in Section 3, for the physical domain see Section 4.

In addition to the framework introduced above, a model-based validation environment is established as shown in Figure 3. It shows the content of both the virtual and physical domains as well as the interfaces between them. The objective is to achieve validated models of the gear mesh. Therefore, the detailed toothing data are calculated by an industry project partner using gear calculation software. These parameters are used as input for the virtual domain as well as manufacturing parameters in the physical domain.

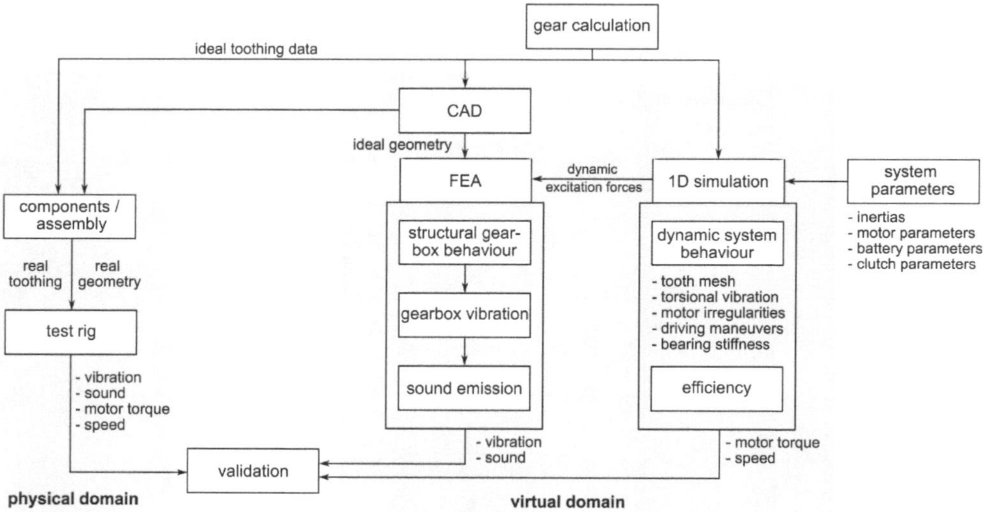

Figure 3. Model implementation.

3. Virtual Model Implementation

To calculate the dynamic system behaviour and the efficiency of the gear pair, a 1D simulation software is used. Subsequently, the dynamic excitation forces at the housing are passed on to a Finite Element Analysis (FEA) where, firstly, the structural behaviour of the ideal gearbox housing geometry is considered. By combining the structural behaviour and the excitation forces from the 1D simulation, the gearbox vibration and the sound emission (both air-borne and structure-borne) can be estimated.

The 1D simulation model contains the shafts (including their masses), helical gears (see Table 1, which includes their masses, inertias, and gearing parameters), and the bearings (including their stiffness and damping coefficients). Figure 4 shows an overview of the implemented system. The dynamic behaviour is predominantly influenced by the gear element. Therein, properties of the tooth contact are defined, which result in irregularities of the output rotational speed. In this model, particularly, the effect of the mesh stiffness on the overall system is examined.

The dynamic equations of the gear system shown in Figure 4 can be written as:

$$[\mathrm{M}]\{\ddot{\mathrm{q}}\} + [\mathrm{D}]\{\dot{\mathrm{q}}\} + [\mathrm{C}]\{\mathrm{q}\} = \{\mathrm{F}(\mathrm{t})\}$$

where M is the matrix consisting of masses and inertias, D is the damping matrix, and C is the stiffness matrix; q is the displacement vector and F the vector including the forces. Periodically varying mesh stiffness leads to the systems' dynamic behaviour. A spring element between two meshing teeth is used to describe the mesh stiffness. The values of the specific tooth stiffness are specified with respect to the normalised meshing length, see Figure 5b.

Table 1. Gear pair data.

Parameter	Symbol	Value
No. of teeth (-)	z_1, z_2	38, 57
Normal module (mm)	m	1.75
Face width (mm)	b	20
Helix angle (°)	β	15
Normal pressure angle (°)	α	20

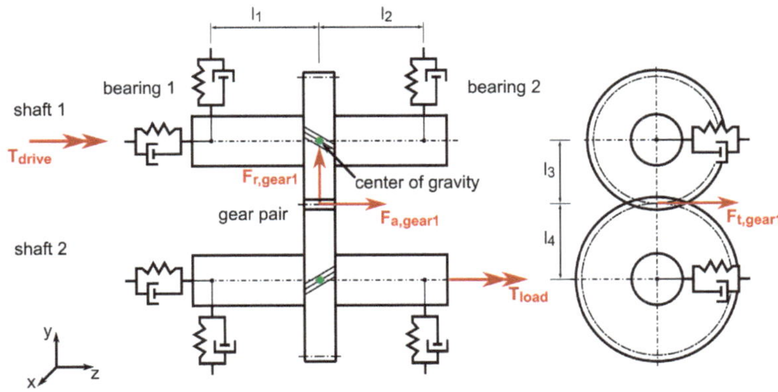

Figure 4. Gear system.

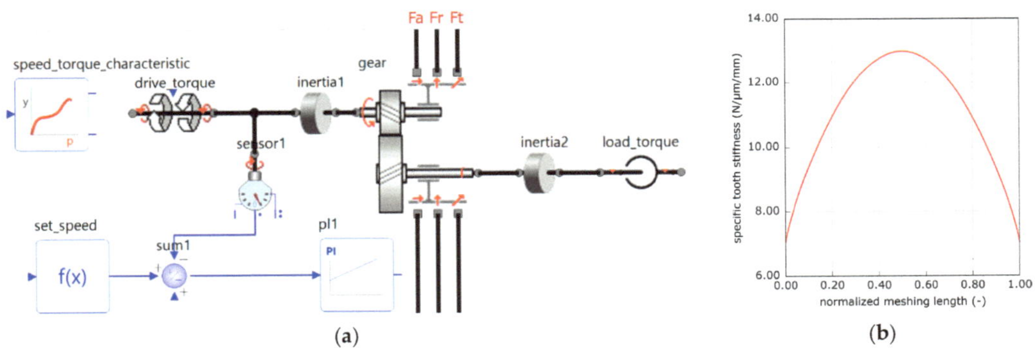

Figure 5. (**a**) Rotational system; (**b**) specific tooth stiffness for the mesh stiffness calculation.

The simulation model is divided into a rotational and a translational domain. In the rotational model an electric motor sets a drive torque. A real electric motor's speed–torque-characteristic is used. With the help of a simple control loop, different load torques can be defined, depending on the speed desired. The gear pair model calculates the gears' dynamic behaviour, including the parameters from the gear calculation. Figure 5a shows the details of the rotational system in the 1D simulation software.

The translational system is used for the calculation of the bearing forces. For this, the tooth forces are provided by the gear pair model. Each shaft is represented by two translational models which contain both the moment and the force equilibrium. To calculate the bearing forces at the gearbox housing in the z- and y-direction, the axial and radial tooth forces are used, see the coordinate system in Figure 4. The second translational model calculates the bearing reaction force in the x-direction using the tangential tooth forces. To consider the bearing stiffness and damping, spring-damper-elements are used. A simulation model detail for gear shaft 1 is shown in Figure 6. The inner forces of the spring-damper models represent the resulting forces in three directions and can be used for the FE analyses described below.

To perform dynamic 3D FE analyses, the project schematic, as shown in Figure 7, is set up. The geometry block contains the solid domain (the gearbox housing) for structural analyses as well as the fluid domain (the acoustic region and air enclosing the gearbox housing) for the acoustic analysis. Based on a modal analysis for the determination of the eigenfrequencies and their shapes, a harmonic (frequency) response analysis with modal superposition is performed. The dynamic bearing loads derived from the 1D simulation are used as input for the harmonic response analyses. The calculated structural behaviour is used as input for the acoustic analyses which result in the acoustic parameters.

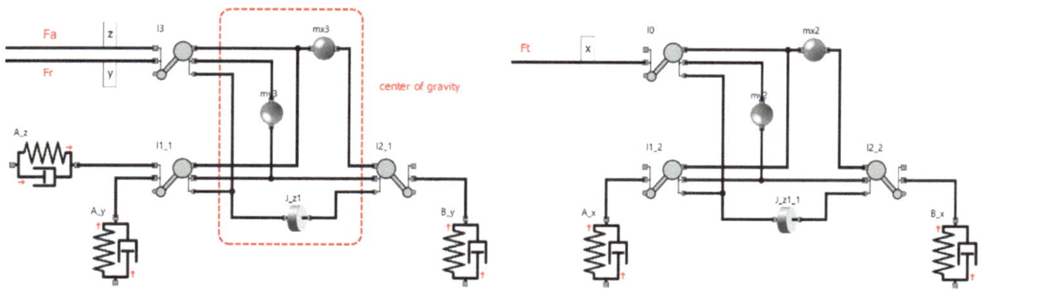

Figure 6. Translational systems for calculating bearing forces in z- and y-direction (**left**), and x-direction (**right**).

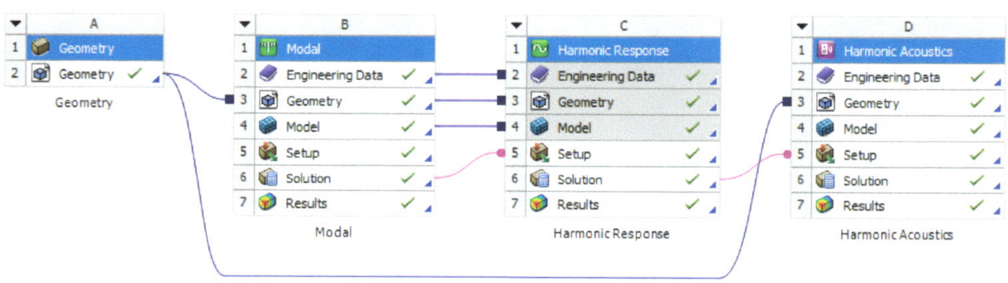

Figure 7. FEA project schematic.

Figure 8 shows a sample result of the harmonic response analysis. In the velocity amplitude spectrum (right) for a specific gearbox housing surface, the eigenfrequencies are indicated by higher amplitudes which are fed by the bearings' dynamic forces. Figure 8a shows an example of the deformation of the gearbox housing at a specific eigenfrequency. To predict the acoustic behaviour of the system (in particular the air-borne sound emission), the acoustic analysis uses the results of the harmonic response analysis.

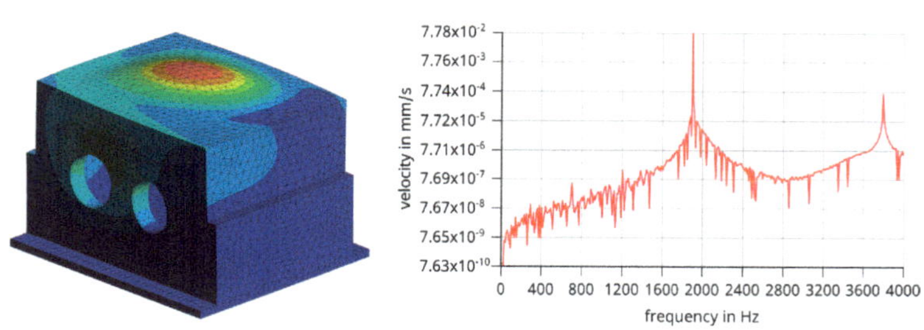

(**a**) (**b**)

Figure 8. (**a**) FEA deformation result of the harmonic response analysis; (**b**) velocity amplitude spectrum.

The acoustic analysis uses the fluid domain. The load is applied by combining the surface velocities calculated using the harmonic response analyses with the interface surface of the fluid (air). The sound parameters in the fluid region are calculated using the acoustic coefficients of radiation. The sound pressure level L_p is used to rate the simulated gear noise. It is calculated as:

$$L_p = 10\,\lg\left(\frac{p}{p_0}\right)^2 = 20\lg\left(\frac{p}{p_0}\right)$$

where p is the root mean square sound pressure and p_0 is the reference sound pressure with a value of 20×10^{-6} N/m^2. This approach has been established and is described in former publications [9,22,23].

By way of example, the sound pressure distribution within the fluid domain at a specific frequency is shown in Figure 9a. The frequency spectrum (Figure 9b) shows the sound pressure level (SPL) at a specific point. These results can be used as a reference for physical tests, such as structure-borne or air-borne sound measurements for validating the implemented virtual models (both 1D and 3D). Additionally, they can be used for further optimisation loops of the gears and the housing.

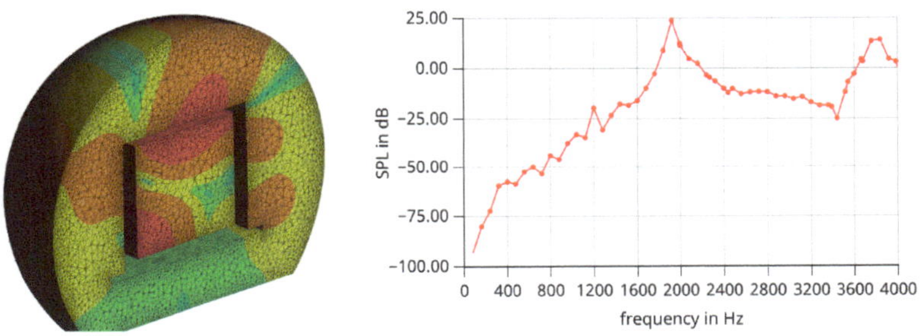

(**a**) (**b**)

Figure 9. (**a**) FEA sound pressure result; (**b**) sound pressure spectrum.

4. Physical Validation Environment and Methods

To validate the results of the virtual models described above, the physical domain is developed according to Figure 3. The objective is to obtain validated virtual models of the gear mesh as well as information about the quality of the acoustic simulation. Since the parameter changes in the gears are in a micron range, a geometrically precise test rig is needed. Due to this, the high rotational speeds (up to 20,000 rpm) and high system frequencies (in particular from the gear mesh), exceptionally high accuracy sensors for torque, speed, and sound pressure (structure-borne and air-borne) are needed. Figure 10 shows a rendering of the test rig. Similarly to the simulation model described in the previous section, the test rig consists of a gear pair which is driven by two electric motors. Both are controlled by a real-time-capable automation software. This now makes the simulation of the rest system according to Section 2 feasible.

Figure 10. Physical validation environment.

The design process took into account both the bearing rollover frequencies and the torsional eigenfrequencies of the entire powertrain. The gearbox housing has a specific structural behaviour as described above. Thus, critical operating points and ideal positions for the placing of accelerometers can be determined. Due to the high influence of the concentricity on the vibration excitation, the gears are connected to the shaft by an optimised cone clamping sleeve. This ensures that the quality of gear alignment following any replacement will remain consistent.

The 3D simulation results suggest that the air-borne sound pressure level is comparatively low. Therefore, the usability of the results must be checked. Appropriate software systems and data acquisition modules for data recording and processing are already available. It is also possible to determine the efficiency of the gear pair enabled by the high precision speed and torque measurement data. This allows for a combined analysis of efficiency and acoustics as well as the interaction between them.

To validate the simulation models which were implemented during the development of the test rig, physical testing must be used. For the validation of the calculated eigenfrequencies and their shapes (gearbox housing), an experimental modal analysis will be applied. Real experimental studies on the rollover frequencies of bearings and natural frequencies of the torsional vibrations will also be performed. In addition, the frequency response of the gearbox housing is measured at certain operating points (e.g., certain speed). In addition, the acoustic simulation will be validated with real measurements. For this purpose, the same points as in the simulation are selected and the structure-borne sound is measured at this location. The hardware required for the physical analyses is available. This includes acceleration sensors for the modal analysis of the gearbox housing as well as high-resolution microphones and structure-borne sound sensors for the acoustic analysis. A dedicated analysis software is also available.

5. Extended Validation Approach

There is an ongoing discussion and expansion of the available validation approaches and frameworks; the both-ends-against-the-middle-approach introduced by the author contributes a new perspective. The focus here is to suggest that validation needs to combine both top-down and bottom-up activities.

The concept of the BEATM-approach is visualised in Figure 11. Starting the product development process with planning the validation activities from the top to the bottom (from vehicle to gear pair, as described in Section 2) allows for the appropriate consideration of the rest system behaviour at each validation layer (compare the top-down planned activities shown in Figure 11). Following this top-down planning of activities, the simultaneous development of virtual validation models as well as the physical validation environment is performed according to Figure 3 in Section 2. The integration of the model itself is mostly a stand-alone activity without a validation layer interface. Hence, a bottom-up approach as shown in Figure 11 also results. However, depending on the time of model implementation, not all the information required for this may be available. Nevertheless, the idea of result transferability from layer to layer should be kept in mind. To achieve this, the validation of all (sub)models implemented is essential. Consequently, it is possible to arrive at validated whole-system models by combining validated sub-models, although the model complexity increases from layer to layer and over the time of development.

Using the method described above, a real frontloading of validation activities within the product development process can be achieved. Working with these validated sub-models guarantees that there is awareness of the customers' requirements over the entire development process, regardless of which validation layer is being worked on. Upon completion of a project, all the implemented and physically validated models can be used for the development of further products with similar requirements. They can be transferred within a product development and reference process framework.

By combining the bottom-up and top-down activities as described above, existing validation approaches may be expanded. Following Figure 11, the content of both the virtual and the physical domain meets at the end of the development and validation time line. At this point, the product is fully developed and validated, and both domains together describe it in its entirety. The approach described herein is new and offers a method to gain a better understanding of the inter-relations between a product's development and its validation.

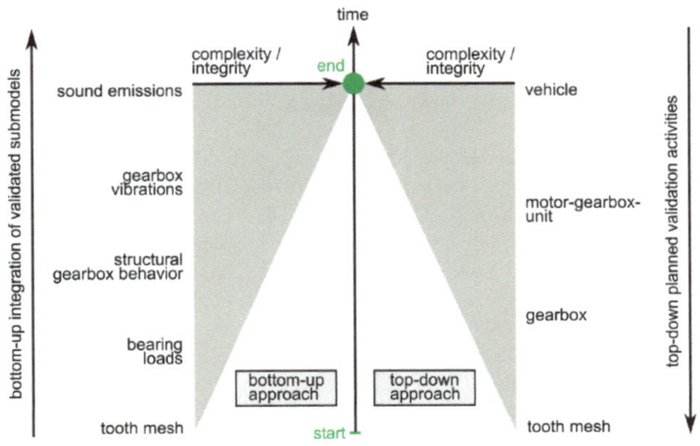

Figure 11. Both-ends-against-the-middle-approach (BEATM).

6. Conclusions

Based on the approaches and frameworks currently available, a physical–virtual coupled validation environment for noise and efficiency studies has been introduced. Existing approaches are extended by using a combination of bottom-up and top-down validation activities.

The virtual domain for a gear pair test is represented by 1D simulation models, which enable the estimation of the dynamic system behaviour of such a gear pair as well as the respective shafts, including the bearing behaviour by stiffness and damping coefficients. Thus, the reaction forces at the housing can be calculated. Subsequently, a 3D FE analysis is performed in which, first, the modal behaviour is computed and, second, the frequency response is estimated using the reaction forces from the 1D simulation as input. Based on the results of the frequency response analysis, respectively, the surface velocities at the gearbox housing, an estimation of the emitted air-borne as well as structure-borne sound can be made through an acoustic analysis.

Controlled by a real-time-capable automation software, a test rig including two electric motors represents the physical domain. The main objective of the test rig operation is the validation of the results from the 1D and 3D simulation models. The test rig behaviour itself needs to be validated, in the future, prior to the start of testing. There are also plans for the extension of the virtual models with regard to efficiency. Finally, the both-ends-against-the-middle-approach remains to be transferred into a reference process model.

Author Contributions: Conceptualization, S.J. and J.S.; methodology, S.J.; software, S.J., J.S. and T.L.; validation, S.J., J.S. and T.L.; formal analysis, S.J., J.S. and T.L.; investigation, S.J., J.S. and T.L.; data curation, S.J., J.S. and T.L.; writing—original draft preparation, S.J., J.S. and T.L.; writing—review and editing, S.J. and T.L.; visualization, S.J., J.S. and T.L.; supervision, S.J.; project administration, S.J.; funding acquisition, S.J. All authors have read and agreed to the published version of the manuscript.

Funding: This research was funded by the Federal Ministry of Education and Research (BMBF) Germany, grant number 13FH527KA9.

Data Availability Statement: Not applicable.

Acknowledgments: The authors are very grateful for the financial support from BMBF (Federal Ministry of Education and Research) of Germany.

Conflicts of Interest: The authors declare no conflict of interest. The funders had no role in the design of the study; in the collection, analyses, or interpretation of data; in the writing of the manuscript; in the decision to publish the results.

References

1. Eghtessad, M.; Meier, T.; Rinderknecht, S.; Küçükay, F. Antriebsstrangoptimierung von Elektrofahrzeugen. *ATZ Automob. Z.* **2015**, *117*, 78–85. [CrossRef]
2. Kwon, K.; Jo, J.; Min, S. Multi-objective gear ratio and shifting pattern optimization of multi-speed transmissions for electric vehicles considering variable transmission efficiency. *Energy* **2021**, *236*, 121–419. [CrossRef]
3. Allendorf, D. Getriebekonzeption auf Basis von Wirkungsgradsimulationen. Ph.D. Thesis, Rheinisch-Westfälische Technische Hochschule, Aachen, Germany, 2018.
4. Albers, A.; Reichert, U.; Ott, S.; Radimersky, A. Software supported development of a Battery Electric Vehicle powertrain considering the efficiency. In Proceedings of the 16th International VDI Congress Drivetrain for Vehicles 2016, Transmissions in Mobile Machine, Friedrichshafen, Germany, 21–22 June 2016; VDI Verlag: Düsseldorf, Germany, 2016. ISBN 978-3-18-092276-8.
5. Esser, A.; Eichenlaub, T.; Schleiffer, J.-E.; Jardin, P.; Rinderknecht, S. Comparative evaluation of powertrain concepts through an eco-impact optimization framework with real driving data. *Optim. Eng.* **2021**, *22*, 1001–1029. [CrossRef]
6. Garambois, P.; Perret-Liaudet, J.; Rigaud, E. NVH robust optimization of gear macro and microgeometries using an efficient tooth contact model. *Mech. Mach. Theory* **2017**, *117*, 78–95. [CrossRef]
7. Geradts, P.; Brecher, C.; Löpenhaus, C.; Kasten, M. Reduction of the tonality of gear noise by application of topography scattering. *Appl. Acoust.* **2019**, *148*, 344–359. [CrossRef]
8. Wang, J.; Yang, J.; Lin, Y.; He, Y. Analytical investigation of profile shifts on the mesh stiffness and dynamic characteristics of spur gears. *Mech. Mach. Theory* **2022**, *167*, 104–529. [CrossRef]
9. Zhang, T.; Shi, D.; Zhuang, Z. Research on vibration and acoustic radiation of planetary gearbox housing. In Proceedings of the Internoise 2014, Melbourne, Australia, 16–19 November 2014.
10. Schweigert, D.; Gwinner, P.; Otto, M.; Stahl, K. Noise and Efficiency Characteristics of High-Rev Transmissions in Electric Vehicles. In Proceedings of the E-Motive—Electric Vehicles Drives, Stuttgart, Germany, 12–13 September 2018.
11. Albers, A. Five Hypotheses about Engineering Processes and their Consequences. In Proceedings of the TMCE 2010, Ancona, Italy, 12–16 April 2010.
12. Jäger, S.; Vogel, S. Validation of a squeeze-film-damper test rig by using multibody cosimulation. *Multibody Syst. Dyn.* **2015**, *34*, 243–257. [CrossRef]
13. Albers, A.; Behrendt, M.; Schroeter, J.; Ott, S.; Klingler, S. X-in-the-Loop: A Framework for Supporting Central Engineering Activities and Contracting Complexity in Product Engineering Processes. In Proceedings of the 19th International Conference on Engineering Design (ICED13), Design for Harmonies, Vol.6: Design Information and Knowledge, Seoul, Republic of Korea, 19–22 August 2013; The Design Society: Glasgow, UK, 2013; pp. 391–400, ISBN 978-1-904670-49-0.
14. Jäger, S. Eine Studie zur Validierung in der Produktentstehung am Beispiel der Entwicklung einer neuartigen Untersuchungsmethode für Quetschöldämpfer = A Study to Validation Activities in Product Development based on the Development of a new Research Method for Squeeze Film Damper. Ph.D. Thesis, Karlsruher Institut für Technologie, Karlsruhe, Germany, 2014.
15. Matros, K. Entwicklung von Hybridantriebssystemen auf Basis des Pull-Prinzips der Validierung und des IPEK-X-in-the-Loop-Ansatzes = Development of Hybrid Powertrain Systems based on the Validation Pull-Principle and the IPEK-X-in-the-Loop-Approach. Ph.D. Thesis, Karlsruher Institut für Technologie, Karlsruhe, Germany, 2016.
16. Utakapan, T.; Kohn, B.; Fromberger, M.; Otto, M.; Stahl, K. Evaluation of gear noise behaviour with application force level. *Forsch. Ing.* **2017**, *81*, 59–64. [CrossRef]
17. Fischer, J.; Behrendt, M.; Lieske, D.; Albers, A. Measurement and analysis of the interior noise and the transfer path of acoustic phenomena into the driver cabin of a battery electric vehicle. In Proceedings of the Internoise 2014, Melbourne, Australia, 16–19 November 2014.
18. Verein Deutscher Ingenieure. *Entwicklungsmethodik für Mechatronische Systeme (VDI 2206): Design Methodology for Mechatronic systems*; VDI: Düsseldorf, Germany, 2004.
19. Albers, A.; Reiss, N.; Bursac, N.; Richter, T. IPeM-Integrated Product Engineering Model in Context of Product Generation Engineering. *Procedia CIRP* **2016**, *50*, 100–105. [CrossRef]
20. Lindemann, U. Models of Design. In *An Anthology of Theories and Models of Design*; Chakrabarti, A., Blessing, L.T.M., Eds.; Springer: London, UK, 2014; ISBN 978-1-4471-6337-4.
21. Geier, M.; Jäger, S.; Stier, C.; Albers, A. Combined real and virtual domain product validation using top-down strategies. In Proceedings of the Vehicle Property Validation, Bad Nauheim, Germany, 19–20 June 2012.
22. Han, J.; Liu, Y.; Yu, S.; Zhao, S.; Ma, H. Acoustic-vibration analysis of the gear-bearing-housing coupled system. *Appl. Acoust.* **2021**, *178*, 108024. [CrossRef]
23. Tang, Z.; Wang, M.; Zhao, M.; Sun, J. Modification and Noise Reduction Design of Gear Transmission System of EMU Based on Generalized Regression Neural Network. *Machines* **2022**, *10*, 157. [CrossRef]

World Electric Vehicle Journal

MDPI

Article

Optimizing Torque Delivery for an Energy-Limited Electric Race Car Using Model Predictive Control

Thomas Maull [1,*] **and Adriano Schommer** [2]

[1] Williams Advanced Engineering, Grove, Wantage OX12 0DQ, UK
[2] Department of Mechanical Engineering, Oxford Brookes University, Oxford OX33 1HX, UK
* Correspondence: tom.maull@wae.com

Abstract: This paper presents a torque controller for the energy optimization of the powertrain of an electric Formula Student race car. Limited battery capacity within electric race car designs requires energy management solutions to minimize lap time while simultaneously controlling and managing the overall energy consumption to finish the race. The energy-managing torque control algorithm developed in this work optimizes the finite onboard energy from the battery pack to reduce lap time and energy consumption when energy deficits occur. The longitudinal dynamics of the vehicle were represented by a linearized first-principles model and validated against a parameterized electric Formula Student race car model in commercial lap time simulation software. A Simulink-based model predictive controller (MPC) architecture was created to balance energy use requirements with optimum lap time. This controller was tested against a hardware-limited and torque-limited system in a constant torque request and a varying torque request scenario. The controller decreased the elapsed time to complete a 150 m straight-line acceleration by 11.4% over the torque-limited solution and 13.5% in a 150 m Formula Student manoeuvre.

Keywords: MPC; Formula Student; torque-limited; energy-limited

Citation: Maull, T.; Schommer, A. Optimizing Torque Delivery for an Energy-Limited Electric Race Car Using Model Predictive Control. *World Electr. Veh. J.* **2022**, *13*, 224. https://doi.org/10.3390/wevj13120224

Academic Editor: Joeri Van Mierlo

Received: 2 September 2022
Accepted: 16 November 2022
Published: 24 November 2022

1. Introduction

In the development of an electric race car, the battery pack sizing is directly related to the ultimate performance of the vehicle, often reflecting a trade-off between vehicle range and mass. As a result, most electric race cars are limited mainly by the energy capacity of the battery rather than the total power capability of the drivetrain. This compromise is especially valid for Formula Student vehicles, which are power-limited to 80 kW by regulations and have a small battery capacity, often of less than 7 kWh. Indeed, preliminary lap time simulations indicate that an energy capacity of approximately 10 kWh would be required to finish the endurance event at the maximum regulatory power level. Thus, the optimal use of this energy can produce significant lap time improvements over simple constant power or torque limitation strategies. Furthermore, while highly skilled drivers can approximate an optimal strategy using lift and coast techniques, Formula Student relies on amateur drivers. Thus, automated torque control and energy-saving methods are preferable.

The need to optimize the energy consumption of racing vehicles powered by an internal combustion engine (ICE) has been long understood, with associated trade-offs among fuel consumption, fuel mass, range, and available power [1]. Numerous approaches have addressed this challenge, including optimising vehicle speed with the driving path [2]. However, most real-time applications of power optimisation have remained mainly independent of the driving path [3–5]. In order to minimise lap time and simultaneously reduce energy consumption, the controller must use energy when it contributes most to acceleration. As drag squares with speed, this occurs at lower velocities [6]. This challenge is exacerbated by drag forces from the aerodynamic packages commonly used in motorsport competitions.

A number of different control approaches could be used for an energy management system, for example proportional integral derivative (PID) control, linear quadratic regulation (LQR), and model predictive control (MPC). PIDs are commonly used in motorsport, industrial, and mobility applications. It is computationally efficient and intuitive to tune in single-input single-output (SISO) applications. However, it is typically unstable or difficult to tune in multiple-input multiple-output (MIMO) plants [7]. LQR uses a fixed horizon to calculate a single optimal solution to the quadratic control problem, described by a cost function. However, this approach only computes a single optimal solution. Furthermore, it poorly adapts to nonlinear conditions, which prevents the ability of the controller to rely on superposition to interpret the system dynamics, thus limiting the performance and dramatically increasing the computational cost of the controller [8]. In addition, because the LQR solution is computed only once, it is susceptible to plant model inaccuracies and does not adapt well to changing conditions [9].

Specifically for motorsport applications, energy management control systems have been reported in the literature using dynamic programming and convex optimization [10–12], nonlinear programming [13], and artificial neural networks (ANNs) [14]. In addition, optimal controls have been applied to motorsports for thermal and current prediction [15] and electrical overloading (EOL) of the battery pack and electric motors for increased vehicle performance [16]. On the other hand, model predictive controller (MPC) approaches have been of significant interest to driverless researchers. They have recently been shown to work in driverless systems at 10–50 ms sample rates, indicating the feasibility of this control methodology for real-time onboard systems [17,18]. Furthermore, MPC has also been used in grid-based energy management systems [19], fast charging methods in battery electric vehicles to reduce lithium plating [20], and to estimate the battery state of power [21,22]. However, real-time MPC-based energy management solutions focused on reducing computational costs are still needed, which can be accomplished through the linearization of models, a reduced number of states, and appropriate quadratic programming solvers [23,24]. In addition, control methodologies have been employed in powertrain and energy management strategies for motorsport applications such as Formula (1) [4,23,25]. However, although MPC approaches have been used in several research areas, the published literature still lacks a predictive algorithm for energy management of purely electric race vehicles.

This paper describes a model-based predictive torque control system for energy management in motorsport applications. It introduces a novel feedback control methodology for energy-limited race vehicles that uses a model-based control algorithm to reduce energy use in an optimal manner when consumption is greater than the target. Furthermore, this real-time algorithm is adaptive to unforeseen changes in race conditions, such as a safety car or driver change, as it repeatedly re-targets the optimal solution [8].

The proposed methodology is compared against a torque-limited approach with the same energy consumption, resulting in an 11.4% improvement in elapsed time to complete the course of a 150 m straight-line acceleration. A representative manoeuvre in a Formula Student track layout was also investigated—the manoeuvre was extracted from a real endurance event at the Michigan, USA, competition. The MPC approach reduced the elapsed time by 13.5%, as well as responding to energy deficit conditions by reducing torque in an optimal manner.

The remainder of this paper is organized as follows. The next section addresses the development of a state-space linearized vehicle model and its verification against a transient model using the commercial software AVL VSM™. Then, the MPC is presented in Section 3 alongside design considerations for the number of prediction and control horizons and the definition of set points and signal weight. Next, results are presented in Section 4 for three scenarios: straight-line acceleration, Formula Student manoeuvre, and a reduced prediction and control horizons scenario to verify the algorithm stability. Finally, this work is summarized and concluded in Section 5.

2. Vehicle Model

Designed to reduce the computational cost of the controller, the plant model is based on a simplified vehicle longitudinal dynamics model using first-principles (Equation (1)). It expresses a lumped mass being accelerated by a thrust force—torque from the electric motors, drive ratio, and tire rolling radius—which is resisted by aerodynamic drag, rolling resistance, and road gradient forces. Equation (2) expresses the power output based on the battery pack voltage v and torque of the motors.

$$\frac{dV}{dt} = \frac{1}{em}\left[\frac{\eta_t T_r i_\theta}{r} - \left\{\frac{1}{2}\rho C_d A V^2 + mgcos(\alpha)f + mgsin(\alpha)\right\}\right] \tag{1}$$

$$\frac{dE}{dt} = v(-T_r(k_t)) \tag{2}$$

where V is the vehicle velocity, E is the energy available in the battery pack, and T_r is the total motor torque—from the four in-hub motors. The remaining parameters are described in Table 1 alongside the values used to populate the model.

Table 1. Vehicle parameters.

Parameter	Value	Units	Description
η_t	0.9	-	Transmission efficiency
e	1.4	-	Rotational mass factor
m	300	(kg)	Vehicle mass
ρ	1.225	(kg/m^3)	Air density
A	2.2	(m^2)	Vehicle frontal area
g	9.81	(m/s^2)	Gravitational acceleration
r	0.203	(m)	Tire rolling radius
k_t	0.26	(A/N.m)	Torque constant
i_θ	15.55	-	Final drive ratio
C_d	0.40	-	Coefficient of drag
α	0	(deg)	Road inclination angle
f	0.015	-	Rolling resistance coefficient

To verify the fidelity of the simplified lumped model in comparison to the real dynamics of the system, a transient model was created using the AVL VSM™ software. The energy consumption was then compared to the lumped model resulting in a root-mean-squared error (RMSE) of 0.008 kWh (Figure 1). The comparison was based on a Formula Student manoeuvre over medium-speed corners from a Formula Student endurance event in Michigan, USA.

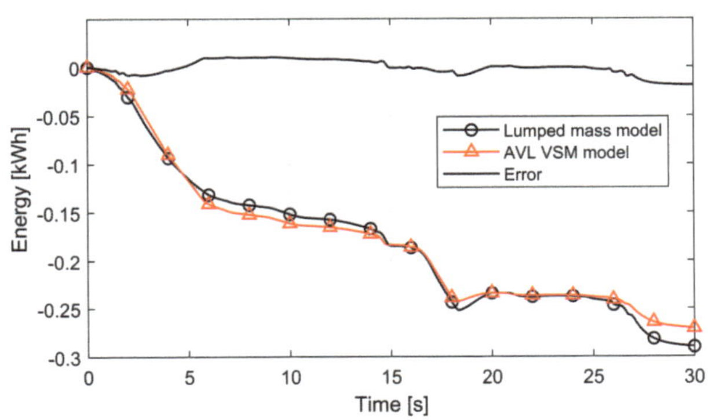

Figure 1. Energy consumption comparison between a transient model developed using the AVL VSM software and a simplified lumped mass model.

Identified Linear Time-Invariant Model

MPC-based control methodologies generally rely on linear time-invariant (LTI) state-space representations of systems [19]. However, real-world dynamical systems are very rarely purely linear [8], and in fact, the lumped mass vehicle model used in this paper is time invariant and nonlinear: as drag forces square with velocity, the outputs remain independent of absolute time. The representation of a LTI assumes the state-space format of Equations (3) and (4).

$$x(k+1) = \boldsymbol{A}x(k) + \boldsymbol{B}u(k) \tag{3}$$

$$y(k) = \boldsymbol{C}x(k) + \boldsymbol{D}u(k) \tag{4}$$

In the first equation, x is the state vector of the system, $\boldsymbol{A}$ is the state matrix, $\boldsymbol{B}$ is the input matrix, and u is the input vector of the plant. In the second equation, the output vector y is a function of the output matrix $\boldsymbol{C}$, the feed-through matrix $\boldsymbol{D}$, and the state and input vectors. The $\boldsymbol{A}$, $\boldsymbol{B}$, $\boldsymbol{C}$, and $\boldsymbol{D}$ matrices are a function of the plant dynamics and were determined through the process of system identification (Appendix A).

With the help of the MathWorks System Identification Toolbox™, a third- and a fourth-order linearized plant were developed. The process consists of evaluating the system response to a given set of inputs and identifying the set of parameters of the state-space model that better approximate the system response compared to the nonlinear plant. Hence, these models are called identified linear time-invariant (IDLTI) models [26]. In this work, three step inputs of magnitudes similar to those expected in real-world conditions were used for the system identification. Figure 2 compares both models with the nonlinear plant.

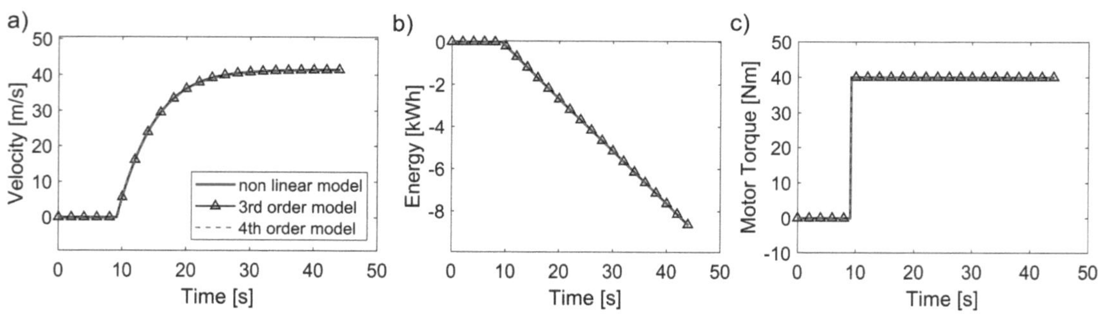

Figure 2. Comparison between 3rd- and 4th-order linearized models with the nonlinear plant. The NRMSEs for all three inputs were greater than 98% for the 3rd-order model and greater 99% for the 4th-order model (Equation (5)).

Because the second derivative of Equation (1) is relatively small, this vehicle model is considered only slightly nonlinear [8]. Even then, the linearization process introduces inaccuracies to the model, and thus, the comparison between the linearized model against the nonlinear plant needs to be evaluated. This was accomplished by a normalized root-mean-squared error (NRMSE) cost function (Equation (5)):

$$fit = 100\left(1 - \frac{||y - \hat{y}||}{||y - mean(y)||}\right) \tag{5}$$

where y is the validation data output, $\hat{y}$ is the output of the model, and fit is a percentage. Values of 100% indicate a perfect fit, whereas a fit of at least greater than 75% is recommended for stable control [7]. The 3rd-order model achieved an NRMSE greater than 98%, whilst the 4th-order linearized model had a 99% fit with the corresponding results of the nonlinear plant. Because of the negligible difference of the model fit, the third-order model was selected to reduce the computational costs.

3. MPC Design

In this work, a dynamic optimization problem was created to control a single-input multiple-output plant. The model predictive controller calculates a minimal cost over a receding prediction horizon using a defined plant model at a repeating time step. It uses two horizon values in its calculation. The first is the prediction horizon P, or number of steps into the future the model predicts and optimizes for. This allows the controller to avoid constraints, as well as make decisions while understanding the long-term implications of the control action, such as in plants with a negative short-term response and positive long-term response or in plants with a long delay period [27]. The second horizon is the control horizon, which predicts the next M control moves in the future. Only the first of these control moves is implemented; the rest are discarded. This is typically a small number, between two and five [7,27]. The MPC algorithm calculates an optimal solution to a quadratic control problem. However, it calculates a minimal cost over a receding prediction horizon using a defined plant model at a repeating time step, as shown in Figure 3 [27].

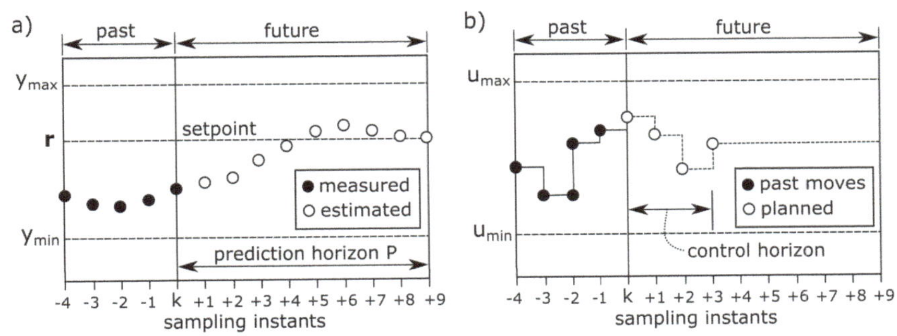

Figure 3. Prediction horizons within boundary limits. Adapted from [27].

The manipulated variables u are held within established limits, as can measured outputs y, allowing the MPC to consider the physical constraints of the system. As the optimal output is recomputed at every time step k, the MPC tends to incur relatively high computational costs compared to the PID and LQR controllers. However, as real-time controllers gain computational power, MPCs have been recently able to be used in high-frequency real-time applications [3,7]. The MPC design is illustrated in Figure 4. The set point block diagram gives the controller the hard limits of the motor torque, u_{max}, and the target velocity set point, r. The MPC then optimizes the energy consumption based on the feedback of the battery energy capacity, torque output, and actual velocity compared to the target velocity. The implementation of the controller was realized using the MathWorks MPC Designer Toolbox™.

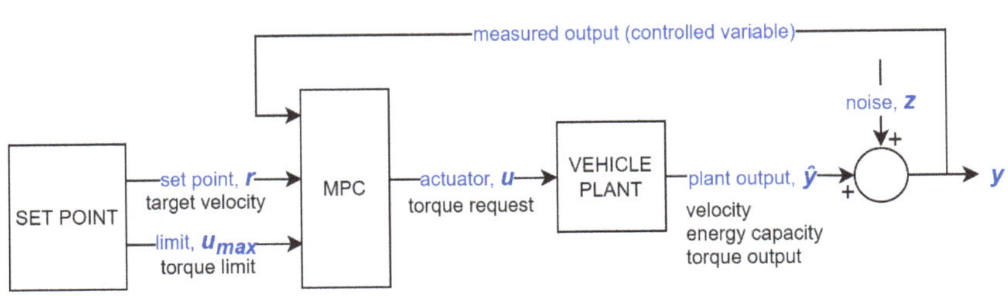

Figure 4. MPC block diagram. Energy consumption is optimized based on the feedback of the battery energy capacity, torque output, and vehicle velocity.

3.1. Horizons

The MPC calculates an optimal solution to a quadratic control problem. The minimal cost is computed over a receding prediction horizon using a defined plant model at a repeating time step. Two horizons are therefore implemented—visualized previously in Figure 3: *prediction* and *control* horizons. In general, longer horizons increase controller stability, but come at a higher computational cost because of the increased number of optimized solutions it computes. If control horizons can be lowered without affecting the controller's response, this is typically taken as an indication that the prediction and control horizon times are adequate [27]. In this work, a *prediction horizon* of ten time steps and a *control horizon* of two time steps were used.

The relation between the cost function and the number of horizons used by the MPC is dictated by how close the evaluation of the cost function is to the design constraints—in this work: energy consumption and target velocity. A longer prediction horizon is needed to find the optimal solution if the evaluation of the cost function is close to those constraints. On the other hand, if the cost function is not reaching those limits, it is possible to reduce the number of the prediction and control horizons without major effects on the optimal solution, which is advantageous to reduce the computational costs. In order to investigate the instabilities caused by reducing the predictive and control horizons, these values were reduced to five and one, respectively, further discussed in the Results Section.

3.2. Set Point and Control Signal Weighting

The MPC uses five set points: torque request (N.m); rate of change of the torque request (N.m/Δt); manoeuvre target velocity (km/h); target energy consumption (kWh); and finally, the torque output (N.m) to meet the energy consumption and target velocity constraints. Each set point is weighted differently in the MPC cost function. Multiple weight schemes were tested by trial and error for this implementation with the help of the MPC Designer Toolbox™. In general, the driver's torque request carries less weight than the velocity term. This is partially because the units used in the torque request are generally larger than those seen in the velocities. The controller attempts to meet the torque request until the cost weighed against that of the other set points grows too high. The energy term carried greater weight because the deviation from the set point occurs over smaller absolute numbers when compared to torque and velocity. This term reduced the torque limits as the energy use exceeded the targets. Similarly, as speed rises over the predetermined set point, the cost increases, and the controller reduces the torque. The set point weighting parameters used in this work are described in Table 2.

Table 2. Set point weighting parameters.

Variable	Weight [1]
Manipulated variable—torque request (N.m)	0
Manipulated variable rate (N.m/Δt)	0.1
Target velocity (km/h)	10
Energy consumption (kW)	13
Torque output (N.m)	3

[1] dimensionless.

4. Results and Discussion

The purpose of the MPC controller presented in this paper is to optimize torque delivery during acceleration events. Two scenarios were investigated: (1) straight line wide-open throttle acceleration event over 150 m, and (2) cornering manoeuvre of positive torque request over 150 m—that is, the torque demand is always positive; no braking or coasting is involved.

In the first scenario, the vehicle accelerates from a low-speed corner onto a straight line, and the length of the straight represents the longest continuous straight a Formula Student car is likely to encounter in an endurance event. In Figure 5, the MPC controller receives an arbitrated constant torque request of 39.8 Nm, which is the maximum torque output of the motors (hardware limit). The system response is then compared between a torque-limited approach with an equivalent energy usage.

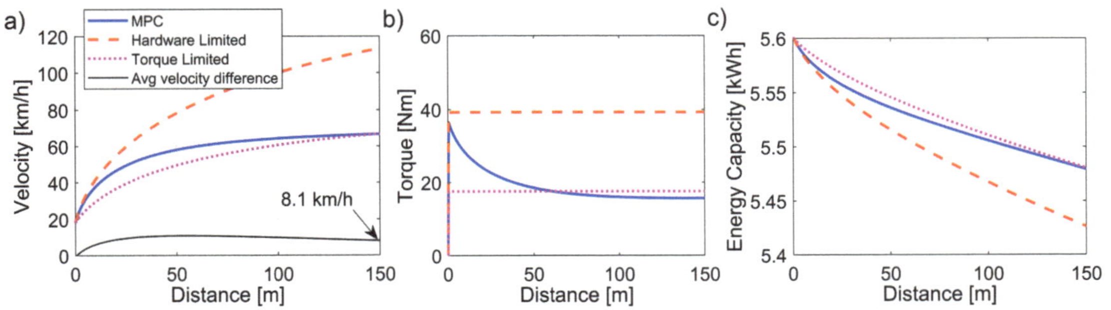

Figure 5. Acceleration torque request scenario. (**a**) Speed profile; (**b**) torque request; (**c**) energy remaining in the battery.

As expected, the MPC torque set point decreased with distance to optimize the energy consumption as shown in Figure 5b). This is explained as the most efficient use of the energy in an acceleration manoeuvre occurs at low speeds, as the thrust force is resisted by the least amount of the drag force, which increases with the square of the velocity ($\frac{1}{2}\rho C_d AV^2$ part of Equation (1)). Compared to the torque-limited approach, an average velocity improvement of 8.1 km/h was observed using the MPC strategy, resulting in a time improvement of 1.31 seconds (11.4%) for the same energy consumed.

In the second scenario, the manoeuvre input was the same torque request used in the vehicle model verification (Figure 1). Figure 6 compares the MPC approach with the torque-limited approach. The MPC results in an average velocity of 42.5 km/h, 5.6 km/h faster than the equivalent energy used by the torque-limited system. This yields a 1.99 second or a 13.5% lap time improvement over the course of 150 m.

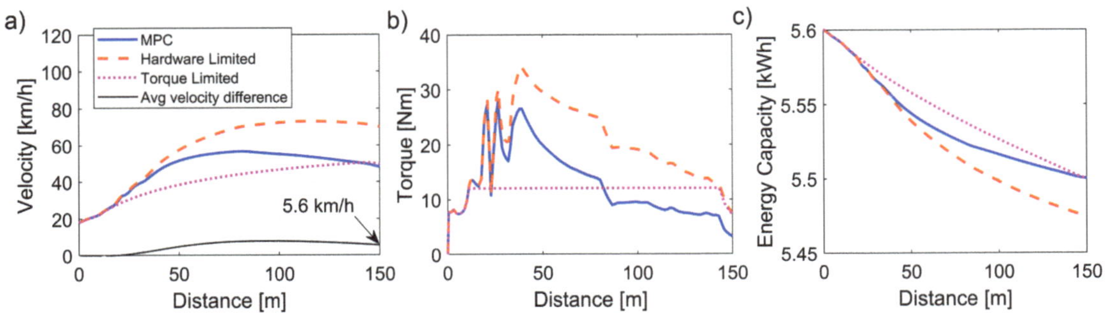

Figure 6. Formula Student manoeuvre torque request scenario. (**a**) Speed profile; (**b**) torque request; (**c**) energy remaining in the battery.

In order to investigate the instabilities caused by reducing the predictive horizon, it was reduced to five and the control horizon reduced to a single time step. The results are shown in Figure 7, while the controller was less able to restrict energy use, no instability or boundary violations in the controller were observed, indicating that, with this controller and plant, reducing the control and prediction horizons is a viable method to lower the computational cost if shorter time steps are desired.

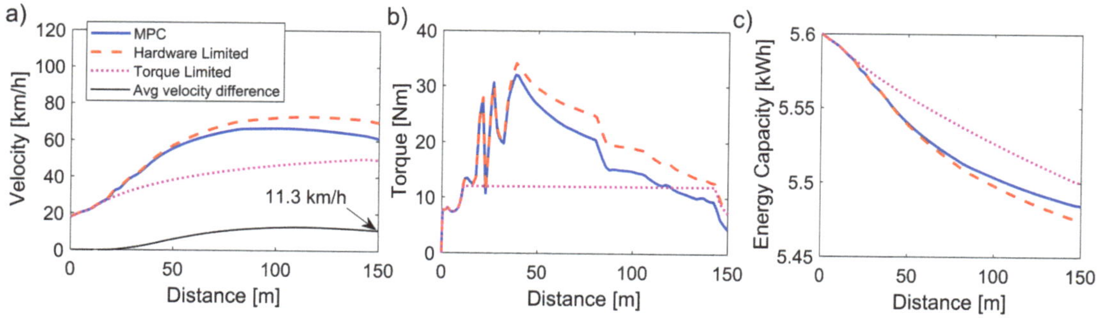

Figure 7. Reduced horizons scenario. (**a**) Speed profile; (**b**) torque request, (**c**) energy remaining in the battery.

5. Conclusions

The MPC torque limiter met the preliminary requirements of the system: it provided significant improvements in lap time and energy use while balancing the immediate torque demand and associated energy use against the overall state of energy in the system. Furthermore, torque safety was maintained by using the hard constraints allowed by the MPC methods. Therefore, model predictive control methods show significant promise for energy-limited race cars, especially Formula Student-class electric vehicles—which are required by the competition rules to be driven by amateur drivers. This controller benefits such drivers since it decreases cockpit workload during competition events, eliminating the need for drivers to use lift and coast techniques to save energy.

Many control architectures are possible to solve a given set of control requirements. While this controller housed the energy management function of the controller within the torque controller, future explorations of this method should consider a low-frequency energy management hypervisor to evaluate the energy state of the system and estimated energy use per lap using a plant model generated from a lap time simulation as suggested by [3]. This offloads the computational cost of energy management to a very low-frequency controller, freeing up computational headroom to increase the frequency of the torque converter. The actuators of the hypervisor can include the weighting, set points, and limitations of the torque controller. The torque controller can then be expanded to include a thermal winding load of the torque request, as well as the battery pack thermal load.

This paper outlined how an MPC-based control algorithm can be implemented in electric motorsport powertrain optimization and provides a starting point for additional control algorithm development in Formula Student vehicles. Furthermore, it addresses the gap identified in the literature where an MPC-based energy management algorithm was applied for a fully electric race car. However, further validations are needed to investigate the robustness of the controller under real-world conditions.

Author Contributions: Conceptualization, T.M.; methodology, T.M.; investigation, T.M.; writing—original draft preparation, T.M. and A.S.; writing—review and editing, T.M. and A.S.; visualization, T.M. and A.S. All authors have read and agreed to the published version of the manuscript.

Funding: This research received no external funding.

Data Availability Statement: Not applicable.

Acknowledgments: Thomas Maull was with Oxford Brookes University at the time this work was conducted. The authors wish to thank Oxford Brookes Racing team members, alumni, faculty, and sponsors for their support and guidance, particularly Gordana Collier and Brady Planden for their supervision. The authors would also like to thank Williams Advanced Engineering for their enthusiastic support and feedback.

Conflicts of Interest: The authors declare no conflict of interest.

Abbreviations

The following abbreviations are used in this manuscript:

FS	Formula Student
ICE	Internal combustion engine
MPC	Model predictive controller
PID	Proportional, integral, derivative
LQR	Linear quadratic regulation
RMSE	Root-mean-squared error
NRMSE	Normalized root-mean-squared error
IDLTI	Identified linear time-invariant
LTI	Linear time-invariant

Appendix A

Linear time-invariant matrices $\boldsymbol{A}$, $\boldsymbol{B}$, $\boldsymbol{C}$, and $\boldsymbol{D}$ from the system identification process:

$$\boldsymbol{A} = \begin{bmatrix} 0 & 1 & 0 & 0 \\ -0.947 & -1.946 & 1.334 & -1.334 \\ 0 & 0 & 0 & 1 \\ -7.735e-08 & 7.868e-08 & 3.296e-05 & 1 \end{bmatrix}$$

$$\boldsymbol{B} = \begin{bmatrix} 0.017 \\ 0.01598 \\ 2.141e-08 \\ -0.0006197 \end{bmatrix}$$

$$\boldsymbol{C} = \begin{bmatrix} 1 & 0 & 0 & 0 \\ 0 & 0 & 1 & 0 \end{bmatrix}$$

$$\boldsymbol{D} = \begin{bmatrix} 0 \\ 0 \end{bmatrix}$$

References

1. Limebeer, D.J.; Rao, A.V. Faster, higher, and greener: Vehicular optimal control. *IEEE Control Syst. Mag.* **2015**, *35*, 36–56.
2. Salazar, M.; Alessandretti, A.; Aguiar, A.P.; Jones, C.N. An energy efficient trajectory tracking controller for car-like vehicles using model predictive control. In Proceedings of the 2015 54th IEEE Conference on Decision and Control (CDC), Osaka, Japan, 15–18 December 2015; pp. 3675–3680.
3. Salazar, M.; Elbert, P.; Ebbesen, S.; Bussi, C.; Onder, C.H. Time-optimal control policy for a hybrid electric race car. *IEEE Trans. Control Syst. Technol.* **2017**, *25*, 1921–1934. [CrossRef]
4. Salazar, M.; Duhr, P.; Balerna, C.; Arzilli, L.; Onder, C.H. Minimum lap time control of hybrid electric race cars in qualifying scenarios. *IEEE Trans. Veh. Technol.* **2019**, *68*, 7296–7308. [CrossRef]
5. Ebbesen, S.; Salazar, M.; Elbert, P.; Bussi, C.; Onder, C.H. Time-optimal control strategies for a hybrid electric race car. *IEEE Trans. Control Syst. Technol.* **2017**, *26*, 233–247. [CrossRef]
6. Gillespie, T.D. *Fundamentals of Vehicle Dynamics*; SAE International: Warrendale, PA, USA, 2021.
7. Khaled, N.; Pattel, B. *Practical Design and Application of Model Predictive Control: MPC for MATLAB® and Simulink® Users*; Butterworth-Heinemann: Oxford, UK, 2018.
8. Brunton, S.L.; Kutz, J.N. *Data-Driven Science and Engineering: Machine Learning, Dynamical Systems, and Control*; Cambridge University Press: Cambridge, UK, 2022.
9. Vivek, K.; Sheta, M.A.; Gumtapure, V. A comparative study of Stanley, LQR and MPC controllers for path tracking application (ADAS/AD). In Proceedings of the 2019 IEEE International Conference on Intelligent Systems and Green Technology (ICISGT), Visakhapatnam, India, 29–30 June 2019; pp. 67–674.
10. Duhr, P.; Christodoulou, G.; Balerna, C.; Salazar, M.; Cerofolini, A.; Onder, C.H. Time-optimal gearshift and energy management strategies for a hybrid electric race car. *Appl. Energy* **2021**, *282*, 115980. [CrossRef]
11. Duhr, P.; Sandeep, A.; Cerofolini, A.; Onder, C.H. Convex Performance Envelope for Minimum Lap Time Energy Management of Race Cars. *IEEE Trans. Veh. Technol.* **2022**, *71*, 8280–8295. [CrossRef]
12. Balerna, C.; Neumann, M.P.; Robuschi, N.; Duhr, P.; Cerofolini, A.; Ravaglioli, V.; Onder, C. Time-optimal low-level control and gearshift strategies for the formula 1 hybrid electric powertrain. *Energies* **2020**, *14*, 171. [CrossRef]

13. Herrmann, T.; Christ, F.; Betz, J.; Lienkamp, M. Energy management strategy for an autonomous electric racecar using optimal control. In Proceedings of the 2019 IEEE Intelligent Transportation Systems Conference (ITSC), Auckland, New Zealand, 27–30 October 2019; pp. 720–725.
14. Liu, X.; Fotouhi, A. Formula-E race strategy development using artificial neural networks and Monte Carlo tree search. *Neural Comput. Appl.* **2020**, *32*, 15191–15207. [CrossRef]
15. Glos, J. Modeling and Control of Electric and Thermal Flows in Fully Electric Vehicles. Ph.D. Thesis, Brno University of Technology, Brno, Czech Republic, 2020.
16. Sedlacek, T.; Odenthal, D.; Wollherr, D. Minimum-time optimal control for battery electric vehicles with four wheel-independent drives considering electrical overloading. *Veh. Syst. Dyn.* **2022**, *60*, 491–515. [CrossRef]
17. Liniger, A.; Domahidi, A.; Morari, M. Optimization-based autonomous racing of 1:43 scale RC cars. *Optim. Control Appl. Methods* **2015**, *36*, 628–647. [CrossRef]
18. Vázquez, J.L.; Brühlmeier, M.; Liniger, A.; Rupenyan, A.; Lygeros, J. Optimization-based hierarchical motion planning for autonomous racing. In Proceedings of the 2020 IEEE/RSJ International Conference on Intelligent Robots and Systems (IROS), Las Vegas, NV, USA, 24 October 2020–24 January 2021; pp. 2397–2403.
19. Lee, J.; Chang, H.J. Explicit model predictive control for linear time-variant systems with application to double-lane-change maneuver. *PLoS ONE* **2018**, *13*, e0208071. [CrossRef] [PubMed]
20. Yin, Y.; Bi, Y.; Hu, Y.; Choe, S.Y. Optimal fast charging method for a large-format lithium-ion battery based on nonlinear model predictive control and reduced order electrochemical model. *J. Electrochem. Soc.* **2021**, *167*, 160559. [CrossRef]
21. Xavier, M.A.; de Souza, A.K.; Plett, G.L.; Trimboli, M.S. A low-cost MPC-based algorithm for battery power limit estimation. In Proceedings of the 2020 American Control Conference (ACC), Denver, CO, USA, 1–3 July 2020; pp. 1161–1166.
22. Xavier, M.A.; de Souza, A.K.; Trimboli, M.S. An LPV-MPC Inspired Battery SOP Estimation Algorithm Using a Coupled Electro-Thermal Model. In Proceedings of the 2021 American Control Conference (ACC), New Orleans, LA, USA, 25–28 May 2021; pp. 4421–4426.
23. Salazar, M.; Balerna, C.; Elbert, P.; Grando, F.P.; Onder, C.H. Real-time control algorithms for a hybrid electric race car using a two-level model predictive control scheme. *IEEE Trans. Veh. Technol.* **2017**, *66*, 10911–10922. [CrossRef]
24. Zhang, F.; Hu, X.; Liu, T.; Xu, K.; Duan, Z.; Pang, H. Computationally efficient energy management for hybrid electric vehicles using model predictive control and vehicle-to-vehicle communication. *IEEE Trans. Veh. Technol.* **2020**, *70*, 237–250. [CrossRef]
25. Limebeer, D.J.; Massaro, M. *Dynamics and Optimal Control of Road Vehicles*; Oxford University Press: Oxford, UK, 2018.
26. Ljung, L. *System Identification Toolbox: User's Guide*; MathWorks: Natick, MA, USA, 2021.
27. Bemporad, A.; Morari, M.; Ricker, N.L. *Model Predictive Control Toolbox User's Guide*; MathWorks: Natick, MA, USA, 2010.

World Electric Vehicle Journal

MDPI

Article

On the Collaborative Use of EV Charging Infrastructures in the Context of Commercial Real Estate

Joela Gauss †, Sascha Gohlke †, Zoltán Nochta *,†

Data-Centric Software Systems (DSS) Research Group, Institute of Applied Research (IAF), Karlsruhe University of Applied Sciences, 76133 Karlsruhe, Germany
* Correspondence: zoltan.nochta@h-ka.de
† These authors contributed equally to this work.

Abstract: Resource sharing in general is a means of solving the problem of infrequent and, thus, inefficient utilization of expensive or scarce resources. In this paper, we present an approach to run shared EV-charging infrastructures in the context of commercial real-estate facilities. Collaborating EV-charger owners thereby create a pool of chargers for shared use. In our work, we consider aspects of economic viability, desirability and technical feasibility as prerequisites of a successful solution. We formally prove that the basic economic potential of the proposed pooling with regard to overall infrastructure utilization is given. In order to operate the shared pool of charging points at a given location, the corresponding management software must fulfil specific requirements. Our prototype implementation that was realized as an extension of the open-source system Open E-Mobility demonstrates the technical feasibility of the sharing idea in a user-friendly way. Questionnaires and personal interviews conducted with owners of small and medium-sized businesses revealed that they would share charging stations if it helped overcome availability bottlenecks, thus improving customer and employee perception.

Keywords: EV charging infrastructure; pooling; sharing; management software; processes

Citation: Gauss, J.; Gohlke, S.; Nochta, Z. On the Collaborative Use of EV Charging Infrastructures in the Context of Commercial Real Estate. *World Electr. Veh. J.* **2022**, *13*, 223. https://doi.org/10.3390/wevj13120223

Academic Editor: Joeri Van Mierlo

Received: 28 September 2022
Accepted: 17 November 2022
Published: 23 November 2022

1. Introduction

Thanks to huge investments made by the automotive industry as well as the political and financial support given by government incentives worldwide, the number of electric vehicles (EV) will continue to rise in the coming years. Therefore, the rapid creation of sufficient EV charging capacity for diverse mobility scenarios is required. For instance, in Germany, by the end of 2021, there were ca. 52,000 publicly accessible charging points (CP) registered [1] serving estimated one million EVs (PHEV and BEV) in the country [2]. The German government's goal is to reach up to 15 million EVs by 2030 [3]. Currently, the installation and operation of *public* EV charging stations—especially in urban areas—appear to be economically unattractive, with the result that the required upscaling effects have so far failed to materialize. It is similarly problematic to scale the number of *private* charging points ("wall boxes"), which are mostly installed in single-family homes in the suburbs: If large numbers of EVs were to be charged at the homes on the same street simultaneously, e.g., during the after-work hours, the local power supply could reach its limits. As a countermeasure, local grid operators can increase grid connection capacity by installing additional cabling, transformers, or by implementing means of remote control to actively manage EV charging. Both options are highly cost intensive as well as having low profit margins. Due to these limitations, in the near future, *semi-public* parking areas and charging stations at office buildings, factories, industrial areas, shopping centers, hotels, etc., might play a significant role in "fuelling" the growing EV population. Such charging facilities primarily power EVs, which are used for business purposes of a company, including, for example, service cars, delivery vans, and shuttles.

At the same time, the charging stations can also cover the electricity demand of privately used vehicles to a huge extent, for example, EVs of employees, customers, visitors, or even neighbours if they are authorized to enter the respective parking area. Thus, companies' semi-public charging infrastructures can help relieve the entire (public) power system [4] and also reduce required investments in public charging infrastructures.

Commercial real-estate facilities and corresponding in- or outdoor parking areas are often used by several tenants simultaneously. For example, the first floor of an office building might house a medical practice and a restaurant, while the upper floors contain offices rented by different corporations. Usually, each tenant has dedicated, reserved car parking lots for its employees, customers, business partners, and visitors. As EVs become increasingly attractive, tenants may want to establish exclusively used EV chargers on their own parking spots. In real-world scenarios, however, the exclusive assignment of EV-charging equipment to single tenants can be challenging due to both economic (high TCO per charging station, low utilization outside of business hours) and technical reasons (insufficient electrical infrastructure in the building, power limitations) [5–9]. The collaborative use of charging stations that are shared among multiple independent tenants of a building might help overcome these problems.

In our approach, collaborating CP owners make their own installed equipment basically available and accessible to each other. The resulting *pool of charging points* helps cover peak demand time intervals of all participating contributors, and it also helps increase the overall utilization of the entire charging infrastructure.

For a better understanding, Figure 1 shows an example scenario for the proposed pooling of CPs in the context of a building with an integrated parking space in the basement. Due to technical limitations, the building can host a maximum of 15 charging points. With e-mobility in mind, the building owner has initially foreseen 15 cable trays for future EV charger installations and installed four charging points before opening the building. Since then, CP 1 has been jointly used by a physician (on weekdays during the daytime) and by a restaurant (in the evening hours and on weekends), while CPs 2, 3 and 4 are rented by a company termed as "Tenant 1".

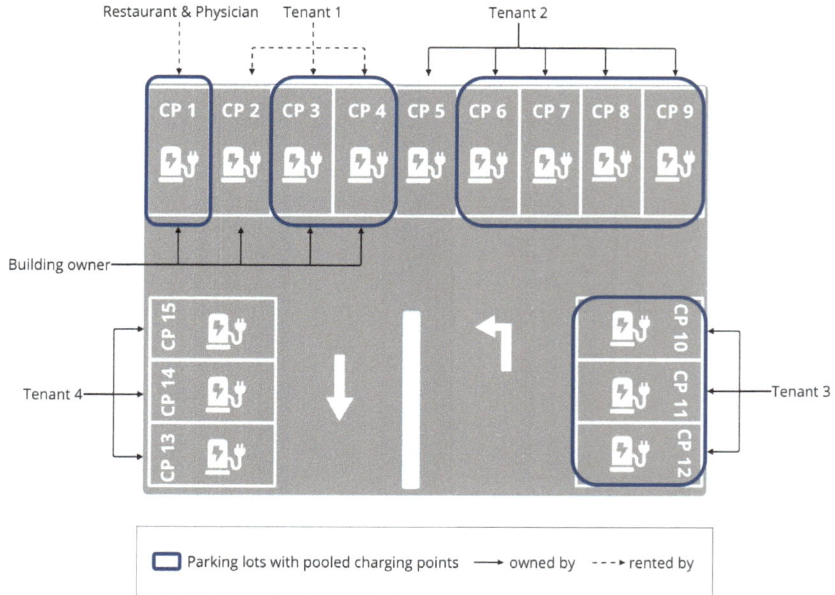

Figure 1. Example CP pooling scenario in an office building's parking area.

Over time, Tenants 2, 3 and 4 had been equipping their parking slots with their own EV chargers because the building owner himself did not want to invest in additional equipment. Meanwhile, the available charging points of single tenants are not sufficient anymore to cover their respective demand. They are faced with availability problems especially at peak times specific to their businesses. Employees, customers, guests and visitors often cannot find an available charging point at the given company's dedicated parking lots, while at the same time chargers of other tenants are unused. To improve the frustrating situation, some tenants joined forces and created a shared pool of charging points. The pool consists of 10 (out of 15 installed) charging points because Tenant 1 and Tenant 2 only partially share their CPs, and Tenant 4 does not participate in the CP pooling initiative at all. The CP pool and related processes are managed with the help of a software system that enables independent participants to individually configure their specific settings and preferences.

This paper is structured as follows: In Section 2, we revisit publications that deal with resource sharing and problems of EV charging. Section 3 summarizes the methodology we have applied in our research project. In Section 4, we explain the benefits of CP pooling and provide a mathematical proof of why sharing partners can expect higher utilization of their charging points. Section 5 introduces main processes, describes identified system requirements, and provides details about the prototype implementation of a software system, which enables the management of pooled CPs at a given location. Section 6 presents the results of a two-part end-user survey, which we conducted with small and medium-sized businesses in a small German town. Finally, in Section 7, conclusions are drawn, and future work is outlined.

2. Related Work

In the area of electric vehicle charging, most of the recent research has focused on solving problems related to power grid constraints and the usage of power at affordable cost. The proposed solutions often combine energy consumption and distribution strategies (e.g., load shifting, peak shaving) with demand response management techniques (e.g., dynamic time-of-use tariffs) to reduce peak demand and optimize the total cost of energy used in enterprise fleets [10–14]. Another approach to reduce the costs is to save energy in a fleet by applying platooning under optimized parameters, such as inter-vehicle gaps [15]. Dynamic pricing systems and related load shifting solutions can help avoid critical electricity demand peaks, but they are not suitable to improve the situation regarding the basic availability of charging points at a specific location. Regarding the issue of availability, previous research has examined the problem of parking spaces and charging stations being blocked by vehicles that are not charging. In [16], a time-based usage fee of charging stations is proposed providing an incentive to drive away after the charging was completed. To automatically detect (unplugged) vehicles that block free parking spaces, the use of proximity sensors is suggested in [17]. Our research and the proposed approach target this gap and help improve the overall availability of charging points through sharing.

In conjunction with the shared use of material goods and immaterial services, the terms *sharing economy* and *collaborative consumption* are widely used in academic literature. According to [18], "Sharing economy consists of the practice of use and share of products and services with two or more individuals, with or without the transfer of ownership, with no material compensation (neither non-monetary compensation) and mediated through social mechanisms". Our CP pooling approach partly fits this definition because pooled charging points are made accessible to other pooling partners without transferring ownership. There is a difference with regard to the pooling agreement, which is negotiated between the participants to regulate the co-usage under fair conditions upfront, including the determination of usage fees as means of compensation.

Usage fees are seen in [18] as a property of collaborative consumption, i.e., "transactions where people coordinate the exchange of goods and services for a fee or other compensation (monetary or non-monetary), where a triadic is existing among a platform provider, peer service provider and a customer, there is no ownership transfer and it is mediated through market mechanisms". However, in our case, the jointly negotiated pooling agreement eliminates the need for involving a platform provider as a mediator. Due to the predetermined fees for using pooled CPs, a mediation through market mechanisms is excluded as well. The term "sharing" is also frequently used in context of ad-hoc and short-term rental services, such as sharing cars or e-scooters in cities [19]. In [20], Belk unmasks and criticizes those rather commercial activities as "pseudo-sharing". In this spirit, we consider our pooling approach as "real sharing" because all involved participants can both provision and consume shared resources under previously negotiated (and therefore fair) conditions. The sharing of expensive and less frequently used machinery is common practice in the agricultural domain [21]. A simple manifestation of it is "mutual aid" between neighboring farmers who help each other out with equipment. Thus, so-called "machinery rings" are established to coordinate the demand-driven cross usage of machinery, for example, to temporarily replace a broken machine. Members of "machinery partnerships" jointly purchase equipment and define operational policies. Our approach can be seen as a mixture of these variants, as neighbouring CP owners help each other out especially to cover EV-charging demand at peak usage hours, while they buy and manage their own equipment without involving a coordinating party.

The concept of allowing others to access privately installed charging points is found in existing products and also in the academic literature. As an example, the Webasto ChargeConnect App [22] allows the owner of a private charging point to grant a time-limited access to guest users. The guest must first register in the system and be authorized by the owner of the CP in question to start the charging process. The use of a third-party platform to coordinate the sharing of private charging points is also discussed in [23,24]. With the help of the demonstrator "CrowdStrom", CP owners provide access to other EV drivers in return for a financial compensation. The price for using a charging point is set on an hourly basis, and authorized EV drivers can reserve advertised charging points before they arrive. This approach can also be characterized as a short-time rental service rather than "real" sharing of charging points among equal peers.

According to survey results, 30–40% of BEV owners use charging facilities at their workplaces, of which 8% exclusively charge at work [25]. In the context of semi-public infrastructures, the development of an operational model, which involves real estate agencies in the installation and operation of charging infrastructures, was considered in [26]. Simulations estimate a significant increase of charging power demand of EVs, based on data captured at shopping centers [27]. Peak electricity demand can be reduced, for instance, by applying dynamic pricing, which makes charging at "peak-times" increasingly unattractive for EV drivers [7]. A broad application of pooling and the resulting increased utilization of charging points could enable basically more drivers to charge their EVs at work. Charging more EVs during the business hours at work can help reduce the power demand at homes, especially during the after-work hours, as advised by Alatise et al. in [4]. The importance of reducing (or shifting) charging demand at home is also underpinned by Gaikwad, who simulated daily charging profiles and predicted substantial transformer overloading, when most cars are charged at the same time [28].

3. Methodology

Our basic working hypothesis is that the shared usage of charging points will only be adopted by the market in the future if the *economic viability, technical feasibility* and *desirability* of the concept and its practical implementation are given [29]. Accordingly, we dealt with all three aspects in parallel in our project:

- **Economic viability:** We created an abstract definition of resource sharing and formally proved the hypothesis that the shared usage of charging points will most likely improve the overall utilization of charging infrastructures. As a consequence, CP pooling helps make related investments more profitable for all involved CP owners.

- **Technical feasibility:** We collected and analysed technical requirements along with non-functional features that charge point operating systems should provide to serve a shared pool of CPs. To prove the technical feasibility of CP pooling, we have decided to implement related requirements as extensions of the open source software system "Open E-Mobility" [30], which is used to operate charging infrastructures in several real-world deployments. Adding new functionality to an existing software system (also known as the brownfield approach) instead of building a dedicated solution (greenfield development) not only helps reduce development costs, but also allows customers to move from exclusively used CPs to shared scenarios in future without changing their existing system. The resulting extended software system was deployed and tested in a prototype environment. The tests covered technical aspects mainly related to the functional correctness of the software system. The implemented prototype system has also been used in five qualitative face-to-face interviews, which we carried out with four retailers and a restaurant owner to gather early feedback from potential end-users.
- **Desirability:** We have collected preferences of potential end-users related to the shared usage of charging points. For this purpose we initially conducted personal interviews and questionnaires with selected business owners in a small German town, consisting of retailers (~30%), service providers (~24%), manufacturers (~10%) and others. The questionnaire was mainly used to find out whether and under which conditions companies would allow others (incl. customers, visitors, employees of other companies, etc.) to access their EV charging equipment.

4. Economic Viability of Sharing Charging Points

The proposed pooling is an approach to bundle charging points that are owned and operated by multiple independent entities (typically businesses). In general, a pool of charging points is not limited to a specific geographic location or context, such as the parking area of a single office building. Instead, a pool may include charging points that are installed at multiple locations of the pooling participants, such as larger companies that usually have offices in different cities or even countries. Establishing a pool increases the basic availability of charging points to individual EV-drivers, e.g., employees, guests and customers of pooling participants, as they gain access to all pooled charging points at their respective location. From a driver's perspective, the creation of a pool of CPs leads to a higher likelihood to find an available charging point at the given location basically at any point of time. In addition to the obvious benefits for EV drivers, the pooling of charging points can also increase the utilization of the overall charging infrastructure. If a particular charging point is exclusively used by only one business, it could remain completely unused outside of that company's regular business hours, i.e., up to 120 h per week. There is also no guarantee that such a bound charging point will be used continuously during the owner's business hours, for example, when the EV-driving employees are on vacation.

The increased utilization of pooled charging points compared to exclusively used equipment can be formally proven as follows:

Let M with $(|M| \in \mathbb{N})$ the set of installed charging points within a charging infrastructure and I with $(|I| \in \mathbb{N})$ the set of CP owners (who contribute to the pool).

M_i denotes the set of charging points of a given owner, whereas $M = \cup_{i \in I} M_i$ $(M_i \cap M_j = \emptyset, i \neq j)$ holds for the entire infrastructure at the facility, meaning that each CP has only one owner.

Let N_i be the set of EV drivers, who are assigned to the charging points of a given CP owner. It is assumed that $N = \cup_{i \in I} N_i$ $(N_i \cap N_j = \emptyset, i \neq j)$ holds because normally an EV driver is an employee (business partner, customer, visitor) of only one company at the same time. $N_i(t) \subseteq N_i$ is the set of EV drivers that are assigned to a given CP owner and require a charging point at time $t \in \mathbb{R}$. Thus, $N(t) = \cup_{i \in I} N_i(t)$ is the set of all EV drivers that look for an available charging point at t. $M_i(t) \subseteq M_i$ denotes the set of charge points of a given owner that are used at time t and $M(t) = \cup_{i \in I} M_i(t)$ is the set of all used CPs at t.

The function $l(t) : \mathbb{R} \to \mathbb{N}, l(t) = |M(t)|$ shows the level of utilization, i.e., the total number of charging points that are used at time $t \in \mathbb{R}$.

Thus, $L = \int_{-\infty}^{\infty} l(t)\ \mathrm{d}t$ stands for the total utilization of the charging infrastructure over time.

$\widetilde{L}$ stands for the utilization of the charging infrastructure in which all CPs are pooled, while $\underset{\sim}{L}$ represents the same CPs' utilization in case each of them is used exclusively by just one CP owner, i.e., without being added to a pool. Accordingly, $\widetilde{M}(t)$ is the number of actually used CPs over time, which are part of the pool, while $\underset{\sim}{M}(t)$ stands for the "non-pooled" case, i.e., no CP owner shares any of its charging points.

Theorem 1.

$$\widetilde{L} \geq \underset{\sim}{L} \tag{1}$$

Proof.

$$\widetilde{L} = \int_{-\infty}^{\infty} \widetilde{l}(t)\ \mathrm{d}t \tag{2}$$

$$= \int_{-\infty}^{\infty} |\widetilde{M}(t)|\ \mathrm{d}t = \int_{-\infty}^{\infty} min\{|M|, |N(t)|\}\ \mathrm{d}t \tag{3}$$

$|\widetilde{M}(t)| = min\{|M|, |N(t)|\}$ holds because, when using a CP pool, the number of actually usable charging points at a given time is basically limited by the total number of pooled charging points of all CP owners:

$$= \int_{-\infty}^{\infty} min\{\sum_{i \in I} |M_i|, \sum_{i \in I} |N_i(t)|\}\ \mathrm{d}t \tag{4}$$

$|M| = \sum_{i \in I} |M_i|$ because considering disjoint charging points $M = \cup_{i \in I} M_i$ holds.
$|N(t)| = \sum_{i \in I} |N_i(t)|$, since assuming a disjoint driver population $N(t) = \cup_{i \in I} N_i(t)$ holds.

$$\geq \int_{-\infty}^{\infty} \sum_{i \in I} min\{|M_i|, |N_i(t)|\}\ \mathrm{d}t \tag{5}$$

$\sum_{i \in I} min\{a_i, b_i\} \leq min\{\sum_{i \in I} a_i, \sum_{i \in I} b_i\}$, is a consequence of $\sum_{i \in I} min\{a_i, b_i\} \leq \sum_{i \in I} a_i$

$$= \int_{-\infty}^{\infty} \sum_{i \in I} |\underset{\sim}{M}_i(t)|\ \mathrm{d}t \tag{6}$$

$|\underset{\sim}{M}_i(t)| = min\{|M_i|, |N_i(t)|\}$ holds because, if charging points are not shared, an EV driver can only occupy an available CP of exactly one CP owner (e.g., her employer) that she is assigned to:

$$= \int_{-\infty}^{\infty} |\underset{\sim}{M}(t)|\ \mathrm{d}t \tag{7}$$

$|\underset{\sim}{M}(t)| = \sum_{i \in I} |\underset{\sim}{M}_i(t)|$ because assuming disjoint charging points $M(t) = \cup_{i \in I} M_i(t)$ holds.

$$= \int_{-\infty}^{\infty} \underset{\sim}{l}(t)\ \mathrm{d}t = \underset{\sim}{L} \tag{8}$$

□

According to the above proven theorem, the sharing of a number of CPs can be beneficial in a given infrastructure. The actual efficiency gains and related cost-savings per participant over time can differ depending on the actual values of M, N, I as well as M_i, N_i for each pooling participant, respectively. Despite the potential for higher usage, it can be assumed that tenants/users of a property will initially try to use exclusively their own self-installed charging points. They will consider sharing as an option, when bottlenecks of the exclusive usage of charging points become noticeable, e.g., frequently reported by frustrated EV drivers. A privileged EV-driver (e.g., a CEO of a company), who can exclusively use a particular parking space and the associated charging station, would not prefer to have that charging station included in the shared pool. However, in exceptional cases, e.g., due to failures, or maintenance work, such users could also benefit from the still functioning CPs of other entities. To ensure that some of the own charging points are always available for own purposes, a CP owner can decide to share only a subset of his charging points (see also the scenario in Figure 1).

5. Technical Feasibility of the Concept

The creation of a CP pool within an already existing, up and running EV-charging environment requires specific technical and non-technical tasks that have to be performed by each contributor. Figure 2 gives an overview of those tasks.

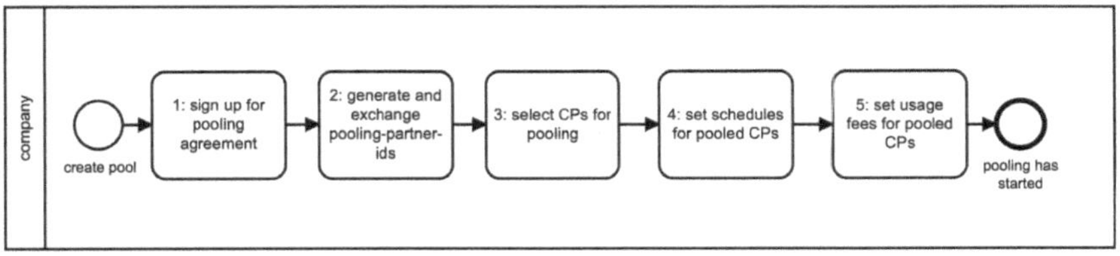

Figure 2. Main steps to initiate a pool of charging points.

As a first step (1), the owners of charging points can jointly negotiate and sign up for an agreement, e.g., in the form of a formal contract. The agreement can state how many or which charging points of a particular owner shall be added to the pool. It can also detail the usage conditions of pooled CPs and the respective parking lots. Examples are time-related aspects (validity period of the agreement, time intervals in which CPs are excluded from pooling), fees for using a pooled CP and the corresponding parking lot, payment modalities, and countermeasures in case of technical problems. In addition, the agreement can regulate operational aspects of the CP pool. For instance, the removal of CPs from the pool must be confirmed by each participant, or temporary access restrictions due to repair or maintenance work must be announced by the given CP owner in advance.

The operative management of pooled CPs can be carried out in a centralized or decentralized way. In a centralized setting, the agreement can explicitly name a dedicated entity (e.g., one of the pooling participants, or a contracted facility manager), who will take over responsibility for managing the CP pool. The managing entity processes all technical change requests concerning the pooled CPs and also helps onboard additional CP owners.

In a rather decentralized approach, the pooling-partners act more autonomously, retain administrative control and self-manage their own charging points. Consequently, they implement and run the CP pool in collaboration. Centralized pooling can reduce the efforts of individual CP owners by bundling all related tasks at one entity. Participating businesses must, of course, cover related costs and make sure that the managing party gets access to all necessary information and related systems. In multi-site environments (of larger companies), it is assumable that the CP pool at each site would be under control of a different (local) entity specific for that site, which can increase the overall complexity.

The main advantages of a decentralized management are increased flexibility and lower operating costs for the pooling participants. It enables each partner to control and monitor its own charging points without further interaction or negotiations with some central party. A continued self-management of processes and related data might also ease the handling of multi-site scenarios. This approach also fits better with the above-mentioned assumption that CP pooling is more of an evolutionary step for companies, which they will take after it has become apparent that their installed capacities are no longer sufficient to meet their demand.

To initially create the pool, in the above-mentioned decentralized way, each participant can generate a unique secret and share it with the pooling partners over a proper trusted channel (see step 2 in Figure 2). There are many possible ways to implement this step. As an example, a CP owner can generate a random number and store it in the form of a QR code, which can be scanned by one or more respective pooling partners during face-to-face meetings via smart phone. The secret code received from a CP owner empowers the receiving pooling partners to co-use the pooled charging points (and only those) of that particular CP owner. For instance, the code can be used as an API key to get access to relevant data in the IT system in a controlled way. In the next step, each pooling partner selects and configures the (subset of its) charging points that shall be part of the CP pool (3), determines associated schedules and time restrictions (4), in case these were agreed on. For example, a particular pooled charging point might not be co-used on weekdays between 8:00 a.m. and 7:00 p.m. Afterwards, the monetary aspects of sharing, i.e., usage fees for each pooled charging point, can be configured (5). The fee for the co-usage could contain a fix component to cover the costs of general operations (maintenance and repair), and a variable part to compensate actual expenses related to EV charging, including cost of electricity, parking fee, taxes, etc.

Following the Business Process Model and Notation (BPMN) standard, the diagram in Figure 3 shows how an EV driver can use a pooled charging point within a semi-public infrastructure. An EV driver, say a restaurant guest, who has never visited the restaurant before, wants to use a charging point close to the restaurant entry. As the restaurant participates in the pooling initiative (refer to Figure 1), the guest can reserve any of the pooled CPs (incl. the parking slot). The reservation can be implemented as part of the usual table reservation procedure, in which the guest's credit card number is also captured. Note that the guest does not need to know, whom the selected charging point actually belongs to. Upon arrival, the guest parks her EV and identifies herself by presenting her credit card at the reserved charging point. In accordance with the available capacity, the local system triggers power scheduling for the particular CP and the charging starts. As the guest actually uses another tenant's charging point, the restaurant receives a notification. After the dinner, the guest accepts the restaurant's friendly offer to take over the costs for using the parking slot and charging the car's battery, so that she only has to pay for the expensive dinner. As a consequence, the restaurant will have to pay the calculated pooling fee to the CP owner as it was stated in the sharing agreement and also configured in the system. Note that, in case the guest would have used the restaurant's own charging point, these costs could be significantly lower (mainly electricity). Thereafter, notifications can be sent to all three involved entities about the related expense claims. Normally, the guest would drive away, i.e., free up the parking slot and the charging point, within a few minutes. As a result of the cost transfer, there will be no additional fees charged on her credit card. Instead, the CP owner will pay the actual electricity costs for the charging (as part of his usual electricity bill), and the restaurant's owner will pay the calculated sharing fee to the CP owner, as a compensation. The latter transaction can take place immediately after the charging process ended. Alternatively, the system may also collect such costs and conduct a periodic clearing between pooling participants. Should the host not agree to take over the costs for its guest's EV charging, the guest would have to pay the respective sharing fee to the actual CP owner, in the example case via a credit card payment transaction. The basic possibility for rejecting the transfer of costs can be helpful to prevent misuse of the charging infrastructure. If the guest has not finished EV charging or blocks the parking lot longer than a certain period of time, the restaurant, as host, may receive a warning about the (meanwhile potentially much higher) sharing fee, so that it can refuse its generous offer. In this case, the EV driver can get a notification message and must pay the bill for the prolonged battery charging. In order to implement such measures against "unfriendly" drivers, additional proximity sensors at the charging points or a camera-based observation of the parking lots may be required.

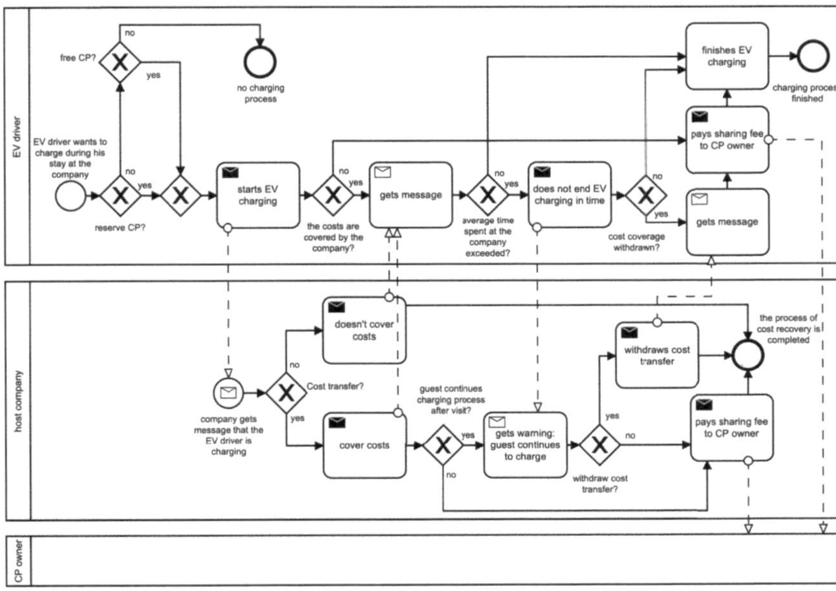

Figure 3. Example process: An external guest EV driver uses a pooled charging point.

The above described process can be significantly simpler, in case the EV driver is a known (permanent) user of the charging infrastructure and is assigned to a particular CP owner, e.g., her employer. The EV driver in this case can possess a personalized ID card. The card can be used for authentication at the company's own charging points and at every other pooled CP. The routing of related costs for "charge at work" to the employer's account can occur without any further interactions and notifications in a fully automated manner.

5.1. System Requirements

In order to run a pool of charging points in semi-public EV charging infrastructures, the corresponding software system must fulfil specific requirements, which we outline here (without claiming completeness):

- **Multi-tenancy:** Provide means of maintaining the data of multiple independent CP owners in one system allowing them to effectively monitor and control their part of the physical infrastructure at a given location in a safe and isolated way.
- **Autonomous administration:** Support the self-management of data about both shared and exclusively used charging points in the context of a given location. Enable CP owners to add/remove charging points to/from the pool, maintain potential time-based restrictions, pricing related information and other optional parameters concerning the shared usage of charging points.
- **Secure access control:** Manage the authentication and authorization of different end-users in case they try to use pooled or non-pooled charging points on site, or when they remotely access the system via an app or web-based user interface. Each CP owner has basically under its control whether and whom he or she might provide access to the charging points, since the overall infrastructure is considered to be "semi-public". An administrator representing a CP owner in the system shall be able to maintain

data that are needed to securely authenticate the internally known, registered end-users, such as related EV-drivers and maintenance staff. The system should allow the assignment of these users to the charging points of the given organization/tenant and manage related permanent credentials, such as personalized RFID badges. In addition, support the CP owner to grant external, previously unknown "guest" users (visitors, customers) temporary access and respective credentials to use the charging points for a limited duration or limited number of charging processes. The CP owner can send some secret (e.g., one-time password) upfront or hand out physical tokens (e.g., RFID cards) to its guests once they arrive. If the charging point supports the technology, previously unknown users can also be identified by using well-established systems, such as credit cards (via the NFC interface). In case an EV driver wants to use a pooled CP that does not belong to his organization/tenant, the system must check the user's presented credential against the authentication data of all tenants that participate in the CP pooling before rejecting the request. Should the user possess a valid credential issued by one of the pooling partners (whether permanent or temporary), the access must be allowed.

- **Reservation:** Offer charging point reservation capability for known, i.e., frequent or permanent users, and also for previously unknown ("guest") users. EV drivers shall be able to select and reserve each available charging point within the pool in addition to the visited host's own CPs. Various ideas and options of implementing a reservation system can be found in [31–33]. In case of yet unknown users, reservation must be coupled with an initial registration procedure in conjunction with the above-mentioned authentication mechanisms.
- **Cost and payment management:** Enable CP owners to maintain fix or variable fees for the co-usage of charging points. At the end of a charging process, ensure that all data items that are needed to calculate relevant expenses are captured correctly, such as consumed power (to calculate electricity costs), duration of charging (to calculate parking fees), etc. Ensure the correct routing of resulting pre-calculated expenses towards the right entities that are involved in the charging process (EV driver, host, actual CP owner) based on predefined rules and configuration. Implement interfaces to respective (external) rating, billing, invoicing, payment and clearing systems to further process actual payments and deliver related information in a secure way.
- **Power management:** Support the optimal utilization of available power capacities within the entire infrastructure [34,35], for example, in conjunction with a local energy management system. The overall system must fulfil the charging demands of EVs, while considering local power limits, total capacity and current state of charge of EV batteries, intended departure times and other relevant parameters. The pooled usage of CPs at a given location should not negatively impact the (previously established) load management system. Once a charging process has started, the power supply for the given charging point has to be managed independently from its membership in the pool.
- **Exception handling:** Implement measures to detect and handle exceptions in case of technical problems (e.g., when a pooled CP is out of order). Implement proper workflows and notifications to deal with the intentional misuse of pooled chargers, such as "long term parking".

5.2. Technical Details

In the context of a facility that hosts and serves multiple businesses, the facility owner may provide tenants with access to an integrated building management system (BMS). Such systems are broadly used today, for example, to control employees' access to floors and offices, book meeting rooms and co-working spaces, report problems, etc. BMS often allow the monitoring of energy flows in the entire facility. Procedures related to EV charging, including the option to self-manage and share own charging points with other co-users of the facility, could be implemented as integrated part of an overarching BMS. Another basic option, which is more common practice today, is to run a specialized system for charging infrastructure management and onboard facility tenants as soon as they install CPs.

To implement our CP-pooling concept and demonstrate its technical feasibility, the open-source system "Open E-Mobility" has been chosen as the technical foundation [30]. "Open E-Mobility" supports the management of EV charging equipment at multiple sites of organizations that operate EV fleets. It has been already deployed in several productive and trial environments. The system can connect to and work with charging stations (both AC and DC) of multiple vendors and provide the operator with real-time status information about ongoing charging processes. Thanks to an integrated smart charging capability, the system can monitor and adapt the use of charging points to optimize energy consumption, while protecting the local grid against overloading.

The software system is designed to be deployed as a scalable cloud application. The high-level architecture is shown on Figure 4. The internal business logic and processes are implemented by the backend server built in NodeJs. The data that are created and managed in the backend persist in the form of document collections in MongoDB. The datasets of multiple CP owners, who are also termed as "tenants" of the system (as this term is commonly used in cloud applications), are maintained in isolated collections. With the help of the front-end server and respective views of the web-based graphical user interface (GUI), each tenant/CP owner can model its EV-charging infrastructure at multiple sites by assigning charging points accordingly. User access to the system and the underlying data of single tenants is controlled by means of roles (e.g., "Admin" or "Basic") that can be assigned to users. Furthermore, a "SuperAdmin" can conduct overarching tasks, including the creation and removal of tenants within the system. The communication between connected charging stations and the backend server is based on the Open Charge Point Protocol (OCPP) using both HTTPS and secure WebSocket as transport options. Each CP owner can maintain information about its users helping to authenticate them at the charging points (e.g., via RFID badges issued by the CP owner) as well as in the web application (via password). In addition, a CP owner can monitor the status of ongoing charging sessions and collect comprehensive logging information about relevant events. End-users can also access the system via a mobile application, for example, to view available charging points at a site, to trigger charging, or to obtain notifications (in form of emails).

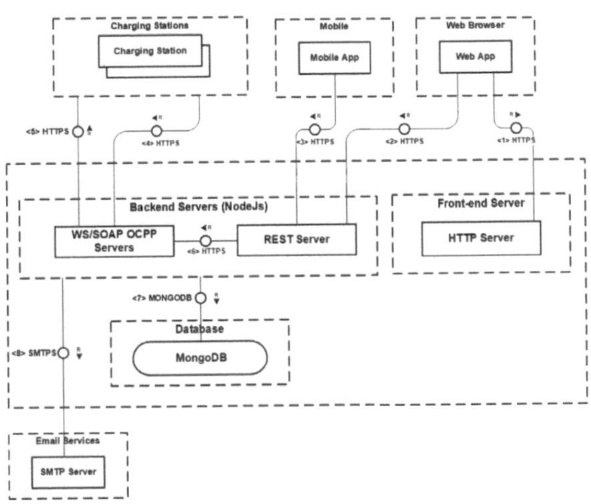

Figure 4. High-level system architecture of "Open E-Mobility". Reprinted with permission from [36]. 2022, SAP Labs France.

In the course of the prototype implementation, the publicly available source code of "Open E-Mobility" (see the current version on GitHub [36]) was extended, by adding required functionalities to the server, creating new and altering existing database objects and by modifying user interfaces.

In order to maintain a pool of charging points, each CP owner manipulates its own local data within the boundaries of its own isolated tenant environment. With other words, the pool is not explicitly created as a system-wide entity that some super-user could maintain. Instead, all information about the pool is stored in the participating tenant's datasets. For that purpose, the previously existing user role "Admin" has been extended by additional activities related to CP pooling that a given CP owner has to perform. Accordingly, a user assigned to this role is allowed to generate a pooling ID (specific per cloud tenant) which is encoded in the form of a QR code. The user can also add pooling partners by scanning QR codes that he received from another CP owners/tenants. In addition, the user can mark charging points as "pooled" and also remove them from the pool. The user is also capable of manually approving (or deny) a requested transfer of costs, if a guest used a pooled charging point. The user role "Basic" has not been enhanced by new software capability related to CP pooling. However, a user in this role (typically an EV driver) will see in her app all charging points of the default-assigned CP owner (employer, business partner, host). In addition, she can also see all other pooled CPs as basically usable charging points. This is achieved by checking the database to see which tenants have actually joined the pool and which of their charging points are flagged as shared. To achieve these steps efficiently, a new document collection is created in each tenant's database in which data about the pooling partners of the respective cloud tenant are stored. With the help of this information, the server can quickly determine which tenants' datasets it should inspect. In case the tenant has no pooling partners, the collection would be empty. In the collection that stores data about the charging stations of the given tenant, the Boolean property "pooled" is added to the documents.

In order to start the charging process at a particular charging point, the authentication request sent by the CP is first checked against the user data of the respective CP owner. In case the user/credential is unknown or not valid, the authentication would normally fail and the charging point would reject to start charging the EV. Following our pooling concept, if the user/credential is unknown to the directly targeted CP owner, the server would check against (potentially all) other tenants' respective data collections whether the presented credential is known and valid and notify the requesting charging point about the result finally. During this procedure, the server uses the above-mentioned mechanism to select and inspect candidate tenant's data. In the current prototype version, all users of the cloud tenant receive access to all pooled charging stations of each pooling partner equally. The enforcement of potential restrictions, such as exclusion time intervals, is not yet supported. For that purpose, further properties could be added to the pooling partner collection scheme.

Concerning the graphical user interface, a new view for the administration of pooling partners was added and assigned to the user role "Admin". The view contains options for adding and removing pooling partners and the generation of a QR code to be exchanged with potential pooling partners. In addition, the already existing detail view of charging points is enhanced by a check-box to add a given charging point to the pool.

In the example screenshot shown in Figure 5, both plugs (termed as *"Connector A"*, *"Connector B"*) of the same charging station (*"CS-ABB-00001"*) are added to the pool called *"Area1"*. To explicitly allow or deny the cost coverage for unregistered guests and to carry out related monetary transactions once the charging has finished, a switch button *"Pay for guest"* as shown in Figure 6 was added to the detail view about ongoing charging sessions.

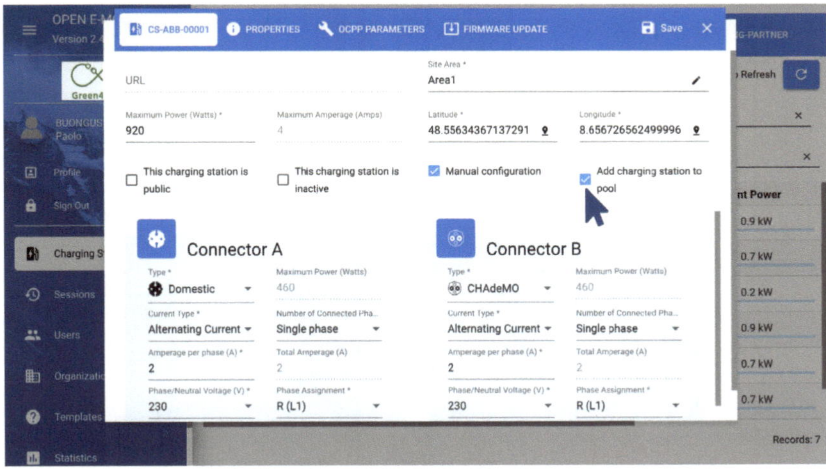

Figure 5. Example screenshot for adding charging points to an existing pool.

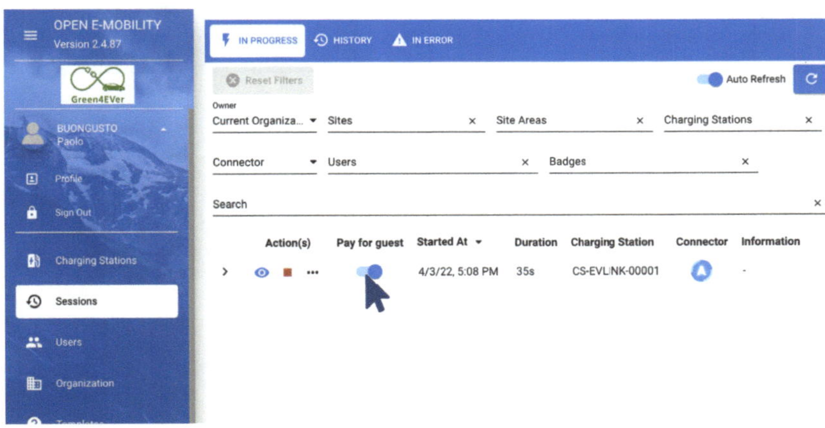

Figure 6. Example screenshot to initiate the transfer of costs of charging in the system.

The switch is visible for the end user together with other information about a running session (see "*IN PROGRESS*" in Figure 6). During the EV charging process, i.e., before the session terminates, an authorized user can turn on the switch. In that case, the backend server can set the respective tenant's account data to which related costs can be invoiced later. In an example, a restaurant owner could use the switch to cover the costs of EV charging, while guests are enjoying their dinner.

Note that the current implementation is not yet capable of preparing and launching actual monetary transactions with payment systems that are otherwise supported by "Open E-Mobility".

6. Desirability Survey with Potential End-Users

In order to capture and understand the preferences of potential end-users with regard to the sharing of charging stations, we reached out to companies in a small German town. In total, 46 companies were actively involved, including retailers (~30%), service providers (~24%), manufacturers (~10%), and others. They responded to an online questionnaire, which contained closed-ended questions asking the respondents to choose from a list of options. In addition to the online survey, we have also conducted personal interviews with four selected retailers and with a restaurant owner. In the following, we summarize the results based on the questionnaire as well as the most important findings of the interviews.

6.1. Results of the Online Survey

One of the main objectives of the survey was to find out under what economic and technical conditions companies would share their existing or yet to be installed charging points. The corresponding questions, response options and the number of respondents, who have chosen the respective option, are shown in Table 1.

Table 1. Summary of company preferences.

Question	Options	#
Would you pay the cost of charging for external guests?	Yes	1
	No	15
	Maybe	27
	No response	3
Whom would you offer EV charging for free?	Business partners	10
	Customers	19
	Employees	22
	Suppliers	1
How should EV drivers interact with pooled CPs?	Guest-card	6
	QR-Code	11
	E-roaming	2
	No preference	27
Which means of reservation would you prefer?	Time based	8
	24h privilege	5
	Equal rights	13
	No preference	20

Considering external guest users, e.g., the customers/visitors of other companies, only one respondent would cover the costs of EV charging (electricity, parking fees, taxes) without further restrictions. One-third of the survey respondents would definitely refuse to pay for external users' consumption.

The willingness to allow the usage of a company's CPs for free is significantly higher, if the respective EV driver has a business relationship to the company (business partner, customer, employee). However, only one respondent considered this option specifically for suppliers.

With regard to user interaction with the pooled CPs, each indicated response option (guest card, QR code, E-roaming) was illustrated and explained by a short process description, incl. the handling of related payments. As shown in Table 1, the majority (27) of respondents have not expressed any preference. The most popular option is to scan a QR code at the charging point in conjunction with a web application that manages further steps. Allowing the usage of roaming platforms (e.g., Gireve or Hubject) via the ID cards of supported e-mobility service providers was preferred by two companies.

Six of the respondents would prefer to hand out guest cards to authenticate their customers at the accessible charging points before the charging process would start.

The results also show that most survey participants (20) had no preference, whether and how EV drivers can reserve pooled charging points.

A relatively high number (13) were in favor of allowing CP reservations for all types of users equally. The scenario termed as "timed-based reservation", which received eight votes, allows CP owners to restrict the reservation and usage of their pooled CPs by others to specific time intervals. Five respondents would keep exclusive reservation rights of their own charging points up to 24 h prior to the intended usage.

Survey results indicate that the shared usage of charging points is basically an interesting application for the involved companies. As survey participants do not prefer a particular technology or method, the charge point management system should offer multiple options to provide CP sharing functionality.

Note that the survey was conducted with a limited number of participants. They represent the business domain in a relatively small German town, where only a few charging points are in operation. Therefore, the responses and opinions should not be considered representative.

6.2. Personal Interviews with Business Owners

To further evaluate the developed concept of pooling charging points with other companies and to gather feedback on the prototype implementation, we carried out qualitative face-to-face interviews with five selected business owners. Thereby, we explained them the basic idea using a simplified version of the example scenario shown in Figure 1 comprising only two tenants (a restaurant and a physician). Note that these example businesses normally have less overlapping opening hours, making CP pooling especially advantageous for them. To gain comparable results from the interviews and reduce bias, a common presentation was used. Nevertheless, the slightly deviating wording of the questions in the individual interview sessions may have influenced the answers given by the respondents. As part of the presentation, we created a pool of CPs by using the software system and asked participants to experiment with the user interface during the interview sessions.

Overall, the concept of sharing charging points is considered useful by the interviewed business owners. In their view, it might be especially feasible for business owners, who know and trust each other, and have common customers. However, two of the interviewees commented that it was not easy for them to empathize with the challenges associated with EV charging because they do not yet own/use EVs themselves. The owner of a restaurant stated the relevance of the actual business location as potential obstacle: His restaurant is located outside of the town, and there are no other companies as potential pooling partners within walking distance. This feedback underpins that CP pooling is in general more beneficial in dense urban areas, where potential pooling partners are in close proximity to each other and the number of charging points per company is rather limited. Nevertheless, the restaurant owner is already looking for a practical solution to make his CPs available for selected guests as an additional service.

Since there are currently only a few charging points operated in town, one of the retailers suggested to rather establish charging points at the publicly accessible parking areas in the city center. In his opinion, it would be more meaningful than equipping the short-term parking spots at the companies' locations. Another retail manager would install charging points as prestige objects for his business and make the CPs exclusively available for his customers and employees. In contrast to this strategy, another interviewee values sharing as a "real bonus" for his customers.

These relatively diverse attitudes and approaches to the shared usage of charging stations might converge in the future. We assume that companies will demand additional charging points (especially at peak times during their specific business hours), if they cannot expand their own infrastructure anymore due to local limitations. Sharing CPs can be highly beneficial for companies with different business/opening hours. It was also confirmed by an owner of a retail store, who already shares expensive parking spots (currently without installed EV chargers) with a nearby restaurant.

The feedback given on the user-friendliness of the prototype system was overall positive. The interviewed business owners particularly highlighted the importance of a simple, user-friendly design and the support of mobile devices (especially for EV-driving employees). They suggested minor improvements concerning the GUI, such as the simple selection of multiple charging points when the CP pool is initialized (instead of adding each CP separately).

Regarding the practical usability of CP pools, interview participants asked several questions that are very valuable for improving the system. Two example questions were: How could a first-time customer know whether or not the cost of charging would be covered by the respective business? How could such customers know which charging points are already reserved? These communication-related problems can be solved, if EV drivers are willing to install and use a specific mobile application that visualizes the status of CPs (e.g., available, occupied, reserved) within the local charging infrastructure. Most business owners consider the usage of such a dedicated app for their customers (with their words) "annoying". In case a business refuses the usage of a provided mobile app, the EV-drivers could receive static information only. For example, signs, stickers, or printed instructions placed at the charging points could indicate whether a CP is part of a pool. Only one interview partner assumes that EV-drivers would be willing to install a specific app for this purpose. One business owner suggests to cooperate with an e-roaming provider that already enables its registered EV drivers to search for and reserve available CPs (at publicly accessible parking/charging locations).

7. Conclusions and Future Work

In this paper, we described main results of the conceptual phase of our ongoing research project. We presented the basic idea to establish and run shared EV-charging infrastructures in the exemplary context of commercial real-estate facilities. Collaborating EV-charger owners, typically companies that are using the same facility (office building, industrial area), jointly create, manage and use a pool of EV chargers. A pooled charger can basically be used by all employees, business partners, customers, guests, etc. of the pooling participants. In our work, we have been considering aspects of economic viability, technical feasibility and desirability in order to create a successful solution.

The economic viability of the solution is of interest for both the solution vendor and the companies that use it. Our related investigation focused more on the users' potential economic benefits. We have shown that CP sharing can help improve the overall utilization of a particular charging infrastructure. As a consequence of a higher utilization, involved companies could reduce investments in EV charging equipment because the chargers of others can substitute self-installed chargers to a certain degree. It is similar for operational costs (inspections, maintenance and repair work) that are in general proportional to the number of installed chargers. From the viewpoint of individual EV drivers, the possibility to use a (larger) shared pool of chargers will increase the likelihood to find an available charging point, e.g., when arriving at work. This is especially to the benefit of the "average" driver, who has no dedicated parking slot with an exclusively accessible charger. To ensure that some of the own charging points are always available for business-critical purposes, e.g., to charge delivery vans, service cars or shuttles, a company can decide to share only a subset of its installed charging points. Applying additional, e.g., temporal restrictions for the shared CPs can also help avoid unwanted situations. The proper selection and optimization of these parameters is a complicated task for companies. It is difficult to predict the demand for available charging stations at a given time, as it depends on the sometimes volatile behavior of their own and external users. We plan to address this problem as part of our future work by running simulations based on random generated and, if available, real datasets, incl. EV departure and arrival times, batteries' state of charge, etc. In addition to improving the infrastructure's utilization, the sharing of CPs can help cover the increased demand for charging capacity caused by exceptional events, such as a wedding party in a restaurant that does not own a sufficient number of chargers to serve all guests.

Despite the above-mentioned advantages, further efforts must be made to validate the profitability and feasibility of CP pooling. It can be yet assumed that in the corporate context the purely economic interest in sharing charging points with other companies would be a less important motivating factor. However, improving customer satisfaction (through the high availability of charging points) can lead to a measurable positive financial impact in specific business areas, such as gastronomy, retail and hotels. National and international regulations, laws and taxation rules can heavily influence whether real-estate investment and management corporations—as building and facility owners—would invest into e-mobility equipment in future. Our current findings show that, at least in Germany, they mostly install no or only a few charging points in new buildings and related parking areas, but allow users/tenants to build and operate own equipment if needed. As of today, it is therefore less likely that a building or facility owner would provide its tenants with access to a software system to manage their own charging points. Without such a commonly used system, however, the creation and operation of a shared pool of CPs can be a challenging task.

In order to explore and understand potential users' preferences, we conducted an online survey with 46 company representatives in a German town. Five of them have also participated in in-depth personal interviews. Results of the survey showed that companies have different opinions and diverse requirements regarding the shared use of EV chargers in future. A technical solution should therefore be as flexible as possible to meet individual needs. Accordingly, we specified and modelled important processes, and formulated technical requirements that a management system should fulfil. The technical feasibility of the concepts has been successfully proven through the development of a software prototype based on the open source system "Open E-Mobility". The required modifications and extensions, especially regarding elements of the existing user interface, led to affordable development efforts. The system was tested with regard to the functional correctness of the software and the user interface. It has also been used in the above-mentioned face-to-face interviews and received positive feedback from the interviewees.

The prototype was so far only used to validate the concept and gather qualitative feedback of potential users for evaluation purposes. As of today, our approach to share charging points between independent business entities has not yet been tested in a real, productive environment. The current lack of quantitative test results gathered in a real-world setup motivates further activities. Together with our project partners, we are preparing a field test in the context of a large office building under realistic conditions. It is planned to actively involve the employees of several companies that are co-using the facility. EV drivers will thereby jointly use a "sandbox" charging infrastructure on site. The tests will run for several months helping to observe usage patterns and to document related energy consumption, infrastructure efficiency, costs and other relevant factors.

Author Contributions: Conceptualization, Z.N.; methodology, J.G., S.G., Z.N.; software, J.G.; validation, J.G., S.G.; formal analysis, J.G., Z.N.; investigation, J.G., S.G.; resources, J.G., S.G.; data curation, J.G., S.G.; writing—original draft preparation, J.G., S.G., Z.N.; writing—review and editing, Z.N.; visualization, J.G., S.G.; supervision, Z.N.; project administration, Z.N.; funding acquisition, Z.N. All authors have read and agreed to the published version of the manuscript.

Funding: As part of the research project "Green4EVer" [37] this work has been funded by the German Federal Ministry for Economic Affairs and Climate Action (BMWK).

Institutional Review Board Statement: Not applicable.

Informed Consent Statement: Informed consent was obtained from all subjects involved in the study.

Data Availability Statement: Not applicable.

Acknowledgments: We especially thank our project partners and colleagues at Art-Invest Real Estate, Webasto, and SAP for their kind support and expertise provided. In addition, we express our gratitude to Timo Kuch for his assistance with the mathematical formulations.

Conflicts of Interest: The authors declare no conflict of interest.

Abbreviations

The following abbreviations are used in this manuscript:

API	Application Programming Interface
BEV	Battery Electric Vehicle
BMS	Building Management System
BPMN	Business Process Model and Notation
CP	Charging Point
EV	Electric Vehicle
GUI	Graphical User Interface
HTTPS	Hypertext Transfer Protocol Secure
NFC	Near Field Communication
OCPP	Open Charge Point Protocol
PHEV	Plug-in Hybrid Electric Vehicle
RFID	Radio Frequency Identification
TCO	Total Cost of Ownership

References

1. Elektromobilität: Öffentliche Ladeinfrastruktur. Available online: https://www.bundesnetzagentur.de/DE/Fachthemen/ElektrizitaetundGas/E-Mobilitaet/start.html (accessed on 23 March 2022).
2. Kraftfahrt-Bundesamt. Anzahl der Elektroautos in Deutschland von 2011 bis 2022. Available online: https://de.statista.com/statistik/daten/studie/265995/umfrage/anzahl-der-elektroautos-in-deutschland/ (accessed on 23 March 2022).
3. Appunn, K. Next, German Government's Key Climate and Energy Plans in 2021 Coalition Agreement. Available online: https://www.cleanenergywire.org/factsheets/future-german-governments-key-climate-and-energy-plans-2021-coalition-treaty (accessed on 7 April 2022).
4. Alatise, O.; Karlsson, A.; Deb, A.; Wu, R.; Ortiz-Gonzalez, J. Expanding EV Charging Capacity in Distribution Networks: A Case Study for Charging EVs at Work. In Proceedings of the CIRED 2021—The 26th International Conference and Exhibition on Electricity Distribution, Online, 20–23 September 2021; IET: Stevenage, UK, 2021; pp. 2945–2949. [CrossRef]
5. Windt, A.; Arnhold, O. *Ladeinfrastruktur Nach 2025/2030: Szenarien für den Markthochlauf*; Nationale Leitstelle Ladeinfrastruktur: Berlin, Germany, 2020.
6. Nicholas, M.; Wappelhorst, S. Regional Charging Infrastructure Requirements in Germany through 2030. 2020. Available online: https://theicct.org/sites/default/files/publications/germany-charging-infrastructure-20201021.pdf (accessed on 7 April 2022).
7. Ghosh, A.; Aggarwal, V. Control of charging of electric vehicles through menu-based pricing. *IEEE Trans. Smart Grid* **2017**, *9*, 5918–5929. [CrossRef]
8. Tao, J.; Huang, D.; Li, D.; Yang, X.; Ling, C. Pricing strategy and charging management for PV-assisted electric vehicle charging station. In Proceedings of the 2018 13th IEEE Conference on Industrial Electronics and Applications (ICIEA), Wuhan, China, 31 May–2 June 2018; IEEE: Piscataway, NJ, USA, 2018; pp. 577–581.
9. Dai, Y.; Qi, Y.; Li, L.; Wang, B.; Gao, H. A dynamic pricing scheme for electric vehicle in photovoltaic charging station based on Stackelberg game considering user satisfaction. *Comput. Ind. Eng.* **2021**, *154*, 107117. [CrossRef]
10. Detzler, S.; Eichhorn, C.; Karnouskos, S. Charging optimization of enterprise electric vehicles for participation in demand response. In Proceedings of the 2015 International Symposium on Smart Electric Distribution Systems and Technologies (EDST), Vienna, Austria, 8–11 September 2015; IEEE: Piscataway, NJ, USA, 2015; pp. 284–289. [CrossRef]
11. Frendo, O.; Karnouskos, S.; Gaertner, N.; Kipouridis, O.; Rehman, K.; Verzano, N. Charging Strategies and Implications for Corporate Electric Vehicle Fleets. In Proceedings of the Proceedings IEEE 16th International Conference on Industrial Informatics (INDIN), Porto, Portugal, 18–20 July 2018; IEEE: Piscataway, NJ, USA, 2018; pp. 466–471. [CrossRef]
12. Frendo, O.; Graf, J.; Gaertner, N.; Stuckenschmidt, H. Data-driven smart charging for heterogeneous electric vehicle fleets. *Energy AI* **2020**, *1*, 100007. [CrossRef]
13. Frendo, O.; Gaertner, N.; Stuckenschmidt, H. Improving Smart Charging Prioritization by Predicting Electric Vehicle Departure Time. *IEEE Trans. Intell. Transp. Syst.* **2021**, *22*, 6646–6653. [CrossRef]
14. Sepetanc, K.; Pandzic, H. A Cluster-Based Model for Charging a Single-Depot Fleet of Electric Vehicles. *IEEE Trans. Smart Grid* **2021**, *12*, 3339–3352. [CrossRef]
15. Caiazzo, B.; Coppola, A.; Petrillo, A.; Santini, S. Distributed Nonlinear Model Predictive Control for Connected Autonomous Electric Vehicles Platoon with Distance-Dependent Air Drag Formulation. *Energies* **2021**, *14*, 5122. [CrossRef]
16. Wolbertus, R.; Gerzon, B. Improving electric vehicle charging station efficiency through pricing. *J. Adv. Transp.* **2018**, *2018*, 4831951. [CrossRef]
17. Nürnberg, M.; Iwan, S. Application of Telematics Solutions for Improvement the Availability of Electric Vehicles Charging Stations. In *Communications in Computer and Information Science, Proceedings of the Development of Transport by Telematics, Jaworze, Poland, 27 February–2 March 2019*; Mikulski, J., Ed.; Springer International Publishing: Cham, Switzerland, 2019; pp. 287–301.

18. Luri Minami, A.; Ramos, C.; Bruscato Bortoluzzo, A. Sharing economy versus collaborative consumption: What drives consumers in the new forms of exchange? *J. Bus. Res.* **2021**, *128*, 124–137. [CrossRef]
19. D'Acierno, L.; Tanzilli, M.; Tescione, C.; Pariota, L.; Di Costanzo, L.; Chiaradonna, S.; Botte, M. Adoption of Micro-Mobility Solutions for Improving Environmental Sustainability: Comparison among Transportation Systems in Urban Contexts. *Sustainability* **2022**, *14*, 7960. [CrossRef]
20. Belk, R. You are what you can access: Sharing and collaborative consumption online. *J. Bus. Res.* **2014**, *67*, 1595–1600. [CrossRef]
21. Cornée, S.; Le Guernic, M.; Rousselière, D. Governing Common-Property Assets: Theory and Evidence from Agriculture. *J. Bus. Ethics* **2020**, *166*, 691–710. [CrossRef]
22. Webasto. Webasto ChargeConnect App Quick Start Guide—EN. Available online: https://dealers.webasto.com/service/files_content.aspx?Files_FileId=27392 (accessed on 7 April 2022).
23. Betzing, J.H.; von Hoffen, M.; Plenter, F.; Chasin, F.; Matzner, M.; Becker, J. One Plug at a Time—Designing a Peer-to-Peer Sharing Service for Charging Electric Vehicles. In Proceedings of the Towards Thought Leadership in Digital Transformation: 13. Internationale Tagung Wirtschaftsinformatik, WI 2017, St. Gallen, Switzerland, 12–15 February 2017.
24. Plenter, F.; Chasin, F.; von Hoffen, M.; Betzing, J.H.; Matzner, M.; Becker, J. Assessment of peer-provider potentials to share private electric vehicle charging stations. *Transp. Res. Part D Transp. Environ.* **2018**, *64*, 178–191. [CrossRef]
25. Lee, J.H.; Chakraborty, D.; Hardman, S.J.; Tal, G. Exploring electric vehicle charging patterns: Mixed usage of charging infrastructure. *Transp. Res. Part D Transp. Environ.* **2020**, *79*, 102249. [CrossRef]
26. Huang, X.; Lin, Y.; Lim, M.K.; Zhou, F.; Ding, R.; Zhang, Z. Evolutionary dynamics of promoting electric vehicle-charging infrastructure based on public–private partnership cooperation. *Energy* **2022**, *239*, 122281. [CrossRef]
27. Simolin, T.; Rauma, K.; Viri, R.; Mäkinen, J.; Rautiainen, A.; Järventausta, P. Charging powers of the electric vehicle fleet: Evolution and implications at commercial charging sites. *Appl. Energy* **2021**, *303*, 117651. [CrossRef]
28. Gaikwad, S.; Mehta, H. Investigation of effects of increasing EV penetration on distribution transformers in Modi Ganpati Area, Pune. In Proceedings of the 2020 IEEE Electric Power and Energy Conference (EPEC), Edmonton, AB, Canada, 9–10 November 2020; IEEE: Piscataway, NJ, USA, 2020; pp. 1–5. [CrossRef]
29. Meinel, C.; Leifer, L.; Plattner, H. *Design Thinking*; Springer: Berlin/Heidelberg, Germany, 2011. [CrossRef]
30. Open E-Mobility. Available online: https://open-e-mobility.fr/ (accessed on 7 April 2022).
31. Basmadjian, R.; Kirpes, B.; Mrkos, J.; Cuchý, M.; Rastegar, S. An interoperable reservation system for public electric vehicle charging stations: A case study in Germany. In Proceedings of the Proceedings of the 1st ACM International Workshop on Technology Enablers and Innovative Applications for Smart Cities and Communities, New York, NY, USA, 13–14 November 2019; pp. 22–29.
32. Basmadjian, R.; Kirpes, B.; Mrkos, J.; Cuchý, M. A Reference Architecture for Interoperable Reservation Systems in Electric Vehicle Charging. *Smart Cities* **2020**, *3*, 1405–1427. [CrossRef]
33. Orcioni, S.; Buccolini, L.; Ricci, A.; Conti, M. Electric Vehicles Charging Reservation Based on OCPP. In Proceedings of the 2018 IEEE International Conference on Environment and Electrical Engineering and 2018 IEEE Industrial and Commercial Power Systems Europe (EEEIC/I CPS Europe), Palermo, Italy, 12–15 June 2018; pp. 1–6. [CrossRef]
34. Zhang, Y.; You, P.; Cai, L. Optimal charging scheduling by pricing for EV charging station with dual charging modes. *IEEE Trans. Intell. Transp. Syst.* **2018**, *20*, 3386–3396. [CrossRef]
35. Huang, J.; Gupta, V.; Huang, Y.F. Scheduling algorithms for PHEV charging in shared parking lots. In Proceedings of the 2012 American Control Conference (ACC), Montreal, QC, Canada, 27–29 June 2012. [CrossRef]
36. Reprinted with Permission from SAP Labs France, Open e-Mobility, 2022. Available online: https://github.com/sap-labs-france/ev-server/blob/master-qa/archi/tam-model.png (accessed on 7 April 2022).
37. Green4EVer. Project Website. Available online: https://www.h-ka.de/en/iaf/green4ever (accessed on 28 September 2022).

World Electric Vehicle Journal

MDPI

Article

Electrification of Vehicle Miles Traveled and Fuel Consumption within the Household Context: A Case Study from California, U.S.A.

Ahmet Mandev [1,*], Frances Sprei [1] and Gil Tal [2]

1 Department of Space, Earth and Environment, Chalmers University of Technology, 412 96 Gothenburg, Sweden
2 Institute of Transportation Studies, University of California, 1590 Tilia Street, Davis, CA 95616, USA
* Correspondence: mandev@chalmers.se

Abstract: Plug-in electric vehicles (PEVs), consisting of battery electric vehicles (BEVs) and plug-in hybrid electric vehicles (PHEVs), combined with the decarbonization of the electricity sector, can significantly help reduce greenhouse gas emissions in the transport sector. This study used empirical data from 287 households with at least one plug-in electric vehicle in California between 2016 and 2020. We estimated electric vehicle miles traveled (eVMT), fuel consumption and utility factor at the household level, i.e., taking into consideration all vehicles. We also studied the effect of household-specific factors—such as frequency of overlaps between vehicles, frequency of charging and frequency of long-distance trips—on eVMT, utility factor and fuel consumption within two-car households. Our results indicate that PHEVs with a range of at least 35 miles have the potential to electrify a similar share of total household miles as some short range BEVs, or can reach up to 70% as much electrification as some long range BEVs and, thus, can play an important role in decarbonizing the transport sector.

Keywords: plug-in electric vehicles (PEVs); eVMT; utility factor; charging behavior

Citation: Mandev, A.; Sprei, F.; Tal, G. Electrification of Vehicle Miles Traveled and Fuel Consumption within the Household Context: A Case Study from California, U.S.A. *World Electr. Veh. J.* **2022**, *13*, 213. https://doi.org/10.3390/wevj13110213

Academic Editor: Joeri Van Mierlo

Received: 7 October 2022
Accepted: 10 November 2022
Published: 15 November 2022

Publisher's Note: MDPI stays neutral with regard to jurisdictional claims in published maps and institutional affiliations.

1. Introduction

1.1. Motivation and Background

The use of plug-in electric vehicles (PEVs), split into two main categories, battery electric vehicles (BEVs) and plug-in hybrid electric vehicles (PHEVs), combined with the decarbonization of the electricity sector, can help reduce greenhouse gas emissions in the transport sector [1–8]. BEVs use an electric drive train for propulsion and their only source of energy is their rechargeable battery packs. PHEVs, on the other hand, combine an electric drive train with a conventional one, which leads to a more complex analysis of fuel economy compared with BEVs and internal combustion engine vehicles (ICEVs) [9]. In order to assess the potential to reduce greenhouse gas emissions from these vehicles, it is important to understand the share and number of miles that can be electrified. Traditionally, this has been performed on a per vehicle basis, however the household context is also important to consider. For example, trips can be shifted between vehicles, increasing or decreasing the electrified miles. However, this has not been previously studied enough, partly due to the difficulty of collecting good empirical data at the household level. Our contribution to the literature and the aim of this paper is to fill this gap by analyzing measures related to electric vehicle miles traveled in the household context. We based our analysis on a case study of 287 households in California.

Utility factor (UF) of a PHEV—defined as the share of electrified vehicle miles traveled (eVMT) within total vehicle miles traveled (VMT)—is the most common metric in analyzing the performance of PHEVs in order to understand the degree to which they provide emission-free travel. Previous literature on the utility factor focuses either solely on the PEV or ignores the other conventional vehicles in the household. The literature on multi-car households with PEVs is also limited due to the difficulty of collecting (and, therefore, lack of) empirical data on multi-car households. See [10–16]. To our knowledge, there has not been any assessment of the UF at the household level—as the share of eVMT within total household miles traveled—that takes into account all vehicles in a household and captures the overall household electrification of miles. The goal of this paper is to fill this gap and assess the UF within the household context, and investigate how household factors impact electrification of vehicle miles traveled and fuel consumption, using an empirical data set. We used data from 650 vehicles which had onboard loggers that recorded driving and charging data for a year, distributed among 287 households where each household had a PEV. Our dataset included BEVs along with PHEVs, and the UF of the BEVs is one by default since all of their traveled miles are electrified. On the household level, UF can be calculated for BEV-households—since each household owns only one PEV, they can be classified as either a BEV or PHEV household—and this enabled us uniquely to compare the UFs of BEV-households and PHEV-households.

The UF of a PHEV encapsulates the eVMT and VMT of that PHEV; similarly, the UF of a household encapsulates the eVMT of the PEV of that household and the VMT of that household. Therefore, in this paper, we use the following metrics regarding electrification of vehicle miles traveled: eVMT of the PEV, VMT of the PEV, VMT of the household, UF of the PHEV, and UF of the household. We define the household context using three categories: (1) PEV technology in the household, (2) household vehicle usage, and (3) ICEVs in the household. For each of these categories, we identify relevant factors and the corresponding variables in our data set. Then, we use descriptive statistics to explain how the most salient factors impact our metrics. This is followed by a regression analysis for each of our metrics where we investigate the statistical significance of the identified factors. In addition, we also analyze the household fuel consumption and perform a regression analysis on how these factors impact fuel consumption in: (1) all households, (2) PHEV households, and (3) BEV households. For our analysis, we focus mostly on two-car households—which is most common in our dataset, with 64% of households being two-car households, and also the largest group of vehicle owners in California—for a more standard comparison between different PEV-type households.

The outline of the paper is as follows: An overview of previous literature regarding the utility factor is presented in Section 1.2. In Section 2, the data and methods are described, followed by the results in Section 3. We close with a discussion and conclusions in Section 4.

1.2. Previous Literature

There are two main approaches to assessing UF in the literature. The first approach is to run simulations based on test-cycles or transportation surveys. In this approach, the UF is calculated under certain assumptions regarding the charging frequency, vehicle characteristics and driver characteristics. It is also common practice to follow the standardized methods from the Society of Automotive Engineers, SAE J2841 and SAE J1711, to calculate the UF [17,18]. Elgowainy et al. [19] used the U.S. National Household Transportation Survey (NHTS) to report on the UFs of PHEVs with an all-electric-range (AER) of 10, 20, 30, 40 and 60 miles. Axsen et al. [20] used survey data from 877 Californian new vehicle buyers and estimated the UF of PHEVs with an AER of 20 and 40 miles under different emission scenarios. Tal et al. [21] used a self-reported web map survey with a sample size of 800 Prius plug-ins and 600 Chevrolet Volts to study charging behavior. Moawad et al. [22] used a test cycle—measured by the U.S. Environmental Protection Agency (EPA), in which they collected driving statistics from 100 drivers in Kansas City for the duration of a day—to estimate the decrease in the consumed fuel of PHEVs with a battery capacity of 4, 8, 12 and 16 kWh, compared with ICEVs. Björnsson et al. [23] used GPS data logged for a representative sample of individual conventional vehicles in private use, to analyze how the utility factor is influenced by the choice of objective function when determining optimal battery sizes for PHEVs. Standardized methods that rely heavily on assumptions were criticized for not accounting for complex scenarios. SAE J2841, for instance, is based on the assumptions that each vehicle starts the day fully charged, does not charge until after the last trip of the day and only charges once a day; it also assumes that PHEVs are driven in the same patterns as national average vehicles [24,25]. Bradley and Quinn [24] investigated how different assumptions in these standards would result in different UF calculations, and they found that UF calculations were very sensitive to assumptions regarding charging behavior, vehicle age and vehicle annual distance driven.

The second approach is to use empirical, real-world data to estimate the UF, which provides an insight into actual travel behavior patterns. Plötz et al. [26] and Plötz et al. [27] used publicly available real-world driving data from two online sources, voltstats.net and spritmonitor.de, to estimate the UFs for several PHEVs in the U.S., Canada and Germany. Ligterink et al. [28] analyzed the fueling and charging data of plug-in vehicles, collected from lease companies in the Netherlands, and estimated the UF for the Dutch plug-in fleet. Davies and Kurani [29] analyzed the data from 25 converted Toyota Prius which had onboard loggers to record driving and charging data; they compared the observed UF with the simulated UF of a scenario that accounted for additional workplace charging, and concluded that additional workplace charging can result in a relatively higher UF. Idaho National Laboratory, in their EV project, analyzed the driving and charging data of 21,600 vehicles in the U.S., obtained from several OEMs, and reported their estimated eVMT and VMT [30]. Nicholas et al. [31] used a subset of the dataset we used in this paper—which contained logger data placed on vehicles for a year—and reported on the UF of Toyota Prius, Ford Energi and Chevrolet Volt. Hao et al. [32] studied the actual UF of seven Chinese cities using PHEV driving big data, concluding that PHEVs with an all-electric-range of over 50 miles would lead to better energy saving and emission reduction potential. Srinivasa Raghavan and Tal [33] and Raghavan and Tal [34] used multi-year longitudinal data from PHEVs in California and studied charging and driving patterns. Plötz et al. [35] analyzed the real world utility factor of over 1000 PHEVs in Germany. Tal et al. [36] analyzed vehicle usage and refueling behavior in California using logger data, surveys and interviews.

In summary, existing studies on the utility factor lack the household context, often ignoring multi-car households and focus solely on the PEV. Here, we fill this gap with an analysis of the UF within the household context, and investigate how household factors impact electrification of vehicle miles traveled and fuel consumption, using empirical data from 287 households with 650 vehicles in total.

2. Data and Methods

The data we used was from the Advanced Plug-in Electric Vehicle Travel and Charging Behavior Project which aims to provide an insight into how plug-in electric vehicles are used on a day to day basis within the household travel context, by placing data loggers in participant households for one year [31]. The project was initiated by the Electric Vehicle Research Center at the University of California. Data was collected from summer 2015 to summer 2020, in California, U.S.A., from 287 households, by placing a monitor in all household vehicles driven more than 1000 miles per year. Participating households were selected, in consultation with the California Air Resource Board, to fit an appropriate sampling of the population. Odometer readings were taken from cars that were driven less than 1000 miles per year. Each household owned only one PEV. Among the 287 households, 20 had a Toyota Plug-in Prius, 45 had a Ford C-Max/Fusion Energi, 25 had a Toyota Prius Prime, 23 had a Chrysler Pacifica, 71 had a Chevrolet Volt, 43 had a Nissan Leaf, 23 had a Chevrolet Bolt, and 37 had a Tesla Model S. Including the conventional vehicles in the households, the dataset has 650 vehicles in total. Nissan Leaf, Chevrolet Bolt and Tesla Model S were the BEVs in the data set, and the rest were PHEVs. The model years for the PEVs in the dataset ranged from 2012 to 2019. The dataset also included an extensive survey with the PEV owners prior to the placement of the monitors.

2.1. Provided Datasets

The raw data collected from the loggers and through the survey was cleaned by the Electric Vehicle Research Center. We used two main sets of data in our analysis: trip data and charging data.

2.1.1. Trip Data

Trip data consisted of each single trip by the logged vehicle. The separating factor between trips was that the car remained at the same position idly with a speed of zero for at least 5 min. The dataset provided information regarding the start time and duration of the trip, the total distance traveled, and fuel consumption during the trip. It also included the electric vehicle miles traveled (eVMT) and gasoline vehicle miles traveled (gVMT) for each single trip. The calculation of eVMT and gVMT were performed by the Electric Vehicle Research Center. eVMT calculation methods do not differ much from each other, as other studies show, e.g., eVMT calculation based on label fuel economy versus vehicle average charge sustaining fuel consumption differed less than 2.5% in a study by the Idaho National Laboratory [30].

Logger information was also available in the dataset with the following variables: logger installation date, logger uninstall date, initial odometer reading and final odometer reading. The dataset also included the following variables related to household identification: household id no., list of PEVs in the household, list of ICEVs in the household, number of drivers in the household, size of the household, number of vehicles in the household and vehicle–driver ratio.

2.1.2. Charging Data

Charging data consisted of each single charging event performed by the logged vehicle. It provided information regarding the start and end times of the charging event, charge levels (either level 1 or level 2), and start and end state of charge (SOC). In addition, the charging dataset also contained information regarding vehicle and household identification and logger information, as in the trip data set. All driving and charging data were annualized prior to analysis.

2.1.3. Other Sources of Information

The range of the PEVs that we used in our analysis was based on the information provided by the U.S. Environmental Protection Agency; which was: 11 miles for Toyota Prius, 19 miles for Ford Energi, 25 miles for Toyota Prius Prime, 33 miles for Chrysler Pacifica, 35–38 miles for Chevrolet Volt, 73–84 miles for Nissan Leaf, 238 miles for Chevrolet Bolt, and 249 miles for Tesla Model S.

2.2. Compiled Dataset for Analysis

From the datasets provided by the Electric Vehicle Research Center, we selected and computed the following variables that we labeled as factors corresponding to the three categories of the household context.

For the category of PEV technology in the household, we selected the variable range and computed the variable frequency of charging. Range was defined as the all-electric range of the plug-in electric vehicle of that household. Frequency of charging was defined as the average number of charging events per day for the PEV of the given household.

Regarding household vehicle usage, we computed the variables frequency of overlaps, frequency of long-distance trips and standard deviation (SD) of daily household VMT. Frequency of overlaps was defined as the percentage of PEV trips that overlapped with any of the ICEV trips in the household. Frequency of long-distance trips was defined as the percentage of single trips (not daily) above 50 miles undertaken by the PEVs. We also computed the frequency of long-distance trips with a threshold of 20, 30 and 40 miles, respectively, in order to see how the definition of a long-distance trip would impact our results.

For the category regarding the ICEVs in the household, we defined the variable MPG (miles per gallon) of the ICEV in the household as the real-life MPG based on actual driving distance and fuel consumption.

We computed the following dependent variables for our analysis: eVMT, VMT of the PEV, VMT of the household, utility factor of the PHEV, utility factor of the household and household fuel consumption in gallons per 100 miles. eVMT was defined as the annualized electric vehicle miles traveled by the PEV in the household. VMT of the PEV was defined as the sum of annualized eVMT and gVMT for the PEV in the household. VMT of the household was the sum of all annualized vehicle miles traveled of all the vehicles in the household. Utility factor of the PHEV was defined as the share of eVMT within the VMT of the PHEV. Utility factor of the household was defined as the share of eVMT within the VMT of the household. In our annualization, we used the first and last recorded trip for that vehicle. Finally, household fuel consumption was defined as the combined fuel consumption of all vehicles in the household in gallons per 100 miles.

All variables that we selected or computed are summarized in Table 1 below.

Table 1. Variables in our compiled dataset.

Dependent Variables	eVMT, VMT of the PEV, VMT of the household, utility factor of the PHEV, utility factor of the household, household fuel consumption.
Independent Variables	Range, frequency of charging, frequency of long distance trips, frequency of overlaps, standard deviation of daily household VMT, MPG of ICEV.

2.3. Methods

We used descriptive and inductive statistical methods, and regression analysis, on our compiled dataset to assess the electrification of vehicle miles traveled and fuel consumption within the household context. We used descriptive statistics to study the household level in more detail, and regression analysis to provide a high-level view and explain overall trends. For the regression analysis, we focused only on two-car households.

Below is our generic regression model, where we used the same independent variables for all:

$$\begin{aligned} Y_i = \beta_0 + & \beta_1 Range + \beta_2 FreqCharging \\ & + \beta_3 FreqLongdistance + \beta_4 FreqOverlaps \\ & + \beta_5 SD\_dailyHHVMT + \beta_6 ICEVMpg + \varepsilon \end{aligned}$$

$i = \{1, \ldots, 8\}$ *where* $Y_1 = eVMT$, $Y_2 =$ *VMT of the PEV*, $Y_3 =$ *VMT of the household*, $Y_4 =$ *Utility factor of the PHEV*, $Y_5 =$ *Utility factor of the households*, $Y_6 =$ *Fuel consumption of households*, $Y_7 =$ *Fuel consumption of PHEV households*, $Y_8 =$ *Fuel consumption of BEV households*

We performed multiple linear regression analysis on $eVMT$, VMT of the PEV and VMT of the household, and logistic regression on the UF of the PHEV, and the UF of the household, since the utility factor was always between 0 and 1. Similarly, we performed a logistic regression on the fuel consumption of all households, because we observed a logarithmic relationship between fuel consumption and all-electric range when all PEVs were included (this can be seen in Figure 3 in Section 3.2). When we observed the relationship between fuel consumption and all-electric-range for PHEV households and BEV households separately, this logarithmic relationship disappeared. Therefore, to investigate further, we performed multiple linear regression analyses on the fuel consumption of PHEV and BEV households, separately.

3. Results

The results section is divided into three parts: first we provide descriptive statistics summarizing the dataset and explain how the most salient factors corresponding to household context categories impact our main metrics. Second, we present the results of the regression analysis for each of our main metrics to estimate the statistical significance of the identified factors. Third, we present the sensitivity analysis for our results.

3.1. Descriptive Statistics

Summary statistics for all two-car households are provided in Table 2 (all households combined) and also in Table A1 (households grouped by PEV-type) in the Appendix A. Frequency of charging was, on average, higher for PHEVs compared with BEVs, ranging from 0.67 per day for Chevrolet Bolts, to 1.1 per day for Ford C-Max/Fusion. The frequency of long-distance trips, in general, was low for all PEVs, except Tesla Model S which went on long-distance trips, on average, almost three to six times more often than the other PHEVs in our dataset.

Table 2. Summary statistics for two-car households.

	Min.	0.25-Quantile	Median	Mean	0.75-Quantile	Max.
All households combined (N = 183)						
Annual household VMT (mi)	7341	16,535	21,200	22,426	27,485	56,521
Daily household VMT (mi)	20.11	45.30	58.08	61.44	75.30	154.85
Household utility factor (%)	0.59%	27.41%	42.99%	41.77%	54.70%	98.35%
PHEV utility factor (%) *	1.27%	31.35%	51.54%	50.15%	69.32%	98.95%
Household fuel consumption (gal/100 mi)	0.04	1.54	2.15	2.12	2.61	5.01
Household tailpipe emissions (gCO_2/km)	2.25	83.72	116.84	115.27	141.55	271.99
ICEV fuel economy (MPG)	11.17	19.52	23.93	26.26	30.85	52.83
PHEV fuel economy (MPG) *	36.64	57.78	86.25	156.80	133.75	4065.52
Frequency of charging (events per day)	0.01	0.61	0.80	0.87	1.13	2.73
Frequency of long-distance trips with PEV (%)	0.00%	0.24%	0.88%	2.53%	2.45%	37.97%
Frequency of overlaps (%)	0.00%	6.22%	10.68%	12.00%	16.61%	51.09%

Note: * Excluding BEV (Leaf, Bolt, Model S) households.

The utility factor of the household allows a comparison between households with different number of cars, and between PHEV and BEV households. As seen in Figure 1, the utility factor of the households had a downward trend when the number of cars increased. In other words, households with more cars electrified a lower share of their total travel, which was expected since these households also drive more in total, as seen in Figure 2. Overall, we observed that there was a general increase in eVMT and UF with all-electric range.

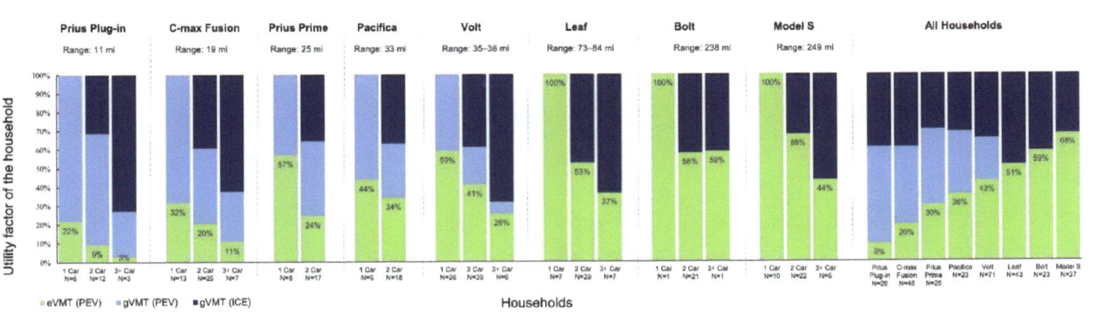

Figure 1. Utility factor of households, categorized by PEV-type and total number of cars in the household.

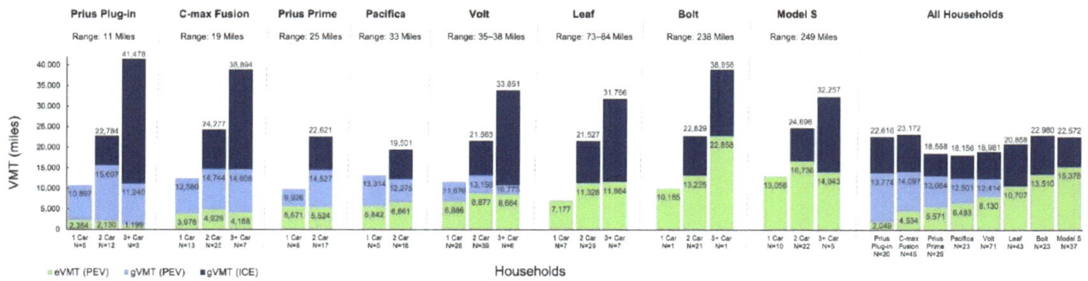

Figure 2. VMT of households, categorized by PEV-type and total number of cars in the household.

Combining same-PEV households and observing their utility factor provided quite interesting results. When we scrutinized the UF of households (Figure 1), we observed that Chevrolet Volt, a PHEV with a range of 35–38 miles, electrified 43% of the total household miles; whereas Nissan Leaf, a BEV with a range of 73–84 miles, was at 51%; Chevrolet Bolt, a BEV with a range of 238 miles, was at 59%; and Tesla Model S, a BEV with a range of 249 miles, was at 68%. This finding shows that, in the context of the whole household, a PHEV such as the Chevrolet Volt, can electrify a similar share of household miles as a low range BEV (Leaf), and can electrify around 70% as many household miles as long range BEVs (Bolt and Model S).

Total annual household VMT among the households were similar, as seen in Figure 2, ranging from 18,156 miles for Pacifica households, to 23,172 miles for C-max Fusion households.

3.2. Regression Analysis

The results of the regression analysis for eVMT, VMT of the PEV, VMT of the household, UF of the PHEV, and UF of the household, are given in Table 3. The results of the regression analysis on household fuel consumption is given in Table 4. Note that both regression analyses were performed only on two-car households.

Table 3. Regression results for how household factors impact vehicle miles and electrification in two-car households.

Dependent	eVMT (mi)		VMT of the PEV (mi)		VMT of the Household (mi)		UF of the PHEV (%)		UF of the Household (%)	
Intercept	−3061.20 (1440.65)	*	2125.20 (1476.15)		3518.39 (2002.98)	.	−3.87 (8.28)		24.88 (4.77)	***
Range (mi)	43.58 (3.77)	***	3.18 (3.87)		1.92 (5.25)		1.82 (0.15)	***	0.17 (0.01)	***
Frequency of charging (events per day)	6023.18 (756.55)	***	5763.80 (775.19)	***	4964.41 (1051.85)	***	17.37 (3.13)	***	15.55 (2.51)	***
Frequency of long-distance trips (%)	307.55 (71.18)	***	519.61 (72.93)	***	274.30 (98.96)	**	−1.31 (0.46)	**	0.16 (0.24)	
Frequency of overlaps (%)	28.24 (39.75)		117.28 (40.73)	**	329.89 (55.27)	***	−0.13 (0.19)		−0.43 (0.13)	**
Standard deviation of daily household VMT (mi)	25.26 (17.71)		79.63 (18.14)	***	197.93 (24.62)	***	−0.26 (0.09)	**	−0.20 (0.06)	***
ICEV fuel economy (MPG)	36.82 (32.56)		−8.90 (33.36)		6.31 (45.27)		0.13 (0.16)		0.11 (0.11)	
N	183		183		183		111 [x]		183	
Multiple R-squared	0.5919		0.5638		0.5517		-		-	
Adjusted R-squared	0.5779		0.5487		0.5362		-		-	
F-statistic	42.07		37.48		35.69		-		-	

Confidence levels *** 99.9%, ** 99%, * 95%, . 90%. Values represent estimates, standard error is given in parentheses.
[x] Excluding BEV (Leaf, Bolt, Model S) households.

Table 4. Regression results for how household factors impact fuel consumption in two-car households.

Dependent	Household Fuel Consumption (gal/100 mi)					
	All Households		*PHEV Households*		*BEV Households*	
Intercept	3.704 (0.174)	***	3.889 (0.238)	***	3.745 (0.339)	***
Range (mi)	−0.004 (0.0004)	***	−0.018 (0.004)	***	−0.003 (0.001)	
Frequency of charging (events per day)	−0.509 (0.091)	***	−0.451 (0.090)	***	−0.753 (0.236)	
Frequency of long-distance trips (%)	−0.023 (0.009)	**	−0.020 (0.013)		−0.023 (0.012)	
Frequency of overlaps (%)	0.021 (0.005)	***	0.020 (0.005)	***	0.019 (0.009)	
Standard deviation of daily household VMT (mi)	0.009 (0.002)	***	0.011 (0.002)	***	0.007 (0.004)	
ICEV fuel economy (MPG)	−0.054 (0.0034)	***	−0.050 (0.005)	***	−0.054 (0.007)	***
N	183		111		72	
Multiple R-squared	-		0.6435		0.6153	
Adjusted R-squared	-		0.6229		0.5787	
F-statistic	-		31.28		16.8	

Confidence levels *** 99.9%, ** 99%, * 95%, . 90%. Values represent estimates, standard error is given in parentheses.

Our results in Table 3 show that range was statistically very significant in electrification of miles, and higher ranges resulted in higher eVMT, UF of the PHEV, and UF of the household. This result confirmed the initial trend observed in the descriptive statistics section, and was also in line with the findings of previous studies [9,24,25].

As seen in Table 3, frequency of charging was also statistically significant, showing up in all main metrics. This suggests that more frequent charging results in higher eVMT, UF of the PHEV and UF of the household. However, it should be noted that the frequency of charging was based on the number of charging events and not the length of these charging events; also, the charging level was not taken into account.

The frequency of long-distance trips was statistically significant for eVMT, VMT of the PEV, VMT of the household, and UF of the PHEV. The results show that more frequent long-distance trips increased eVMT and VMT of PEV as expected, but decreased the UF of the PHEV. This suggests that the decrease in the UF of the PHEV can be explained by the increase in gVMT of the PHEV, meaning a lower fuel economy as long-distance trips become more frequent. Plötz, Funke and Jochem [9] and Mandev et al. [37] also reached the same conclusion in their papers on the impact of daily and annual driving on fuel economy, where they concluded that the tendency for long-distance trips decreased the UF of a PHEV, and PHEV fuel economy.

As seen in Table 3, the frequency of overlaps was statistically significant for the VMT of the household and the UF of the household. As expected, more overlaps between PEVs and ICEVs resulted in higher ICEV usage which increased the gVMT and VMT of the household and, thus, lowered the UF of the household.

In Table 4, we observe that the range and frequency of charging was significant in reducing household fuel consumption in PHEV households, but not in BEV households. Table 4 also shows that for every 10-mile increase in range for PHEV households, household fuel consumption and household tailpipe emissions were reduced by 0.2 gal/100 mi and 17.5 gCO_2/mi (10.9 gCO_2/km), respectively. Conversion of fuel consumption to tailpipe emission was performed by using U.S. Environmental Protection Agency values [38].

In Figure 3, we show the household fuel consumption vs range for two-car households categorized by PEV-type, with a logistic regression curve (marked as B in the figure) and also box plots with the distribution of household fuel consumption. We observed only a small difference between long range BEV households and short range BEV households. The decrease in fuel consumption was smaller in long range BEV households than what was intuitively expected, compared with short range BEV and long range PHEV households. In two-car households with one PEV, BEVs did not reduce household fuel consumption as significantly, compared with long range PHEVs; and consequently, some long range PHEVs reduced household fuel consumption to a similar degree as BEVs.

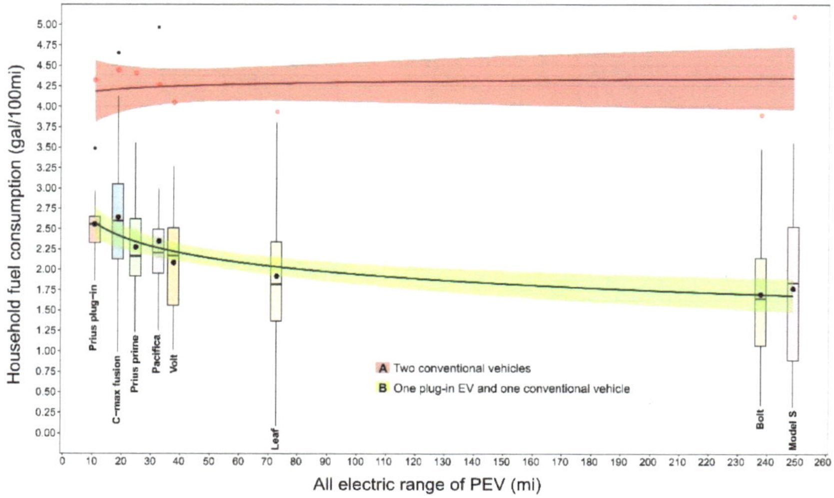

Figure 3. Household fuel consumption vs. range of PEV in two-car households.

Tesla model S households had the lowest ICEV MPG, as given in Table A1 in the Appendix A. This behavior of owning low MPG cars increased the overall household fuel consumption, as seen in Figure 3. This was not observed for Bolt, even though these two cars have a similar range.

Figure 3 also shows a logistic regression curve (marked as A in the figure) for the household fuel consumption if the households had two of the same ICEV instead of one PEV and one ICEV. (We assumed a hypothetical scenario that the second ICEV would be the same as the one already owned, where in reality it could have a higher or lower MPG.) The difference between curve A and B visualizes how much fuel consumption is hypothetically saved through owning a PEV instead of an ICEV. We do not expect the difference in fuel consumption between households with a PHEV or a BEV to change much in the future because long range PHEVs such as Volt can have a higher fuel economy than a new ICEV. Volt production has ended, but similar long range PHEVs could have the potential to further lower household fuel consumption. The new EPA target of 1.9 ga/100 mi (166 gCO_2/mi) [39] for 2023 and EU target of 1.75 ga/100 mi (153 gCO_2/mi) [40] during 2020-24 for newly sold passenger cars, means that buying two new conventional passenger cars would end up consuming the same amount of fuel by the household as having a long range BEV and an older conventional car with lower MPG. To really achieve a decrease in fuel consumption, households will need to shift to two PEVs and no ICEV.

3.3. Sensitivity Analysis

In our results, we presented fewer variables than identified in the data and methods section to label factors corresponding to the categories of the household context. Some of the variables we considered included the number of drivers, commute distance, size of ICEVs in the household, size of the household, whether a charger was present at work, and the age of the PEV users. We first ran our regression models with those variables included, we estimated variance inflation factors to check for multicollinearity, and ran likelihood ratio tests to see if our models were better off without those variables. Removing those variables made our model more robust, so we decided to omit them. The regression models presented in this paper had variance inflation factors lower than 2 for all variables.

For the frequency of long-distance trips, we tested several different variations to make our model more robust. We considered trips over 50 miles in our models, but we also set the threshold at 20, 30 and 40 miles. A higher threshold increased the statistical significance of the variable; thus, we decided to use 50 miles as the threshold for long-distance trips.

4. Discussion and Conclusions

As for all datasets, our dataset has its limitations. First, the sample size of our data was limited to 287 households, of which 183 were two-car households. This had implications both for the regression analysis and the representativeness of our data. From a regression point of view, a larger dataset might have allowed us to include more variables. On the other hand, the strengths of the dataset were that it was collected for the length of a year, which control for the learning phase of a new car, seasonality and other external effects; it also contained several variables and included all vehicles in the households, which is quite unique. There is often a trade-off in empirical data sets between level of detail, length of measurement and number of entities (households, individuals, or vehicles) that are measured.

Our study was regionally bound to California, and the households included could be considered early adopters, with higher education levels and income. Of the households in our dataset, 76% had an annual income of over USD 100,000 compared with the USD 78,672 median income in California in 2020 [41]. Based on the household survey, 86% of the adults had at least a college degree or higher. We recognize that this might have created a bias towards more conscious driving and charging behavior. Our results might have differed slightly if a larger and geographically more diverse population sample had been used. However, California is one of the leading EV markets, and at the time of the data collection for this study, California was in the early adopter phase regarding EV adoption [42]; therefore, we considered these limitations inevitable.

In 2020, 30% of households in California owned one vehicle, 37% owned two vehicles, 26% owned three or more vehicles, and 7% owned no vehicles [43]. This shows that two-car households are the dominant group in the population sample, with every two out of five households being two-car households. This was in line with our dataset, where the majority were also two-car households. Therefore, investigating how two-car households behave is vital to understanding the trends in the general population, and this further strengthened our focus on two-car households in this paper.

Analysis of PHEVs, regarding their electrification potential if conducted without taking into account the household context, lacks the tools to assess how they are used (1) in combination with ICEVs and (2) in comparison to households that have BEVs instead of PHEVs. For instance, Chevrolet Volt as a stand-alone PHEV can electrify, on average, 70% of its miles driven; a two-car household with a Chevrolet Volt and a conventional vehicle can electrify, on average, 43% of the total household miles; this amounts to 63% of the electrification potential of a two-car household with a Tesla Model S, and 73% of the electrification potential of a two-car household with a Chevrolet Bolt.

In this paper, we aimed to analyze how factors within the household context impact electrification of miles and fuel consumption. The results indicated that a PHEV with a higher all-electric-range resulted in more electrification and, thus, a higher UF for both the PHEV and the household. Furthermore, our results showed that—considering the UF of the household—a PHEV such as the Chevrolet Volt, with half the range of a BEV such as the Nissan Leaf, can electrify a similar share of miles, and can electrify around 70% as much household miles as long-range BEVs (Bolt and Model S) within the household context. Our results also indicated that more frequent charging resulted in higher electrification of miles and a higher UF for both the PHEV and the household. However, it should be noted that we did not take charging duration into account.

The results showed that, for two-car households, every 10-mile increase in range for PHEV households reduced the household fuel consumption and household tailpipe emissions by 0.2 gal/100 mi and 17.5 gCO_2/mi (10.9 gCO_2/km), respectively. In addition, our results indicated that in two-car households with one PEV, BEVs did not reduce household fuel consumption as expected (compared with long range PHEVs) and some long range PHEVs reduced household fuel consumption to a similar degree. Aside from the impact of all-electric-range in reducing fuel consumption, our results also showed how significantly having a PHEV or BEV, instead of an ICEV, can reduce household fuel consumption and related tailpipe emissions. This aspect of replacing an ICEV with a PEV is especially important to understand, especially for future households with multiple PEVs. For instance, in a two-car household with one BEV and one ICEV, replacing the ICEV with a PHEV could significantly increase electrification of vehicle miles and reduce fuel consumption. Two-car households with long range BEVs such as the Chevrolet Bolt and Tesla Model S in our dataset had a household utility factor of 58% to 68%, on average. If the ICEVs in those households were replaced by efficient long-range PHEVs such as the Chevrolet Volt (with a mean utility factor of 70%, see Table A1), their total household electrification could increase by a further 20% to 30%, and reach almost 90%. On the other hand, in future households with two BEVs, the challenge lies with how to tackle long-distance trips. We observed in our dataset that Chevrolet Bolt and Tesla Model S have very different frequencies of long-distance trips. Bolt averaged 0.73% for the frequency of long-distance trips (over 50 miles) which was significantly lower than the PHEVs in our dataset (which ranged from 1.35% to 3.17%, on average); Tesla Model S, however, had an average frequency of long-distance trips of 8.75%, which was three to six times higher than PHEVs (see Table A1). This difference between the two long range BEVs is likely due to the difference in the availability of fast charging opportunities, with Tesla supercharger stations being quite prevalent in California. Contrary to the general assumption that BEVs would have lower frequencies of long-distance trips than PHEVs, Tesla Model S users in our dataset showed that, with the right charging infrastructure, long-distance trips can stop being an obstacle. This finding strengthens the transition to future households with multiple BEVs, which can make the goal of reaching 100% electrification in the household a reality. Our analysis in this paper, therefore, provides a basis for understanding future households with multiple PEVs.

In this paper, we found that PHEVs with more than a 35 mile all-electric-range when used by private households in the U.S., can be an effective tool to decarbonize the transport sector. However, further research is needed to see how much this electrification potential changes from country to country, e.g., how much this is impacted by electricity and gasoline prices. To reach 100% electrification in the household, a switch to households with multiple BEVs is necessary. Understanding how households with multiple BEVs drive and charge is an important topic of future research, to be able to understand the overall environmental benefits of these vehicles. Further research is needed to see how multiple BEVs in a household can be integrated into the grid, utilizing their batteries as storage units and using vehicle-to-grid technologies, e.g., to address peak demand and irregularity of renewable energy sources [44,45]. One potential drawback to using electric vehicle batteries to store energy is the loss of energy and reduced efficiency arising from double conversion (grid-to-vehicle and vehicle-to-grid). This suggests the feasibility of vehicle-to-grid technologies as an area of future research, and future developments in battery technology can help address this issue in coming years. Future developments in charging infrastructure, such as e-road technologies, are also an important area of future research [46–48]. Households with multiple BEVs utilizing e-road technologies has the potential to remove long-distance trips from being an obstacle to reaching 100% electrification in the transport sector.

In conclusion, our results provide an insight into the electrification of vehicle miles traveled within the household context, which is rarely taken into consideration. The implication for policy makers is that PHEVs with a range of at least 35 miles have the potential to electrify a similar share of total household miles as some short range BEVs, or can reach up to 70% as much electrification as some long range BEVs and, thus, can play an important role in decarbonizing the transport sector.

Author Contributions: Conceptualization, A.M., F.S. and G.T.; data curation, G.T.; formal analysis, A.M.; methodology, A.M., F.S. and G.T.; software, A.M.; visualization, A.M.; writing—original draft, A.M.; writing—review and editing, A.M., F.S. and G.T. All authors have read and agreed to the published version of the manuscript.

Funding: This research received no external funding.

Data Availability Statement: Restrictions apply to the availability of these data. Data was obtained from the Electric Vehicle Research Center at the Institute of Transportation Studies, University of California, Davis and are available only with their permission.

Acknowledgments: This research was supported by the Swedish Electromobility Centre. The data for analysis were provided by the Electric Vehicle Research Center at the Institute of Transportation Studies, University of California, Davis.

Conflicts of Interest: The authors declare no conflict interest.

Appendix A

Table A1. Summary statistics of two-car households categorized by PEV-type.

	Min.	0.25-Quantile	Median	Mean	0.75-Quantile	Max.
Prius plug-in households (N = 12) [PHEV]						
Annual household VMT (mi)	7984	12,708	23,923	22,784	28,628	43,070
Daily household VMT (mi)	21.87	34.82	65.54	62.42	78.43	118.00
Household utility factor (%)	2.26%	7.10%	10.39%	10.97%	11.21%	31.62%
PHEV utility factor (%)	3.19%	9.98%	14.17%	16.57%	18.57%	49.10%
Household fuel consumption (gal/100 mi)	1.84	2.32	2.55	2.55	2.65	3.50
Household tailpipe emissions (gCO_2/km)	100.10	126.00	138.80	138.80	143.80	190.30
ICEV fuel economy (MPG)	16.00	18.05	23.60	25.64	27.13	50.23
PHEV fuel economy (MPG)	45.66	53.63	56.57	59.53	59.44	102.14
Frequency of charging (per day)	0.13	0.74	0.97	0.92	1.24	1.64
Frequency of long-distance trips with PEV (%)	0.11%	0.20%	0.67%	3.17%	2.94%	16.70%
Frequency of overlaps (%)	0.11%	4.74%	9.29%	9.63%	13.79%	20.08%
C-max fusion households (N = 25) [PHEV]						
Annual household VMT (mi)	9280	17,081	24,515	24,277	31,197	50,807
Daily household VMT (mi)	25.43	46.80	67.16	66.51	85.47	139.20
Household utility factor (%)	0.59%	10.45%	22.90%	23.52%	31.48%	62.95%
PHEV utility factor (%)	1.27%	20.12%	37.09%	38.07%	50.83%	98.95%
Household fuel consumption (gal/100 mi)	0.96	2.11	2.60	2.64	3.06	4.69
Household tailpipe emissions (gCO_2/km)	52.29	114.91	141.11	143.43	166.17	254.81
ICEV fuel economy (MPG)	13.54	18.67	22.54	25.31	30.25	49.94
PHEV fuel economy (MPG)	36.64	48.06	61.55	234.64	89.35	4065.52
Frequency of charging (per day)	0.01	0.67	0.99	1.10	1.42	2.73
Frequency of long-distance trips with PEV (%)	0.00%	0.30%	1.31%	2.31%	3.10%	8.38%
Frequency of overlaps (%)	0.00%	8.97%	12.20%	13.79%	19.33%	51.09%
Prius prime households (N = 17) [PHEV]						
Annual household VMT (mi)	8220	18,624	23,165	22,621	27,805	40,374
Daily household VMT (mi)	22.52	51.03	63.47	61.97	76.18	110.61
Household utility factor (%)	4.06%	20.01%	27.61%	26.63%	34.66%	43.44%
PHEV utility factor (%)	5.73%	34.68%	42.27%	40.91%	52.67%	65.91%
Household fuel consumption (gal/100 mi)	1.21	1.90	2.16	2.27	2.62	3.57
Household tailpipe emissions (gCO_2/km)	65.49	103.41	117.10	123.12	142.45	194.21
ICEV fuel economy (MPG)	13.94	20.40	21.41	24.97	24.88	48.44
PHEV fuel economy (MPG)	51.77	86.25	95.57	105.72	103.98	222.10
Frequency of charging (per day)	0.14	0.61	0.88	0.82	0.96	1.41
Frequency of long-distance trips with PEV (%)	0.00%	0.42%	0.73%	1.35%	2.18%	4.67%
Frequency of overlaps (%)	2.83%	7.62%	10.40%	11.77%	14.08%	28.62%

Table A1. *Cont.*

	Min.	0.25-Quantile	Median	Mean	0.75-Quantile	Max.
Pacifica households (N = 18) [PHEV]						
Annual household VMT (mi)	7341	14,426	20,720	19,501	22,718	30,299
Daily household VMT (mi)	20.11	39.52	56.77	53.43	62.24	83.01
Household utility factor (%)	16.88%	23.42%	32.40%	34.49%	43.37%	55.94%
PHEV utility factor (%)	26.48%	44.04%	55.15%	55.13%	65.12%	86.21%
Household fuel consumption (gal/100 mi)	1.22	1.94	2.20	2.34	2.49	5.01
Household tailpipe emissions (gCO_2/km)	66.53	105.26	119.28	127.19	135.23	271.99
ICEV fuel economy (MPG)	11.17	20.44	23.26	26.70	31.00	50.40
PHEV fuel economy (MPG)	40.26	56.50	65.33	89.24	118.06	227.14
Frequency of charging (per day)	0.36	0.69	0.95	1.04	1.39	1.78
Frequency of long-distance trips with PEV (%)	0.00%	0.45%	1.38%	1.59%	2.23%	5.63%
Frequency of overlaps (%)	0.89%	5.17%	9.57%	10.28%	13.95%	24.65%
Volt households (N = 39) [PHEV]						
Annual household VMT (mi)	7406	16,455	19,825	21,563	26,757	37,197
Daily household VMT (mi)	20.29	45.08	54.31	59.08	73.31	101.91
Household utility factor (%)	8.25%	35.26%	42.99%	43.00%	53.15%	66.66%
PHEV utility factor (%)	18.74%	58.99%	71.81%	69.96%	85.12%	96.29%
Household fuel consumption (gal/100 mi)	0.99	1.54	2.16	2.07	2.51	3.28
Household tailpipe emissions (gCO_2/km)	53.68	83.55	117.48	112.60	136.12	178.17
ICEV fuel economy (MPG)	14.60	20.00	25.10	26.61	30.38	43.29
PHEV fuel economy (MPG)	41.95	88.53	134.72	190.29	241.98	916.66
Frequency of charging (per day)	0.14	0.63	0.83	0.89	1.11	1.77
Frequency of long-distance trips with PEV (%)	0.00%	0.51%	1.23%	2.39%	2.34%	26.48%
Frequency of overlaps (%)	1.51%	6.27%	11.40%	11.65%	14.12%	33.99%
Leaf households (N = 29) [BEV]						
Annual household VMT (mi)	8947	16,602	21,657	21,527	26,137	34,203
Daily household VMT (mi)	24.51	45.48	59.33	58.98	71.61	93.71
Household utility factor (%)	19.82%	44.23%	49.65%	51.88%	54.94%	98.35%
Household fuel consumption (gal/100 mi)	0.04	1.34	1.80	1.90	2.34	3.82
Household tailpipe emissions (gCO_2/km)	2.25	73.07	97.95	103.46	126.89	207.52
ICEV fuel economy (MPG)	13.62	21.20	27.00	28.79	34.25	52.83
Frequency of charging (per day)	0.23	0.67	0.70	0.79	0.87	1.53
Frequency of long-distance trips with PEV (%)	0.00%	0.00%	0.10%	0.62%	0.44%	6.46%
Frequency of overlaps (%)	0.00%	7.43%	9.84%	10.75%	13.27%	23.92%
Bolt households (N = 21) [BEV]						
Annual household VMT (mi)	7930	17,020	19,953	22,829	28,929	38,190
Daily household VMT (mi)	21.73	46.63	54.67	62.55	79.26	104.63
Household utility factor (%)	37.36%	50.94%	54.54%	58.06%	67.84%	88.14%
Household fuel consumption (gal/100 mi)	0.30	1.05	1.63	1.68	2.13	3.50
Household tailpipe emissions (gCO_2/km)	16.48	56.78	88.47	91.34	115.92	189.96
ICEV fuel economy (MPG)	13.94	19.84	26.42	28.32	35.00	44.00
Frequency of charging (per day)	0.17	0.38	0.60	0.67	0.93	1.68
Frequency of long-distance trips with PEV (%)	0.00%	0.15%	0.36%	0.73%	0.79%	4.38%
Frequency of overlaps (%)	0.00%	3.88%	12.26%	11.41%	18.99%	22.73%
Model S households (N = 22) [BEV]						
Annual household VMT (mi)	12,149	17,043	20,793	24,696	29,811	56,521
Daily household VMT (mi)	33.29	46.69	56.97	67.66	81.67	154.85
Household utility factor (%)	46.22%	56.35%	63.29%	65.88%	74.27%	89.42%
Household fuel consumption (gal/100 mi)	0.28	0.86	1.83	1.80	2.53	3.58
Household tailpipe emissions (gCO_2/km)	15.15	46.82	99.31	97.76	137.20	194.26
ICEV fuel economy (MPG)	13.10	16.00	18.14	22.38	24.96	49.11
Frequency of charging (per day)	0.22	0.54	0.69	0.72	0.85	1.39
Frequency of long-distance trips with PEV (%)	0.81%	2.01%	3.85%	8.75%	8.19%	37.97%
Frequency of overlaps (%)	3.23%	8.14%	14.14%	15.80%	19.16%	37.54%

Note: Please note that the household utility factors here show the distribution among households and is not aggregated as in Figures 1 and 2 and, thus, are slightly different.

References

1. Duvall, M.; Knipping, E.; Alexander, M. *Environmental Assessment of Plug-in Hybrid Electric Vehicles*; Electric Power Research Institute: Palo Alto, CA, USA, 2007.
2. Kromer, M.A.; Bandivadekar, A.; Evans, C. Long-term greenhouse gas emission and petroleum reduction goals: Evolutionary pathways for the light-duty vehicle sector. *Energy* **2010**, *35*, 387–397. [CrossRef]
3. Stephan, C.H.; Sullivan, J. Environmental and Energy Implications of Plug-In Hybrid-Electric Vehicles. *Environ. Sci. Technol.* **2008**, *42*, 1185–1190. [CrossRef] [PubMed]
4. Yang, C.; McCollum, D.; McCarthy, R.; Leighty, W. Meeting an 80% reduction in greenhouse gas emissions from transportation by 2050: A case study in California. *Transp. Res. Part D Transp. Environ.* **2009**, *14*, 147–156. [CrossRef]
5. Girardi, P.; Gargiulo, A.; Brambilla, P.C. A comparative LCA of an electric vehicle and an internal combustion engine vehicle using the appropriate power mix: The Italian case study. *Int. J. Life Cycle Assess.* **2015**, *20*, 1127–1142. [CrossRef]
6. Shi, S.; Zhang, H.; Yang, W.; Zhang, Q.; Wang, X. A life-cycle assessment of battery electric and internal combustion engine vehicles: A case in Hebei Province, China. *J. Clean. Prod.* **2019**, *228*, 606–618. [CrossRef]
7. Wang, N.; Tang, G. A Review on Environmental Efficiency Evaluation of New Energy Vehicles Using Life Cycle Analysis. *Sustainability* **2022**, *14*, 3371. [CrossRef]
8. Xia, X.; Li, P.; Xia, Z.; Wu, R.; Cheng, Y. Life cycle carbon footprint of electric vehicles in different countries: A review. *Sep. Purif. Technol.* **2022**, *301*, 122063. [CrossRef]
9. Plötz, P.; Funke, S.Á.; Jochem, P. The impact of daily and annual driving on fuel economy and CO_2 emissions of plug-in hybrid electric vehicles. *Transp. Res. Part A Policy Pract.* **2018**, *118*, 331–340. [CrossRef]
10. Khan, M.; Kockelman, K.M. Predicting the market potential of plug-in electric vehicles using multiday GPS data. *Energy Policy* **2012**, *46*, 225–233. [CrossRef]
11. Tamor, M.A.; Milačić, M. Electric vehicles in multi-vehicle households. *Transp. Res. Part C Emerg. Technol.* **2015**, *56*, 52–60. [CrossRef]
12. Jakobsson, N.; Gnann, T.; Plötz, P.; Sprei, F.; Karlsson, S. Are multi-car households better suited for battery electric vehicles? Driving patterns and economics in Sweden and Germany. *Transp. Res. Part C Emerg. Technol.* **2016**, *65*, 1–15. [CrossRef]
13. Karlsson, S. Utilization of battery-electric vehicles in two-car households: Empirical insights from Gothenburg Sweden. *Transp. Res. Part C Emerg. Technol.* **2020**, *120*, 102818. [CrossRef]
14. Srinivasa Raghavan, S.; Tal, G. Behavioral and technology implications of electromobility on household travel emissions. *Transp. Res. Part D Transp. Environ.* **2021**, *94*, 102792. [CrossRef]
15. Jakobsson, N.; Sprei, F.; Karlsson, S. How do users adapt to a short-range battery electric vehicle in a two-car household? Results from a trial in Sweden. *Transp. Res. Interdiscip. Perspect.* **2022**, *15*, 100661. [CrossRef]
16. Chakraborty, D.; Hardman, S.; Tal, G. Integrating plug-in electric vehicles (PEVs) into household fleets- factors influencing miles traveled by PEV owners in California. *Travel Behav. Soc.* **2022**, *26*, 67–83. [CrossRef]
17. *J2841_201009*; Utility Factor Definitions for Plug-In Hybrid Electric Vehicles Using Travel Survey Data. SAE International: Warrendale, PA, USA, 2010.
18. *J1711_201006*; Recommended Practice for Measuring the Exhaust Emissions and Fuel Economy of Hybrid-Electric Vehicles, Including Plug-in Hybrid Vehicles. SAE International: Warrendale, PA, USA, 2010.
19. Elgowainy, A.; Burnham, A.; Wang, M.; Molburg, J.; Rousseau, A. Well-To-Wheels Energy Use and Greenhouse Gas Emissions of Plug-in Hybrid Electric Vehicles. *SAE Int. J. Fuels Lubr.* **2009**, *2*, 627–644. [CrossRef]
20. Axsen, J.; Kurani, K.S.; McCarthy, R.; Yang, C. Plug-in hybrid vehicle GHG impacts in California: Integrating consumer-informed recharge profiles with an electricity-dispatch model. *Energy Policy* **2011**, *39*, 1617–1629. [CrossRef]
21. Tal, G.; Nicholas, M.A.; Davies, J.; Woodjack, J. Charging Behavior Impacts on Electric Vehicle Miles Traveled: Who is Not Plugging In? *Transp. Res. Rec.* **2014**, *2454*, 53–60. [CrossRef]
22. Moawad, A.; Singh, G.; Hagspiel, S.; Fellah, M.; Rousseau, A. Impact of Real World Drive Cycles on PHEV Fuel Efficiency and Cost for Different Powertrain and Battery Characteristics. *World Electr. Veh. J.* **2009**, *3*, 186–195. [CrossRef]
23. Björnsson, L.-H.; Karlsson, S.; Sprei, F. Objective functions for plug-in hybrid electric vehicle battery range optimization and possible effects on the vehicle fleet. *Transp. Res. Part C Emerg. Technol.* **2018**, *86*, 655–669. [CrossRef]
24. Bradley, T.H.; Quinn, C.W. Analysis of plug-in hybrid electric vehicle utility factors. *J. Power Sources* **2010**, *195*, 5399–5408. [CrossRef]
25. Paffumi, E.; De Gennaro, M.; Martini, G. Alternative utility factor versus the SAE J2841 standard method for PHEV and BEV applications. *Transp. Policy* **2018**, *68*, 80–97. [CrossRef]
26. Plötz, P.; Funke, S.A.; Jochem, P.; Wietschel, M. CO_2 Mitigation Potential of Plug-in Hybrid Electric Vehicles larger than expected. *Sci. Rep.* **2017**, *7*, 16493. [CrossRef] [PubMed]
27. Plötz, P.; Funke, S.Á.; Jochem, P. Empirical Fuel Consumption and CO2 Emissions of Plug-In Hybrid Electric Vehicles. *J. Ind. Ecol.* **2018**, *22*, 773–784. [CrossRef]
28. Ligterink, N.E.; Smokers, R.T.M.; Bolech, M. Fuel-electricity mix and efficiency in Dutch plug-in and range-extender vehicles on the road. In Proceedings of the 2013 World Electric Vehicle Symposium and Exhibition (EVS27), Barcelona, Spain, 17–20 November 2013; pp. 1–4.

29. Davies, J.; Kurani, K.S. Moving from assumption to observation: Implications for energy and emissions impacts of plug-in hybrid electric vehicles. *Energy Policy* **2013**, *62*, 550–560. [CrossRef]
30. Carlson, B. *Electric Vehicle Mile Traveled (eVMT): On-Road Results and Analysis*; Idaho National Laboratory: Idaho Falls, ID, USA, 2015.
31. Nicholas, M.A.; Tal, G.; Turrentine, T.S. *Advanced Plug-in Electric Vehicle Travel and Charging Behavior Interim Report*; UC Davis Institute of Transportation Studies: Davis, CA, USA, 2017.
32. Hao, X.; Yuan, Y.; Wang, H.; Ouyang, M. Plug-in hybrid electric vehicle utility factor in China cities: Influencing factors, empirical research, and energy and environmental application. *eTransportation* **2021**, *10*, 100138. [CrossRef]
33. Srinivasa Raghavan, S.; Tal, G. Influence of User Preferences on the Revealed Utility Factor of Plug-In Hybrid Electric Vehicles. *World Electr. Veh. J.* **2020**, *11*, 6. [CrossRef]
34. Raghavan, S.S.; Tal, G. Plug-in hybrid electric vehicle observed utility factor: Why the observed electrification performance differ from expectations. *Int. J. Sustain. Transp.* **2022**, *16*, 105–136. [CrossRef]
35. Plötz, P.; Moll, C.; Bieker, G.; Mock, P. From lab-to-road: Real-world fuel consumption and CO₂ emissions of plug-in hybrid electric vehicles. *Environ. Res. Lett.* **2021**, *16*, 054078. [CrossRef]
36. Tal, G.; Karanam, V.C.; Favetti, M.P.; Sutton, K.M.; Ogunmayin, J.M.; Raghavan, S.S.; Nitta, C.; Chakraborty, D.; Davis, A.; Garas, D. *Emerging Technology Zero Emission Vehicle Household Travel and Refueling Behavior*; UC Davis: Plug-In Hybrid & Electric Vehicle Research Center: Davis, CA, USA, 2021.
37. Mandev, A.; Plötz, P.; Sprei, F.; Tal, G. Empirical charging behavior of plug-in hybrid electric vehicles. *Appl. Energy* **2022**, *321*, 119293. [CrossRef]
38. EPA. *Emission Facts: Average Carbon Dioxide Emissions Resulting from Gasoline and Diesel Fuel*; Environmental Protection Agency: Washington, DC, USA, 2005.
39. EPA. *Revised 2023 and Later Model Year Light-Duty Vehicle Greenhouse Gas Emissions Standards*; Environmental Protection Agency: Washington, DC, USA, 2021.
40. European Environment Agency. *Regulation (EU) 2019/631 of the European Parliament and of the Council*; European Environment Agency: Copenhagen, Denmark, 2019.
41. UC Bureau. Median Household Income (in 2020 Dollars), 2016–2020. Available online: https://www.census.gov/quickfacts/fact/table/CA# (accessed on 19 September 2022).
42. Slowik, P.; Lutsey, N. *Evolution of Incentives to Sustain the Transition to a Global Electric Vehicle Fleet*; International Council on Clean Transportation: Washington, DC, USA, 2016.
43. UC Bureau. American Community Survey, Selected Housing Characteristics. Available online: https://data.census.gov/cedsci/table?tid=ACSDP5Y2020.DP04&g=0400000US06&hidePreview=true (accessed on 19 September 2022).
44. Xu, N.Z.; Chung, C.Y. Reliability Evaluation of Distribution Systems Including Vehicle-to-Home and Vehicle-to-Grid. *IEEE Trans. Power Syst.* **2016**, *31*, 759–768. [CrossRef]
45. Inala, K.P.; Sah, B.; Kumar, P.; Bose, S.K. Impact of V2G Communication on Grid Node Voltage at Charging Station in a Smart Grid Scenario. *IEEE Syst. J.* **2021**, *15*, 3749–3758. [CrossRef]
46. Coban, H.H.; Rehman, A.; Mohamed, A. Analyzing the Societal Cost of Electric Roads Compared to Batteries and Oil for All Forms of Road Transport. *Energies* **2022**, *15*, 1925. [CrossRef]
47. Shoman, W.; Karlsson, S.; Yeh, S. Benefits of an Electric Road System for Battery Electric Vehicles. *World Electr. Veh. J.* **2022**, *13*, 197. [CrossRef]
48. Morfeldt, J.; Shoman, W.; Johansson, D.J.A.; Yeh, S.; Karlsson, S. If Electric Cars Are Good for Reducing Emissions, They Could Be Even Better with Electric Roads. *Environ. Sci. Technol.* **2022**, *56*, 9593–9603. [CrossRef] [PubMed]

World Electric Vehicle Journal

MDPI

Article

The "Semiconductor Crisis" as a Result of the COVID-19 Pandemic and Impacts on the Automotive Industry and Its Supply Chains

Benjamin Frieske [1,*] and Sylvia Stieler [2]

1 Institute of Vehicle Concepts, German Aerospace Center (DLR), Pfaffenwaldring 38-40, 70569 Stuttgart, Germany
2 IMU Institut GmbH, Hasenbergstraße 49, 70176 Stuttgart, Germany
* Correspondence: benjamin.frieske@dlr.de; Tel.: +49-(0)711-6862-623

Abstract: In the first half of 2020, the coronavirus pandemic led to a drastic slump in the automotive industry, which was replaced by a surprisingly rapid growth in demand in the fall of 2020, and consequently led to the current shortages in microelectronic products. The prospect of an equally rapid economic recovery in the automotive industry is still threatened by supply bottlenecks for raw materials and key components, foremost for semiconductors. The so-called 'semiconductor crises' show exemplarily the overlapping of short-term supply chain disruptions with long-term structural features of the semiconductor industry. The combination of both is preventing that the supply situation in the automotive industry will improve quickly. First in this paper, the reasons for and respective effects of the crisis on the automotive industry are investigated by a quantitative market analysis. Second, specific strategic measures and options of automotive Original Equipment Manufacturers (OEM) and suppliers in Germany to cope with the situation and increase resilience in future supply chains are described by the means of qualitative expert interviews. By that, the paper helps in understanding the actual situation in the automotive industry, on the one hand, and contributes to the field of strategic supply chain and risk management with a focus on practical implications on the other hand. The results aim to support political stakeholders as well as small and medium sized enterprises to prepare themselves for future developments in the automotive market and changes in manufacturer–supplier relationships due to the transformation to new powertrain technologies and digitization.

Keywords: market development; automotive industry; strategy; semiconductor; supply chain; COVID-19

Citation: Frieske, B.; Stieler, S. The "Semiconductor Crisis" as a Result of the COVID-19 Pandemic and Impacts on the Automotive Industry and Its Supply Chains. *World Electr. Veh. J.* **2022**, *13*, 189. https://doi.org/10.3390/wevj13100189

Academic Editor: Joeri Van Mierlo

Received: 4 July 2022
Accepted: 9 October 2022
Published: 16 October 2022

1. Introduction

As a direct consequence of the coronavirus pandemic, both demand and sales in the global passenger car market collapsed. The COVID-19-related closures of production plants led to global production losses for automotive manufacturers; in the first and second quarters of 2020, automotive production fell by a total of around 9.6% (approx. 7.7 million vehicles) and OEMs reduced their capacity planning requirements for supplier parts accordingly, including electronic components and semiconductors [1]. At the end of 2020, however, demand rose again surprisingly quickly, driven, in particular, by a recovery of the Chinese automotive market and by the higher sales of electrified vehicles associated with the German innovation premium. By the end of the year, passenger car sales in China alone were around 500,000 vehicles higher than the figures forecast in the fall of 2020 [2].

At the same time, demand for consumer electronics (e.g., smartphones, game consoles, TVs) increased during the coronavirus pandemic, which led to a change in prioritization and a shift in capacity of semiconductor manufacturers towards IT and consumer electronics customers. As a consequence, bottlenecks in the production and supply of electronic components for the automotive industry occurred, first and foremost among suppliers in the fourth quarter of 2020. In the meantime, and still ongoing in 2022, however, it is not only the suppliers who are affected, but almost all automotive manufacturers, who have had to reduce or (temporarily) completely stop the production of vehicles due to a lack of components and parts.

The use of semiconductors in vehicles will continue to increase due to electromobility and autonomous driving. Therefore, the stability of the supply chain is of high strategic importance for both the German and the European automotive industries.

The goal of this paper is to analyze the reasons for the ongoing shortage in semiconductor supply in the automotive industry, its impacts on automotive OEMs and suppliers, as well as to show how automotive companies in Germany develop strategies to cope with the situation and increase resilience in future supply chains.

2. Methodology

The analyses described in this paper were conducted using a combination of quantitative market analysis to (1) illustrate the causes and impacts of the crisis on the automotive sector and its companies, on the one hand, and qualitative expert interviews to (2) derive short- and long-term strategies for dealing with the crisis and increasing the resilience of supply chains in the future, on the other hand.

For the former (1), particular reference was made to news articles in the automotive industry, production and industry data, as well as annual reports of the companies affected at all stages of the value chain (OEM, Tier 1, Tier 2, Tier 3). It builds on the findings of existing market studies and enhances them with a special focus on the automotive and semiconductor industry and its dependencies in Germany [3–11]. It is accompanied by a description of the specifics and peculiarities of supplier relationships and value chains in the German automotive industry.

For the latter (2), and complementary to the market analysis, the study is based on 26 guided expert interviews conducted between July 2020 and September 2021. The interviewees primarily hold executive positions (including CEOs, heads of purchasing, heads of logistics, plant managers, production, supplier management) in the German automotive and mechanical engineering industries and cover all relevant stages of value creation in the automotive sector. In total, the interviews covered 3 OEMs, 4 system suppliers (Tier 1), 6 Tier 2 suppliers, 6 Tier 3 suppliers, 3 service providers, as well as 2 associations and 2 trade unions. They were conducted by telephone or video call, recorded, transcribed, anonymized, summarized and relevant statements were clustered according to the structure of the interview guide. The interview guide consisted of 7 main sections covering all relevant topics to assess the effects of the semiconductor crisis in the respective institutions: 1. Impacts of COVID-19/lockdowns/shortages on the company; 2. impact on production sales and labor volume; 3. specific supply chain disruptions and causes; 4. measures to handle the specific disruptions; 5. options to avoid future disruptions; 6. changes in future purchasing/procurement strategies; 7. expectations regarding transformation and electrification of the powertrain.

Results on how companies dealt with the semiconductor shortages and how they will try to increase supply chain resilience in the future were processed and compared in anonymized form. These will be described in Section 4 for the areas of operational measures (short-term) and strategic options (long-term). By that, this paper contributes significantly to the field of strategic management of supply chain risks with a special focus on the current practical implementation of solutions using the example of the German automotive industry. These results are unique because comparable studies focusing on practical solutions and options in the German automotive industry based on in-depth interviews do not exist in the current literature.

3. Results

In this section, the main results of the market study on the reasons for the semiconductor crisis (Section 3.1) and impacts on the automotive industry (Section 3.2) will be described and subsequently supplemented by the results of the expert interviews on (Section 3.3) options and strategies for German automotive companies to cope with the crisis and increase resilience in future supply chains.

3.1. Reasons for the Semiconductor Crisis

Various factors have led to the current supply bottlenecks, with both acute incidents and longer-term conditions having an impact on production volumes and the availability of electronic components for the automotive industry. The incidents that had an acute impact at the beginning of the crisis include:

- A fire at a semiconductor factory owned by chipmaker AKM in Nobeoka City, Japan:

In October 2020, a fire at a semiconductor plant of the Japanese chip manufacturer AKM, a subsidiary of Asahi Kasei K.K., led to the complete closure of production lines at the plant in Nobeoka City for about 12 months [12]. The electronic components produced there are used, among others, in audio and navigation systems for the automotive industry.

- A cold snap in Austin, Texas, North America (USA):

In February 2021, an arctic winter storm led to the collapse of the power supply and the closure of semiconductor factories in the US state of Texas. Among others, plants of NXP Semiconductors N.V., one of the largest suppliers to the automotive industry, Samsung Group and Infineon Technologies AG were affected [13–15]. Production was restricted for approximately 1 to 1.5 months.

- A fire at a semiconductor plant owned by chipmaker Renesas in Naka, Japan:

In March 2021, a fire at a semiconductor plant of the Japanese chip manufacturer Renesas Electronics K.K. in Naka led to a halt in the production of wafer components in particular [16].

The longer-term factors are rooted in the structure of the semiconductor market and the semiconductor industry itself, as well as in complex, elaborate production processes for manufacturing the chips. These more structural conditions mean that the problem is difficult for the automotive industry to solve in the near future. They will be described in detail in the following.

3.1.1. Structure of the Semiconductor Market and Automotive Share

The current structure of the semiconductor market works against the possibilities of a short-term increase in production capacities for customers in the automotive industry, as it represents only a small share of sales compared with other segments (such as communications). In 2020, for example, they accounted for only around 11% of global sales, while the communications and data technology segments accounted for almost 65% of total sales, amounting to EUR 352 billion [3].

China is now the world's largest sales market for semiconductor applications, with a market share of 35% in 2020. The development of market shares for the world regions of China, the Americas, Europe and Asia/Pacific from the year 2000 is shown in the following Figure 1:

Europe and Japan have an overall market share for semiconductors of less than 10% worldwide, and furthermore, the demand is mainly automotive-driven: the automotive industry accounted for 37% of semiconductor sales in Europe and 28% in Japan in 2019. By comparison, other segments dominate demand in the USA (with 10% automotive share) and China (7%). In the USA, for example, the "Computers" segment is in the lead with a share of 38%, in China the "Communications" segment is in the lead with 42% [4].

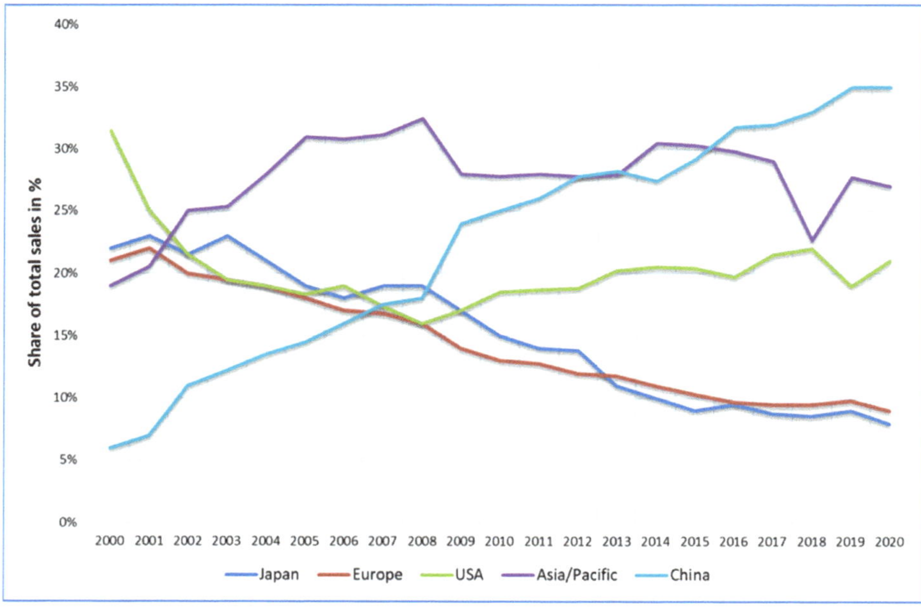

Figure 1. Development of market shares in semiconductor sales, 2000–2020 [3,4].

The shares of the individual segments in the total sales of the semiconductor market in comparison of the regions "World" and "Europe" are shown again graphically in the following Figure 2—the strong automotive focus for Europe (shown in purple) is clearly visible:

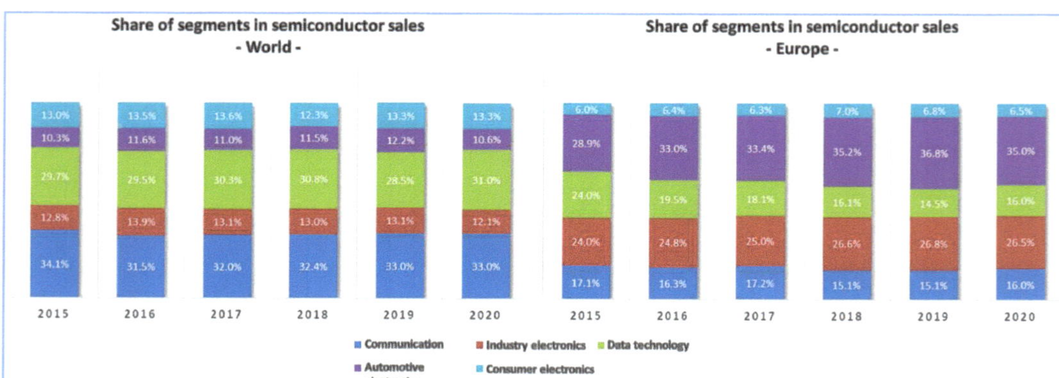

Figure 2. Share of individual segments in semiconductor sales in comparison with the world and Europe, 2015–2020 [3,4].

As an example, for eight of the largest semiconductor manufacturers, the respective relevance of the automotive business in terms of sales is shown below, whereby only NXP (44%), Infineon (42%) and Renesas (48%) have significant sales shares in the automotive sector. The other manufacturers, Texas Instruments (20%), Qualcomm (4%), AMD/ATI (1%) and NVIDIA (6%) primarily serve customers in the communications and consumer segments, which therefore buy much higher volumes overall and also provide chip manufacturers with greater profit margins. Even the world's largest chip manufacturer, TSMC

(Taiwan), has an automotive share of sales of only 3% (annual reports of NXP, Infineon, Renesas, Texas Instruments, Qualcomm, AMD, NVIDIA, TSMC, 2020). The market and negotiating position of the automotive industry for semiconductor products is correspondingly lower compared to other industries.

In the area of semiconductor components for automated and autonomous driving, the German manufacturers Audi, BMW, Mercedes and VW rely mainly on cooperation with the U.S. companies Qualcomm and Texas Instruments, and also in part on NVIDIA. In the case of electronic components for processing signal and information data, the German OEMs are supplied by Texas Instruments, BMW and Volkswagen, also by NXP (Netherlands), and Audi, also by Renesas (Japan). For the supply of power semiconductors for vehicle control, all German OEMs rely primarily on cooperation with Infineon (Germany) (Annual reports Audi, BMW, Infineon, Mercedes, NXP, Qualcomm, Renesas, Texas Instruments, VW, 2019).

3.1.2. Structure of the Semiconductor Industry and Production Capacities

The structure of the semiconductor industry itself, as well as the complex production process, also counteract any short-term easing of the supply situation in the automotive industry. Companies in the semiconductor industry can generally be divided into contract manufacturers with production capacities (so-called foundries) and companies without their own manufacturing and production facilities (so-called fabless) [5].

The latter focus their activities, in particular, on the areas of development of superior functions and designs (e.g., circuits), but not on the development and construction of investment-intensive production facilities. The so-called fabless companies include NVIDIA (USA), AMD/ATI (USA) and Qualcomm (USA), all of which—as described above—do not have a sales focus on the automotive business. Nevertheless, they also develop products that are a prerequisite for the realization of automated and autonomous driving functions. The main customer of these companies is TSMC in Taiwan, the world's leading company in terms of production capacity only in the 200 mm wafer sector [5]. However, TSMC also has a very small automotive share of sales in comparison.

The so-called foundries have their own production capacities for the manufacture of semiconductor products, which are characterized by high investment sums in the construction of highly automated production lines and can only operate economically competitively with large production volumes and quantities—often in three-shift operation and 24/7. The cost of setting up suitable production facilities amounts to several billion dollars [17]. The foundries include manufacturers such as TSMC (Taiwan), Texas Instruments (USA), Infineon (Germany), NXP (Netherlands) and Renesas (Japan). These producers focus mainly on the manufacture of chips for processing signal and information data (e.g., network controllers and navigation) or for vehicle control. Most of the other producing companies are located in Asia, e.g., Samsung (Korea), UMC (Taiwan), SMIC (China), TowerJazz Panasonic Semiconductor (Japan), VIS (Taiwan), PSMC (Taiwan), Hua Hong (China) and DB HiTek (Korea).

Production facilities at Chinese locations are also leading in the production of so-called wafers (the basic material of electronic components, primarily made of silicon) required for semiconductor components in a comparison of monthly production capacities of 200 mm variants: with approx. 5.6 million units and a share of 20%, ahead of Taiwan with approx. 5.3 million units (19%), South Korea and Japan with approx. 4.8 million units each (17%), the USA with 3.1 million units (11%) and Europe with 2.2 million units (8%). Fabs in these six world regions represent a total of 92% of global wafer production (approx. 28 million units per month). Experts expect this share to shift even further towards Chinese production facilities by 2024, with simultaneous growth in global output volume per year of approx. 5% to then approx. 36 million units per month [3].

In an analysis by headquarters of the leading semiconductor companies, however, U.S. manufacturers dominated the global market in 2019, with a 51% share of sales, followed by South Korea (19%), Europe and Japan (10% each), Taiwan (6%) and China (5%), as shown in Figure 3. Over a period of around 10 years, US companies have consistently accounted for a high share of sales (between 48% and 54%), similarly Europe and Taiwan with, however, much lower shares of sales (between 12% and 9% and 4% and 7%, respectively). Japan has seen a sharp decline over the same period since 2001, from 28% to just 10% in 2019, while at the same time South Korean companies increased their sales share from 6% to 24% in 2018. In 2019, however, this fell back to 19%. Chinese companies also increased their share of sales from 0% in 2001 to 5% in 2019 [3].

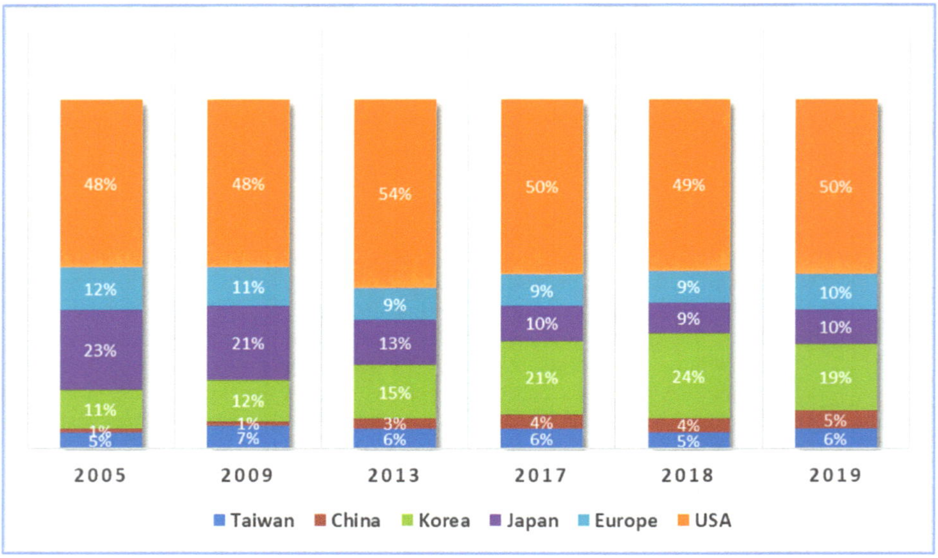

Figure 3. Development of market shares of sales by company headquarters [3].

3.1.3. Complex Production Processes and Supply Chains

The manufacture of semiconductor components is subject to a complex and elaborate production process, which also counteracts any rapid easing of the supply situation in the automotive industry. A total of around 5 to 8 months elapse between the order and the delivery of the component. The production of the wafer takes the most time, at around 3 months.

The raw material silicon required for this is separated from the quartz rock at high temperatures via a chemical reaction (so-called raw silicon) and freed from foreign substances (such as iron, aluminum, phosphorus). Subsequently, the arrangement of the atomic lattice structure is homogenized, and a so-called single crystal is produced from the silicon substrate or "pulled" from the silicon melt (so-called ingot).

These single-crystalline rods are processed into wafers in various steps, which are then coated and polished to form extremely flat surfaces, into which the circuit structures are burned on different levels in the nanometer range, for example using high-energy lasers (lithography). Finally, the individual chips are separated from the wafer, e.g., by sawing.

Wafers are produced in different sizes from 100 mm to 300 mm, whereby an increase in the diameter allows a higher production throughput and thus, lower production costs. By increasing the diameter from 200 mm to 300 mm, for example, the number of chips per wafer can be doubled. If the diameter is increased further (to 450 mm, for example), problems arise due to more complex processing steps and possible deformation.

The following Figure 4 schematically illustrates the production process and supply chain from ordering to delivery:

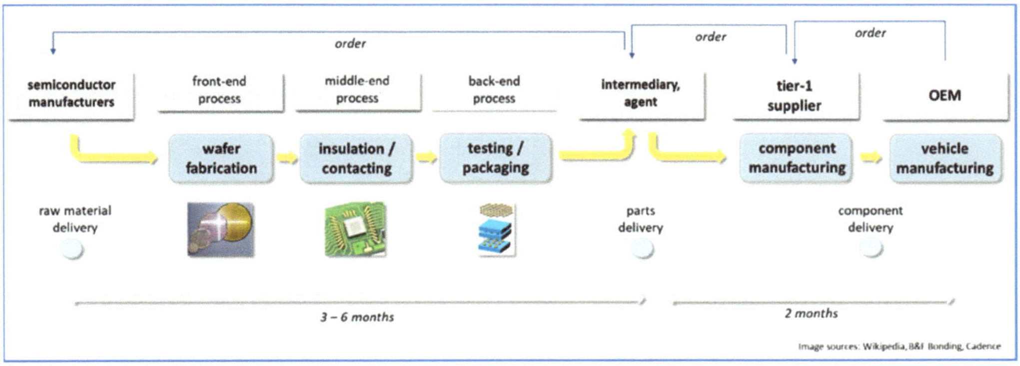

Figure 4. Production and supply chain for semiconductor components [18].

3.1.4. Increasing Importance of Microelectronics in the Automotive Industry

The demand and market for semiconductor elements will continue to grow in the automotive sector in the future. This is due to the higher requirements for digital and networked functions in the vehicle compared to today's conventionally powered vehicles, increasing levels of automation, the use of driver assistance systems and the electrification of the powertrain [6].

The value shares for semiconductor components thus increase from approx. EUR 330 to EUR 690 material value for plug-in hybrids or purely battery-electric vehicles against the background of electrified drive components alone. Automated and/or autonomous driving functions will require additional semiconductor elements with a pure material value of up to approx. EUR 1,030 [19]. According to Roland Berger, the cost of the semiconductor end products in a premium vehicle with a combustion engine is already around EUR 2,500 today and will rise to EUR 5,900 by 2025 in a passenger car with semi-autonomous driving functions [6]. The growing market for electrified and automated vehicles will further increase sales opportunities for semiconductors in the future. Semiconductor manufacturers are also responding to this by investing in the expansion of additional production capacities. As a rule, the construction phase for a semiconductor fab is expected to take around 3 years and cost up to EUR 20 billion (for state-of-the-art production in the 5 nm range).

3.2. *Impacts on the Automotive Industry and its Supply Chains*

In this section, the specifics of the supply chains in the automotive industry will be described in general, reaching from the extraction of raw materials to the delivery of products to end customers, before exemplary impacts of the semiconductor shortages on OEMs and their production capabilities are shown.

3.2.1. Characteristics of Automotive Supply Chains

The so-called "supplier–manufacturer–customer network" in the automotive industry extends around the entire world, especially for raw materials and for some intermediate products and components. To coordinate this network, OEMs and suppliers have established a supply chain management system to secure and optimize product, information and cash flows in a competitive environment. This supply chain management is defined as "the establishment and management of integrated logistics chains [...] across the entire value creation process, starting from the extraction of raw materials through the refinement stages to the end consumer" [20].

A distinction can be made between the so-called basic supply chain and the extended supply chain [21]. The former comprises the manufacturing company, its direct suppliers (Tier 1) and direct customers—in the automotive industry, this can be the retail trade, but in some cases, it can also be bulk buyers such as car rental companies or logistics companies, or direct sales to end customers. The extended supply chain expands the management of the supply chain to include additional value-added stages up to suppliers active in the fields of raw material processing. In the course of the change in value creation strategies of automotive manufacturers in the transformation to electrified powertrains, this concept is even being extended by OEMs to the level of raw material extraction.

Today's supply chain in the automotive industry consists of up to five internal and ten external stages [22]. Automotive manufacturers are integrators at the top of a hierarchical supplier system characterized by a division of labor between OEMs and suppliers, and by a high degree of modularity and flexibility in the selection and use of suppliers.

OEMs and Tier 1 suppliers influence the structure of their respective suppliers by limiting the number of direct suppliers. This structure, described by the image of the "supplier pyramid", is based on the fact that manufacturers and a few globally active suppliers are at the top, followed by development service providers, contract manufacturers, just-in-sequence and just-in-time suppliers, and logistics service providers as Tier 2 to Tier n suppliers, depending on the complexity of the input products and the share of product and process innovations. At the lowest level (e.g., Tier 3), highly standardized parts, e.g., screws, are usually manufactured, while at the next higher level (e.g., Tier 2), components such as rotors, stators, valves or cylinders are manufactured. At the highest level of system suppliers (e.g., Tier 1), components are integrated into modules and systems, e.g., battery modules, steering systems, and then delivered to the automaker for final assembly within the vehicle. The position in the supplier pyramid and the dependency on individual customers or OEMs decisively determines the dependency of the company and thus the scope of its own corporate strategy; with their buying power, OEMs hold central positions here [23]. The close interaction between OEMs and suppliers has emerged during a long phase of industry growth. It is considered an important element of the premium and quality strategy of German OEMs, has been maintained over the long term despite high cost and price pressure and can be described as a production knowledge-based innovation cluster [24].

The image of the supplier pyramid also describes a model gradation of suppliers according to the complexity of the preliminary product. Since the 1980s, suppliers in the automotive industry have been taking on ever larger shares of the value chain, including more complex tasks and parts. As a result, they also have greater obligations when it comes to financing upstream products and materials. For OEMs, reducing their own depth of value added and outsourcing to suppliers was and is a lever for maximizing earnings.

The internationalized value chain, which is based on a high division of labor, places high demands on companies to ensure trouble-free supply. Starting from the original purchasing or procurement, OEMs and Tier 1 suppliers have now established a supplier or supply chain management system for this purpose. Whereas the initial focus of supplier relationships was primarily on the question of whether primary products or components could be manufactured in-house or purchased more cheaply ("make or buy"), suppliers are now audited/certified by OEMs and production processes, approved, advised on optimizing production processes and supported financially or by other means in the event of threats to the supply chain.

OEMs have long been shifting value creation to their suppliers in order to be able to focus on global markets and the interfaces with their customers [22]. Due to their financial strength, they continue to maintain a central position and exert a strong influence on contract design with suppliers. Nevertheless, the image of a strictly hierarchical supply chain or supplier pyramid is changing; at least in part, value networks are emerging instead, in which companies can focus better on their core competencies and act more flexibly. Suppliers can improve their position in the value chain through more strategic partnerships with OEMs and Tier 1 suppliers. The shift to electromobility, with new supply relationships, and in some cases new suppliers, is accelerating these changes. However, this option is limited, on the one hand, by the fact that mastering complexity along global supply chains will continue to be a key challenge for OEMs and Tier 1 suppliers in the future. On the other hand, OEMs are taking back at least some of their production shares to compensate for the loss of value creation in electric vehicles.

3.2.2. Exemplary Impacts of the Semiconductor Shortages on the Automotive Industry

An exemplary overview of impacts of the semiconductor shortages on the automotive industry in terms of sales and production figures are shown below, with a special focus on the effects for single OEMs and suppliers [25]:

- Audi had to send part of its workforce to short-time work due to a shortage of electronic components at its Ingolstadt and Neckarsulm plants (A4 and A5).
- BMW suffered short-term production stoppages at its plants in Regensburg (1 Series, 2 Series, X1, X2), Leipzig (i3) and Oxford, England (Mini) due to a lack of parts deliveries.
- Daimler had to initiate a production stop at the Sindelfingen (E-Class), Bremen (C-Class, GLC) and Rastatt plants, as well as at the Kecskemet plant in Hungary (A-Class, B-Class, GLA); the respective employees were moved to part-time working for a limited period. The supply of electronic components is being managed in such a way that they can be used in high-margin models (such as the S-Class or EQS) on a prioritized basis and these can continue to be produced.
- Ford had to stop production at the Cologne plant for several weeks, affecting around 5000 of the 15,000 employees there. Production was also reduced or completely interrupted at the plants in Saarlouis (Focus), Gölcük in Turkey, Craiova in Romania, Valencia and other plants in the USA. The company expects the supply bottlenecks to reduce profits by USD 1 billion to USD 2.5 billion in 2021.
- Regarding Volkswagen, the plants in Wolfsburg (Golf), Emden (Passat), Mexico (Jetta, Tiguan) and Slovakia (Bratislava, mainly SUV models) have been affected. Across the Group, approximately 100,000 fewer vehicles could be produced in the first quarter of 2021 due to the semiconductor bottlenecks. As a consequence, Volkswagen is responding by establishing its own relationships with semiconductor producers in order to be able to negotiate capacities directly and manage them better. Inventories are also to be increased for special chips in the future.
- Furthermore, Opel, Tesla, Toyota, Nissan, Hyundai, Jaguar Land Rover, Peugeot and Renault, among others, also reported problems due to missing components and supply bottlenecks in 2021.

Overall, supply bottlenecks have led to a drastic reduction in production volumes in 2021 and even beyond. In May 2021, the management consultancy Alix Partners assumed that around 3.9 million fewer vehicles, worth about EUR 90 billion, were to be produced worldwide in 2021 [1]. Also, LMC Automotive expected that the lower availability of chips would lead to a drop in production of approx. 1.3 to 2.2 million passenger cars, which would correspond to a decline of approx. 1.5% to 2.5%, with an expected total volume of around 88 million units for 2021 [2]. In reality, the number of vehicles not produced because of the semiconductor crisis added up to about 9.7 million light-duty vehicles in 2021 alone [26]. Experts expect the supply shortage to continue in 2022 and beyond [27].

3.3. *Strategies and Options to Cope with Shortages*

In this section, company strategies to deal with the disruptions in supply chains will be analyzed and described in detail. With its high dependence on exports, and the structures of a diversified, internationalized division of labor in global value networks that have been created over decades, the German automotive industry appears to be particularly vulnerable to external disruptions in the supply chain. In the course of the study, this thesis was discussed in expert discussions and in-depth interviews. The aim was to identify the effects of the supply chain disruption on the affected companies in the automotive industry and to identify measures for overcoming them. The latter include both operational, rather short-term measures, arising directly from the COVID-19 crisis, as well as strategic options and consequences with a rather longer-term perspective. The latter are aimed at generally increasing resilience in relation to external disruptions to production and supply chains.

3.3.1. Operational Measures (Short-Term)

Short-term and operational measures to deal with the disruptions caused by the COVID-19 pandemic and the shortage in semiconductor components could be clustered into the following five topics: "Use of short-time work and closure of production plants", "Adjustment of production processes", "Adjustment of inventories", "Establishment of central emergency teams and task forces" and "Search for alternative suppliers". The mentions relating to "Short-time work" and "Adjustment of production processes" were the most frequent. In the case of multiple mentions, the area "Establishment of central emergency teams and task forces", in particular, was highlighted as a very relevant measure, both in relation to the company itself and in relation to the higher-level customer. The "Search for alternative suppliers" was mentioned only once.

Use of short-time work and closure of production plants: 66% of the interviewees were affected by the lockdown and the associated halt of production activities. They mainly used labor market policy instruments as a result of the lockdown, with part-time work in particular.

Adaptation of production processes: 53% adapted their production processes as a result of the pandemic, in particular to be able to implement safety measures to protect against infection. Where the minimum distance of 1.5 m could not be maintained (e.g., on production lines), plexiglass walls were built up, masks were made compulsory and disinfection facilities were installed. Overall, the implementation of these measures resulted in an increase in infection protection, but at the same time led to productivity losses for the companies. Unexpectedly, hygiene articles such as masks have turned out to be "critical goods" and have presented companies with procurement problems

Adjustment of inventories: 60% mentioned adjusting inventories as a tactical measure to deal with the pandemic and customers' often unclear demands and call-offs. In this context, inventory levels were both increased and decreased. On the one hand, in the early phase and before the (partial) lockdown, production was continued to a large extent despite unclear customer demand, or deliveries were accepted and inventories increased in order to be able to continue to deliver with high volumes and at short notice in the event of an improvement in the coronavirus situation. Secondly, in the later course of the crisis, existing inventories were reduced again in order to be able to manufacture and deliver the company's own products. This meant that the company was less dependent on upstream products that were no longer supplied or only supplied in small quantities. Adjustments to inventories were mainly made for products identified as "critical", such as electronic parts or cable harnesses.

Establishment of centralized emergency response teams and task forces: The establishment of task forces was named by 64% of the interviewees as a short-term operational measure for dealing with the crisis. In this context, the interviewees were referring, in particular, to difficulties in general communication with customers in terms of the reliability and predictability of requirements and orders. They emphasized that customers, in particular, were able to quickly set up task forces and establish them at the interfaces of the supply chain, which have their own central logistics in a strong position and a distinct, established supplier management. Reference was also made here to those companies that were affected by Fukushima in 2011, where they were able to gain experience in setting up emergency task forces.

In addition to the organizational establishment of centralized emergency response teams and task forces, digital technologies for the exchange of information and, in particular, the tracking of the flow of goods along the value chain were also mentioned by 23% of the interviewees. This includes blockchain technology, which is intended to ensure a secure flow of information and is being tested by the automotive industry in initial pilot projects. Here, however, it is questionable whether cross-company data exchange along the entire supply chain is actually feasible. On the one hand, the emerging data volume is currently unmanageable, and on the other hand, the companies must agree to the data transfer. It is hard to imagine the semiconductor industry granting what it considers an insignificant market share appropriate access here.

Search for alternative suppliers: Only one interviewee mentioned the search for alternative suppliers as an operational measure to deal with the crisis. The majority of the companies were able to maintain their own production capacities on the basis of existing supplier relationships and/or by reducing inventories until production was able to restart again.

3.3.2. Strategic Options (Long-Term)

In the first weeks of the lockdown, broken supply chains were discussed as the cause of the production halt in the automotive industry. To avoid and/or better manage this problem in the future, a greater return to local production was called for as a way to contain the pandemic and stabilize the economy, according to broader public discussion. However, in the expert interviews, the opposite position was stated for the automotive industry: there will be no fundamental change in international division of labor or associated global supply chains. Instead, a modification of purchasing strategies is more likely. Overall, four fields of action could be identified in the question of long-term strategic fields of action for stabilizing supply chains in crisis situations: "Increase stockholding for critical components", "Strengthen dual sourcing and flexible shares", "Support local supply chains" and "Monitoring of the reliability of political action". The responses were equally distributed across all interviews. In the case of multiple mentions, the area "Reliability of political action" was identified as most relevant option.

Increase inventory levels for critical components: 40% of the interviewees named an increase in inventories for critical components as a reasonable strategic option. Vulnerability to disruptions was already evident in 2011 during the Fukushima crisis but affected not as many products there. The COVID-19 pandemic with its global impact endangered production in the entire supply industry. Therefore, in addition to the identification of critical components and suppliers, higher stock levels for single parts were set up. Since warehousing requires corresponding areas, systems and staff, activities always have to be weighed against disadvantages in cost and efficiency in the automotive industry. For this reason, larger warehousing—if at all—would only be practicable for individual components that are considered to be particularly critical.

Strengthen dual sourcing and flexible shares: Another 40% stated that strengthening dual sourcing and flexible shares with suppliers is being discussed as a strategic option for the future. Dependence on one supplier, especially a foreign supplier, has already revealed weaknesses in supply chains during Fukushima and even more so in the pandemic. In the interests of resilient supply chains, consideration is therefore being given to purchasing from two suppliers instead of one for selected components and to a limited extent. As a result of the global disruption, splitting with a low-cost supplier, e.g., in Asia, and a European or German supplier is being considered here. However, this will only lead to a small return of production, if at all. This is because, when split between the suppliers, the significantly larger volume would still be sourced from the lower-cost foreign countries; the experts cited proportions of 70 to 80% here.

Supporting local supply chains: 33% mentioned a stronger focus on local supply chains as a possible strategic option, and as critical for the resilience of future value networks. In many cases, it is not only supplier relationships in Germany that are rated as "geographically local", but also European ones on a larger scale. As a result of the pandemic, the need to adapt purchasing strategies in the interests of greater stability is seen. Here, the company experts see an advantage in greater geographical proximity, and thus, a focus of the supply chains in the European countries or in the future also in the (North) African countries. At the same time, the experts have pointed out that sourcing components and parts from European locations is associated with higher costs compared with sourcing from Asian production. Therefore, the advantages of regionalized supply chains must be weighed against higher costs. This raises the question of how far the current impressions from the pandemic will extend into the future, or whether purchasing decisions will soon again be made primarily on the basis of cost efficiency.

Monitoring the reliability of political action: When asked about possible political action, the experts focused mainly on reliability, and thus, on maintaining social and economic stability. This topic was mentioned most frequently among the long-term measures (47%). Three levels of action should be distinguished here: the municipal level or that of the federal state, the national level and the pan-European level. With its decisions on how to deal with the pandemic, the political arena provides the essential framework conditions under which economic activity, such as the company's production, the purchase of inputs and the sale of its own products, is possible in the first place. Only the most effective possible limitation of the pandemic protects against restrictions on the company's own production. The interviewees certainly acknowledged the great difficulties that politicians faced with the first lockdown decision in German history. Nevertheless, it is helpful for companies at the local level to have reliable and comparable regulations across different counties and cities. This was not always the case in the first lockdown in March/April 2020. For companies in the automotive industry, which are heavily dependent on foreign trade, this demand for politically regulated stable framework conditions also applies to other states with which they maintain economic relations. After all, limiting the pandemic can protect against a lockdown there as well. However, this change in supply chains is associated with disadvantages in terms of cost efficiency and, depending on the corporate strategy, is not feasible for everyone. In return, companies expect stability within the European economic area and agreements on border crossings for business purposes. Here, however, the expectations of companies in the automotive industry and the machine tool industry differ: while the automotive industry was primarily concerned with the supply of components and parts from foreign production sites, for the machine tool industry, the greatest difficulty was primarily in the supply of machines and in the commissioning and repair of machines abroad.

4. Conclusions

Despite previous crises, the COVID-19 pandemic and the subsequent shortage in semiconductor components proved to be a shock to the economy and the automobile industry: for the first time the entire world and almost all economic sectors were affected, and supply chain disruptions will continue to negatively impact production capacities even in and beyond 2022. The production plant lockdowns and semiconductor shortages, however, were not the sole trigger of the slump in demand that has been evident in the automotive industry since 2019 and in mechanical engineering since 2018. Already in the summer of 2020—after the initial lockdown and improvement in the infection situation—the pandemic took a back seat to the demand crisis and questions of transformation toward new technologies of electrification and digitalization.

Our results contribute to the field of strategic management of supply chain risks, with a special focus on the current practical implementation of solutions using the example of the German automotive industry on the one hand and to the practical handling of supply chain risks in the German automotive industry based on the current semiconductor shortages on the other hand. Our findings are unique because comparable studies focusing on practical solutions and options in the German automotive industry based on in-depth interviews do not exist in the current literature.

They clearly show that the European and German automotive industries have little influence on the stability of semiconductor supply chains due to their small market share, but are heavily affected in terms of production capacities. In this respect, approaches for new purchasing and logistics strategies to make the supply chains more resistant to crises in the long term are being discussed in automotive companies. Operationally, increased warehousing and dual sourcing—especially for critical components—could become more important in the future [28]. Strengthening local (European) value networks for critical components is also being considered by OEMs and suppliers. However, both are described more as "incremental improvements" than as a fundamental change in strategy, and both lead to a more holistic view of supplier relationships in the sense of more systematic management of supply chains, product and supplier risks. Research limitations of our study are based on the quantity of expert interviews held and the quality of the statements therein. The research design of this study has therefore been selected in such a way that each stage of the value chain in the automotive industry (OEM tier 1, tier 2, tier 3) and different perspectives in the supplier–manufacturer alliance could be fully covered by the expert interviews conducted.

The implementation of these strategies, their impact and their potential to improve the handling of supply chain risks in the German automotive industry should be investigated in further research and also be compared to specifics in other world regions, such as France, the United States of America, Japan, South Korea and China. Building up on our findings, the aim of these studies could be to (1) identify best practice examples for dealing with potential shortages of critical materials and components on the basis of further analyses and (2) to expand the research scope and also include change potentials for value creation networks and supply chains resulting from the transformation of the automotive industry towards new powertrain technologies, digitalization and CO_2-neutrality.

Author Contributions: Conceptualization, B.F. and S.S.; methodology, B.F. and S.S.; validation, B.F. and S.S.; formal analysis, B.F. and S.S.; investigation, B.F. and S.S.; resources, B.F. and S.S.; data curation, B.F. and S.S.; writing—original draft preparation, B.F.; writing—review and editing, S.S.; visualization, B.F.; supervision, S.S.; project administration, B.F.; funding acquisition, B.F. All authors have read and agreed to the published version of the manuscript.

Funding: This research was funded by e-mobil BW GmbH, Landesagentur für neue Mobilitätslösungen und Automotive Baden-Württemberg.

Data Availability Statement: Not applicable.

Conflicts of Interest: The authors declare no conflict of interest. The funders had no role in the design of the study; in the collection, analyses, or interpretation of data; in the writing of the manuscript; or in the decision to publish the results. Sylvia Stieler is employee of the IMU Institut GmbH. The paper reflects the views of the scientists, and not the company.

References

1. Alix Partners. Der Chipmangel Führt Weltweit zu 3,9 Millionen Weniger Produzierten Fahrzeugen im Jahr 2021. Available online: https://www.alixpartners.com/media-center/press-releases/prognose-chipmangel-in-automobilproduktion-2021/ (accessed on 30 June 2021).
2. LBBW Research. *Produktionsausfälle Durch Fehlende CHIPS*; LBBW Research: Stuttgart, Germany, 2021.
3. ZVEI. *Entwicklung der Halbleiterindustrie 2019*; Zentralverband der Elektro- und Digitalindustrie e.V.: Frankfurt am Main, Germany, 2019.
4. ZVEI. *Entwicklung der Halbleiterindustrie 2020*; Zentralverband der Elektro- und Digitalindustrie e.V.: Frankfurt am Main, Germany, 2020.
5. IC Insights. *Global Wafer Capacity 2021–2025—Detailed Analysis and Forecast of the IC Industry's Wafer Fab Capacity*; IC Insights Inc.: Scottsdale, AZ, USA, 2021.
6. Roland Berger. *Computer on Wheels/Disruption in Automotive Electronics and Semiconductors*; Roland Berger GmbH: Munich, Germany, 2020.
7. Roland Berger. *Steering through the Semiconductor Crisis*; Roland Berger GmbH: Munich, Germany, 2021.
8. McKinsey & Company. Coping with the Auto-Semiconductor Shortage: Strategies for Success. Available online: https://www.mckinsey.com/industries/automotive-and-assembly/our-insights/coping-with-the-auto-semiconductor-shortage-strategies-for-success (accessed on 30 June 2021).
9. McKinsey & Company. Semiconductor Shortage: How the Automotive Industry Can Succeed. Available online: https://www.mckinsey.com/industries/semiconductors/our-insights/semiconductor-shortage-how-the-automotive-industry-can-succeed (accessed on 30 June 2021).
10. IHS Markit. Semiconductor Shortage Update: Nearly One Million Vehicles Delayed. Available online: https://ihsmarkit.com/research-analysis/semiconductor-shortage-update-one-million-vehicles-delayed.html (accessed on 30 June 2021).
11. Deloitte. My Kingdom for a Chip: The Semiconductor Shortage Extends into 2022. Available online: https://www2.deloitte.com/xe/en/insights/industry/technology/technology-media-and-telecom-predictions/2022/semiconductor-chip-shortage.html (accessed on 18 December 2021).

12. AKM. Situation Regarding Semiconductor Plant Fire and Product Supply. Available online: https://www.akm.com/kr/en/about-us/news/information/20210122-information/ (accessed on 30 June 2021).
13. NXP. NXP Resumes Operations at Austin, Texas Facilities Following Weather-Related Shutdown and Pro-Vides Revenue Update. Available online: https://media.nxp.com/news-releases/news-release-details/nxp-resumes-operations-austin-texas-facilities-following-weather/ (accessed on 30 June 2021).
14. BusinessKorea. Samsung's U.S. Plant Shutdown Feared to Continue until Mid-April. Available online: http://www.businesskorea.co.kr/news/articleView.html?idxno=61405 (accessed on 30 June 2021).
15. Infineon. Infineon Re-Ramps Production in Austin, Texas, and Provides Update on Customer Impact; Pre-Shutdown Output Level Expected in June 2021. Available online: https://www.infineon.com/cms/en/about-infineon/press/press-releases/2021/INFXX202103-054.html (accessed on 30 June 2021).
16. Renesas, Notice Regarding the Semiconductor Manufacturing Factory (Naka Factory) Fire. Available online: https://www.renesas.com/us/en/about/press-room/notice-regarding-semiconductor-manufacturing-factory-naka-factory-fire, (accessed on 30 June 2021).
17. Puffer, W. *Technisch-ökonomische Effizienzbetrachtungen für die Halbleiterfertigung*; Technische Universität München: Munich, Germany, 2007.
18. e-mobil BW. *Zukunftsfähige Lieferketten und neue Wertschöpfungsstrukturen in der Automobilindustrie*; e-mobil BW GmbH: Stuttgart, Germany, 2022.
19. Infineon. *Fourth Quarter FY 2020—Quarterly Update*; Infineon Technologies AG: Neubiberg, Germany, 2020.
20. Gabler Wirtschaftslexikon. Supply Chain Management. Available online: https://wirtschaftslexikon.gabler.de/definition/supply-chain-management-scm-49361/version-272597 (accessed on 21 September 2021).
21. Muchna, C.; Brandenburg, H.; Fottner, J.; Gutermuth, J. *Grundlagen der Logistik*; Begriffe, Strukturen und Prozesse: Wiesbaden, Germany, 2018.
22. Proff, H. *Neue Dimensionen der Mobilität. Technische und betriebswirtschaftliche Aspekte*; Springer Fachmedien Wiesbaden GmbH: Wiesbaden, Germany, 2020.
23. Bratzel, S.; Retterath, G.; Hauke, N. *Automobilzulieferer in Bewegung: Strategische Herausforderungen für Mittelständische Unternehmen in Einem Turbulenten Umfeld*; Nomos Verlag: Baden-Baden, Germany, 2015.
24. e-mobil BW. *Strukturstudie BWe Mobil 2019—Transformation Durch Elektromobilität und Perspektiven der Digitalisierung*; e-mobil BW GmbH: Stuttgart, Germany, 2019.
25. Automobil-Produktion. Alle Infos zur Halbleiterkrise in der Autoindustrie. Available online: https://www.automobil-produktion.de/hersteller/wirtschaft/autoindustrie-leidet-unter-halbleiter-engpaessen-241.html (accessed on 19 April 2022).
26. IHS Markit. *Supply Chain Crisis Update: Semiconductor Focus*; HIS Markit Ltd.: London, UK, 2022.
27. Elektroniknet. Investitionen für IC-Produktionskapazitäten Steigen. Available online: https://www.elektroniknet.de/halbleiter/investitionen-fueric-produktionskapazitaeten-steigen.179378.html (accessed on 30 June 2021).
28. EC—European Commission. *Strategic Dependencies and Capacities—Updating the 2020 New Industrial Strategy: Building a Stronger Single Market for Europe's Recovery*; European Commission Working Staff Document; European Commission: Brussels, Belgium, 2021.

World Electric Vehicle Journal

MDPI

Article

Empirical Analysis of the User Needs and the Business Models in the Norwegian Charging Infrastructure Ecosystem

Erik Figenbaum *, Paal Brevik Wangsness, Astrid Helene Amundsen and Vibeke Milch

Institute of Transport Economics, Gaustadalléen 21, N-0349 Oslo, Norway
* Correspondence: efi@toi.no; Tel.: +47-90-57-33-98

Abstract: The Norwegian charging infrastructure ecosystem was investigated from a user perspective by (1) developing knowledge of end-user experiences with public charging, (2) mapping BEV owners and future owner's user-friendliness needs and the extent to which these needs are met, (3) pointing at potential user-friendliness improvements, (4) mapping the charging infrastructure ecosystem and business models, and (5) developing scenarios for the future system development and the impact on charging infrastructure user-friendliness. The article draws on the literature, a BEV (battery electric vehicle) and ICEV (internal combustion engine vehicle) owner survey, 15 BEV owner interviews, 21 charging infrastructure actor interviews, and open information sources on the charger actors. The unregulated charging system evolved into a complex web of actors that developed their own charging networks following their individually sensible business models, which in sum led to serious user-friendliness issues. To gain access to all chargers, users need to interact with up to 20–30 apps and 13 payment systems, which comes on top of different plug types, power levels, and charger interfaces. Some actors support roaming, while others oppose it. OEMs want users to interface with chargers through the navigation system. In the future, the system will become even more complex and less user friendly as more actors join unless, e.g., consolidation, regulation, or independent network orchestrators reduce the complexity.

Keywords: charging; infrastructure; business models; user experience; user-friendliness

Citation: Figenbaum, E.; Wangsness, P.B.; Amundsen, A.H.; Milch, V. Empirical Analysis of the User Needs and the Business Models in the Norwegian Charging Infrastructure Ecosystem. *World Electr. Veh. J.* **2022**, *13*, 185. https://doi.org/10.3390/wevj13100185

Academic Editor: Joeri Van Mierlo

Received: 8 August 2022
Accepted: 27 September 2022
Published: 3 October 2022

1. Introduction

This article regarding battery electric vehicle (BEV) charging infrastructure in Norway builds on two papers the authors presented at the World Electric Vehicle Symposium in Oslo 13–15 June 2022 titled: "Demand for standardization of public fast charger stations—user perspectives" [1] and "The Norwegian Charging Infrastructure Ecosystem" [2], as well as a report with an overview of the Norwegian charging infrastructure ecosystem [3]. These documents were developed as part of a research and innovation action that focused on finding ways to improve the user-friendliness of the overall Norwegian charging infrastructure ecosystem. The articles cover user needs for and experiences with public charging infrastructure and the business perspectives of the charging infrastructure through 2021. The hypothesis is that the way the market was developed and has become organized has led to unsatisfactory user experiences. The focus of the study is on user-friendliness issues related to the access to and use of charging infrastructure, not on the number of chargers deployed. The study does not cover the Tesla Supercharger network.

The aim of the study was to (1) develop knowledge concerning end-user experiences of using public charging services in Norway, (2) map BEV owners and future owner's charging needs in terms of the user-friendliness of charging services, (3) map the charging infrastructure ecosystem and the various applied business models, (4) identify to what extent user needs are met within the current charging infrastructure ecosystem, (5) to point out potential user-friendliness improvements, and (6) establish scenarios for the future system development and their impact on the user-friendliness of the charging infrastructure and the actors' business models to provide guidance on how the system can be made to function better.

The article contributes to the global research on charging infrastructure by shedding light on the gap between BEV user needs and what an unregulated charging infrastructure ecosystem actually provides in the world's largest per capita BEV market. Other countries with rapidly growing BEV fleets will need guidance for developing charging infrastructure, as do actors developing charging infrastructure for heavy trucks. Both groups can learn from the Norwegian experience.

The article starts off in Section 2 with a presentation of the BEV and charging infrastructure status in Norway and an overview of the current knowledge of BEV owners need for and interaction with charging infrastructure. Section 3 presents an overview of the research on business models for charging. The materials and methods used is the present study are presented in Section 4, followed by the results in Section 5 and the discussion and the conclusion in Section 6.

2. BEV Ownership and Charging Infrastructure Use in Norway

2.1. BEV Ownership in Norway

The Norwegian charging infrastructure ecosystem case is particularly interesting as Norway is the world leader in BEV adoption per capita [4]. Over 460,000 BEVs were on the road at the end of 2021, which constituted 16% of the total passenger car fleet [5]. The BEV market share reached 54% in 2020, 65% in 2021 [6], and 79% in the first half of 2022 [7]. All other countries are far behind these fleet and market shares. This leading role is the result of three decades of active BEV policies with large incentives [4,8,9], market activities, industrial innovation [10], and charging infrastructure deployment. The parliament's ambitious goal since 2017 has been that all new passenger cars sold in Norway shall be zero-emission from 2025 [11]. This target was a back-casting of the required transport sector contribution to reduce the national greenhouse gas emissions (GHG) by 40% by 2030, which Norway committed to in the 2016 Paris Agreement NDC [12]. These developments put pressure on the build out of fast chargers and other public and private charging infrastructure so that the charging barrier to BEV adoption could be reduced.

2.2. BEV Charging in Norway

In 2000, there were about 3000 BEV owners in Norway [10]. They charged their BEVs mainly at home using domestic outdoor sockets and occasionally at existing publicly available outdoor sockets outside offices, hotels, or cafés [10]. Fast charging was not possible until after 2010. An unregulated fast-growing charging infrastructure ecosystem has since 2010 emerged as a BEV enabler. However, this ecosystem built out by actors following their individual strategies and business models is not providing the expected user experience to many BEV users, which may somewhat deter future BEV growth. Charging infrastructure needs to work more effortlessly as BEV buyers progress from being mainly technology-interested multi-vehicle households to being more like vehicle owners in general [13,14].

BEV owners' preferred charging locations are still primarily at home, then at work and in other public locations [13], a pattern also found in other countries according to a review of the global literature on BEV charging by Hardman et al. [15]. Over 90% of Norwegian BEV owners charge at home [13]. Most households can install a wall box charging solution [13] as they have access to a private parking spot. Flat owners must rely on installation of shared charging systems in the buildings' common parking facilities. Such installations are supported economically by several counties and municipalities [16]. Many of the early adaptors changed to BEVs from internal combustion engine vehicles (ICEVs) due to environmental/technological interest and economic incentives [13,17,18]. When "everyone" is driving BEVs, they will demand more or less the same "seamlessness" at the public charging stations as at petrol stations. This will require an increased standardization, both regarding charging and payment, as well as a better handling of queues. This will be even more important in the future when more car owners without access to private charging facilities make the transition to BEVs.

Fast chargers along major roads and in cities enable longer trips and solve everyday range issues [16]. About 75% of BEV owners in three Norwegian counties reported having used fast chargers when driving long distances [19]. Another Norwegian survey found that the average non-Tesla BEV driver used public fast chargers about 19 times per year and obtained about 5% of the energy used by their BEV from fast chargers [13,20]. Yet, 40% used public fast chargers as rarely as 1–3 times per year. Long distance driving remains one of the final barriers to BEV adoption [13,16] in single-car households. Most BEVs are, however, used by multi-vehicle households that can use an ICEV for long distance driving. BEV models with a longer range can enable long distance trips, but many BEV owners dread taking their BEV for long trips [13,19]. The fast charge network expansion has helped, but there are still several user-friendliness issues that need to be addressed. Users want easy access to public chargers, but user surveys show that charging queues can be an issue on peak travel days/times [13], chargers can be out of order, payment could be easier, and users do not want to use a myriad of apps [13,19] to access chargers.

There were in early 2022 about 4000 public fast charging points in Norway, within a steadily expanding network [21]. There are around 10 major operators in the market with more than 280 charging stations (including Tesla), as well as 20–25 operators with fewer stations and charging points [21]. Unlike petrol stations, where pumps and user interfaces are rather standardized, fast charging in Norway is characterized by different charging systems, power levels, user interfaces, plugs, access and payment solutions, and (up to 2022) also a variety of pricing schemes. Consequently, BEV users must navigate in a complex charging landscape where they interact with a variety of systems and solutions that makes charging less streamlined and more time consuming than filling petrol, especially for those who seldom use public fast chargers. Very few public fast chargers have bank terminals, so users must pay via apps, operator-specific payment cards, SMS messaging, or other means. To avoid extra costs associated with SMS messaging, customers must register their credit card information in each charging operator's system. When registered, the customer identification can operate the charger via an app or an RFID (radio-frequency identification) card. Few actors have opened for roaming, which would have made it easier for users to use the same authentication and payment solution across operators, making the BEV user experience more streamlined. The networks of slow chargers have similar payment systems, access, roaming, and user interface issues.

3. Business Models for Charging Stations

The functioning of the Norwegian charging infrastructure ecosystem and how it is influenced by the business models of its actors has not been investigated previously. There is substantial research on business models for charging infrastructure that focuses on the ability to develop sustainable income. Yet, it seems that research on how charging infrastructure business models can deliver the charging services that users need and prefer has been neglected in Norway and elsewhere. The gap between the services delivered and the needs and wants of users can become a threat to the sustainability of the charging infrastructure ecosystem in the long run and a barrier to increasing the adoption rate of BEVs.

International studies point out that for the BEV charging market to be economically sustainable in the long run, there is a need for businesses that generate value for the end customer, but also a sufficient return on the huge investments made in the charging infrastructure hardware and software. Well-suited business models are needed to achieve this. There is a large amount of public discussion about how the economics of public charging infrastructure depends on the applied business models, but not so much in the scientific literature, according to Zhang et al. [22]. Greene et al. [23] points out that as the charging market appears now, there is uncertainty about what are well-suited business models for public charging, partly because the utilization rate is still generally low, and that the roles of the public and private sectors are not well defined. This conclusion is also supported by de Rubens et al. [24] and Van der Kam et al. [25], who also point out that the difficulties in finding profitable business models are one of the biggest challenges for the roll-out of public charging infrastructure. Studies such as those by the Boston Consulting Group [26], Deloitte [27], Helmus and Van den Hoed [28], and Schroeder and Traber [29] point out that it is common for businesses to struggle to find profitability in investments and operations in public charging infrastructure, especially fast-charging infrastructure. This situation is related to the fundamental fact that the costs of investment and operation are high relative to the size of the initial customer base, i.e., BEV users, and their willingness to pay. A Norwegian charging deployment strategy developed in 2012 [30,31] therefore found that fast chargers should be deployed at locations where other revenue streams could be pursued, e.g., kiosks, restaurants, or shopping centers. BCG [32] found that lack of government guidelines and requirements have resulted in a chaotic ecosystem for charging in several countries.

The ecosystem for EV charging covers many different sectors. Within each sector, there is a myriad of products and services to deliver to different customer segments. Different companies can provide products and services both within a segment of a sector and across segments and sectors. In other words, there are thousands of possible business models based on combinations of these aspects. A classical illustration of the enormous range of possible business models can be found in the morphological boxes in Kley et al. [33] that are used to show the potential for new business models in the expanded ecosystem for electric cars, batteries, charging infrastructure, and the energy system.

4. Materials and Methods

The analysis in this article is mainly qualitative, building on different types of data sources in order to obtain a broad overview, as well as to go in-depth on certain aspects of the current charging infrastructure user needs and to understand the functioning of the current ecosystem. How well the charging infrastructure ecosystem accommodates these user needs and the reasons for any discrepancies are then investigated for the current situation and in future scenarios.

4.1. Methods Used to Investigate User Needs

4.1.1. Literature Review

The literature on user needs and interaction with charging infrastructure in Norway were analyzed to provide the background for the analysis. The literature has consisted mainly of peer-reviewed scientific articles, but also research and consultancy reports and government documents. The literature review focused on studies of the situation in Norway. Norway is so far ahead of other countries in terms of BEV adoption that results obtained elsewhere have limited relevance to the current Norwegian conditions.

4.1.2. Vehicle User Survey

An online survey was carried out in June 2021. The participants were recruited from the NAF (Norwegian Automobile Federation) member base through a special invitation sent out to 10,000 NAF members, of which 1237 responded. A total of 76% of these were BEV owners, 18% were ICEV owners, 5% were PHEV owners, and 1% used car sharing services. ICEV owners represent here the potential future BEV owners that have not yet made the transition. The mean age of the respondents was 54 years, and the majority were men. The survey contained questions about car ownership and driving behavior. BEV owners were asked about charging behavior and views on public charging infrastructure in Norway, and more specifically about payment methods, pricing, experiences pertaining to charging queues, user-friendliness, and preferred charging station facilities. ICEV owners were asked about their views on BEVs and asked to state their imagined preferences regarding number of charging stops and acceptable charging time if they were to become BEV owners.

4.1.3. BEV Owner Interviews

A total of 15 semi-structured anonymous interviews were conducted with BEV owners in 2021. In order to gain insight into different user groups experiences and views regarding their use of charging infrastructure, a strategic sampling strategy was utilized. The target was to reflect variations in BEV user groups and be representative of different types of BEV users with different housing situations, car usage patterns, and charging behavior. Participants were recruited from the NAF member base, from members of the car sharing service "Bilkollektivet", and from contacts within our existing professional network. Nine of the participants were male, and six were female. The respondents' ages ranged from 20 to 70 years. Interviews were carried out virtually via Microsoft Teams or over telephone. The interviews lasted between 30 and 80 min. Upon completion, each interview was transcribed, and a thematic analysis was performed to identify salient themes in the data reflecting BEV users experiences.

4.2. Methods Used to Investigate the Functioning of the Charging Infrastructure Ecosystem

4.2.1. Literature Review

The international literature on business models for charging was analyzed as background for understanding the motives of the charging infrastructure actors and the potentials and preconditions for developing sustainable businesses for the different types of actors. Peer-reviewed studies, consultancy reports, and government documents were applied. No literature covering this topic was available specifically for Norway.

4.2.2. Open Data Collection

Open data on the Norwegian charging infrastructure ecosystem and its actors were systematically collected to be able to understand how the overall charging ecosystem functions, i.e., who are the actors, what do they do, and who interacts with whom and how do they interact? These open data include data and information from the web pages of the actors and their partners, as well as news articles and other open documents.

4.2.3. Charging Infrastructure Ecosystem Value Chain Analysis

The BEV charging ecosystem actors were systemized as a value chain for the purpose of the analysis in this article, which is consistent with the approach in reports by ADL [34], the Boston Consulting Group [26], Capgemini [35], Deloitte [27], and PwC [36]. Inspiration was drawn from these when illustrating the value chain in this article. Supporting functions from government and membership organizations were included alongside the value chain, as illustrated in Figure 1.

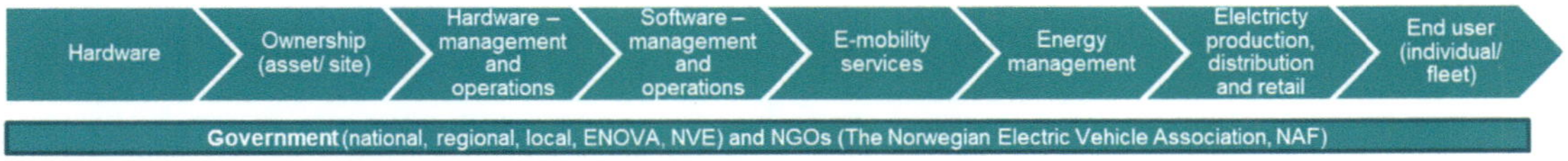

Figure 1. Illustration of the value chain for electric car charging.

4.2.4. Charging Ecosystem Actor Interaction Mapping

The ecosystem was also systemized as a functional map that shows the monetary and co-operative interactions between the different actors and actor types, e.g., CPOs, EMSPs, land and facility owners, MaaS actors, RFID card issuers, software suppliers, and owners of common parking facilities (public facilities as well as private parking facilities for flat owners). The map provides a holistic view of the Norwegian charging infrastructure ecosystem. It was built up by analyzing data from reports, press articles, and open sources.

4.2.5. Semi-Structured Interviews with Charging Infrastructure Actors

A total of 21 semi-structured anonymous interviews were carried out via Microsoft Teams. To cover the entire ecosystem as defined in Figure 1, the following actors were interviewed: charge point operators (CPOs), electromobility service providers (EMSPs), mobility as a service (MaaS) companies, hardware and software (HW/SW) suppliers, landowners, BEV producers (OEMs) represented by their importers, energy companies, and government agencies. Some of the actors had activities in several parts of the ecosystem. The interviews were carried out during the period of June–October 2021 and were transcribed afterwards.

4.3. Charging Infrastructure Scenarios

On the basis of the outcome of an internal workshop, 10 potential scenarios for the future charging infrastructure landscape were developed and discussed. To span out the potential future of the ecosystem in a manageable setting, the authors used their best judgment to reduce the number of scenarios to the five most likely scenarios. These final five scenarios were then further discussed in terms of their ability to meet future user needs and their business model implications.

5. Results

5.1. User Needs

5.1.1. User-Friendliness

Respondents from the survey and participants in the interviews reported experiencing several user-friendliness issues when using public fast chargers. Many of the issues BEV users experience can be traced back to the diversity of service providers in the market for fast charging. Each of these have developed their own charging offering on the basis of their own business model. The charging stations vary greatly in terms of layout, user interfaces, payment solutions, and charger activation, which comes on top of the different types of plugs, the cables with different lengths and locations, and different charging power levels associated with the BEV technology and chargers in general. Consequently, several of the respondents were dissatisfied with public fast charging services and emphasized the need for standardization of all public charging stations. Quotes from interview participants:

> *"There are many different systems in terms of physical plugs. What kind of plug is this? What power can I get out of this one? What is most up to date? Is it the same standard on a Tesla? After a while, you start to understand". Informant 6*

> *"Often I see people that are not accustomed to driving an electric vehicle, trying to use the charger that I'm already using, thinking they can use the other cable (there are two options, one that I can't remember the name of and CSS). Visible information on the charger should have been available where it says that you can only use one cable at a time". Informant 8*

> A quote from a survey participant is as follows: *"Charging of a BEV should be just as simple as filling petrol. Drive to the station, insert the cable, and pay with a bank/credit card. I have never experienced that this succession differs between different petrol stations. It should be equal to use all public charging stations; it should be simple."*.

To avoid having to learn the operation and payment system of other operators, 55% of the respondents in our survey stated that they deliberately used the same charger operator each time they use public chargers, a fact that could be mistaken for customer loyalty. Almost half of the respondent stated that they had bought a BEV with large batteries to avoid having to use public chargers. This development could threaten the long-term profitability of fast charging. It can also lead to larger than necessary batteries in Norwegian BEVs, which would have a negative ecological impact.

5.1.2. Pricing and Payment Method

The operators of the fast-charging stations in Norway had different prices and pricing models up to 2022. During 2022, all major actors changed to pricing per kWh. The survey and interviews were carried out in 2021, and pricing was then a major user concern. The survey and interview results on this topic are presented here, as this experience will be important to other countries to learn from. Up to the end of 2021, some CPOs and EMSPs charged for kWh, some per minute to avoid users charging slowly past 80% battery state-of-charge, and some used a combination of these solutions.

Many found paying by the minute unfair because the important part is how much kWh you receive. Most of the respondents found it difficult to understand in advance how much the charging would cost. This was evident in the survey as well as in the interviews. About 70% of the BEV owners in the survey either partly or fully agreed to the statement *"it is difficult to understand what the price of using public fast chargers will be"*. A majority within both the group using public fast chargers 1–3 times a year and the group using public chargers more than 20 times each year agreed on the difficulty of understanding the price. Price information was often lacking or difficult to understand. In the interviews, several of the participants drew parallels to fueling on petrol stations, where pricing information tends to be clear and easy to understand. In the case of public fast chargers, BEV users were often not provided with price information until the charging had finished.

Quotes from interview participants:

> *"What annoys me is the price. The taximeter starts before the charging has begun and it ends a couple of seconds after the charging is finished. The Norwegian Metrology Service should conduct controls on fast charger stations, like they do on gas stations, so we are not taken advantage of. It boils down to seconds". Informant 4*

> *"The problem is lacking information about how fast I'm charging; how much I'm charging and how much it costs. These three things tend to be hidden at most charging stations. It's normally when you have finished charging that you know how much you've spent, unless you've brought a calculator and do the math yourself". Informant 9*

> *"It's like going to a petrol station where they've covered all the prices with white tags. You can't see the price before after you've charged. That really sucks". Informant 5*

The following are quotes from survey participants:

> *"It is hopeless that precise pricing is not available, like at petrol stations. I would prefer information about actual costs and received kWh during the charging session. Ideally, I could indicate the wanted charging percentage at the beginning, and the charging session would automatically stop when this percentage is achieved".*

> *"I want to be able to pay by credit/debit card, get information on the price before I start the charging, and receive information about the actual costs after the charging session is ended".*

More than 50% of the BEV owners found the payment method at public fast charging stations inconvenient (Figure 2) according to the survey. Figure 2 shows the level of dissatisfaction/inconvenience with the payment systems, depending on how often they used public fast chargers during a year. Many state that they wished paying with credit/bank card was an option (this is currently only tested out at a small percentage of the fast charger stations in Norway). BEV owners who seldom use public fast chargers find the payment method slightly more inconvenient than the ones that use public chargers more often. Even among BEV owners who use the chargers more than 20 times a year, the level of dissatisfaction is relatively high.

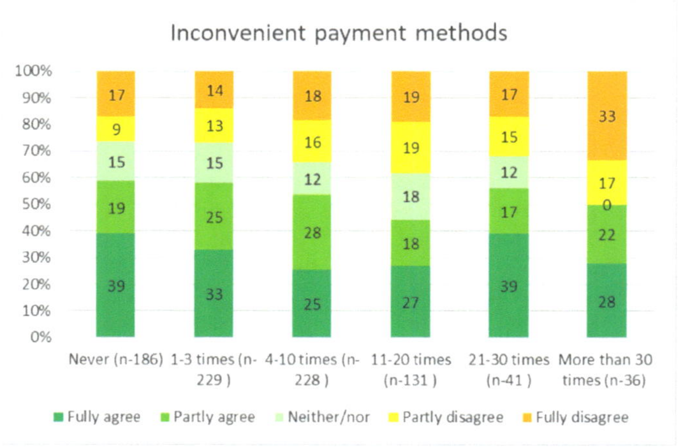

Figure 2. Distribution showing BEV owners' level of agreement to the statement *"I find payment solutions at fast-charging stations inconvenient"*, according to the number of times the respondents have used public fast chargers the last 12 months (n-851).

The average BEV driver in the user survey had downloaded three different charger operator apps. About 75% agree to the statement *"It is inconvenient when charging operators operate with different apps"*. The perceived inconvenience was fairly independent of the number of times the respondent used public fast chargers (Figure 3). There were some variations between age groups, but this could be sample related. It was a tendency that those who seldom (three times or less per year) or often (30 times or more) used public chargers were somewhat more annoyed.

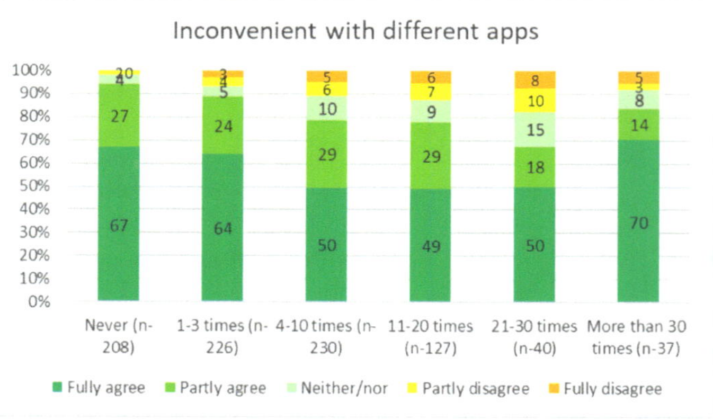

Figure 3. Distribution showing BEV owners' level of agreement to the statement *"It is inconvenient when charging operators operate with different apps"*, on the basis of the number of times the respondents used public fast chargers the last 12 months (n-868).

The following are quotes from interview participants:

"Having all these apps is confusing, because if I'm planning a trip, I can't just use Google Maps and find all the different charger services there. My experience is that I'm unable to get a full overview, and that I'm forced to use each app individually to locate charging stations, and get information about the terms and so forth, and then I'm not even sure if I have the right app". Informant 10

"An opportunity for simpler payment would be nice. To be able to pay with credit card (...). An elderly person who uses fast chargers on rare occasions would want to do it as simple as possible. If this was better, I think the development in terms of the number of electric vehicles would go much faster". Informant 5

The following is a quote from a survey participant:

"It should have been a common method for payment at all public charging stations, without having to download an app to the phone. Should be possible to pay with debit or credit card. Not everybody is comfortable with apps, especially the elderly".

5.1.3. Information and On-Site Services

Findings from the survey and the interviews suggest that BEV owners desire better information about the availability and operational status of chargers, i.e., information about out-of-order charging points and available charging points at the different charging stations. Moreover, several informants wished for road signs indicating the location of charging stations along main roads. For long-distance drivers (driving 100 km+), access to the same amenities at fast charging stations as they are used to from petrol stations are highly valued when using public chargers. Amenities mentioned include basic facilities such as toilets or benches, but also service options and the possibility to buy food and beverages. Lighting conditions on site were also frequently mentioned. Enough lighting is important to be able to read the information on the charger and to be able to locate and insert the correct cable. The fact that most charging stations are built without roofing or other means of shelter is a topic of great frustration for many BEV owners. Being under a roof when using public chargers, especially during the winter, is appreciated when it is raining/snowing. Many BEV owners pointed out the need for a system for handling charging queues, which has become an increasing problem, especially during the seasonal vacation times, public holidays, and on weekends.

Figure 4 shows how BEV owners consider the importance of different factors when choosing a charging station. Location and payment methods were considered the most important factors, but price and amenities were also important. About 75% agree that amenities at fast charger stations are important (the question did not differ between the charging stations located in the cities or along main roads connecting cities). The availability of amenities was more important at stations typically used by long distance travelers.

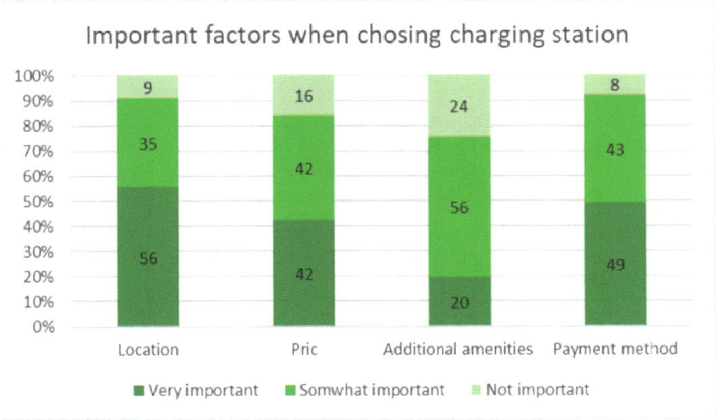

Figure 4. Important factors when choosing public charging station. BEV owners; *n*-675.

The following are quotes from interview participants:

"I normally charge where I can use toilets, buy ice-cream, coffee etc. That is much more OK than (charging stations) where there are just two chargers in the middle of

> *nowhere. It's nice to be able to do something other than just sit in the car. I would not let it determine my path, but along the route I am going, I would choose charging stations with extra service options. This is more important than if I save 15 kroner (NOK, NOK 1 = EUR 0.1) on charging". Informant 7*

> *"I'm not very fond of chargers that are in the middle of nowhere. It must be facilities that allows you to do something there while charging. Gas station, toilet, etc.". Informant 8*

> *"I have experienced that the there are no available charging points when arriving at a charging station. It's difficult to know how long others plan to charge and where to park in the meantime. How does the queuing system work? Are you supposed to drive to the next charging point? (...) I get that all charging points are busy sometimes, but the fact that you're not provided with any information regarding when it will be available, creates a sense of uncertainty. There should be a function on the charger where you could plot in 'I need to have 50 km left so I get home, how long does it take?'". Informant 5*

> The following are quotes from survey participants:

> *"When it comes to amenities, the type of amenities is not so important, if it is a grocery store, a kiosk or a café. The important thing is that there is something. The best option would be if the chargers were placed at petrol station because they are usually open 24/7 and have toilet facilities".*

> *"A form of booking/queue handling system should be available, to avoid arguments and "pointed elbows" on popular weekends".*

These user opinions about location and amenities points at a need to involve facility owners, fuel and energy stations, and roadside cafés and shops in the development of a nation's charging infrastructure, as pointed out already in 2012 in a proposed strategy for charging infrastructure deployment [30,31]. Such actors have indeed been important in developing the Norwegian charging infrastructure ecosystem, as seen in the next section.

5.2. The Norwegian Charging Infrastructure Ecosystem Actor Landscape

Turning to the supply side, the charging infrastructure market consists of a number of actors from the public sector and many different parts of the private sector that have built up a vivid charging market since 2010. These actors play different roles in the value chain, as well as in making the charging market work. The most important roles in the charging infrastructure ecosystem are outlined in Table 1.

Table 1. Actor types and roles in the Norwegian charging infrastructure ecosystem.

Actor Type	Role
CPOs (charge point operators)	Installs, owns, and operates charging station hardware and software standalone or in co-operation with land or facility owners. They want as many users as possible to use their chargers and offer access through roaming networks and EMSPs.
CPOs with EMSPs, closed	These CPOs install, own, and operate charging station hardware and software standalone or co-operating with land/facility owners, including EMSP functions. Their networks provide complete customer experiences and secure the full charging revenue stream.
CPOs with EMSPs, open	As above, but their network is open for roaming through other EMSPs.
EMSPs	Connect BEV users with a CPO charger and starts and stops the charging and organizes payment. They take a percentage of what the user pays. The rest goes to the CPO.
Roaming facilitators	Integrates several CPO (and EMSP) networks into one network and takes a percentage of the transaction cost. Often international.
Producers of charging hardware	They manufacture charging hardware for one or more of the various charging segments; slow charging (home, workplace, destination, or public charging) or fast charging.

Table 1. *Cont.*

Actor Type	Role
Facility and landowners	Cooperate with CPOs to install chargers to attract customers to their facilities. Large actors can leverage good locations to get chargers installed also in the less attractive locations, while CPOs can include such chargers into their national network.
Municipalities and counties	Property owner that may cooperate with CPOs to install chargers that serve those that cannot charge at home. Municipalities and counties may also provide economic support and facilitate installation of fast chargers on municipal land and in other locations.
Parking facility owner	Have chargers in their facilities, i.e., public parking, or parking facilities of flat owners. They often buy turn-key solutions from suppliers.
OEMs, their Norwegian importers	Sell BEVs. Want to deliver good experiences for their buyers. Some will roll-out plug and charge capabilities in 2022, and some have their own EMSP integrated into the vehicle navigation systems via roaming platforms or bi-lateral agreements with CPOs. The latter is pursued by some integrated CPO/EMSPs.
RFID card issuers	Want to give their customers and members easier charging experiences in the absence of roaming solutions. RFID cards allow for identification and payment if the user registers the card number and payment details with the CPOs they use.
BEV fleet operators	Car rental and car sharing companies and leasing services have customers that need and want easy access to charging services.
Maps, route planners	Shows the location and type of publicly available chargers, irrespective of who is the CPO in online and app-based map/route planners.
Software providers	Develop software required to provide charging services, such CPO/EMSP back- and frontend systems, as well as database services with location data and technical characteristics.
Electricity producers and retailers	Sell electricity used for BEV charging. Some of them are owners of CPOs, and/or sell home and destination charging hardware. Some retailers bundle rebate for charging together with the household electricity supply.
Supporting businesses	Electricians, entrepreneurs, payment system providers, etc.
Structural elements	Laws and regulations, authorities, support schemes, etc.

The core actors in the charging infrastructure ecosystem are the CPOs (Mer, Recharge, BKK/Eviny, Lyse, Ionity, Kople, CircleK, Tesla, and others), some of which have an integrated EMSP service, such as Mer, Eviny, CircleK, Kople, and Tesla (non-Tesla BEV charging), or offer such services via a broader roaming service (Fortum Charge and Drive). The roaming EMSPs Elton, Kople, and Elbilforeningen/Ladeklubben (EV Association) connect BEV drivers with CPOs via an app or an RFID card and handle payment at connected CPO networks so that their user gets one bill. RFID cards work as identification, but payment information must be linked to the RFID card number within each CPO/EMSP system/app when they do not support roaming. International roaming actors (Hubject, Digital Charging Solutions, Plugsurfing, Shell NewMotion) aggregate CPO networks into international charging networks, facilitate payment between actors, and are accessible for EMSPs. They have a limited foothold in Norway.

Landowners, i.e., municipalities (Oslo, Bergen, Trondheim, Stavanger, etc.) and other public property owners (Avinor, BaneNor) rent out land to put chargers on. Fast-food chains (McDonalds, Burger King), shopping centers (Steen and Strøm, Olav Thon Gruppen), grocery chains (Norges-gruppen, Coop, Rema 1000) and other store chains (Ikea), hotel chains (Olav Thon Gruppen), and fuel stations (CircleK, and others) install chargers at their facilities, often in cooperation with CPOs, in order to attract customers to their facilities.

OEMs provide BEV owners access to chargers through their proprietary charging ecosystems (VW We charge, Mercedes Me Charge, Kia Charge, etc.) and enable interaction with charging infrastructure via the vehicle navigation system. They are aided by international software and roaming actors such as Digital Charging Solutions GMBH and enter into bilateral agreements with Norwegian CPOs to obtain access to their charger networks. These OEMs' strategy is to offer as many chargers as possible through their EMSPs, a strategy opposing that of the integrated CPO/EMSPs that wants to own the customer. For the future, OEMs such as VW, Audi, and BMW will pursue plug and charge in co-operation with Ionity (OEM owned ultra-fast charger CPO).

EV fleet operators (imove, Bilkollektivet, Bildeleringen, MoveAbout, Vy, Hyre, Lease-plan, Easly, Hertz, Avis, and others) rent out BEVs over longer or shorter periods but often leave it to the user to find out how and where to charge. Peer-to-peer charging is explored by some actors (Plugshare, Cloudcharge, ChargeBnB).

Map services such as A Better Route Planner, Google Maps, Ladestasjoner.no, Elbil.no, Opplysningen1881, and others display charger maps and offer route planners. They use the Nobil database of public charging infrastructure. Norwegian and international companies such as Driivz, Current, Virta, and others provide back-office and white label charging infrastructure management software.

The energy retail sector (Fjordkraft, Hafslund Eco, BKK, and others) sells the electricity used for charging. Some also provide charging solutions that allow users to shift charging to off-peak hours (Tibber). Some electricity grid operators install normal and fast chargers locally, often in co-operation with larger CPOs. Finally, there are a number of actors that develop (Zaptec), sell (Fjordkraft), and install chargers (Smartly/lyse, BKK, Kople, etc.) in public parking facilities (Apcoa, Easypark), at workplaces, in apartment buildings (Ohmia charging, Hafslund, Mer), and in private houses (CircleK). Many of these actors are also involved elsewhere in the charging infrastructure ecosystem, for instance, as CPOs and energy retailers.

As the actors within these actor types pursued their individually sound, and in some cases, opposing business models, a complex web of charging infrastructure actors and solutions emerged. While it may make sense for an individual actor to oppose roaming and "own the customer", the result is a form of "tragedy of the commons" [37] in which the overall system does not take sufficient care of user needs for easy access to and use of the total public charging infrastructure. This is illustrated by the fact that BEV owners in 2022 needed more than 20 apps, e.g., Mer, Bilkraft (Eviny), Fortum Charge and Drive, Circle K charge, Ionity, Kople, E.on drive, Ishavsveien, Vattenfall InCharge, Cloudcharge, Shell Recharge, Elton, Ladeklubben, Apcoa Flow, EasyPark, Bergen parkering, Avinor, Smart charge, Charge365, Plugsurfing, Tesla (from 31 January 2022), and Ladeklubben (from February 2022) to obtain access to all chargers in Norway. The average user has about three apps installed, and 50–70% of surveyed BEV users thought that it is inconvenient to have so many apps (see Section 4.1). Users may also encounter 13 different payment methods [3], including apps, RFID cards, SMS payment, web page payments accessed manually or via QR code scanning, plug and charge, tap to charge, instant payment solution apps (the Norwegian payment solution Vipps) or QR codes, variants of subscription services, payment integration with vehicle navigation systems, and OEMs' EMSPs. The first CPO that offered bank card payment opened their first station in 2021, but only a few stations have this option, which is highly desired by the users. A large share of the surveyed BEV owners found that the multitude of different payment methods at the different public fast charging stations are inconvenient. BCG [32] found similar issues with apps and payment system complexity in other countries.

The main actors in the ecosystem are positioned in Figure 5 in relation to each other, and lines show with whom and how they interact through, e.g., roaming transactions, generation of traffic, location cooperation, or other forms of cooperation. The figure shows where the cooperation between actors in different sectors of the ecosystem is most and least extensive, and it provides a good illustration of how something as seemingly simple as charging BEVs involves a very complex business landscape. One of the reasons for the added complexity is that some actors allow roaming while others do not. In fact, several of the largest actors have an opposing strategy to roaming. They want to "own the customer", as they expect such a strategy to be most profitable. If all CPOs had allowed for roaming, the relationship could be depicted by one line going from each CPO to the roaming hubs, with lines going out to all EMSPs. Instead, there is large variation in the number of lines going from a CPO to the various EMSPs and roaming hubs. Those that do not allow roaming have entered into a number of bi-lateral contracts that provide some users external access to these otherwise closed networks. From the end users' point of view, the lack of roaming means less interoperability between CPOs, which causes friction if they want to switch between different CPOs and makes the user experience less streamlined.

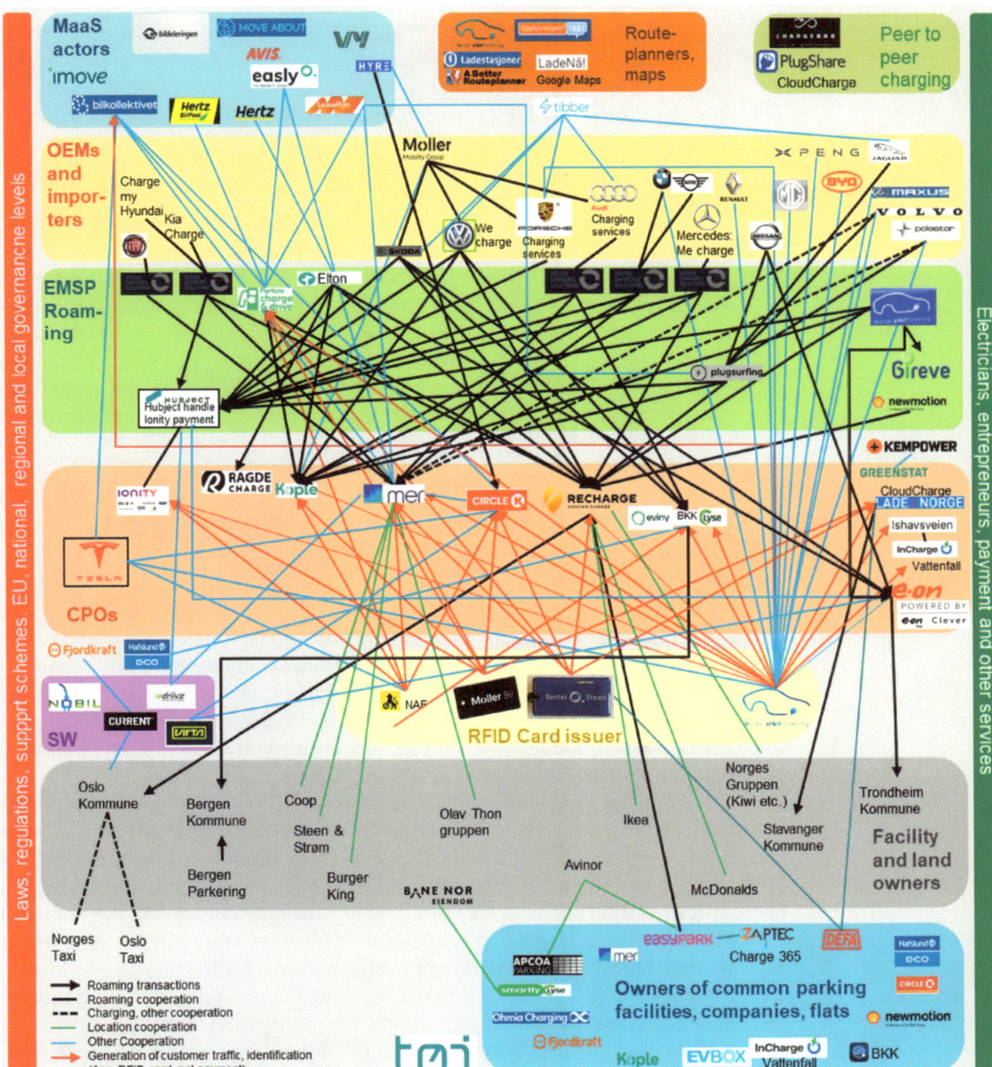

Figure 5. The Norwegian charging infrastructure ecosystem. Roaming transactions: thick black line with arrow; roaming co-operation: thick black line; other charging co-operation: stippled thick black line; location co-operation: green line; other co-operation: blue line; generation of customer traffic by co-operation on customer identification (RFID card issuers, etc.): red line with arrow. Source: the authors.

5.3. Promising Business Models in the Norwegian Charging Infrastructure Ecosystem

Wangsness and Figenbaum [3] found that more than 40 business models within the seven parts of the EV charging value chain shown in Figure 1 are currently practiced in Norway. They provided a qualitative evaluation of these business models according to their profitability, scalability, and contribution to the public charging infrastructure, which were the parameters also used by Spöttle et al. [38]. Seven business models were highlighted as particularly interesting, with typical actor examples in parentheses:

1. Integrated location owner + CPO + EMSP + charging system supplier for housing cooperatives/condominiums, private households and companies, semi-closed but cooperates with other CPOs (Circle K).
2. Integrated CPO + EMSP + charging system supplier for apartment buildings, companies + pop-up charging, open for roaming and also a provider of CPO services under contract for a number of municipalities/counties (Kople).
3. Roaming EMSP with map services (Elton).
4. Provider of platform solutions for CPOs, EMSPs, installers, electricity producers, and grid companies, which enable operation and management, smart charging, and eventually V2G (current).
5. Electricity supplier EMSP (without roaming) + supplier of hardware and software solutions for charging at homes, housing cooperatives/condominiums, workplaces, and destinations (Fortum Charge and Drive).
6. Integrated CPO + EMSP + charging system supplier for apartment buildings and companies, semi-closed. Can also be an asset owner, and in some cases a site owner (Mer and BKK).
7. Manufacturer of hardware and software—all segments (up to 22 kW AC) (Easee and Zaptec). These actors mainly produce chargers for home and parking facilities.

The first six business models are discussed further below as they are most relevant for the user experience associated with charging away from home.

It is highlighted in several places in the reviewed literature, e.g., Pagani et al. [39], PwC [36], Schroeder and Traber [29], and Zhang et al. [22], that much of the profit potential in the charging market will come from being able to combine several revenue streams. Circle K is doing just that. Their energy stations make money from BEVs charging there, where they act as both CPO and EMSP for several of the charging stations. With larger volumes to their energy stations, they draw greater revenue from the retail part of the business (food etc.). In addition, they sell hardware, software, and support solutions to homes, for apartment buildings, property developers, and workplaces.

Kople also has a business model that stands out. They act as a "turnkey provider" for apartment buildings, workplaces, and destinations (hotels, shops, etc.) and an "end-to-end integrator" as owner, CPO, and EMSP for its charging network with normal, fast, and ultra-fast chargers. Their charging network is also open for roaming, which means that they contribute to a larger charging network for their customers and other networks' customers.

Although the reviewed literature does not often refer to aggregators/roaming hubs, they are sometimes highlighted as a promising business model, for instance by the Boston Consulting Group [26] and Capgemini [35]. Elton's business model is interesting in this respect. After a few months of developing the solution, it became possible to use their platform and app to roam between major players such as Ionity, Recharge, and Kople in both Norway and Sweden, with the possibility to search for the fastest and cheapest charging option in the app.

Vehicle-to-grid (V2G) services are considered to have a very high revenue potential according to ADL [34] and Bland et al. [40]. In the Norwegian ecosystem, Current stands out. In addition to providing platform solutions to CPOs, EMSPs, installers, power producers, and grid companies, as well as enabling operation and management of charging services, the Current platform also supports smart charging and V2G functionality. Current provides software solutions to several stages of the value chain and several charging segments that support roaming through open standards (where they also offer certification), smart charging, and V2G.

Fortum Charge and Drive is also an interesting business model as a large EMSP that is also an electricity retailer that sells home charging equipment. Fortum is thus close to having an "end-to-end energy" business model with multiple revenue streams, but without having to take the risk with the investment and operation of hardware, which a CPO must do. Fortum Charge and Drive was an integrated CPO/EMSP until the CPO part was spun off as "Recharge" in 2020.

It will be interesting to follow the business model of more closed CPOs that do not offer roaming in general but cooperate with some OEMs and see if they manage to grow while retaining a firm grip on their customers. In the Norwegian ecosystem, Mer and Eviny (previously BKK) stand out. If they manage to deliver a large enough stand-alone network to their customers and ensure loyalty and bring in new customers through their large presence, it is possible that they will be more profitable than if they had chosen a more open strategy with roaming. Their current strategy also enables bilateral agreements that open the network to specific groups.

The position of these actors in the value chain is shown in Table 2, together with a selection of other Norwegian actors.

Table 2. Selected actors positioning in the charging infrastructure value chain (see Figure 1).

	Segment	Hardware Production	Hardware Sales	Ownership (Site Asset)	Hardware Management and Operation	Software Management and Operation	E-Mobility Services	Energy, Powert Management	Electricity Production, Distribution and Retail	Retail Sale, other Income Sources	End User (Individual, Fleet), End User NGO
MER	Public—Fast chargers										
	Public—Normal										
	Workplace—Normal										
	Home—Flat owner										
BKK	Public—Fast chargers										
	Public—Normal										
	Workplace—Normal										
	Home—Flat owner										
	Home—Detached house										
CircleK	Public—Fast chargers										
	Public—Normal										
	Workplace—Normal										
	Home—Flat owner										
	Home—Detached house										
Kople	Public—Fast chargers										
	Public—Normal										
	Workplace—Normal										
	Home—Flat owner										
Recharge	Public—Fast chargers										
	Public—Normal										
	Workplace—Normal										
	Home—Flat owner										
Ionity	Public—Fast chargers										

Table 2. *Cont.*

	Segment	Hardware Production	Hardware Sales	Ownership (Site Asset)	Hardware Management and Operation	Software Management and Operation	E-Mobility Services	Energy, Powert Management	Electricity Production, Distribution and Retail	Retail Sale, other Income Sources	End User (Individual, Fleet), End User NGO
E.ON, Powered by E.ON Drive & Clever	Public—Fast chargers			■	■	■	■				
	Public—Normal			■	■	■	■				
	Workplace—Normal			■	■	■		■			
	Home—Flat owner			■	■	■		■			
Defa, Lade i Norge	Public—Normal	■		■	■	■	■				
	Workplace—Normal	■		■	■	■					
	Home—Flat owner	■		■	■	■					
	Home—Detached house	■									
Elton	Public—Fast chargers						■				
	Public—Normal						■				
Current	Software platform		■	■	■	■	■				
Tibber	Electricity provider							■	■		
Fjordkraft	Home—Flat owner		■					■	■		
Easypark	Public—Normal			■			■				
NAF	Automobile owners Ass.										■
EV Association	Public—Fast chargers						■				■
	Public—Normal						■				■
Oslo Municipality	Public—Fast chargers			■							
	Public—Normal			■							
	Taxis—Fast chargers			■							
ZAPTEC	Public—Normal	■	■								
	Workplace—Normal	■	■								
	Home—Flat owner	■	■								
	Home—Detached house	■	■								
EASEE	Public—Normal	■									
	Workplace—Normal	■									
	Home—Flat owner	■									
	Home—Detached house	■									
Bilkollektivet	Car sharing										■
Avinor	Airport owner			■							
McDonalds				■							

5.4. Scenarios for the Future

Time will tell which business models will deliver on long-term profitability and scalability while meeting user needs. The ecosystem for BEV charging is developing rapidly and there is great uncertainty about what this ecosystem will look like in, e.g., 5–10 years. Differences in the relative strength of different economic drivers (e.g., technological developments, regulation, or customer loyalty to incumbents) will lead to different futures for this ecosystem and different results for the end-user experience. In such instances, it is useful to construct a set of scenarios for the different directions the ecosystem can take. A total of 10 different scenarios were identified during and following an internal project workshop. All these scenarios are described and discussed in more detail in a report by Wangsness and Figenbaum [3]. For the purpose of this article, the 10 scenarios are summarized briefly below:

1. Business-as-usual (BAU) with many and probably increasing number of players. Voluntary roaming is limited, and none of the CPO networks are large enough by themselves so that BEV drivers can get by with only one user interface on a long trip to a new destination. New actors joining will lead to increased user complexity by adding more apps and payment solutions to the overall system. The user needs for simple and efficient charging analyzed in Section 5.1 will thus not be met.
2. Consolidation into large, vertically integrated semi-closed CPOs. The least profitable CPOs have dropped out of the market or merged with the large remaining ones that grow alongside the BEV market. They tend to have a large horizontal (multiple charging customer segments) and vertical coverage (e.g., energy retail and energy stations with convenience products). This scenario will function better for users than the BAU. The number of actors will be reduced, and each actor will have a larger network. Given their size, some users may only need to use one network.
3. Interaction and roaming platforms connect charging infrastructure players with charging customers. Interaction and roaming platforms gain enough volume so that it becomes hard for the "stubborn" CPOs and EMSPs to "lock in" their customers by refusing roaming. This scenario will allow users to virtually interact with only one app to gain access to multiple CPOs networks, as Elton is today on a small scale. This development would improve the user-friendliness of charging but can be more costly for the user as the platform will add a small cost to the transaction.
4. "The people want to roam freely!" Roaming becomes the industry standard. Preferences of both charging customers for maximal roaming possibilities and site owners for maximal volumes will favor CPOs that agree to roaming and put pressure on the CPOs that follow a more closed strategy. Charging will be more easily accessible for users but may be slightly more costly due to the roaming charges.
5. Car manufacturer (OEM)-controlled future with charging integrated into the car's navigation system. OEMs will directly or indirectly provide EMSP services through the cars' navigation system, as a part of their expanded service portfolio. Such services will increase in relative importance, as revenue from vehicle maintenance services will decline. OEMs will have leverage to obtain seamless access to CPO networks and even be a force pressuring for expanded roaming possibilities. This scenario will significantly improve the user-friendliness of charging.
6. Plug and charge—the cars identify themselves automatically and payment is seamless. Some OEMs and CPOs will lead the way towards increased use of plug and charge in order to compete with the customer experience of Tesla, which already offers this service. Plug and charge will make it much easier to charge where it is enabled, but may not work on all chargers, at least in a long switchover period. The potential for improved user-friendliness is very large, but in an early transition period it may add complexity by becoming yet another solution for identification and payment.
7. Regulation requires a splitting into separate CPOs and eMSPs. In order to ensure more competition and less "lock-in" of customers, regulators require in this scenario that CPOs and EMSPs are separate companies and require CPOs to provide fair access to all EMSPs. This scenario will be similar for users, like scenario 4. The difference is how it comes about.
8. BEVs obtain a longer range that allows charging mainly at home and at destinations. Fast charging will in this scenario be of less relative importance compared to home and destination charging, which is much cheaper. There will be a strong growth in destination charging, "become-your-own-CPO-solutions", and peer-to-peer charging solutions. For users, this will be a simplification as charging issues will be encountered less frequently and long-distance driving will be easier.
9. The EU sets the standard for ensuring seamless charging across national borders in Europe (including Norway). The regulation will make ad hoc payment simpler, likely with bank/credit cards, as well as more transparent, and it will allow BEV users to be anonymous. It will reduce the need for EMSPs. It also reduces CPOs' ability to "lock

in" customers, but other user-friendliness issues such as different plugs, power levels, and charging interfaces will remain.

10. The technology giants take over. Tech giants (e.g., Google and Apple) have become the dominant EMSPs as they provide services through products the charging customer already uses, i.e., the smart phone. These tech giants have payment solutions and ample amounts of data for prediction of real-time traffic conditions to fine tune their services already. OEMs will have strong incentives to give the tech giants' platforms access to their navigation systems. This scenario can provide BEV owners with services not available to ICEV owners, such as real-time info about the likelihood of queues when needing to charge on the basis of data they already use for travel time prediction.

The five most likely scenarios (the authors' assessment) were analyzed further, as shown in Figure 6, and differentiated across the degree of consolidation in the market and whether the market development is steered from charging actors or external actors. The customer experience compared to BAU will improve to a variable degree in these scenarios. It can be useful to have a mix of main scenarios A and B in mind when making a strategy in the BEV charging infrastructure ecosystem, as these scenarios exemplify outcomes of the battle between the large integrated CPOs/EMSPs and OEMs regarding the control of the charging market. Both of these types of actors need to be aware of the snowball effect of an increasing number of players and platforms offering e-roaming connected together by network orchestrators and interaction platforms as in scenario E. Tech giants can also be in the lead if they gain access to the navigation systems/vehicle displays as in scenario C. Better customer experiences can also be achieved through regulation of the market as in scenario D. EMSPs can in scenario D become less needed if fair pricing is required for drop-in customers and if unified payment systems that do not require user registration become mandatory. Authorities may need to pursue such regulations if the market actors themselves cannot establish an overall acceptable customer experience, through voluntary roaming, co-operation, or consolidation.

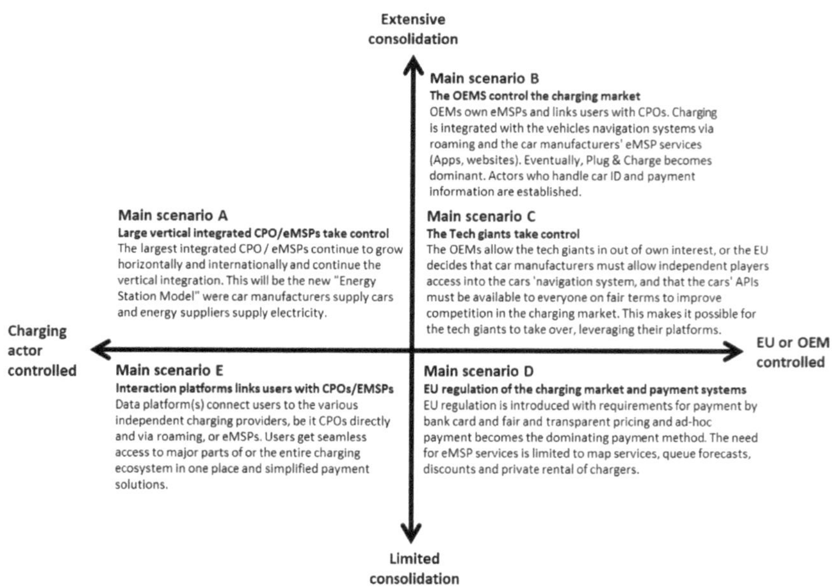

Figure 6. Main scenarios for the future of the Norwegian charging market. Source: the authors.

6. Discussion and Conclusions

The user needs for and experiences with charging infrastructure, the motivations and strategies of the business actors that delivers these services to users, and the ability of the overall charging infrastructure ecosystem to meet the user needs were investigated. Combined, the analyses enable an understanding of how the Norwegian charging infrastructure system evolved, how it now works, and how it is likely to evolve in the future with and without interventions and how this will or will not meet user needs.

6.1. Considering Developments up to 2021

The Norwegian BEV market has been built up organically over 30 years with incentives for purchase and use of BEVs [4,8,9], which created a large and competitive market. The same applies to the deployment of public charging infrastructure, which started from 2009. Enova, Transnova, and counties and municipalities have provided economic support for charging infrastructure deployment, requiring only a minimum of technical standards and that at least one common payment system was available, without specifying which. Within the terms of these public support schemes and tenders, actors pursued their business models, competed with other actors, and developed their own payment and user interface solutions from scratch. This development started when BEVs had small batteries and each charging session had a low value. Low transaction cost and basic user interface systems were therefore prioritized over user-friendliness. Many different payment solutions were pursued, and each actor developed their own app for access and payment. While this approach likely enabled installation of more chargers early on, the large number of individual actors with a variety of incompatible strategies, solutions, payment systems, and apps has become a major challenge for BEV users [21,41,42], as also seen in other countries [32]. The total charging infrastructure network has expanded significantly, i.e., as the sum of all the CPO networks, but the overall system has not become intuitive or simple to use for BEV owners. Major reasons for the lack of user-friendliness are that roaming is only offered by a few actors and that credit/debit card payment is not available apart from at some test stations. To try to make charging easier, various patches have been introduced such as the use of RFID cards for identification at charging stations and various map services providing an overview of the total offering. Large actors have refused to open their charging networks for roaming as they want to keep their customers exclusively in their networks, believing that this will generate more profit for them. The way the market functions is thus a result of how individual actors have followed their individually sensible business models and used their market power for their own benefit. This is an example of the classic "tragedy of the commons" dilemma.

Survey and interview results shows that many BEV users as a result of these developments experience several issues when using public fast chargers. The lack of user-friendliness and the lack of standardized payment systems across operators are important disadvantages that may become a barrier to future BEV adoption. According to a survey by the Norwegian EV Association, 88% of current BEV owners want a simpler universal system that is independent of which company that operates the charging station [41]. More than 70% of BEV owners want to pay for charging with a credit/debit card [41]. Findings from the present study are consistent with these results.

Recent years have shown consistent improvements to the fast charger network, particularly in terms of availability [13,16]. However, findings from the present study suggest that there is still a long way to go in terms of user-friendliness. As BEV market shares increase and BEV buyers are becoming the majority car buyers [4], the need for a more user-friendly experience will become pressing. Standardization can increase the overall user-friendliness by

1. Standardizing the payment method. The same type(s) of payment methods should be available at all stations. Preferably, it is one common app for all charging stations, as well as easy options for drop-in clients (debit/credit card option).

2. Standardizing the pricing system. The price should be clearly marked, and the pricing system should be easily comparable between different charging stations (without having to download an app).
3. Standardizing the charging operation (user-friendliness). Available user information, placing of charging cables to fit all vehicle models, method for starting the charging, etc. Universal design at fast charging stations (easy use independent of nationality, technological skills, or disabilities).
4. Standardizing the level of amenities at fast charging stations along the major highway network, such as, for instance, toilets, light, garbage bins, road signs, roof, etc.

There has been no progress on the standardization of payment systems, which could solve many of the issues that BEV users are experiencing today. Access to a universal app for use of and payment for chargers would for instance have solved the issue of having to install many apps and interface with a multitude of payment systems. It would particularly simplify identification and payment for those using public chargers occasionally. Moreover, a universal app could also provide users with accurate and simpler price information, regardless of service provider. Large actors have refused to provide external actors access to their networks, which so far has made a universal user-friendly solution impossible. Elton and the EV Association's "Ladeklubben" have developed such solutions on a small scale with a handful of CPOs that support roaming. Another improvement would be to make payment with credit/debit card standard. This payment option is only tested out in a handful locations, but findings from the current and other studies [21,41,42] imply that most BEV users would like to have this option. The EU [43] is considering standardizing this payment method at public charging stations. For charging at 50 kW and more, the EU's preliminary plan is to require payment card readers at all charging stations by 2027 (or a contactless solution). Other methods to simplify the existing authentication and payment methods discussed in Norway [21] are to use the toll road AutoPass chip or a type of ANPR (automatic number plate recognition) system.

It was up until 2022 difficult for BEV owners to obtain an overview of the price of charging. Different operators used different pricing systems, i.e., paying for minutes, paying for kWh, or a combination of both. The price of charging also varied depending if you were a drop-in customer or a registered customer of the operator. Drop-in customers typically paid two times more with SMS messaging than those registered as customers and using an app or RFID card [21]. The pricing system became more standardized during 2022. All actors now price charging in NOK per kWh charged, which is what the users have wanted, and the added cost of charging for unregistered customers has been reduced.

Standardization of the charging access, which is important for the user-friendliness of charging, has not progressed much apart from better and more accessible user information [21]. There are also issues regarding universal access [21]. The chargers must according to Norwegian law be accessible for people with disabilities, which requires more space around the vehicle and charger for wheelchair users. Such requirements can be difficult to meet as the chargers were often installed in narrow parking spaces and because of the need to install barriers to protect cables and chargers from being hit by vehicles [44]. The operators have focused on expanding the number of chargers, at the expense of universal access [44,45], but a standard for universal charging access is being developed [21]. In Norway, many of the BEV owners only use public chargers 1-3 times a year [13,19,20], which makes it important that the user information at the charging stations is available (in both Norwegian and English) and is easy to understand for inexperienced users. The EU [43] is planning to include regulation for accessible user information, traffic signs, and information on prices in the updated version of Directive 2014/94/EU.

Customers generally want the same amenities at charging stations as at petrol stations. For long-distance travelers, the following amenities are especially important [19,21,30]: toilets, cover/roof protecting for rain/snow, access to a food vendor or a café, resting areas, garbage bins, 24/7 service, and access to water for window washing. Most fast chargers have therefore been installed in existing parking spots [44,45] at fuel stations, cafés, restaurants, food stores, and shopping centers. Having such amenities available is not just a question of providing a better user experience, it also enables more revenue streams related to the vehicle charging that could add positively to the overall return on investment in charging infrastructure. These locations also have electricity available already, which can reduce the cost of installing chargers.

6.2. Future Developments and Furter Research

The direction the charging infrastructure ecosystem will evolve in if no regulation is put in place is not clear. Business as usual will likely increase the overall complexity as new actors will join with different degrees of cooperation and interoperability with existing providers of charging solutions. It would likely lead to even more apps and varieties of access and payment systems and be less desirable for users wishing to obtain simpler access to chargers. Business as usual is unlikely to continue long into the future. Consolidation of CPOs and EMSPs into larger and fewer actors is a scenario that can reduce the user complexity, but users may still need to establish customer relations with several operators to obtain full access everywhere. OEMs have a key role by controlling the vehicle navigation system and may thus be able to force the charging infrastructure actors to realign. In other scenarios, integrated CPO and EMSPs may need to be split into separate entities through regulation to make it possible for OEM EMSPs and independent EMSPs to offer a large share of the chargers through their service. This has advantages in terms of transparency and simplicity under normal operation. Furthermore, tech giants such as Google or Apple may disrupt the charging infrastructure market by linking their map services that have real-time traffic flow data, to charger status data and access control, and integrate that with their existing payment systems (e.g., Google pay, Apple pay), in order to offer an advanced EMSP service with real-time queue predictions and routing options to less used chargers.

Charging infrastructure for heavy-duty trucks will also be an important development for the EV charging ecosystem. Developers of such infrastructure and government support agencies should learn from the Norwegian experiences with BEV charging infrastructure. They should, for instance, agree on one common standard, and roaming should be a requirement from the outset. The charging transactions will be much larger than for BEVs due to the larger batteries (currently about 2-6 times larger than the largest BEVs). The added cost of roaming charges will be small compared to the transaction value, so there is little reason not to plan for and demand roaming. Moreover, the value of time for a trucking operation is a great amount higher than for an average passenger car driver, implying high demand for smooth charging operations with low time use. Truck charging infrastructure will also need to be utilized fully, due to the large surface areas and high-power levels required for truck charging. The truck infrastructure should be separate from that of passenger cars and light commercial vehicles so that trucks do not block several chargers when charging at the infrastructure put in place for BEVs. This is already an issue with trucks with CCS charge inlets [46]. CharIn is working on the MegaWatt charging system (MCS) for trucks, and a prototype of the connector design allowing for up to 3.7 MW charging was demonstrated at the 35th World Electric Vehicle Symposium in Oslo in June 2022 [47].

There are many ways to extend the research on the charging infrastructure users, actors, and ecosystem. Further research can, for instance, expand the analysis of the user needs with real charger transaction data, and of the ecosystem with financial data from the businesses operating in the ecosystem. Even if financial data at the desired level of detail are not available, official (more aggregated) financial data may over time say something about the magnitudes and development of importance within the ecosystem. Since this ecosystem is a rapidly evolving, it is of great interest to follow, document, and analyze the changes taking place. This will enable a better understanding of the ecosystem dynamics and make it possible to learn from the success factors and pitfalls in the transition to a fully electrified transport sector.

6.3. Caveats

The ecosystem for EV charging is complex. The qualitative assessments in this article have provided a holistic overview of charging of BEVs in Norway but are subject to uncertainty. The data for this article were collected until the end of 2021, so the article does not consider developments in 2022. Some important developments in Norway during 2022 were that

- Tesla opened on 31 January a total of 15 locations and 315 charge points for non-Tesla owners, which was expanded to 58 locations and 812 charge points on 3 May. Non-Tesla owners must, however, use yet another app to gain access, and not all BEVs can use Tesla chargers due to the positioning of the charge inlet and the length of the cables.
- The EV Association launched their roaming EMSP service "Ladeklubben" (Charging club) on 17 February. A blue RFID card provides access to charging at Recharge, Kople, Ionity, E.on, and Powered by E.on, providing access to 275,000 chargers across Europe.
- The popular Norwegian instant payment service "Vipps" became more common as Mer and Eviny also took it into use during 2022.
- Pricing became standardized during 2022. All the large CPOs and EMSPs now charge in NOK per kWh.

Author Contributions: Conceptualization, E.F., P.B.W., A.H.A. and V.M.; methodology, E.F., P.B.W., A.H.A. and V.M.; survey software, QuenchTec; validation, E.F., P.B.W., A.H.A. and V.M.; formal analysis, E.F., P.B.W., A.H.A. and V.M.; investigation, E.F., P.B.W., A.H.A. and V.M.; resources, E.F.; survey data curation, A.H.A.; writing—original draft preparation, E.F., P.B.W., A.H.A. and V.M.; writing—review and editing, E.F., P.B.W., A.H.A. and V.M.; visualization, E.F., P.B.W., A.H.A. and V.M.; supervision, E.F.; project administration, E.F.; funding acquisition, E.F. All authors have read and agreed to the published version of the manuscript.

Funding: The research was funded by the Research Council of Norway, through the Spot-on project (grant no. 321090), with a small contribution from the MoZEES FME (grant no. 257653).

Institutional Review Board Statement: The study was conducted in accordance with the Declaration of Helsinki, and approved by the Ethics Committee of the Norwegian Agency for Shared Services in Education and Research (protocol code NSD 976704, 4 May 2021).

Informed Consent Statement: Informed consent was obtained from all the anonymous subjects involved in the study.

Data Availability Statement: Not applicable.

Acknowledgments: The authors wish to thank the partners of the Spot-on research and innovation project, e.g., Powerzeek, Bouvet, NAF, Q-Free, Know-it, Kluge, and ITS-Norway, for their constructive comments and co-operation on interviews with the charging infrastructure actors and BEV owners.

Conflicts of Interest: The authors declare no conflict of interest.

References

1. Amundsen, A.H.; Milch, V. Demand for standardization of public fast charger stations—User perspectives. In Proceedings of the 35th International Electric Vehicle Symposium and Exhibition (EVS35), Oslo, Norway, 11–15 June 2022.
2. Figenbaum, E.; Wangsness, P.B. The Norwegian Charging Infrastructure Ecosystem. In Proceedings of the 35th International Electric Vehicle Symposium and Exhibition (EVS35), Oslo, Norway, 11–15 June 2022.
3. Wangsness, P.B.; Figenbaum, E. The Charging Market—Complex and Dysfunctional or Future-Oriented? TOI Report (in Norwegian) 1867/2022. Institute of Transport Economics, September 2022. Available online: https://www.toi.no/publications/the-charging-market-complex-and-dysfunctional-or-future-oriented-article37371-29.html (accessed on 27 September 2022).
4. Figenbaum, E. Norway—The World Leader in BEV Adoption. In *Who's Driving Electric Cars. Lecture Notes in Mobility*; Contestabile, M., Tal, G., Turrentine, T., Eds.; Springer: Cham, Switzerland, 2020. [CrossRef]
5. IEA. Norway (Electromobility Status 2022). In The Electric Drive Charges Ahead. IEA HEVTCP Annual Report 2022. 2022. Available online: https://ieahev.org/wp-content/uploads/2022/05/DIGITAL-HEVTCP_2022_Annual_Report_Final-with-Cover.pdf (accessed on 1 August 2022).

6. Statistics Norway, to Av Tre Nye Personbiler er Elbiler. Available online: https://www.ssb.no/transport-og-reiseliv/landtransport/statistikk/bilparken/artikler/to-av-tre-nye-personbiler-er-elbiler (accessed on 25 April 2022).
7. OFVAS. Car Sales Statistics. Bilsalget i JUNI 2022 (og Første Halvår 2022). OFV AS. 2022. Available online: https://ofv.no/bilsalget/bilsalget-i-juni-2021-2-2 (accessed on 1 August 2022).
8. Figenbaum, E. Retrospective Total Cost of Ownership analysis of Battery Electric Vehicles in Norway. *Transp. Res. Part D* **2022**, *105*, 105103246. [CrossRef]
9. Figenbaum, E. Perspectives on Norway's supercharged electric vehicle policy. *Environ. Innov. Soc. Transit.* **2017**, *25*, 14–34. Available online: http://www.sciencedirect.com/science/article/pii/S2210422416301162 (accessed on 22 March 2022). [CrossRef]
10. Figenbaum, E. The 1990 to 2020 Technology Innovation System supporting Norway's BEV revolution. Working Paper. 19 March 2022. Available online: https://papers.ssrn.com/sol3/papers.cfm?abstract_id=4061401 (accessed on 10 August 2022).
11. Ministry of Transport. National Transport Plan 2022–2033. Oslo. 2021. Available online: https://www.regjeringen.no/en/topics/transport-and-communications/content-2021/national-transport-plan-20222033/id2866098/ (accessed on 10 January 2022).
12. Government. UNFCCC Paris Agreement. Norway's First NDC 2016. Norway's Nationally Determined Contribution to the Paris Agreement. 20 June 2016. Available online: https://unfccc.int/sites/default/files/NorwayINDC.pdf (accessed on 27 September 2022).
13. Figenbaum, E.; Nordbakke, S. Battery Electric Vehicle User Experiences in Norway's Maturing Market. TØI Report 1719/2019. Available online: https://www.toi.no/publications/battery-electric-vehicle-user-experiences-in-norway-s-maturing-market-article35709-29.html?deviceAdjustmentDone=1 (accessed on 10 May 2022).
14. Fevang, E.; Figenbaum, E.L.; Halse, A.H.; Hauge, K.E.; Johansen, B.J.; Raaum, O. 2021. Who goes electric? The anatomy of electric car ownership in Norway. *Transp. Res. Part D* **2021**, *92*, 102727. [CrossRef]
15. Hardman, S.; Jenn, A.; Tal, G.; Axsen, J.; Beard, G.; Daina, N.; Figenbaum, E.; Jakobsson, N.; Jochem, P.; Kinnear, N.; et al. A review of consumer preferences of and interactions with electric vehicle recharging infrastructure. *Transp. Res. Part D* **2018**, *62*, 508–523. Available online: https://www.sciencedirect.com/science/article/pii/S1361920918301330 (accessed on 27 September 2022). [CrossRef]
16. Figenbaum, E. Electromobility status in Norway: Mastering Long Distances—The Last Hurdle to Mass Adoption. TØI Report 1627/2018. Institute of Transport Economics. 2018. Available online: https://www.toi.no/publications/electromobility-status-in-norway-mastering-long-distances-the-last-hurdle-to-mass-adoption-article34903-29.html (accessed on 27 September 2022).
17. Bjerkan, K.Y.; Nørbech, T.E.; Nordtømme, M.E. Incentives for Promoting Battery Electric Vehicle (BEV) Adoption in Norway. *Transp. Res. Part D* **2016**, *43*, 169–180. [CrossRef]
18. Figenbaum, E.; Kolbenstvedt, M. Learning from Norwegian Battery Electric and Plug-In Hybrid Vehicle Users. Results from a Survey of Vehicle Owners. TØI Report 1492/2016. Institute of Transport Economics. Available online: https://www.toi.no/publications/learning-from-norwegian-battery-electric-and-plug-in-hybrid-vehicle-users-results-from-a-survey-of-vehicle-owners-article33869-29.html (accessed on 27 September 2022).
19. Ydersbond, I.M.; Amundsen, A.H. Fast Charging and Long-Distance Driving by Electric Cars in Inland Norway, ISBN 978-82-480-2148-3, TOI Report 1775, Oslo, Institute of Transport Economics. 2020. Available online: https://www.toi.no/publications/fast-charging-and-long-distance-driving-by-electric-cars-in-inland-norway-article36311-29.html (accessed on 27 September 2022).
20. Figenbaum, E. Battery electric fast charging—Evidence from Norwegian market. *World Electr. Veh. J.* **2020**, *11*, 38. [CrossRef]
21. Norwegian Environment Agency & Norwegian Public Road Association. Kunnskapsgrunnlag om Hurtigladeinfrastruktur for Veitransport. Oslo. 2022. Available online: https://www.regjeringen.no/contentassets/a07ef2d3142344989dfddc75f5a92365/kunnskapsgrunnlag_1mars.pdf (accessed on 27 September 2022).
22. Zhang, Q.; Li, H.; Zhu, L.; Campana, P.E.; Lu, H.; Wallin, F.; Sun, Q. Factors influencing the economics of public charging infrastructures for EV—A review. *Renew. Sustain. Energy Rev.* **2018**, *94*, 500–509. Available online: https://www.sciencedirect.com/science/article/abs/pii/S136403211830460X (accessed on 27 September 2022). [CrossRef]
23. Greene, D.L.; Kontou, E.; Borlaug, B.; Brooker, A.; Muratori, M. Public charging infrastructure for plug-in electric vehicles: What is it worth? *Transp. Res. Part D Transp. Environ.* **2020**, *78*, 102182. Available online: https://www.sciencedirect.com/science/article/abs/pii/S1361920919305309 (accessed on 27 September 2022). [CrossRef]
24. de Rubens, G.Z.; Noel, L.; Kester, J.; Sovacool, B.K. The market case for electric mobility: Investigating electric vehicle business models for mass adoption. *Energy* **2020**, *194*, 116841. Available online: https://www.sciencedirect.com/science/article/abs/pii/S0360544219325368 (accessed on 27 September 2022). [CrossRef]
25. van der Kam, M.; van Sark, W.; Alkemade, F. Multiple roads ahead: How charging behavior can guide charging infrastructure roll-out policy. *Transp. Res. Part D Transp. Environ.* **2020**, *85*, 102452. Available online: https://www.sciencedirect.com/science/article/pii/S1361920920306398 (accessed on 10 August 2022). [CrossRef]
26. Boston Consulting Group. Winning the Battle in the EV Charging Ecosystem. BCG. 2021. Available online: https://www.bcg.com/publications/2021/the-evolution-of-charging-infrastructures-for-electric-vehicles (accessed on 9 June 2021).
27. Deloitte. Hurry up and … Wait: The Opportunities Around Electric Vehicle Charge Points in the UK. 2019. Available online: https://www2.deloitte.com/content/dam/Deloitte/uk/Documents/energy-resources/deloitte-uk-Electric-Vehicles-uk.pdf (accessed on 21 January 2021).
28. Helmus, J.; Van den Hoed, R. Key performance indicators of charging infrastructure. *World Electr. Veh. J.* **2016**, *8*, 733–741. Available online: https://www.mdpi.com/2032-6653/8/4/733 (accessed on 10 August 2022). [CrossRef]

29. Schroeder, A.; Traber, T. The economics of fast charging infrastructure for electric vehicles. *Energy Policy* **2012**, *43*, 136–144. Available online: https://www.sciencedirect.com/science/article/abs/pii/S0301421511010470 (accessed on 10 August 2022). [CrossRef]
30. Pöyry 2012a. Strategi og Kriteriesett for Utplassering av Hurtigladere (del 1). Utarbeidet for Transnova og Statens Vegvesen. Pöyry Rapport R-2012-007. Available online: https://docplayer.me/1090511-Strategi-og-kriteriesett-for-utplassering-av-hurtigladere-del-1-utarbeidet-for-transnova-og-statens-vegvesen-r-2012-007.html (accessed on 10 August 2022).
31. Pöyry 2012b. Alternative Forretningsmodeller for Etablering av Hurtigladestasjoner—Del 2. Available online: https://www.osti.gov/etdeweb/servlets/purl/22000205 (accessed on 10 August 2022).
32. How Governments Can Solve the EV Charging Dilemma. Boston Consulting Group (BCG). 12 October 2021. Available online: https://mkt-bcg-com-public-pdfs.s3.amazonaws.com/prod/electric-vehicle-charging-station-infrastructure-plan-for-governments.pdf (accessed on 10 August 2022).
33. Kley, F.; Lerch, C.; Dallinger, D. New business models for electric cars—A holistic approach. *Energy Policy* **2011**, *39*, 3392–3403. Available online: https://www.sciencedirect.com/science/article/abs/pii/S0301421511002163 (accessed on 10 August 2022). [CrossRef]
34. ADL. Electric Vehicle Charging in Europe: What Will be the Winning Business Models in a Rising Multibillion Euro Market? 2021. Available online: https://www.adlittle.com/en/insights/viewpoints/electric-vehicle-charging-uk-and-europe (accessed on 29 September 2022).
35. Capgemini. Key Factors Defining. The E-Mobility of Tomorrow: A Focus on the EV Charging Infrastructure Ecosystem and Emerging Business Models. 2019. Available online: https://www.capgemini.com/wp-content/uploads/2019/02/Capgemini-Invent-EV-charging-points.pdf (accessed on 27 September 2022).
36. PwC. Powering Ahead! Making Sense of Business Models in Electric Vehicle Charging. 2018. Available online: https://www.pwc.co.uk/power-utilities/assets/powering-ahead-ev-charging-infrastructure.pdf (accessed on 21 January 2021).
37. Hardin, G. Science. New Series, Volume 162, No. 3859 1968. pp. 1243–1248, American Association for the Advancement of Science. Available online: http://web.mit.edu/2.813/www/readings/HardinCommons.pdf (accessed on 27 September 2022).
38. Spöttle, M.; Jörling, K.; Schimmel, M.; Staats, M.; Grizzel, L.; Jerram, L.; Drier, W.; Gartner, J. Research for TRAN Committee-Charging Infrastructure for Electric Road Vehicles. European Parliament. 2018. Available online: https://www.europarl.europa.eu/thinktank/en/document/IPOL_STU(2018)617470 (accessed on 2 July 2021).
39. Pagani, M.; Korosec, W.; Chokani, N.; Abhari, R.S. User behaviour and electric vehicle charging infrastructure: An agent-based model assessment. *Appl. Energy* **2019**, *254*, 113680. Available online: https://www.sciencedirect.com/science/article/abs/pii/S0306261919313674 (accessed on 27 September 2022). [CrossRef]
40. Bland, R.; Gao, W.; Noffsinger, J.; Siccardo, G. Charging Electric-Vehicle Fleets: How to Seize the Emerging Opportunity. McKinsey & Company. 2020. Available online: https://www.mckinsey.com/~{}/media/McKinsey/Business%20Functions/Sustainability/Our%20Insights/Charging%20electric%20vehicle%20fleets%20How%20to%20seize%20the%20emerging%20opportunity/Charging-electric-vehicle-fleets-how-to-seize-the-emerging-opportunity-FINAL.pdf (accessed on 28 April 2021).
41. Thronsen, M. Ny Undersøkelse: Nær Halvparten Forstår Ikke Prismodellene for Hurtiglading. Article in Norwegian, The Norwegian EV Association, Oslo, 2021-06-16. Available online: https://elbil.no/ny-undersokelse-naer-halvparten-forstar-ikke-prismodellene-for-hurtiglading/ (accessed on 10 August 2022).
42. Norwegian Automobile Federation (NAF). Skal Alle Med—Noe Må Skje, Oslo. 2021. Available online: https://res.cloudinary.com/nafmedier/image/upload/v1625087243/Elbil/Rapporter/NAF_Laderapport_2021.pdf (accessed on 27 September 2022).
43. European Commission. Proposal for a Regulation of the European Parliament and of the Council on the Deployment of Alternative Fuel Infrastructure and Repealing Directive 2014/94/EU of the European Parliament and the Council. Brussels, 14.7.2021. COM(2021) 559 Final. Available online: https://eur-lex.europa.eu/resource.html?uri=cellar:dbb134db-e575-11eb-a1a5-01aa75ed71a1.0001.02/DOC_1&format=PDF (accessed on 29 September 2022).
44. TU 2022a.—Etter Vår Vurdering er Dette Brudd på Loven. TU 022. February 2022. Available online: https://www.tek.no/nyheter/nyhet/i/BjnKPe/etter-vaar-vurdering-er-dette-brudd-paa-loven (accessed on 6 July 2022).
45. TU 2022b.—Rullestolbrukerne må Slite for å få Ladet. TU 18. January 2022. Available online: https://www.tek.no/nyheter/nyhet/i/MLJgnM/rullestolbrukerne-maa-slite-for-aa-faa-ladet (accessed on 6 July 2022).
46. Pinchasik, D.; Figenbaum, E.; Hovi, I.B.; Amundsen, A.H. Grønn Lastebiltransport? Teknologistatus, Kostnader og Brukererfaringer (Green Trucking? Technology Status, Costs, User Experiences). TØI Rapport 1855/2021 (in Norwegian). 2021. Available online: https://www.toi.no/publications/green-trucking-technology-status-costs-user-experiences-article37226-29.html (accessed on 10 August 2022).
47. CharIn 2022. Megawatt Charging System (MCS). Available online: https://www.charin.global/technology/mcs/ (accessed on 7 July 2022).

World Electric Vehicle Journal

MDPI

Article

Potentials of Light Electric Vehicles for Climate Protection by Substituting Passenger Car Trips

Simone Ehrenberger [1,*], Isheeka Dasgupta [1], Mascha Brost [1], Laura Gebhardt [2] and Robert Seiffert [2]

[1] DLR Institute of Vehicle Concepts, Pfaffenwaldring 38–40, 70569 Stuttgart, Germany
[2] DLR Institute of Transportation Research, Rudower Chaussee 7, 12489 Berlin, Germany
* Correspondence: simone.ehrenberger@dlr.de

Abstract: For the transformation of the mobility sector, small and light electric vehicles (LEV) show great promise, owing to their efficiency and low vehicle weight resulting in low energy consumption and lower greenhouse gas emissions per driven kilometer. The presented study focuses on the theoretical potential of substitutability of passenger car trips in Germany by varied LEVs based on the "Mobilität in Deutschland 2017" ("Mobility in Germany 2017") dataset, for the year 2030. A detailed approach for identifying substitutable car trips was developed, reflecting age, trip purpose, number of passengers, and other decision criteria. By conducting a life cycle assessment of the considered LEVs and passenger cars, potential emission savings were analyzed. In the considered baseline scenario, it is found that emissions could be reduced by 44 % with 50% of passenger car mileage being substituted by LEVs. This study, thereby, gives way to further research on LEVs, and would urge both policy makers and general users to steer towards comprehensive measures that encourage a switch from cars to LEVs.

Keywords: car trip substitution; greenhouse gas emissions; life cycle assessment; small and light electric vehicles

Citation: Ehrenberger, S.; Dasgupta, I.; Brost, M.; Gebhardt, L.; Seiffert, R. Potentials of Light Electric Vehicles for Climate Protection by Substituting Passenger Car Trips. *World Electr. Veh. J.* **2022**, *13*, 183. https://doi.org/10.3390/wevj13100183

Academic Editor: Joeri Van Mierlo

Received: 16 August 2022
Accepted: 27 September 2022
Published: 2 October 2022

1. Introduction

With 16% of worldwide emissions emerging from transport-related activities [1], transforming transportation with new vehicle concepts and mobility modes play a vital role in reaching target climate goals and reduction in emissions of noise and pollutants, and improve living conditions. Light electric vehicles (LEV) show great promise in this regard owing to their efficiency and low ratio of vehicle weight to payload, resulting in low energy consumption and, thus, lower greenhouse gas emissions per driven kilometer. Not only are the emissions significantly lower than those of cars with internal combustion engines, but also the comparison with EVs shows a significant advantage of LEVs in terms of both energy consumption in operation and resource requirements for production [2]. When situated in a multi and inter modal environment, they have a certain potential to replace car trips and, thus, reduce emissions [3].

Today's market offers a large variety of LEV models, from electric scooters and e-bikes up to two-, three-, and four-wheelers that fall under categories L1e to L7e according to EU regulation 168/2013 [4]. LEVs address diverse target groups and applications and, accordingly, differ regarding design and technical features. Common characteristics of all LEVs are that they are battery−electric and have a much lower weight and energy consumption per kilometer travelled compared to passenger cars with electric or other drive systems. The study analyses the following selection of LEV categories: e-scooters, e-bikes, and "e-bike+" that allow transport of cargo or children, speed pedelecs (L1e), mopeds (L1e), motorcycles (L3e), and three types of microcars differing in maximum speed and number of wheels (L5e, L6e, L7e). Section 2.1 provides detailed information on each category.

In order to quantify the theoretical potential of emission reduction, this study models a scenario in 2030, in which a major modal shift away from trips with full-sized passenger cars to LEVs has taken place. For the analysis, today's mobility patterns are maintained, with only a few trips still made by car, such as very long trips or trips with many occupants. This setting was chosen to avoid complex influences affecting the potential of LEVs, in the spirit of a ceteris paribus. For future studies, it would be interesting to investigate more complex scenarios involving different forms of mobility such as public transport or long-distance trains in combination with LEVs to gain differentiated insights into their potential, as well as possible adverse effects. The study uses Germany as an example of a national savings potential, as cars play an important role there, both as a means of mobility and as a sector of the economy.

2. Materials and Methods

2.1. LEV Properties

For the potential analyses and calculation of life cycle emissions, technical characteristics per evaluated vehicle categories are required, such as the maximum design speed, which determines the use on highways. In addition, values are needed that are derived on the basis of technical properties, as they do not result directly from this. This is the case for the maximum trip length for which a trip is considered to be substitutable. The so-called "relevant trip length" defines a trip length that is well-drivable with a specific LEV category on the basis of the literature and expert assessments. It is shorter than the technical electric range.

Most parameters such as maximum design speed, weight, battery capacity, technical electrical range, and seating capacity are based on exemplary models for the analyses. The data were taken from manufacturer information and sales websites, as far as available, or calculated on the basis of technical data. The energy consumption per kilometer is either taken from the sources described, or calculated with electric range and battery capacity if not provided. As vehicles differ considerably regarding maximum design speed, there are no standardized test cycles for assessing the energy consumption per kilometer that apply to all categories. This is considered tolerable, as the trip duration is also not necessarily kept constant for the replaced trips. The lack of standardized driving cycles, different handling of manufacturer specifications, and the high influence of user behavior complicate the estimation of energy consumption. In contrast to the technical parameters, which are based on today's technological status, for modelling a scenario for the year 2030, an increased lifetime mileage compared to today's vehicle mileage is assumed, resulting from expected technological progress. Although it is likely that other technical parameters also improve, these were left at the current level, in order to make a conservative estimate. Table 1 provides an overview of the values used for the further potential analysis.

Table 1. LEV properties (see [5]).

	Max. Speed (km/h)	Technical Electric Range (km)	Battery Capacity (kWh)	Weight (kg)	Energy Consumption (kWh/100 km)	Lifetime Mileage (km)
E-scooter	20	65	0.6	20	0.8	16,000
E-bike	25	120	0.4	25	0.3	50,000
E-bike+	25	70	0.4	51	0.6	50,000
Speed pedelec	25	70	1.2	29	1.7	70,000
Moped	45	100	2.7	100	2.7	70,000
Motorcycle	120	130	8.5	231	7.7	100,000
Microcar45	45	110	6.1	440	5.5	70,000
Microcar90	90	200	14.4	571	7.2	160,000
Microcar125	125	256	25	454	10.0	160,000

The maximum design speed indicates the speed used within the analyses. According to EU regulation 168/2013 [5], for the categories evaluated in the study, it is limited to 45 km/h for light two-wheel vehicles (L2e−B) and for microcars (L6e), to 90 km/h for microcars of category L7e (depending on sub-categories), and is not limited for vehicles belonging to category L3e or L5e.

For each category that was used for analyzing the trip substitution potential and for assessing life cycle emissions, an exemplary LEV model that is available (soon) on the market serves as basis for definition of the technical parameters. The selection process of LEV categories and respective exemplary models for deriving LEV properties aimed for (1) covering a broad spectrum of properties and capabilities in order to satisfy a full range of use cases and, thus, offering a high trip substitution potential, (2) capturing the differences between LEV categories and models, (3) limiting the number of selections to maintain an achievable modelling effort leading to comprehensible and memorable results, and (4) choosing models with low expected emission values, e.g., having a battery capacity corresponding to a required range, and not significantly higher. According to these aims, the technical parameters defined for each category do not represent average values of models available on the market or sold units in each category, but offer values for each category that are realistic on the one hand (as based on existing models) and oriented towards sustainability on the other hand, which does not necessarily correspond to average values of each category. The sustainability aspect refers in particular to choosing vehicles with rather low weight and battery capacity. A certain exception is the exemplary model of the L5e category, as here a model with three seats and a relatively high maximum design speed was aimed for, and no vehicle was found that combines these characteristics with a relatively small battery capacity.

2.2. Trip Substitution

This study uses data from the German national travel survey "Mobilität in Deutschland 2017" (MiD) [6], which surveyed 156,000 households, resulting in a dataset containing 960,000 trips. The dataset itself provides weighting and extrapolation factors that can be used to calculate the daily mobility pattern of the entire German population [7]. Based on this data, this study analyses the theoretical substitutability of car trips by LEVs of various categories considering their constraints. Real-world, market-ready LEVs were taken as reference for using technical properties of 9 types of LEVs, ranging from e-scooters and speed bikes to microcars, as described in Section 2.1. Their constraints regarding relevant travel distance, occupants, trip purposes, weather, age of driver, impairments, and street category were compared with the characteristics of the surveyed car trips. The relevant travel distance describes the distance that is assumed to be comfortably driven with an LEV per day and is lower than the technical electrical range. The number of persons reported for each trip is compared with the seating configuration of the LEVs, considering also if adults or children are travelling, which affects the suitability of the categories e-bike+, moped, and motorcycle. Regarding trip purposes, some trips dedicated to shopping and accompaniment of other persons, as well as professional trips, are excluded and considered as unsuitable for LEVs. The MiD data also provide information on weather conditions. Heavy rain, snowfall, or icy roads were defined as not usable with two-wheeled vehicles and vehicles without weather protection. The use of two-wheeled vehicles was also limited by the age of the driver, introducing a maximum age of 70 for these kind of vehicles for the analysis. Walking impairments are another exclusion criterion for the use of two-wheeled LEVs. Otherwise, it was assumed that people who can use a passenger car can also take a microcar. Regarding the use of different road categories, the maximum design speed of the respective LEVs was taken into account. This determines whether a vehicle is allowed to use motorways in Germany, for which the vehicle must be able to travel at least 60 km/h. For example, microcars have a higher relevant travel distance and are appropriate for most street types and all-weather conditions and, therefore, offer a broader range of suitability than e-bikes (Table 2). Only if all constraints that characterize a trip are fulfilled is an LEV classified as suitable for this trip.

Table 2. Trips substitution criteria (see [5]).

<table>
<tr><th></th><th>Relevant Travel Distance (Round Trip, km)</th><th>Number of Occupants</th><th>Street Category</th><th>Range of Age of Driver (Years)</th><th>Weather Conditions</th><th>Impairments (Suitability)</th></tr>
<tr><td>E-scooter</td><td>8</td><td rowspan="2">1</td><td rowspan="5">excl. highway</td><td rowspan="6">18–70</td><td rowspan="6">All, without heavy rain, snowfall, or icy roads</td><td rowspan="6">none</td></tr>
<tr><td>E-bike</td><td>30</td></tr>
<tr><td>E-bike+</td><td>30</td><td>1 + 3 children</td></tr>
<tr><td>Speed pedelec</td><td>60</td><td>1</td></tr>
<tr><td>Moped</td><td>60</td><td rowspan="2">2 (excl. children)</td></tr>
<tr><td>Motorcycle</td><td>90</td><td>all</td></tr>
<tr><td>Microcar45</td><td>80</td><td rowspan="2">2</td><td>excl. highway</td><td rowspan="3">18–99</td><td rowspan="3">all</td><td rowspan="3">Walking impairment</td></tr>
<tr><td>Microcar90</td><td>140</td><td rowspan="2">all</td></tr>
<tr><td>Microcar125</td><td>140</td><td>3</td></tr>
</table>

Since everyday trips are seldom independent of each other but usually belong to trip chains of an individual, substitutability was not only tested for individual trips, but also on the level of the associated trip chain. As the trips surveyed in the MiD dataset mostly start and end at the same location, they can be regarded as round trips. With this approach, daily car trips in Germany that could be substituted by an LEV were determined. Finally, identified trips were translated into substituted vehicle km for one day according to the trip single lengths reported in the MiD dataset.

2.3. Life Cycle Assessment and Emission Reduction Potential

Quantifying emissions savings by LEVs versus conventional passenger cars both during vehicle use phase and production requires a life cycle analysis (LCA) comparison. Scarce data and research for new LEV types and models for components and vehicle use behavior presents a challenge for analyzing potential emission reduction. Based on an analysis of currently available LEVs and prototypes, a representative set of technical characteristics was defined for every LEV category (see Table 1). Range and battery capacity represent upper boundaries of current vehicles, as we assume the technical characteristics to be representative for the year 2030 to which we refer in our analysis.

The LCA analysis of LEVs is based on modelling the production and vehicle use greenhouse gas (CO_{2eq}) emissions of e-scooters, e-bicycles, e-mopeds, and small 4 wheelers. Models using the material and energy flow software Umberto and the ecoinvent 3.7 [8] database are created using material flow details from the available literature. The data for vehicle characteristics and material composition of the vehicles are taken from various data sources. For e-scooters, several authors analyzed the production and use, and data for this study are taken from [9,10]. In case of e-bikes, we used data and the bill of materials from [11]. For the e-mopeds and motorcycles, we referred to [12,13]. For microcars, less information on representative vehicles of these classes is available. Therefore, we assumed that the generic microcars that we consider in our study have a similar material composition as a small electric vehicle, which is available in [14]. We adapted all vehicle material data to the technical characteristics of the generic LEVs in Table 1. While aluminium is the dominant material for e-scooters, the share of steel increases with the vehicle size (Figure 1) The lithium ion battery is assumed to be a NMC type for all vehicles. The production of these batteries is adapted from the production process in the GREET model [15] and combined with background data of the ecoinvent database. Lifetime mileage (see Table 1) plays a crucial role in the emission reduction potential, and care is taken by accounting for uncertainty emerging from the literature-derived vehicle use, as well as suitable battery and material properties.

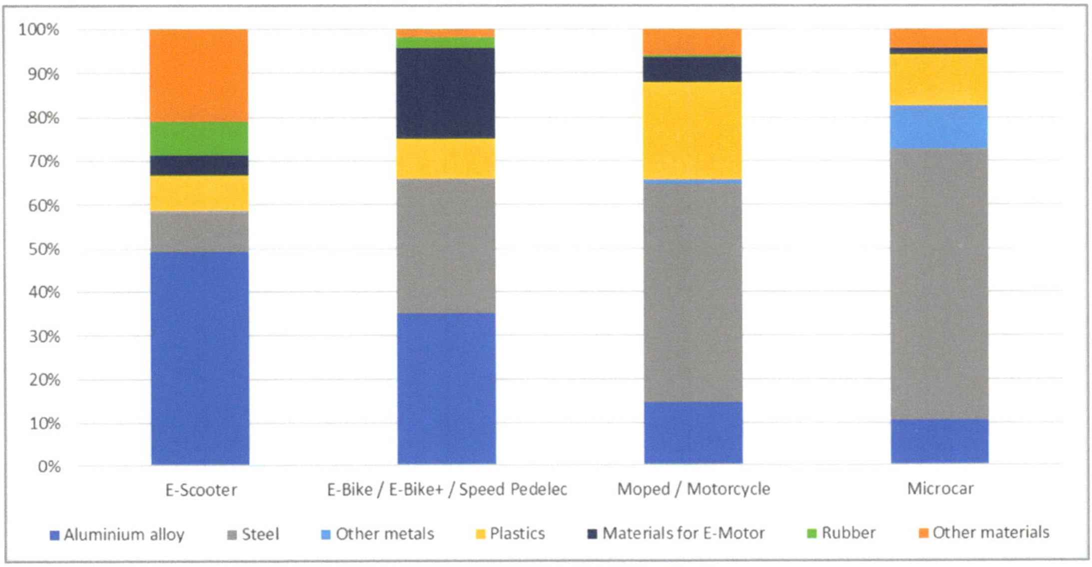

Figure 1. Weight share of vehicle materials and components.

The production of the passenger cars was calculated according to data from the ecoinvent 3.7 database [8]. We distinguish mid-class passenger cars with different drive trains, such as gasoline or diesel vehicles, cars using compressed natural gas (CNG), and hybrid and plug-in hybrid electric vehicles, as well as battery electric vehicles (BEV). The vehicle lifetime mileage for all drive train types is assumed to be 200,000 km. Emissions for the vehicle operation were taken from the handbook of emission factors [16]. Such emissions from the use of conventional fuels in internal combustion engines do not occur for purely electric vehicles, that is, all LEVs and BEVs. The electricity mix for the vehicle use is taken from [17]. The report describes pathways for the electricity system in Germany until 2035. From this, we derived an electricity production mix for the year 2030 with a renewable share of 74%. Conventional electricity production in this scenario consists mainly of natural gas and a smaller share of coal and lignite. A second scenario having a higher share of renewable energies (76%) and significantly lower share of lignite is derived from this report, and used for a sensitivity analysis.

In order to calculate the overall emission-saving potential for the passenger car substitution, we analyzed which LEVs are suitable for the trips identified from MiD (see Section 2.2), and chose the LEV with the lowest CO_{2eq} emissions as a substitute for the passenger car. The passenger car emissions are averaged according to a reference fleet for 2030 [18] containing a BEV share of 35% in the new vehicle fleet and 14% in the stock. Using this information, the difference in CO_{2eq} emissions between the average car and a LEV trip (emissions per km times trip length, 2030 scenario, Germany) is calculated using the following formula:

$$Emission\ Reduction\ Potential\ (\%) = \frac{\sum_i l_i * \left(E_c - E_{LEV_{lowest,i}}\right)}{\sum_i l_i * E_c} * 100 \quad (1)$$

where i = trip in MiD dataset; l = trip length; E_c = stock-weighted LCA emission of passenger car; and $E_{LEV\ lowest}$ = LCA emission of least emitting substitutable LEV for given trip.

As shown in Equation (1), when estimating the emission reduction potential from LEV substitution, the least LCA-based emission-producing LEV was selected when multiple LEVs satisfied a trip's constraints. This value is based on trips made on a representative day as reported in Section 2.2. Finally, the daily trips are aggregated and upscaled for one year.

3. Results

3.1. Potential of Substituting Car Trips and Mileage

Figure 2 shows the maximum potential of mileage substitution by the considered LEVs with a baseline range of these vehicles. A maximum of 76% of all trips accounted by the dataset can be substituted, which accounts for 50% of the mileage driven in Germany (Figure 2).

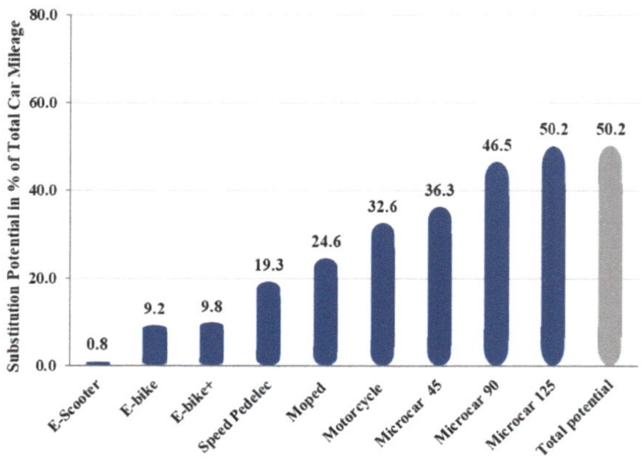

Figure 2. Maximum mileage substitution potential as a percentage of passenger car trips.

The difference between the trip substitution potential and the substitution potential in terms of car mileage is mainly based on the fact that a significant proportion of car mileage in Germany results from long distance car trips (see [7]), which could not be substituted by LEVs due to their limited range. Due to their different characteristics, there is a considerable difference in the car mileage that each LEV could substitute. While the e-scooter, as the simplest LEV model in these calculations, could only substitute under 1% of all car mileage in Germany, e-bikes already could substitute up to almost 20% of car mileage. Two-wheeled LEV models of the category moped and motorcycle offer a substitution potential that ranges from around 25 to 33% of car mileage. With the four-wheeled microcars, LEV models offer characteristics quite similar to small ordinary combustion engine cars, which is also reflected in the results: up to around 50% of all car mileage in Germany could be substituted by these types of LEVs. Since the microcar 125 offers the highest range and the most space for passengers among the LEV models, its substitution potential is equal to the total potential that results from the sum of all car trips that can be substituted by the LEV models.

3.2. Emission Reduction Potential

Battery size and capacity are found to play a decisive role in the overall greenhouse gas emissions. Emissions from battery and materials are directly related to battery capacity and vehicle weight. Consequently, microcars and motorcycles cause considerably more greenhouse gas emissions than smaller two–wheeler LEVs (Figure 3). The microcars with the largest batteries (microcar 125 km/h) have around 16 times higher greenhouse gas emissions in their production stage than an e-bike. More than half of these microcar emissions stem from the production of the traction battery. Comparing the technical electric range of more than 250 km possible with this battery and the assumed relevant travel distance of 140 km per day, an optimization of battery size in relation to the vehicle use bears some emission reduction potential.

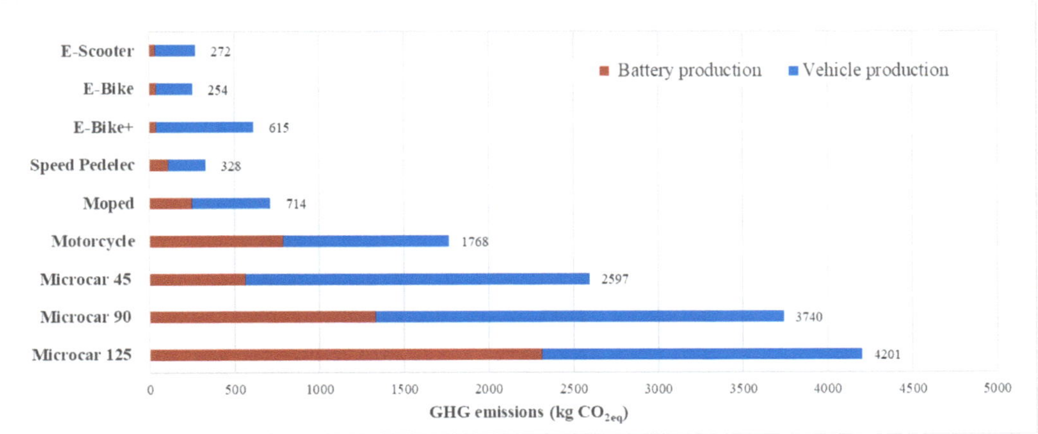

Figure 3. LEV production emission for baseline scenario.

For a per-kilometer greenhouse gas emission comparison with passenger cars, lifetime mileages of the vehicles are highly relevant, as they reflect the different technical characteristics, as well as the different utilities, of the vehicle types. Due to the higher lifetime mileages of the microcar 90 km/h and microcar 125 km/h, the greenhouse gas emissions on a per km basis are closer to the smaller LEVs, which have a lower lifetime mileage (Figure 4).

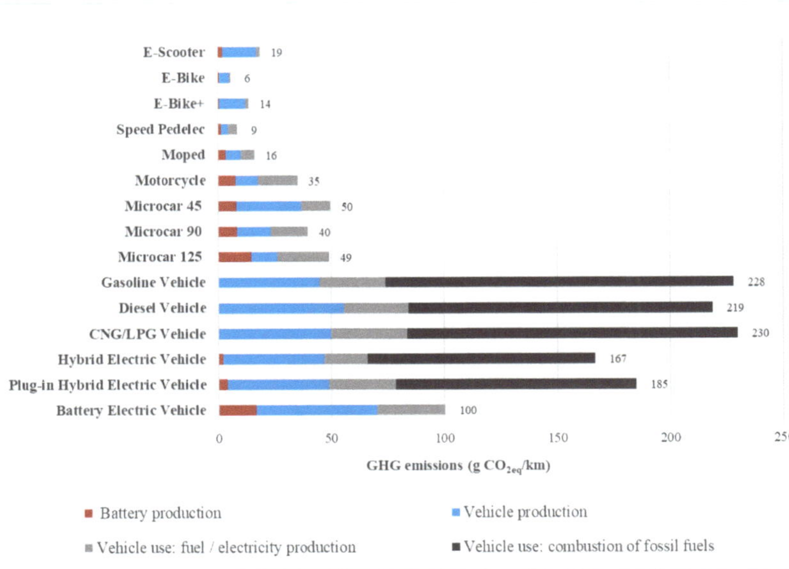

Figure 4. LCA emissions per kilometer of LEVs and passenger cars for baseline scenario.

For the vehicle-use stage, apart from the direct emissions of fuel combustion, the indirect emissions of the traction electricity supply are decisive for the results. The greenhouse gas emissions for the baseline energy mix of 2030 amount to around 230 g CO_{2eq} per kWh. Compared to the emissions from fossil fuel combustion and production, the emissions from the traction electricity supply lead to significantly lower emissions in the case of electric vehicles. This applies for BEVs in comparison to other drive trains for the passenger car, and also for the LEVs. Due to their lower energy consumption for operation, the LEVs in the worst case still cause only half of the greenhouse gas emission emissions of a BEV, and less than a quarter of conventional gasoline or diesel passenger cars (Figure 4).

When analyzing the substitutable trips with low-emission LEVs, the average greenhouse emissions of the LEVs (substituted mileage weighted average) is 24 g CO_{2eq}/km including vehicle production. This is only 12% of the replaced passenger car greenhouse gas emissions of 203 g CO_{2eq}/km. The average passenger car emissions represent the vehicle stock composition as described in Section 2.3. This difference results in a high emission reduction potential. With the identified mileage substitution potential of 50% (see Section 3.1), an overall saving of 44% would be achieved compared to the use of passenger cars on these trips. In absolute numbers, this amounts to 157 kilo tons CO_{2eq} per day compared to 356 kilo tons CO_{2eq} per day. This is equivalent to a reduction of 58 Mio tons CO_{2eq} per year (Figure 5). Overall, all LEV types contribute to this emission reduction potential with a slightly higher potential for small two-wheelers compared to microcars, mopeds, and motorcycles (Figure 5). This can be understood with the fact that even with similar passenger car mileage substitution for e-scooters, e-bikes, and speed pedelecs when compared to microcars, they save more emissions on LCA and tank-to-wheel basis (Figure 5), given their lower LCA emissions per kilometer (Figure 4).

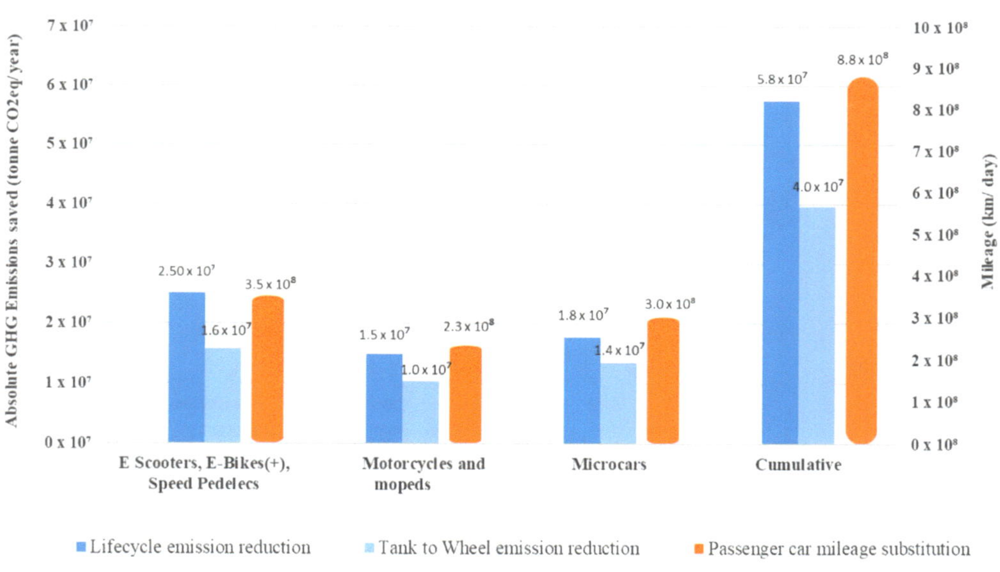

Figure 5. Emission reduction potential by LEV categories ("Lifecycle emissions" include vehicle production, fuel/electricity production, and vehicle use).

When evaluating the reduction potential that is allocated to the transport sector, only emissions on a tank-to-wheel basis would be considered. The analysis of the emission reduction potential with the tank-to-wheel system boundary leads to savings of 40 million tons CO_{2eq} per year (Figure 5), with e-bikes having the highest contribution.

3.3. Sensitivity Analysis

A sensitivity analysis was performed with respect to influencing parameters such as vehicle range, lifetime mileage, and energy mix possibilities. Figure 6 shows the variation in the LCA-based emission reduction potential by substituting emission optimum LEVs for passenger car trips. The variation is shown as a function of a common range multiplier applied to the range of all LEVs, which is the relevant travel distance in Table 2 when considered for trip substitution and an energy mix carbon intensity multiplier that affects the emissions generated during LEV vehicle use. The alternative energy mix is related to around 140 g CO_{2eq} per kWh compared to 230 g CO_{2eq} per kWh of the baseline electricity mix. For the sensitivity analysis, a range between these two electricity mixes is used. The multipliers are applied to the baseline vehicle scenario parameters as explained in Tables 1 and 2. This results in the carbon intensity for the electricity mix in the sensitivity analysis vary from 138 to 322 g CO_{2eq} per kWh.

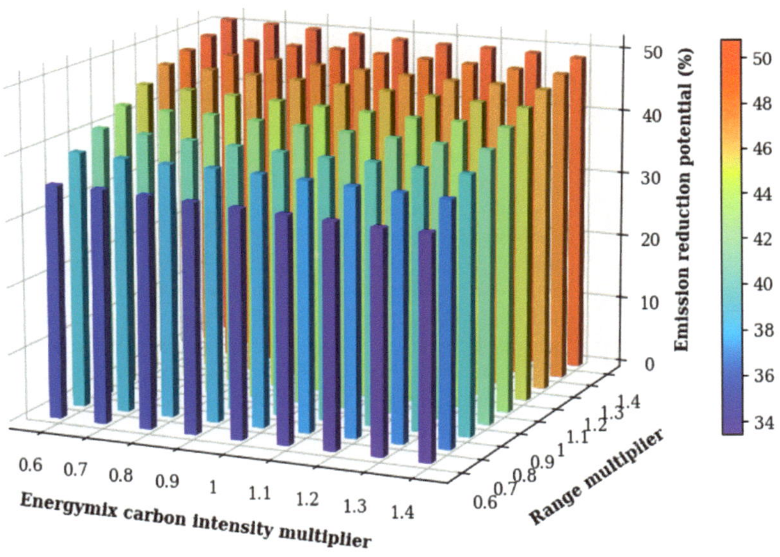

Figure 6. Sensitivity of the results to the carbon intensity of the energy mix (varies from 138 to 322 g CO_{2eq} per kWh) and to the relevant travel distance (varies from 0.6 to 1.4 times the baseline values for each LEV in Table 2).

As can be seen, the emission reduction potential varies from 33% to 51% of the total passenger car and motorcycle emissions calculated from the MiD dataset when both multipliers are varied from 0.6 to 1.4 times the baseline scenario value. The gradient of the resulting surface is more aligned towards the range multiplier direction, i.e., higher percentage change in emission reduction potential by change in range multiplier as more trips can be substituted by lower emitting LEVs. The effect of this seems larger than reducing the carbon intensity of the energy mix to as low as 60% of the assumed base scenario. On an average across this variable space, with a 10% increase in range, the emission reduction potential increases by approximately 2%, whereas for a 10% decrease in carbon intensity of the energy mix, the emission reduction potential increases by approximately 0.3%. It should be noted that rate of change of emission reduction potential is not constant with respect to the range multiplier, since it is also a function of the trip distribution in the MiD dataset and the parameters of the LEVs considered for substitution.

4. Discussion and Conclusions

The analysis quantifies a significant theoretical potential for decreasing greenhouse gas emissions with LEVs, and is a first step that sets the ground for future scientific work evaluating approaches and obstacles to realizing the emissions reduction potential of LEVs. It, thereby, gives way to further research on LEVs, and urges both policy makers and general users to steer towards comprehensive measures that to encourage a switch from cars to LEVs. Looking at today's trends in the vehicle market in Germany, such as the increasing sales of SUVs, it seems obvious that the identified potential can only be realized if extensive measures are taken, and even then only in parts. High profit margins make it attractive for manufacturers to sell large, heavy, and expensive vehicles, which could be one of many reasons why hardly any LEVs are found in the portfolios of major manufacturers. However, also on the buyer side, there are currently still many obstacles to the market success of LEVs, which need to be illuminated in further studies.

The potential for trip substitution and reduction in greenhouse gas emissions might seem high at first glance, however, given that 60% of the car mileage in Germany results from trips under 50 km, and some LEVs can cover even longer trips, the result with 50% mileage replacement potential and respective emission reduction is plausible. Conservative assumptions were made regarding technical properties, substitutability of trips, and replaced vehicles. Evaluation of data from the German Federal Motor Transport Authority on new battery electric cars registered in Germany in the recent years show a trend towards batteries with higher capacity than today. As long as battery technology does not improve significantly regarding production-related greenhouse gas emissions, the life cycle impacts of electric cars would tend to worsen, and the relative potential of LEVs to reduce greenhouse gases would improve. Furthermore, it would be interesting to expand the scope of the study and analyze the extent to which LEVs can support modal shift and multimodal mobility. For instance, long car trips could be partly covered by using a privately owned or shared EV in combination with long-distance trains, thereby, further reducing emissions. A mix of mobility solutions is necessary for future sustainable mobility, as LEVs are no stand-alone solution for substituting car trips and reducing car ownership. Extensive sustainability effects would be achieved with a holistic mobility approach, integrating LEVs in a system together with a high share of public transport, active modes, low private care ownership, attractive vehicle-sharing offers, and other sustainable mobility components such as reducing the number of trips by approaches such as the city of short distances.

The analyses quantify as a first step of theoretical potential to substitute car trips and, thereby, reduce greenhouse gas emissions, not considering aspects of user acceptance, or possible changes in mobility behavior, infrastructure, or political framework conditions. Future research should, therefore, evaluate paths towards greater use of LEVs, with specific measures to promote a sustainable and safe use of LEVs, and investigate social aspects, such as user acceptance and preferences in mode choice, as well as technological aspects such as improved vehicle safety.

Author Contributions: Conceptualization, S.E., M.B. and L.G..; methodology, S.E., I.D., M.B., L.G. and R.S.; software, S.E., I.D., L.G. and R.S.; validation, I.D., M.B. and L.G.; formal analysis, I.D. and R.S.; investigation, S.E., I.D. and M.B.; resources, S.E. and L.G.; data curation, I.D.; writing—original draft preparation, S.E., I.D., M.B., L.G. and R.S; writing—review and editing, S.E., I.D. and M.B.; visualization, I.D. and R.S.; supervision, S.E. and M.B.; project administration, M.B.; funding acquisition, M.B. All authors have read and agreed to the published version of the manuscript.

Funding: This research was funded by LEVA EU.

Data Availability Statement: Not applicable.

Acknowledgments: We thank the members of the scientific panel of this project for the constructive comments and fruitful discussion. We further thank our colleagues Robert Hahn and Stephan Schmid for supporting us with a critical review of our results.

Conflicts of Interest: The authors declare no conflict of interest.

References

1. IEA. Global CO_2 Emissions by Sector. Available online: https://www.iea.org/data-and-statistics/charts/global-co2-emissions-by-sector-2019 (accessed on 22 December 2021).
2. Brost, M.; Ewert, A.; Schmid, S.; Eisenmann, C.; Gruber, J.; Klauenberg, J.; Stieler, S. Elektrische Klein− und Leichtfahrzeuge: Chancen und Potenziale für Baden−Württemberg (Small and Light Electric Vehicles: Opportunities and Potential for Baden−Württemberg). Project Report. 2019. Available online: https://elib.dlr.de/133048/1/LEV_e$-$mobil_BW_Leichtfahrzeug_Studie.pdf (accessed on 3 March 2022).
3. Gebhardt, L.; Ehrenberger, S.; Wolf, C.; Cyganski, R. Can shared E−scooters reduce CO_2 emissions by substituting car trips in Germany? *Transp. Res. Part D Transp. Environ.* **2011**, *109*, 103328. [CrossRef]
4. Regulation (EU) No 168/2013 of the European Parliament and of the Council of 15 January 2013 on the Approval and Market Surveillance of Two− or Three−Wheel Vehicles and Quadricycles (Text with EEA Relevance). Available online: https://eur$-$lex.europa.eu/legal$-$content/EN/TXT/?uri=CELEX%3A02013R0168$-$20201114 (accessed on 12 April 2022).
5. Brost, M.; Gebhardt, L.; Ehrenberger, S.; Dasgupta, I.; Hahn, R. The Potential of Light Electric Vehicles for Climate Protection through Substitution for Passenger Car Trips−Germany as a Case Study. Project Report 2022. Available online: https://elib.dlr.de/186036/1/2022$-$03$-$24_LEV4Climate_DLR_LEVA$-$EU_report.pdf (accessed on 24 May 2022).
6. Federal Ministry of Transport and Digital Infrastructure (BMVI). Mobillitaet in Deutschland (Mobility in Germany). Available online: http://www.mobilitaet$-$in$-$deutschland.de/publikationen2017.html (accessed on 22 December 2021).
7. Nobis, C.; Köhler, K. *Mobilität in Deutschland–MiD Nutzerhandbuch (User Guide), Study by infas, DLR, IVT and infas 360 (FE−Nr. 70.904/15)*; Federal Minister of Transport and Digital Infrastructure: Bonn/Berlin, Germany, 2018; Available online: www.mobilitaet$-$in$-$deutschland.de (accessed on 23 September 2021).
8. Wernet, G.; Bauer, C.; Steubing, B.; Reinhard, J.; Moreno-Ruiz, E.; Weidema, B. The Ecoinvent Database Version 3 (Part I): Overview and Methodology. *Int. J. Life Cycle Assess.* **2016**, *21*, 1218–1230. Available online: www.ecoinvent.org (accessed on 23 September 2021). [CrossRef]
9. Severengiz, S.; Finke, S.; Schelte, N.; Wendt, N. Life Cycle Assessment on the Mobility Service E−Scooter Sharing. In Proceedings of the 2020 IEEE European Technology and Engineering Management Summit, E−TEMS 2020. Dortmund, Germany, 5–7 March 2020.
10. Moreau, H.; Meux, L.J.; Zeller, V.; D'Ans, P.; Ruwet, C.; Achten, W.M.J. Dockless e−scooter: A green solution for mobility? Comparative case study between dockless e−scooters, displaced transport, and personal e−scooters. *Sustainability* **2020**, *12*, 1803. [CrossRef]
11. Bonilla-Alicea, R.J.; Watson, B.C.; Shen, Z.; Tamayo, L.; Telenko, C. Life cycle assessment to quantify the impact of technology improvements in bike−sharing systems. *J. Ind. Ecol.* **2020**, *24*, 138–148. [CrossRef]
12. Kerdlap, P.; Gheewala, S.H. Electric Motorcycles in Thailand—A Life Cycle Perspective. *J. Ind. Ecol.* **2016**, *20*, 1399–1411. [CrossRef]
13. Cox, B.L.; Mutel, C.L. The environmental and cost performance of current and future motorcycles. *Appl. Energy* **2018**, *212*, 1013–1024. [CrossRef]
14. Helmers, E.; Dietz, J.; Weiss, M. Sensitivity Analysis in the Life−Cycle Assessment of Electric vs. Combustion Engine Cars under Approximate Real−World Conditions. *Sustainability* **2020**, *12*, 1241. [CrossRef]
15. Dai, Q.; Kelly, J.; Dunn, J.; Benavides, P. Update of Bill−of−Materials and Cathode Materials Production for Lithium−Ion Batteries in the GREET® Model, Memo 2018. Available online: www.greet.es.anl.gov (accessed on 23 September 2021).
16. INFRAS. *Handbook Emission Factors for Road Transport (HBEFA)*; Bundesamt für Umwelt BAFU: Bern, Switzerland; Umweltbundesamt GmbH: Vienna, Austria; Umweltbundesamt: Dessau, Germany; Swedish Transport Adiministration: Borlänge, Sweden; Norwegian Environment Agency: Oslo, Norway; Joint Research Centre: Ispra, Italy, 2017.
17. Drees, T.; Medert, H.; Meinecke, M.; Halici, C. Netzentwicklungsplan Strom 2035 (2021), Zweiter Entwurf | Übertragungsnetzbetreiber CC−BY−4.0. 2021. Available online: https://www.netzentwicklungsplan.de/sites/default/files/paragraphs-files/NEP_2035_V2021_2_Entwurf_Teil1.pdf (accessed on 18 January 2022).
18. Kopernikus−Projekt Ariadne, Ariadne−Report: Deutschland auf dem Weg zur Klimaneutralität 2045−Szenarien und Pfade im Modellvergleich. 2021. Available online: https://ariadneprojekt.de/publikation/deutschland-auf-dem-weg-zur-klimaneutralitat-2045-szenarienreport/ (accessed on 2 February 2022).

World Electric Vehicle Journal

MDPI

Article

Impact of Electric Vehicle Charging Synchronization on the Urban Medium Voltage Power Distribution Network of Frederiksberg

Tim Unterluggauer [1,*], F. Hipolito [2], Sergey Klyapovskiy [1,†] and Peter Bach Andersen [1]

1 Department of Wind and Energy Systems, Technical University of Denmark, 2800 Kongens Lyngby, Denmark
2 Department of Management Engineering, Technical University of Denmark, 2800 Kongens Lyngby, Denmark
* Correspondence: timun@dtu.dk
† Current address: Utiligize ApS, Gammel Kongevej 1, 1610 Copenhagen, Denmark.

Abstract: The uptake of electric vehicles (EVs) may pose a challenge to power distribution networks (PDNs). While smart charging can be deployed to relieve stress on the grid, user-centric smart charging strategies could also exacerbate peak power demand due to synchronization when optimizing charging with regard to different objectives, such as charging costs. In this paper, we assess the charging demand emerging from a large fleet of EVs, with models for the decision to charge and distribution of the steady-state state-of-charge (SoC). These are applied to the municipality of Frederiksberg, Denmark, using data from the Danish national travel survey. Home and workplace charging are mapped to the urban 10 kV medium voltage PDN of Frederiksberg considering different charging behaviors and degrees of synchronization. Results indicate that the likelihood of severe congestion in the power distribution network is low and that it can be attributed to rare scenarios in which high synchronization is observed, particularly when maintaining the normal steady-state demand. Despite the low likelihood, preventive measures should be devised to mitigate such scenarios, especially if additional high-power consumers are connected.

Keywords: electric vehicles; charging synchronization; grid impact; medium voltage; urban power distribution network

Citation: Unterluggauer, T.; Hipolito, F.; Klyapovskiy, S.; Andersen, P.B. Impact of Electric Vehicle Charging Synchronization on the Urban Medium Voltage Power Distribution Network of Frederiksberg. *World Electr. Veh. J.* **2022**, *13*, 182. https://doi.org/10.3390/wevj13100182

Academic Editor: Joeri Van Mierlo

Received: 1 September 2022
Accepted: 27 September 2022
Published: 30 September 2022

1. Introduction

Electrification of road transport, particularly personal vehicle utilization, has been identified as an important means to address the global challenge of reducing carbon emissions [1]. The growth of the EV market share has stimulated extensive research in the analysis of the impact of uncontrolled charging on both low [2,3] and medium voltage [4,5] PDNs. Works include the impact analysis for different grid topologies, e.g., rural, suburban or urban PDNs [6,7], for different countries [2,8–10] or commonly used test systems [11–13] due to a lack of data. The use of smart charging to mitigate this risk on the PDN has also been largely covered in the literature [14,15]. While several definitions of smart charging exist, optimally smart charging should consider both the power systems' and vehicle users' benefits [16].

However, recent studies also show that the coincidence factor (CF) of uncontrolled charging with increasing fleet size and charging power is expected to be rather low, typically less than 25% [17,18], and thus might not pose the primary risk of overloading PDNs. The CF is defined as the ratio of the simultaneous maximum demand of a group of consumers (EV demand for the purpose of this work) to the sum of their individual maximum demands within a specified period. In contrast, little attention has been dedicated to the analysis of the potential negative impacts of a large-scale simultaneous start of charging stemming from different optimization strategies or the unforeseen use of the system, such as user-centric smart charging strategies that do not consider network constraints and the large-scale response of EV users to distribution network tariffs [11,19,20] or the malicious manipulation of charging control resulting from cyberattacks [21,22].

The first encompasses the risks posed by the synchronization of charging to external signals, such as electricity spot price and carbon emission fluctuations, or in its simplest form, to distribution network tariff variations. While several Danish companies offer such smart charging services [23,24], the existing literature [11,19] indicates that cost minimization could be a driving force for undesirable synchronization effects in home charging. Furthermore, network tariffs aim at incentivizing customers to shift demand away from peak load times. One example is the recently announced new tariff system of the Danish distribution system operator (DSO) Radius [25], where prices for households during peak load periods in winter will be roughly nine times higher than during low load periods. While tariffs can be a means to relieve stress on the grid, several pilot projects related to network tariffs also point towards the potential adverse effects of high rebound effects causing secondary peaks [26–29] due to undesirable synchronization at the intersection of two pricing periods.

The interplay of network tariffs and electricity price-based smart charging schemes (based on 24 h forecasts) could increase the CF, stemming from EV users deciding to charge on the given day. In addition, price-based smart charging could potentially increase the CF even further when considering multi-day forecasts in the optimization, thus rising the potential for simultaneous charging of a large fleet of EVs on the same day at the same time. Despite the potential for large impacts, little attention has been dedicated to quantifying and analyzing such impact on PDNs [14,20,30,31].

We address this knowledge gap in the paper by analyzing the potential impact of different degrees of charging simultaneity by comparing several EV charging scenarios. Recent literature credits home and workplace charging with the greatest potential for EV flexibility due to long idle times [32]. Therefore, we focus our analysis on these two charging locations where smart charging might be pursued most frequently, with the risk of adverse synchronization effects if PDN constraints are not taken into account. Furthermore, this paper focuses its analysis on the municipality of Frederiksberg, the inner part of the metropolitan area of the city of Copenhagen (Denmark). Due to its high population density and heterogeneous combination of charging demand, it is a good case study for assessing the impact of EV adoption on the power grid in the urban context. This municipality is currently part of a comprehensive project, FUSE [33], dedicated to analyzing the impacts of the electrification of private road transport, both from the perspective of transport demand and from the power grid supply standpoint. By combining real data from user behavior, baseline power consumption in the PDN and its topology, we offer a practical approach to analyze power grid utilization in real-world scenarios. We focus the analysis on the medium voltage PDN as part of a two-step approach. First, we access the boundaries of the impact on the medium voltage PDN. Second, we proceed with the analysis of the impact on the low voltage grid in future work. This approach addresses the lack of real-world case studies that simultaneously consider the transportation and power distribution perspectives [34] and aims at assisting DSOs to understand the potential impact of high charging synchronization in urban areas.

The main contributions of the paper are:

- Providing a large-scale case study based on real data for modelling the PDN and EV demand,
- Making use of novel steady-state SoC and decision-to-charge models to be able to simulate and analyze different charging patterns,
- Providing a novel and straightforward charging allocation methodology to map EV charging events to the PDN,
- Addressing the lack of studies devoted to analyzing the grid impact of charging synchronization which might become more prominent in the future when EV users will follow the same user-centric smart charging objective.

The subsequent parts of this paper are organized as follows: Section 2 provides an overview of the methodology introduced in this paper; Sections 3 and 4 present and discuss the key findings; finally, conclusions are given in Section 5.

2. Methodology

An overview of the methodology of this paper is illustrated in Figure 1 and consists of five steps, to be discussed in the following. The different steps are highlighted by large rectangles ranging from light to dark green.

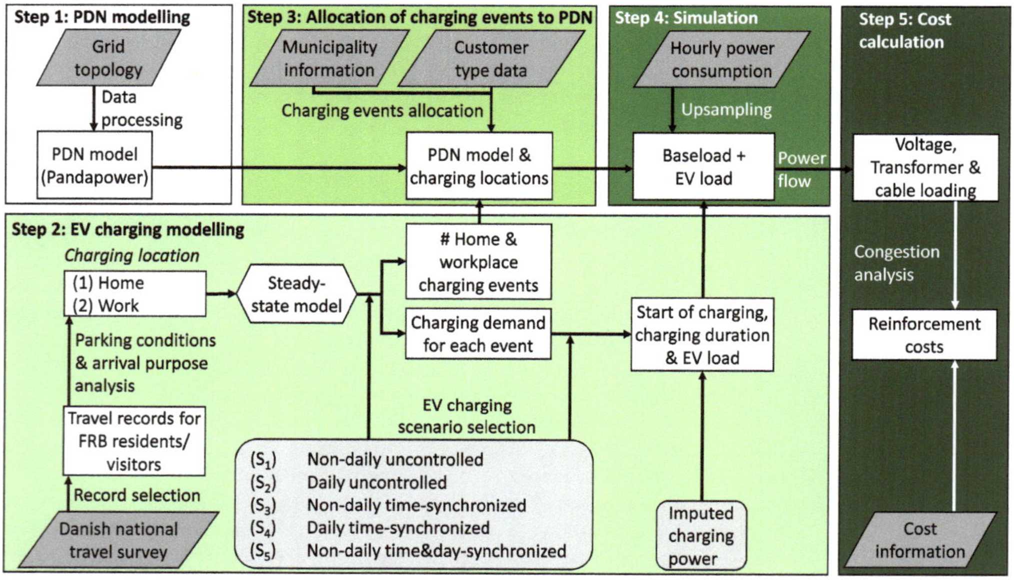

Figure 1. Methodology of the paper to determine the impact of different charging strategies on the power distribution network of Frederiksberg (FRB). While dark-gray-shaded rhomboids depict the input data, gray-shaded shapes represent the external parameter. The light-gray-shaded hexagon represents the external models [35] used in this work. White-shaded rectangles illustrate results obtained in each step of the process.

2.1. Power Distribution Network Modelling

The modelling software to analyze the PDN in Frederiksberg makes use of the Python package Pandapower. This package allows for a straightforward representation of the 10 kV PDN of Frederiksberg to simulate the power flow in the respective infrastructure. For the purpose of this paper, we focus the analysis on the 10/0.4 kV transformers and the respective feeding underground cables. The power grid network topology is defined based on data collected from the DSO for this area, Radius, which provided the location and characteristics of its transformers and respective cables. In normal operation, the PDN of Frederiksberg is divided into three independent networks labelled FRBi with $i = \{1, 2, 3\}$. Each PDN has its own main station 30/10 kV coupled to multiple feeders connecting the respective transformers. Additional connections between the PDNs exist, which can be deployed as tie-lines to connect the transformers in contingency scenarios. In this work, we only consider normal operation in which the three PDNs are operated independently. A brief description of the PDNs' main characteristics is given in Table 1, and the layout is illustrated in Figure 2.

Table 1. Summary of the power distribution network infrastructure in Frederiksberg. For each FRBi, there are N^i_{tr} transformers connected to N^i_c cables covering a total length L^i_c.

FRBi	N^i_{tr}	N^i_c	L^i_c (km)
1	152	222	93.4
2	98	157	60.0
3	34	61	19.9

Figure 2. Visualization of the urban power distribution network of Frederiksberg. The grid topology is shown on the footprint of the municipality, with gray lines indicating roads and walking paths and buildings represented in light blue. The 10/0.4 kV transformers $i = \{1, 2, 3\}$ are illustrated by red, blue, and black solid dots. Medium voltage underground cables for all PDNs are shown in green.

2.2. EV Charging Modelling

The evaluation of the daily EV demand is based on two pillars. The first is travel records (private car trips only) from the Danish national travel survey in the years 2006–2019 [36]. The second is the models for the decision to charge and steady-state SoC distributions introduced in [35], where the decision to charge is represented by a simple parametric model that takes as inputs only the SoC level and relative daily range, i.e., the ratio of daily-driven distance to the vehicle's range. Moreover, this model can also be used to infer an initial steady-state distribution of SoC levels that is consistent with real-world utilization, thus preventing either under or overcharging due to poor representation of the initial SoC levels.

From the travel survey, we collected all travel records for private car utilization that visited Frederiksberg at least once, which defined the donor pool of representative travel records. Since the travel survey samples a small fraction of the population yearly, each record includes a calibration factor λ that ensures a representative picture of the population in each year. We used this calibration factor to generate an expanded pool of travel records, such that each donor record spawns $\lfloor \lambda \rfloor$ records in the expanded pool. In this process, we utilized origin–destination traffic matrices to introduce random variation to the location of each activity in the expanded pool of records, while still preserving the traffic patterns. By making use of the information available, we characterized the records by home residency and workplace municipalities, the parking conditions and the type of day. The travel survey does not indicate the specific day for each record, rather it identifies the respective type of day. Normal weekdays, defined as Monday to Thursday where the next day is not a holiday or special day, comprise the bulk of the travel records, approximately 57.4% of all records. The next largest subset of records concerns Fridays or weekdays prior to holidays, accounting for approximately 14.6%. The remaining groups contain even smaller shares of the donor pool, making these susceptible to non-representative variations from outlier records. Therefore, we restricted our analysis to travel records from the largest group.

In addition, we leveraged the data on parking conditions at home and work to identify the potential access to home and workplace charging, which we used to categorize the records into non-overlapping subgroups, as depicted in Figure 3. Given the scarcity of records for EVs in the travel survey, we used records from vehicles with conventional drive-trains. For each record, we attributed a battery capacity based on a review of new EV specifications, assuming a fleet with a mean capacity of 68 kWh with a standard deviation of 18 kWh and set the mean rate of conversion for all records at $\eta = 0.2$ kWh/km.

The models introduced in [35] were then used to determine the required energy to meet travel requirements $\delta\varepsilon$, the probability of charging at a given day p_d, as well as the mean interval between charging events d_p.

2.2.1. Classification of Travel Records

The distribution of the charging events was determined by a hierarchical filtering process, based on the parking conditions at home and work, as well as the travel diaries details, as depicted in Figure 3. To begin with, we segregated the records according to the residency municipality to identify the share of charging events from residents and the demand emerging from visitors. The visitors comprise any person that at any point in their travel record drives into the selected municipality. Then, we started the hierarchical filtering process. First, we assessed the parking conditions at home as indicated in the travel survey [36]. For those with reliable access to parking at their premises (i.e., in carport/garage or in the front yard/driveway) or on/next to the property (i.e., reserved, always or normally space parking), we inferred that home charging will be available to such drivers. If the travel records indicate that the vehicle returns home on that day, we attributed the demand from this record to *home* charging. Second, remaining records were then screened for good parking conditions at work. Those include a permanent parking space provided by the employer, as well as free parking, defined as always or normally available. By the same token, we attributed *workplace* charging if the travel records indicated a visit to the workplace on that day. Since the workplace municipality can be different from the residency, we accounted for the contribution inside and outside Frederiksberg. The remaining demand was assigned to charging using the public infrastructure and was exempt from our impact analysis within this work. It is worth noting that we are likely overestimating the penetration of home and workplace charging by assuming that everyone who has conditions to have a charger will install one. Hence, the results shown in this paper should be interpreted as upper bounds to the share of charging demand met at home and the workplace.

2.2.2. Charging Scenarios and Energy Demand

Even though the market share of EVs in Denmark remains low so far [37], for the purpose of this paper, we considered an ambitious scenario of full electrification of the private passenger car fleet of the year 2020, i.e., 24,252 cars [38]. Assuming a homogeneous distribution of EVs at the national level and present-day traffic patterns, we should expect in this scenario that the total number of EVs transiting in Frederiksberg daily reaches $N_t = N_r + N_v$ = 59,070, where N_r = 24,252 and N_v = 34,818 are the number of EVs from residents and visitors, respectively. To ensure a representative load profile, we randomly sampled travel records for each day in the power flow simulation, collecting N_r residents and N_v visitors. We collected travel records for 40 weekdays of peak loading to be discussed in more detail in Section 2.4.

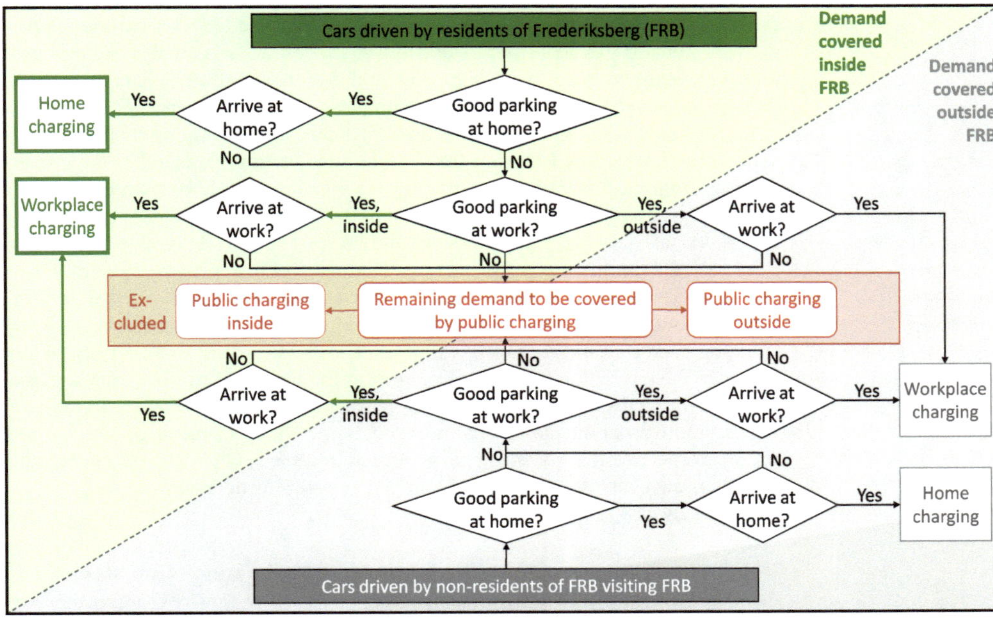

Figure 3. Representation of the segmentation of travel records according to residency and workplace municipality, as well as the parking conditions at the respective locations.

Based on this set of records, we ran five base scenarios S_i for EV charging to analyze the impact on the PDN, namely:

- (S_1) uncontrolled non-daily charging,
- (S_2) uncontrolled daily charging,
- (S_3) time-synchronized non-daily charging,
- (S_4) time-synchronized daily charging, and
- (S_5) time-and day-synchronized non-daily charging.

The first and second scenarios concern unconstrained charging, in which no management is imposed on the charging process and the EVs charge upon arrival. S_1 sets a baseline for the charging demand, in which drivers charge $\delta\varepsilon$ with a mean interval of d_p days, according to the decision model and steady-state demand introduced in [35]. In this scenario, only a subset of EVs charge on a given day, as determined by the aforementioned decision model. It is worth noting that most EVs will charge on a non-daily basis, yet a small portion could require daily charging. Hence, we refer to the charging pattern determined by the decision model as non-daily charging behavior. In contrast, S_2 probes the impact of satisfying the energy demand on a daily basis, i.e., charging $\delta\varepsilon/d_p$ every day. In this scenario, we forced all EVs to charge their daily demand independent of the decision model. The impact of high charging simultaneity was addressed in the remaining scenarios. Scenarios S_3 and S_4 follow the charging patterns of scenarios S_1 and S_2, respectively, but the charging process is controlled and only allowed to start, if altogether possible, after the coordination signal time. Both scenarios illustrate intraday synchronization (labelled as T.S. for time-synchronization), where the charging starts simultaneously for the EVs that will charge on the given day. In brief, scenarios $S3$ and $S4$ concern different levels of demand. In $S3$, the demand is governed by the steady-state model, which leads to a fraction of the total EVs charging each day to cover their multi-day energy demand. In contrast, $S4$ concerns a scenario where each EV charges daily the amount required to satisfy its daily demand. For large fleets, the aggregated energy demand in both cases should be similar, but distributed over a much larger number of vehicles in S4 than in S3. Conversely, the duration of charge events is considerably smaller in S4 than in S3. Hence, S4 represents the potential peak power demand. Finally, S_5 simulates an extreme demand case as a combination of S_3 and S_4. This scenario should be understood as a worst-case scenario which explores peak power and peak energy demand in a single day. In this scenario, all EVs charge their steady-state demand $\delta\varepsilon$ as in S_3 (i.e., the demand accumulated since the last charge) on the same day as in S_4, thus representing the extreme case of synchronizing charging on the same day at the same time (labelled as T.D.S. for time- and day-synchronization). For large fleets of EVs driving different distances every day, such a scenario should be extremely unlikely, as it would require a commensurate alignment of the steady-state charging cycle of all vehicles.

In scenarios S_1 and S_3, the number of charging events is determined by the probability p_d of each EV charging on a given day, as defined in [35]. For the purpose of this paper, we drew a random number from a uniform distribution in the interval $r \in [0,1]$ and defined the acceptance criteria for the decision to charge as $r \leq p_d$. Consequently, the number of charging events in said scenarios was stochastic in nature and smaller than the number of events in the remaining scenarios. Conversely, in the remaining scenarios, we observed as many charging events as records matching the filtering criteria depicted in Figure 3.

A summary of the maximum number of home N_H and workplace N_W charging events and the respective charged energy is presented in Table 2, where the total charging demand for j charging events at the location X for S_1, S_3 and S_5 is defined as $E_X = \sum_j \delta\epsilon_j$ and for S_2 and S_4 as $E_X = \sum_j \delta\epsilon_j / d_p$.

Table 2. Maximum number of charging events and respective energy demand per day during the simulation time span. Results are segregated by home and workplace charging for each scenario under consideration.

	S_1	S_2	S_3	S_4	S_5
N_H	2793	12,686	2793	12,686	12,686
N_W	867	3089	867	3089	3089
E_H (MWh)	68.4	81.2	68.4	81.2	350.5
E_W (MWh)	21.2	26.1	21.2	26.1	83.4

Having identified the EV demand and number of charging events for each scenario, we will now focus our attention on the simulation of the charging events.

2.2.3. Simulation of Charging Events

In this paper, we are interested in the power grid impact of normal AC charging. Recently released EV models are converging towards supporting three-phase 11 kW AC charging, in line with the most common configuration of normal charging stations in Denmark. We frame our analysis assuming a large EV fleet capable of fully utilizing this type of infrastructure. Hence, throughout this paper we consider the charging power to be 11 kW, regardless of the scenario under analysis and the EV's SoC. Therefore, the charging duration is determined by dividing the required energy by the constant charging power.

Given the daily variation of demand on the PDN, we consider different scenarios for charging control according to a broadcasting control signal or a common objective. For home charging, we consider two sub-scenarios, namely *a* and *b*. In sub-scenario *a*, the control signal delays the start of charging until midnight, i.e., 24:00 (00:00 of the next day), whereas sub-scenario *b* delays charging until 18:00. Scenario *a* was devised to illustrate the impact of charging synchronization due to a simultaneous response to price variations, such as to distribution network tariffs where the low price period starts at midnight [25]. Conversely, scenario *b* was chosen to illustrate the impact of EV charging during the peak loading period. Although such a scenario is rather unlikely, it could occur when controlling charging to provide ancillary services to the transmission system or through malicious manipulation of charging as a result of cyberattacks. During working hours, we consider only one synchronization control signal that takes place at 10:00 and affects only vehicles charging at the workplace. Therefore, the start of charging at a location X is defined as $t_{s,X} = \max(t_{c,X}, t_{a,X})$, where $t_{c,X}$ is the synchronization control signal time (for uncontrolled charging in S_1 and S_2 we set $t_{c,X} = 00{:}00$) and $t_{a,X}$ is the arrival time at the location. When travel records include multiple visits to home or workplace, we select the arrival time as follows: for home charging we consider the last arrival, whereas for workplace we take the first. Vehicles visiting the home location during the day, but not terminating the day at home, are exempted from the synchronization and start charging immediately upon arrival.

A set of labels was introduced to distinguish home from workplace charging strategies, namely H_i and W_i, as the sub-scenarios a and b only apply to home charging. A summary of the configuration of each scenario can be found in Table 3. Charging events were simulated over a period of 24 h in 15 min resolution starting at 05:00 on the given day to be able to both capture early charging at work and late charging at home without compromising capturing the majority of EV demand throughout one day. Furthermore, it is worth noting that some records encompass parking durations that are insufficient to fully charge the EV. While scarce and mostly affecting workplace charging, our simulations force vehicles to complete charging.

Table 3. Summary of the characteristics of each charging scenario S_i, including sub-scenarios a and b, as well as H_i and W_i. The non-daily (n.d.) charging demand $\delta\varepsilon$ is defined as the required energy according to the steady-state model, whereas the daily demand is $\delta\varepsilon / d_p$ as introduced in the definition of S_i.

	S_1 H_1, W_1	S_2 H_2, W_2	S_{3a} H_{3a}, W_3	S_{3b} H_{3b}, W_3	S_{4a} H_{4a}, W_4	S_{4b} H_{4b}, W_4	S_{5a} H_{5a}, W_5	S_{5b} H_{5b}, W_5
pattern	n.d.	daily	n.d.	n.d.	daily	daily	n.d.	n.d.
control	none	none	T.S.	T.S.	T.S.	T.S.	T.D.S.	T.D.S.
$t_{c,H}$	none	none	24:00	18:00	24:00	18:00	24:00	18:00
$t_{c,W}$	none	none	10:00	10:00	10:00	10:00	10:00	10:00

2.3. Allocation of Charging Events to the Power Distribution Network

Having modelled the PDN topology as discussed in Section 2.1 with its tripartite nature, we now have to allocate the charging events to the different transformers. The PDNs under consideration contain multiple classes of transformers, ranging from public to private (owned by consumers) and finally reserve. For the purpose of our simulations, we only allocated charging events to transformers from the public infrastructure. The remaining transformers were accounted for in the simulation, but no additional EV load was added. The distribution of charging events also accounted for the type of housing associated with home charging, such that private home charging was mapped to single-family infrastructure (i.e., detached, semi-detached or terraced houses) and shared home charging was associated with denser housing infrastructure, e.g., apartment blocks. The allocation procedure was based on an analysis of the topology of Frederiksberg municipality, with the aim to identify the areas that are primarily dedicated to workplace parking and private residential housing areas, as illustrated in Figure 4.

This procedure can be summarized in four broad steps, as shown in Figure 5.

Figure 4. Visualization of the municipality of Frederiksberg, with gray lines indicating roads and walking paths, and buildings represented in light blue. The identified residential housing areas with predominantly single-family houses are highlighted with a yellow shade, and the primary workplace parking areas are represented in a purple shade.

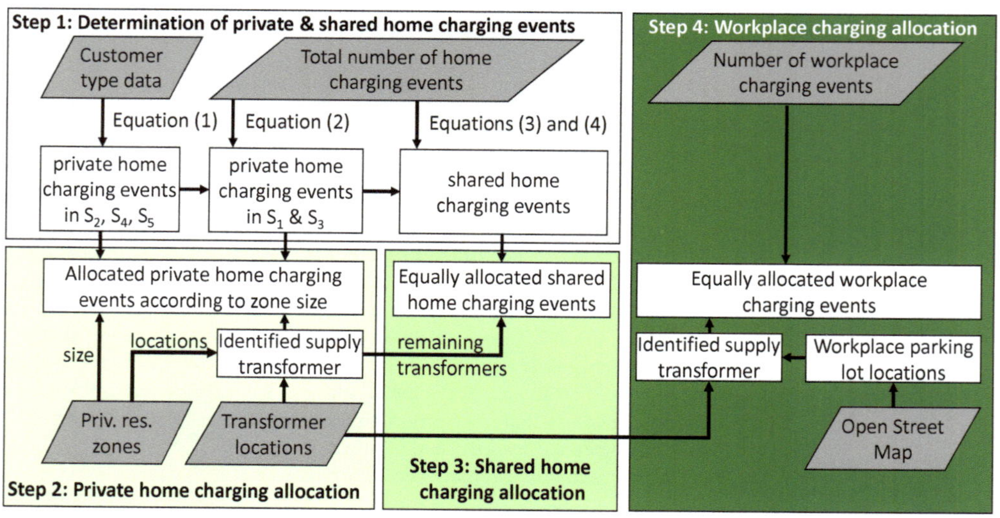

Figure 5. Charging event allocation methodology. Dark-gray-shaded rhomboids indicate the input data, and white-shaded rectangles illustrate the results obtained in each step of the process.

First, we divided home charging events according to the type of charging infrastructure used, i.e., private or shared. Based on customer type data provided by the DSO, we inferred that the number of daily private home charging events $N_{H,p}$ during one day in S_2, S_4

and S_5 equates roughly to 80% of the total number of private houses H_p in Frederiksberg (detached, semi-detached or terraced), as described in Equation (1).

$$N_{H,p}^{S_{2,4,5}} = H_p * 0.8. \tag{1}$$

for S_1 and S_3, the number of private home charging events needed to be reduced to account for the fact that fewer EVs will charge on the given day. To achieve such a task, as shown in Equation (2), the number of private home charging events was multiplied with the ratio between the total number of home charging events in S_1/S_3 and the total number of home charging events in $S_2/S_4/S_5$.

$$N_{H,p}^{S_{1,3}} = N_{H,p}^{S_{2,4,5}} * N_H^{S_{1,3}} / N_H^{S_{2,4,5}}. \tag{2}$$

the number of shared home charging events $N_{H,sh}$ was then calculated as the difference between total and private home charging events accordingly, namely

$$N_{H,sh}^{S_{2,4,5}} = N_H^{S_{2,4,5}} - N_{H,p}^{S_{2,4,5}}, \tag{3}$$

$$N_{H,sh}^{S_{1,3}} = N_H^{S_{1,3}} - N_{H,p}^{S_{1,3}}. \tag{4}$$

In the second step, we considered the allocation of private home charging events. The respective charging events are distributed to the transformers that supply residential zones with predominantly single-family houses. This attribution is based on the overlap of single-family housing zones (including a vicinity buffer) and the location of the transformers, with the number of events being proportional to the relative size of the respective housing zone. The third step concerned the allocation of shared home charging. Shared home charging events were distributed uniformly over the remaining transformers, i.e., those that do not supply single-family housing zones. The fourth step handled the allocation of the load from workplace charging. It began with the identification of transformers that can supply this load, based on the locations of workplace parking lots near company or institution offices and other intensive labor locations. This part of the procedure relied on public data collected from OpenStreetMap. Workplace charging events were then allocated to the nearest transformers using the shortest path between the transformer and respective location, according to the real road and public walking paths in Frederiksberg. Finally, it is important to note that the allocation procedure has to be repeated for each simulation scenario and simulated day, as the number of charging events is dependent on the configuration of each scenario and sampled records.

A summary of the maximum number of charging events allocated to each of the three PDNs is depicted in Table 4. In addition, we observed that the amount of charging events allocated to a single transformer in scenarios S_1 and S_3 were in the following ranges: 2 to 25 for private home, 11 to 12 for shared home and 18 to 19 for workplace charging events. In scenarios S_2, S_4 and S_5, the number of events increased considerably, but private homes continued to exhibit the largest interval and shared and workplace exhibited minimal variation as in the other scenarios. The respective ranges read: 11 to 108 private home, 51 to 52 shared home and 65 to 66 workplace charging events.

Table 4. Summary of charging events allocation in the FRB*i* networks. We list the maximum number of events allocated to a private home $n_{H,p}^{(i)}$, a shared home $n_{H,sh}^{(i)}$ and workplace $n_W^{(i)}$ charging for each scenario under consideration.

	$n_{H,p}^{(1)}$	$n_{H,sh}^{(1)}$	$n_W^{(1)}$	$n_{H,p}^{(2)}$	$n_{H,sh}^{(2)}$	$n_W^{(2)}$	$n_{H,p}^{(3)}$	$n_{H,sh}^{(3)}$	$n_W^{(3)}$
S_1, S_{3a}–S_{3b}	275	1312	480	21	821	313	11	353	74
S_2, S_{4a}–S_{5b}	1244	5953	1710	95	3734	1116	54	1606	263

2.4. Power Flow Simulation and Reinforcement Cost Calculation

The power flow simulation is a computationally demanding problem to solve. Since the PDN needs to be properly dimensioned for the days of highest loading, we made use of the provided power consumption data for 2020 to select all days in which at least one transformer experienced its maximum loading. Based on this criterion, we identified a total number of 67 days of peak load, which fell within the periods January–March and August–December. Furthermore, as discussed in Section 2.2, we considered only normal weekdays. By this token, the size of the identified pool was further reduced to 40. By focusing on this reduced sample, we were able to effectively reduce the computational load and analyze multiple EV charging scenarios.

To be able to assess the EV impact on the PDNs, both in terms of voltage deviations and transformer and cable overloading, the provided baseload and estimated EV load were aggregated for each transformer within the networks. The baseload data for each transformer is only available in hourly resolution. Hence, we made use of cubic interpolation to infer the load in a 15 min timescale. Moreover, as previously mentioned, the EV demand was modelled for a period of 24 h starting at 05:00 on the given day of peak load. Thus, the baseload of the PDNs was chosen accordingly within the same period. While the baseload varies according to the day, the estimated EV load also exhibits variation for each day due to the sampling, as discussed in Section 2.2. It is worth noting that the reactive power of EV charging is neglected within this work. Finally, the Newton–Raphson method was used for the power flow simulation of each selected day in 15 min resolution.

To assess the reinforcements cost associated with the overloading of transformer and cables observed in our simulations, we assumed a typical cost of 133 k€/km for cable and 21.64 k€ for transformer replacements, as used in Refs. [39,40].

3. Results

Having discussed the methodology of the paper, this section is dedicated to the main results of our work. Here, we address the findings for both the EV and aggregated demand, as well as the power flow simulation.

3.1. EV Demand and Aggregated Power Consumption

To begin with, we focus on the estimated EV demand and aggregated power consumption for Frederiksberg. This estimate serves as an input to the power flow simulations. Figure 6 illustrates the aggregated EV demand over the course of 24 h and associated CF. The CF was calculated for each type of charging location separately as the ratio between the simultaneous maximum charging demand of EVs charging at home or work, respectively, and the sum of their individual maximum charging demand. Therefore, the CF can be understood as the share of EVs charging at the same time in relation to the total number of EVs with the opportunity to charge at home or work, respectively.

By comparing home charging scenarios *a* and *b*, it can be seen that charging synchronization at midnight led to a significant higher peak load compared to synchronizing at 18:00, as the majority of EVs arrived at home and charged at the same time. For scenario *a*, a maximum CF of 94.8% is observed. The CF decreases significantly to 58.7% in scenario *b*, with a similar variation in the maximum peak power demand, respectively, 131.7 and 81.6 MW. Furthermore, the time of synchronization also has an impact on the magnitude of experienced peak load in S_4 and S_5. While synchronization at midnight led to similar peak loads in both scenarios, the peak load in S_5 was notably higher when considering synchronization time at 18:00 due to longer charging durations.

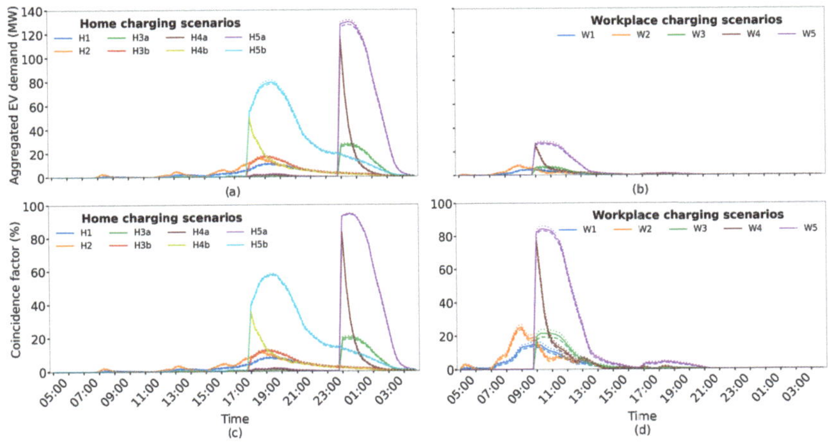

Figure 6. EV demand simulation results. The aggregated EV demand over 24 h is illustrated for (**a**) home and (**b**) workplace charging scenarios. Furthermore, the concomitant coincidence factor is illustrated for (**c**) home and (**d**) workplace charging scenarios. Average, minimum and maximum values for the 40 demand profiles are illustrated by solid, dashed and dotted lines, respectively.

A summary of the time and scale of the maximum peak load and maximum CF is provided in Table 5. Comparing the results to the existing literature, the CF in H_1 and H_{4a} is similar to results in [18], which estimates the CF of a fleet of 10,000 EVs to be 9% for non-daily uncontrolled charging and 83% for daily price-responsive charging starting at 22:00.

Table 5. Summary of the aggregated EV peak power demand for each charging scenario, decomposed according to home and workplace contributions. We list its magnitude $P_{\max}$, time $t_{\max}$ and coincidence factor $CF_{\max}$.

	$P_{\max}$ (MW)	$t_{\max}$	$CF_{\max}$ (%)
H_1	11.6	19:15	8.5
H_2	16.9	18:30	12.4
H_{3a}	28.6	00:30	21.0
H_{3b}	18.0	19:00	13.1
H_{4a}	116.8	24:00	83.7
H_{4b}	50.4	18:00	36.5
H_{5a}	131.7	00:30	94.8
H_{5b}	81.6	19:30	58.7
W_1	5.3	10:00	16.8
W_2	9.1	09:00	27.0
W_3	7.6	10:15	23.9
W_4	26.6	10:00	79.8
W_5	28.9	10:15	86.2

The aggregated active power demand for Frederiksberg is illustrated in Figure 7. For scenarios S_1–S_3, an increase in peak load of less than 32% was experienced. In contrast, the peak load increased by roughly 183%, 100%, 215% and 157% in S_{4a}, S_{4b}, S_{5a} and S_{5b}, respectively. Thus, charging synchronization of a large fleet of EVs at midnight could cause higher concerns compared to the synchronization at 18:00 where the EV load coincides with the baseload of the system.

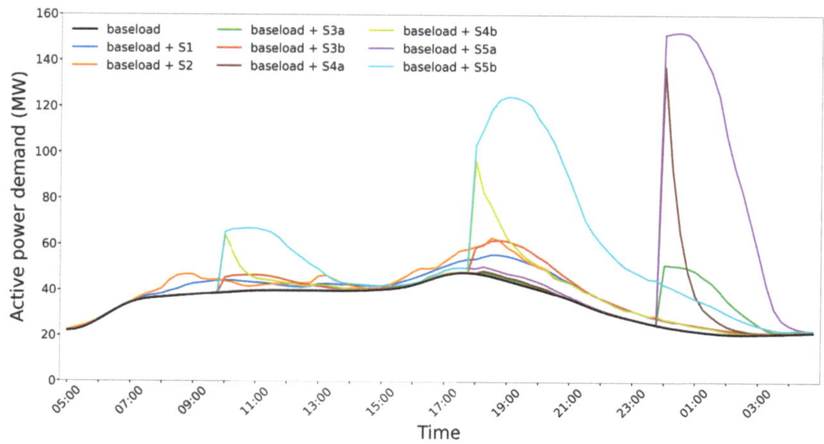

Figure 7. Aggregated active power demand of all three power distribution networks in Frederiksberg, illustrated from 05:00 on 6 January 2020 to 04:45 on 7 January 2020. While the black plot indicates the aggregated baseload, the aggregated demand (comprising both baseload and EV load) is illustrated by different colors for each charging scenario.

3.2. Power Distribution Network Impact and Reinforcement Costs

The EV impact on the PDNs of Frederiksberg was evaluated with respect to bus voltages and the loading of transformers and underground cables. The replacement of a PDN's components is a time-consuming process, and thus it is important for DSO planners to look at components that could potentially face congestion in the future to be able to make timely decisions regarding network reinforcements. To analyze the loading of components, two thresholds are considered, namely 75% and 100%. A summary of the total number of transformers and cables loaded above 75% and 100%, as well as the respective cable length and the estimated reinforcement costs, is shown in Table 6. Voltage violations were not experienced in any of the simulated scenarios and are thus not further addressed in this section. Comparing the loading of transformers and cables, it can be seen that transformer overloading was more prominent. For scenarios S_1–S_{3b}, no cable overloads were recorded, and a maximum of two transformers experienced overload situations. For the extreme scenarios S_4 and S_5, a notable number of transformers and cables experienced congestion. Compared to home charging synchronization at 18:00 (*b*), synchronization at midnight (*a*) exerted a significantly higher impact on the PDNs, both in terms of transformer and cable overloading, indicated by a significant increase in reinforcement costs.

Table 6. Summary of reinforcement costs required to match the increased load in each scenario S_i. We indicate the number of transformers $n_{tr} \equiv n^1_{tr}, n^2_{tr}, n^3_{tr}$, cables $n_c \equiv n^1_c, n^2_c, n^3_c$ and total cable length $l_c \equiv l^1_c, l^2_c, l^3_c$ (km) per PDN in Frederiksberg that exceed 75% and 100% of the nominal capacity. The last column contains an estimate for the total costs C_T (k€) associated with the respective reinforcement.

Load (%)	n_{tr} >75	>100	n_c >75	>100	l_c (km) >75	>100	C_T (k€) >75	>100
S_1	5, 1, 0	1, 0, 0	0, 0, 0	0, 0, 0	0, 0, 0	0, 0, 0	130	22
S_2	12, 7, 0	1, 1, 0	0, 0, 0	0, 0, 0	0, 0, 0	0, 0, 0	411	43
S_{3a}	6, 2, 0	1, 0, 0	0, 0, 0	0, 0, 0	0, 0, 0	0, 0, 0	173	22
S_{3b}	7, 2, 0	1, 0, 0	0, 0, 0	0, 0, 0	0, 0, 0	0, 0, 0	195	22
S_{4a}	120, 69, 32	59, 40, 9	32, 9, 0	16, 3, 0	16.5, 5.3, 0	8, 2.1, 0	7676	3680
S_{4b}	74, 52, 14	34, 21, 4	12, 2, 0	2, 0, 0	5.1, 1.3, 0	0.4, 0, 0	3878	1333
S_{5a}	122, 69, 32	86, 67, 25	34, 13, 0	19, 4, 0	17.6, 7.1, 0	9.2, 3.3, 0	8103	5509
S_{5b}	112, 69, 32	54, 38, 8	21, 9, 0	11, 2, 0	10.1, 5.3, 0	4.9, 1.3, 0	6656	2983

To showcase the impact of home and workplace charging, we illustrate the loading of four different transformers in Frederiksberg over the course of 24 h starting at 05:00 on 6 January 2020 in Figure 8. Here, we consider four different examples, starting with no overloading in any scenario (subplot (a)). Examples of overloading caused by either home or workplace charging are shown in subplot (c) and (d). Finally, an example of both workplace and home charging causing overloading is depicted in subplot (b). Thus, even though the home charging demand is estimated to be significant higher than the workplace charging demand, synchronized charging at work also contributes to potential congestion.

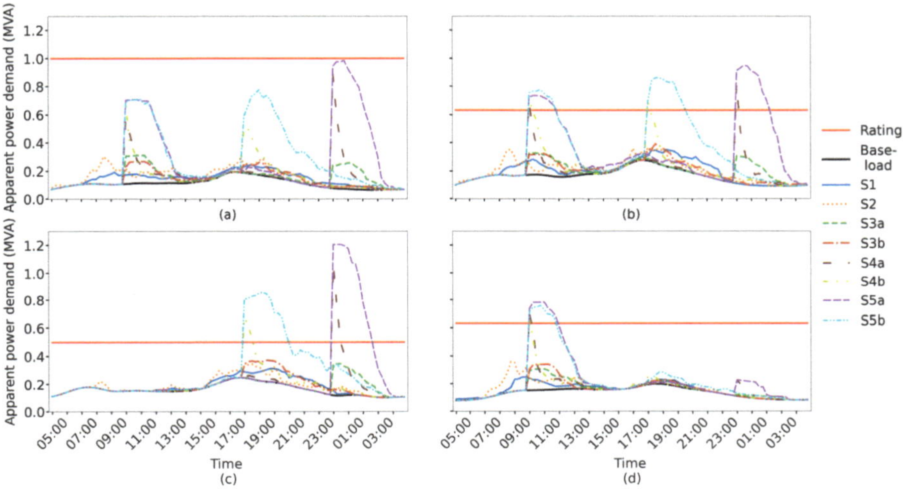

Figure 8. Transformer loading illustrated for a period of 24 h starting at 05:00 on 6 January 2020. Four different 10/0.4 kV transformers are presented, illustrating a: (**a**) no-overload scenario; and congestion scenarios caused by (**b**) both work and home charging; (**c**) solely home charging; and (**d**) solely workplace charging. Transformer ratings and the baseload are indicated by red and black lines. Colored lines in different shapes indicate the aggregated apparent power demand (baseload + EV demand) in each scenario.

4. Discussion

Results indicate that even if full EV penetration is reached, uncontrolled charging is not likely to cause any significant challenge to the urban PDNs of Frederiksberg. Furthermore, even in the case of time-synchronized charging, the grid impact is negligible, assuming a non-daily charging behavior of EVs. However, high synchronization of charging caused by smart charging objectives that do not consider PDN constraints could potentially lead to severe congestion within the PDN when the number of EVs charging on a given day is high, i.e., assuming daily charging patterns or the synchronization of charging on the same day. Those worst-case scenarios, represented by S_4 and S_5 (and respective sub-scenarios), while rare and unlikely, should be taken into consideration. Charging operators should be incentivized or mandated to implement algorithms that avoid such scenarios. Smart charging schemes considering PDN constraints that were not explored within this paper could be used as dampening mechanisms to curb excessive demand surges, e.g., by spreading charging over low price periods. Grid-aware smart charging could not only avoid the occurrence of such extreme scenarios, but also mitigate the small impact experienced in the other scenarios.

Furthermore, one important aspect to mention is the potential mutually interacting relationship between electricity prices and large-scale EV charging. Within this work, any potential implication of EV charging on the spot market is neglected because those implications are in large part uncertain. For the purpose of the paper, we consider that

EV users will also be able to exploit any form of price variation in the future. Another important facet is the fact that full electrification of private passenger transport is still a long way off, and most PDN equipment will have to be renewed before then, as it reaches the end of its service life. Nonetheless, the results obtained in this work could provide valuable insights for planning and prioritizing such reinforcements of the grid in the near future.

Moreover, several factors not taken into account within this work could have a significant impact on the results, which can either reduce or aggravate the load on the PDN. Concerning the former, the scenarios introduced in this paper represent extremes both in terms of charging patterns and control, as well as in terms of the availability of home and workplace charging. Therefore, results should be taken as illustrative upper bounds for the expected impact on the grid. Regarding the charging pattern, each charging scenario assumes the same charging behavior for all EV users. However, in reality, charging behavior is complex in nature, involves multiple patterns (e.g., daily vs. non-daily), is subject to different control strategies (e.g., uncontrolled vs. smart-charging) and different implementations of controls (i.e., the control will vary between charging operators), and variable charging power is likely a critical element. On top of this, future market products offered by charging operators could add another dimension of complexity by offering potential financial benefits to the customers. Regarding charger installation, we assume throughout this paper that all EV owners with access to good parking conditions, either at home or work, will have access to a charger, which could lead to overestimating the number of users with access to charging at those locations. Furthermore, workplace and housing associations are prime candidates for the utilization of load sharing solutions in their charging infrastructure, which should naturally dampen high-demand surges. Last but not least, constraints in the low voltage PDN may also dampen the impact on the medium voltage PDN, i.e., bottlenecks might materialize on the low voltage level before any congestion is experienced on the medium voltage level.

In contrast to the abovementioned factors that will likely dampen demand, we now delve into four factors that can aggravate demand. First, our simulations are based on present-day numbers of vehicles and driven distance, which fails to account for the likely growth in the number of private cars, as expected by the Danish authorities [41]. Second, given the constraints associated with travel records, we assessed the impact of EV charging on the grid based only on normal weekdays data, which covers the days when high synchronization is most likely to occur. Yet, it is important to note that high demand could also occur during weekends or holidays. Third, the impact of public charging is not considered in the present study. Even in the least impactful scenario of evenly distributed utilization of public charging, this will add additional load to the power grid. The fourth factor concerns the spatial dispersion of charging infrastructure and respective connections to the PDN. As previously discussed, the likelihood of global synchronization of demand on the entire grid is rather low. Yet, congestion could also arise from excessive concentration of charging in areas with a limited number of transformers and reduced cable connections, leading to hot spots of demand that could locally overload the power grid.

5. Conclusions

This paper probes the impact of full electrification of private vehicle utilization on the urban medium voltage PDN of Frederiksberg (Denmark), both in terms of transformer and cable loading, as well as voltage deviations. Making use of a Danish travel survey data, we estimated the future EV demand for home and workplace charging, that will have to be served by the PDN. To identify potential congestion in the PDN, we devised five different charging strategies comprising different charging patterns and degrees of control. We simulated the impact of each charging scenario on top of the present base load, selecting 40 weekdays of peak loading within the system. Three of the five charging scenarios focused on a highly simultaneous start of charging to account for a potential high level of charging synchronization in the future caused by user-centric smart charging objectives, such as cost minimization, that do not consider PDN related constraints.

The results indicate that transformer overloading is the primary concern for the analyzed PDN, with cable overloading being less prominent and no detected voltage violations. Moreover, uncontrolled charging does not pose a massive concern for the PDN, mostly due to a low CF of charging, which does not exceed 13% and 27% at home or work, respectively. Furthermore, even with high charging synchronization, no severe impacts are to be expected when EVs are not charged on a daily basis given a low CF of less than 24% for both types of charging locations. However, rare events involving either time-synchronization for daily charging patterns or day- and time-synchronization for non-daily charging patterns could lead to a massive increase in CF of up to 95%. Such events could lead to severe congestion within the PDN, resulting in overloading risk of up to 63% of transformers and 11% of cables. Finally, due to a higher coincidence factor of home charging at midnight, synchronization during the night could potentially pose a bigger challenge to PDNs compared to the synchronization at peak loading times in the evening.

Future work involves the modelling of a more realistic mix of charging patterns and control to account for the diversity of EV user behavior and charging preferences and the inclusion of public charging to analyze the impact on the PDN. Our future work will also focus on analyzing the impact on the low voltage PDN, which may be subject to more significant challenges from electric vehicle synchronization, especially with regard to home charging.

Author Contributions: Conceptualization, T.U. and S.K.; methodology, T.U., S.K. and F.H.; software, T.U., S.K. and F.H.; validation, T.U. and F.H.; formal analysis, T.U.; investigation, T.U., S.K. and F.H.; resources, P.B.A.; data curation, T.U., S.K. and F.H.; writing—original draft preparation, T.U.; writing—review and editing, T.U., F.H. and P.B.A.; visualization, T.U.; supervision, P.B.A.; project administration, T.U.; funding acquisition, P.B.A. All authors have read and agreed to the published version of the manuscript.

Funding: The work in this paper has been supported by the research project FUSE (EUDP grant nr. 64020-1092). Website: http://www.fuse-project.dk.

Data Availability Statement: Restrictions apply to the availability of these data. Data was obtained from DTU and can be requested at https://www.cta.man.dtu.dk/english/national-travel-survey/access_to_tu-data. Furthermore, this work uses data provided by the DSO Radius (https://radiuselnet.dk/). This data is not publicly available due to legal restrictions.

Conflicts of Interest: The authors declare no conflict of interest.

Abbreviations

The following abbreviations are used in this manuscript:

CF	Coincidence factor
DSO	Distribution system operator
EV	Electric vehicle
FRB	Frederiksberg
PDN	Power distribution network
SoC	State-of-charge

References

1. Wolfram, P.; Lutsey, N. *Electric Vehicles: Literature Review of Technology Costs and Carbon Emissions*; The International Council on Clean Transportation: Washington, DC, USA, 2016; pp. 1–23.
2. Calearo, L.; Thingvad, A.; Suzuki, K.; Marinelli, M. Grid Loading Due to EV Charging Profiles Based on Pseudo-Real Driving Pattern and User Behavior. *IEEE Trans. Transp. Electrif.* **2019**, *5*, 683–694. [CrossRef]
3. Haider, S.; Schegner, P. Simulating the Impacts of Uncontrolled Electric Vehicle Charging in Low Voltage Grids. *Energies* **2021**, *14*, 2330. [CrossRef]
4. Jones, C.B.; Lave, M.; Vining, W.; Garcia, B.M. Uncontrolled Electric Vehicle Charging Impacts on Distribution Electric Power Systems with Primarily Residential, Commercial or Industrial Loads. *Energies* **2021**, *14*, 1688. [CrossRef]

5. Bitencourt, L.d.A.; Borba, B.S.; Dias, B.H.; Maciel, R.S.; Dias, D.H.; Oliveira, L.W. Electric vehicles charging optimization to improve the insertion considering the Brazilian regulatory framework. *Energy Storage* **2019**, *1*, e76. [CrossRef]
6. Yu, Y.; Shekhar, A.; Mouli, G.C.R.; Bauer, P.; Refa, N.; Bernards, R. Impact of uncontrolled charging with mass deployment of electric vehicles on low voltage distribution networks. In Proceedings of the 2020 IEEE Transportation Electrification Conference & Expo (ITEC), Chicago, IL, USA, 23–26 June 2020; pp. 766–772.
7. Stiasny, J.; Zufferey, T.; Pareschi, G.; Toffanin, D.; Hug, G.; Boulouchos, K. Sensitivity analysis of electric vehicle impact on low-voltage distribution grids. *Electr. Power Syst. Res.* **2021**, *191*, 106696. [CrossRef]
8. Neaimeh, M.; Wardle, R.; Jenkins, A.M.; Yi, J.; Hill, G.; Lyons, P.F.; Hübner, Y.; Blythe, P.T.; Taylor, P.C. A probabilistic approach to combining smart meter and electric vehicle charging data to investigate distribution network impacts. *Appl. Energy* **2015**, *157*, 688–698. [CrossRef]
9. Procopiou, A.T.; Quirós-Tortós, J.; Ochoa, L.F. HPC-based probabilistic analysis of LV networks with EVs: Impacts and control. *IEEE Trans. Smart Grid* **2016**, *8*, 1479–1487. [CrossRef]
10. Yu, Y.; Reihs, D.; Wagh, S.; Shekhar, A.; Stahleder, D.; Mouli, G.R.C.; Lehfuss, F.; Bauer, P. Data-Driven Study of Low Voltage Distribution Grid Behaviour With Increasing Electric Vehicle Penetration. *IEEE Access* **2022**, *10*, 6053–6070. [CrossRef]
11. Humayd, A.S.B.; Bhattacharya, K. A Novel Framework for Evaluating Maximum PEV Penetration Into Distribution Systems. *IEEE Trans. Smart Grid* **2018**, *9*, 2741–2751. [CrossRef]
12. Boglou, V.; Karavas, C.S.; Arvanitis, K.; Karlis, A. A fuzzy energy management strategy for the coordination of electric vehicle charging in low voltage distribution grids. *Energies* **2020**, *13*, 3709. [CrossRef]
13. Sun, W.; Neumann, F.; Harrison, G.P. Robust Scheduling of Electric Vehicle Charging in LV Distribution Networks Under Uncertainty. *IEEE Trans. Ind. Appl.* **2020**, *56*, 5785–5795. [CrossRef]
14. Veldman, E.; Verzijlbergh, R.A. Distribution grid impacts of smart electric vehicle charging from different perspectives. *IEEE Trans. Smart Grid* **2014**, *6*, 333–342. [CrossRef]
15. Crozier, C.; Morstyn, T.; McCulloch, M. The opportunity for smart charging to mitigate the impact of electric vehicles on transmission and distribution systems. *Appl. Energy* **2020**, *268*, 114973. [CrossRef]
16. IRENA. Electric Vehicle Smart Charging—Innovation Landscape Brief. Technical Report. 2019. Available online: https://irena.org/-/media/Files/IRENA/Agency/Publication/2019/Sep/IRENA_EV_smart_charging_2019.pdf?la=en&hash=E77FAB7422226D29931E8469698C709EFC13EDB2 (accessed on 15 September 2022).
17. Bollerslev, J.; Andersen, P.B.; Jensen, T.V.; Marinelli, M.; Thingvad, A.; Calearo, L.; Weckesser, T. Coincidence Factors for Domestic EV Charging From Driving and Plug-In Behavior. *IEEE Trans. Transp. Electrif.* **2022**, *8*, 808–819. [CrossRef]
18. Gonzalez Venegas, F.; Petit, M.; Perez, Y. Plug-in behavior of electric vehicles users: Insights from a large-scale trial and impacts for grid integration studies. *eTransportation* **2021**, *10*, 100131. [CrossRef]
19. Verzijlbergh, R.A.; De Vries, L.J.; Lukszo, Z. Renewable energy sources and responsive demand. Do we need congestion management in the distribution grid? *IEEE Trans. Power Syst.* **2014**, *29*, 2119–2128. [CrossRef]
20. Jones, C.B.; Vining, W.; Lave, M.; Haines, T.; Neuman, C.; Bennett, J.; Scoffield, D.R. Impact of Electric Vehicle customer response to Time-of-Use rates on distribution power grids. *Energy Rep.* **2022**, *8*, 8225–8235. [CrossRef]
21. Sayed, M.A.; Atallah, R.; Assi, C.; Debbabi, M. Electric vehicle attack impact on power grid operation. *Int. J. Electr. Power Energy Syst.* **2022**, *137*, 107784. [CrossRef]
22. Soykan, E.U.; Bagriyanik, M.; Soykan, G. Disrupting the power grid via EV charging: The impact of the SMS Phishing attacks. *Sustain. Energy, Grids Netw.* **2021**, *26*, 100477. [CrossRef]
23. Monta ApS. Skær Ned på Din Elregning og CO2 Udslip. Available online: https://monta.com/dk/hjemmeladning/ (accessed on 16 September 2022).
24. True Energy. Smart Charge with True Energy. Available online: https://www.trueenergy.io/ev-charging/ (accessed on 16 September 2022).
25. Radius. De Højere Elpriser Medfører æNdringer i Nettariffen. Available online: https://radiuselnet.dk/elnetkunder/tariffer-og-netabonnement/prisaendringer/#:~:text=P%C3%A5%20grund%20af%20de%20h%C3%B8je,skemaet%20l%C3%A6ngere%20nede%20p%C3%A5%20siden (accessed on 16 September 2022).
26. GEOTAB Energy. Charge the North: Phase Two—Charging Infrastructure with Dynamic and Distributed Price Signals for Load Shaping—Final Report. Technical Report. 2021. Available online: https://5vtj648dfk323byvjb7k1e9w-wpengine.netdna-ssl.com/wp-content/uploads/2021/07/geotab-charge-the-north-report-2-2021-DS03521.pdf (accessed on 15 September 2022).
27. Alexeenko, P.; Bitar, E. Achieving Reliable Coordination of Residential Plug-in Electric Vehicle Charging: A Pilot Study. *arXiv* **2021**, arXiv:2112.04559.
28. van Bokhoven, P.; Gardien, L.; Klapwijk, P.; Refa, N.; Berende, M.; van Zante, A.; Heinen, J.W.; Kats, R. Charge Management of Electric Vehicles at Home—Testing Smart Charging with a Home Energy Management System. Technical Report. 2020. Available online: https://elaad.nl/wp-content/uploads/2022/08/rapport_charge-management-of-electric-vehicles-at-home.pdf (accessed on 14 September 2022).
29. Esther Dudek, K.; Platt, N.S. Electric Nation Customer Trial Final Report. Technical Report. 2019. Available online: https://eatechnology.com/media/girhcnsc/electric-nation-customer-trial-report.pdf (accessed on 14 September 2022).

30. Kröger, D.; Rauma, K.; Spina, A.; Rehtanz, C. Scheduled charging of electric vehicles and the increase of hosting capacity by a stationary energy storage. In Proceedings of the 25th International Conference on Electricity Distribution (CIRED 2019), Madrid, Spain, 3–6 June 2019.
31. Mahat, P.; Handl, M.; Kanstrup, K.R.; Lozano, A.P.; Sleimovits, A. Price based electric vehicle charging. In Proceedings of the 2012 IEEE Power and Energy Society General Meeting, San Diego, CA, USA, 22–26 July 2012; pp. 1–8.
32. Sadeghianpourhamami, N.; Refa, N.; Strobbe, M.; Develder, C. Quantitive analysis of electric vehicle flexibility: A data-driven approach. *Int. J. Electr. Power Energy Syst.* **2018**, *95*, 451–462. [CrossRef]
33. FUSE—Frederiksberg Urban Smart Electromobility. Available online: http://www.fuse-project.dk (accessed on 31 March 2022).
34. Unterluggauer, T.; Rich, J.; Andersen, P.B.; Hashemi, S. Electric vehicle charging infrastructure planning for integrated transportation and power distribution networks: A review. *eTransportation* **2022**, *12*, 100163. [CrossRef]
35. Hipolito, F.; Vandet, C.; Rich, J. Charging, steady-state SoC and energy storage distributions for EV fleets. *Appl. Energy* **2022**, *317*, 119065. [CrossRef]
36. Baescu, O.; Christiansen, H. *The Danish National Travel Survey Annual Statistical Report TU0619v2*; Technical Report; DTU Management: Copenhagen, Denmark, 2020.
37. Danmarks Statistik. BIL54: Bestand af Motorkøretøjer Efter Brugerforhold, Køretøjstype, Område, Drivmiddel og tid. Available online: https://www.statbank.dk/statbank5a/SelectVarVal/Define.asp?MainTable=BIL54&PLanguage=0&PXSId=0&wsid=cftree (accessed on 31 March 2022).
38. Danmarks Statistik. BIL707: Bestanden af Køretøjer pr 1 Januar Efter Område og Køretøjstype. Available online: https://www.statistikbanken.dk/bil707 (accessed on 31 March 2022).
39. Klyapovskiy, S.; You, S.; Michiorri, A.; Kariniotakis, G.; Bindner, H.W. Incorporating flexibility options into distribution grid reinforcement planning: A techno-economic framework approach. *Appl. Energy* **2019**, *254*, 113662. [CrossRef]
40. Klyapovskiy, S.; You, S.; Cai, H.; Bindner, H.W. Incorporate flexibility in distribution grid planning through a framework solution. *Int. J. Electr. Power Energy Syst.* **2019**, *111*, 66–78. [CrossRef]
41. Finansministeriet. *Grøn skattereform Aftale om Grøn skattereform*; Technical Report; Finansministeriet: Copenhagen, Denmark, 2020.

World Electric Vehicle Journal

Article

Novel Battery Module Design for Increased Resource Efficiency

Simon Schmidt [1,*], Jan Clausen [1], Robin van der Auwera [1], Oliver Klapp [1], Rico Schmerler [2], David Löffler [2], Maximilian Jakob Werner [3] and Lukas Block [4]

1 Fraunhofer-Institut für Fertigungstechnik und Angewandte Materialforschung, Wienerstr. 12, 28359 Bremen, Germany
2 Fraunhofer-Institut für Werkzeugmaschinen und Umformtechnik, Reichenhainer Straße 88, 09126 Chemnitz, Germany
3 Institut für Arbeitswissenschaft und Technologiemanagement, University of Stuttgart, Allmandring 35, 70569 Stuttgart, Germany
4 Fraunhofer-Institut für Arbeitswirtschaft und Organisation, Nobelstraße 12, 70569 Stuttgart, Germany
* Correspondence: simon.schmidt@ifam.fraunhofer.de

Abstract: The work presented focuses on a material efficient, modular design of a battery module for vehicle applications. Furthermore, the possibility of disassembly of individual components was considered. The constructive design focused on the combination of cast aluminum components, lightweight composites panels, and aluminum-foam phase-change material (PCM) composites. This led to an innovative battery module, which was finally implemented on a demonstrator level. The required cooling power of the module could be reduced by approx. 20% compared to conventional battery module setups. Furthermore, the constructive design of the module and the use of a "debonding-on-demand" technology enabled significantly faster disassembly. Die to the combination of these advantages and the possibility to give individual parts of the module a second life for new modules, the module shows a high resource efficiency as well as high CO_2 savings potential.

Keywords: design for recycling; scalable modular design; casting technology; thermal management; adhesive bonding

Citation: Schmidt, S.; Clausen, J.; van der Auwera, R.; Klapp, O.; Schmerler, R.; Löffler, D.; Werner, M.J.; Block, L. Novel Battery Module Design for Increased Resource Efficiency. *World Electr. Veh. J.* **2022**, *13*, 177. https://doi.org/10.3390/wevj13100177

Academic Editor: Joeri Van Mierlo

Received: 15 August 2022
Accepted: 19 September 2022
Published: 23 September 2022

1. Introduction

Efficiency and sustainability are crucial aspects of electric vehicle design. On the one hand, electric vehicles are seen as key factors in reducing the environmental impact of road traffic. On the other hand, the technology is not yet fully perfected. Electric vehicle production requires several critical or rare resources itself. Thus, increasing the efficiency of electric vehicles in order to reduce their use of resources and to ensure sustainability contribution is fundamental. Thereby, the battery is a significant factor and focal point, causing up to a third of the total weight and cost [1]. Currently, range and capacity, respectively, are unique selling points of a battery electric vehicle (BEV) even though most private trips are quite short and most commercial trips are well-known in advance [2]. In contrast, the battery's size is critical for the economic and ecological use of natural resources (see, e.g., [3]). Thus, battery size and capacity should be scalable to the use case. For instance, public transit operators planning to electrify their fleet would not be limited to the small number of BEV models' battery variants offered but could order individual configurations with a specific capacity, e.g., considering the topography, the share of urban and interurban routes, average travel speeds, etc., in their respective business area.

Accordingly, high flexibility in automotive battery design is important to remain responsive to differences in individual use cases such as private use, ride pooling, or commercial applications [4]. A new battery design needs to be developed that is scalable in terms of space, weight, and the usage of natural resources [5]. The battery needs to be adaptable and upgradable throughout its lifecycle phases, repairable when necessary [6], and disassembled at the end of its lifetime. In this context, innovative life-cycle engineering approaches to support engineers in evaluating their design decisions with respect to requirements of individual, varying life cycle paths of a product are gaining relevance [7]. This is especially valid with regard to the increasing spread of circular economy (CE) strategies and the focus they place on products' later-stage life cycle phases, which is owed to the aim of decoupling value-creating activities, e.g., production, from the consumption of finite natural resources by replacing the 'end-of-life' concept with alternatives such as the reuse, refurbishment, or recycling of product components (see e.g., [8])

The current battery designs provide variants that are selected prior to manufacturing of the car. This is neither resource-efficient nor flexible during the batteries' lifetime. Thus, a battery system with multiple self-sufficient modules that can be flexibly plugged together and that are expandable and upgradable during their life cycle must be developed. In the case of the public transit operator mentioned above, the battery capacity of vehicles could be flexibly adapted to changing operational requirements, caused by new business areas or other external influences.

To address these issues, the scope of the presented work was to develop an innovative battery module concept focusing on sustainability goals in terms of the efficient use of space and material. On a higher level, emphasis was placed on a particularly modular concept with single-battery modules that can be flexibly assembled to obtain a certain desired capacity instead of a non-variable battery pack with fixed modules. A bolt joint allows for easy disassembly of individual modules in order to replace, refurbish, or even reuse modules when needed. On a lower level, CE goals are addressed by reducing the weight, the number of parts and variants, and the production effort. Therefore, the lightweight housing concept (i) consists of two halves that are common parts and (ii) integrates the function of temperature control via integrated active cooling. As an extension to conventional pouch-cell battery modules, an active and close-to-contour liquid cooling system is located directly in the battery module itself. The integration of active liquid cooling in series production has so far been limited to installation in the base plate of the battery box and has not yet been transferred to the level of the battery modules. Examples of this are the use of the Volkswagen Group's MEB (German: "Modularer E-Antrieb Baukasten"), Hyundai/Kia's E-GMP ("electric-Global Modular Platform"), and PSA's platform eVMP ("electric Vehicle Modular Platform") [9].

2. Materials and Methods

The developed battery module is a feasibility study for the technological demonstration of a sustainable product approach. Topics of a circular economy approach, such as the recyclability of individual sub-elements, the separation of different materials used in order to be able to carry out sorted recycling, or the replacement of damaged sub-components are to be achieved through the disassembly of the battery module.

The battery system of today's battery electric vehicles is structured according to three levels: the battery pack, the module, and the cell level. Many battery structures are based on the use of Li-Ion pouch cells, which are combined as a cell stack in battery modules. These battery modules are located in a battery box, combining the modules in a modular or partially modular structure.

These levels each serve different functions. The battery modules serve to bundle the cells for electrical interconnection and under fire-protection aspects, while the tasks of cooling and structural load-bearing are adopted in the battery box.

This structure offers a lot of optimization potential. For a future modular battery system design, a self-sufficient deployment strategy of the individual battery modules without the previous strong dependence on a load-bearing frame and the cooling structure of the battery box is advantageous.

2.1. Structural Design

The formulation of the technological claim and the objective in comparison to conventional battery structures was followed by the derivation of a specific design approach within the framework of the project. In the process, a modular construction was strongly emphasized. The battery module consists of two metallic housing halves. For reasons of lightweight construction, the flat areas of the housing are represented by carbon-fiber-reinforced polymer (CFRP) structures (Kaneka Aerospace LLC, Benica, CA, USA, Typ: KP BZ9704 Fabric), which are bonded to the housing halves by adhesive bonding. The interconnection side is realized by a phase-change material (PCM)foam cover, which is screw-connected to the housing. All individual parts, including the joining processes, are designed to achieve an efficient construction of the battery module and to allow for the easy dismantling of the module in order to repair individual battery modules or to reuse intact individual parts in the sense of a sensible recycling economy.

The individual modules are designed in such a way that several modules can be positioned next to each other. The modules can be interchanged as required as all modules are of the same design. This offers particular advantages in terms of assembly and for the flexible design, expansion, or reduction in the battery modules used in a vehicle.

The designed castings take an essential role in the module's construction. Both the fastening between the modules is provided through them and an integrated tempering structure, which enables internal tempering of the module. The casting was designed so that the base and top are identical parts in order to simplify the production processes and the number of variants.

The side and rear panels are made of lightweight CFRP based on a temperature-resistant polybenzoxazine matrix. The box-shaped geometry of the module housing allows for the efficient use of flat sheet material. Adhesive bonding joins the cast housing and the lightweight walls. The use of a pressure-sensitive adhesive (PSA) allows for a simple bonding process on the one hand, and on the other hand makes it possible to dismantle the module housing by debonding on demand. Here, too, the maximum reusability of the individual components of the battery module housing is ensured.

Contrary to the focus on joining high-temperature-resistant CFRP with correspondingly temperature-resistant adhesives or by means of a thermal joining process, it became apparent in the course of the project that the joining processes considered there were not applicable for joining a battery module housing, in which battery cells and power electronics were already pre-installed, with a material bond using the above-mentioned processes. Both of the joining processes considered would have involved heating of the joining partners to well over 100 °C. To ensure the functionality and safety of the battery cells, they may be exposed to a maximum temperature of 60 °C. Therefore, it was necessary to use a low-temperature joining method that had been developed. The aim was to join the housing parts, which are made of cast aluminum, by means of the CF-BZ laminate described above. This was to avoid holes in the laminate, which on the one hand would have a strength-reducing effect or on the other hand would have to be sealed against gas leakage in the event of fire.

The sandwich lid made of closed-cell aluminum foam in the module front combines the structural functions, such as module stiffening and power electronics mounting with thermal functionality via the infiltrated PCM. In addition, a bumper made of aluminum foam is attached vertically to the outside of the lid. This bumper contains the module connectors and acts as a crash absorber in case of impacts from the modules front side. The lid is connected with the casted halves by screws, sealing the module via a gasket. This allows for simple disassembly and the use of the components for a second life.

An exploded CAD (Catia V5, Dassault Systemes, Vélizy-Villacoublay, France) view is shown in Figure 1 and can be compared with the assembled battery module with stacked battery pouch cells in Figure 1.

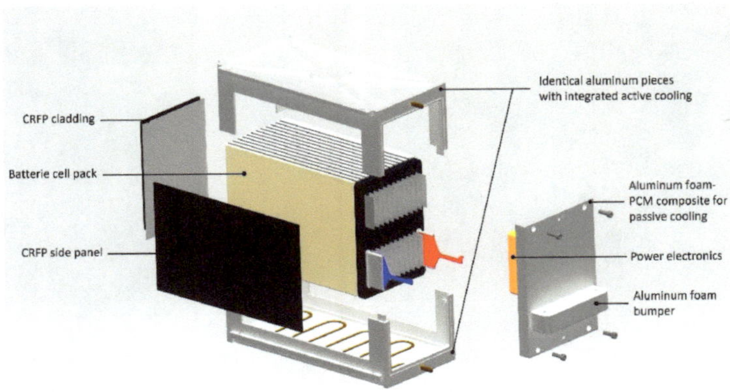

Figure 1. Exploded CAD view of the developed battery module and its components.

2.2. Thermal Design

There are two electrical components in the module that require a thermal management: the battery cells and the power electronics (PE).

The battery cells were tempered by active cooling through the fluid channels integrated in the module halves. The water-glycol fluid on the top flowed in the opposite direction to the one in the bottom. The defined maximum temperature for the battery cell was defined as 40 °C to prevent aging and damages. The assumed load case was a 2C charging scenario from 20% to 80% state of charge (SoC). The thermal loss of the cells was calculated by taking into account the measured internal cell resistance.

In comparison to the actively cooled cells, a passive temperature control system was design for the PE, sitting on the backside of the lid. PCM was integrated into this structural sandwich component to act as a thermal inertia buffering heat losses and flattening thermal load peaks. PCM allows for the storage or release of large amounts of thermal energy in a reversible process during its phase change, in this case from solid to liquid or vice versa. Due to corrosion issues, paraffin was used, which enables a wide range of adjustable melting range and is physiologically harmless [10]. By integrating the paraffin into the aluminum foam, the very low thermal conductivity of 0.2 W/mK can be increased significantly by more than a factor of 80 [11], which enables the rapid thermal loading and unloading of the PE system connected through a thin thermal conductive interface. The boundary condition for the maximum temperature of PE in combination with a balancing load case was defined with 50 °C. The assumed heat power loss was 60 W for a 10 min load case at an ambient temperature of 20 °C.

3. Results

Manufacturing and Joining Technology—Realization of the Demonstrator

The battery module housing was manufactured using low-pressure sand casting due to the small number of components to be produced. By using 3D-printed sand molds, the desired scale of the prototype series can thus be produced economically.

Figure 2 shows a sand core package with dimensions of approx. 400 × 400 mm and a weight of approx. 15 kg. The core package is constructed in two parts. The positioning of a copper cooling channel to be integrated in the casting is realized by the use of core blocks and thus fixed between the two core package halves. In this way, the castings were produced as symmetrical identical parts, so that two of the manufactured components each represent a closed module housing. The castings were produced on a low-pressure casting system within the Fraunhofer Center Wolfsburg at the Open Hybrid LabFactory.

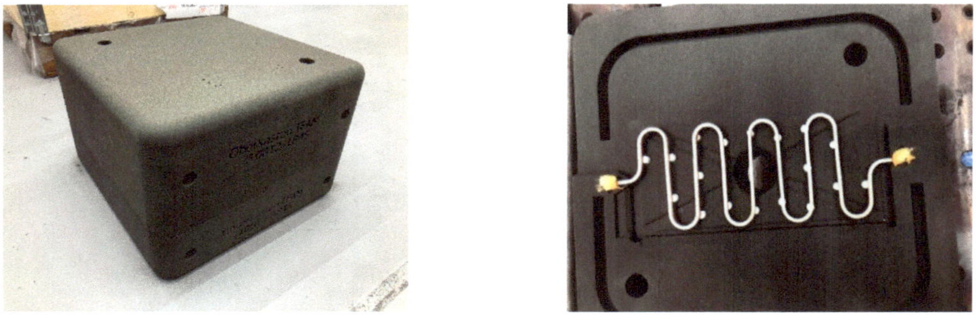

(**a**) (**b**)

Figure 2. Core package made of 3D-printed sand, in two parts. (**a**) Top and bottom box assembled; (**b**) copper cooling channel positioned in the bottom box with the help of core blocks.

By embedding square copper tubes with external dimensions of 4 × 4 mm and a wall thickness of 0.5 mm, it was possible to provide active cooling close to the contour, directly in the floor or ceiling panel. Copper was chosen as the material for the tempering structures due to its superior thermal conductivity. The rod material was hand-formed into the meandering shape and cleaned with a solvent and sandblasted for better bonding before casting.

To investigate the dimensional accuracy of the castings and the fit of the upper and lower halves, a scan was performed using fringe light projection. Figure 3 shows the three-dimensional measurement of a finished cast component. Good dimensional accuracy of the internal geometry was achieved. The 2.44 mm deformation of a housing leg shown here, which is due to the mechanical demolding process, was corrected by mechanical straightening. The castings were designed in such a way that, apart from the removal of gates and overflows at the upright corners of the housing half, no mechanical finishing is required. This also takes into account the basic concept of resource-saving production and the use of this battery module.

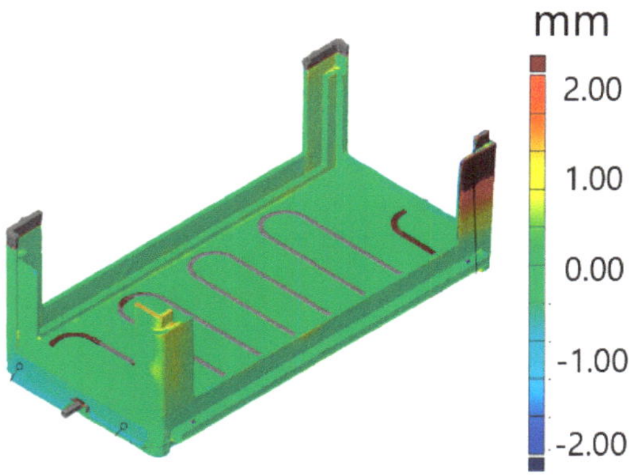

Figure 3. Three-dimensional measurement of the real components by means of fringe light projection and comparison with the CAD target geometry.

For the thermal investigations, the maximum temperature vs. time curves have been simulated for different PCM melting temperatures and sandwich thicknesses, see Figure 4. The red curve represents the temperature increase for the PE placed on a 1 mm aluminum sheet. The nomenclature of the aluminum foam sandwiches (AAS, German: Aluminium-Schaum mit Aluminium-Decklage) is the total thickness as the first number, followed by the cover sheet thicknesses.

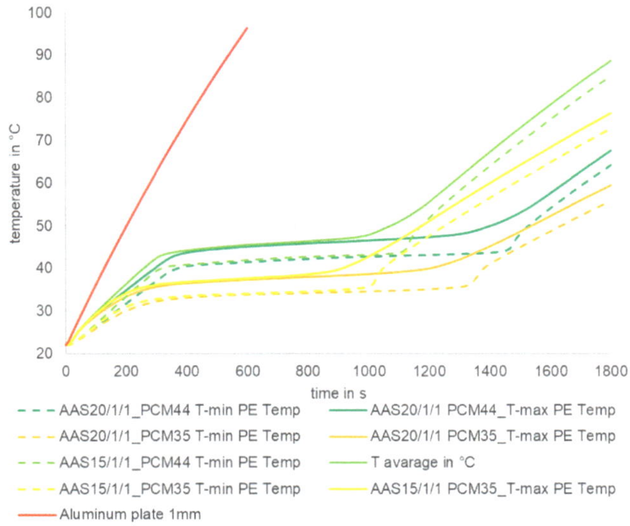

Figure 4. Simulated temperature curves of aluminum foam sandwiches (AAS) infiltrated with PCM and aluminum plate for the defined thermal PE load case.

From the simulation results, it can be concluded that all sandwich structures with PCM setups result in significantly lower maximum temperatures than the reference 1 mm aluminum plate, due to the thermal inertia. For the scenario of a 60 s applied load and a sandwich thickness of 15 mm, the maximum temperature for PCM with a melting area around 35 °C (PCM35) could be decreased by 61%, and for PCM with a melting range of 44 °C (PCM44) it could be decreased by 53%. Moreover, the graph shows that for the given scenario, PCM35 results in lower temperatures than PCM44. Furthermore, it can be seen that the total sandwich thickness has a very small influence on the maximum temperature, before reaching the heat storage capacity of the composite. A thicker sandwich, consequently containing a higher amount of PCM, shifts the temperature rise after the PCM melting towards longer times. The temperature differences between maximum and minimum are below 5 K, which characterizes the high thermal conductivity and thus the rapid thermal loading of the composite.

Based on these results a setup of AAS15/1/1, which means a sandwich with a total thickness of 15 mm and 1 mm-thick cover sheets, infiltration with PCM35 was chosen for the battery module.

Figure 5 shows the simulation result for the operating module load case. The maximum cell temperature remains below 38 °C, and the maximum PE temperature does not exceed 50 °C. Further thermal design investigations that were examined on the fire behavior of the used materials and their impact on thermal runway/propagation will be part of additional publications.

For the front lid sandwich structures made of closed cell aluminum foam, covered aluminum sheets were produced via a powder metallurgical foaming route. The first setup of AAS15/1/1 for the base plate and a second setup of AAS25/2/2 for the bumper, joined vertically on the base plate, were manufactured. After placing inserts for joining the lid with the module frames and sealing the sandwich along the edges, PCM was infiltrated with pressure support. The closed cell foam, which consists of cell rods as well as cell walls, has micro and macro cracks in its wall structure, enabling the foam to be infiltrated with the PCM. Finally, the infiltration ports were sealed.

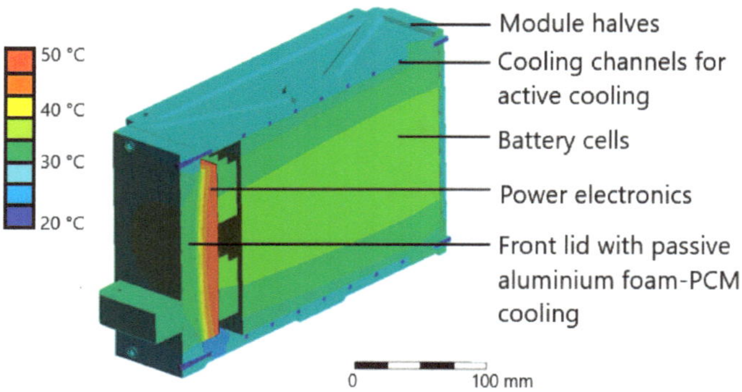

Figure 5. Sectional view of temperature distribution within the module.

To meet the bonding requirements, as explained in chapter 2, an adhesive bonding concept was developed, which is shown in Figure 6. For this purpose, a temperature-resistant adhesive tape (DuploCOLL® 930 F, Lohmann, Neuwied, Germany, temperature range −40 to 125 °C) was applied all around the pure resin sides of the CFRP side walls over a width of 11 mm. The side walls are positioned relative to the pouch cell package (Figure 6, top left) and, using two pressure plates, the pouch cell package is subsequently compressed (top center) to such an extent that it can be inserted into the first half of the housing (top right). The second half of the housing is positioned in a next step (bottom left and center). Once compression is complete, which causes the pouch cell package to expand and exerts the necessary joining pressure, the adhesive tape is pressed against the inner adhesive flanges of the housing parts, thus joining the two housing parts via the two side walls (Figure 6, bottom right).

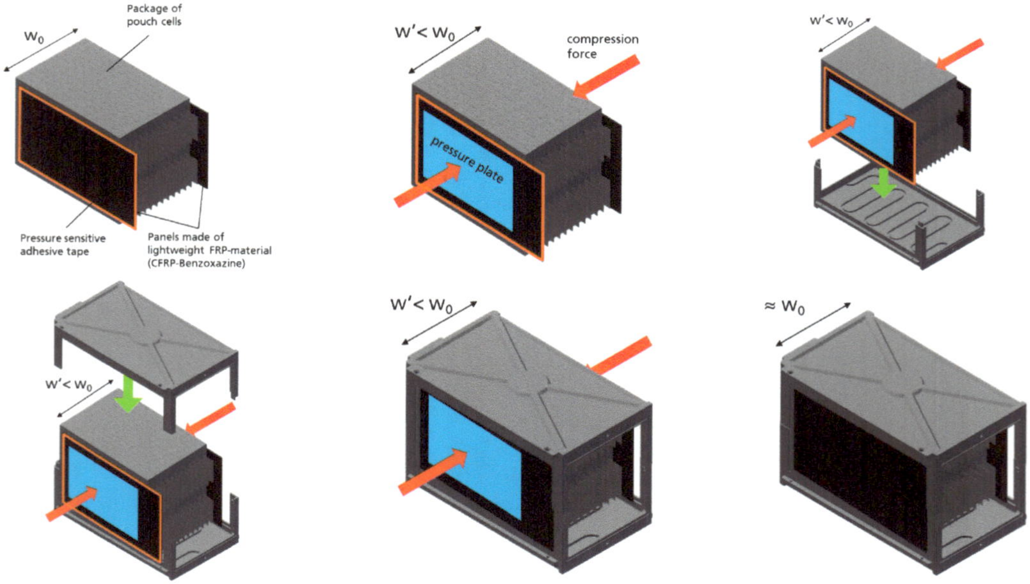

Figure 6. Concept for joining the battery module housing parts.

On the rear-end face of the already joined battery module housing, a further CFRP lightweight panel was connected from the outside onto the adhesive flange, also using the above-mentioned adhesive tape. Among other things, this wall also acts as a shear field to limit the bending up of the housing struts under the load case "internal pressure as a result of swelling of the punched cell stack" (Figure 7).

Figure 7. Load case "internal pressure as a result of swelling of the punch cell stack during loading".

The device used to join the side panels with the two battery module housing parts is shown in Figure 8 on the left. A joined battery module is shown in the figure on the right. Figure 9 shows the assembly process of the demonstrator modules. The finally assembled battery module can be seen in Figure 10.

(a)

(b)

Figure 8. (**a**) Device for inserting the battery cells into the battery module housing halves and joining them and joined battery module housing. (**b**) A dummy battery-cell stack made of wooden elements of the same thickness was used for testing.

Dismantling of the lateral lightweight walls (e.g., at the end of use of the battery module to remove the pouch cells) can be carried out with the above-mentioned device. Since adhesive tapes generally have only limited adhesion, they can be removed from the substrates by means of peeling stresses in Mode I at a sufficiently low removal speed and, if necessary, with temperature support. In the case of the tape-joined CFRP lightweight sidewalls of the battery module housing, the device developed for joining the walls also enables a sufficiently high head tensile load to be applied, which locally causes a peeling stress on the tape so that it detaches from one of the joining partners and the FRP sheets can thus be separated from the cast Al structure again. The lightweight CFRP panel joined from the outside on the face side must first be dismantled by being levered out.

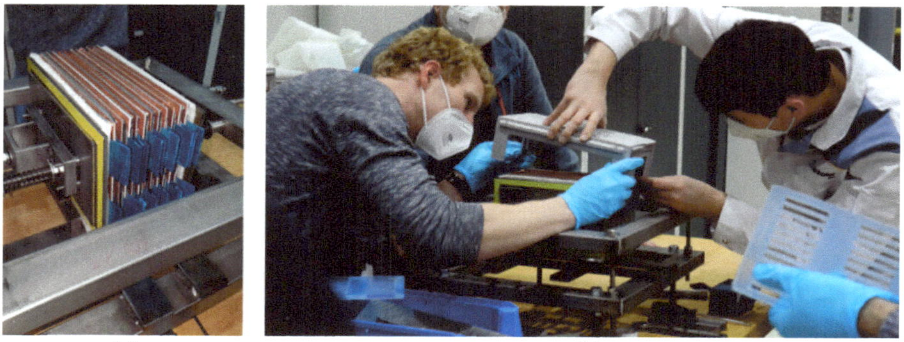

(**a**) (**b**)

Figure 9. Demonstrator assembly: (**a**) positioning of the cell stack in the joining device; (**b**) assembly of the module shells.

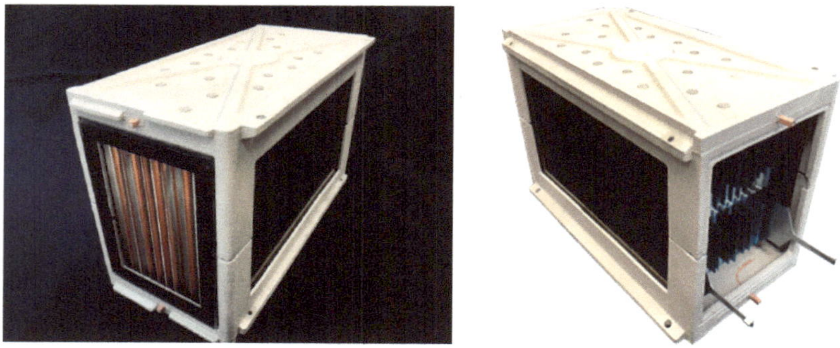

Figure 10. Demonstrator of the battery module. **Left**: rear view with window to provide insight into the module. **Right**: front view with stacked battery pouch cells and without cover.

4. Discussion

Vision and Further Goals

The presented battery module (Figure 10) forms the basis for a novel, modular design for vehicle batteries. The design makes it possible to dispense with an additional battery box in the future as all relevant functions are integrated within each individual module. In addition, the new battery module was designed to withstand structural loads. This brings several benefits in terms of CE and the sustainability of future BEV design. During early life-cycle phases, especially production, the number of parts as well as the number of resources used can be reduced. Later on, users of the vehicle can flexibly configure the battery capacity according to their individual needs in order to scale down unnecessary weight or material consumption and increase efficiency.

A potential application of this flexible and modular battery design is for vehicle classes such as light commercial vehicles, people movers, or conventional buses/minibuses that regularly travel on fixed routes with a plannable necessary range. The modular design allows for the battery to be efficiently tailored to the size of the vehicle in question. If the range requirement changes in the course of the vehicle life cycle, battery modules can be added or removed with little effort thanks to the modular design. This enables the efficient utilization of resources.

In future investigations, the modular battery will be optimized with respect to the car's critical body load cases using finite elements methods. Furthermore, subsequent work shall consider alternative bonding and debonding methods for the battery module's constituents to support different CE strategies with regard to extending the respective life cycles of these parts. As a rather visionary idea, it might be interesting to analyze the integration of the battery module concept into a swappable battery concept that is still pursued by several industry players from China (see e.g., [12–14]). Following a product-service system approach, providers could offer a business model that allows users to configure their vehicle with a battery capacity sufficient for their daily needs and to flexibly rent additional battery modules at a swapping station when needed, e.g., for a long-distance ride.

5. Conclusions

The developed module design shows significant potential for further follow-up of the concept. Some promising technologies were summarized based on the demonstrator presented and show potential for future use in battery module housings. This is especially the case for the basic design approach, which allows for easy module assembly and disassembly.

The technologies demonstrated on the demonstrator, such as "debonding-on-demand" CFRP for applications with high fire protection requirements as well as EMC protection requirements, passive cooling using a PCM foam, the use and design of an aluminum cast housing without additional post-processing, and the considered suitability for the re-use of the individual parts, entail interesting approaches.

In further investigations, open questions still need to be addressed with regard to actual achievable loads (e.g., crash), the vehicle integration of module packages, and the design of the interfaces between the individual modules for electrical contact and media supply. Furthermore, there is still the potential for a weight reduction in the housing through, for example, an optimized design or adaptations in the casting process.

Author Contributions: Conceptualization, methodology, formal analysis, resources, and writing—original draft preparation, review, and editing: S.S., J.C., R.v.d.A., O.K., R.S., D.L., M.J.W. and L.B. All authors have read and agreed to the published version of the manuscript.

Funding: This research was funded by German Federal Ministry of Education and Research (BMBF), grant number L1FHG42421; the authors are thankful.

Institutional Review Board Statement: Not applicable.

Informed Consent Statement: Not applicable.

Data Availability Statement: Not applicable.

Conflicts of Interest: The authors declare no conflict of interest.

References

1. König, A.; Nicoletti, L.; Schröder, D.; Wolff, S.; Waclaw, A.; Lienkamp, M. An Overview of Parameter and Cost for Battery Electric Vehicles. *Electr. Veh. J.* **2021**, *12*, 21. [CrossRef]
2. McGuckin, N.; Fucci, A. *Summary of Travel Trends: 2017 National Household Travel Survey*; Daniel, E., Jenkins, P.E., Eds.; Federal Highway Administration: Washington, DC, USA; US Department of Transportation: Washington, DC, USA, 2018; p. 20590. Available online: https://nhts.ornl.gov/assets/2017_nhts_summary_travel_trends.pdf (accessed on 27 April 2022).
3. Bastin, Z.; Bhattacharya, S.; Kumar, A. International Electric-Vehicle Consumer Survey, Are Battery Electric Vehicles Here to Stay? Automotive & Industrial. 2019. Available online: https://www.alixpartners.de/media/13453/ap-electric-vehicle-consumer-study-2019.pdf (accessed on 27 April 2022).
4. Kostorz, N.; Fraedrich, E.; Kagerbauer, M. Usage and User Characteristics—Insights from MOIA, Europe's Largest Ridepooling Service. *Sustainability* **2021**, *13*, 958. [CrossRef]
5. Block, L.; Werner, M.; Mikoschek, M.; Stegmüller, S. Developing Technology Strategies for Flexible Automotive Products and Processes. In *Advances in Automotive Production Technology—Theory and Application. ARENA2036*; Weißgraeber, P., Heieck, F., Ackermann, C., Eds.; Springer: Berlin/Heidelberg, Germany, 2021. [CrossRef]
6. Füller, K.H.; Schäfer, J.; Zehnpfennig, L.; Budaker, B.; Hossfeld, M.; Keilhoff, D. FlexCar-A Technology Platform Concept for the Future Cyberphysical Vehicle at the Research Campus ARENA2036. In Proceedings of the Werkstoff Plus Auto-9, Tagung für neue Fahrzeug- und Werkstoffkonzepte, Stuttgart, Germany, 20–21 February 2019.

7. Block, L.; Werner, M.; Spindler, H.; Schneider, B. A variability Model for Individual Life Cycle Paths in Life Cycle Engineering. In Proceedings of the FAPC 2022, Future Automotive Production Conference FAPC 2022, Wolfsburg, Germany, 17–18 May 2022; Dröder, E., Herrmann, C., Vietor, T., Eds.; pp. S1–S10.
8. Kirchherr, J.; Denise, R.; Hekkert, M. Conceptualizing the circular economy: An analysis of 114 definitions. *Recour. Conserv. Recycl.* **2017**, *127*, 221–232. [CrossRef]
9. Ganesh, S.; Venkatesan, S. Evolution of flexible modular electric vehicle platforms among automotive industry and its influence on battery integration. *Int. J. Veh. Struct. Syst.* **2021**, *13*, 361–366. [CrossRef]
10. Mehling, H.; Cabeza, L.F. *Heat and Cold Storage with PCM: An Up to Date Introduction into Basics and Applications*; Springer: Berlin/Heidelberg, Germany, 2008.
11. Schmerler, R.; Hipke, T.; Drebenstredt, C. SMC-Aluminum lightweight battery case design with switchable temperature control functionality. In Proceedings of the Automotive Circle-Conference for Battery Systems in Car Body Engineering, Bad Nauheim, Germany, 26 October 2021.
12. NIO. Battery Swapping. Available online: https://www.nio.com/blog/current-state-ev-battery-swapping (accessed on 27 April 2022).
13. CATL Battery Swap Solution. Available online: https://www.catl.com/en/news/856.html (accessed on 27 April 2022).
14. Geely Battery Swapping Tech. Available online: http://zgh.com/media-center/story/8-reasons-why-geelys-battery-swapping-tech-is-the-future-for-evs/?lang=en (accessed on 27 April 2022).

World Electric Vehicle Journal

Article

Green Public Procurement for Accelerating the Transition towards Sustainable Freight Transport

Hampus Karlsson [1,*], Solveig Meland [1], Kristin Ystmark Bjerkan [1], Astrid Bjørgen [1], Nina Møllerstuen Bjørge [2] and Sahar Babri [1]

1 SINTEF Community, NO-7465 Trondheim, Norway
2 SINTEF Digital, NO-7465 Trondheim, Norway
* Correspondence: hampus.karlsson@sintef.no; Tel.: +47-97648672

Abstract: Requests for emission reduction in the freight transport sector will be more intense in the coming years. One possible strategy to reduce emissions from freight transport is through utilising zero emission vehicles, which requires substantial investments both by transporters and by authorities. This paper examines how green public procurement (GPP) can be used to push the market in an environmentally sustainable direction. For this purpose, interviews with both public authorities and freight service providers are conducted. The results show that GPP is considered a useful tool for public authorities to both boost the uptake of zero emission vehicles and to share the investment costs with freight service providers. However, our study shows that there are differences between small and large municipalities. Moreover, to succeed with GPP, public authorities must prioritise such tasks in their daily routines through political decisions and strategies. Additionally, barriers related to financial possibilities are crucial to handle, as public support schemes are important to reduce costs for all involved stakeholders. Altogether, our paper shows that with the right tools and willingness among both public and private stakeholders, GPP can contribute to the use of more environmentally friendly solutions in the freight transport sector.

Keywords: freight; green public procurement; transport

Citation: Karlsson, H.; Meland, S.; Ystmark Bjerkan, K.; Bjørgen, A.; Bjørge, N.M.; Babri, S. Green Public Procurement for Accelerating the Transition towards Sustainable Freight Transport. *World Electr. Veh. J.* **2022**, *13*, 173. https://doi.org/10.3390/wevj13090173

Academic Editor: Joeri Van Mierlo

Received: 5 August 2022
Accepted: 13 September 2022
Published: 16 September 2022

1. Introduction

Transforming an entire mobility system in a more environmentally sustainable direction requires attention to the many different domains that transport systems comprise. Transitioning personal mobility has already been well explored by scholars in regard to electric cars [1–3], while the hard-to-abate shipping sector has been increasingly recognised as a paramount research objective [4–6]. One transport domain still insufficiently recognised in research on sustainability transitions is road-based freight transport. Despite accelerated technological innovation within freight transport [7–9], it has yet to be explored as a transition site. This paper aims to contribute to enhance knowledge in this area by turning to the Norwegian freight transport sector.

The freight transport sector is characterised by a wide range of companies and operators, both international and domestic, targeting wider or more niche parts of the logistics market, and applying a variety of business models and contracts. On a national level in Norway, road transport in 2020 represented 17% of greenhouse gas (GHG) emissions. Heavy vehicles and vans accounted for 49% of these emissions. Other negative side effects of freight transport are traffic noise, congestion, air pollutions and traffic accidents, which are closely related to vehicle milage travelled [10]. Like any other transport sector, freight transport poses a series of climate and environmental challenges, particularly in cities, that have increasingly compelled local authorities to incorporate freight into climate and planning strategies [11]. However, freight service providers are mainly aware of their own emission, and several large actors have incorporated their own goals for low or zero emissions operations within their self-defined time horizon.

The European unions' (EU) White Paper [12] and the following EU's working document entitled "A call to action on Urban logistics" [13] highlights the need to be aware of logistics and freight transport, thus being a part of public planning processes [14].

Green public procurement (GPP) is an instrument used by public actors to purchase goods and services with a lower environmental impact in their lifetime than goods and services otherwise bought. Green public procurement in the EU is not mandatory, but with several policies, the EU seeks to stimulate for the use of green procurement. In 2016, they launched the Handbook of GPP, which may be applied as a guide for buying green [15]. In Norway, however, green public procurement is one of several strategic instruments for reducing emissions and achieving Norway's environmental goals [16].

In the European context, public procurements have increased in the last 20 years, and now constitute 14% of the EU's gross domestic product (GDP) [17]. In Norway, public authorities spend approximately 600 billion NOK annually on public procurements, and the products and services procured by public authorities represent 16% of the national total emissions [16].

As procurers of large volumes of goods and services, public authorities are able to move the priorities of commercial freight service providers in a greener and more sustainable direction. Although ambitions have been defined at both an international and national level, procurement processes on regional and local levels play an important role in fulfilling the ambitions for sustainability transitions. Along Oslo's Climate strategy [18], the municipality has developed a procurement strategy, as they want to use public procurement as a powerful tool to accelerate the green shift, climate goals and circular economic thinking in accordance with ambitions of being an emission free city [19]. This requires procurement processes that strengthen the competitiveness of climate- and environmentally friendly solutions with a small environmental footprint, high expected quality, long service life and good recycling opportunities [20].

Norwegian green public procurement is anchored in national legislation (e.g., [21]) that obliges public administrations to establish procurement practices that enhance sustainability; for instance, by requiring specific technological solutions or setting specific emission limits for the products or services that are procured. Although green public procurement is increasingly emphasised as a climate measure, The Office of the Auditor General in Norway has concluded that current public procurement practices are not contributing enough to diminish GHG emissions and to provide the use of climate friendly solutions. They also point to a lack of holistic approaches for procurement, which are needed to take sufficient environmental actions [17].

The paper delves into the governance of sustainability transitions in freight transport by studying practices with green public procurement. Specifically, this paper explores preliminary experiences with green public procurement in freight transport and discusses how such practices should be developed to accelerate sustainability transitions in freight transport. The study focuses on procurements where freight transport is a result of the procurement, but not the main purpose. To do so, we study practices and perspectives among Norwegian freight service procurers as well as freight service providers through qualitative interviews and nation-wide surveys.

In the following section, we introduce green public procurement as a governance instrument for progressing sustainable technologies in freight and provide an overview of existing knowledge about green public procurement in sustainability transitions. We then account for the methods and data used to explore the research questions in Section 3. We present the results in Section 4, following a discussion based on the results in Section 5. Finally, we conclude in Section 6.

2. Literature Review

2.1. Transition through Green Public Procurement

The governance of sustainability transitions is an increasingly important research topic, as the pace of transition depends on the degrees to which transitions are managed or incentivised [22]. Within the field of sustainability transitions, Transition Management has been particularly prominent in devising governance strategies for steering and accelerating transitions in different sectors [23,24]. Transition Management is a "prescriptive framework" that allows for creating "space for short-term innovation and develop long-term sustainability visions linked to desired societal transitions" [25]. Specifically, Transition Management points to particular governance activities (i.e., strategic, tactical, operational, reflexive) that allow for dealing with the complexity of sustainability problems and the societies that they are enmeshed in. Within the framework of Transition Management, green public procurement (GPP) could be an example of tactical governance activities, which intervene into established structures such as rules and regulations, routines, and institutions.

Within the scholarly field of sustainability transitions, green public procurement could be considered an expression of Strategic Niche Management. This field of research tends to focus on two ways in which emerging technologies and innovations can diffuse and become competitive in existing markets or become dominant in new niche markets. Scholars of Strategic Niche Management are thus concerned with how emergent innovations could be nurtured and matured in protective spaces that shield them from market competition until they are sufficiently competitive [26]. Thus, green public procurement could also be a fruitful strategy for accelerating transitions when following ideas of Strategic Niche Management. Green public procurement could nurture niche innovations by creating markets for specific technologies or for any technologies that reduce emissions. Green public procurement is therefore an example on how sustainable transitions could be steered on low-level governance.

The potential for stimulating sustainable innovations and technological changes by public procurement is high in the transport and logistics sector. The sector is undergoing rapid changes due to digitalization in the supply chain and forcing changes in the individual behavioural pattern such as increased online shopping, home delivery services, and new consumer patterns. It is therefore a golden opportunity to link green public procurement to the transport sector and identify its possibilities and challenges as a tool towards achieving CO_2-free city logistics by 2030 [13,18,27].

Ref. [28] has investigated the success factors in implementing GPP and identified four criteria to succeed with the implementation of GPP: (1) consistent and operational policy goals; (2) nation-wide campaigns for GPP; (3) ethics, professionalism, capacity, and knowledge among employers; and (4) systems for checks and balances among actors in the entire purchasing process. Ref. [29] has examined GPP in Swedish municipalities and observed that most previous measures to curb emissions from transport in urban areas focussed on passenger transport, and that urban freight has almost been forgotten. Ref. [29] also mentions that there are exceptions, and that knowledge from the few studied municipalities might be transferred to other municipalities to accelerate the transition towards more sustainable freight transport. Ref. [30] identifies significant differences in GPP uptake between countries and national levels: the countries working most with GPP are characterised by a large governmental sector, and municipalities and local administrations are more prone to work with GPP than authorities on EU and national levels.

Ref. [31] has developed a method for GPP, which was tested on public procurers of transport in Sweden. This method is a participatory method, seeking to enhance the procurer's understanding and knowledge about practical needs. Their results emphasise that the participatory method and improved insight for the procurer's contribute to broad environmental assessments, enhancing green procurements and to a more equitable assessment of alternative technologies. With Rotterdam and The Hague as examples, [32] studied public procurement as a driver for more sustainable freight transport. They state that it is hard to determine the exact number of trips that are generated from procurements, and that procurements are first applied as an administrative activity, and not as an activity to meet policy.

2.2. Green Vehicles in Freight

When choosing a vehicle, transporters consider criteria such as the size of the vehicle, its capacity, the range, fuel usage, etc. such that they match with their requirements. With the new technological advances, use of battery electric vehicles is supported as a measure to decarbonise freight transport, especially in urban areas in which transport is characterised by short distances, many stops and low driving speed [33].

Studies show that despite the high investment cost of battery electric freight vehicles, they are cost competitive with conventional diesel vehicles [34]. However, in some cases, transporters need to implement measures such as reducing the scope of service, mixing the fleet with conventional vehicles, using opportunity charging, etc., to increase the profitability of EVs as a compensation to high investment costs [35,36]. Moreover, external factors such as the availability of charging stations on public spaces [37] or exemption from paying road-toll or parking fees [38] are also recognised as influential factors in promoting the use of BEVs in urban freight transport.

Norway has been successful in introducing policies for boosting the uptake of EVs in passenger transport. However, the incentives such as financial support schemes administered by the public agency Enova are less comprehensive for the BEVs in freight. Exemption from value added tax and discounts for registration taxes are already offered for fossil fuel vehicles [39] and support for purchasing the vehicles of establishing charging infrastructure covers a subtle part of the investment costs [40].

Considering the limited options to offer as financial support for freight transporters in Norway, the role of Green Public Procurement is highlighted as a tool to accelerate the transition towards more sustainable freight transport. However, the barriers for effectively using the tool to promote zero emission vehicles into freight system is not well studied in the literature.

This paper aims to fill this research gap by investigating the potentials of green public procurement from both procurers and transporters points of view. It highlights the important factors to succeed with the GPP as a tool for transitioning towards more environmentally sustainable freight transport and how it is perceived both by public officials and transporters by providing empirical data.

3. Materials and Methods

This study is based on a combination of qualitative (interviews) and quantitative (survey) methods. Qualitative methods were used to shed light on the topic and gain deeper insight into whether and how environmental criteria are used in public procurements within the Norwegian municipalities today. Interviews were chosen because they allow us to interact with the informant and explore the reasoning behind their answers. Quantitative methods, here represented by surveys, allow us to examine if specific findings from the interviews are also viable in a broader context.

3.1. Interviews

Table 1 shows the number of interviews and respondents in the respective survey used in the further analyses. Two of the municipalities are regional centres and two are smaller municipalities. The interviewees from public procurers were staff responsible for procurement processes in the respective municipality. Two of the three interviewees from freight service providers operate in the business-to-business market, while the third also delivers to private households. The informants were persons with responsibilities for vehicle fleets, daily businesses, and submitting tenders for public procurements.

Table 1. Overview of data collection.

	Public Procurers	Freight Service Providers
Interviews	N = 4	N = 3
Survey	N = 71	N = 220

Interview guides were developed based on national documents [41,42] and the Public Procurement Act [43]. The interview guide for municipalities revolves around the following topics: (1) strategy and collaboration, (2) experience with public procurements, and (3) public support schemes and national goals. The interviews with freight service providers highlighted the following themes: (1) technology, (2) logistic solutions and (3) policy to explore drivers and barriers impacting the freight operators in the transition of urban freight. The interviews were conducted as digital meetings

The results from the interviews were further used as an input to design survey questions, as well as a basis for analysis in this study.

3.2. Surveys

Two separate surveys were designed and distributed among potential participants: (1) survey for public procurers distributed among municipalities' staff, and (2) survey for freight transport providers distributed among transporters. The design and distribution of the two surveys are briefly explained in this section.

The survey for municipalities is structured according to the same three main topics as in the interviews, while the questions are reformulated to fit into the format of a survey. Findings from the interviews were used to create response categories. The survey was distributed via e-mail to all 356 Norwegian municipalities with a request to forward it to administrative personnel responsible for or experienced with procurements. A reminder was sent out to non-response municipalities after two weeks. In total, we received 71 individual responses from 61 different municipalities, implying that the request had been forwarded to more than one person in some municipalities, and both had responded. However, as the roles and the answers from the same-municipality respondents are not identical, all responses are included in the analysis. There was no record of more than two answers from identical municipalities. Although only 46 participants responded to all the questions in the survey, all 71 responses were included in the analysis since those who were not able to complete the survey also provided useful information about why they did not include requirements for low or zero emission solutions.

Using the classification of municipalities by population size from Statistic Norway [44], and comparing to national statistics for 2022, we found that the smallest municipalities were somewhat underrepresented, and the largest somewhat overrepresented in the survey sample, while the remaining categories were proportionally represented (Table 2). The informants represent a range of roles related to public procurement (Table 3).

The survey among freight service providers was distributed to the members of the Norwegian association for lorry owners (NLF), targeting freight service providers at different levels. In the survey, we asked whether they submit tenders for public procurements, together with other questions about public procurements, how they relate to it, and their opinion on its effects on sustainable transitions. The survey was initially distributed to over 3000 members of the NLF, of which 220 responded.

Table 2. Categorization of included municipalities.

Population Size	Survey Sample		Norwegian Municipalities 2022 [45]
	N	%	%
Under 2000 inhabitants	8	12	22
2000–4999 inhabitants	18	26	27
5000–9999 inhabitants	13	19	20
10,000–19,999 inhabitants	10	15	13
20,000–49,999 inhabitants	11	16	12
50,000 or more inhabitants	9	13	6
Missing	2		
Total	71	100	100

Table 3. Respondent's role in municipality.

Role in Municipality	N	%
Responsible for procurement strategies	27	38
Advisor to procurers	20	28
Procurer	19	27
Other role related to procurement	5	7
Total	71	100

Roughly one third (39%) of the respondents provide services for the public sector, and one out of four (27%) are personally involved in tenders for the public sector (Table 4). The results presented in this paper are based on responses from companies which do provide services to the public sector (N = 87).

Table 4. Respondent's role in freight company.

Are You Involved in Tenders for Public Procurement?	N	%
Yes, I respond to procurement tenders	60	27
Yes, but someone else in our company is responsible	27	12
No, we do not provide services to the public sector	133	60
Total	220	100

4. Results

The results are presented in three sub sections. The first section presents existing requirements on public procurements and how they are currently used. The second section gives an overview of the municipalities' perspectives and their requirements on public procurements. The last section describes freight service providers' perspectives and their obligations for public procurements.

4.1. Existing Low and Zero Emission Requirements in Freight Transport

The in-depth interviews and comments in the survey show that there are differences, both in the types of requirements and how requirements are used in practice in the municipalities' procurement process. The most common requirement, related to transport in Norwegian public procurements for goods and service, is for the type of propulsion technology in the vehicles. One reason for this might be its simplicity to require and control, compared to requiring a specific emission level, which is the second most common requirement. Some municipalities also put restrictions on specific types of deliveries: for example, by demanding smaller vehicles, consolidation, or deliveries outside of rush hour. However, all such requirements are considered more demanding, both for the public procurer to control and the freight service provider to document.

There are also variations in which part of the transport leg such requirements apply. Although many goods are produced outside of Norway and therefore include transport activities abroad, the transport-related requirements are mainly only applied for the last mile in Norway. This might also be explained by the tendency to ask for requirements which are easier to monitor, as the transport activities outside of Norway are not under the control of a Norwegian distributor and there might be limited alternatives with low or zero emission options in the market abroad.

The third difference is whether the application of requirements depends on the value of the tender. The results are presented in Table 5.

Table 5. Municipalities practice in terms of threshold values for applying of requirements.

Are Environmental Requirements for Transport Included in Your Procurement Processes?	Municipalities by Population Size						Sum across Municipalities
	Under 2000 Inhabitants	2000–4999 Inhabitants	5000–9999 Inhabitants	10,000–19,999 Inhabitants	20,000–49,999 Inhabitants	50,000 ≤ Inhabitants	
Yes, always	13%	6%	31%	40%	36%	33%	25%
Yes, for purchases over 100,000 NOK	0%	17%	15%	30%	18%	44%	20%
Yes, for purchases over 500,000 NOK	13%	22%	15%	0%	9%	11%	13%
Yes, for purchases over 1 million NOK	0%	6%	0%	10%	18%	11%	7%
No, never	75%	50%	38%	20%	18%	0%	35%
N of respondents	8	18	13	10	11	9	71

The response categories (specifically values NOK 100,000 and NOK 500,000) are based on answers in the interviews. Although these limits are also viable for other municipalities, the majority using requirements always include them (Table 5). The motivation for applying value thresholds for including requirements is related both to the possibilities for monitoring what is delivered, and how large the emissions from the transport in that procurement is. The Public Procurement Acts says requirements should be in proportion to both costs and emissions.

4.2. The Perspective of Municipalities

The results are cross tabulated with the size of municipalities to see if there are any differences that may depend on the number of inhabitants in each municipality.

Observing the statistics in Table 6 reveals a relationship between adoption of strategies and number of inhabitants in the municipality. Municipalities with a large population are more probable to have an implemented strategy for promoting green public procurement (GPP). Political decisions seem to be more important than the Public Procurement Act from 2017. An important insight from the interviews is that in smaller municipalities, both political will and personal interest in the administrations play more important roles in success with GPP for freight transport.

Table 6. Which municipalities has a strategy and why.

Does Your Organisation Have a Strategy to Increase Share of Environmentally Friendly Transport in Procurement?	Municipalities by Population Size						Sum across Municipalities
	Under 2000 Inhabitants	2000–4999 Inhabitants	5000–9999 Inhabitants	10,000–19,999 Inhabitants	20,000–49,999 Inhabitants	50,000 ≤ Inhabitants	
Yes, as a result of a political decision	25%	18%	46%	50%	36%	78%	40%
Yes, according to the legislation for Public Procurements from 2017	13%	24%	15%	30%	27%	11%	21%
Yes, of other reasons	13%	12%	8%	20%	27%	11%	15%
No, the organization missing such a strategy	50%	41%	23%	0%	9%	0%	22%
Don't know	0%	6%	8%	0%	0%	0%	3%
N of respondents	8	17	13	10	11	9	68

Municipalities were also asked what kind of commodities and services they included as requirements related to sustainable transport. The responses show that requirements apply for all kind of goods, but are more common for office supplies, furniture, consumables, and food.

As shown in Table 7, dialogue with freight service providers is the most common way to gain knowledge about low emission solutions or delivery alternatives. Responses summarised in Table 7 are based on a multiple-response question, therefore the total sum of responses is higher than the number of respondents. Percentages show the share of the respondents choosing each of the alternative responses.

Table 7. How municipalities gain knowledge.

Do You Acquire Information about Low Emission Solutions or Delivery Alternatives? (Multiple Responses Possible)	Municipalities by Population Size						
	Under 2000 Inhabitants	2000–4999 Inhabitants	5000–9999 Inhabitants	10,000–19,999 Inhabitants	20,000–49,999 Inhabitants	≥50,000 Inhabitants	Total
Yes, have employees who are constantly working to map the market	0%	0%	0%	0%	27%	56%	12%
Yes, in dialogue with freight service providers/vehicle manufactures	50%	33%	54%	80%	91%	56%	58%
Yes, participate in networks at local level (neighbouring municipalities)	13%	17%	31%	40%	18%	0%	20%
Yes, participate in networks at regional level	38%	17%	38%	40%	18%	33%	29%
Yes, participate in networks at national level	0%	6%	0%	10%	9%	33%	9%
Yes, in other ways	13%	11%	8%	0%	0%	0%	6%
No, do not actively seek knowledge	13%	39%	23%	0%	9%	0%	17%
N of respondents	8	18	13	10	11	9	69

The technological maturity for low and zero emission vehicles varies between large and small vehicles [46]. The technology for small vehicles is mature enough to compete with conventional vehicles regarding range and payload. For heavy vehicles, however, the new technologies could currently only be competitive within urban areas. Considering the significantly higher purchase price compared to vehicles with internal combustion engines, it is reasonable to expect that requirements for low or zero emission vehicles would negatively affect the tender price. Results presented in Table 8, however, show that the municipalities do not clearly acknowledge this. A large share of the respondents from municipalities state that they do not know the effect on prices, and the share stating that the price remains unchanged is almost the same the share stating that the price increases due to environmental requirements in tenders.

Table 8. Procurers' assessment of consequences for offered price in procurements.

Is the Offered Price Affected by a Requirement for Low Emission Transport?	N	%
Yes, the price will be higher	8	17
Yes, the price will be lower	1	2
No, the price remains unchanged	10	22
Don't know	27	59
Total	46	100

Different public support schemes to cover extra costs have been launched such as *Klimasats* by the Norwegian Environment Agency or support to infrastructure by the local municipality. This is because there is a difference in price for fossil fuelled vehicles and low or zero emission vehicles. The new Public Procurements Act also tries to speed up the transition in sustainable solutions when it asks public authorities to always examine possibilities to include requirements for low or zero emission solutions. For such support schemes to be useful, they must be communicated to potential users, e.g., freight service providers. According to the responding procurers, the knowledge level among smaller actors must be enhanced (Table 9).

Table 9. Procurers' assessment of bidders' knowledge about legislation and public support schemes.

How Do You Experience Level of Knowledge of Legislation and Support Schemes among Bidders?	N	%
Overall good	5	11
Good among larger freight service providers	10	22
Bad	10	22
Don't know	21	46
Total	46	100

It is worth mentioning that Tables 8 and 9 are based on answers from respondents who include requirements for low or zero emission transport in their procurements. Therefore, the number of analysed respondents is lower than Tables 5–7.

4.3. The Perspective of Freight Service Providers (FSP)

When it comes to what affects the freight industry's acquisition of vehicles running on alternative fuels, environmental requirements in public procurement are assessed to be significant, although not the most important factor (Table 10).

Table 10. Freight service providers' (FSP) assessment of what affects the company's acquisition of alternative fuel-vehicles.

How Important Are These Issues for Your Company's Choice of Vehicles with Alternative Fuels?	Of Very Little Importance	Of Little Importance	Important	Very Important	N
Support for purchase of vehicles	16%	2%	10%	72%	83
Improved infrastructure for alternative fuels	27%	2%	13%	58%	83
Reduced road tolls	20%	6%	22%	52%	83
Requirements in public procurement	30%	10%	22%	39%	83
Priority at loading bays	51%	8%	12%	29%	83
Requirements from the private market	51%	11%	16%	23%	83
Access to streets/areas otherwise off limits for vehicles	53%	18%	10%	19%	83
Access to public transport lanes	49%	20%	11%	19%	83
Other	73%	14%	7%	5%	83

As shown in Table 11, findings from interviews with freight service providers indicate that an orientation towards public clients seems to both affect their experiences with procurements processes and their internal goals for low- and zero emission vehicles. Further findings from the interviews are embedded in discussions where it is appropriate.

Respondents directly involved in providing bids for GPP tend to consider the environmental requirements to represent a challenge for the industry with negative effects on price and quality of services offered, although they also tend to agree that such requirements have a positive effect on sustainable transition (Table 12).

Table 11. Findings from interviews with freight service providers.

FSP	Experience with Public Procurement	Goal
A	Experience a growing interest for FSP opinion/knowledge regarding requirements. Poor evaluation and monitoring of requirements.	Zero emission city logistics in 2026
B	PP is time-consuming; therefore, they concentrate on the private market.	Euro VI fleet
C	Experience a growing interest for FSP opinion/knowledge regarding requirements. Poor evaluation and monitoring of requirements.	Fossil free vehicles and buildings in 2025

Table 12. FSP's assessment of consequences of environmental requirements in public procurements.

How Do You Consider the Following Statements about Environmental Requirements Included in Public Procurement Processes?	Strongly Disagree	Disagree	Agree	Strongly Agree	N
The tenders include requirements which support sustainable transition	20%	20%	49%	11%	45
The requirements have a negative effect on price and quality of services	7%	18%	29%	47%	45
It is challenging for freight operators to meet the requirements	7%	4%	36%	53%	45

Findings from the in-depth interviews with freight service providers indicate that the largest drawback with low or zero emission vehicles requirements is how it affects the operational routines, and, consequently, the price and quality of deliveries. In addition, freight service providers are experiencing a growing interest for market dialogues with public procurers and are mainly positive to share their knowledge about possibilities and barriers in the market.

The interviews also include statements from the industry about the importance of follow-up routines after a tender has been awarded. Since offered solutions are part of the competitive basis in the procurement, it is important to make sure that the winner delivers according to the tender. If not, other bidders are able to deliver a more cost-effective solution. From the experience of the interviewees, today's follow-up practice has the potential to be improved.

5. Discussion

As findings from the surveys indicate, a large proportion of Norwegian municipalities are already working with green public procurement (GPP). Today's practices of GPP involving freight services have been presented above and the results indicate potential for further refining of the practices. Part two of our research question is «how should GPP practices be developed further to accelerate sustainability transitions in freight transport? This section starts with a brief overview of the most important issues recognised through the analyses as a basis for the following discussion, which also relates to other studies within the same field.

According to the results, factors such as political decisions, strategies, requirements, capacity/budget, competence, and methods for monitoring/follow-up are the most prominent issues in public procurements of goods and services for accelerating the transition towards a more sustainable freight transport. These issues are important for both public procurers and freight service providers, but in different ways.

Political decisions and strategies are identified as the most prominent reasons for municipalities to work with GPP in this sample. However, there are differences between the larger and smaller municipalities, where the smaller tend to be lagging behind regarding the implementation of GPP. One reason for not implementing the GPP in smaller municipalities could be related to the lack of competence and personal interest. Statements from interviews point to personal interest as often being essential when prioritizing different tasks if public procurements is one of several responsibilities of the respondent. In contrast, the larger municipalities can work with GPP in a more professional way, since they often have designated procurement departments, in line with studies where professionalism and knowledge are mentioned as success criteria in GPP [28].

It is worth noting that a large administration is not mandatory to succeed with GPP. One possibility for empowering the smaller municipalities and enhancing their involvement in GPP could be to increase their access to resources or more efficient use of existing resources, e.g., through inter-municipal collaboration in procurement processes. Some of the municipalities already use networks on a local and/or regional level as a source for acquiring information about low emission solutions and delivery alternatives. Enhanced collaboration through such networks could help smaller municipalities to increase their competency for implementation of GPP. As understood in [16,29], enhancing knowledge through transfer it between actors is important, especially in smaller municipalities. Easy access to or already acquired competence about GPP probably facilitates its inclusion in daily routines. Capacity can also be related to access of vehicles running on alternative fuels. This is mentioned by both freight service providers and small municipalities. Access to appropriate vehicles and corresponding fuelling/charging stations is crucial for succeeding with transitioning of the freight transport.

In addition, our results highlight the importance of dialogue between the freight industry and municipalities for enhancing procurers' knowledge about new solutions and alternatives. At the same time, the procurers indicate that there is room for improvement when it comes to the freight service providers' knowledge of legislation and support schemes. Utilising the dialogue between procurers and the freight industry to provide suppliers with information about legislation and support schemes could be worth exploring.

An active strategy is not merely important for municipalities. Setting strategies and aiming to achieve them requires generating requirements that both encourage and force freight service providers to develop solutions for meeting the requirements. Although the results here have shown that economic incentives are the most important parameter for freight service providers, findings from the in-depth interviews show that the freight service providers working for the public sector have the most ambitious goals for delivering low or zero emission solutions. This may not only be due to the orientation in the market affecting internal strategies, but also due to internal budgets and fleet sizes which reflect the capacity of the companies in terms of both financial and administrational capacity. To become an early adopter freight service provider, they need financial resources, just as municipalities. Requirements for low or zero emission vehicles may therefore favour larger freight service providers already in possession of adequate vehicles. Investments on new technologies or dealing with the risks involved for ordering new vehicles are easier to face for larger freight providers.

Thus, GPP could be combined with public support schemes to secure that both small and large municipalities and freight service providers can engage in the green shift. Public support schemes covering some of the extra expenses often associated with the introduction of new technologies are examples of strategic niche management, which has proved to be a useful tool in transition management based on both our study and previous findings [23,26]. However, many small municipalities state that the funding from current public support schemes is scarce and that it often includes cumbersome application processes. So, to accelerate the transition, and reduce the bias for larger actors, easier application processes, and more financial support to cover for both risks and extra work would be beneficial.

Otherwise, less competitions may contribute to a negative effect on price and quality when fewer bids are delivered, something freight service providers already have reported.

Finally, proper methodologies for monitoring the compliance of requirements are needed to ensure fair competition. It is important from a competitive point of view that the winning freight service provider deliver what is promised. The development of standardised monitoring methodologies may facilitate the engagement of smaller enterprises and municipalities as it reduces the need for follow-up resources. Moreover, increased use of value-based inclusion of environmental requirements, as many municipalities already apply, can also be a measure to accelerate more sustainable transport solutions associated with the larger purchases of goods and services.

The results presented in this paper confide to a Norwegian context, which is characterised by a large public sector, and therefore might not be transferable to other contexts without modification. However, this study sheds light on how municipalities and freight providers perceive GPP, and the success factors related to it. The results reveal that despite highlighting the topic by National legislation, local willingness is essential to effectively include GPP in public procurement. Similarly, there exist tools and networks in a wider perspective, such as the EU, though mixed national contexts. This study presents a useful approach to enhance knowledge about the topic, which can be further used in other contexts.

6. Conclusions

This study aimed to gain knowledge on GPP, its practices in Norwegian contexts and how it is perceived from both public procurers and freight transport providers. The results show that Norwegian municipalities in all sizes have already started to implement requirements for low and zero emission transport in their procurement of goods and services. However, there are still some barriers to overcome, especially in the smaller municipalities. Building and utilising networks to enhance and disseminate knowledge and resources among municipalities and between municipalities and the freight service operators should be further explored. Future research can probably contribute to enhance knowledge on how to establish attractive networks for smaller actors, providing them with relevant and practical knowledge about GPP. Moreover, carrying out research on GPP in countries with a smaller public sector will also be of huge relevance, since Scandinavian countries are characterized by a larger public sector than many others.

Finally, several barriers are directly or indirectly related to financial resources. Therefore, extended public support schemes adapted to the needs of the smaller actors, both private and public, should be considered to avoid only larger freight service providers being able to compete on procurements, including requirements for low or zero emission solutions. Other relevant measures include simplifying application routines to reduce the work related to the process of applying for financial support for extra expenses related to low or zero emission transport.

Author Contributions: The authors confirm contribution to the paper as follows: conceptualization, H.K., S.M., K.Y.B., A.B., N.M.B. and S.B.; methodology, H.K., S.M., K.Y.B., N.M.B. and S.B.; validation, H.K., S.M., A.B. and N.M.B.; formal analysis, H.K., S.M., S.B. and N.M.B.; investigation, H.K., S.M., A.B. and N.M.B.; writing—original draft preparation, H.K., S.M., K.Y.B., A.B. and N.M.B.; writing—review and editing, H.K., S.M., S.B., K.Y.B., A.B. and N.M.B.; project administration, S.M. and K.Y.B.; funding acquisition, S.M. and K.Y.B. All authors have read and agreed to the published version of the manuscript.

Funding: The authors wish to thankfully acknowledge the Norwegian Research Council for financing this research through the MoZEES research centre, grant number 257653 and Ntrans research centre, grant number 296205. We further wish to thank all interviewees for providing their perspectives.

Data Availability Statement: Not applicable.

Acknowledgments: We wish to thank all interviewees and respondents for providing their perspectives.

Conflicts of Interest: The authors declare no conflict of interest. The funders had no role in the design of the study; in the collection, analyses, or interpretation of data; in the writing of the manuscript; or in the decision to publish the results.

References

1. Figenbaum, E. Perspectives on Norway's supercharged electric vehicle policy. *Environ. Innov. Soc. Transit.* **2017**, *25*, 14–34. [CrossRef]
2. Köhler, J.; Turnheim, B.; Hodson, M. Low carbon transitions pathways in mobility: Applying the MLP in a combined case study and simulation bridging analysis of passenger transport in the Netherlands. *Technol. Forecast. Soc. Chang.* **2018**, *151*, 119314. [CrossRef]
3. Nilsson, M.; Nykvist, B. Governing the electric vehicle transition—Near term interventions to support a green energy economy. *Appl. Energy* **2016**, *179*, 1360–1371. [CrossRef]
4. Bergek, A.; Bjørgum, Ø.; Hansen, T.; Hanson, J.; Steen, M. Sustainability transitions in coastal shipping: The role of regime segmentation. *Transp. Res. Interdiscip. Perspect.* **2021**, *12*, 100497. [CrossRef]
5. Pettit, S.; Wells, P.; Haider, J.; Abouarghoub, W. Revisiting history: Can shipping achieve a second socio-technical transition for carbon emissions reduction? *Transp. Res. Part D Transp. Environ.* **2018**, *58*, 292–307. [CrossRef]
6. Tvedten, I.; Bauer, S. Retrofitting towards a greener marine shipping future: Reassembling ship fuels and liquefied natural gas in Norway. *Energy Res. Soc. Sci.* **2021**, *86*, 102423. [CrossRef]
7. Jaller, M.; Otay, I. Evaluating Sustainable Vehicle Technologies for Freight Transportation Using Spherical Fuzzy AHP and TOPSIS. *Intell. Fuzzy Tech. Smart Innov. Solut.* **2021**, *1197*, 118–126. [CrossRef]
8. Meyer, T.; von der Gracht, H.A.; Hartmann, E. Technology foresight for sustainable road freight transportation: Insights from a global real-time Delphi study. *Futur. Foresight Sci.* **2021**, *4*, e2101. [CrossRef]
9. Strale, M. Sustainable urban logistics: What are we talking about? *Transp. Res. Part A Policy Pract.* **2019**, *130*, 745–751. [CrossRef]
10. Browne, M.; Allen, J.; Nemoto, T.; Patier, D.; Visser, J. Reducing Social and Environmental Impacts of Urban Freight Transport: A Review of Some Major Cities. *Procedia Soc. Behav. Sci.* **2012**, *39*, 19–33. [CrossRef]
11. Cui, J.; Dodson, J.; Hall, P. Planning for Urban Freight Transport: An Overview. *Transp. Rev.* **2015**, *35*, 583–598. [CrossRef]
12. European Commission. *Roadmap to a Single European Transport Area*; European Commission: Brussels, Belgium, 2011; Available online: https://eur-lex.europa.eu/legal-content/EN/TXT/PDF/?uri=CELEX:52011DC0144&from=EN (accessed on 25 March 2022).
13. European Commission. *A Call to Action on Urban Logistics*; European Commission: Brussels, Belgium, 2013; Available online: https://smartcities.at/wp-content/uploads/sites/3/A-call-to-action-on-urban-logistics.pdf (accessed on 25 March 2022).
14. European Commission. *Topic Guide: Planning for More Resilient and Robust Urban Mobility*; European Commission: Brussels, Belgium, 2021; Available online: https://www.mobiel21.be/assets/documents/sump_topic-guide_planning_for_more_resilient_and_robust_urban_mobility-gecomprimeerd.pdf (accessed on 25 March 2022).
15. European Commission. Directorate General for the Environment and ICLEI—Local Governments for Sustainability. In *Buying Green: A Handbook on Green Public Procurement*, 3rd ed.; Publications Office: Luxembourg, 2016.
16. Riksrevisjonen. *Riksrevisjonens Undersøkelse av Grønne Offentlige Anskaffelser*; Riksrevisjonen: Oslo, Norway, 2022; Available online: https://www.riksrevisjonen.no/globalassets/rapporter/no-2021-2022/gronne-offentlige-anskaffelser.pdf (accessed on 25 March 2022).
17. Tátrai, T.; Diófási-Kovács, O. European Green Deal—The way to Circular Public Procurement. *ERA Forum* **2021**, *22*, 523–539. [CrossRef]
18. Oslo Municipality. *Climate Strategy for Oslo towards 2030*; Oslo Municiaplity: Oslo, Norway, 2020; Available online: https://www.klimaoslo.no/2020/06/10/oslos-new-climate-strategy/ (accessed on 24 April 2022).
19. Oslo Municipality. *Procurement Strategy Byrådssak 1104/17*; Oslo Municipality: Oslo, Norway, 2017; Available online: https://www.oslo.kommune.no/getfile.php/13254871-1510047808/Tjenester%20og%20tilbud/Politikk%20og%20administrasjon/Anskaffelser/Anskaffelsesstrategi%202017%20-.pdf (accessed on 29 April 2022).
20. Oslo Municipality. *Standard Climate and Environmental Requirements for Transport for the Municipality*; Byrådssak 1123/19; Oslo Municipality: Oslo, Norway, 2019; Available online: https://tjenester.oslo.kommune.no/ekstern/einnsyn-fillager/filtjeneste/fil?virksomhet=976819837&filnavn=byr%2F2019%2Fbr2%2F2019055303-2184743.pdf (accessed on 29 April 2022).
21. Agency for Public Management and Egovernment. Grønne Anskaffelser Hva Er Det Og Hvorfor Er Det Viktig. Available online: https://www.difi.no/blogg/2017/08/gronne-oentligeanskaelser-hva-er-det-og-hvorfor-er-det-viktig (accessed on 7 March 2019).
22. Sovacool, B.K. How long will it take? Conceptualizing the temporal dynamics of energy transitions. *Energy Res. Soc. Sci.* **2016**, *13*, 202–215. [CrossRef]
23. Kemp, R.; Avelino, F.; Bressers, N. Transition management as a model for sustainable mobility. *Eur. Transp.* **2011**, *47*, 25–46.
24. Loorbach, D.; Frantzeskaki, N.; Huffenreuter, R.L. Transition Management: Taking Stock from Governance Experimentation. *J. Corp. Citizsh.* **2015**, *2015*, 48–66. [CrossRef]

25. Loorbach, D. Transition Management for Sustainable Development: A Prescriptive, Complexity-Based Governance Framework. *Governance* **2010**, *23*, 161–183. [CrossRef]
26. Kemp, R.; Schot, J.; Hoogma, R. Regime shifts to sustainability through processes of niche formation: The approach of strategic niche management. *Technol. Anal. Strat. Manag.* **1998**, *10*, 175–198. [CrossRef]
27. Ministry of Transport and Communication. *The National Transport Plan (2018–2029)*; Meld. St. Nr. 33; Ministry of Transport and Communication: Oslo, Norway, 2017.
28. Wang, C.; Qiao, Y.; Li, X. A systems approach for green public procurement implementation. *J. Public Procure* **2020**, *20*, 287–311. [CrossRef]
29. Björklund, M.; Gustafsson, S. Toward sustainability with the coordinated freight distribution of municipal goods. *J. Clean. Prod.* **2015**, *98*, 194–204. [CrossRef]
30. Rosell, J. Getting the green light on green public procurement: Macro and meso determinants. *J. Clean. Prod.* **2020**, *279*, 123710. [CrossRef]
31. Lindfors, A.; Ammenberg, J. Using national environmental objectives in green public procurement: Method development and application on transport procurement in Sweden. *J. Clean. Prod.* **2020**, *280*, 124821. [CrossRef]
32. Quak, H.; Nesterova, N.; Kok, R. Public procurement as driver for more sustainable urban freight transport. *Transp. Res. Procedia* **2019**, *39*, 428–439. [CrossRef]
33. Oliveira, C.M.D.; Albergaria De Mello Bandeira, R.; Vasconcelos Goes, G.; Schmitz Gonçalves, D.N.; D'Agosto, M.D.A. Sustainable vehicles-based alternatives in last mile distribution of urban freight transport: A systematic literature review. *Sustainability* **2017**, *9*, 1324. [CrossRef]
34. Jaller, M.; Pineda, L.; Ambrose, H.; Kendall, A. Empirical analysis of the role of incentives in zero-emission last-mile deliveries in California. *J. Clean. Prod.* **2021**, *317*, 128353. [CrossRef]
35. Taefi, T.T.; Kreutzfeldt, J.; Held, T.; Fink, A. Strategies to Increase the Profitability of Electric Vehicles in Urban Freight Transport. In *E-Mobility in Europe: Trends and Good Practice*; Leal Filho, W., Kotter, R., Eds.; Springer International Publishing: Cham, Switzerland, 2015; pp. 367–388. [CrossRef]
36. Teoh, T.; Kunze, O.; Teo, C.-C.; Wong, Y.D. Decarbonisation of Urban Freight Transport Using Electric Vehicles and Opportunity Charging. *Sustainability* **2018**, *10*, 3258. [CrossRef]
37. Ehrler, V.; Hebes, P. Electromobility for City Logistics—The Solution to Urban Transport Collapse? An Analysis Beyond Theory. *Procedia Soc. Behav. Sci.* **2012**, *48*, 786–795. [CrossRef]
38. Juan, A.A.; Mendez, C.A.; Faulin, J.; De Armas, J.; Grasman, S.E. Electric Vehicles in Logistics and Transportation: A Survey on Emerging Environmental, Strategic, and Operational Challenges. *Energies* **2016**, *9*, 86. [CrossRef]
39. Figenbaum, E. Can battery electric light commercial vehicles work for craftsmen and service enterprises? *Energy Policy* **2018**, *120*, 58–72. [CrossRef]
40. Enova Landtransport. Available online: https://www.enova.no/bedrift/landtransport/ (accessed on 29 March 2022).
41. Direktoratet for forvaltning og økonomistyring. Modenhet i Offentlige Anskaffelser. Available online: https://anskaffelser.no/sites/default/files/modenhetsrapport_v4_2021.pdf (accessed on 7 April 2022).
42. Norwegian Government Agency for Financial Management. *Nullutslippstransport i Leveranser Til Det Offentlige*; Norwegian Government Agency for Financial Management: Oslo, Norway. Available online: https://sintef.sharepoint.com/teams/work-10833/Delte%20Dokumenter/RA5%20Use%20Case%203%20TRANSPORT/2021%20Sampublisering%20Green%20Public%20Procurement/Litteratur/Miljodirektoratet%20(2020)%20Nullutslippstransport%20i%20leveranser.pdf (accessed on 7 April 2022).
43. Ministry of Trade, Industry and Fisheries. Forskrift om Offentlige Anskaffelser. Available online: https://lovdata.no/dokument/SF/forskrift/2016-08-12-974?q=forskrift%20om%20offentlige%20anskaffelser (accessed on 4 April 2022).
44. Statistics Norway. Standard for Klassifisering av Kommuner Etter Innbyggertall. Available online: https://www.ssb.no/klass/klassifikasjoner/115 (accessed on 25 March 2022).
45. Statistics Norway. Areal og Befolkning i Kommuner, Fylker og Hele Landet. Available online: https://www.ssb.no/system/ (accessed on 29 April 2022).
46. Jordbakke, G.N.; Amundsen, A.; Sundvor, I.; Figenbaum, E.; Hovi, I.B. Technological Maturity Level and Market Introduction Timeline of Zero-Emission Heavy-Duty Vehicles. Available online: https://www.toi.no/getfile.php/1348918-1538579785/Publikasjoner/T%C3%98I%20rapporter/2018/1655-2018/1655-2018-elektronisk.pdf (accessed on 7 April 2022).

World Electric Vehicle Journal

MDPI

Article

Synergy of Unidirectional and Bidirectional Smart Charging of Electric Vehicles for Frequency Containment Reserve Power Provision

Jonas Schlund [1,*], Reinhard German [2] and Marco Pruckner [3]

1 Ampcontrol.io, New York, NY 10001, USA
2 Computer Networks and Communication Systems, University of Erlangen, 91058 Erlangen, Germany
3 Modeling and Simulation, Institute of Computer Science, University of Würzburg, 97074 Würzburg, Germany
* Correspondence: jonas@ampcontrol.io

Abstract: Besides the integration of renewable energies, electric vehicles pose an additional challenge to modern power grids. However, electric vehicles can also be a flexibility source and contribute to the power system stability. Today, the power system still heavily relies on conventional technologies to stay stable. In order to operate a future power system based on renewable energies only, we need to understand the flexibility potential of assets such as electric vehicles and become able to use their flexibility. In this paper, we analyzed how vast amounts of coordinated charging processes can be used to provide frequency containment reserve power, one of the most important ancillary services for system stability. Therefore, we used an extensive simulation model of a virtual power plant of millions of electric vehicles. The model considers not only technical components but also the stochastic behavior of electric vehicle drivers based on real data. Our results show that, in 2030, electric vehicles have the potential to serve the whole frequency containment reserve power market in Germany. We differentiate between using unidirectional and bidirectional chargers. Bidirectional chargers have a larger potential but also result in unwanted battery degradation. Unidirectional chargers are more constrained in terms of flexibility, but do not lead to additional battery degradation. We conclude that using a mix of both can combine the advantages of both worlds. Thereby, average private cars can provide the service without any notable additional battery degradation and achieve yearly earnings between EUR 200 and EUR 500, depending on the volatile market prices. Commercial vehicles have an even higher potential, as the results increase with vehicle utilization and consumption.

Keywords: smart charging; electric vehicles; simulation; ancillary services; smart grid

Citation: Schlund, J.; German, R; Pruckner, M. Synergy of Unidirectional and Bidirectional Smart Charging of Electric Vehicles for Frequency Containment Reserve Power Provision. *World Electr. Veh. J.* **2022**, *13*, 168. https://doi.org/10.3390/wevj13090168

Academic Editor: Joeri Van Mierlo

Received: 12 July 2022
Accepted: 1 September 2022
Published: 2 September 2022

1. Introduction

The increasing renewable infeed and the aging infrastructure pose major challenges to the operation of EPSs (electrical power systems) all over the globe. In addition, we have recently seen an increased trend in new consumers, such as EVs (electric vehicles), which starts to have a considerable impact on the EPS [1] as well. Forecasts for the worldwide EV stock in 2030 amount to 250 million EVs based on the EV30@30 scenario [2]. If the charging processes of EVs are not coordinated in the future, they will result in a considerable demand peak increase in the evening hours and amplify the problem [3]. Thus, supplying EVs via the EPS brings new challenges [4], but also new opportunities as possible flexibility sources. Due to the high parking duration of privately owned vehicles, EV batteries represent a great flexibility that can be used to support the EPS. For example, renewable energies can be better integrated or ancillary services that were previously provided by conventional power plants can be provided by aggregating many EVs.

A future EPS with highly volatile, uncertain RES (renewable energy sources) and an inflexible demand will not work without extensive grid expansion, large scale storage or other flexibility sources. In this context, flexibility in an EPS is defined as the ability of assets to take different courses of action at a given point in time and thereby provide a service to third parties [5]. However, the energy transition thus far mostly focuses on producing and distributing enough RES, and the system stability itself still depends heavily on conventional technologies.

In the age of digitization, we are able to use ICT (information and communication technology) to pool and actively control the charging processes of many EVs in a VPP (virtual power plant). This has the advantage of charging processes being able to be postponed without restricting the mobility of the user while preserving the electrical grid. The largest part of the charging events will happen at private EVSE (electric vehicle supply equipment) and will be needed for short distance travels [3]. Such EV-charging processes are highly flexible loads [1,6] and are thus theoretically a good source for system-stabilizing ancillary services [7,8]. The tremendous flexibility potential of coordinating charging processes of a large fleet of EVs is also shown by Strobel et al. [9].

In this paper, we analyzed in depth how the smart charging of EVs can contribute to providing FCR (frequency containment reserve) power. For a case study, the German FCR power market serves as an example. FCR power is the fastest and one of the most economically interesting ancillary services in Europe [10]. For the analysis, we utilized a comprehensive and flexible simulation framework that consists of several components, such as a mobility behavior model, flexibility model, and a VPP component. For the case study, we differentiated between unidirectional and bidirectional EVSE and considered stochastic driving behavior based on real data [11,12]. Whereas unidirectional EVSE uses cheaper hardware and does not contribute to additional battery degradation, bidirectional EVSE can provide more flexibility using V2G (vehicle-to-grid). Note that, by ramping the load up and down, it is also possible to provide a bidirectional flexibility service with unidirectional EVSE; it is just more constrained than using V2G-capable EVSE. Subsequently, we aim to combine the advantages of both approaches and provide an analysis of the synergy potential of a mixed approach with only partially V2G-capable EVSE.

The paper is organized as follows: Section 2 presents related work in the field of smart charging of EVs and ancillary service provision. In Section 3, we introduce our methodology by introducing fundamentals about FCR power and our simulation framework. Section 4 discusses results about different cases with regard to uni- and bidirectional charging, as well as a short economical evaluation. We conclude the paper with a short summary and an outlook on future work in Section 5.

2. Related Work

The consideration of smart charging algorithms and the additional use of EV batteries for ancillary service provision has become increasingly important in recent years. As EVs can be seen as a flexible resource, it is important to have an effective and sufficient charging infrastructure. A good overview on the energy and charging infrastructure requirements are given by Thingvad [13]. Literature pertaining to the additional use of EVs for the EPS is often focused on different applications. For instance, Englberger et al. [14] investigated several applications for EV multi-use, such as self-consumption increase, peak shaving, frequency regulation and spot market trading, from a techno-economic point of view. The fleet sizes vary between 1 and 150 vehicles. Results show that the stacking of several services is very profitable with additional cash flows of up to EUR 2224 per EV per year. In this context, the provision of FCR plays an important role and contributes significantly to the annual cash flow. In addition, Tsagkaroulis et al. [15] studied the provision of ancillary services based on real driving data from over 7000 Nissan LEAFs. Based on real data, the authors consider the driving time, distance and parking time at different locations, which constrains the availability of each EV. Under full knowledge of future driving consumption and future grid frequency, an optimization problem is formulated for maximizing the profit of each EV. The profit spreads from EUR 51 to EUR 1654 per year and highly depends on the individual user profiles. In the study by Moncecchi et al. [16], the authors investigated an EV parking garage providing ancillary services in the Italian balancing market. The idea is to adapt the charging rate of the EV to obtain an income from participating on the balancing market. The results show that a large flexibility power can be offered with a high level of reliability.

Due to the increase in EVs as well as other consumers, such as heat pumps, some publications are also focused on the combined delivery of FCR from a fleet of EVs and heat pumps. In the study by Meesenburg et al. [17], the authors present a combined approach of EVs and a large-scale heat pump to exploit the synergies between both systems. The results show that an additional profit from capacity and power price payments can be generated that could not be achieved by the EV or the heat pump alone.

In order to provide ancillary services with a fleet of EVs, the quantification of the flexibility potential and the resulting smart control of charging processes is essential. With regard to the quantification and methodological description of the flexibility potential, numerous approaches can be found in the literature. A review can be found in the dissertation of Schlund [12]. In particular, the FlexAbility concept [18], which is also used in this work, provides a method for modeling the flexibility availability of decentralized electrical loads (e.g., charging processes of EVs). Similar approaches can also be found in [19,20]. Knowledge of the flexibility of charging processes can be used to shift them within the plug-in duration under the usage of smart charging algorithms.

There are many different approaches to the smart charging of EVs that have been published in recent years. For this purpose, many different approaches, such as linear programming, mixed-integer linear programming, model predictive control and machine learning techniques, are used. The optimization goals are a cost reduction from the customer's point of view, a minimization of greenhouse gas emissions, and the best possible integration of renewable energies into the grid. For instance, in the study by Hussain et al. [21], the authors developed a two-layer decentralized charging approach for residential EVs based on fuzzy data fusion. Therefore, they presented a fuzzy integer linear program and showed that the charging costs for EVs can be reduced while guaranteeing their required energy by determining the optimal charging schedule. A optimization-based coordination is also presented by Spitzer et al. [22]. They analyzed the impact of uncoordinated as well as optimization-based coordination strategies on a low-voltage grid. For the optimization-based coordination, they defined three strategies: a cost-optimized, a valley-filling-optimized and a greenhouse-gas-emission-optimized strategy. Depending on the objective, the costs can be reduced by more than 50% and the greenhouse gas emissions by around 40%. Tuchnitz et al. [23] presented a charging coordination system based on reinforcement learning. In contrast to optimization-based charging strategies, general parameters such as arrival times or the parking duration do not have to be known beforehand. The results show that the developed method works very well as all EVs have enough energy for their next trip.

To sum up, there is a large amount of literature available for the ancillary service provision with EVs and the smart control of charging processes. Nevertheless, we identified some research gaps, as some papers omit a deeper comparison of unidirectional and bidirectional charging for the provision of FCR power. Our paper has the following main goals:

- We investigate how unidirectional and bidirectional smart charging can contribute to FCR power provision.
- We are the first to propose and analyze the advantages of a synergistic operation with uni- and bidirectional charging for FCR power provision
- We present a comprehensive and flexible simulation framework.
- We conduct several sensitivity analyses.

3. Methodology

3.1. Frequency Containment Reserve Power

In the European EPS, the nominal power frequency f_0 is 50 Hz and may only vary within ±0.2 Hz. Therefore, generation and consumption in the whole synchronous EPS need to stay in balance. In order to guarantee this balance, several standardized flexibility services exist. In essence, those services either ramp up or down large amounts of power within certain time constraints. Those services are tendered by the TSO (transmission system operator), which, in Germany, are 50 Hertz Transmission, Amprion, Tennet TSO and TransnetBW.

The FCR power is the flexibility service of highest quality, and the full tendered amount of power needs to be available within 30 seconds. The service is traditionally provided by conventional power plants and, in the recent past, the market became dominated by new players—mostly stationary battery storage systems. In this paper, we analyzed how EVs can theoretically contribute to the service using a simulation model. Thus, we mimicked the requirements from FCR power by equal power manipulations of large amounts of charging processes without constraining the user experience of the vehicle owners. In the following, we summarize the mechanisms of FCR power that are constraining the aggregated power of charging processes when providing the service with a VPP of EVs.

The unit providing the service needs to be able to provide the full reserve power (positive and negative) for at least 15 min at any given time during the contract period [24,25]. The period to be covered per incident is below 15 min [24,25]. The activation begins when the power frequency lies outside the dead band, e.g., of ±10 mHz in the synchronous grid of Continental Europe [26,27]. Then, the FCR power provision increases linearly with an increasing power frequency deviation until a full power provision of the whole contracted power P_{contr} at a power frequency deviation of 200 mHz [26,27].

Thereby, the power difference ΔP to the baseline power of an asset providing FCR power needs to follow the frequency f as defined in Equation (1) [28].

$$\Delta P(f) = \begin{cases} P_{\text{contr}} & \forall f < 49.8\text{ Hz} \\ P_{\text{contr}} \cdot \frac{f_0 - f}{0.2\text{ Hz}} & \forall 49.8\text{ Hz} \leq f \leq 50.2\text{ Hz} \\ -P_{\text{contr}} & \forall f > 50.2\text{ Hz} \end{cases} \quad (1)$$

The major challenges to providing FCR power are the requirements of a fast activation time and the required power reservation over a quarter hour [24,25]. The provision is fully symmetric, i.e., an asset providing the service needs to be able to ramp up and down the generation accordingly. The technical feasibility of controlling charging processes of commercially available EVs with a fast enough response time has already been validated by [29]. Thus, the major constraint is the ability of a VPP to reserve power and guarantee the availability over 15 min.

In order to differentiate such an aggregation of flexible loads from a VPP (that typically also includes generation), we will use the term VFP (virtual flexibility plant) in the following. When providing the service with a VPP, the operation follows Equation (1) (counting power generation positively). With a VFP, we are unable to generate power. Thus, we count power consumption positively and follow Equation (2) to provide an equivalent service.

$$\Delta P(f) = \begin{cases} 0 & \forall f < 49.8\text{ Hz} \\ 2 \cdot P_{\text{contr}} \cdot \frac{f - 49.8\text{ Hz}}{0.4\text{ Hz}} & \forall 49.8\text{ Hz} \leq f \leq 50.2\text{ Hz} \\ 2 \cdot P_{\text{contr}} & \forall f > 50.2\text{ Hz} \end{cases} \quad (2)$$

3.2. Simulation Model

We used the discrete events simulation model from [12] to simulate EV fleets of any size. It uses up to 50,000 simulated EV instances that are able to represent up to millions of EVs. The model was implemented in AnyLogic [30], a simulation software based on Java 8 [31]. It uses the framework i7-AnyEnergy [32,33] to implement efficient interfaces between different model components. The model structure is summarized in Figure 1.

The model includes a stochastic component for the social behavior of vehicle owners based on [11]. Behavior in this context means at what time EVs depart or arrive at certain chargers, to which location they drive, how far and fast they drive and how much energy the trips consume. It is important to use realistic driving behavior for our analysis as the behavior essentially poses constraints on the availability of vehicles at the chargers and on the energy amounts that need to be charged. While all this behavior is sampled randomly for individual EV instances, we assumed that the resulting constraints are known to the central VFP controller.

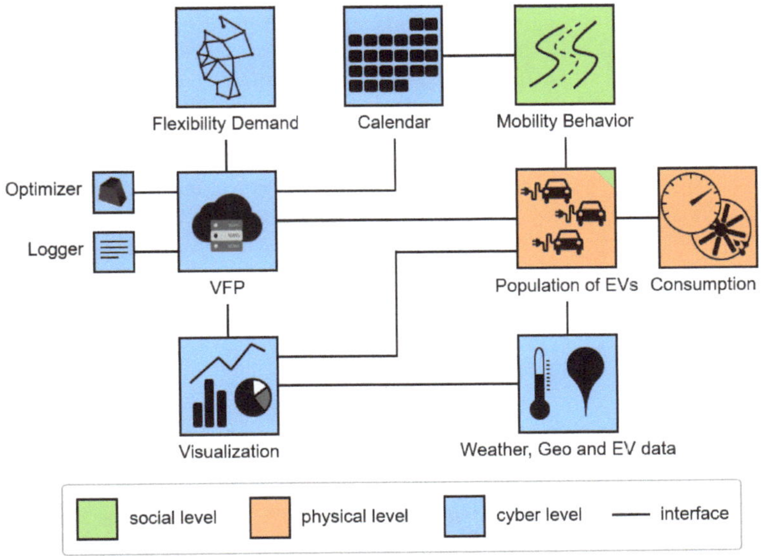

Figure 1. Overall simulation model structure: objects of the different levels and interfaces [12].

The used stochastic mobility behavior model was based on a Bayesian network. Its key assumptions are that EVs are only adopted at mass scale if users do not have to change their behavior and that they mainly charge where they park. It models the mobility patterns, including a correct overnight stay behavior, without any magic rules or numbers, and just based on the input data from [11]. Thereby, the directed acyclic graph structure ensures that the most important variable inter-dependencies are adhered to.

The model describes the daily mobility activity of a vehicle with individual trips between abstract locations (such as *Work*, *Business*, *Leisure*, *Home*, *Education* or *Shopping*). A trip chain then describes the full mobility behavior over a day. It includes all individual constraints of each trip (such as *arrival time*, *departure time*, *duration*, *speed*, *stay time*, *distance* or *purpose*). It can represent patterns of different user groups (*behaviorally homogeneous groups*, *mobility groups* or *age groups*) on different days (the different days of the week and holidays) and locations (different regional types from urban to rural and federal states in Germany). A Python re-implementation of the mobility model component is publicly available on GitHub (https://github.com/jsschl/ev_mobility_model, accessed on 2 September 2022).

On the physical level, we used empirical models of the technical components [12,34], such as battery, on-board charger, HVAC (heating, ventilation, air conditioning) and other consumers, to represent the consumption while driving and the losses while charging. Each simulated EV instance obtains trip chains assigned by the mobility model and then drives accordingly, and is assumed to be available for smart charging whenever it is located at a charger. Thereby, the main charging locations can be varied for different shares between expected amounts at home and workplaces [35].

On the cyber layer, we modeled the cloud-based IT system that is in charge of smart charging control. The VFP connects to the chargers, aggregates the data and combines them with different third party data integration, such as vehicle telematics, weather or geographic data. It is connected to an optimizer that can use different smart charging strategies and algorithms [12] to realize the smart charging application. The model includes a visualization component for fast prototyping and experimentation.

For the use case of FCR power provision, the flexibility demand component represents the requirement of a guaranteed reservation of the full contracted power over 15 min. The VFP uses the FlexAbility model from [12,18] based on the time flexibility [36] to exactly quantify the possible power reservation over an arbitrary time horizon, e.g., 15 min for FCR power provision, at any given point in time. Thus, in order to quantify the potential of FCR power provision with unidirectional chargers, we simulated replications of whole years of operation of a fleet of 5000 EVs and quantified the reservation capability of the overall EV fleet.

Thereby, we firstly considered a baseline scenario with distributions of typical car users and average EV type distributions based on the registration figures in Germany [37,38] and the technical data of the most common EV types [39]. Secondly, we varied certain parameters, such as the battery size of the vehicles, to determine the sensitivity of the results. In the last step, we derived results for mixed fleets with a share of V2G-capable chargers and analyzed the synergy.

4. Results

4.1. Unidirectional Charging

Figure 2 summarizes the sensitivity of key results to key parameters for a fleet size of 5000 EVs in Germany with unidirectional charging only. Note that the points on the y-axes represent simulation results based on the baseline parameterization. As explained above, this baseline parameterization represents the current distribution of technical, social and spatial parameters for EVs in Germany today. The lines on the graphs represent scenarios, where all parameters are kept at the baseline parameterization apart from the parameter that is varied on the x-axis.

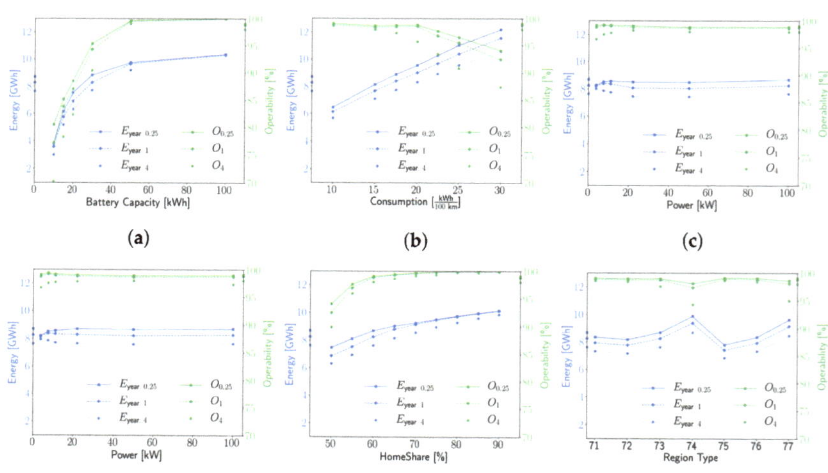

Figure 2. Sensitivity analysis of $E_{\text{year},x}$ (blue) and O_x (green) for key parameters with the default parameterization marked on the y-axes [12]. (**a**) Battery capacity. (**b**) Consumption. (**c**) Max. EV charging power. (**d**) Max. EVSE power. (**e**) Charging at home vs. work. (**f**) Regional type [40].

The results are linearly scalable for large fleet sizes (i.e., larger than 1000 EVs) as the stochastic behavior levels out [12]. The energy flexibility $E_{\text{year},\,x}$ (blue) describes the total power over time horizon x in hours, i.e., energy, that can be reserved over the year. The operability O_x (green) describes the share of the year in which the VFP is able to provide the service. Note that the points on the blue and green y-axes represent the according values for the mixed parameterization in Germany based on the current figures, whereas the points in the graphs and the interpolated lines describe the sensitivity to the given parameter. For these points, the parameters of interest were all set to the given value for all EVs in the simulation.

For the current requirement of $x = 0.25$ in Germany, an operability of over 99% is reached and, on average, a vehicle can contribute 1.75 MWh of reserved bidirectional energy flexibility. For larger time horizons x, these values drop slightly. Without V2G, it is thus possible to provide roughly 200 W per average private car (including the constraint of a full bidirectional reservation over 15 min) over the whole year.

We observe the highest sensitivity to the battery capacity, the consumption and the regional type. The sensitivity to the battery capacity in Figure 2a saturates at large capacities and, interestingly, the current mixed parameterization reaches results that are already close to the maximum. Thus, a fleet with different kinds of battery sizes seems to have synergistic effects.

For the consumption in Figure 2b, we observe that the potential generally increases the larger the consumption is. The reason for this is that the service is generally upper bounded by the ability to decrease the load. A higher consumption means a higher average load and thus also more potential to decrease the load. However, for larger values than 20 kWh per 100 km, the operability decreases significantly as large amounts of charge become necessary, and the reduction in the load is no longer possible throughout certain time periods.

The maximum EV and EVSE charging powers in Figure 2c,d do not show significant sensitivities. This means that the maximum charging powers are large enough in any case and do not pose a constraint to the investigated service. However, as shown in Figure 2e, the location of the main charging stations has a significant effect. It is generally favorable to include a larger share of users that primarily charge at home. Lastly, Figure 2f shows that the potential is generally higher in town and village areas with larger distances (types 74 and 77 [40]). This has the same reasons as described above for the consumption. Other investigated parameters, such as the weather year or the federal state, did not show a notable sensitivity. Even the plug-in behavior of the users does not influence the results considerably, as long as the operational strategy of the VFP is adapted accordingly.

With the knowledge of the sensitivity, an aggregator can compose an optimized fleet for the service. The result improves considerably if the aggregator targets users that live in towns and villages in urban regions and users that have EVs with larger batteries (50 kWh) and a slightly above average consumption ($0.2\ \frac{\text{kWh}}{100\ \text{km}}$). In addition, the aggregator should aim at 90% of users that are primarily charging at home. In this specific case, the yearly 0.25 h energy flexibility increases to 2.94 MWh per EV (+69% in comparison to the default case) at an operability of one. This results in an average possible provision of 336 W of FCR power per EV during the whole year.

4.2. Uni- and Bidirectional Charging

Including a certain amount of bidirectional V2G-capable EVSE can improve these results considerably without actually having to discharge any notable amounts during operation. The major constraint for FCR power provision is the reservation and not the actual operation. This becomes clear when observing how the actual power frequency is typically distributed. Figure 3 visualizes a typical power frequency distribution measured second-by-second over a whole year. The power frequency varies in both directions from the nominal power frequency, i.e., a dispatch in both directions is necessary. The measurements follow a Gaussian distribution and the most extreme deviations are considerably below ± 100 mHz. A very large proportion of the observations are even within the deadband. In general, the actual observations are only within a very narrow area around f_0, which means that a full provision of the contracted power is never necessary.

If we operate a VFP with unidirectional charging only, the VFP is idle, i.e., neither charging nor discharging, at a frequency of 49.8 Hz (see Equation (2)). Thus, in the whole frequency band of 49.8 Hz to 50.2 Hz, the VFP always charges. It charges more power or less power depending on the frequency, but never discharges the EV.

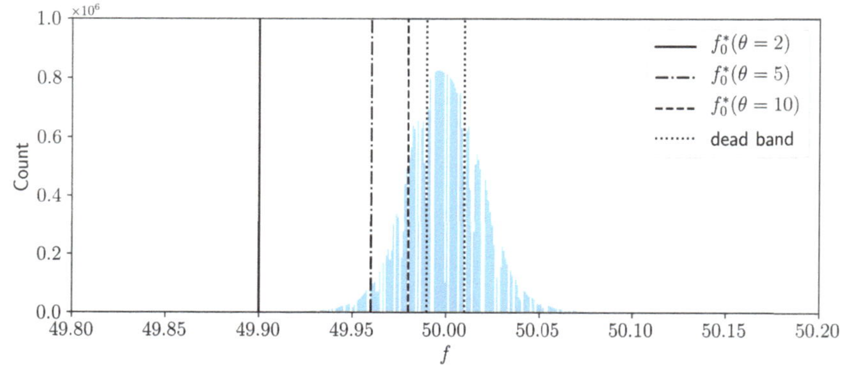

Figure 3. Power–frequency histogram based on second-by-second measurements from [41].

With a VPP of bidirectional chargers only, we can operate symmetrically around f_0 according to Equation (1). This means that the VPP is idle at f_0 (=50 Hz). It charges in the upper half of the frequency spectrum and discharges in the lower half. As the distribution is mostly symmetric, we also roughly need to discharge half of the time.

When we combine unidirectional and bidirectional charging, we can shift this frequency with an idle operation between 49.8 Hz and f_0. This way, it is possible to contract higher amounts of power than with unidirectional charging only, as parts of the flexibility spectrum are covered by discharging. We call the frequency at which we operate in idle mode f_0^*. For instance, with a configuration of f_0^* = 49.9 Hz, the spectrum below 49.9 Hz is covered by discharging and the spectrum above is covered by charging. However, there are typically no or almost no observations below 49.9 Hz. Thus, we can use the discharging capability of V2G for reservation purposes but actually operate in charging mode all or most of the time. This is interesting as additional battery degradation only occurs if EVs are actually discharged.

When we assume that (a) the baseline charging operation is, on average, 200 W per EV and that (b) we are able to contract ten times this amount for FCR power provision when the EVSE has V2G capability (P_{contr} = 2 kW), a discharge is only necessary if the power frequency is at least 20 mHz below the nominal frequency f_0 of 50 Hz. The reason for this is that, in the given scenario, the VFP has a baseline consumption of 200 W and thus only needs to discharge when the frequency f is lower or equal to 49.98 Hz (see Equation (1)).

The over-subscription θ describes the ratio of contracted power P_{contr} to the average possible contracted power only with G2V (grid-to-vehicle) $P_{\text{contr}}^{\text{G2V}}$. It is defined in Equation (3). In the given example above, $P_{\text{contr}}^{\text{G2V}}$ is 200 W per EV, P_{contr} is 2 kW per EV and the over-subscription θ is ten. As defined in Equation (4), the frequency $f_0^*(\theta)$ is then the minimal frequency that does not yet result in a discharge of the VFP. In the example above, $f_0^*(10)$ is 49.98 Hz.

$$\theta = \frac{P_{\text{contr}}}{P_{\text{contr}}^{\text{G2V}}} \tag{3}$$

$$f_0^*(\theta) = f_0 - \frac{0.2\ \text{Hz}}{\theta} \tag{4}$$

Thus, the frequency $f_0^*(\theta)$ describes at which frequency the overall VFP is in idle operation, i.e., neither charging nor discharging. Thereby, θ is the factor of how much more power is contracted in comparison to the scenario without V2G. Without V2G, $f_0^*(\theta = 1)$ needs to be 49.8 Hz as the system needs to be able to provide a full power reduction without being able to discharge. With a share of 10 % and, respectively, 40 % V2G, this frequency can be shifted further to the right and a θ of 2 (without any discharging in operation) and, respectively, 5 (with just very rare occasions of discharging) can be reached.

4.3. Economical Evaluation

Overall, the expected amount of 6.2 million EVs in Germany in 2030 [42] has, in any scenario, the theoretical potential to supply the whole FCR power market of 562 MW [43]. The current coalition contract of the German government even aims at 15 million EVs in 2030 [44], which means an even larger potential. The total costs for this market amounted to EUR 64.5 million in 2018 [25]. Nonetheless, an economic evaluation is difficult due to the recent price volatility, where we have seen prices per MW and week between EUR 1000 and EUR 10,000 [45]. Assuming prices between EUR 2000 and EUR 5000, a V2G share of 35%, which results in a θ of 4.46, and a provision of 1.5 kW per EV, an average EV can achieve earnings between EUR 200 and EUR 500. In this scenario, the necessary average discharge in operation only results in additional battery degradation equivalent to driving 40 km per year, which can be neglected. In comparison, previous research [46] concluded with an additional equivalent aging of driving 1573 km per year and a vehicle for the exact same use case, with bidirectional charging that operates symmetrically around f_0.

5. Conclusions

In conclusion, providing FCR power with EVs generally has a high potential from a technical perspective. The economical feasibility is uncertain but looks promising, with potential yearly earnings between EUR 200 and EUR 500. This can become even more interesting in the future, especially with the recent price increases on the market.

Using the synergy between unidirectional EVSE and V2G-capable EVSE combines the advantages of both technologies, and larger amounts of flexibility can be offered without causing additional charging cycles and battery degradation through discharging during operation. The discharging capability is mostly used for reservation purposes for very rare events, and most of the operation is covered by charging only.

Lastly, follow-up studies with a focus on different commercial vehicles are promising, as the potential for FCR power provision increases considerably with the consumption and utilization of the vehicles. Additionally, weaker stochastic effects in commercial setups simplify real-world applications.

Author Contributions: Conceptualization, J.S., R.G. and M.P.; methodology, J.S. and M.P.; software, J.S.; validation, J.S. and R.G.; formal analysis, J.S.; data curation, J.S.; writing—original draft preparation, J.S.; writing—review and editing, M.P.; visualization, J.S.; supervision, R.G. and M.P.; project administration, J.S. All authors have read and agreed to the published version of the manuscript.

Funding: This research received no external funding.

Institutional Review Board Statement: Not applicable.

Informed Consent Statement: Not applicable.

Data Availability Statement: The used mobility model is available on GitHub: https://github.com/jsschl/ev_mobility_model (accessed on 2 September 2022).

Acknowledgments: We want to thank our former students Felix Posner, Ronny Steinert and Leo Strobel, whose research and contributions made it possible to perform this study.

Conflicts of Interest: The authors declare no conflict of interest.

Abbreviations

The following abbreviations are used in this manuscript:

EPS	electrical power system
EV	electric vehicle
EVSE	electric vehicle supply equipment
FCR	frequency containment reserve
G2V	grid-to-vehicle
HVAC	heating, ventilation, air conditioning
ICT	information and communication technology
RES	renewable energy sources
TSO	transmission system operator
V2G	vehicle-to-grid
VFP	virtual flexibility plant
VPP	virtual power plant

References

1. Flammini, M.G.; Prettico, G.; Julea, A.; Fulli, G.; Mazza, A.; Chicco, G. Statistical characterisation of the real transaction data gathered from electric vehicle charging stations. *Electr. Power Syst. Res.* **2019**, *166*, 136–150. [CrossRef]
2. International Energy Agency. *Global EV Outlook 2019—Scaling-Up the Transition to Electric Mobility*; International Energy Agency: Paris, France, 2019.
3. Ebner, M.; Fattler, S.; Ganz, K. *Kurzstudie Elektromobilität. (Short Study Electric Mobility)*; Forschungsstelle für Energiewirtschaft e.V.: Munich, Germany, 2019.
4. Mu, Y.; Wu, J.; Jenkins, N.; Jia, H.; Wang, C. A Spatial Temporal model for grid impact analysis of plug-in electric vehicles. *Appl. Energy* **2014**, *114*, 456–465. [CrossRef]
5. Lehmann, N.; Kraft, E.; Duepmeier, C.; Mauser, I.; Förderer, K.; Sauer, D. Definition von Flexibilität in einem zellulär geprägten Energiesystem (Definition of Flexibility in a Cellular Energy System). In Proceedings of the Zukünftige Stromnetze 2019, Berlin, Germany, 30–31 January 2019; Conexio: Pforzheim, Germany, 2019; pp. 459–469.
6. Gottwalt, S.; Gärttner, J.; Schmeck, H.; Weinhardt, C. Modeling and Valuation of Residential Demand Flexibility for Renewable Energy Integration. *IEEE Trans. Smart Grid* **2017**, *8*, 2565–2574. [CrossRef]
7. Develder, C.; Sadeghianpourhamami, N.; Strobbe, M.; Refa, N. Quantifying flexibility in EV charging as DR potential: Analysis of two real-world data sets. In Proceedings of the 2016 IEEE International Conference on Smart Grid Communications (SmartGridComm), Sydney, Australia, 6–9 November 2016; pp. 600–605. [CrossRef]
8. Sadeghianpourhamami, N.; Refa, N.; Strobbe, M.; Develder, C. Quantitive analysis of electric vehicle flexibility: A data-driven approach. *Int. J. Electr. Power Energy Syst.* **2018**, *95*, 451–462. [CrossRef]
9. Strobel, L.; Schlund, J.; Pruckner, M. Joint analysis of regional and national power system impacts of electric vehicles—A case study for Germany on the county level in 2030. *Appl. Energy* **2022**, *315*, 118945. [CrossRef]
10. Englberger, S.; Hesse, H.; Hanselmann, N.; Jossen, A. SimSES Multi-Use: A simulation tool for multiple storage system applications. In Proceedings of the 2019 16th International Conference on the European Energy Market (EEM), Ljubljana, Slovenia, 18–20 September 2019; pp. 1–5. [CrossRef]
11. Infas; DLR; IVT; Infas 360. *Mobilität in Deutschland 2017 (Mobility in Germany 2017)*; Bundesministerium für Verkehr und digitale Infrastruktur: Bonn, Germany, 2018.
12. Schlund, J. Electric Vehicle Charging Flexibility for Ancillary Services in the German Electrical Power System. Ph.D. Thesis, Friedrich-Alexander-Universität Erlangen-Nürnberg (FAU), Erlangen, Germany, 2021.
13. Thingvad, A. The Role of Electric Vehicles in Global Power Systems. Ph.D. Thesis, Technical University of Denmark, Lyngby, Denmark, 2021.
14. Englberger, S.; Abo Gamra, K.; Tepe, B.; Schreiber, M.; Jossen, A.; Hesse, H. Electric vehicle multi-use: Optimizing multiple value streams using mobile storage systems in a vehicle-to-grid context. *Appl. Energy* **2021**, *304*, 117862. [CrossRef]
15. Tsagkaroulis, P.; Thingvad, A.; Marinelli, M.; Suzuki, K. Optimal Scheduling of Electric Vehicles for Ancillary Service Provision with Real Driving Data. In Proceedings of the 2021 56th International Universities Power Engineering Conference (UPEC), Middlesbrough, UK, 31 August–3 September 2021; pp. 1–6. [CrossRef]
16. Moncecchi, M.; Rancilio, G.; Dimovski, A.; Bovera, F. Smart Charging Algorithm for Flexibility Provision with Electric Vehicle Fleets. In Proceedings of the 2021 IEEE 15th International Conference on Compatibility, Power Electronics and Power Engineering (CPE-POWERENG), Florence, Italy, 14–16 July 2021; pp. 1–8. [CrossRef]
17. Meesenburg, W.; Thingvad, A.; Elmegaard, B.; Marinelli, M. Combined provision of primary frequency regulation from Vehicle-to-Grid (V2G) capable electric vehicles and community-scale heat pump. *Sustain. Energy Grids Netw.* **2020**, *23*, 100382. [CrossRef]

18. Schlund, J.; Pruckner, M.; German, R. FlexAbility—Modeling and Maximizing the Bidirectional Flexibility Availability of Unidirectional Charging of Large Pools of Electric Vehicles. In Proceedings of the Eleventh ACM International Conference on Future Energy Systems: e-Energy'20, Virtual Event, Australia, 22–26 June 2020; Association for Computing Machinery: New York, NY, USA, 2020; pp. 121–132. [CrossRef]
19. Neupane, B.; Siksnys, L.; Pedersen, T.B.; Hagensby, R.; Aftab, M.; Eck, B.; Fusco, F.; Gormally, R.; Purcell, M.; Tirupathi, S.; et al. GOFLEX: Extracting, Aggregating and Trading Flexibility Based on FlexOffers for 500+ Prosumers in 3 European Cities [Operational Systems Paper]. In Proceedings of the Thirteenth ACM International Conference on Future Energy Systems: e-Energy'22, Online, 28 June–1 July 2022; Association for Computing Machinery: New York, NY, USA, 2022; pp. 361–373.
20. Lilliu, F.; Pedersen, T.B.; Šikšnys, L.; Neupane, B. *Uncertain Flexoffers, a Scalable, Uncertainty-Aware Model for Energy Flexibility: e-Energy'22, Online, 28 June–1 July 2022*; Association for Computing Machinery: New York, NY, USA, 2022; pp. 448–449.
21. Hussain, S.; Thakur, S.; Shukla, S.; Breslin, J.G.; Jan, Q.; Khan, F.; Kim, Y.S. A two-layer decentralized charging approach for residential electric vehicles based on fuzzy data fusion. *J. King Saud Univ.-Comput. Inf. Sci.* 2022, *in press*. [CrossRef]
22. Spitzer, M.; Schlund, J.; Apostolaki-Iosifidou, E.; Pruckner, M. Optimized Integration of Electric Vehicles in Low Voltage Distribution Grids. *Energies* **2019**, *12*, 4059. [CrossRef]
23. Tuchnitz, F.; Ebell, N.; Schlund, J.; Pruckner, M. Development and evaluation of a smart charging strategy for an electric vehicle fleet based on reinforcement learning. *Appl. Energy* **2021**, *285*, 116382. [CrossRef]
24. German Transmission System Operators. General Information on Control Reserve—Technical Aspects. Available online: https://www.regelleistung.net/ext/static/technical?lang=en (accessed on 20 October 2020).
25. Bundesnetzagentur (German Federal Network Agency). *Monitoringbericht 2019 (Monitoring Report 2019)*; German Federal Network Agency: Bonn, Germany, 2020.
26. Biernacka, I. *Transmission System Operations and Control*; Lecture; Friedrich-Alexander-Universität Erlangen-Nürnberg (FAU): Erlangen, Germany, 2014.
27. NEXT Kraftwerke. Was Ist Primärregelleistung (What Is Frequency Containment Reserve Power)? Available online: https://www.next-kraftwerke.de/wissen/primaerreserve-primaerregelleistung (accessed on 20 October 2020).
28. Schlund, J.; Steber, D.; Bazan, P.; German, R. Increasing the efficiency of a virtual battery storage providing frequency containment reserve power by applying a clustering algorithm. In Proceedings of the 2017 IEEE Innovative Smart Grid Technologies—Asia (ISGT-Asia), Auckland, New Zealand, 4–7 December 2017; pp. 1–8. [CrossRef]
29. Marinelli, M.; Martinenas, S.; Knezovic, K.; Andersen, P.B. Validating a centralized approach to primary frequency control with series-produced electric vehicles. *J. Energy Storage* **2016**, *7*, 63–73. [CrossRef]
30. Borshchev, A.; Brailsford, S.; Churilov, L.; Dangerfield, B. Multi-method modelling: AnyLogic. In *Discrete-Event Simulation and System Dynamics for Management Decision Making*; Wiley Online Library: Hoboken, NJ, USA, 2014; pp. 248–279.
31. Krüger, G.; Hansen, H. *Java-Programmierung—Das Handbuch zu Java 8*; O'Reilly: Heidelberg, Germany, 2014.
32. Bazan, P.; Luchscheider, P.; German, R. Rapid Modeling and Simulation of Hybrid Energy Networks. In Proceedings of the 2015 SmartER Europe Conference, Essen, Germany, 10–12 February 2015; pp. 47–54.
33. Bazan, P. Hybrid Simulation of Smart Energy Systems. Ph.D. Thesis, Friedrich-Alexander-Universität Erlangen-Nürnberg (FAU), Erlangen, Germany, 2017.
34. Schlund, J.; Betzin, C.; Wolfschmidt, H.; Veerashekar, K.; Luther, M. Investigation, modeling and simulation of redox-flow, lithium-ion and lead-acid battery systems for home storage applications. In Proceedings of the 11th International Renewable Energy Storage Conference (IRES 2017), Düsseldorf, Germany, 14–16 March 2017.
35. Engel, H.; Hensley, R.; Knupfer, S.; Sahdev, S. *Charging Ahead: Electric Vehicle Infrastructure Demand*; McKinsey Center for Future Mobility: Düsseldorf, Germany, 2018.
36. Lee, Z.J.; Li, T.; Low, S.H. ACN-Data: Analysis and Applications of an Open EV Charging Dataset. In Proceedings of the Tenth ACM International Conference on Future Energy Systems: e-Energy'19, Phoenix, AZ, USA, 25–28 June 2019; Association for Computing Machinery: New York, NY, USA, 2019; pp. 139–149. [CrossRef]
37. Kraftfahrt Bundesamt (German Federal Motor Transport Authority). Bestand nach Zulassungsbezirken (Stock According to Registration Districts). Available online: https://www.kba.de/DE/Statistik/Produktkatalog/produkte/Fahrzeuge/fz1_b_uebersicht.html (accessed on 10 December 2020).
38. Kraftfahrt Bundesamt (German Federal Motor Transport Authority). Bestand nach Marken, Hersteller (Stock by Brand, Manufacturer). Available online: https://www.kba.de/DE/Statistik/Fahrzeuge/Bestand/MarkenHersteller/marken_hersteller_node.html (accessed on 10 December 2020).
39. Posner, F. *Techno-Ökonomische Analyse zur Reduzierung von Engpassmanagement mit Elektrofahrzeugen (Techno-Economical Analysis of Reducing Congestion Management with Electric Vehicles)*; Friedrich-Alexander-Universität Erlangen-Nürnberg (FAU): Erlangen, Germany, 2020.
40. Bundesministerium für Verkehr und Digitale Infrastruktur (Federal Ministry of Transport and Digital Infrastructure). Regionalstatistische Raumtypologie—RegioStaR (Regional Statistical Spatial Typology). Available online: https://www.bmvi.de/SharedDocs/DE/Artikel/G/regionalstatistische-raumtypologie.html (accessed on 10 December 2020).

41. Gobmaier, T. Netzfrequenzmessung—Messdaten des Jahres 2015 (Power frequency Measurement—Measurement Data of the Year 2015). Available online: https://www.netzfrequenzmessung.de (accessed on 25 January 2021).
42. Deloitte. *Elektromobilität in Deutschland—Marktentwicklung bis 2030 und Handlungsempfehlungen (Electromobility in Germany—Market Development until 2030 and Recommendations for Action)*; Deloitte: Munich, Germany, 2020.
43. German Transmission System Operators. regelleistung.net—Data Center. Available online: https://www.regelleistung.net/apps/datacenter/tenders/ (accessed on 20 January 2021).
44. Reuters. German Transport Minister Reverses from 15 mln Electric Vehicles Goal. Available online: https://www.reuters.com/world/europe/german-transport-minister-reverses-15-mln-electric-vehicles-goal-2022-01-17/ (accessed on 12 July 2022).
45. German Transmission System Operators. Leistungspreise (Power Prices). Available online: https://www.regelleistung-online.de/prl/leistungspreise/ (accessed on 21 December 2021).
46. Schlund, J.; Steinert, R.; Pruckner, M. Coordinating E-Mobility Charging for Frequency Containment Reserve Power Provision. In Proceedings of the Ninth International Conference on Future Energy Systems: e-Energy'18, Karlsruhe, Germany, 12–15 June 2018; ACM: New York, NY, USA, 2018; pp. 556–563. [CrossRef]

World Electric Vehicle Journal

MDPI

Article

Where to Charge Electric Trucks in Europe—Modelling a Charging Infrastructure Network

Daniel Speth *, Verena Sauter and Patrick Plötz

Fraunhofer Institute for Systems and Innovation Research ISI, Breslauer Strasse 48, 76139 Karlsruhe, Germany
* Correspondence: daniel.speth@isi.fraunhofer.de; Tel.: +49-721-6809-243

Abstract: Heavy-duty trucks account for 27% of the European greenhouse gas emissions in the transport sector. To decarbonize road freight transport, the European Union plans to build a fast charging network for trucks. This paper presents two scenarios, covering European highways with charging stations at regular intervals every 50 or 100 km along the most important highways. For each location, the required number of charging points at 15% battery electric trucking is calculated individually using queueing theory. A third scenario takes into account the infrastructure ramp-up in 2025 and assumes a share of 5% battery electric trucking in a network with a 100 km distance. We define a network of 660 (100 km distance) or 1468 stations (50 km distance). Depending on the scenario and the individual station, the projected number of charging points per station varies between 1 and 18 in 2030. The results give a first insight into what a fast charging infrastructure for trucks in Europe might look like. In particular, we show that large charging stations with more than ten charging points could be necessary in the next few years. This knowledge might help to design future charging infrastructure for electric road freight transport.

Keywords: battery electric truck; charging infrastructure; heavy-duty trucks; infrastructure; network modelling; queueing theory

Citation: Speth, D.; Sauter, V.; Plötz, P. Where to Charge Electric Trucks in Europe—Modelling a Charging Infrastructure Network. *World Electr. Veh. J.* **2022**, *13*, 162. https://doi.org/10.3390/wevj13090162

Academic Editor: Joeri Van Mierlo

Received: 25 July 2022
Accepted: 23 August 2022
Published: 28 August 2022

1. Introduction

Road transport causes approximately one quarter of current European greenhouse gas (GHG) emissions. In turn, heavy-duty trucks and buses account for 27% of road transport GHG emissions [1]. European Union (EU) legislation requires CO_2 emissions from newly registered heavy-duty vehicles to be reduced by 30% in 2030 compared to current levels [2]. Analyses show that this is only possible when using zero emission vehicles (ZEV), i.e., electric or hydrogen powered trucks [3]. When comparing different options to electrify heavy-duty trucks (e.g., fast charging, overhead catenary or battery swap), fast charging seems to be the best option in the short term, mainly due to a comparatively scalable infrastructure [4]. Current research shows that charging infrastructure is essential for the diffusion of battery electric trucks (BET) [5,6]. Consequently, the EU plans to install a fast charging infrastructure for BET [7].

According to [8,9], charging infrastructure planning models can be divided into three groups: node-based, path-based, and tour-based models.

Prominent representatives for node-based models are p-median models, where p facilities are positioned at nodes so that the demand of neighboring nodes can be fulfilled with a minimum distance travelled [9]. The approach is based on [10], who used it to determine the optimal position of police stations in a road network. A further development is the set-covering problem, which ensures that every demand can be served. Ref. [11] originally used this approach to position emergency service facilities. This approach can also be used for the positioning of charging locations for electric vehicles [12]. In node-based models, a charging station covers a certain area or a certain part of a road. Therefore, typically a quite dense charging network with many charging points is modelled.

The coverage approach described in [13] represents a special case. Unlike the previous approaches, charging locations are placed at regular intervals and sized based on the traffic volume determined for the neighboring nodes. Therefore, the focus is not on minimizing the number of charging locations but on sizing them. Ref. [14] shows a coverage approach to design an infrastructure for battery electric cars in Germany. Ref. [13] transferred the approach to BET in Germany. Like all node-based models, the coverage approach benefits from low data requirements [9]. Only the traffic volume at the nodes is required [13].

In contrast, path-based models rely on traffic flows within a network and try to cover a maximum of passing traffic with a minimum of stations. The Flow Capturing Location Model (FCLM), the first subgroup of path-based models, was introduced by [15]. For example, [16] using a FCLM to position 27 charging stations for battery electric cars in the city of Barcelona would serve 92% of the considered flows. Ref. [17] introduced the Flow Refueling Location Model (FRLM) that also considers multiple stops for one path. The general idea of placing locations such that a maximum of origin–destination paths known in advance can be supplied with a fixed number of stations remains the same. FRLM has been used to model charging infrastructure for battery electric cars in the USA [18] or in Europe [19]. Ref. [20] transferred this approach to hydrogen powered trucks and combined the model with a capacity restriction to avoid unrealistic large stations. However, due to the high computational effort, restrictions are usually necessary. For example, [18] clustered 4486 regions to 196 regions. Ref. [19] reduced the problem by ignoring flows with less than 5000 vehicles per year.

Finally, a tour-based model considers individual driving profiles and locates charging stations such that they fit the driving profiles. While the level of detail increases from the node-based to the path-based to the tour-based models, the demands on the input data also increase. For the node-based approach, data from local traffic counts are sufficient. Path-based models require origin–destination relations. The tour-based models typically rely on journey logs. For example, [21] using driving trip data to model slow-charging infrastructure for cars in the city of Columbus (OH, USA). For a deeper comparison of the models, refer to [8,9].

Regarding infrastructure modelling for trucks, initial publications already exist. Ref. [22] used GPS data from eight million vehicle trips and implemented a path-based model with individual trip data for short-haul BET in South East Queensland. The vehicle trips were clustered to 13,456 origin–destination paths and 116 possible charging locations were identified. Using up to 10 optimally positioned realized charging locations, 85 charging networks were modeled and a network coverage of up to 93% was derived. Also using a path-based model, [20] modelled a hydrogen refueling infrastructure for trucks in Germany. The analysis was based on 2655 origin–destination paths. As a result, 100 refueling stations can serve Germany (13,000 km highway) without any capacity restrictions. Depending on the number of hydrogen trucks, a capacity restriction, e.g., a maximum daily capacity per refueling station, can significantly increase the number of refueling stations. In contrast, [13] extended the coverage approach for battery electric cars from [14] and designed a public fast charging infrastructure for BET in Germany. Depending on the specified distance, 142 to 267 charging stations were identified. At 15% BET trucking, these charging stations are equipped with 2 to 13 charging points per station.

As shown, previous analyses usually consider geographically small areas and/or are based on significant simplifications. To the best of the authors' knowledge, there is no publication that presents a European fast charging infrastructure for BET. Since the EU is currently debating a Europe-wide charging network for trucks [7], it is highly important to develop an idea of what such a network could look like in the next few years.

Therefore, the aim of this paper is to design a public fast charging network for BET throughout Europe for the mid-term future (e.g., 2030) based on transport flows. We limit the analysis to the EU, extending to Great Britain, Switzerland and Norway (EU27+3). Since the proposal for the Alternative Fuel Infrastructure Regulation (AFIR) [7] suggests charging infrastructure at regular intervals along the European highway network, we follow [13] and apply a node-based approach and combine it with a queuing model. The queueing model determines the number of necessary charging points for a given share of BET for each charging station.

This paper differs from existing studies in several aspects. First, the geographical coverage includes the entire EU as well as Great Britain, Switzerland and Norway. Publicly available data is used for this purpose. Second, the paper follows the logic of the AFIR proposal [7] and models charging infrastructure at regular intervals. Third, the approach extends beyond location identification and focuses on the sizing of the locations using queueing theory. Fourth, the scenarios in this paper show various options and their implications for a potential short- to medium-term public infrastructure for trucks.

The paper is structured as follows: First, we present relevant input data and the methodological approach. Afterwards, Section 3 contains the results. Finally, we discuss some critical aspects in Section 4 and conclude our work in Section 5.

2. Materials and Methods

2.1. Materials and Scenarios

To model a charging infrastructure for Europe, we require the traffic volumes on the European highway network. Since homogeneous traffic counts for all European countries are rarely available and the standardization of country-specific data would be very burdensome, modelled European traffic flows [23] serve as our basis. The dataset represents an update of the truck traffic flows published in the European Transport policy Information System (ETIS) project [24] and contains projections between NUTS3 (Nomenclature des unités territoriales statistiques. Level 3: small regions, large cities)-regions in Europe. The dataset contains only transports between NUTS3-regions. Inner regional traffic is not included. However, inner regional traffic is not relevant for public fast charging infrastructure on the highway network. In the following, we refer to the truck traffic flows in the dataset as long-haul traffic. The underlying road network contains 17,435 nodes that are connected with 18,447 edges. In some sections, we add additional nodes to ensure a maximum distance of 10 km between two nodes. The nodes serve—simplified—as potential station locations in this paper. For each edge and node, the number of vehicles that pass them per year in 2010, 2019 and 2030 is known. In the following, we refer to those edges as subsections of one road. Figure 1 shows the traffic volume in Europe in 2030 according to [23]. Within this paper, we filter the updated ETIS dataset and focus on the international E-road network in Europe.

Additionally, we derive a cumulative annual mileage of 162,397 million km in 2019 and 215,042 million km in 2030 for long-haul traffic on roads in the EU, including Switzerland, Norway and Great Britain from [23]. Simplified, we assume 188,719 million km in 2025.

As shown in [13], and based on the automated traffic census in Germany [25], we assume that a maximum of 6% of the daily charging events happen in the most trafficked hour of the day. For this hour, the average waiting time should not exceed 5 min, according to experts from the automotive industry [26]. Thus, the charging process of 30 min can be carried out within the mandatory break of 45 min after 4.5 h of driving, including a 10 min buffer. Therefore, a recharge for approximately 300 km is required each time [13].

We assume that 25% of the charging events occur on public fast charging infrastructure. Survey data [27] show that about half of the heavy-duty vehicles drive less than 500 km per day [4]. For these vehicles, we assume that they are almost exclusively charged at the depot. For the other vehicles, we assume that half of their charging events take place at the origin or destination depot. This means that, on average, half of the BET use public infrastructure and, for these trucks, every second charging process takes place publicly. However, this assumption comes with a high level of uncertainty. As shown in [13], the required number of charging points increases approximately linearly with the share of public charging events. Table 1 sums up the most important parameters.

Figure 1. Traffic volume in Europe 2030. Own illustration based on [23]. Background: OpenStreetMap.

Table 1. Input parameters for infrastructure calculation.

Parameter	Abbreviation	Value	Reference
Cumulative annual mileage	$AM_{HDV,EU27+3}$	188,719 Mio. km (2025) 215,042 Mio. km (2030)	Own calculation, based on [23]
Range in 4.5 h	$range_{BET}$	300 km	[13]
Share of public charging	CE_{public}	25%	Own estimation, based on [13,26]
Average charging time		30 min	[13]
Average waiting time	W_q	5 min	[13]
Share of daily charging events in most trafficked hour		6%	[13]

In the following, we will focus on three scenarios: First, we will design a startup network for 2025. Within the startup network, we assume 5% of the annual mileage being electrified ($BET_{share} = 0.05$), following [3] and expert opinion [26]. Additionally, we assume a distance of $d_{avg} = 100$ km between the charging locations. This is slightly more than the EU's proposal with 60 km [7]. Second, we will design an expansion network for 2030 that will densify the startup network to 50 km, by putting additional stations between the stations of the startup network. The BET_{share} will grow to 15%. Third, we will design a widemeshed network for 2030 with a distance of 100 km and a BET_{share} of 15%. Table 2 sums up the most important information. Figure 2b illustrates the relationships between the three scenarios.

Table 2. Scenario definition.

Scenario	Targeted Year	d_{avg}	BET_{share}
Startup	2025	100 km	5%
Expansion	2030	50 km	15%
Widemeshed	2030	100 km	15%

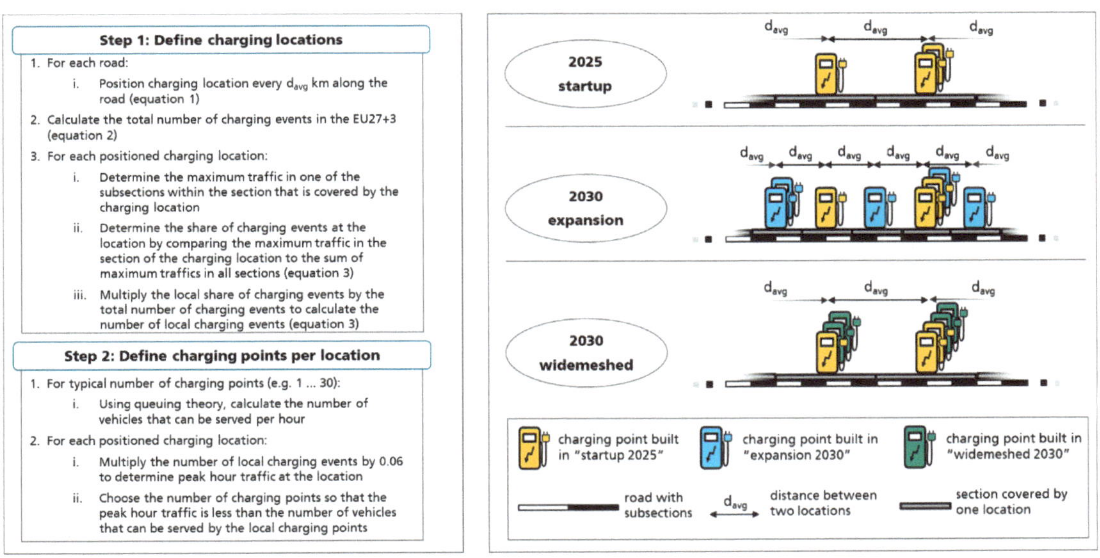

(**a**) (**b**)

Figure 2. Methodological procedure (**a**) and simplified draft (**b**) for the calculation of the charging networks.

2.2. Methods

The methodological procedure is divided into two steps: First, the charging locations and the number of charging events at every location are determined. Second, the number of charging points for every location is calculated. An overview of the methodological procedure is given in Figure 2a. The steps are described in more detail in Sections 2.2.1 and 2.2.2. For more information, compare [13].

2.2.1. Determine Charging Locations

According to the coverage approach described in [13], every single road in the E-road network graph is traversed successively according to a predefined scheme. The road network from [23] serves as our basis. Within the E-road network, there are two major groups: odd numbers indicate roads that run from north to south, whereas even road numbers run from east to west. Each of these roads is processed one after the other in ascending order, along the previously defined direction of travel. Every node in the network is a potential charging location. Locations are positioned at regular intervals. Equation (1) shows this approach, where CL_L is a bivariate variable indicating whether infrastructure is built in L or not. $d_{CL,L}$ indicates the distance between the last positioned charging location and location L. d_{avg} defines the distance between two charging stations in the network.

$$CL_L = \begin{cases} 1, \ if \ d_{CL,L} \geq d_{avg} \\ 0, \ else \end{cases} \tag{1}$$

Afterwards, the total number of daily public charging events in the EU27+3 CE_{EU27+3} is calculated as follows:

$$CE_{EU27+3} = \frac{BET_{share} * (AM_{HDV,EU27+3}/313)}{range_{BET}} * CE_{public} \tag{2}$$

BET_{share} stands for the share of BET on the total cumulative annual mileage $AM_{HDV,EU27+3}$ of all heavy-duty trucks. The annual mileage is divided by 313 to derive daily mileage, excluding Sundays. $range_{BET}$ refers to the range that a truck can cover in 4.5 h of driving. This corresponds to the maximum driving time before a mandatory break is required. Finally, we multiply this with the share of charging events on public infrastructure CE_{public}.

Finally, the expected daily public charging events have to be allocated to individual charging locations. For this purpose, the maximum traffic volume in the area in front of and behind the location is calculated and compared with the total maximum traffic volume of all locations. The number of trucks in both directions is considered together. The 04_network_edges dataset described in [23] serves as the basis for the subsection-by-subsection traffic volume. Equation (3) describes the calculation of the daily charging events at each realized charging locations. $MAX_{CL_{i-0.5}}^{CL_{i+0.5}}(TV_j)$ describes the maximum traffic volume of all of subsection j on half the distance between the realized charging location i and the realized station before this location ($CL_{i-0.5}$) and half the distance to the subsequent location ($CL_{i+0.5}$). The individual maximum traffic volume is set in relation to the sum of all maximum traffic volumes of all realized stations.

$$CE_{CL_i} = CE_{EU27+3} * \frac{MAX_{CL_{i-0.5}}^{CL_{i+0.5}}(TV_j)}{\sum_{CL} MAX_{CL_{i-0.5}}^{CL_{i+0.5}}(TV_j)} \tag{3}$$

2.2.2. Dimension Charging Locations

The calculation of the number of charging points per location is based on queueing theory. For the peak hour, we assume 6% of the daily charging events of one location, as described above. The system is designed for this size. We stick to the Kendall notation $(A/S/c/d/k/m)$, to define the queueing system. To define the arrival process A, we assume Poisson-distributed arrivals [28], with the average arrival rate $\lambda = CE_{CL_i} * 6\%$. The inter-arrival times are therefore exponentially distributed (Markovian Distribution M). This means A = M. With regard to the service process S, [28] show that a General distribution G with normally distributed service times fits quite well. The average number of customers served per period is defined by μ. For example, an average charging time of 30 min results in an average service rate μ = 2 trucks/hour. The number of service units c—charging points—shall be calculated. For D, k, and m, we assume the default values. The queue's discipline d follows the First-In-First-Out principle. This means that the trucks are served in the order of their arrival. The number of customers waiting in the queue k is assumed to be infinite. The same applies to the number of customers in total. Therefore, we define an M/G/c queueing system. Since the exact solution for the mean waiting time of M/G/c systems is not known, the mean waiting time is approximated, according to [29]:

$$W_q^{M|G|c} = \frac{C^2+1}{2} W_q^{M|M|c} \tag{4}$$

C is defined as the variation coefficient of the distribution of the service times, i.e., the standard deviation (5 min) divided by the mean value of the service time distribution (30 min). This formula is used with the waiting time of the original M/M/c system, given in Equation (5):

$$W_q^{M|M|c} = \frac{1}{1-\rho}\frac{1}{c\mu}\frac{(c\,\rho)^c}{c!}\left((1-\rho)\sum_{n=0}^{c-1}\frac{(c\rho)^n}{n!}+\frac{(c\rho)^c}{c!}\right)^{-1} \; with\; p = \frac{\lambda}{c\mu} \tag{5}$$

Finally, we calculate the maximum average arrival rate λ that allows for an average waiting time of 5 min for all possible numbers of charging points c. For each location, we compare the local average arrival rate $\lambda = CE_{CL_i} * 6\%$ to the arrival rates with different c. Afterwards, we can choose the number of charging points c for each location so that the average waiting time is less than 5 min.

3. Results

Figures 3 and 4 show the distribution of the charging locations in the EU27+3. For the startup network and the widemeshed network with a distance of 100 km, we receive 660 charging locations. For the expansion network with a 50 km distance, we count 1486 charging locations. There is more than a doubling since very short sections as well as peripheral areas also receive charging locations in the closer network.

With regard to the geographical distribution of the charging points, there is a concentration on Central Europe (e.g., France and Germany). The traffic generated by the ports in the Netherlands, Belgium and Germany is of great interest for the dimensioning of charging locations. The surrounding countries (e.g., Norway, Sweden, Finland, Greece, Italy and Spain) are equipped with smaller locations that cover the whole area.

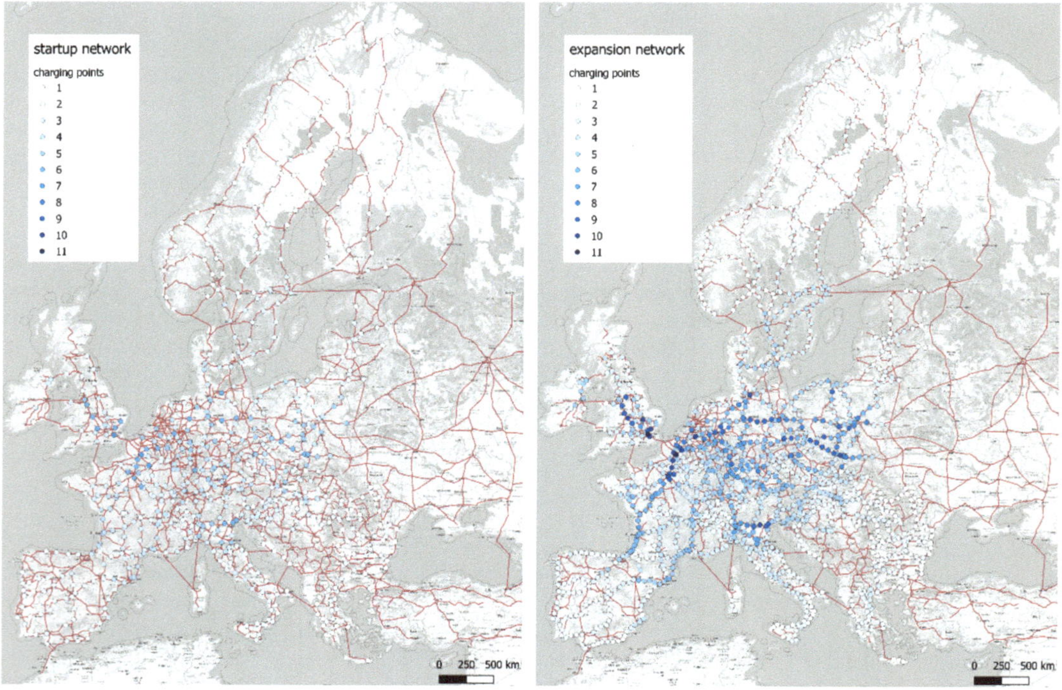

(a) (b)

Figure 3. Location of 660 charging locations in the startup network (**a**) and 1468 charging locations in the expansion network (**b**). Background: OpenStreetMap.

The startup network consists of 1697 charging points at 660 locations. This means that an average charging location includes 2–3 charging points in 2025 (mean = 2.57, median = 2). Assuming a tripling of electric traffic (BET_{share} 5% versus 15%) and a densification of the network from 100 km to 50 km, the average charging location still has three charging points (mean = 3.25, median = 3). However, the largest station contains 11 charging points instead of seven. We refer to this scenario as expansion network. In total, 4778 charging points are needed in this scenario. Figure 5a shows the change in the required number of charging points from the startup network to the expansion network. In less than 2% of all charging locations, between two and four charging points are removed. These are typically locations nearby areas with high traffic, such as ports, where a new charging location is opened closer to the high traffic location. If the network is not densified, 3679 charging points are required, as shown in the widemeshed network. However, the individual charging locations will be significantly larger. An average charging location is equipped with 5–6 charging points (mean = 5.57, median = 5). The biggest charging location consists of 18 charging points. Figure 5b shows the distribution of charging points among the individual charging locations for all scenarios.

Since our model relies on assumptions, Figure 6 shows a sensitivity analysis for BET_{share}, CE_{public}, and $range_{BET}$. We varied the parameters by $+/-50\%$. An increase in BET_{share} or CE_{public} by 50% increases the number of charging points from 4778 to 6211. This means the number of charging points increases by 30%. The same effect can be observed when reducing BET_{share} or CE_{public}. A reduction of $range_{BET}$ by 50% leads to an increase of charging points from 4778 to 7.581. An increase of $range_{BET}$ by 50% reduces the number of charging points to 3813.

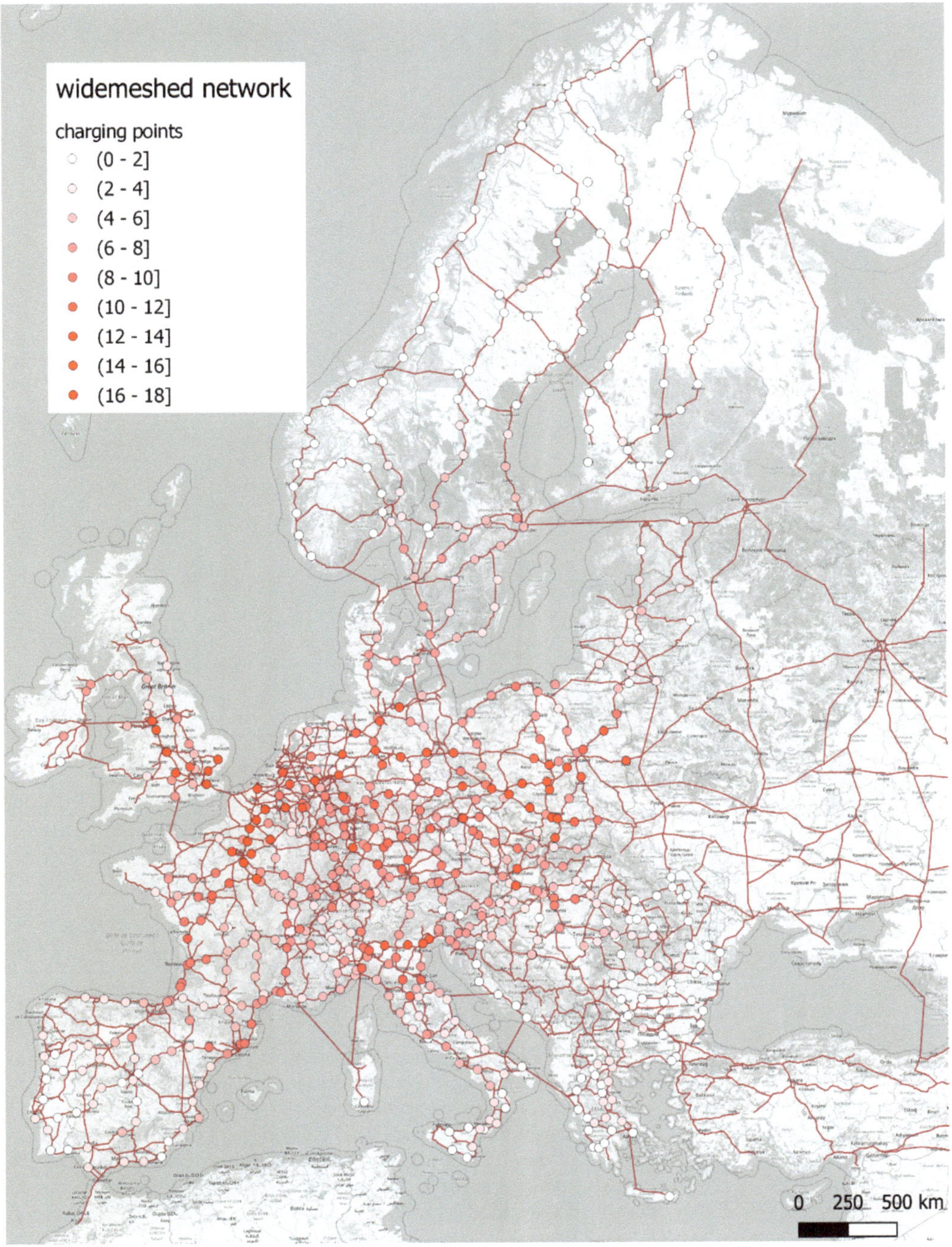

Figure 4. Location of 660 charging locations in the widemeshed network. Background: OpenStreetMap.

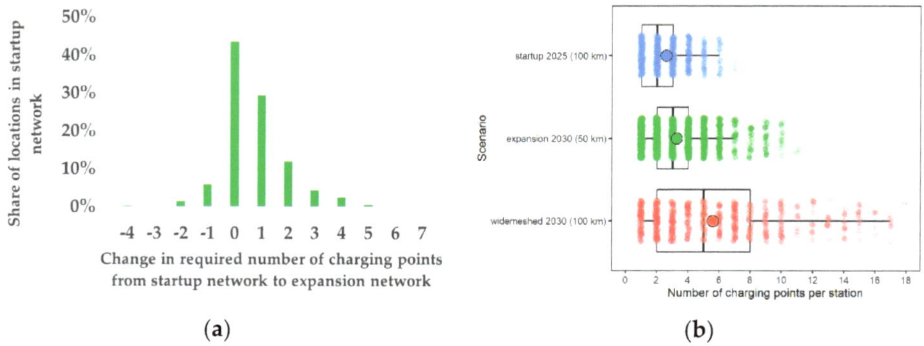

Figure 5. (**a**) Change in number of charging points from startup network to expansion network. (**b**) Boxplot of charging points per location for all scenarios.

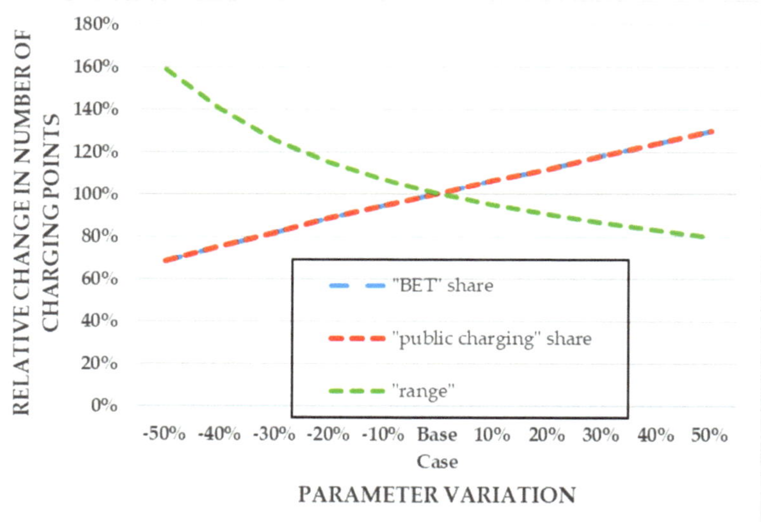

Figure 6. Sensitivity analysis for the expansion network.

4. Discussion

4.1. Data Limitations

The results shown in this paper rely on different assumptions and data sources. In particular, assumptions regarding the share of battery electric trucking and the vehicle range are currently based on expert estimates. As shown in Figure 6, the number of charging points relies on these assumptions. The effects of under or overestimation are remarkable but are not of a different order of magnitude. Therefore, the assumptions should be further investigated and validated in the future. Coupling the infrastructure ramp-up with a market diffusion model for trucks could also improve the reliability compared to our predefined market diffusion scenarios.

In contrast to previous work [13], the traffic volumes in this paper are based on synthetic road freight transport flow data [23]. On the one hand, this significantly improves the results, since regional traffic that is unlikely to use public fast charging infrastructure is not included in the dataset. On the other hand, the dataset itself relies on significant simplifications, e.g., in terms of resolution, scaling, and non-consideration of multi-stop-tours. From a scientific perspective, additional data sources, such as driver logbooks, should be integrated to better reflect actual driving behavior. From a planner's perspective, activities at local rest areas should be observed to validate the model results.

While [13] uses the entire German highway network as potential locations for charging stations, we restrict our analysis to the E-road network. This reduces the number of charging stations from 267 [13] to 154 in the 50-km network (expansion network) and from 142 [13] to 67 in the 100-km network (widemeshed network) in Germany. Given the same assumptions in both models, the number of charging points in total remains almost identical for Germany. This shows the quality of the synthetic dataset [23] compared to the real traffic count data used in [13]. As in the comparison between the expansion network and the widemeshed network, it also becomes clear that the average number of charging points per station depends on the underlying network distance and density.

4.2. Model Limitations

In addition to the data limitations, the model itself also comes with some limitations. First, the location selection does not take into account the suitability of the location for a charging area. Aspects such as parking area availability or the power grid connection are not part of the analysis. The locations are intended as representatives for the particular highway section, not as a defined location. However, the model gives a good impression of the general distribution of charging locations as well as the total number of charging points required. As part of the model development, location details could be integrated up to a certain level in the future. However, a planner will evaluate the local conditions in the targeted area in detail.

Second, as a node-based model, the coverage approach does not consider traffic flows. The origin–destination paths contained in the synthetic dataset are therefore converted into traffic volumes per subsection of a road. Vehicles are counted several times for different subsections they pass. It is plausible that vehicles recharge evenly distributed throughout the road network according to the local traffic volume. However, special effects may occur. As an example, the underlying dataset contains extensive port-hinterland-traffic. This traffic leads to a high traffic volume locally. As shuttle transports, these vehicles are probably charged at private depots. The model tends to overestimate the required public charging infrastructure in these sections. The assumption that charging processes are distributed equally to traffic volumes should be verified in the future, for example with driving logbooks or with data from parked vehicles. To overcome the underlying problem, the use of path- or tour-based models can also improve the results if the computation time allows it without too much simplification.

Third, the differentiation between public fast charging in the mandatory break and public slow charging is highly simplified and can only be modeled as part of CE_{public}. At this point, path-based or tour-based models are better suited. However, these models are associated with significantly higher computational effort and a high demand on data availability.

5. Conclusions

In conclusion, a dense fast-charging infrastructure for heavy-duty trucks is essential for the successful electrification of road freight transport. The EU is currently defining the legal framework to create a corresponding network. This paper gives a first insight into what a fast-charging network in Europe with charging stations at regular intervals might look like. Our results indicate a demand of approximately 700 to 1500 charging locations, with up to 4800 charging points in total within the next few years. In addition, the paper shows that even in an early stage of market, large locations with more than ten charging points are required. The size of the charging stations depends on the distance between the individual stations. In the first years of battery electric trucking, a dense network—e.g., the expansion network with 50 km distance between the charging stations—can avoid large stations with significantly more than ten charging points. Given the EU's targets for the

CO_2 reduction of newly sold vehicles in 2030, technical issues should be clarified, path decisions should be made and construction projects should be initiated.

From a scientific perspective, we have shown a first approach to estimate the required charging infrastructure even for large areas. Future research should focus on using path- or tour-based models to analyze the future needs for public truck charging infrastructure. This would provide an even more detailed view on how to locate and dimension individual charging stations.

Author Contributions: Conceptualization, D.S. and P.P.; methodology, D.S., V.S. and P.P.; software, D.S. and V.S.; validation, D.S. and P.P.; formal analysis, D.S.; investigation, D.S.; data curation, D.S. and V.S.; writing—original draft preparation, D.S.; writing—review and editing, P.P.; visualization, D.S.; supervision, P.P.; project administration, P.P.; funding acquisition, P.P. All authors have read and agreed to the published version of the manuscript.

Funding: The works in this paper were supported by EU STORM project funded from the European Union's Horizon 2020 research and innovation program under grant agreement No 101006700. The Federal Ministry of Transport and Digital Infrastructure (BMVI) Germany funded this research within the project HOLA (FKZ 03EMF0404A).

Data Availability Statement: Not applicable.

Conflicts of Interest: The authors declare no conflict of interest. The funders had no role in the design of the study; in the collection, analyses, or interpretation of data; in the writing of the manuscript; or in the decision to publish the results.

References

1. Eurostat. Greenhouse Gas Emissions by Source Sector (Source: EEA). Available online: https://ec.europa.eu/eurostat/databrowser/view/ENV_AIR_GGE__custom_1533603/default/table?lang=en (accessed on 8 November 2021).
2. European Union. Regulation (EU) 2019/1242 of the European Parliament and of the Council of 20 June 2019 Setting CO_2 Emission Performance Standards for New Heavy-Duty Vehicles and Amending Regulations (EC) No 595/2009 and (EU) 2018/956 of the European Parliament and of the Council and Council Directive 96/53/EC. Available online: https://eur-lex.europa.eu/legal-content/EN/TXT/PDF/?uri=CELEX:32019R1242&from=EN (accessed on 22 August 2022).
3. Breed, A.K.; Speth, D.; Plötz, P. CO_2 fleet regulation and the future market diffusion of zero-emission trucks in Europe. *Energy Policy* **2021**, *159*, 112640. [CrossRef]
4. Speth, D.; Funke, S.Á. Comparing Options to Electrify Heavy-Duty Vehicles: Findings of German Pilot Projects. *WEVJ* **2021**, *12*, 67. [CrossRef]
5. Nykvist, B.; Olsson, O. The feasibility of heavy battery electric trucks. *Joule* **2021**, *5*, 901–913. [CrossRef]
6. Transport & Environment. *Unlocking Electric Trucking in the EU: Recharging in Cities: Electrification of Urban and Regional Deliveries*; Transport & Environment: Brussels, Belgium, 2020.
7. European Commission. Proposal for a Regulation of the European Parliament and of the Council on the Deployment of Alternative Fuels Infrastructure, and Repealing Directive 2014/94/EU of the European Parliament and of the Council. Available online: https://eur-lex.europa.eu/resource.html?uri=cellar:dbb134db-e575-11eb-a1a5-01aa75ed71a1.0001.02/DOC_1&format=PDF (accessed on 22 August 2022).
8. Metais, M.O.; Jouini, O.; Perez, Y.; Berrada, J.; Suomalainen, E. Too much or not enough? Planning electric vehicle charging infrastructure: A review of modeling options. *Renew. Sustain. Energy Rev.* **2022**, *153*, 111719. [CrossRef]
9. Deb, S.; Tammi, K.; Kalita, K.; Mahanta, P. Review of recent trends in charging infrastructure planning for electric vehicles. *WIREs Energy Environ.* **2018**, *7*, e306. [CrossRef]
10. Hakimi, S.L. Optimum Locations of Switching Centers and the Absolute Centers and Medians of a Graph. *Oper. Res.* **1964**, *12*, 450–459. [CrossRef]
11. Toregas, C.; Swain, R.; ReVelle, C.; Bergman, L. The Location of Emergency Service Facilities. *Oper. Res.* **1971**, *19*, 1363–1373. [CrossRef]
12. Gopalakrishnan, R.; Biswas, A.; Lightwala, A.; Vasudevan, S.; Dutta, P.; Tripathi, A. Demand Prediction and Placement Optimization for Electric Vehicle Charging Stations: In Proceedings of the Twenty-Fifth International Joint Conference on Artificial Intelligence (IJCAI-16). Available online: https://www.ijcai.org/Proceedings/16/Papers/442.pdf (accessed on 22 August 2022).
13. Speth, D.; Plötz, P.; Funke, S.; Vallarella, E. Public fast charging infrastructure for battery electric trucks–a model-based network for Germany. *Environ. Res. Infrastruct. Sustain.* **2022**, *2*, 025004. [CrossRef]
14. Funke, S.; Plötz, P. A Techno-Economic Analysis of Fast Charging needs in Germany for Different Ranges of Battery Electric Vehicles. In Proceedings of the European Battery, Hybrid and Fuel Cell Electric Vehicle Congress, Geneva, Switzerland, 17 March 2017.
15. Hodgson, M.J. A Flow-Capturing Location-Allocation Model. *Geogr. Anal.* **1990**, *22*, 270–279. [CrossRef]

16. Cruz-Zambrano, M.; Corchero, C.; Igualada-Gonzalez, L.; Bernardo, V. Optimal location of fast charging stations in Barcelona: A flow-capturing approach. In Proceedings of the 2013 10th International Conference on the European Energy Market (EEM 2013), Stockholm, Sweden, 27–29 May 2013; IEEE: Piscataway, NJ, USA, 2013; pp. 1–6, ISBN 978-1-4799-2008-2.
17. Kuby, M.; Lim, S. The flow-refueling location problem for alternative-fuel vehicles. *Socio-Econ. Plan. Sci.* **2005**, *39*, 125–145. [CrossRef]
18. He, Y.; Kockelman, K.M.; Perrine, K.A. Optimal locations of U.S. fast charging stations for long-distance trip completion by battery electric vehicles. *J. Clean. Prod.* **2019**, *214*, 452–461. [CrossRef]
19. Jochem, P.; Szimba, E.; Reuter-Oppermann, M. How many fast-charging stations do we need along European highways? *Transp. Res. Part D Transport Environ.* **2019**, *73*, 120–129. [CrossRef]
20. Rose, P.K.; Nugroho, R.; Gnann, T.; Plötz, P.; Wietschel, M.; Reuter-Oppermann, M. Optimal development of alternative fuel station networks considering node capacity restrictions. *Transp. Res. Part D Transport Environ.* **2020**, *78*, 102189. [CrossRef]
21. Xi, X.; Sioshansi, R.; Marano, V. Simulation–optimization model for location of a public electric vehicle charging infrastructure. *Transp. Res. Part D Transport Environ.* **2013**, *22*, 60–69. [CrossRef]
22. Whitehead, J.; Whitehead, J.; Kane, M.; Zheng, Z. Exploring public charging infrastructure requirements for short-haul electric trucks. *Int. J. Sustain. Transp.* **2021**, 1–17. [CrossRef]
23. Speth, D.; Sauter, V.; Plötz, P.; Signer, T. Synthetic European road freight transport flow data. *Data Brief.* **2022**, *40*, 107786. [CrossRef] [PubMed]
24. Szimba, E.; Kraft, M.; Ihrig, J.; Schimke, A.; Schnell, O.; Kawabata, Y.; Newton, S.; Breemersch, T.; Versteegh, R.; van Meijeren, J.; et al. ETISplus Database Content and Methodology: ETISplus Deliverable D6. In Proceedings of the project co-funded by the European Commission under the 7th Framework Programme, Zoetermeer, The Netherlands, 1 February 2013. [CrossRef]
25. BAST. Automatische Zählstellen auf Autobahnen und Bundesstraßen: Stundendaten. Available online: https://www.bast.de/BASt_2017/DE/Verkehrstechnik/Fachthemen/v2-verkehrszaehlung/Stundenwerte.html (accessed on 14 December 2019).
26. HoLa. HoLa-Hochleistungsladen Lkw-Fernverkehr: Construction, Operation and Accompanying Research for the First Megawatt Charging Systems for Trucks in Europe. Available online: https://www.hochleistungsladen-lkw.de/hola-en/ (accessed on 6 October 2021).
27. KiD. *Kraftfahrzeugverkehr in Deutschland 2010*; Schlussbericht: Braunschweig, Germany; Flensburg, The Netherlands, 2010.
28. Gnann, T.; Funke, S.; Jakobsson, N.; Plötz, P.; Sprei, F.; Bennehag, A. Fast charging infrastructure for electric vehicles: Today's situation and future needs. *Transp. Res. Part D Transport Environ.* **2018**, *62*, 314–329. [CrossRef]
29. Funke, S.Á. *Techno-ökonomische Gesamtbewertung heterogener Maßnahmen zur Verlängerung der Tagesreichweite von batterieelektrischen Fahrzeugen: Dissertation an der Universität Kassel zur Erlangung des akademischen Grades eines Doktors der Ingenieurwissenschaften (Dr.-Ing.) im Fachbereich 16 Elektrotechnik und Informatik*; Universitätsbibliothek Kassel: Kassel, Germany, 2018; Available online: https://d-nb.info/1158692242/34 (accessed on 22 August 2022).

World Electric Vehicle Journal

MDPI

Article

Technical Feasibility of Heavy-Duty Battery-Electric Trucks for Urban and Regional Delivery in Germany—A Real-World Case Study

Steffen Link [1,2,*] and Patrick Plötz [1]

1 Fraunhofer Institute for Systems and Innovation Research ISI, Breslauer Str. 48, 76139 Karlsruhe, Germany
2 Karlsruhe Institute of Technology, Kaiserstraße 12, 76131 Karlsruhe, Germany
* Correspondence: steffen.link@isi.fraunhofer.de

Abstract: Cutting greenhouse gas emissions to comply with the Paris Agreement is challenging for road freight. While heavy-duty battery-electric trucks (BET) promise tremendous and immediate reduction potential, literature increasingly confirms technical feasibility in general, and several manufacturers launched BET models. However, their real-world application is still being questioned by fleet owners due to the limited range or payload penalties. Thus, our case study aims to assess the technical feasibility of urban and regional delivery in Germany based on real-world and per-vehicle operational data that feed into an energy simulation with Monte-Carlo modeling. Our results demonstrate the importance of vehicle-specific examination for the right battery capacity that ideally matches the vehicle's operating profile. We find that full electrification may be most accessible for 18-t and 26-t rigid solo trucks, soon followed by tractor-trailers, while truck-trailers turn out as most challenging. With up to 600 kWh battery capacity available in all truck classes, we find nearly 40% of all transport performance and 60% of all diesel trucks may be replaced with BET—while already 400 kWh is sufficient for half of all trucks. Additional measures such as intermediate charging and adjusted and more flexible truck-tour allocation may significantly accelerate electrification.

Keywords: technical feasibility; heavy-duty; battery-electric trucks; energy simulation; Monte-Carlo simulation; real-world tour data; battery size

Citation: Link, S.; Plötz, P. Technical Feasibility of Heavy-Duty Battery-Electric Trucks for Urban and Regional Delivery in Germany—A Real-World Case Study. *World Electr. Veh. J.* **2022**, *13*, 161. https://doi.org/10.3390/wevj13090161

Academic Editor: Joeri Van Mierlo

Received: 5 August 2022
Accepted: 18 August 2022
Published: 27 August 2022

1. Introduction

A broad consensus has been reached that cutting greenhouse gas (GHG) emissions rapidly and eventually reaching climate neutrality by 2050 is essential to comply with the Paris Climate Agreement, i.e., to keep the global mean temperature below 1.5 °C, and thus mitigate the anthropogenic climate change. Today, despite their minor significance in the total vehicle fleet, heavy-duty vehicles contribute about 8% of the total EU GHG emissions [1]. While several technological pathways for zero-emission trucks exist, heavy-duty battery-electric trucks (BET) benefit from the technological experience and recent battery innovations—i.e., costs, volumetric and gravimetric energy density, and fast charging capability [2–4]—and, thus, short-term large-scale availability [5]. The increasing European manufacturer commitment toward BET further accentuates this shift [6].

The literature increasingly confirms technical feasibility in general. While several studies imply a great potential for urban and regional delivery with a daily mileage lower than 400 km [3], most recent studies even see long-haul transport close to a threshold where BETs become feasible [3,4,7]. Despite this commitment and literature-proofed feasibility, truck fleet owners are still questioning the technical feasibility of BETs for their application in light of limited vehicle range, insufficient public charging infrastructure, and payload restrictions [8,9]. This individual reservation demands a shift from generalized feasibility assessments based on synthetic operating schedules [10], fleet analyses [11,12], or standardized driving profiles and generic use patterns [2–4,7,13]. Thus, we aim to assess the technical feasibility of BETs with a comprehensive case study by using real-world and per-vehicle operational data. This answers which shares within the truck fleet may be replaced by BET today and which technological or operational factors may boost or limit electrification.

The paper is structured as follows: First, we introduce the data and describe our methods to evaluate the technical feasibility in Section 2. Section 3 continues with data evaluations and presents our results. Section 4 discusses our results and their representativeness. We close with conclusions and recommendations in Section 5.

2. Data and Methodology

2.1. Operational Vehicle Data

We use tour data from the food retail industry that is provided by commercial tour scheduling software (CSV format) for two depots in the northeast region of Germany. Our sample covers one month from 2021, roughly 9500 individual tours, 543 retail stores, around 1 million km, and 224 different trucks operating within 220 km around Berlin. These are refrigerated trucks with a gross vehicle weight (GVW) of 18 or 26 tons as rigid trucks, truck-trailer combinations, and tractor-semitrailer combinations. The data cover the planned temporal sequence, routing, and payload information. Each tour starts and ends at either one of both depots. Based on timestamp information, we chain these individual tours to more than 4000 daily tours. While Depot1 primarily supplies Berlin and partially the metropolitan area, Depot2 additionally supplies the entire northeast region. Figure 1 shows both depot locations within the northeast region. Evaluations and statistics are provided in Section 3.1.

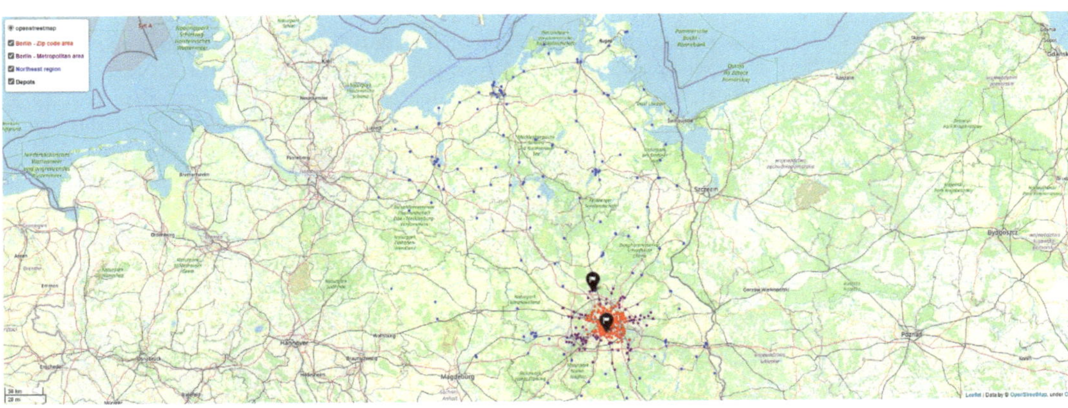

Figure 1. Data sample—Maps of the northeast region of Germany. Points represent individual retail stores; Red: Retail stores in the Berlin zip code area (share: 50%); Purple: Berlin metropolitan area (40 km) (share: 20%); Blue: Other retail stores (share: 30%)—own illustration based on leaflet.

2.2. Methodology

2.2.1. Energy Simulation

The technical feasibility involves a tour-specific energy simulation for each truck. First, we use the simplified mathematical-physical vehicle model by [14] to account for vehicle dynamics and energy losses related to aerodynamic drag forces, frictional forces, and inertial forces (Equation (1)). Second, we incorporate energy demand from both accessories and refrigeration as well as restrictions to the depth of discharge (DoD) and minimum residual range requirements (Equation (2)). Parameter and values are shown in Tables 1 and 2.

$$E_{driving} = \left[\begin{array}{c} \frac{\left(\frac{1}{2} \cdot \rho \cdot C_D \cdot A \cdot v_{rms}^3 + c_{rr} \cdot m_T \cdot g \cdot v_{av} + \partial_\alpha \cdot m_T \cdot g \cdot v_{av}\right)}{\eta_{bw}} + \\ \partial_{Reku} \cdot m_T \cdot a_{av} \cdot v_{av} \cdot \left(\frac{1}{\eta_{btw}} - \eta_{btw} \cdot \eta_{brk}\right) \end{array} \right] \cdot \frac{D}{v_{av}} \quad (1)$$

$$E_{req} = \frac{E_{Driving} + P_{Aux} \cdot t_{Driving} + P_{Cool} \cdot (t_{Driving} + t_{Stopp}) + E_{Residual}}{\eta_{DoD}} \quad (2)$$

$E_{driving}$ represents the net battery capacity [kWh] required to overcome all driving resistances. Major parameters are vehicle drag coefficient (C_D), vehicle frontal area (A), mean vehicle velocity (v_{av}), root-mean-square velocity (v_{rms}), mean vehicle acceleration (a_{av}), tire rolling resistance coefficient (c_{rr}), trip distance (D), and total vehicle mass (m_T). Truck specifications such as aerodynamics, tire rolling resistance, and diesel chassis curb weight are based on empirical data [14] and aggregated per truck class—lower quantile (Q25), median (Q50), and upper quantile(Q75). Tour-specific parameters such as payload, mean vehicle velocity, trip distance, or timestamps for driving and stopping are taken from the tour scheduling software. Parameter (∂_α) approximates the average road gradient per tour that is calculated based on truck routing software [15]. This involves a piecewise linearization of the reconstructed tour (500 m steps) and the calculation of distance-weighted quantiles (Q25 and Q75) as minimum and maximum values. We adopt the most likely value from [16]. The root-mean-square velocity is calculated using the Steiner–König–Huygens theorem, including crosswind influence. The mean vehicle acceleration is approximated by cycle-specific values based on 18 different American driving cycles [17]. Since urban driving usually features higher dynamics due to stop-and-go traffic, traffic lights, or planned stops per distance than regional deliveries, we distinguish between both use cases. η_{BTW} denotes the battery-to-wheels efficiency and summarizes battery discharge efficiency $(\eta_{Battery})$ and drivetrain efficiency $(\eta_{Drivetrain})$. The proportion of recoverable energy is specified via $\partial_{Reku} \cdot \eta_{brk}$ accounts for additional braking losses. m_T represents the total truck weight, comprising truck curb weight, potential trailer curb weight, and tour-specific payload [kg]. Here we apply a top-down approach starting from diesel truck values, suppose a common vehicle chassis, and then calculate the possible BET curb weight. Therefore, we subtract all major diesel powertrain-related components such as the ICE, gearbox, and fuel tank from the diesel curb weight (m_{Curb_D}). Afterward, we add major electric powertrain components such as motor and battery. For truck-trailers and tractor-trailers, we add the trailer curb weight. We close by adding the tour-specific payload weight.

E_{req} represents the gross battery capacity [kWh] required to complete any tour successfully. We approximate additional per-vehicle energy demand from accessories (pneumatics, hydraulics, heating and air conditioning (HVAC), on-board power grid) by using mean power consumption (P_{Aux}) per driving time $(t_{Driving})$—following [18,19]. Likewise, the energy demand for refrigeration (P_{Cool}) is calculated. However, continuous refrigeration is needed even during stationary periods at the retail stores (t_{Stopp}). The mean power consumption—generally highly dependent on the temperature delta, the cooling volume, and the frequency of opening and closing—is calculated based on the ATP/DIN 8959. The temperature difference is assumed to be 25 °C. The customer-specific safety margin corresponds to about 20–30 km of residual range.

Main uncertainties result from the energy consumption formula (Equations (1) and (2)) based on the simplified vehicle model, uncertain technical vehicle parameters, and variable and dynamic real-world operating conditions. We follow [14] and cast all major parameters using individual PERT distributions to increase robustness instead of running sensitivity analyses of selected parameters afterward. Minimum, most likely, and maximum values are specified based on empirical data, literature values, or assumed to spread ± 20%. Finally, we perform a standard Monte Carlo simulation for each tour (n = 100). Tables 1 and 2 indicate the chosen approach and parameter spreads. We perform all calculations on a standard Lenovo notebook with i7-8565U @1.8 GHz and 16 GB RAM.

Table 1. Truck-class-specific simulation parameters. Ranges indicate the PERT distribution's minimum, most likely, and maximum values. Individual parameters are constant values.

Parameter	Unit	18-t Rigid	26-t Rigid	Truck-Trailer	Tractor-Trailer	Source
m_{Curb_D}	[kg]	5761–6475–7125	8239–8679–9073	8239–8679–9073	5761–6475–7125	Q25-Q50-Q75 [14]
$m_{Trailer}$	[kg]	-	-	6500 ± 20%	8500 ± 20%	derived from [18]
$C_D{\cdot}A$	[m^2]	5.559–5.698–5.837	5.463–5.997–5.737	6.557–7.839–9.179	5.559–5.698–5.837	Q25-Q50-Q75 [14]
c_{rr}	[N/kN]	5.5–5.7–6.9	5.0–5.6–6.8	5.0–5.6–6.8	4.9–5.1–6.5	Q25-Q50-Q75 [14]
P_{Aux}	[kW]	2.97 ± 20%	3.39 ± 20%	4.32 ± 20%	4.11 ± 20%	[18,19]
P_{Cool}	[kW]	3.11 ± 20%	3.11 ± 20%	5.90 ± 20%	5.14 ± 20%	ATP/DIN 8959
P_{Motor}	[kW]	200–228–265	265–323–350	265–323–350	331–355–368	Q25-Q50-Q75 [14]
v_{Std}	[m/s]	0.413	0.417	0.744	0.677	[-]

Table 2. Other simulation parameters. Ranges indicate the PERT distribution's minimum, most likely, and maximum values. Individual parameters are constant values.

Parameter	Unit	Value/Value Range	Source
η_{DoD}	[%]	90% ± 5 %	[13]
ρ_{Bat}	[Wh/kg]	150–175–225	[13,16,20]
∂_{Reku}	[%]	50% ± 10%	[16]
a_{av}	[m/s^2]	Urban: 0.331 ± 20%, Regional: 0.160 ± 20%	Q25 and Q75 [17]
η_{BTW}	[%]	$\eta_{BTW} = \eta_{Battery}$ (95% ± 2.5%)$\cdot\eta_{Drivetrain}$ (90% ± 2.5%)	[13,16]
η_{brk}	[%]	97%	[16]
v_{RMS}	[m/s]	$= \sqrt{v_{av}^2 + v_{std}^2} + v_{Wind}$ (3 ± 20%)	Based on [16] and VECTO [21]
$E_{Residual}$	[kWh]	30	Own assumption
p	[kg/m^3]	1.15–1.225–1.3	Own assumption
m_{Emot}	[kg/kW]	1.43	[22]
m_{DPT}	[kg]	$= m_{Gearbox}$ (300 kg) + m_{Tank} (108 kg) = 408 kg	Based on [2]
m_{ICE}	[kg/kW]	3.3	[23]
m_{PL}	[kg]	Base value from the truck schedule (± 20%)	
$P_{Charge,Dep}$	[kW]	$\in \{50, 150, 250, 350, 450, 1000\}$	
$P_{Charge,CR}$	[kW]	150	
$r_{PlanVSActual}$	[%]	75% ±10%	Own assumption
η_{NCP}	[%]	68.1% (184/270)–82% (164/200)–92.6% (250/270)	[24]

2.2.2. Assumptions and Premises

There is no tour optimization, truck re-allocation, or adjusted scheduling for all calculations. Tours are presumed to be exactly as of February 2021 so that potential BET would mimic the existing diesel truck schedule. There is no opportunity for public charging during mandatory driving breaks, as these coincide with the stops at customer retail stores. Thus, private or semi-public charging infrastructures may be the most promising possibility to avoid time losses.

2.2.3. Feasibility Calculations and Scenarios

We simulate gross battery capacities between 100 and 800 kWh in 50 kWh increments per truck and compare those against the required gross battery capacity (E_{req}). This happens for each tour. We include higher permitted GVW limits for BETs in Germany, i.e., up to 2 tons depending on truck class [25]. If E_{req} is lower than the simulated battery capacity and the permitted GVW is not exceeded, this simulation run is labeled as "technically feasible". If at least 95% (i.e., 95 out of 100) of all simulation runs are labeled as "technically feasible", we label this tour as "technically feasible". If all tours per vehicle are labeled "technically feasible", we affirm the BET replacement for this truck and denote the simulated gross battery capacity. This approach is rather restrictive, as just one daily tour might negate BET replaceability.

We investigate three scenarios in total. Our base scenario (S0) assumes that vehicles depart from the depot fully charged in the morning, and installed gross battery capacities must be sufficient throughout the day. Additionally, we investigate the effect of private charging opportunities to extend vehicle coverage. Thus, we integrate potential intermediate depot-charging within the depot premises (S1) at the beginning of each tour that might happen directly at the cargo terminals during vehicle commodity loading ($t_{Loading}$). Additionally, we investigate the effect of charging opportunities at individual customer retail stores (S2), where the stop time (t_{Stopp}) at the local cargo terminals might be used for charging. We still assume vehicles depart from the depot fully charged in the morning. Vehicles recharge without time losses in both scenarios, and the tour schedule is maintained. We approximate the state of charge (SoC) evolution throughout any tour based on consumed energy (driving, accessories, and cooling) per traveled distance. For depot charging, we consider six different peak charging powers ($P_{Charge,Dep}$). For individual retail stores, we explore the effect of 150 kW peak charging power ($P_{Charge,CR}$) at any retail store. The latter might align with a joint passenger car charging infrastructure deployment at local customer parking lots. We use an average charging power across the whole SoC corridor based on peak charging power, empirical findings on the average net charging power η_{NCP} from passenger cars, and a 2C charging rate limit. Parameters are included in Table 2.

3. Results

3.1. Operating Patterns: Daily Mileage and Timestamps

The daily operating distances are visualized in Figure 2. The visualization splits per truck class and depot location and involves aggregated boxplots and individual values as scatter. The daily operating distance typically ranges from 46 to 105 km (25% and 75% quantile) for Depot1 and from 143 to 384 km for Depot2. Across both depots, we typically find 1–5 daily tours per truck. Each tour usually serves 1–4 customer retail stores. Rigid trucks (18 t and 26 t) focus on Berlin and the metropolitan area, while truck-trailer and tractor-semitrailer combinations also supply the entire northeast region. Daily mileages in the urban and metropolitan deliveries are usually less than 200 to 300 km, while from 500 to 700 km may be a typical upper limit for regional deliveries. However, over 800 to 1200 km are possible in multi-shift and cross-daily operations.

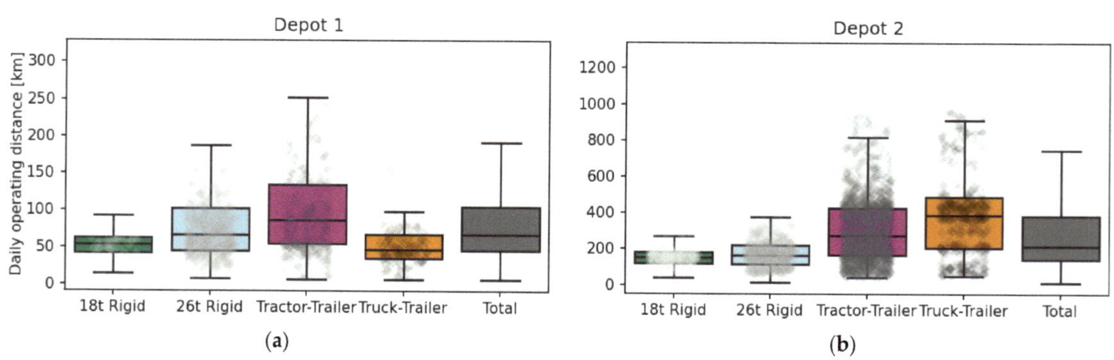

Figure 2. Evaluation of daily operating distances per truck class; (**a**) Depot1; (**b**) Depot2; Sample points are scattered. Boxplots include the lower quartile, median, upper quartile, and whiskers (±1.5 IQR); Colors indicate different truck classes: 18-t rigid (green), 26-t rigid (blue), tractor-trailers (purple), truck-trailers (orange), and total (grey)—own illustration.

The corresponding daily operating times are visualized in Figure 3. Vehicle scheduling specifies four timestamps, from vehicle loading at the cargo terminals within the depots ($t_{Loading}$), driving time ($t_{Driving}$), stop time at customer retail stores (t_{Stopp}), and eventually vehicle unloading at the cargo terminals to complete one single tour. The visualization aggregates depot locations and visualizes each truck class separately. While vehicle loading typically takes 60–80 min (lower and upper quartile) for rigid solo trucks, 90–110 min for tractor trailers, and 110–120 min for truck-trailers. Similar to the daily mileage, truck trailers and tractor trailers exhibit higher driving times and wider span compared to rigid solo trucks. Stops at customer retail stores typically last between 40–60 min across all truck classes, with over 80 min occurring as well, especially for larger truck classes. Eventual unloading takes only 15–22 min across all truck classes. As mentioned earlier, additional breaks such as the mandatory 4.5 h driving break are not scheduled as these are covered at customer stops.

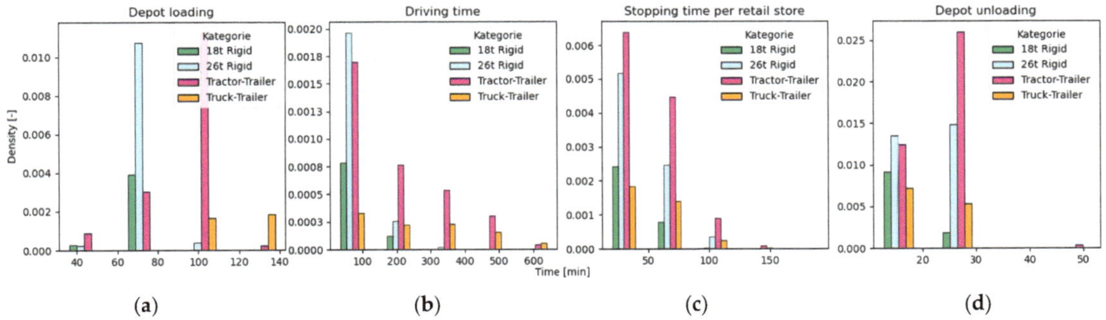

Figure 3. Evaluation of vehicle operating times per tour across both depots and per truck class; (**a**) Vehicle loading at the depot; (**b**) Driving time; (**c**) stopping time per retail store; (**d**) Vehicle unloading at the depot—own illustration.

To derive net charging times from $t_{Loading}$ and t_{Stopp}, we need to approximate potential time losses for waiting, vehicle docking, or connecting to the charging station. Here, we refer to the difference between our tour data's planned and actual data. We find certain deviations between planned and actual data so that specified planned timestamps are undercut, and only 65% to 85% of the specified planned times are usually obtainable. Thus, we multiply $t_{Loading}$ and t_{Stopp} and reduce the possible charging time by this fraction ($r_{PlanVSActual}$). For the solo trucks, only 39–68 min would be available during vehicle loading at the depot (instead of the 60–80 min). In addition, we set 10 min as the minimum net charging time to ensure practical implementation.

3.2. Simulation Results: Specific Energy Consumption

Figure 4 shows our simulation results for the specific energy consumption. The left side visualizes the energy consumption as a density plot per truck class over different battery capacities to illustrate the spread and influence of larger batteries. Note that the density plots extrapolate beyond simulated battery capacities. The right side shows the sample-weighted boxplot per truck class and aggregated over all battery capacities, whereas outliers are removed. All plots comprise only feasible tour-battery combinations. The sample-weighted median spans from 1.01 kWh/km for the 18-t rigid truck, 1.14 kWh/km for the 26-t rigid truck, and 1.52 kWh/km for tractor-trailers, to 1.66 kWh/km for the truck-trailer combinations. The higher variation for tractor-trailers and truck-trailer is striking and may be attributed to broader territorial coverage and thus operating patterns. Depending on the truck class and total operating times, around 0.1 to 0.4 kWh/km on average may be attributed to accessories and refrigeration.

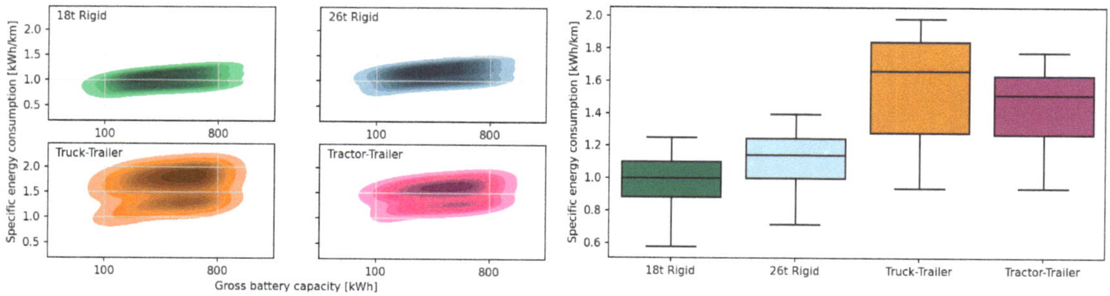

(a) (b)

Figure 4. Trip-averaged specific energy consumption [kWh/km] per truck class. (**a**) Density plot over battery capacity (**b**) Sample-weighted boxplot per truck class. Boxplots indicate the lower quartile, median, upper quartile, and whiskers (±1.5 IQR); Colors indicate different truck classes: 18-t rigid (green), 26-t rigid (blue), tractor-trailers (purple),and truck-trailers (orange)—own illustration.

3.3. Scenario 0: Technical Feasibility without Intermediate Charging

The effect of different battery capacities on BET feasibility aggregated per truck class is visualized in Figure 5. The y-axis (CDF—cumulative density function) indicates the share of feasible trucks that would cope with this or less gross battery capacity throughout each vehicle's daily tours. We remark that we require all daily tours per vehicle to be technically feasible to affirm its BET replaceability.

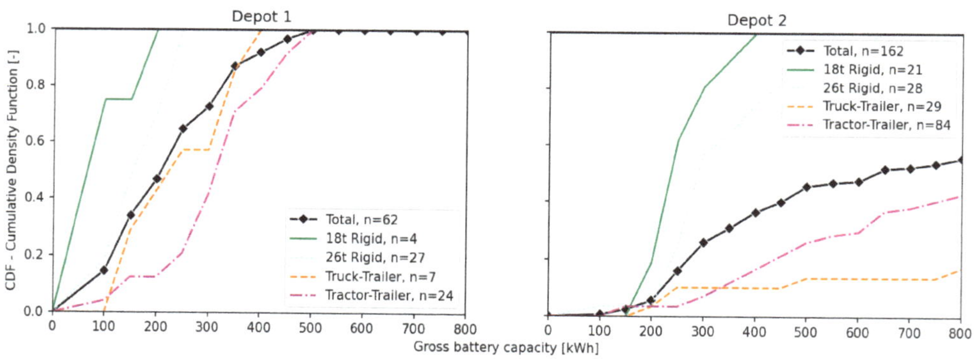

(a) (b)

Figure 5. BET feasibility per truck class as CDF over gross battery capacity; (**a**) Depot1; (**b**) Depot2—own illustration.

For Depot1, the technical feasibility is high given that primarily urban and partially metropolitan operating areas and daily operating distances are usually below 300 km. Battery capacities from 100 to 200 kWh for 18-t rigid trucks and 100 to 250 kWh for 26-t rigid trucks are sufficient for full electrification. Truck-trailers (400 kWh) and tractor-trailers (550 kWh) require slightly larger batteries. The highest gains are reached between 250 and 350 kWh. Up to 400 kWh is sufficient to affirm 80% BET replaceability for the whole fleet. If we had required all simulation runs per tour to be technically feasible instead of 95%, we would have increased the capacities for full electrification to 450 kWh for 26-t rigid trucks and 550 kWh for truck-trailers, and 650 kWh for tractor-trailers. Up to 400 kWh would then be sufficient to affirm 75% BET replaceability for the whole fleet. The GVW limits may be exceeded for rigid trucks at around 450 kWh and 650 kWh for tractor-trailers. We conclude that around 400 kWh may already be sufficient to replace over three-quarters of the truck fleet, while around 600 kWh may be sufficient to replace the entire one.

For Depot2, the technical feasibility is significantly lower given the greater operating area in the northeast region, so larger batteries are required. Battery capacities from 100 to 400 kWh for 18-t rigid trucks and 200 to 650 kWh for 26 t rigid trucks are sufficient for full electrification. The vast majority (80%) of rigid trucks require 200 to 350 kWh (18-t rigid) or 200–450 kWh (26-t rigid). For truck-trailers and tractor-trailers, full electrification even fails. Around one-third of all tractor-trailers may be replaced with up to 600 kWh. With up to 800 kWh, only 18% of truck-trailers and 42% of tractor-trailers may be replaced. Overall, 37% of the total fleet may be electrified with up to 400 kWh, 48% with 600 kWh, and 56% with 800 kWh. If we had released our 95% feasibility threshold per tour, fleet replaceability for Depot2 would have dropped to 29% (400 kWh), 42% (600 kWh), or 50% (800 kWh).

If we consolidate both truck fleets, we find 52% BET replaceability with 400 kWh, 62% with 600 kWh, and 68% with 800 kWh. If we had required all simulation runs per tour to be technically feasible, the total fleet replaceability would have dropped to 44% (400 kWh), 57% (600 kWh), or 63% (800 kWh). Thus, given the available and announced battery capacities across all truck classes, we conservatively conclude that at least half of the total truck fleet is suitable for ad-hoc electrification.

The following section discards aggregating tours to trucks and focuses solely on the feasibility of individual tours to answer how many of the 4000+ daily tours would be feasible if examined separately. Through this, we want to approximate the effect of a modified tour planning (sequence and vehicle assignment). Daily trip chains remain untouched, and tours are still presumed to be exactly as of February 2021. The effect is visualized in Figure 6. Here we distinguish between simple tour count and the associated transport performance (i.e., ton-kilometers) per tour as a common metric in transport statistics. The y-axis indicates the share of feasible daily tours or electrifiable transport performance with this or less gross battery capacity.

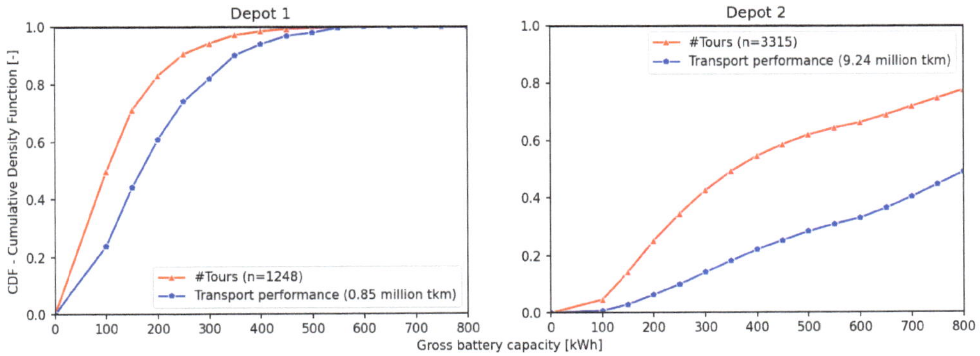

(a) (b)

Figure 6. BET feasibility on tour- and tkm-level as CDF over gross battery capacity; (**a**) Depot1; (**b**) Depot2—own illustration.

For Depot1, battery capacities between 200 and 300 kWh are sufficient to electrify 83 to 94% of all tours. Up to 400 kWh are sufficient to electrify almost all tours (99%) and, as the previous analysis has shown, 550 kWh are required for full electrification. Feasibility in terms of transport performance is lower, most likely because this adds information in terms of short and light-loaded tours versus long and heavily loaded tours instead of equal counting. However, 400 kWh is still sufficient to electrify 94% of all ton-kilometers. If we had required all simulation runs per tour to be technically feasible, the effect would have been negligible as almost 98% of all tours, or 92% of all ton-kilometers, respectively, would have been feasible with up to 400 kWh.

For Depot2, over one-half of all tours may be electrified with about 400 kWh and up to 66% with 600 kWh. Up to 78% may be possible with 800 kWh. The difference between tour count and transport performance is considerably bigger than for Depot1. Feasibility is around 23% with up to 400 kWh, 33% with up to 600 kWh, and 49% with 800 kWh.

If we consolidate both depots, we find 65% of all tours feasible with up to 400 kWh, 75% with 600 kWh, and 84% with 800 kWh. Equivalently in terms of transport performance, this is 29% with up to 400 kWh, 39% with 600 kWh, and 53% with 800 kWh. Thus, given the available and announced battery capacities across truck classes, we conservatively conclude that roughly two-thirds of all tours and roughly one-third of transport performance may be feasible for ad-hoc electrification.

To summarize, our base scenario highlights two main findings: (1) There is no one battery capacity per truck class, even within one fleet. Thus, a vehicle-specific examination for the right battery capacity that ideally matches the vehicle's operating profile is crucial to avoid over- or undersized battery capacities. (2) If the tour-truck allocation is neglected, the tour feasibility is significantly higher than on the truck level. Most likely, only a few unfeasible tours may be the crunch. This implies certain potential by re-allocating daily tours within the truck fleet.

3.4. Scenario 1: Technical Feasibility with Intermediate Depot Charging

For our scenario evaluation, we aggregate both depots. Results are visualized in Figure 7, where the color scale indicates BET replaceability. The median (50% threshold) is given in white. We limit to truck-level evaluation across all simulated gross battery capacities, as findings from our base scenario regarding tours and transport performance react similarly. The y-axis indicates the charging power ($P_{Charge,Dep}$). Thus, the x-axis matches the base scenario.

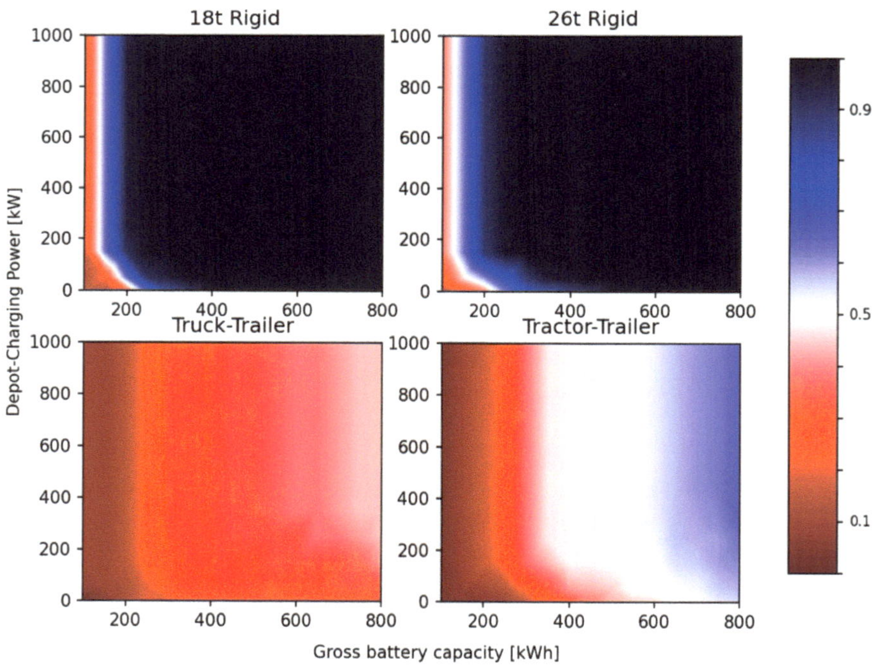

Figure 7. BET feasibility per truck class—scenario 1. Variations for battery capacity [kWh] and depot-charging power [kW]; color bar indicates technical feasibility from 0% to 100%—own illustration.

Intuitively, higher charging power leads to higher feasibility with smaller batteries. However, a substantial effect saturates beyond 350 kW, and we see higher sensitivity towards installed battery capacity rather than charging power. While there is hardly any effect on rigid solo trucks above 200 kWh, depot charging particularly boosts BET replaceability for truck-trailers and tractor-trailers. While truck-trailers do not even reach the 50% threshold, tractor-trailers may reach up to 66%. The broad 50% corridor at 400 to 600 kWh is striking here. This shows the trade-off between installed battery capacity and intermediate depot charging paired with higher charging power, yet the replaceability effect on the fleet remains equal.

Considering all 224 trucks, with up to 600 kWh maximum gross battery capacity and 350 kW available depot fast charging, we may replace nearly 70% of the whole fleet with BET, and almost half the transport performance may be electrified (i.e., +10% compared to the base scenario).

3.5. Scenario 2: Technical Feasibility with Intermediate Charging at the Depot and Customer Retail Stores

Figure 8 visualizes the results of our second scenario and the layout follows the previous logic from Figure 7. However, as introduced in the methodology section, we additionally explore the effect of charging opportunities at customer retail stores ($P_{Charge,CR}$ = 150 kW).

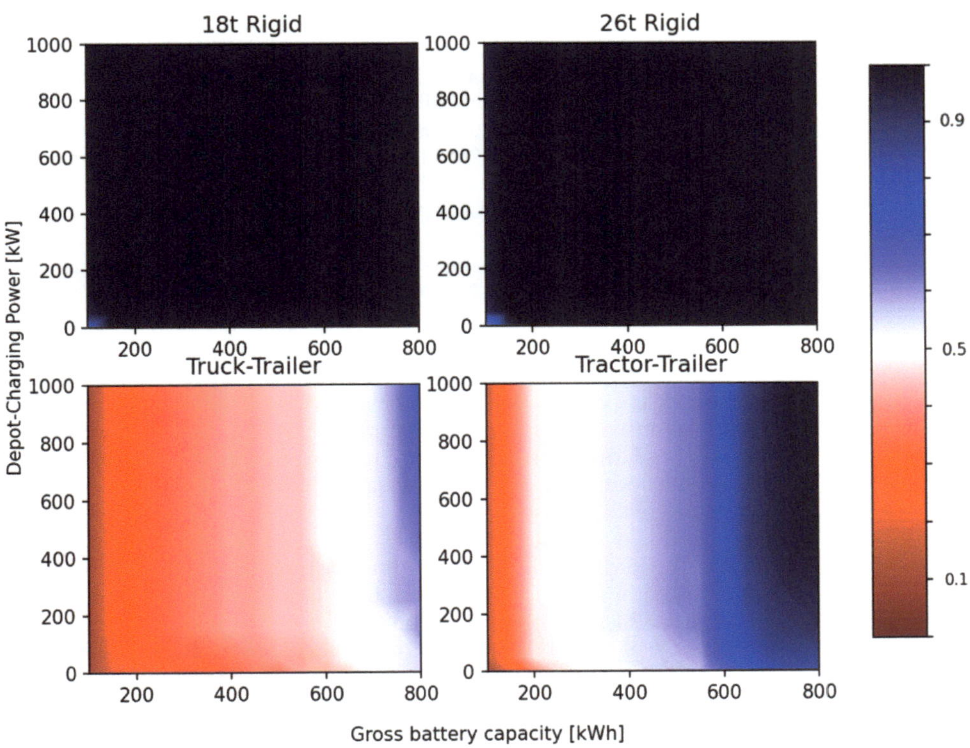

Figure 8. BET feasibility per truck class—scenario 2. Variations for battery capacity [kWh] and depot-charging power [kW] plus charging at retail stores (fixed, 150 kW; color bar indicates technical feasibility from 0% to 100% —own illustration.

The additional charging at customer retail stores may enable full electrification for solo trucks (18 t and 26 t rigid) and tractor-trailers, while significantly smaller battery capacities are needed. This effect is most apparent in the 50% corridor for tractor-trailers, which decreases to 200–400 kWh instead of 400–600 kWh. The substantial effect of higher depot charging power already saturates beyond 150 kW. Truck-trailers show the lowest feasibility, and full electrification fails as only 72% may be reached.

Considering all 224 trucks, with up to 600 kWh maximum gross battery capacity, 350 kW available depot fast charging, and 150 kW fast charging at retail stores, we may replace over three-quarters of the whole fleet with BET and roughly 70% of the transport performance may be electrified (i.e., +30% compared to the base scenario).

4. Discussion

The following discussion includes tour data, representativeness, energy consumption modeling, charging assumptions, and battery aging.

(1) Tours and vehicle allocation are presumed to be exactly as of February 2021 so that potential BET would mimic the existing diesel truck schedule rather than using a tour schedule optimized for BET (e.g., allocate energy-consuming tours to available trucks with highest SoC). Plus, daily tour chains are untouched.

(2) We acknowledge that our food retail case study may not represent the entire European or even German distribution logistics and road freight transport since each industry has its unique characteristic usage patterns and constraints. However, we analyzed four relevant truck classes covering about 87% of the German N3 truck stock [25]. The studied region (i.e., Berlin, Mecklenburg-Western Pomerania, and Brandenburg) covers a relevant share of German land area (15%) and total population (9%). Plus, distribution logistics for the food retail industry covers up to a quarter of the annual transport performance in Germany (2019: 498 billion tkm) [26]. However, our sample covers less than 1%. Our calculated annual mileage is typically from 15,000 km to 124,000 km (10% and 90% quantile), with a mean value of 56,000 km and a median of 42,000 km. Our annual mileage is lower than official statistics [27] and driving data surveys [28]. However, pure long-haul transport is missing in our data but included in the others.

(3) Uncertainties for our simulated energy consumption and battery sizes result from the simplified vehicle model, parameter values from standard driving cycles, underlying generic BET technical specifications, and neglected dynamics (i.e., vehicle-related, speed profile, or ambient conditions). Since exact time- or distance-based vehicle speed profiles are missing from such tour data, we refrain from setting up a dynamic component-based simulation, such as used by [2]. Since we define the SoC as a function of traveled distance, we ignore any fluctuations that may limit real-world feasibility, nor did we simulate different seasonal effects. Even though we performed a Monte-Carlo simulation to increase robustness, there are no dynamics within one trip, and we may have incorporated only certain variations and irregularities. However, our results, taking the tractor-trailer as an example, are consistent with other studies indicating ranges between 1.2 and 1.9 kWh/km [2,3,7,10], which underlines our results.

(4) For convenience, we assume that charging is available at all retail stores. Plus, all cargo terminals for depot loading are equipped with charging infrastructure. Thus, we assume charging availability at any time. While this mirrors a full rollout, it seems intuitive that charging infrastructure may not be built at all cargo terminals or cannot be built at all retail store locations due to different constraints (e.g., costs, available space, or grid connection). Different optimization approaches may be used to determine the most relevant locations or the optimal number of equipped cargo terminals.

(5) We neglect battery aging effects (i.e., cyclic and calendar). Typically, the battery state-of-health (SoH) would decrease to 70–80% toward an ending truck service life. This impacts technical feasibility, assuming that the truck-tour allocation remains identical throughout the service life. In contrast, assuming more variable vehicle planning, the latest trucks might master the more difficult daily tours given a typically ongoing truck fleet renewal, while older trucks perform on easier daily tours (i.e., SoH-based tour allocation). To approximate aging, one might choose the next higher battery increment (+50 kWh) when affirming the feasible battery capacity threshold per truck.

5. Conclusions and Recommendations

Our case study quantifies the technical feasibility of BET for urban and regional delivery in Germany, covering over 9000 real-world tours, over 540 customers, and more than 200 heavy trucks from four different truck classes. Even though trucks are operating within only 220 km around Berlin, we see 200–300 km as daily mileage in urban delivery and 500–700 km as typical for regional deliveries, while even higher distances are possible in multi-shift and cross-daily operations. The opportunity for public charging at service areas and parking lots during the mandatory driving breaks may be limited for those delivery types, an effect possibly not limited to this case study and industry. Thus, private or semi-public charging infrastructure may be most promising to avoid time losses and stick to current schedules.

We find high potential for BET feasibility even if we mirror the existing diesel truck operating schedule. With up to 600 kWh and no additional charging infrastructure, we reach 39% of electrified transport performance and may replace nearly 60% of all trucks. Full electrification may be most accessible for rigid solo trucks (18 t and 26 t) and tractor-trailers. Truck trailers remain a concern. We find no one battery capacity per truck class but high heterogeneities, even within one fleet. Thus, we recommend that fleet owners and shippers carefully evaluate the modular battery capacities offered by the manufacturers to find the most suitable capacity per truck to avoid over- or undersized batteries. However, this might also limit flexible and universal vehicle planning.

Interim charging options at the depot (S1) and at individual retail stores (S2) boost feasibility and the share of electrified tkm. However, we find an overall higher sensitivity to battery capacity given long individual journeys than additional depot charging. In any case, overnight charging at the depot is crucial. Intermediate charging offers little added value for urban delivery, and the effect is greater for regional deliveries. The same applies to rigid solo trucks versus truck-trailers as well as tractor-trailers. The individual effect of each scenario measure is larger than the combined effect and, thus, should be balanced against each other. However, both may be necessary in selected cases to achieve full electrification. Apart from intermediate charging, we highlight the potential through tour optimization, truck re-allocation, and adjusted tour schedules. These imply significant potential without additional structural measures and costs (note: neglecting planning costs).

Given our findings, representativeness, and the literature-proofed general feasibility, we recommend all fleet owners and shippers to start examining their transition to climate-friendly commercial vehicles. Recent studies already indicate cost-effective operations of BET as of today or in the near future [29,30], currently especially in light of the notable subsidies for vehicles and charging infrastructure (e.g. in Germany: 80% refund for extra costs compared to a comparable diesel-powered truck as well as 80% of infrastructure-related expenses [31]). We emphasize the necessity of finding the right battery capacity per truck by analyzing its operational patterns and ad-hoc potential through tour optimization, adjusted and variable truck-tour allocation (i.e., SoC- and SoH-based), or even mixed fleet considerations with BETs and diesel trucks. Further practical BET research should focus on more case studies from other relevant industries, highlight custom pitfalls in daily operations, and enhance techno-economic evaluations.

Author Contributions: S.L.: Conceptualization, Methodology, Software, Writing—Original Draft. S.L. is the first author. P.P.: Methodology, Supervision, Writing—Review & Editing. All authors have read and agreed to the published version of the manuscript.

Funding: This work was sponsored by the Federal Ministry of Education and Research Germany within the project "BetterBat" under grant number 03XP0362C.

Institutional Review Board Statement: Not applicable.

Informed Consent Statement: Not applicable.

Data Availability Statement: Not applicable.

Acknowledgments: This research extends the ZeroEmissionDeliveries—Berlin project funded by the European Climate Foundation (ECF) and managed by Transport & Environment (T&E) Germany. The authors thank the REWE Group for kindly providing the data and supporting their work.

Conflicts of Interest: The authors declare no conflict of interest.

Abbreviations

BET	Heavy-duty battery-electric trucks
GHG	Greenhouse gas
GVW	Gross vehicle weight
SoC	State of charge (battery)
SoH	State of health (battery)
tkm	ton-kilometers

References

1. European Commission. *EU Transport in Figures: Statistical Pocketbook 2020*; European Commision: Luxembourg, 2020. Available online: https://op.europa.eu/en/publication-detail/-/publication/da0cd68e-1fdd-11eb-b57e-01aa75ed71a1 (accessed on 1 August 2022).
2. Mareev, I.; Becker, J.; Sauer, D. Battery Dimensioning and Life Cycle Costs Analysis for a Heavy-Duty Truck Considering the Requirements of Long-Haul Transportation. *Energies* **2018**, *11*, 55. [CrossRef]
3. Phadke, A.; Khandekar, A.; Abhyankar, N.; Wooley, D.; Rajagopal, D. *Why Regional and Long-Haul Trucks are Primed for Electrification Now*; Lawrence Berkeley National Laboratory: Berkeley, CA, USA, 2021. Available online: https://escholarship.org/uc/item/3kj8s12f (accessed on 1 August 2022).
4. Nykvist, B.; Olsson, O. The feasibility of heavy battery electric trucks. *Joule* **2021**, *5*, 901–913. [CrossRef]
5. International Energy Agency. *Global EV Outlook 2021: Accelerating Ambitions Despite the Pandemic*; International Energy Agency: Paris, France, 2021. Available online: https://iea.blob.core.windows.net/assets/ed5f4484-f556-4110-8c5c-4ede8bcba637/GlobalEVOutlook2021.pdf (accessed on 1 August 2022).
6. Transport and Environment. *Easy Ride: Why the EU Truck CO2 Targets Are Unfit for the 2020s*; Transport and Environment: Brussels, Belgium, 2021.
7. Basma, H.; Beys, Y.; Rodriguez, F. *Battery Electric Tractor-Trailers in the European Union: A Vehicle Technology Analysis*; ICCT: Berlin, Germany, 2021. Available online: https://theicct.org/sites/default/files/publications/eu-tractor-trailers-analysis-aug21-2.pdf (accessed on 1 August 2022).
8. Zayer, E.; Gnamm, J.; Strempel, K.; Di Loreto, G.; Johansson, M. *European Truck Market Outlook 2022*; Bain & Company: Munich, Germany, 2022.
9. Youngeun, B.; Suman Kumar, M.; Craig, R.R.; Stephen, G.R. Factors influencing alternative fuel adoption decisions in heavy-duty vehicle fleets. *Transp. Res. Part D Transp. Environ.* **2022**, *102*, 103150. [CrossRef]
10. Hildermeier, J.; Jahn, A.; Rodriguez, F. *Electrifiying EU City Logistics: An Analysis of Energy Demand and Charging Cost*; ICCT: Berlin, Germany, 2020. Available online: https://theicct.org/sites/default/files/publications/EU-logistics-electrification-fv-202011.pdf (accessed on 1 August 2022).
11. Liimatainen, H.; van Vliet, O.; Aplyn, D. The potential of electric trucks—An international commodity-level analysis. *Appl. Energy* **2019**, *236*, 804–814. [CrossRef]
12. Çabukoglu, E.; Georges, G.; Küng, L.; Pareschi, G.; Boulouchos, K. Battery electric propulsion: An option for heavy-duty vehicles? Results from a Swiss case-study. *Transp. Res. Part C Emerg. Technol.* **2018**, *88*, 107–123. [CrossRef]
13. Mauler, L.; Dahrendorf, L.; Duffner, F.; Winter, M.; Leker, J. Cost-effective technology choice in a decarbonized and diversified long-haul truck transportation sector: A U.S. case study. *J. Energy Storage* **2022**, *46*, 103891. [CrossRef]
14. European Environment Agency. *Monitoring of CO2 Emissions from Heavy-Duty Vehicles: Flattened HDV 2019–20 Data*; European Environment Agency: Copenhagen, Denmark, 2021.
15. HERE Truck Navigation. Available online: https://developer.here.com/documentation/android-premium/dev_guide/topics/routing-truck.html (accessed on 1 August 2022).
16. Sripad, S.; Viswanathan, V. Performance Metrics Required of Next-Generation Batteries to Make a Practical Electric Semi Truck. *ACS Energy Lett.* **2017**, *2*, 1669–1673. [CrossRef]
17. NREL DriveCAT. Chassis Dynamometer Drive Cycles. Available online: www.nrel.gov/transportation/drive-cycle-tool (accessed on 30 September 2021).
18. Delgado, O.; Rodriguez, F.; Muncrief, R. *Fuel Efficiency Technology in European Heavy-Duty Vehicles: Baseline and Potential for the 2020–2030 Timeframe*; ICCT: Berlin, Germany, 2017.
19. Gao, Z.; Lin, Z.; Davis, S.C.; Birky, A.K. Quantitative Evaluation of MD/HD Vehicle Electrification using Statistical Data. *Transp. Res. Rec.* **2018**, *2672*, 109–121. [CrossRef]
20. Löbberding, H.; Wessel, S.; Offermanns, C.; Kehrer, M.; Rother, J.; Heimes, H.; Kampker, A. From Cell to Battery System in BEVs: Analysis of System Packing Efficiency and Cell Types. *World Electr. Veh. J.* **2020**, *11*, 77. [CrossRef]

21. European Commission. *Vehicle Energy Consumption Calculation TOOL (VECTO)*; European Commission: Luxembourg, 2019.
22. Doppelbauer, M. *Grundlagen der Elektromobilität: Technik, Praxis, Energie und Umwelt*; Springer Professional: Wiesbaden, Germnay, 2020; ISBN 978-3-658-29730-5.
23. Van Basshuysen, R.; Schäfer, F. *Handbuch Verbrennungsmotor: Grundlagen, Komponenten, Systeme, Perspektiven*, 7th ed.; Springer Professional: Wiesbaden, Germany, 2015.
24. P3 automotive GmbH. *P3 Charging Index—Update 2021: Comparison of the Fast Charging Capability of Different Electric Vehicles from an User Perspective*; P3 Group GmbH: Stuttgart, Germany, 2021.
25. Federal Transport Authority (KBA). *Bestand an Kraftfahrzeugen und Kraftfahrzeuganhängern am 1. Januar 2021 Nach Bundesländern und Fahrzeugklassen Absolut: FZ 27*; KBA: Flensburg, Germany, 2021. Available online: https://www.kba.de/DE/Statistik/Produktkatalog/produkte/Fahrzeuge/fz27_b_uebersicht.html (accessed on 27 September 2021).
26. Göckeler, K.; Hacker, F.; Mottschall, M.; Blanck, R.; Görz, W.; Kasten, P.; Bernecker, T.; Heinzelmann, J. *Status Quo und Perspektiven Alternativer Antriebstechnologien für den Schweren Straßengüterverkehr: Erster Teilbericht des Forschungs-und Dialogvorhabens „StratES: Strategie für die Elektrifizierung des Straßengüterverkehr"*; Öko-Institut e.V.: Berlin, Germany, 2020.
27. Federal Transport Authority (KBA). Verkehr in Kilometern (VK): Zeitreihe Jahre 2014 bis 2020. Available online: https://www.kba.de/DE/Statistik/Kraftverkehr/VerkehrKilometer/vk_inlaenderfahrleistung/vk_inlaenderfahrleistung_node.html (accessed on 30 September 2021).
28. Wermuth, M.; Neef, C.; Wirth, R.; Hanitz, I.; Löhner, H.; Hautzinger, H.; Stock, W.; Pfeiffer, M.; Fuchs, M.; Lenz, B.; et al. *Mobilitätstudie „Kraftfahrzeugverkehr in Deutschland 2010" (KiD 2010) 70.0829/2008*; WVI, IVT, DLR, KBA: Braunschweig, Germany, 2012. Available online: https://www.bmvi.de/SharedDocs/DE/Anlage/G/kid-2010.pdf?__blob=publicationFile (accessed on 2 May 2021).
29. Basma, H.; Saboori, A.; Rodriguez, F. *Total Cost of Ownership for Tractor-Trailers in Europe: Battery Electric versus Diesel*; ICCT: Berlin, Germany, 2021.
30. Basma, H.; Rodriguez, F.; Hildermeier, J.; Jahn, A. *Electrifying Last-Mile Delivery: A Total Cost of Ownership Comparison of Battery-Electric and Diesel Trucks in Europe*; ICCT: Berlin, Germany, 2022.
31. Federal Ministry for Digital and Transport. *Richtlinie Über die Förderung von Leichten und Schweren Nutzfahrzeugen Mit Alternativen, Klimaschonenden Antrieben und Dazugehöriger Tank-und Ladeinfrastruktur: Richtlinie KsNI*; Federal Ministry for Digital and Transport: Berlin, Germany, 2021.

World Electric Vehicle Journal

MDPI

Article

Parametric Optimisation of a Direct Liquid Cooling–Based Prototype for Electric Vehicles Focused on Pouch-Type Battery Cells

Manex Larrañaga-Ezeiza [1,2,*], Gorka Vertiz Navarro [1], Igor Galarza Garmendia [1], Peru Fernandez Arroiabe [2], Manex Martinez-Aguirre [2] and Joanes Berasategi Arostegui [2]

1 CIDETEC, Basque Research and Technology Alliance (BRTA), Po. Miramón 196, 20014 Donostia-San Sebastián, Spain
2 Mechanical and Industrial Production Department, Faculty of Engineering, Mondragon Unibersitatea, Loramendi 4, 20500 Arrasate-Mondragón, Spain
* Correspondence: mlarranaga@cidetec.es

Abstract: In this work, a numerical optimisation process is applied to improve the fluid dynamical aspect of an innovative direct liquid cooling strategy for lithium-ion–based HEV/EV. First, the thermofluidic numerical model of the battery cell defined by means of CFD computational tools was validated with experimental tests. Then, a comparison between different flow patterns was developed to analyse the influence of the fluid distribution geometry. Finally, a parametric multi-objective optimisation process was implemented arranged by a two-level full factorial design. Considering as input variables the height of the fluid, the number of cooling channels, the number of distributors, and the flow rate, the optimal relationship between the thermal performance of the battery cell, the volumetric energy density of the system, and the power consumption of the strategy was obtained. As a result, the energy density of the system was maximised, and the power consumption was reduced while keeping the cell temperature within the optimal range.

Keywords: lithium-ion; battery; thermal management; CFD; direct liquid cooling; energy density

Citation: Larrañaga-Ezeiza, M.; Vertiz Navarro, G.; Galarza Garmendia, I.; Fernandez Arroiabe, P.; Martinez-Aguirre, M.; Berasategi Arostegui, J. Parametric Optimisation of a Direct Liquid Cooling–Based Prototype for Electric Vehicles Focused on Pouch-Type Battery Cells. *World Electr. Veh. J.* **2022**, *13*, 149. https://doi.org/10.3390/wevj13080149

Academic Editor: Michael Fowler

Received: 12 July 2022
Accepted: 29 July 2022
Published: 9 August 2022

1. Introduction

Aiming to reduce CO_2 emissions in densely populated areas, in recent years HEV/EV vehicles have been one of the global benchmark topics in the mobility sector [1]. Considering batteries as the energy source, these technologies enable 100% zero-emission operation. Today, due to their characteristic large density, high discharge capacity, and low maintenance, lithium-ion–based batteries are the reference energy storage technology in the electric vehicle sector.

Although this energy storage technology has high electrical performance, the proper operation of these types of systems is influenced by the temperature at which they operate. The thermal management of these systems is therefore essential. To avoid premature degradation of the energy storage system and prevent risky situations, the optimal working range of lithium-ion batteries is between 15 and 40 °C [2,3], maintaining a cell level temperature uniformity of 0–5 °C. These temperature ranges ensure the best compromise between technical performance and system safety [4].

To control the working temperature of the battery system, in recent years different cooling strategies have been considered. These strategies are generally classified as air cooling, liquid cooling (LC), phase change materials (PCM), heat pipes (HP), and thermoelectric coolers (TEC) [5]. Among these technologies, owing to the thermal conductivity and heat capacity of the implemented fluids, strategies based on liquid cooling (LC) predominate the thermal management of the electric vehicle sector [6]. LC strategies can be divided into two main groups, indirect liquid cooling (ILC) and direct liquid cooling (DLC). Today, ILC is the leading strategy in the field of electric vehicles. However, the fluids used in ILC are electrically conductive, a nature that avoids direct contact with the battery cell. This characteristic decreases the cooling capacity of the strategy compromising the performance of the system [7,8]. Consequently, DLC has been the subject of many studies in recent years [9–13]. DLC strategy uses fluids with high dielectric strength that enable direct contact between the heat generation source (battery cell) and the refrigeration fluid, improving the thermal management efficiency of the system [14].

Recently, some studies have been carried out to analyse the features that can optimise the performance of DLC system. These types of optimisation processes are developed in the numerical environment using multi-physics models to characterise the real working conditions of the system. An example of this type of analysis is the study that Y. Fan et al. [15] developed with a battery module based on 32 cylindrical cells with a DLC strategy. Using a full-immersed casuistry as a reference, the authors analysed the effect of partial cooling by reducing the height of the fluid channel. Results showed that the thermal behaviour of the battery module was worsened by decreasing the height of the fluid channel. However, the authors highlighted that the power consumption of the auxiliary system and the energy density of the battery module were improved. Considering other design parameters, in the development process of a novel direct liquid immersion cooling strategy, M.-Y. Lee et al. [16] analysed the effect of the cell spacing on the system performance. Using a battery module of 14 pouch-type cells of 20 Ah as a reference, the results of the study demonstrated that cell temperatures and pressure drop of the system improves increasing cell spacing. However, the authors mentioned that in a practical scenario, the lowest cell spacing is preferred due to space limitations. Finally, X. Ju et al. [17] performed a geometrical optimisation analysis using a manifold immersion cooling structure for a 51 Ah prismatic battery module. Defining a uniform heat generation on the battery, the structure of the manifold was parametrised to analyse the influence of the fluid line geometry on the thermal and fluid dynamical performance of the strategy. As a result, at the expense of an increment on the pressure drop, a higher and more homogeneous heat transfer coefficient was led on the surface of the battery cells decreasing maximum temperatures and thermal heterogeneities of the system.

Based on the above literature review, we appreciate the importance of analysing the influence of the design parameters to improve the performance of the cooling strategy and, thus, develop a robust, effective, and economical design. To date, there have been few optimisation studies that consider direct liquid cooling strategies as a reference, and the ones based on pouch-type cells, where the thermal characterization becomes more complicated, are in the minority. Furthermore, most of the research has focused on the independent analysis of the influence of the parameters.

Therefore, in this article, a numerical design optimisation of a direct liquid cooling prototype is proposed, defining a specific design of experiments methodology (DoE) to study the interaction of the parameters. The prototype was developed as a partial DLC to decrease the impact of the fluid weight on the energy density of the systems, and it was focused on a large-scale lithium-ion 60 Ah NMC pouch-type cell, a type of cell with high energy to mass ratio and packaging efficiency [18]. To improve the performance of the reference prototype, an optimisation process based on the following steps was undertaken. First, using the numerical CFD environment, the battery cell simulation model was characterised and validated with experimental tests. Then, a comparison between different flow patterns was developed to analyse the influence of the fluid distribution geometry. Finally, a parametric multi-objective optimisation process was implemented to obtain the optimal relationship between the thermal performance of the battery cell, the volumetric energy density of the system, and the power consumption of the strategy.

2. Numerical Model Description

The numerical model is defined to optimize the cell level thermofluidic performance before the scale-up to a module level prototype. To this end, Ansys Fluent 21R2 was used to develop the numerical models. This software enables a coupled analysis, solving the fluid dynamics and the heat transfer considering non-uniform and transient heat generation of the battery cell.

2.1. Physical Model

The partial cooling concept developed in this work is focused on the cooling effect of the cell surface, enabling a direct thermal management of the battery cell. To develop this strategy, a prismatisation component was designed. This component facilitates the internal circulation of the cooling fluid creating a space between the cell contact surface and the component's internal surface, as is shown in Figure 1.

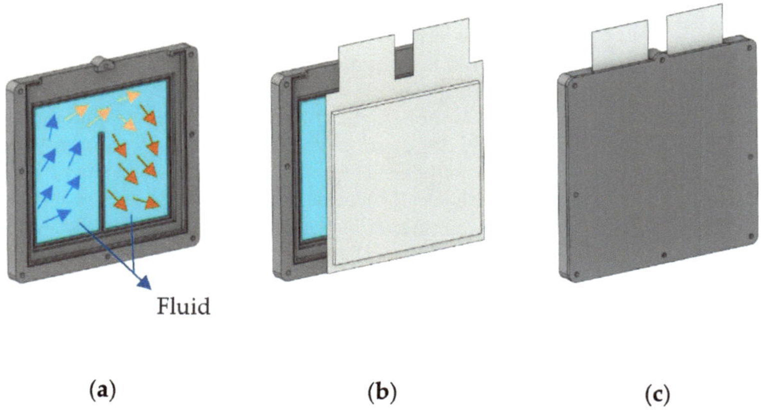

Figure 1. Reference model (**a**) flow distribution, (**b**) principal components, and (**c**) assembly.

2.2. Simulation Model Geometry

To replicate the experimental setup, the principal components defined in the simulation model were the following: the battery cell, the fluid, and the clamps (Figure 2). Clamps were implemented because of their influence on the thermal behaviour of the battery cell. These components have almost ten times the mass of the battery tabs and were used to develop the electrical connexion of the tests.

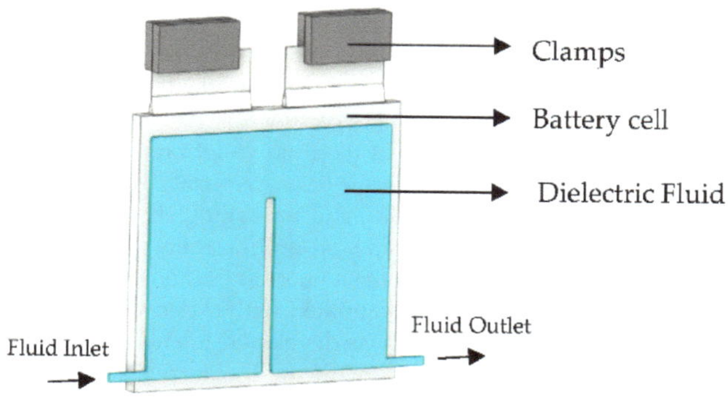

Figure 2. Principal components defined in the simulation model.

The battery cell implemented in the model is a large-scale 60 Ah NMC pouch-type battery cell. As the battery cell, the dielectric fluid implemented on the model mirrors the characteristics of the commercial dielectric fluid based on mineral oil. The thermal and electrical characteristics of each component are presented in Table 1.

Table 1. Properties of the battery cell, the clamps, and the dielectric fluid, for 25 °C.

Property	Battery Cell	Clamps	Dielectric Fluid
Material	Battery cell	Aluminium	Mineral oil
Kinematic Viscosity (mm^2/s)	-	-	4.3
Heat Capacity (J/kgK)	1306	871	2130
Thermal conductivity (W/mK)	x,y: 17.9/z: 0.65	202.4	0.135
Density (kg/m^3)	2183	2719	774
Resistivity (MΩm)	-	-	>5 × 10^6

2.3. Battery Modelling

The heat generation of a lithium-ion battery can be expressed by the simplified equation of Newman $Q_{gen} = Q_{irr} + Q_{rev}$. Where Q_{gen} is the total heat generation, Q_{irr} is the irreversible heat, and Q_{rev} is the reversible heat. The irreversible heat generation, also referred as Joule effect, is represented by the equation $Q_{irr} = I^2 R_{int}$. Where I is the current (A) and R_{int} the internal resistance of the cell (Ω), which depends on the temperature (T) and the state of charge (SOC). On the other hand, the reversible heat generation is developed by the insertion and disinsertion of lithium ions into the electrodes (anode and cathode). This generation is based on the electrochemical reactions that take place when discharging and charging the lithium-ion cells, reactions that create a variation in the entropic level of the system. This generation is represented by equation $Q_{rev} = IT\, dE_{OCV}/dT$. Where I is the current (A), T is the temperature (K), and dE_{OCV}/dT is the variation of the open circuit voltage with the temperature (V/K).

To characterise the heat generation of the battery cell numerically, an electrothermal model was implemented to mirror the thermal response of the battery cell properly. This model calculates the heat generation of the cell considering the instantaneous temperature (T) and the state of charge (SOC) level. Moreover, this model characterises the influence of the current density increase on the thermal behaviour of the battery cell at high C-rate working conditions.

In this aspect, to characterise the battery cell, the equivalent circuit model (ECM) was selected on the battery module of Ansys Fluent 21R2. This model is widely used because of its simplicity and effectiveness [19]. Using experimental hybrid pulse power characterization (HPPC) testing data, the parameters of the ECM were calculated. This characterization mirrors only the irreversible heat generation of the battery cell. Therefore, to characterize the endothermic behaviour of the battery cell, the entropic factor of the reversible heat generation was implemented by user-defined functions (UDFs). This factor was experimentally characterised analysing the open circuit voltage (OCV) variation with the temperature (dE_{OCV}/dT).

The battery cell heat generation model was validated using experimental voltage and temperature information of a 1C discharge test (Figure 3). The simulation model was initialised at the same SOC (100%) and temperature level (25 °C) of the experimental test. The discharge process was followed until the SOC level of 20% was reached. To characterise the interaction between the climatic chamber and the prototype, an equivalent heat transfer coefficient of 25 W/m^2K was defined on the boundary surfaces of the battery cell [20,21], which was experimentally validated.

As it is depicted in Figure 3a, the numerical correlation of the voltage performs adequately below an error of 1%. Considering results from the temperature of the cell surface (Figure 3b), it is observed that the simulation results adequately replicate the distribution and evolution of the experimental curves with a maximum error of 0.5 °C.

Pulse profiles in a SOC range of 40–50% were implemented to analyse the influence of the geometrical parameters of the fluid channel on the thermal performance, energy density, and power consumption of the proposed strategy. These tests were developed until the thermal stabilization of the battery cell was achieved to avoid dynamics effects during the tests. Therefore, the heat generation of the battery cell model in charging and discharging process was analysed between a SOC range of 80–20%. In this range both charging and discharging processes have a similar heat generation. As presented in

Figure 4, the heat generation of the simulation model adequately characterises the shape and the heat generation range of the charging and discharging processes. It is observed that in the charging process the simulation model has a deviation from the experimental results. However, considering that the dynamics of the cell are well characterised, the electrical response is correlated, and the pulse tests are carried out at intermediate SOC values, it is concluded that the numerical model presents a high reliability to predict trends in the parametric optimisation process.

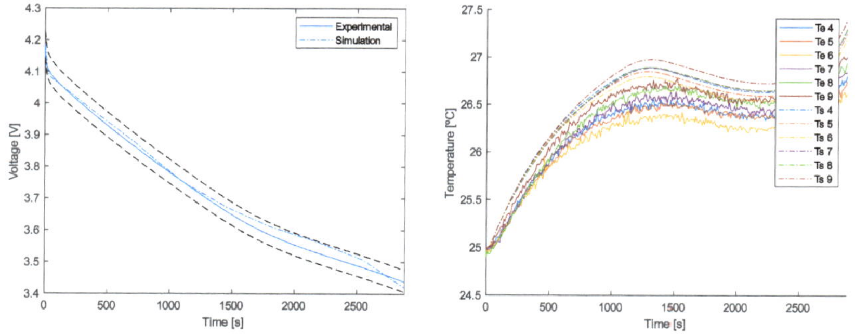

(a) **(b)**

Figure 3. Battery cell model validation with experimental (**a**) voltage and (**b**) surface temperature distribution in a range of 100–20% SOC of a 1C discharge test.

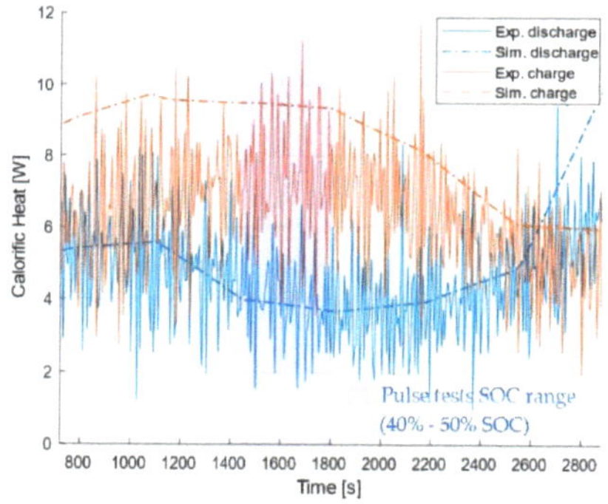

Figure 4. Battery cell model heat generation validation with experimental heat generation information (80–20% SOC range).

2.4. Mesh Independence

To define proper mesh for accurate results, a mesh independence study was developed using as a reference the particular case of the U-shape flow pattern design analysed on the geometrical comparison section (Section 3). To analyse the stabilization of the maximum temperature of the battery cell and the pressure drop of the cooling strategy, a sensitivity study was developed using seven models with different grid cell numbers. All cases were analysed defining constant inlet fluid flow and temperature at 0.4 L/min and 25 °C, consecutively. The flow rate was calculated using as a reference the characteristics of an indirect liquid cooling strategy–based cold plate [22]. The laminar model was defined to characterise the flow development because the Reynolds number was lower than 2300 in all case studies. To characterise the cell heat generation, a 1C pulse profile was implemented. Owing the prototype prismatisation casing, the interaction of the cell body with the climatic chamber environment was omitted. Therefore, an adiabatic definition was imposed on the battery cell surface that had no contact with the fluid. On the tab zone and on clamps an equivalent heat transfer coefficient of 25 W/m^{2}K was set. As Figure 5 presents, both maximum temperature and pressure drop curves are stabilised with a mesh of 4,778,025 elements. The results variation maintains below 1% when the elements where increased to 8743816. Thus, it was concluded that the results are independent for the model of 4,778,025 elements [23,24].

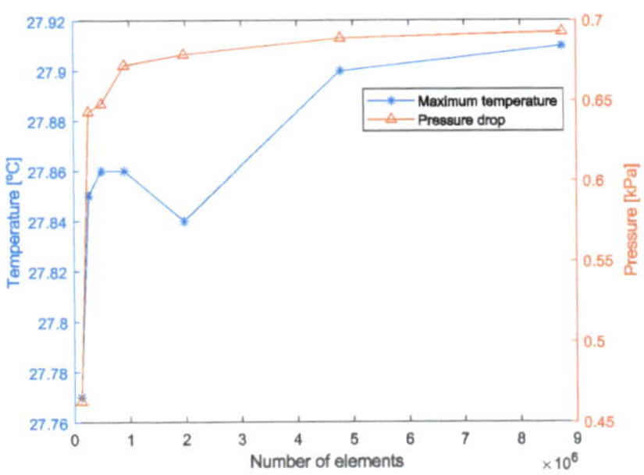

Figure 5. Mesh independence analysis with the maximum temperature of the battery cell (°C) and pressure drop (kPa).

3. Flow Pattern Design Selection

Before optimising the geometrical aspect of the prototype flow pattern, a design selection process was specified. This process was aimed at selecting among different flow pattern geometries analysed in the bibliography [23]. Defining the same working conditions, a comparison between the four main designs presented in Figure 6 was developed: U-shape, convex, honeycomb and airfoil. The objective of this study was to examine the impact of the geometry on the flow rate distribution and the resultant thermal performance and power consumption of each of the prototype.

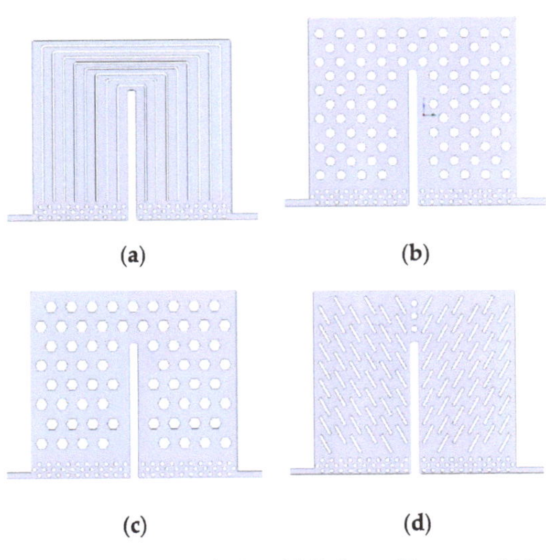

Figure 6. Flow pattern designs (**a**) U-shape, (**b**) convex, (**c**) honeycomb, and (**d**) airfoil.

To make the designs comparable, the same surface contact areas between the battery cell and flow patterns were defined: 0.026 m^2. This was a conditioning parameter to define the same heat absorption capacity with the same fluid volume in all flow pattern designs.

3.1. Boundary Conditions

As a variable of the study, simulations at a flow rate of 0.1, 0.2, 0.4, 0.8, and 1.6 L/min were developed for each flow pattern design. This study of the flow rate range makes it possible to draw the curves for each case and, thus, to develop the comparison between the flow pattern designs in a wide working range. The inlet temperature of the fluid and the boundary temperature that mirrors the climatic chamber influence were defined at 25 °C. The battery cell working condition was defined with a 1C pulse test. These tests were developed until the thermal stabilization of the battery cell was achieved; thus, dynamics effects during the tests were avoided to analyse more clearly the influence of the parameters under study.

3.2. Results and Discussion

To analyse how the flow pattern geometry influences the performance of the battery cell and the auxiliary system, battery cell maximum temperature (T_{max}), temperature homogeneity (ΔT), and the pumping power consumption related to the auxiliary system (P_h) were defined as the output variables. Figure 7 presents the steady state results of each case at flow rates of 0.1, 0.2, 0.4, 0.8, and 1.6 L/min.

As it is depicted in Figure 7a,b, at high flow rates, the airfoil design shows the lowest battery cell temperature and heterogeneity. However, at low flow rates the U-shape model presents the best thermal performance. Although the flow rate is greatly reduced, due to the channels of the U-shape design, the fluid is forced to flow all over the surface of the cell, avoiding the appearance of hot spots, and provides better homogeneity. For their part, considering the power consumption impact presented in Figure 7c, the convex and honeycomb models present the lowest pressure drop impact because the pattern design enables a greater freedom to the flow than the other design. Nevertheless, this freedom hinders the flow control and facilitates the development of hot spots. Consequently, both models show the highest maximum temperature and heterogeneity values.

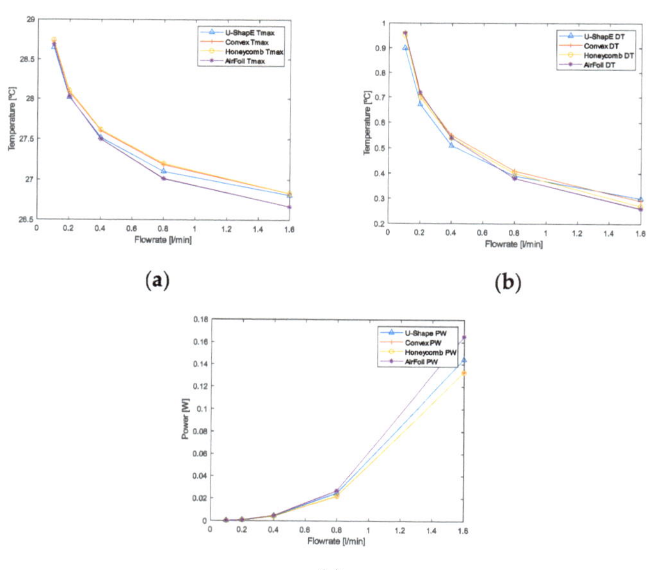

Figure 7. Geometry comparison results of (**a**) battery cell maximum temperature (T_{max}), (**b**) temperature homogeneity (ΔT), and (**c**) the pumping power consumption related to the auxiliary system (P_h) at different flow rate.

To further analyse each design, Figure 8 shows the surface contact temperature between the fluid and the battery cell at a flow rate of 0.4 L/min. As it is presented, the U-shape model provides the most homogenous temperature distribution in the battery cell surface.

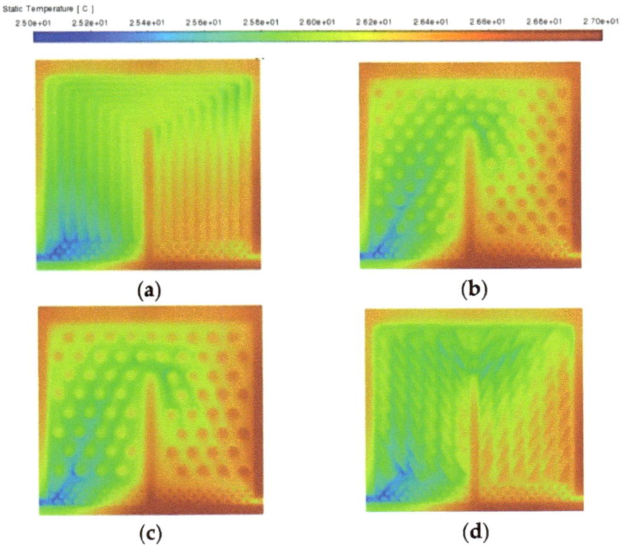

Figure 8. Battery cell surface temperature distribution at 0.4 L/min of flow rate for (**a**) U-shape, (**b**) convex, (**c**) honeycomb, and (**d**) airfoil designs.

The objective of the prototype is to work with low flow rates to decrease the impact of the cooling strategy on the power consumption of the auxiliary system. Therefore, it is concluded that the U-shape design is the most appropriate one to develop the geometrical optimisation process.

4. Design Optimisation

4.1. Parametrisation of the Geometry

To develop the multi-objective optimisation of the selected U-shape flow pattern, four parameters were selected including the height of the fluid (H_f), the number of cooling channels (N_c), the number of inlet and outlet distributors (N_d), and the flow rate (Q). Figure 9 presents each parameter definition.

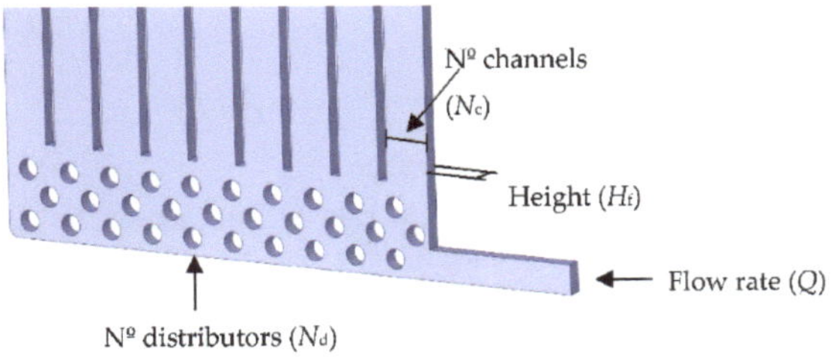

Figure 9. Parameters implemented in the optimisation process.

The height of the fluid was defined between 1 mm and 3 mm to study the impact of the fluid volume reduction on the energy density and the power consumption of the system. The number of cooling channels and distributors were defined between 3 and 9 channels and 10 and 30 distributors, respectively, to analyse the influence of those components and to spread the fluid in a homogeneous way through all the cooling space. This definition was based on the cell surface temperature analysis presented in Figure 10. For each case the cooling channels and the distributors were omitted and the temperature of the cell surface analysed. Considering Figure 10, results show that the lack of these components creates a heterogeneous fluid distribution developing hot spot areas on the battery cell surface. Finally, inlet flow rate was defined as variable of study to analyse how impacts this input the relationship between the thermal performance of the battery system and the power consumption of the auxiliary system.

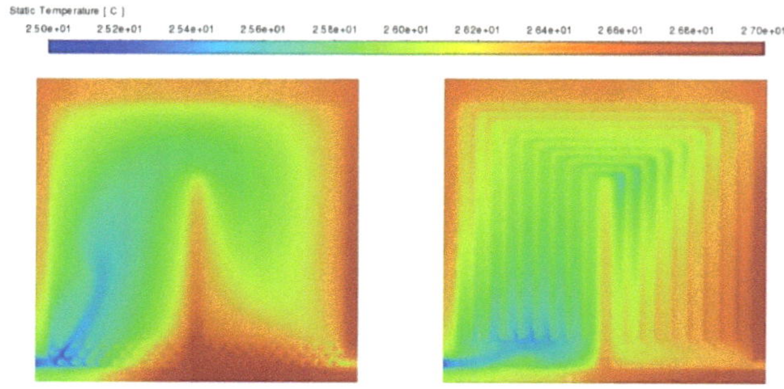

(**a**) (**b**)

Figure 10. Battery cell surface temperature distribution for (**a**) a cooling design without channels and (**b**) a cooling design without distributors.

4.2. Output Variables

To evaluate the performance of each case, four objective output variables were defined, including maximum temperature (T_{max}), temperature homogeneity (ΔT), system volumetric energy density (VED), and system power consumption (P_{h}). As it is presented in Figure 11, the first two variables (T_{max} and ΔT) were related to the cell level design to analyse how the strategy controls the thermal behaviour of the battery system. Then, scaling up to a battery module of 24 battery cells, the impact of each cooling strategy on the system volumetric energy density (VED) and system power consumption (P_{h}) were analysed.

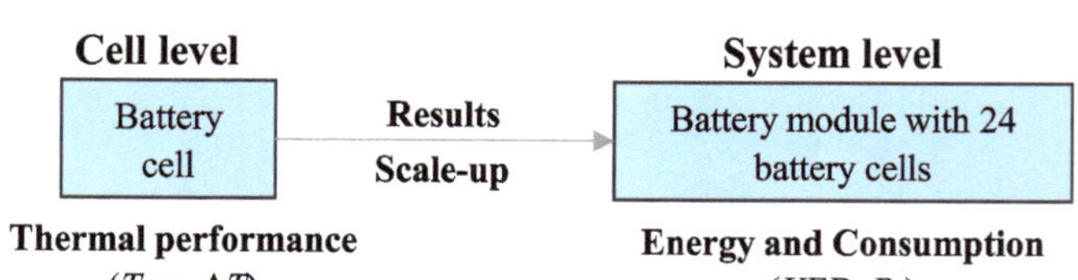

Figure 11. Output variable analysis level explanation.

4.3. Optimisation Process Definition

To evaluate how the flow rate definition and the geometrical parameters influence the performance of the strategy, a two-level full factorial matrix was defined. This is a numerical analysis method which considers all the possible variations between the defined factors. Hence, considering four variables of study, a simulation process with 16 simulation cases was defined. With the proposed analysis, all the interactions between H_{f}, N_{c}, N_{d}, and Q were determined. To investigate the importance of the defined variables on the output results of the full factorial design, analysis of variance (ANOVA) was employed. This is, a statical model that evaluates the importance of the defined factors by comparing the response variable means at the defined factor levels. To properly develop the regression equations, all variations up to the second order of interaction between the defined factors were first analysed. Then, the most relevant interactions were selected by defining a significance level of 0.05. Once the most influential factors conditioning the output variable were selected, the factorial design was analysed again, and the ANOVA results were extracted. With this information, the regression coefficients were calculated, and regression equations were developed for each output variable. To analyse the reliability of these regression model, values of R^2 (Adequate), R^2 (Predicted), and R^2 (Adjusted) were examined, and the residual normal plots were presented. After the validation, the optimum case was selected considering the composite desirability function, where the optimum case selection process guideline was defined assigning specific weight and importance values for each of the proposed output variables.

4.4. Results and Analysis

The corresponding results of the two-level full factorial design are presented in Table 2. Among the results, cell level T_{max} and ΔT, and module level VED and Ph output variable results can be observed for all the possible variations between H_f, N_c, N_d, and Q.

Table 2. The corresponding simulation cases of the two-level full factorial design and the output variable results of each case.

N° Simulations	H_f (mm)	N_c	N_d	Q (L/min)	Cell Level		Module Level	
					T_{max} (°C)	ΔT (°C)	VED (Wh/L)	P_h (W)
1	3	9	30	0.40	27.52	0.51	248.70	0.109
2	1	9	30	0.40	27.00	0.37	279.78	1.455
3	3	3	30	0.40	27.51	0.56	248.70	0.095
4	1	3	30	0.40	26.98	0.38	279.78	1.301
5	3	9	10	0.40	27.56	0.47	248.70	0.076
6	1	9	10	0.40	27.02	0.35	279.78	1.162
7	3	3	10	0.40	27.51	0.49	248.70	0.069
8	1	3	10	0.40	26.98	0.36	279.78	1.060
9	3	9	30	0.13	28.36	0.78	248.70	0.009
10	1	9	30	0.13	27.73	0.68	279.78	0.131
11	3	3	30	0.13	28.31	0.84	248.70	0.007
12	1	3	30	0.13	27.7	0.71	279.78	0.118
13	3	9	10	0.13	28.37	0.73	248.70	0.006
14	1	9	10	0.13	27.72	0.65	279.78	0.115
15	3	3	10	0.13	28.27	0.76	248.70	0.005
16	1	3	10	0.13	27.69	0.67	279.78	0.103

To obtain the coefficients of the regression model, ANOVA was used. As it was mentioned in the optimisation procedure section, these coefficients were evaluated using the most relevant interactions effects between the defined factors. Therefore, the coefficients defined for the regression equations were the ones that presents a significance level below 0.05. Considering this guideline, the following regression equations were obtained for the output variables of cell maximum temperature (T_{max}), cell surface temperature homogeneity (ΔT), system volumetric energy density (VED), and system power consumption (P_h):

$$T_{max} = 27.6 + 0.28\, H_f - 0.38\, Q \quad (1)$$

$$\Delta T = 0.58 + 0.06\, H_f - 0.014\, N_c + 0.022\, N_d - 0.146\, Q + 0.008\, H_f\, N_d + 0.01\, H_f\, Q \quad (2)$$

$$VED = 264.24 - 15.54\, H_f \quad (3)$$

$$P_h = 0.364 - 0.316\, H_f + 0.039\, N_d + 0.302\, Q - 0.031\, H_f\, N_d - 0.261 H_f\, Q + 0.035\, N_d\, Q \quad (4)$$

To analyse the reliability of the regression model, values of R^2 (Adequate), R^2 (Predicted), and R^2 (Adjusted) were analysed. These are significance indexes that present the adequate quality, the quality of the predicted regression models, and the quality of the models after adjustment, respectively. Considering results from Table 3, index values agree for T_{max}, ΔT, VED, and P_h. Index values for VED were 100% in all R^2 cases. This result means that there is not variability on results, as the volumetric energy density *VED* is the only one proportional to a single parameter (H_f). Therefore, the quality and the correlation of these values demonstrates the reliability of the regression models.

Table 3. Values of R^2 (Adequate), R^2 (Predicted), and R^2 (Adjusted) for the output variables T_{max}, ΔT, *VED*, and P_h.

	R^2 (Adequate)	R^2 (Predicted)	R^2 (Adjusted)
T_{max}	90.32%	85.34%	88.84%
ΔT	99.68%	99.00%	99.47%
VED	100%	100%	100%
P_h	99.59%	98.37%	99.23%

Apart from R^2 index values, it is important to make sure that the results of the residuals for each response are consistent. Figure 12 presents the normal residual plots for each output variable. As it is shown the residuals are normally distributed, which means that there is a good agreement between the predicted and actual values. Figure 12c corroborates the interpretation made with R^2 index values for the volumetric energy density *VED*. Therefore, it was confirmed that *VED* was uniquely and exclusively proportional to H_f. These results demonstrated that the proposed regression models could predict adequately the variability of the output parameters. The reliability of the regression models is therefore justified.

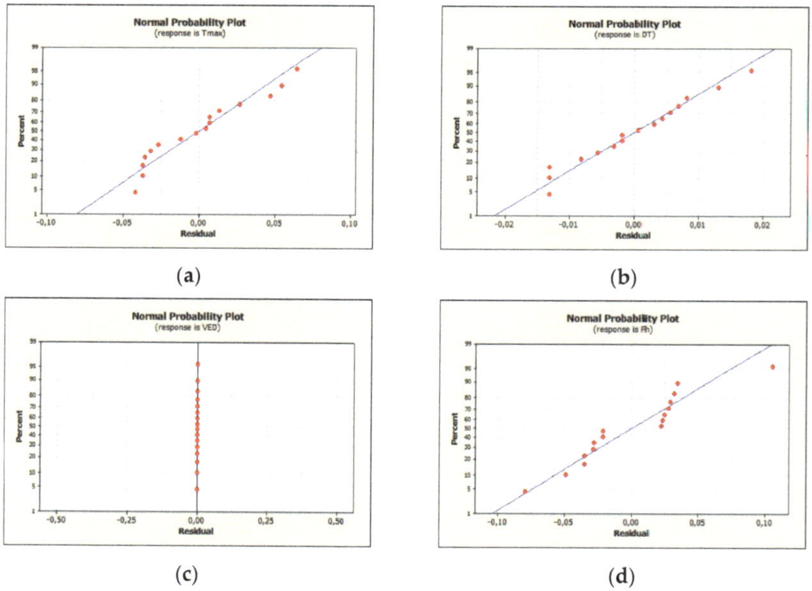

Figure 12. Residual normal plots for the output variables of (**a**) T_{max}, (**b**) ΔT, (**c**) *VED*, and (**d**) P_h.

After validating the regression models, the principal influences of the optimisation parameters on the defined output variables were analysed. As it is presented in the regression equation of cell maximum temperature (Equation (1)), the height of the fluid channel (H_f) and the inlet flow rate definitions (Q) were the main parameters that influences the maximum temperature of the cell body. This means that the influence of the number of channels (N_c) and distributors (N_d) for T_{max} is negligible. However, the impact of those input parameters is appreciated on the variability of the temperature homogeneity of the battery cell surface (Equation (2)). With a higher N_c number, the temperature difference of the cell surface increases. In contrast, the effect of N_d is the opposite. Therefore, the results recommended to decrease the number of distributors and increase the number of channels to improve the temperature homogeneity of the battery cell surface.

As mentioned above, the volumetric density (Equation (3)) is influenced only by the parameter H_f. Therefore, analysing the results, the objective of developing systems with high volumetric energy density will be achieved by reducing this parameter H_f as much as possible. In this case, the minimum value of 1 mm for H_f is the one that gives the best results of volumetric energy density.

Finally, to achieve the objective of decreasing the power consumption of the system, the regression equation (Equation (4)) presents that all the parameters influence it. However, analysing the estimated coefficients for P_h, it is observed that H_f, Q, and the interaction between them (H_f Q) have more impact than the N_d and related interactions. The power consumption of the system is based on the pressure drop defined by the fluid; therefore, H_f and Q influence is coherent. The pressure drop is related in a quadratic way by the velocity of the fluid [25]. Thus, to decrease the power consumption of the system lower fluid velocity profiles were recommended. Studying the influence of the parameters N_c and N_d it can be observed that the rise in the number of channels and distributors increases the power consumption of the system. Therefore, it is concluded that the quantity of these components should be minimised for achieving the desired fluid distribution to maintain the lowest impact on the power consumption of the system.

Once the reliability of the regression models was validated, and the principal influences of the optimisation parameters on the defined output variables were analysed, the composite desirability function for T_{max}, ΔT, VED, and P_h was implemented. This function calculates, according to the desirability values defined for each output variable, the optimal case within the range of the proposed two-level full factorial model results.

To examine results that ensure the best performance of the battery cell, first, a composite desirability function was developed with the maximum desirability values defined for T_{max} and ΔT. In this case, the optimum values to ensure the best thermal performance were $H_f = 1$ mm, $N_c = 9$, $N_d = 10$, and $Q = 0.4$ L/min. This configuration maintains the highest volumetric energy density at the expense of a considerable increase in the P_h output parameter. Therefore, VED and P_h were implemented with a lower weight impact and the same importance on the composite desirability function to develop a multi-objective optimisation that minimises the power consumption of the strategy and maximises the volumetric energy density of the system. For this case, the corresponding values of the design variables were $H_f = 1$ mm, $N_c = 9$, $N_d = 10$, and $Q = 0.13$ L/min. Maintaining the maximum volumetric energy density of 279.7 Wh/L, this configuration implies an increase in the maximum temperature from 27.02 °C to 27.72 °C, and the temperature difference was increased by 0.3 °C. However, the power consumption of the system was decreased by 90% from 1.16 W to 0.11 W.

5. Conclusions

In this work, a parametric optimisation of a direct liquid cooling strategy is proposed for a large-scale lithium-ion pouch-type cell as a reference. First, the simulation model was developed and validated with experimental results. Moreover, a non-uniform heat generation was defined in the battery model to adequately represent the temperature distribution of the large-scale pouch-type cell. Then, a comparison between different flow pattern designs was developed to analyse the influence of the fluid distribution geometry. Finally, a parametric optimisation process was implemented to obtain the optimal relationship between the thermal performance of the battery cell, the volumetric energy density of the system, and the power consumption of the strategy. The analysis carried out in the present work gave rise to the following conclusions:

- At flow rates below 0.4 L/min, the flow distribution channels defined on the U-shape design improve the fluid dynamical aspect of the cooling strategy, maintaining the highest thermal performance of the battery cell without increasing the power consumption. It was therefore selected to develop the parametric optimisation process.

- Developed surrogate models presented that the most critical parameters that influence the output variable results were the height of the fluid channel (H_f) and the flowrate definition (Q), which are directly related to the fluid local velocity.
- The number of channels (N_c) increases the power consumption of the system (P_h) while decreasing the thermal heterogeneity of the battery cell (ΔT). Therefore, it is recommended to decrease the number of channels remaining the thermal distribution of the cell within the optimal range.
- The number of distributors (N_d) increases the power consumption of the system (P_h) and the thermal heterogeneity of the battery cell (ΔT). However, a minimum number of components to adequately distribute the inflow and outflow are necessary, thus avoiding hot spots in the system.
- The proposed parametric optimisation defined the optimum design of the DLC strategy that ensures the optimal relationship among T_{max}, ΔT, *VED*, and P_h. The corresponding values of the design parameters were H_f = 1 mm, N_c = 9, N_d = 10, and Q = 0.13 L/min. This design case maintains T_{max} at 27.72 and ΔT at 0.65 with the maximum *VED* value and reducing P_h by 90%.

Author Contributions: Conceptualization, M.L.-E. and G.V.N.; methodology, M.L.-E. and G.V.N.; software, I.G.G. and P.F.A.; validation, M.L.-E., I.G.G. and P.F.A.; formal analysis, M.L.-E.; investigation, M.L.-E.; resources, M.L.-E.; data curation, M.L.-E., G.V.N. and J.B.A.; writing—original draft preparation, M.L.-E.; writing—review and editing, M.M.-A., J.B.A. and P.F.A.; visualization, M.L.-E.; supervision, G.V.N., M.M.-A., J.B.A. and P.F.A.; project administration, M.M.-A. All authors have read and agreed to the published version of the manuscript.

Funding: This research did not receive any specific grant from funding agencies in the public, commercial, or not-for-profit sectors.

Institutional Review Board Statement: Not applicable.

Informed Consent Statement: Not applicable.

Data Availability Statement: Not applicable.

Conflicts of Interest: The authors declare no conflict of interest.

Abbreviations

The following abbreviations are used in this manuscript:

ANOVA	analysis of variance
CFD	computational fluid dynamics
DLC	direct liquid cooling
DoE	design of experiments
ECM	equivalent circuit model
EV	electric vehicle
HEV	hybrid electric vehicle
HP	heat pipes
HPPC	hybrid pulse power characterization
ILC	indirect liquid cooling
LC	liquid cooling
NMC	nickel manganese cobalt
OCV	open circuit voltage
PCM	phase change material
SOC	state of charge
TEC	thermoelectric material
UDF	user-defined functions
VED	volumetric energy density

References

1. International Energy Agency. Global EV Outlook 2020. *Glob. EV Outlook* **2020**, *191*, 116565.
2. Choudhari, V.; Dhoble, A.; Sathe, T. A review on effect of heat generation and various thermal management systems for lithium ion battery used for electric vehicle. *J. Energy Storage* **2020**, *32*, 101729. [CrossRef]
3. Han, X.; Lu, L.; Zheng, Y.; Feng, X.; Li, Z.; Li, J.; Ouyang, M. A review on the key issues of the lithium ion battery degradation among the whole life cycle. *eTransportation* **2019**, *1*, 100005. [CrossRef]
4. Patil, M.S.; Seo, J.-H.; Panchal, S.; Jee, S.-W.; Lee, M.-Y. Investigation on thermal performance of water-cooled Li-ion pouch cell and pack at high discharge rate with U-turn type microchannel cold plate. *Int. J. Heat Mass Transf.* **2020**, *155*, 119728. [CrossRef]
5. Kim, J.; Oh, J.; Lee, H. Review on battery thermal management system for electric vehicles. *J. Appl. Therm. Eng.* **2019**, *149*, 192–212. [CrossRef]
6. Rouaud, C. Automotive Thermal Management, Online Conference. 19.11.2020. Available online: https://lp.bcf-events.com/automotive-thermal-management-online-conference-2020/ (accessed on 19 November 2020).
7. Chung, Y.; Kim, M.S. Thermal analysis and pack level design of battery thermal management system with liquid cooling for electric vehicles. *Energy Convers. Manag.* **2019**, *196*, 105–116. [CrossRef]
8. Yang, C.; Cao, L. The Role of Interfacial Thermal Resistance in Li-Ion Battery Thermal Management. In Proceedings of the ASME 2019 International Technical Conference and Exhibition on Packaging and Integration of Electronic and Photonic Microsystems, Anaheim, CA, USA, 7–9 October 2019. [CrossRef]
9. 3M. 3M Novec 7000 Engineered Fluid Product Information. 2005, pp. 1–6. Available online: http://multimedia.3m.com/mws/mediawebserver?66666UuZjcFSLXTtlXftMxMVEVuQEcuZgVs6EVs6E666666-- (accessed on 16 April 2020).
10. XING Mobility Dynamic Power on Demand. 2022. Available online: https://www.xingmobility.com/assets/pdf/XingMobility_DynamicPower_OnDemand.pdf (accessed on 3 April 2020).
11. Battery, K. Kreisel Battery Pack. 2022. Available online: https://www.kreiselelectric.com/ (accessed on 10 March 2020).
12. Park, S.; Jung, D. Battery cell arrangement and heat transfer fluid effects on the parasitic power consumption and the cell temperature distribution in a hybrid electric vehicle. *J. Power Sources* **2013**, *227*, 191–198. [CrossRef]
13. Sundin, D.W.; Sponholtz, S. Thermal Management of Li-Ion Batteries with Single-Phase Liquid Immersion Cooling. *IEEE Open J. Veh. Technol.* **2020**, *1*, 82–92. [CrossRef]
14. Dubey, P.; Pulugundla, G.; Srouji, A. Direct Comparison of Immersion and Cold-Plate Based Cooling for Automotive Li-Ion Battery Modules. *Energies* **2021**, *14*, 1259. [CrossRef]
15. Tan, X.; Lyu, P.; Fan, Y.; Rao, J.; Ouyang, K. Numerical investigation of the direct liquid cooling of a fast-charging lithium-ion battery pack in hydrofluoroether. *Appl. Therm. Eng.* **2021**, *196*, 117279. [CrossRef]
16. Patil, M.S.; Seo, J.-H.; Lee, M.-Y. A novel dielectric fluid immersion cooling technology for Li-ion battery thermal management. *Energy Convers. Manag.* **2020**, *229*, 113715. [CrossRef]
17. Le, Q.; Shi, Q.; Liu, Q.; Yao, X.; Ju, X.; Xu, C. Numerical investigation on manifold immersion cooling scheme for lithium ion battery thermal management application. *Int. J. Heat Mass Transf.* **2022**, *190*, 122750. [CrossRef]
18. Battery University. Battery University—BU-301a: Types of Battery Cells. 2021. Available online: https://batteryuniversity.com/article/bu-301a-types-of-battery-cells (accessed on 25 February 2020).
19. Tran, M.-K.; Mevawala, A.; Panchal, S.; Raahemifar, K.; Fowler, M.; Fraser, R. Effect of integrating the hysteresis component to the equivalent circuit model of Lithium-ion battery for dynamic and non-dynamic applications. *J. Energy Storage* **2020**, *32*, 101785. [CrossRef]
20. Li, Y.; Chen, M.; Bai, F.; Song, W.; Feng, Z. Thermal equilibrium characteristic of large-size lithium-ion pouch battery: Resting time between charge and discharge. *Energy Procedia* **2019**, *158*, 2623–2630. [CrossRef]
21. Wu, X.; Lv, S.; Chen, J. Determination of the Optimum Heat Transfer Coefficient and Temperature Rise Analysis for a Lithium-Ion Battery under the Conditions of Harbin City Bus Driving Cycles. *Energies* **2017**, *10*, 1723. [CrossRef]
22. Larrañaga, M. Design and implementation of a Direct Liquid Cooling strategy for HEV/EV focus on a pouch type cell body refrigeration. In Proceedings of the Thermal Management Systems Symposium (21TMSS), Online, 12–14 October 2021.
23. Li, M.; Wang, J.; Guo, Q.; Li, Y.; Xue, Q.; Qin, G. Numerical Analysis of Cooling Plates with Different Structures for Electric Vehicle Battery Thermal Management Systems. *J. Energy Eng.* **2020**, *146*, 04020037. [CrossRef]
24. Panchal, S.; Khasow, R.; Dincer, I.; Agelin-Chaab, M.; Fraser, R.; Fowler, M. Numerical modeling and experimental investigation of a prismatic battery subjected to water cooling. *Numer. Heat Transfer Part A Appl.* **2017**, *71*, 626–637. [CrossRef]
25. Brown, G.O. The History of the Darcy-Weisbach Equation for Pipe Flow Resistance. *Proc. Environ. Water Resour. Hist.* **2002**, *40650*, 34–43. [CrossRef]

World Electric Vehicle Journal

MDPI

Article

Characterisation of Norwegian Battery Electric Vehicle Owners by Level of Adoption

Nina Møllerstuen Bjørge [1], Odd André Hjelkrem [2,*] and Sahar Babri [1]

1 SINTEF, 7465 Trondheim, Norway
2 SINTEF Energy Research, 7465 Trondheim, Norway
* Correspondence: oddandre.hjelkrem@sintef.no

Abstract: In this paper we investigate differences between groups of Norwegian electric vehicle owners, sorted by their adoption level. The grouping is based on adoption theory and the share of battery electric vehicles in new car sales numbers. We investigate Norwegian adopters' preferences, values, and motivations for choosing a battery electric vehicle. The main data source is a yearly survey between 2015 and 2020 amongst Norwegian electric vehicle drivers. The motivation of the study is to reveal different choices by the adopter groups, contributing to policy recommendations and incentives for other countries. However, the Norwegian case might be a special one, having economic advantages which many other countries do not have access to. We assess the validity of the results and policy recommendations by analysing the results of a survey amongst the Nordic countries on investment choices concerning battery electric vehicles.

Keywords: policy; market development; user behaviour

Citation: Bjørge, N.M.; Hjelkrem, O.A.; Babri, S. Characterisation of Norwegian Battery Electric Vehicle Owners by Level of Adoption. *World Electr. Veh. J.* **2022**, *13*, 150. https://doi.org/10.3390/wevj13080150

Academic Editor: Joeri Van Mierlo

Received: 13 July 2022
Accepted: 5 August 2022
Published: 9 August 2022

1. Introduction

As a response to local pollution and aims to reduce greenhouse gas emissions in the atmosphere, many countries have applied extensive measures in different sectors. Targeting the transport sector, Norway has focused on electromobility and has introduced and upheld beneficial local and national policies, making it convenient and affordable to own electric vehicles [1]. This accelerated the sales and the adoption of battery electric vehicles (BEVs), as data on new vehicle registrations in 2022 (year to date) show that 83.7% of new registrations are BEVs [2].

Multiple incentives were introduced in Norway throughout different periods of the adoption phases. The added value taxes have traditionally been high, and already in 1991, the BEV users were exempted from this tax. The exemption led to considerable price reduction and accelerated the import of BEVs onto the Norwegian market [1]. In the late 1990s, exemption from toll road fees were introduced, and after a test period of free parking, this was fully established in 1998 [1]. These measures contributed to expanding the share of BEVs onto the Norwegian market, starting the diffusion of the early market phase. In this phase, the innovation process also took place in Norway. Even though international car producers dominated the market in the long run, the technology development was important for policy incentives. The exemption from VAT and BEV access to the bus lanes were pointed out to be of the most important incentives provided in the early market phase. Numbers from the former Norwegian EV-producer, Th!nk, shows that the demand for BEVs increased considerably after the adoption of these policies [1]. In 2009, the bus lane access became a permanent incentive, and all BEVs could also use ferry connections without cost [3].

Two key documents were launched in 2012: "The Climate Message" and "The Climate settlement" [4]. The documents pointed out clear goals for the transport sector and put a maximum limit of greenhouse gas emission for new cars in 2020. The limit should not exceed an average of 85 g of CO_2 per km. "The Climate Message" also emphasised several measures to make the Norwegian car park more climate friendly, such as utilising low emission technology for taxis, facilitating the use of electric vehicles and establishing new electrification infrastructure. These measures contributed to the spread of BEVs in other sectors as well, enforcing the spread of technology onto freight operators and other transporters.

The measures above have contributed to decreasing the barriers to purchase BEVs and have been important for the increasing adoption in Norway. A report from 2020 suggests that BEV owners have become more similar to internal combustion engine vehicle (ICEV) owners. They have also become richer, emphasising economic incentives when purchasing a BEV. Moreover, there are specific trends to own a BEV as a second car [5]. Gradually, the state authorities have reversed many of these exemptions; ferry fees and toll road tickets where reintroduced in 2018, and the bus lane access became more restricted. It is therefore likely that there are some differences between people who purchased BEVs in the early market phase, compared to those purchasing it after the incentives became more prominent, and later removed. Identifying and discussing these differences is the main aim of this paper.

Previous Literature on BEV Adoption

The diffusion phase of BEVs in Norway started already in the 1970s, and the period from 1990 to 2009 is the "technology niche creation phase" characterised by an almost non-existent market. As stated above, several policies were implemented during this period with the main intent to nurture the BEV industry and to create a market [6]. The BEV users in Norway have since evolved from a typical innovator and early adopter group in 2014–2015 to an early majority in 2018. During this evolution, the BEV users have become more akin to ICEV users [5].

Several factors seem to affect people to adopt BEVs, and multiple studies have been done on Norwegian adopters due to the high uptake of BEVs in Norway compared to other countries. A study on local electromobility policies in Oslo and Bergen found that the policies contributed to an increase in the relative advantage of BEVs and therefore spiked the sales of BEVs in these cities [7]; in particular, policies that expanded the charging infrastructure and gave access to bus lanes for BEVs. A study from 2016 [8] reports that in small communities such as Norway, pricing incentives are the most efficient measure for encouraging people to adopt BEVs. This result was confirmed in another study [3], which highlights the exemption of both purchase tax and VAT as the most important incentive for BEV adoption in Norway. The people responding to reduced BEV costs are often highly educated, with lower income, and living in the third largest city of Norway, Trondheim [3]. Users that are concerned with incentives being removed are often over-emphasising the importance of the incentives itself [3].

Multiple studies on electromobility in other countries than Norway also investigate criteria for users to adopt BEVs. A study from 2017 [9] identified barriers for adopting BEVs in China through surveys. Experienced drivers emphasised battery and purchase cost as an important barrier for adopting BEVs. An increase in the amount of public charging infrastructure, installing new charging infrastructure as well as purchase subsidies were also measures emphasised in the literature. The cost considerations being the most powerful incentive in China was confirmed in another study in 2019 [10]. Also in China, it was found that only 33.6% of the respondents not owning a car were interested in purchasing BEV [9]. Of the respondents already owning a vehicle, 46.8% were interested in purchasing a BEV as a second family car. The Chinese BEV population has now developed to an early majority. However, by comparing China and the early adopter staged South Korea, [11] investigated motives behind BEV purchases in the two countries. The environmental concern was higher in China than in South Korea, where the economical motives were the most important for buying a BEV [11]. The Chinese early majority also had a stronger level of symbolic perception towards their vehicle and were sensitive to the fact that driving a BEV would contribute to an image of being environmentally conscious and responsible. However, the Chinese respondents were less satisfied in general, and emphasized lack of charging points and too high purchasing price. The Korean BEV users were more satisfied, mainly due to the large cost exemptions.

A comparison between Chinese and US BEV users from 2015 [12] focused on the difference in willingness to pay, and found that users from the US were willing to pay higher prices for BEVs than Chinese users. The study also suggests that the Chinese respondents are more receptive to BEVs than American respondents regardless of the subsidies. This is due to key distinctions of the Chinese market, which is not prominent in the US. About 2/3 of the Chinese adopters are first time car buyers who typically have less experience with both gasoline and plug-in vehicle technology [12].

Several studies have looked into specific traits of BEV users and what they emphasise as motivating factors for adopting BEVs. A Norwegian study [5] state that most people owning BEVs are characterised by high income and adds that very few have a minority background. Their study also confirms that BEV owners usually live in large cities and suburbs. Households owning one BEV are often families consisting of couples with kids or households consisting of singles [5]. In this group, the age span between 25 to 44 years is dominating. Couples, with or without kids, are found to be most frequent in the group owning two BEVs [5]. A qualitative study [13] investigated BEV users through in-depth interviews and found that the BEV users were a heterogenous group, which for either economical or practical purposes, adopted a BEV. They also emphasised to be environmentally friendly expressing pride of their vehicle. In a German study [14], the early adopters of BEVs in Germany were found to be middle-aged men living in suburbs and rural areas. Another study of Danish and Swedish BEV users found that males with high levels of income and education were predominant [15]. For regions without strong policies for BEV uptake, it was found that BEV users were characterised by high income, high technological competence, home ownership and green political preferences [16]. Regarding changes in attitudes over time as BEVs become more widespread in a market, it was found in [10] that people with more experience with BEVs were also more willing to adopt them. The more people who are in favour of BEV technology and aware of environmental issues, the more others are willing to adopt them [17].

Several studies have used theoretical frameworks to classify groups of BEV adopters. In [18], two distinct groups of early adopters are identified, which are high-end and low-end adopters, with differences in opinions, preferences, and sociodemographic properties. As high-end adopters had a more positive view on future BEV ownership, it was suggested that policies should target low-end BEVs. A study of EV adopters in China showed that a specific type of adopter, lead users, are important for successful adoption of BEVs due to their expertise and experience [19].

It is apparent from the presented literature that there are differences between BEV adopters, but the studies are almost exclusively focusing on early adopters. Hence, there is a gap in the literature on the classification of BEV adopters beyond the early adopter stage. The main aim of this paper will be to contribute to this field by characterising how Norwegian BEV users have developed from the early marked phase until today, and which incentives they emphasise in different adoption phases.

It is important to investigate Norwegian BEV adopter groups to gain more insight on adopter choices. This knowledge may contribute to recommend policies and incentives for other countries. In this study we address this issue by investigating how BEV users have changed from 2015 to 2020, depending on how long they have owned a BEV. Based on previous literature, it is reasonable to assume that there will be distinct differences between a user adopting a BEV in 2015, versus a user adopting it in 2020. We therefore aim to investigate these research questions:

Which incentives were emphasized by first movers and early adopters when purchasing BEVs in Norway?

How can the adoption experiences from Norway be used in policy making for further BEV acceleration in other European countries?

First, we investigate data from a yearly survey amongst Norwegian BEV users between 2015 and 2020 to see what influenced different stages of first movers, early adopters and early majority when adopting BEVs. The second part is to investigate numbers from European countries (EU + EFTA + UK + Turkey), to see what stage each country is in, depending on the theory.

2. Theory of Diffusion of Innovation

The increase in market share for BEVs in the Norwegian market, demands for an investigation of different type of user groups. In [20], Roger's theory of diffusion of innovation is presented. The theory is based on different characteristics related to groups that adopt new innovations at different times, which are presented in short in this chapter.

The very first segment of the population to adopt are innovative in nature and are referred to as first movers. They are characterised as a deviation from the social system, having high technological competence. The first movers might be closely related to the innovation process, and a prerequisite to be a first mover is control over substantial financial resources and the means by which to cope with non-profitable investments. The innovator also plays a crucial role in the diffusion process. An early adopter is someone that starts using the innovation earlier than the rest of the participants in the social system. Rogers describe an early adopter as an opinion leader that is quite respected by his or her peers. This person often ends up as a role model to whom later adopters turn for advice. Further on, Rogers describes early adopters as more educated, having higher social status and are in general more wealthy than later adopters. They are well integrated into the social system, trust science and cope well with risk taking.

In the social system among the first movers and the early adopters, the early majority, late majority, and the laggards also exist. The early majority are the most numerous groups of adopters and deliberate for a while before deciding on adopting a new technology. They are the last group to adopt before the other half of the social system. The late majority adopts a technology later than the rest and are in general sceptic towards new innovations. The laggards are the last adopters and have their point of reference in the past. Comparing socioeconomic characteristics between early adopters and late adopters, Roger points to a small difference in age. Early adopters have higher years of formal education and are more eager to change, coping well with risks and uncertainties.

The stages of diffusion of innovation are shown in Figure 1, which groups the population according to percentages of the population that sequentially adopt an innovation. The first 2.5% of the population that adopts an innovation are known as first movers. The second group are the early adopters and consist of the next 13.5% of adopters in the population. The segment of the population which is within the adoption interval of 16% to 50% are regarded as early majority. The late majority are in the range of 50% to 84%, while the last 16% that adopts an innovation are called laggards. In our study we limit the social system to new vehicle buyers and define the stages of adoption in line with the percentage of BEVs sold per year compared to total new vehicle sales.

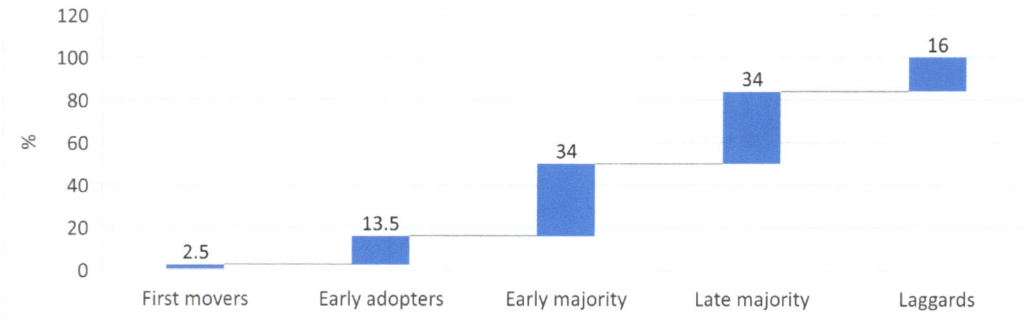

Figure 1. The sequence of adopter stages from Roger's "Diffusion of Innovation" [20], and the size of each group in terms of percentage of the population.

3. Data and Methods

The main data source in this study is a survey conducted by the Norwegian EV association between 2015 and 2020. This is a yearly survey amongst members of the EV association, and the number of respondents is in the order of 10,000 each year. Most of the questions are repeated each year.

The reasons for choosing data between 2015 and 2020 are based on our theory, "Diffusion of innovation" and numbers on new BEV registration obtained from European Alternative Fuel Associations (EAFO) [2]. The numbers from Norway from the year 2011 to 2020 are shown in Table 1.

Table 1. New registrations of passenger BEVs in Norway. Source: [2].

First Movers		Early Adopters				Early Majority			Late Majority
2011	2012	2013	2014	2015	2016	2017	2018	2019	2020
1.5%	3.1%	5.8%	12.6%	17.1%	15.7%	20.8%	31.2%	42.3%	54.3%

According to our application of Roger's classification theory, the Norwegian new registrations of BEVs already went into the phase of early adopters in 2012. The early majority stage was reached in 2015, and the late majority stage in 2020.

With data from 2015, it will be possible to extract data from respondents who obtained a BEV four years earlier, i.e., in 2011. This gives us important insight into first mover reasons for obtaining a BEV. Applying data from 2015 and investigating respondents that obtained BEV two or more years prior, we will see the important factors for BEV adoption in the early adopter stage. By applying data from 2020 and investigating people that obtained a BEV two or more years earlier, we will gain insights into early majority reasons for adopting BEV.

We chose newly registered passenger cars because we know that these are cars that most likely are in everyday use. If we chose to investigate the total amount of BEVs in the car park, it might be large parts of a country's total car park that are old vehicles being used less frequently.

The surveys have been distributed mostly through the BEV associations channels, and it is therefore reason to believe that the results will be generalisable to the populations of BEV users.

In addition to the Norwegian survey described above, the Nordic EV barometer is conducted yearly. This is a narrow survey focusing only on the next car purchase of the respondent. A representative sample of the populations in Denmark, Finland, Iceland, Norway and Sweden are interviewed through a web-based questionnaire.

The method applied in this paper includes a descriptive analysis of the data from the survey conducted by the Norwegian EV Association, with the theory of diffusion of innovation as a framework for classification of BEV users. This is further discussed in a European setting using new car registration data from EAFO and the Nordic EV barometer to discuss the main barriers for higher EV adoption on a European level.

4. Results

In this chapter, we present differences between the defined adoption groups, with results from both the Norwegian and Nordic survey. Some of the demographic differences between the groups are shown in Figure 2. Here, we see that the first movers are mainly men, older, have higher education levels, and are more often in smaller households compared to the early adopters and early majority. The share of females increases further as the adoption groups evolve. The education level decreases throughout the adoption groups. Usually, first movers are in households of 1–2 and 3–4 people.

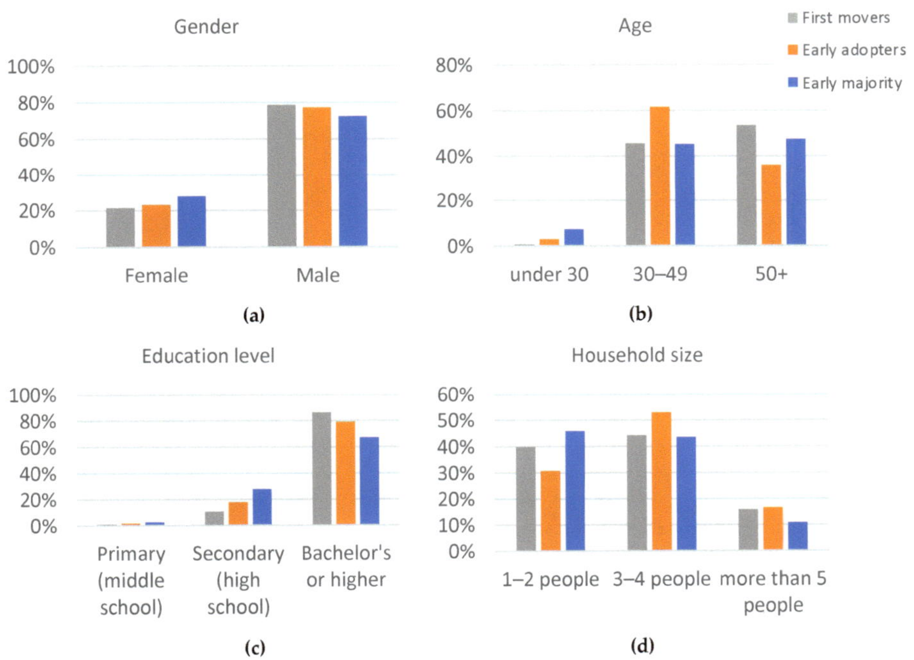

Figure 2. The distribution of sociographic differences across the three adopter groups in percent: First movers (grey), Early adopters (orange), Early majority (blue). (**a**): Gender distribution. (**b**): Age distribution. (**c**): Education distribution. (**d**): Household size distribution.

In Figure 3, results about vehicle choice, travel mode choice and purchase inspiration are presented. Regarding car purchasing choice, this question is also asked in the Nordic EV barometer. However, the latter is sampled in a representative part of the population, while the sample in Figure 3a are members of the EV association. We see that the limited sample—theoretically representing adopters up to and including the early majority—states that only 2–4% would have bought a petrol or gasoline car. In the Nordic survey, the equivalent numbers are 26% (Norway), 34% (Sweden), 30% (Denmark), 45% (Finland) and 39% (Iceland).

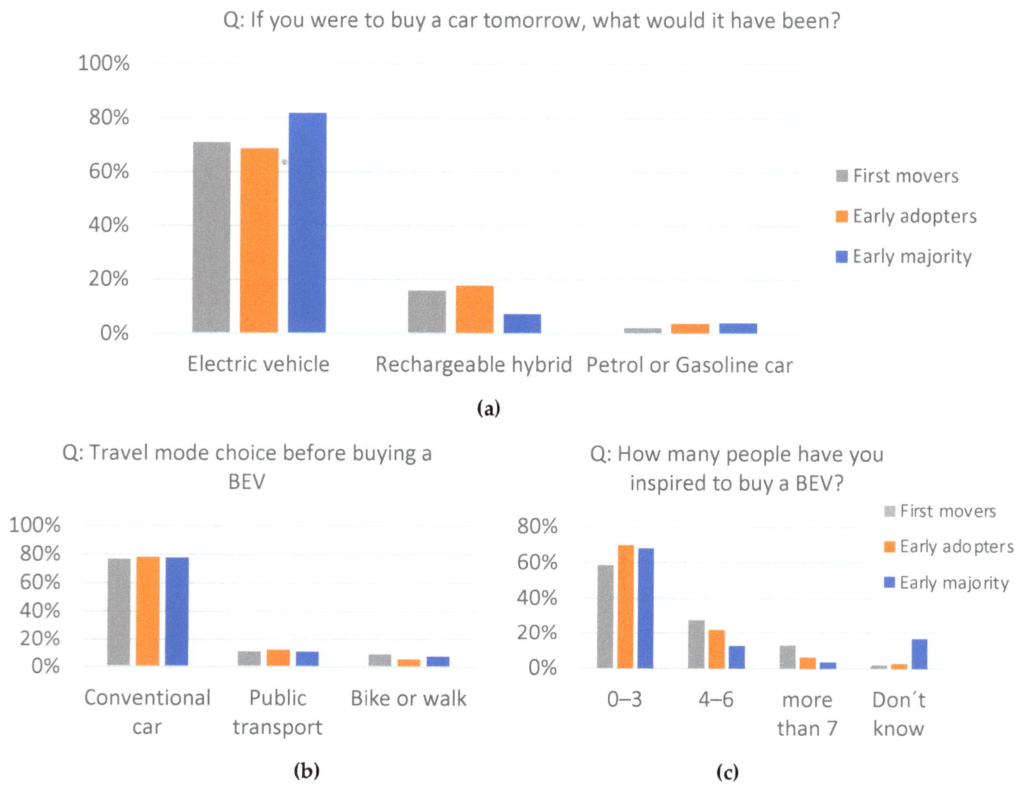

Figure 3. The share of responses for the three adopter groups regarding questions about their future choice of vehicle (**a**), previous travel mode (**b**) and the amount of people they have inspired to buy a BEV (**c**).

In Figure 3b, the main travel mode choice before buying a BEV is investigated. The results indicate that early adopters had a slightly higher preference for cars than other modes compared to the first movers and early majority. Regarding inspiring other people to buy a BEV, Figure 3b shows that the first movers and early majority state to a higher degree than the early majority that they have been influential in other people's choices.

When looking at the drivers for buying a BEV, as shown in Figure 4, we see that the first movers emphasized climate friendliness, while the early majority emphasized low operating costs. Regarding the most important information source used before buying a BEV, the first movers found information about BEVs in media before purchasing. However, the early majority got information to a higher degree from friends, family and from web pages.

As shown in Table 2, each adopter group consists of responses sampled over several years. To see if there is a change over time, a specific response for one of the adopter groups is investigated. In Figure 5, the differences in opinions by early adopters between 2015 and 2017 regarding drivers for buying a BEV are presented. Regarding lower operating costs, the opinions are the same for each year. For the other responses, the change between years is within a difference of two percentage points, in contrast to the maximum difference between adopter groups to the same responses of five percentage points.

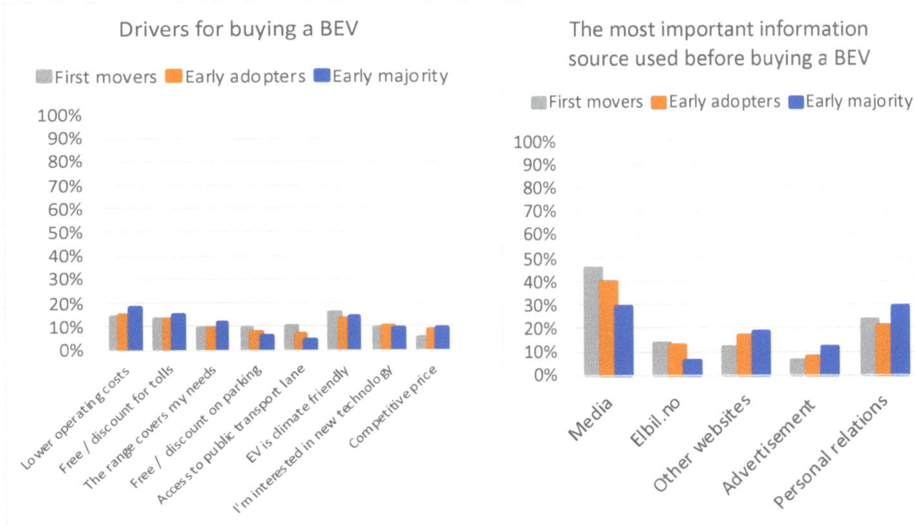

(a) (b)

Figure 4. The share of responses for the three adopter groups regarding drivers for buying a BEV (**a**) and important information source when buying a BEV (**b**). Elbil.no is the website of the Norwegian EV association.

Table 2. The classification of adopters in the surveys and number of respondents per year.

Group	Definition of Each Group from the Dataset	No. of Observations
Early majority	Dataset 2020: Those who has owned BEV for 1–4 years. Dataset 2019: Those who has owned BEV for 0–4 year Dataset 2018: Those who has owned BEV for 0–3 year Dataset 2017: Those who has owned BEV for 0–2 year	6784 7997 6766 7183
Early adopters	Dataset 2018: Those who has owned BEV for 4–5 years Dataset 2017: Those who has owned BEV for 3–5 years Dataset 2016: Those who has owned BEV for 2–4 years Dataset 2015: Those who has owned BEV for 1–3 years	718 1547 914 3107
First movers	Dataset 2016: Those who has owned BEV for more than 5 years Dataset 2015: Those who has owned BEV for more than 4 years	139 423

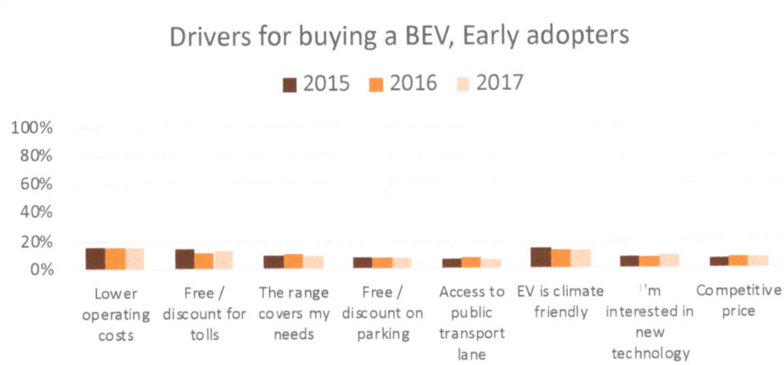

Figure 5. The share of responses between 2015 and 2017 regarding drivers for buying a BEV by early adopters.

European Adoption Stages

In Figure 6, the innovation stages in European countries are shown, split into the four different categories first movers, early adopters, early and late majority. As for the first movers, there is a predominance of Southern and Eastern European countries. In the early adopter group, we find that the Northern European countries are dominating. The countries that have come furthest in BEV adoption are Iceland, Norway and the Netherlands. A clear distinction between Western and Eastern Europe is apparent in the graphic representation of the numbers.

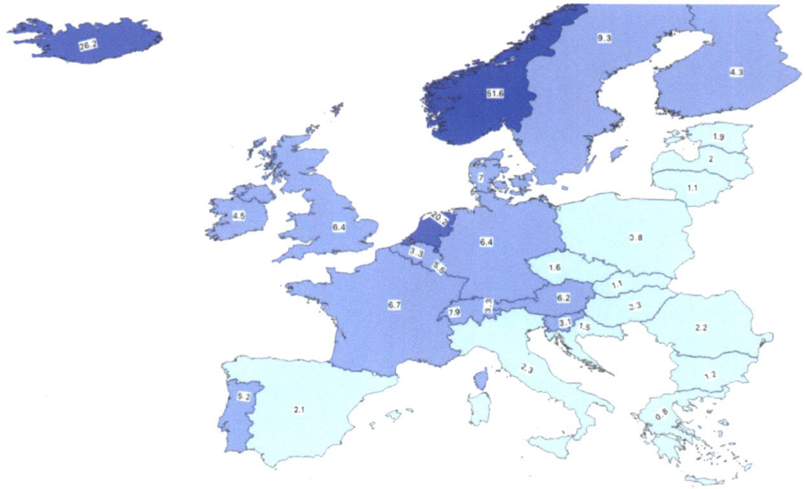

Figure 6. BEV Distribution across Europe. Dark blue: Late majority (50–84%). Blue: Early majority (16–50%). Light blue: Early adopters (2.5–16%). Turquoise: First movers (0–2.5%).

5. Discussion

This study aims to improve the understandings of BEV adopters in Norway and drivers for different adoption phases. Based on the Diffusion of Innovation theory, people that adopt new technologies have different characteristics. Based on this, they should also emphasise different kinds of incentives. Knowing the emphasis of each adopter group in Norway may contribute with valuable knowledge about which incentives and policies to implement in other countries, aiming to accelerate electromobility.

The application of the theory should not be done without consideration to the representativeness of the sample. Therefore, it is important to emphasize that the sample in the Norwegian survey consists of people already driving a BEV. The results are generally in line with the expectations based on the presented literature review. In addition, as shown in Figure 5, there are some small variations over time in the responses, but it seems that the early adopter group is relatively consistent compared to the differences between adopter groups in Figure 4a. The two most frequent responses to important drivers for the early adopters, lower operating costs, and BEV climate friendliness, are observed at the same level for the 2015, 2016 and 2017 surveys. Although there might be individuals participating in the survey repeatedly each year, these results increase robustness in the application of the innovation theory.

5.1. Diffusion of Innovation Theory and BEV Users in Norway

The Diffusion of Innovation theory draws on both socioeconomic and personal traits in each adopter group. According to the theory, there are small differences in age between the early adopters and the later adopters. Our results, as shown in Figure 2, do not align completely with this. Instead, the early adopters have a higher number of people between 30 and 49 years old than the rest of the adopter groups. Our results point to the education level among people adopting technology in the early stages being higher than later adopters. This aligns with the theory stating that early adopters are more likely to have higher education than late adopters. First movers are also specified as highly technologically competent, which also might be related to the amount of higher educated respondents in this group.

There are several reasons for purchasing a BEV, and the driver for purchasing varies between each adopter group. In our results, the first movers emphasise on being climate friendly, the access to public transport, and environmental discount on parking in public parking lanes. Overall, the first movers are emphasising on economic incentives to a smaller degree than the early adopters and early majority. The theory establishes that first movers often are in a robust financial situation since they are investing in a technology very early in the innovation process. Because of this, the first movers do not seem to emphasise on economic incentives as much as the later adopter groups. This is visible in our results and might imply a smaller emphasis on economic incentives for this group, since they are relatively affluent.

The first movers emphasise on being climate friendly more than the later adopters when purchasing a BEV. The diffusion of innovation theory points to first movers emphasising on knowledge and science to a high degree, therefore acquiring much knowledge before choosing to invest. This might be seen in relation to their BEV adoption and their concern regarding climate change and environmental research.

The theory points to early adopters as opinion leaders, which implies their being used as reference points for providing information for later adopters. In our results, the first movers and the early adopters get most of their information from the media, while the early majority favours consulting other people that own a BEV, such as family or friends.

5.2. Transferability to European Countries

Our results show differences in how incentives are emphasized by adopter groups. In this section, we will attempt to draw some main learning points from our adopter study with transferability to other countries in Europe in mind.

Electric mobility is relatively new for large parts of Eastern and Southern Europe regarding the small share of BEVs in new car sales. According to the implementation of adoption theory presented in this paper, these markets are in the first mover phase. To accelerate and make it a desired means of transport, the evidence from our result might give important insights for policy makers in the respective countries. As per our results, it was not necessarily a case of economic incentives that drove the adoption for the first movers, since they already hold substantial amounts of money. This group seemed more driven by exclusive accesses and being recognised as environmentally friendly when choosing a BEV. Therefore, exclusive access to, for example, bus lanes or restricted parking areas might apply to the first mover group in Eastern and Southern Europe. This group also emphasises the environmental aspects of BEV technology. The first movers tend to rely on science and knowledge in their decision. The authorities in the respective countries can, based on this, run information campaigns with information about BEV and their impact on emission reduction and other environmental benefits.

For the countries in the early adopter phase, which is mostly North Western Europe, another set of incentives could be more efficient. We see that price reductions are more efficient for this group. However, it is worth mentioning that Norway is in a special position regarding both general wealth and taxes on fossil fuel cars, which is not necessarily transferable to other countries. Therefore, any incentive which increases the economic benefit of car ownership in favour of BEVs should be prioritised.

These kinds of incentives should also be continued when entering the early majority phase to continue the adoption and accelerating the BEV share. In this stage, purchasing price and operating costs are even more emphasised than by first movers and early adopters, suggesting a higher priority should be given to economic incentives compared to other faciliatory incentives.

6. Conclusions

This study has related the theory of diffusion of innovation on BEV users to classify them in different stages of adoption based on the percentage of new cars sold being BEVs. By applying this classification on respondents in surveys conducted by the Norwegian EV association, we investigated differences between adopter groups to see if the characteristics of each group coincide with adoption theory. The majority of the results, and especially gender distribution, educational level, environmental opinions and influencing power, are as expected with regards to adoption theory.

With the Norwegian BEV users already being in the late majority phase, large parts of Europe are still in the first mover and early adopter stage. While the first movers are driven by, e.g., environmental reasons, it seems to be necessary with economic incentives to accelerate the transition to other adoption stages to increase the BEV share in general.

Author Contributions: Conceptualization, N.M.B. and O.A.H.; methodology, N.M.B., O.A.H. and S.B.; formal analysis, N.M.B. and S.B.; investigation, N.M.B. and S.B.; data curation, S.B.; writing—original draft preparation, N.M.B., O.A.H. and S.B.; writing—review and editing, N.M.B., O.A.H. and S.B.; supervision, O.A.H.; project administration, O.A.H.; funding acquisition, O.A.H. All authors have read and agreed to the published version of the manuscript.

Funding: This research was supported by the project "Grid and Charging Infrastructure of the Future" (FuChar). The FuChar project is funded by the Research Council of Norway and industry partners under the ENERGIX programme, with project number 295133/E20.

Institutional Review Board Statement: Not applicable.

Informed Consent Statement: Not applicable.

Data Availability Statement: Data from the survey "Elbilisten" is not publicly available and was made available for this study by the Norwegian EV Association. Data from the Nordic EV barometer is available at https://www.nordicenergy.org/project/nordic-ev-barometer-2021/ (accessed on 6 July 2022).

Conflicts of Interest: The authors declare no conflict of interest.

References

1. Figenbaum, E.; Kolbenstvedt, M. *Electromobility in Norway-Experiences and Opportunities with Electric Vehicles*; Institute of Transport Economics (TOI): Oslo, Norway, 2013; Report 1281/2013.
2. European Alternative Fuels Observatory-Vehicles and Fleet. Available online: https://alternative-fuels-observatory.ec.europa.eu/transport-mode/road/european-union-eu27/vehicles-and-fleet (accessed on 6 July 2022).
3. Bjerkan, K.Y.; Nørbech, T.E.; Nordtømme, M.E. Incentives for promoting Battery Electric Vehicle (BEV) adoption in Norway. *Transp. Res. D Transp. Environ.* **2016**, *43*, 169–180. [CrossRef]
4. Norwegian Climate Policy. Available online: https://www.regjeringen.no/no/dokumenter/meld-st-21-2011-2012/id679374/ (accessed on 28 April 2022).
5. Fevang, E.; Figenbaum, E.; Fridstrøm, L.; Halse, A.H.; Hauge, K.E.; Johansen, B.G.; Raaum, O. *Who Goes Electric? Characteristics of Electric Car Ownership in Norway 2011-2017*; Institute of Transport Economics (TOI): Oslo, Norway, 2020; Report 1780/2020.
6. Skjølsvold, T.M.; Ryghaug, M. Temporal echoes and cross-geography policy effects: Multiple levels of transition governance and the electric vehicle breakthrough. *Environ. Innov. Soc. Transit.* **2016**, *35*, 232–240. [CrossRef]
7. Bjerkan, K.Y.; Bjørge, N.M.; Babri, S. Transforming socio-technical configurations through creative destruction: Local policy, electric vehicle diffusion, and city governance in Norway. *Energy Res. Soc. Sci.* **2021**, *82*, 102294. [CrossRef]
8. Mersky, A.C.; Sprei, F.; Samaras, C.; Qian, Z.S. Effectiveness of incentives on electric vehicle adoption in Norway. *Transp. Res. D Transp. Environ.* **2016**, *46*, 56–68. [CrossRef]
9. She, Z.Y.; Sun, Q.; Ma, J.J.; Xie, B.C. What are the barriers to widespread adoption of battery electric vehicles? A survey of public perception in Tianjin, China. *Transp. Policy* **2017**, *56*, 29–40. [CrossRef]
10. Sovacool, B.K.; Abrahamse, W.; Zhang, L.; Ren, J. Pleasure or profit? Surveying the purchasing intentions of potential electric vehicle adopters in China. *Transp. Res. A Policy Pract.* **2019**, *124*, 69–81. [CrossRef]
11. Chu, W.; Im, M.; Song, M.R.; Park, J. Psychological and behavioral factors affecting electric vehicle adoption and satisfaction: A comparative study of early adopters in China and Korea. *Transp. Res. D Transp. Environ.* **2019**, *76*, 1–18. [CrossRef]

12. Helveston, J.P.; Liu, Y.; Feit, E.M.D.; Fuchs, E.; Klampfl, E.; Michalek, J.J. Will subsidies drive electric vehicle adoption? Measuring consumer preferences in the U.S. and China. *Transp. Res. A Policy Pract.* **2015**, *73*, 96–112. [CrossRef]
13. Søraa, R.A.; Anfinsen, M.; Foulds, C.; Korsnes, M.; Lagesen, V.; Robison, R.; Ryghaug, M. Diversifying diversity: Inclusive engagement, intersectionality, and gender identity in a European Social Sciences and Humanities Energy research project. *Energy Res. Soc. Sci.* **2020**, 62. [CrossRef]
14. Plötz, P.; Schneider, U.; Globisch, J.; Dütschke, E. Who will buy electric vehicles? Identifying early adopters in Germany. *Transp. Res. Part A Policy Pract.* **2014**, *67*, 96–109. [CrossRef]
15. Haustein, S.; Jensen, A.F. Factors of electric vehicle adoption: A comparison of conventional and electric car users based on an extended theory of planned behavior. *Int. J. Sustain. Transp.* **2018**, *12*, 484–496. [CrossRef]
16. Brückmann, G.; Willibald, F.; Blanco, V. Battery Electric Vehicle adoption in regions without strong policies. *Transp. Res. Part D Transp. Environ.* **2021**, *90*, 102615. [CrossRef]
17. Bas, J.; Zou, Z.; Cirillo, C. An interpretable machine learning approach to understanding the impacts of attitudinal and ridesourcing factors on electric vehicle adoption. *Transp. Lett.* **2021**, 1–12. [CrossRef]
18. Hardman, S.; Shiu, E.; Steinberger-Wilckens, R. Comparing high-end and low-end early adopters of battery electric vehicles. *Transp. Res. Part A Policy Pract.* **2016**, *88*, 40–57. [CrossRef]
19. Wang, N.; Huang, Y.; Fu, Y.; Chen, L. Does lead userness matter for electric vehicle adoption? An integrated perspective of social capital and domain-specific innovativeness. *J. Consum. Behav.* **2022**. [CrossRef]
20. Rogers, E.M. *Diffusion of Innovations*, 5th ed.; Free Press: New York, NY, USA, 2003.

World Electric Vehicle Journal

MDPI

Article

Experimental Validation of an Automated Approach for Estimating the Efficiency and Heat Balance of Gearboxes Based on an Electrified Heavy Commercial Vehicle Axle

Roland Uerlich *, Sven Köller, Gordon Witham, Theo Koch and Lutz Eckstein

Institute for Automotive Engineering (ika), Steinbachstr. 7, 52074 Aachen, Germany; sven.koeller@ika.rwth-aachen.de (S.K.); gordon.witham@ika.rwth-aachen.de (G.W.); theo.koch@rwth-aachen.de (T.K.); office@ika.rwth-aachen.de (L.E.)

* Correspondence: roland.uerlich@ika.rwth-aachen.de

Abstract: Freight transport accounts for about half of all distances travelled in Europe. Therefore, freight transport is one of the decisive factors for reducing greenhouse gases and air pollutants. For this reason, the electrification of road freight transport is being promoted as part of the project "BEV Goes eHighway—[BEE]". The data basis for the modelling used in this project is an electric drive axle for a heavy commercial vehicle, which was developed in the "Concept-ELV^2" project. Based on the results of the previous project, the methodological tools that were developed are presented in this paper. These allow a wide range of possible powertrain topologies to be considered at the concept stage of development based on an estimation of future system characteristics. For this purpose, the components are automatically designed taking into account the mutual influence of the requirements and are evaluated in the context of the holistic system. This publication focuses on the efficiency and thermal evaluation of the transmission stages of the addressed electric drive units and validates the developed models using a pototypically designed electric commercial vehicle axle.

Keywords: electric powertrain; heavy duty; holistic powertrain design; modular topology

Citation: Uerlich, R.; Köller, S.; Witham, G.; Koch, T.; Eckstein, L. Experimental Validation of an Automated Approach for Estimating the Efficiency and Heat Balance of Gearboxes Based on an Electrified Heavy Commercial Vehicle Axle. *World Electr. Veh. J.* **2022**, *13*, 142. https://doi.org/10.3390/wevj13080142

Academic Editor: Zonghai Chen

Received: 28 June 2022
Accepted: 27 July 2022
Published: 2 August 2022

1. Introduction

For the evaluation of commercial vehicle-related emissions, a standardized Europe-wide tool was developed in 2018 in the form of VECTO (Vehicle Energy Consumption Calculation Tool). As is already customary for the passenger car segment with the WLTP (Worldwide harmonized Light vehicles Test Procedure), this tool defines reference cycles, test procedures and their boundary conditions, as well as the recording of emission values in real road traffic [1]. Since the segment of heavy-duty commercial vehicles covered by VECTO comprises fewer units but a significantly higher number of variants, component tests are used more frequently in this procedure [2]. These also enable less common vehicle variants to be evaluated concerning their emissions without an individual test procedure using standardized longitudinal dynamic simulation models. In this approach, transverse dynamic effects are neglected, as they have only a small impact on vehicle efficiency.

In the current certification processes, battery-electric vehicles are considered zero-emission vehicles both in the commercial and passenger car sectors, as they do not exhaust any local CO_2 emissions. Thus, even a small number of these vehicles has a significant impact on the fleet footprint of a vehicle manufacturer. However, this approach neglects the preceding process of energy generation, which also disregards the efficiency of the energy use of the respective vehicles. As a result, this factor, which is crucial for the success of the technology change, is only of secondary importance [3].

Consequently, the current legislative situation is similar to the situation in the entire commercial vehicle sector before the introduction of the VECTO, as no standardized approaches are currently defined to assess the efficiency of a commercial electric vehicle. As this topic has the previously described significant influence on the success of the technology and is gaining in importance due to the new regulatory environment, the first approaches for a supplement to the VECTO have already been developed. These are currently being implemented in a binding form [4,5].

A key aspect of the increased applicability of electric powertrains in the commercial vehicle segment is the enhancement of the usable driving range of the vehicles [6]. Due to the considerable battery capacities to be expected in this segment and the associated charging times, this aspect is gaining in importance compared to the passenger car segment with respect to economic viability. In this context, besides the consideration of the battery characteristics, the efficiency of the drivetrain is of great importance, as it has a significant influence on the achievable range of the vehicles. Due to the changed requirements compared to conventional powertrains, simpler powertrain and gearbox structures are currently being implemented. Especially in the case of gearboxes, only a few gear stages are currently common to meet the driving requirements. This results in the possibility to automate the design of these systems and thus cover a wide range of possible solutions. However, this automated process is also subject to divergent requirements with respect to efficiency and costs [7].

The automation of the design process also has the advantage that the transition from conventional to electric drives is underway, and manufacturers currently only have a limited pool of experience in this segment. Therefore, methods must be provided that allow an early estimation of the system properties. In this paper, therefore, thermal analysis of the gear stages is carried out in addition to the evaluation of the system efficiency. This aspect has gained importance due to the proximity of the electric motor and gearbox [8]. For the thermal modelling of gearboxes, approaches exist according to [9–11]. However, these do not link the efficiency evaluation with the thermal evaluation and are also not automated; therefore, the user constantly has to build a thermal analysis model manually for each gearbox. Thus, covering a large solution space is only possible with enormous manual effort.

2. General Methodology

In order to meet the challenges of electrifying powertrains, a tool was expanded and applied within the 'Concept-ELV^2' project that was developed at the Institute for Automotive Engineering (ika) for the synthesis and design of powertrains [12]. The holistic approach chosen starts with the basic parameters of the vehicle and derives component-centered requirements using longitudinal dynamic vehicle models. This systematic assessment of the individual component requirements enables the optimization of each individual component and the simultaneous modelling of the interdependencies between those components. One reason is that each powertrain component's individual optimum does not often lead to the optimum within the entire powertrain [13]. The conceptual design process of ika is visualized in Figure 1 in the form of a V-diagram.

The process of generating a suitable powertrain is subdivided into several sub-models, starting with the longitudinal dynamics model of the vehicle as described earlier. Taking into account the desired driving requirements, the specifications of the powertrain can be derived using this typical simualtive approach. Depending on the selected vehicle and application, different levels of detail of longitudinal dynamics models are used. In the following step, the automated design of the powertrain begins. Since initially only the global requirements of the powertrain are available, their dependencies on each other are implemented in the form of a parameter set. The underlying holistic approach is presented in [13]. It considers the effects of the scaling of a component on the requirements of the other components of the drivetrain and thus addresses the direct interaction of the component dimensioning in the overall system. In the following, the design of the gearbox is described in detail. The approach implemented within the holistic methodology for the dimensioning of the electric machine is described in detail in [14].

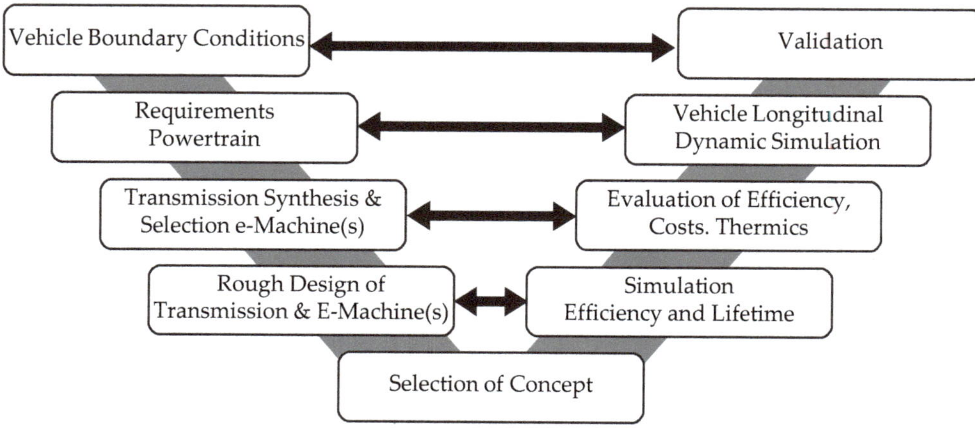

Figure 1. V-model of the ika design methodology for electric powertrains.

For the design of the gearbox, each parameter set of the holistic model is processed individually. In this subordinate step of the design process, possible drive topologies are designed algorithm-based on these requirements. Predefined criteria are used for evaluation, including the system's complexity, the achievable driving performance and the number of system elements. The most promising topologies according to these criteria are then further detailed. The shaft-bearing system is first defined in this detailing, and an initial design of the basic gearing parameters is carried out. This enables a component-based efficiency calculation of the concepts whereby a well-founded second evaluation and prioritization of the remaining topologies can be carried out. Based on this efficiency simulation, a thermal node model of the gearbox is created automatically. This model assigns the losses calculated for each component to the thermal nodes of the respective components and links them to each other according to the gearbox configuration. In addition, a simplified gearbox housing is generated to calculate heat dissipation to the environment. The gear oil, which plays a decisive role in heat transfer, is also considered in this method. An approach based on particle simulations was developed to integrate the oil into the thermal model presented in the publications by Köller et al. and Uerlich et al. [15,16]. Since this paper focuses on validating the models using measurement data, we refer to the mentioned publications for a detailed elaboration of the modelling process.

2.1. Requirement Specification Based on VECTO

The urban delivery cycle of VECTO was chosen as the basis for this design process in the exemplary application for the 'Concept-ELV[2]' project. As there is currently no modification of this cycle for battery electric vehicles integrated in VETCO, the cycle of conventionally powered vehicles was used for the evaluation. This driving cycle is shown in Figure 2 [1].

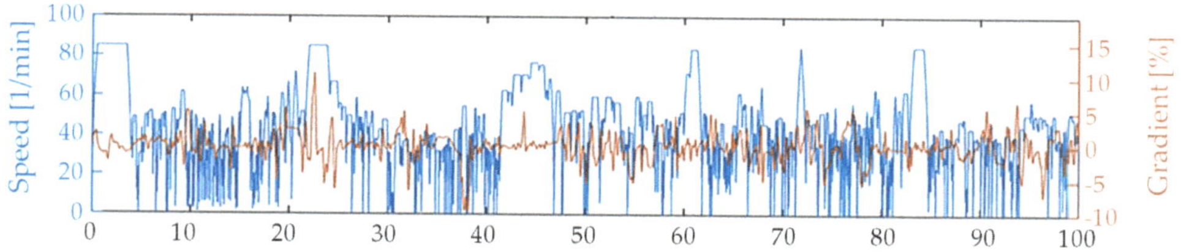

Figure 2. Urban delivery cycle of VECTO according to [1].

The diagram shows the required speeds in blue and the gradients in orange. It can be seen that the urban delivery cycle involves comparatively few constant-speed sections, but rather continual acceleration and braking demands. In addition, a wide range of road gradients is taken into account. This results in a set of requirements which, due to the homogeneous distribution of driving conditions within the cycle, delivers significantly more complex requirements with respect to overall vehicle efficiency.

2.2. Methodical Approach for the Automated Identification of Suitable Powertrain Topology and Gearbox Design

The processing of these requirements is carried out according to a method presented in detail in the work of Köller et al. [17]. The method introduced there differs from other state-of-the-art methods in its holistic approach to identifying solutions and subsequent optimization. In addition, the degree of automation sets it apart from other prior art processes and enables a large solution space of possible topologies to be included in the approach. The three essential design steps of topology synthesis, initial gear dimensioning and evolutionary combination are carried out within the method.

In the topology synthesis, all possible transmission topologies are generated that comply with the driving requirements and the restrictions of the solution space, such as the maximum number of gear stages, required torque and maximum speed. For these topologies, an estimation of the characteristic gearing parameters is conducted already in the first development step, allowing prioritization of the topologies according to their fitness. For this purpose, the criteria complexity, costs and efficiency are assessed based on the limited number of available parameters. The second step is followed by an initial dimensioning of the gear elements for the most promising variants. This analytical process is based on ISO 21771 and ISO 6336 [18,19]. Through the analytical approach, a large number of gear variants are considered and optimized in the final step of the method with the help of a genetic algorithm. This algorithm weighs and combines the solution variants concerning system mass, safety, lifetime and efficiency. The user determines the importance of these criteria according to the application's requirements. In the context of the application within the 'Concept ELV^{2}' project, 10^9 gearbox variants were initially generated with this algorithm, and about 400 variants were dimensioned in a second step. A selection of these variants and the prototypically realized variant is shown in Figure 3.

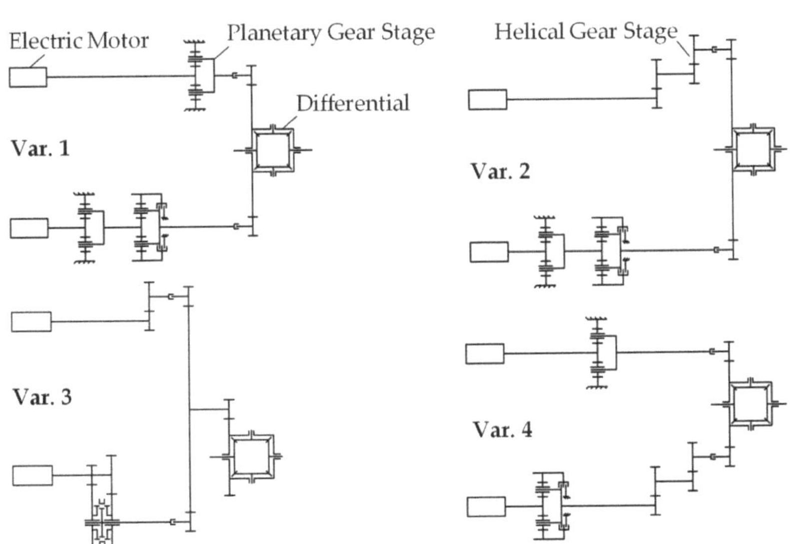

Figure 3. Exemplary selection of considered gearbox topologies according to [16,17].

The selection of gearbox topologies shown here provides small insight into the bandwidth. In all transmission topologies, two electric machines were set as a boundary condition, as these should ensure more efficient coverage of the widely varying load requirements of a heavy commercial vehicle. In variant 1, mainly planetary stages are used to realize the transmission, whereas in variant 3, only spur gear stages are used. Variant 2 and variant 4 represent different combinations of these gearing elements, so the advantages of both element types can generate an advantage with respect to service life or efficiency in one system. In addition to the automated weighting, the selection of the realized topology was based on the aspects of manufacturability and research-relevant system designs. Thus, a stepped planet was designed for electric machine one and a shiftable planetary stage with decoupled through-drive was designed for electric machine two. The topology selected for the prototypical implementation and its CAD design is shown in Figure 4. A detailed discussion of the synthesis process can be found in [17].

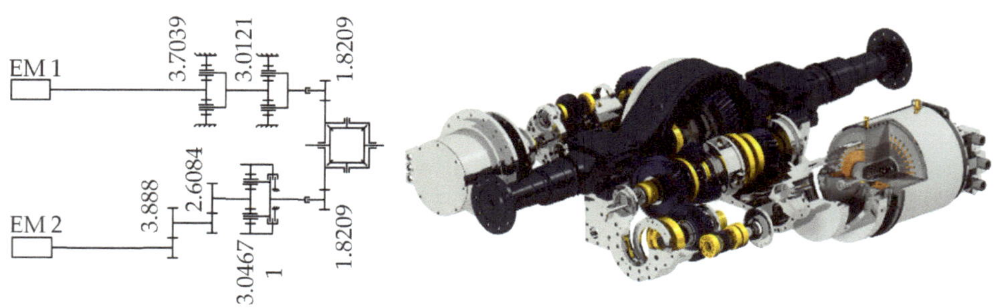

Figure 4. Prototypical realised gearbox topology including ratios and CAD rendering of the prototypical electric drive axle [15,16].

2.3. Method of Iterative Efficiency Simulation

An iterative calculation of the component losses is subsequently carried out for the generated and top-rated gearbox topologies for in-depth evaluation of the concepts. Based on the initially defined overall structure, it is necessary to determine the shaft-bearing system and to specify the bearing positions as well as the gear configuration. The general sequence of the process is divided into the modelling of the gearbox component bearings, shafts and gears, followed by linking those to each other with elements such as gear pairs and couplings. After the complete modelling of the mentioned components, the system of equations is set up automatically for each topology treated. In these, the speed relationships and the force and torque flows within the system are taken into account. In the first solution step, this equation system is solved without considering the system losses. These are then determined iteratively in a subsequent calculation step. The resulting losses lead to a load reduction for the following elements in the torque flow. Hence the iterative consideration is necessary. This process was presented comprehensively in the publication by Kieninger et al. and in further publications by Köller et al. and Uerlich et al. and is therefore not explained in depth here [13,15,17]. The results of this calculation approach for the designed topology of the 'Concept-ELV^2' project are shown in Figure 5 for the switchable transmission part of electric machine two (EM2).

The addition of the planetary stage causes a reduction in the overall efficiency of about one percent over the entire map. This influence is comprehensible due to the increased number of tooth contacts generated by the addition of the planetary stage and corresponds to the expectations based on known literature values [20]. Drag losses are neglected in the loss calculation, as no approach for the early stage of the gearbox design could be identified that would allow a feasible estimation. The gear losses calculated for each component are used in the following process step for the thermal assessment of the gearbox concept.

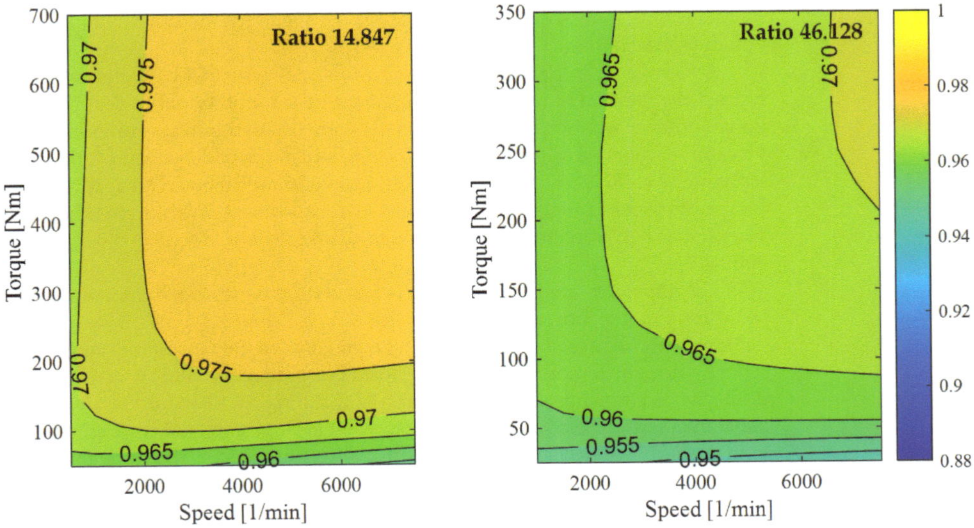

Figure 5. Simulative efficiency maps of the switchable gear stage of EM2 from Concept ELV[2].

2.4. Thermal Modelling Approach

Within ISO/TR 14179-2 [21], an approach for calculating the gearbox temperature is introduced, which relates the power loss of the gearbox to the outer surface of the housing. With this method, a thermal equilibrium state is calculated for the gearbox. This approach is not possible to identify thermally critical system parts due to the simplified procedure. The procedure also provides only limited information, as the standard already pointed out that a deviation of 14% from measured values is to be expected. To be able to make a more precise prediction, a thermal network is generated for the gearbox at this point in the ika method, which includes at least three nodes for each loss-generating gearbox component. An exemplary network of a gearbox section is shown in Figure 6. In this figure, the meshing of the bearings, gears, shaft, seal, housing, oil and environment can be traced on a gearbox section.

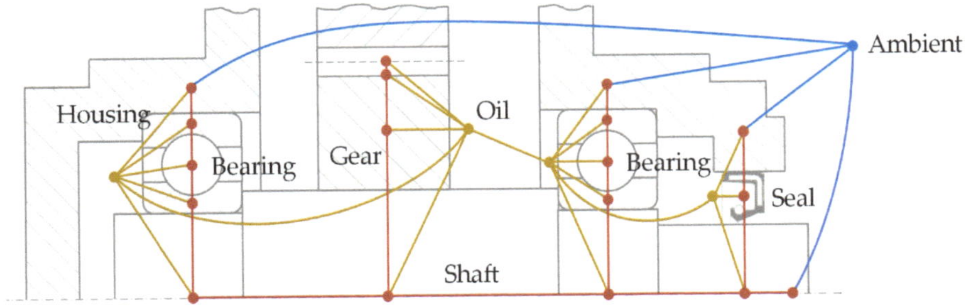

Figure 6. Exemplary representation of the thermal node model for a gearbox section.

The thermal modelling approach of a gear using thermal networks is already established. Such an approach already used by Blok [22] to model a single-stage FZG gear test rig. This modelling aimed to obtain feedback on the thermal behavior of the tooth contact, the lubricant film in the tooth contact and the tooth flank. This modelling approach was not validated by measurement in his work and used a too-coarse mesh with five accounts. In Funck's [23] work, this approach was refined and validated by measurement. However, a deviation of 25% between the measured value and the model was determined there.

A significant improvement in the results was achieved by Geiger [9], who further refined the thermal network and divided the gear into thermal nodes tooth flank, tooth body and gear wheel. Similarly, the shafts were divided into several nodes. The gear oil was still represented as one node at this point, which can be considered sufficient due to the structure of an FZG test rig. Additionally, his work implemented an approach that enables temperature calculation under transient load conditions. With these adjustments, good compliance between the model and the measurements was achieved, with a deviation of about 10%.

The approach used in this paper is based on the previously mentioned approaches but differs in the chosen application. Instead of a single-stage FZG test rig, multi-stage vehicle gearboxes are modelled. These vary not only in the number of stages but also in the housing design. A more closely fitting form is assumed for vehicle transmissions instead of the cuboid housing shape used in the FZG test rig. In this approach, an automatically generated housing is created based on the shaft arrangement and corresponds to a detailed housing only as an initial approximation. The interior of the housing, which is also modelled in this way and filled with an oil–air mixture during operation, is subdivided into further sub-elements and is represented with a thermal node for each. This subdivision enables thermally critical points within the gearbox to be identified based on the oil node temperatures. The gears in the model consist of the node flank, tooth and gear body, and the bearings are represented by the node inner ring, rollers and outer ring. The shafts are divided into smaller shaft segments. The calculation procedure is divided into four essential steps: the generation of the thermal network, the time-step calculation, the temperature calculation and the evaluation. These steps and their details are shown in a flow chart in Figure 7.

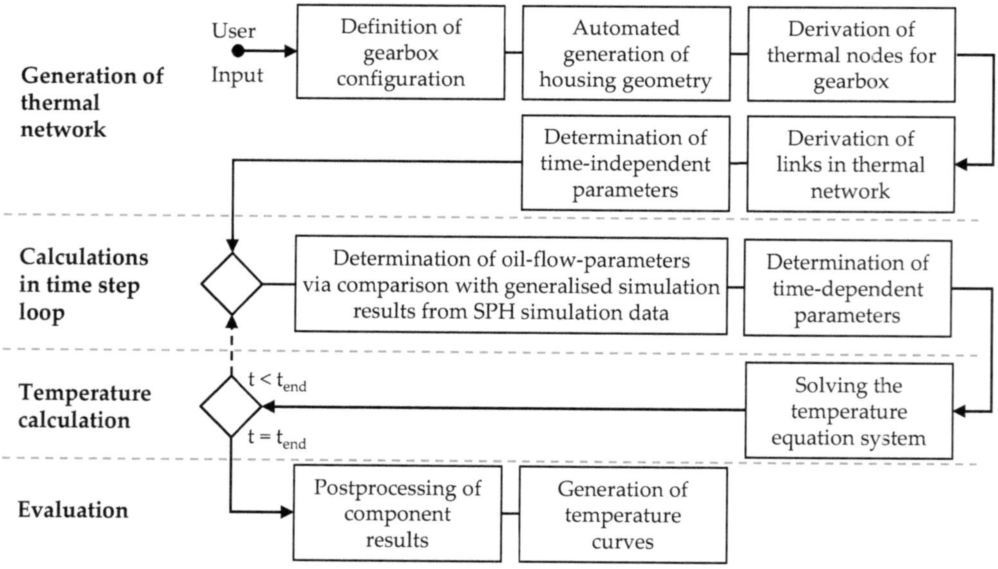

Figure 7. Stepwise process of model generation and calculation of the thermal network.

First, the gearbox components are divided into single nodes, and the required data are specified and determined. After automatically detecting all links between the nodes, the time-independent parameters are calculated. In step two, the time-dependent parameters and conductivity values are determined. Moreover, the flow state of oil is calculated based on look-up tables from SPH simulations. The next step is solving the equation system for the component temperatures for the respective time step using the calculated parameters before. In the final step, the results of the component-by-component calculation are combined and evaluated. The detailed description of the configuration of the thermal simulation model is given in [24].

3. Results

The measurement results of the prototype gear unit are presented in the following and are used to validate the simulation results. For the measurement, the gear unit was mounted between test bench machines. These emulated the electric motor as the drive unit on the one hand and the output on the wheel on the other. The mechanical efficiency was determined using a torque measurement sensor at the gearbox input and a torque measurement sensor at the gearbox output. The representation of these measuring points on the prototype can be seen in Figure 8.

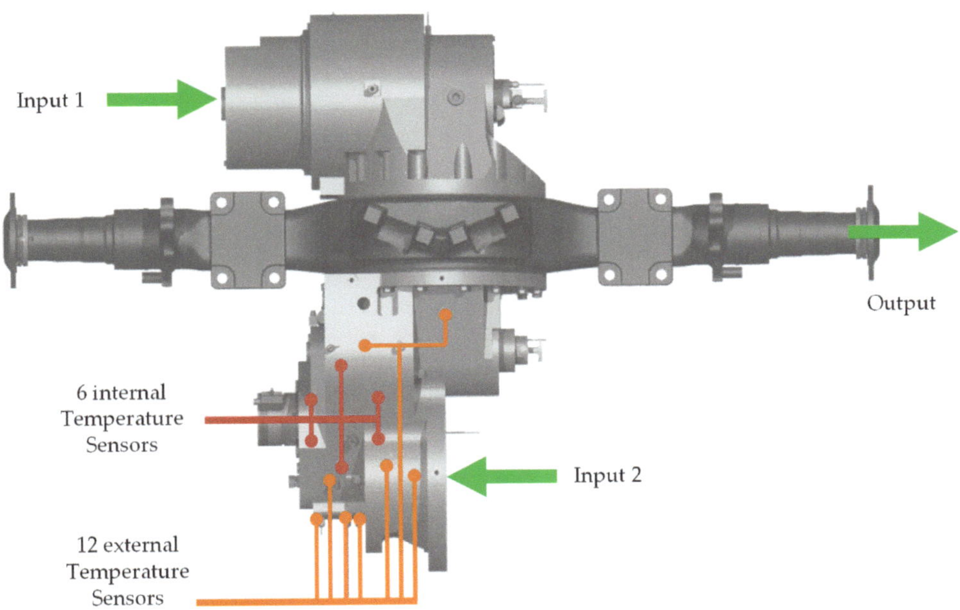

Figure 8. Measurement setup of the prototype for measuring the mechanical efficiency as well as the temperatures of 12 stationary and 6 rotating gearbox components.

For the validation of the efficiency calculation presented in this paper, only the measurement of the gearbox part of EM2 was carried out, thus the measurement of the mechanical performance of input 2 compared to the output. In addition, for the validation of the simulated thermal balance, the temperature was measured on twelve fixed gearbox components, such as bearing outer rings, seals, housing and the oil inlet and outlet. In addition, the temperatures of six rotating gear components were recorded. These were each recorded as a pair; therefore, three different component temperatures are available for comparison with the simulation. These are located in the tooth flank, a shaft segment and a bearing inner ring.

3.1. Efficiency Results

The gearbox efficiency maps of Figure 9 were recorded with this measurement setup. These maps show the system's efficiency concerning the input speed and the input torque. Both maps show gaps in the upper speed and torque range compared to the simulation. These ranges have not been measured since the driving requirements do not demand them of the vehicle and could not be covered using the emulated electric motor. Furthermore, these areas were not considered in the in-depth design of the gearbox, as this would lead to a massive over-dimensioning of the unit compared to the requirements.

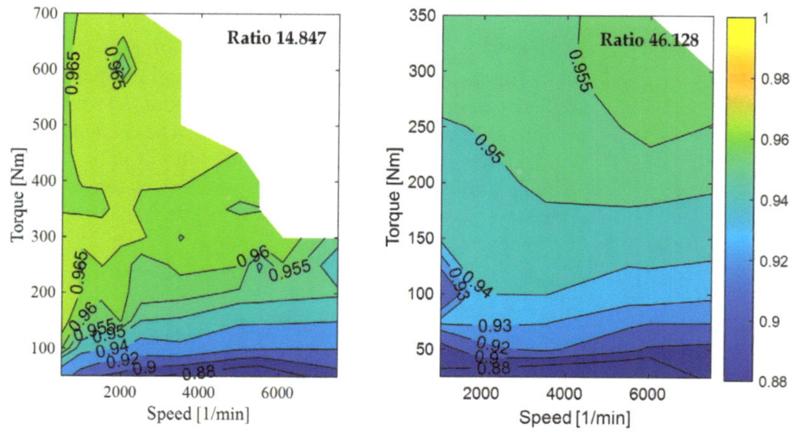

Figure 9. Measured efficiency maps of the prototype electric drive axle.

Compared to the simulation, there is deterioration in the efficiency of one to two percent in both gears in the range above 150 Nm input torque and over the entire speed range. Below this input torque, this deviation increases for the gear with the transmission ratio of 14.847 to five percent at 100 Nm and up to ten percent in torque-free operation. This is the same as for the gear with a ratio of 46.128, but a five percent deviation only occurs at an input torque of 75 Nm and a ten percent deviation during towing operation. These slight deviations can be attributed to the changed speeds caused by the different gear ratios, especially since the load-independent losses significantly influence the efficiency in these load ranges. Thus, the comparatively high deviation of the measurement results in the lower efficiency map range, which is largely due to the lack of the modelling of the drag losses, the implementation of which is currently still a current research topic point.

3.2. Thermal Modelling and Measurement Results

Alongside the efficiency measurement of the gearbox, a thermal characterization was also carried out. For this purpose, the load cases listed in Table 1 were recorded in a steady state. Each of these load cases was held for 30 min. Then, the gearbox was cooled down to its initial state; therefore, the measurement of the following load case could be carried out with approximately the same initial state. However, slight deviations due to different initial temperatures of the individual cannot be avoided due to the test setup.

Table 1. Overview of load cases that were carried out on the test bench.

Gear Ratio	Rot. Speed [1/min]	Torque [Nm]	Oil Temp. [°C]	Gear Ratio	Rot. Speed [1/min]	Torque [Nm]	Oil Temp. [°C]
14.847	2500	100	40, 50, 60	46.128	2500	100	50
	2500	200	40, 50, 60		2500	200	40, 50
	2500	400	50		2500	300	50
	5000	100	50		7500	100	50
	5000	200	40, 50, 60		7500	200	40, 50, 60
	5000	400	40, 50, 60		7500	300	40, 50
	7500	100	50				
	7500	200	40, 50, 60				
	7500	300	50				

Input torques of 100, 200 and 300 or 400 Nm were run in both gears, depending on the total power applied. For the gear with a ratio of 14.847, input speeds of 2500, 5000 and 7500 rpm were used to represent the wide range of driving speeds of this gear. In the gear with the ratio of 46.128, which was designed for the starting maneuvers, only 2500 and 7500 rpm were used as engine speeds, as this gear only covers a comparatively lower speed range. The inlet of the injected oil was always kept at a constant 50 °C; therefore, its influence as the primary heat transporter could be specifically demonstrated. For this purpose, the injection temperatures were changed to 40 °C and 60 °C at individual load points. This also allowed the influence on the inertia of the thermal system to be traced.

Figure 10 shows the result of such a thermal characterization for the load case of 5000 1/min input speed 400 Nm input torque and 50 °C oil injection temperature in the gear with the transmission ratio of 14.847 and with its simulation superimposed. In this illustration, the three thermal accounts of a bearing inner ring, a gear tooth and a housing segment for the second gearbox shaft are shown. When designing the prototype, the integration of temperature sensors in the rotating gearbox elements for their thermal measurement was feasible in this shaft due to the gearbox layout.

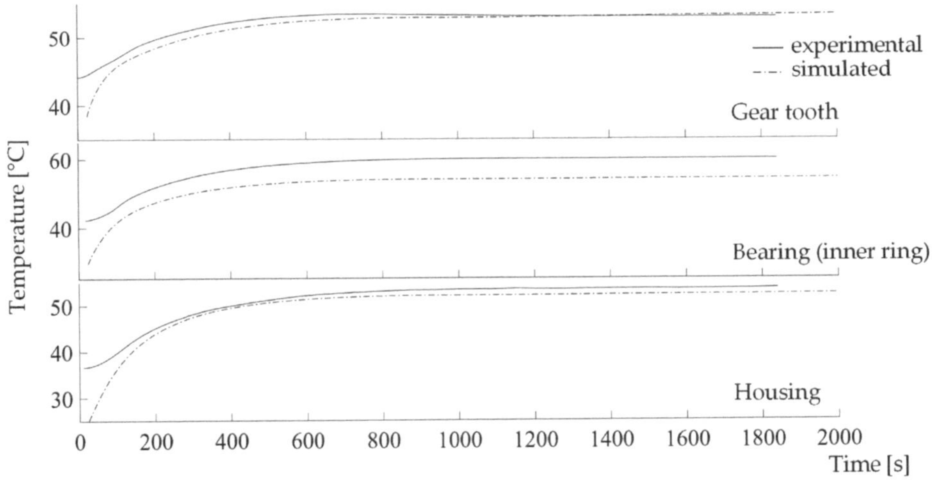

Figure 10. Comparison of temperature curves for a gear tooth, an inner ring of a bearing and a housing segment.

Although the end temperature deviation of the bearing is about ten degrees celsius, the simulated temperatures of the gear tooth and the housing segment show good accordance with the measured temperatures. Since the gearbox was not always cooled down to the ambient temperature between test runs, there are more significant temperature differences for $0\ \text{s} \leq 100\ \text{s}$. To derive a statement about the thermal inertia of the system despite this transient, the gradient of the temperature increase was introduced as an evaluation variable and is shown in Figure 11.

However, since the first 100 s of the measured values are needed for the system to stabilize, the time when the system has already reached 50% of the final temperature is selected as the starting point. Starting from this point, the gradients for 35% and 85% of the remaining 50% are determined according to Equations (1)–(6). These are evaluated as characteristic values in the following.

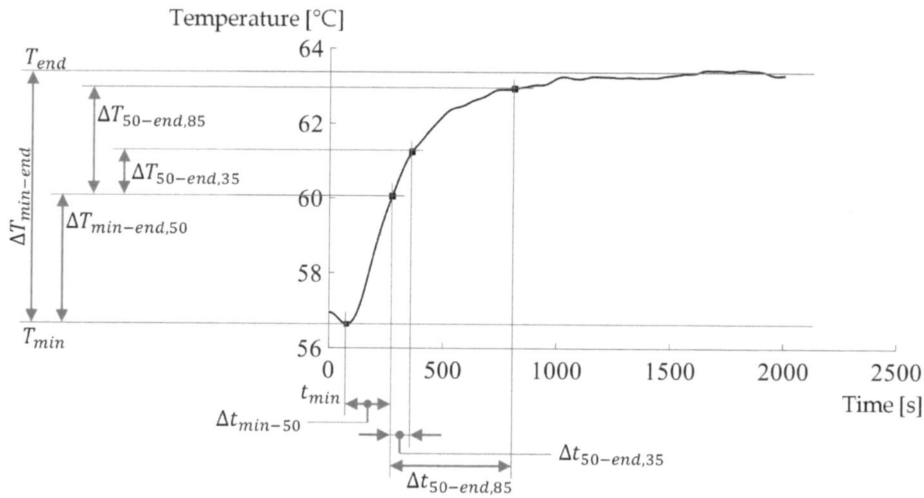

Figure 11. Key temperature and time parameters.

An average value of the last five percent of the test time is chosen as the final temperature for this calculation to not give too much weight to any measurement deviations of a single measuring point. The minimum system temperature results in the initial temperature delta according to Equation (1).

$$\Delta T_{min-end} = T_{end} - T_{min} \quad (1)$$

Subsequently, the final temperature differences are determined:

$$\Delta T_{min-end,50} = 0.5 \cdot (T_{end} - T_{min}) \quad (2)$$

$$\Delta T_{50-end,35} = 0.35 \cdot \Delta T_{min-end,50} \quad (3)$$

$$\Delta T_{50-end,85} = 0.85 \cdot \Delta T_{min-end,50} \quad (4)$$

The characteristic time intervals are then calculated by:

$$\Delta t_{50-end,35} = t(T_{min} + \Delta T_{min-end,50} + \Delta T_{50-end,35}) - t(T_{min} + \Delta T_{min-end,50}) \quad (5)$$

$$\Delta t_{50-end,85} = t(T_{min} + \Delta T_{min-end,50} + \Delta T_{50-end,85}) - t(T_{min} + \Delta T_{min-end,50}) \quad (6)$$

The key temperature parameter and the key time parameters are equally calculated for the simulation. For example, if the measured temperature course is steeper than the simulated, the key time intervals are shorter for the measurements than for the simulations. Figure 12 illustrates the comparison of the final temperatures between the simulation and the experiment in the left-hand diagram, and the right-hand graph shows the inertia based on the time interval for reaching $\Delta t_{50-end,35}$ and $\Delta t_{50-end,85}$.

For the ideal case of precise temperature prediction, all points in the diagram on the left would lie on the origin line drawn in. The two additional straight lines in the diagram indicate a deviation of ten percent. If points within the diagram lie above the origin diagonal, the simulation temperatures are higher than the measured temperatures, and if the points lie below the straight line, the simulated temperatures are lower than the measured temperatures. The graph shows that the majority of the determined values are within a ten percent deviation. This corresponds to a deviation of between 1.5 and 4.5 °C at the temperature level considered. This value is to be emphasised as a positive result, especially in comparison with the currently established approach of ISO TR 14179, since ISO-TR-14179 provides one system and one housing temperature as a result, and deviations of up to 10 °C are specified in the ISO as the expected accuracy. The developed model thus allows the prediction of internal component temperatures with higher accuracy and thus represents a significant improvement compared to established approaches.

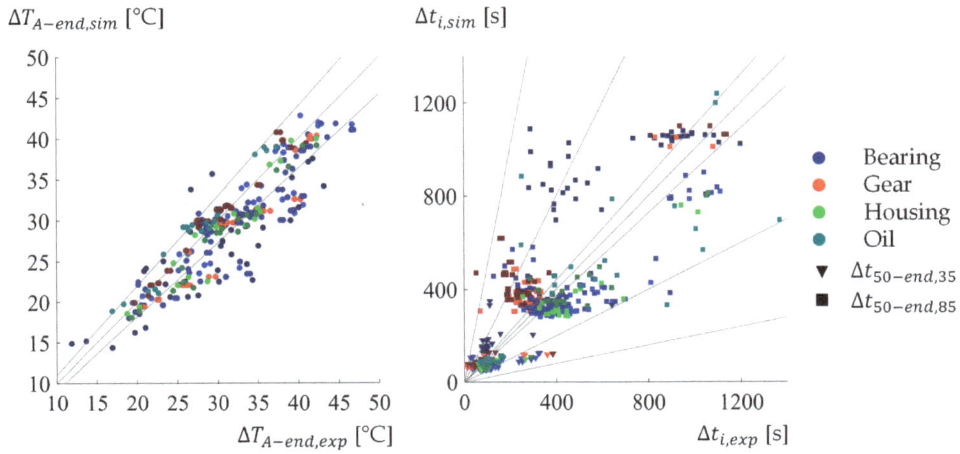

Figure 12. Comparison of simulated key parameters versus measured key parameters for all load cases in Table 1.

Figure 12 also shows that the simulated temperatures tend to be lower than the measured ones. On the one hand, this is due to the detailed design of the housing and its surrounding airflow and, on the other hand, to the already proven deviations of the loss modelling. A further source could be the thermocouples used, which have a relatively high measuring deviation compared to the measuring range with a tolerance of 1.5 K.

The evaluation of the system inertia in the right-hand diagram of Figure 12 is not part of the approach of ISO-TR-14179 and thus represents a new evaluation parameter for this type of system evaluation. It turns out that the inertia shows a wider deviation. On the one hand, this is due to the chosen modelling approach, as simplifications in the component geometry and oil distribution were assumed. On the other hand, only a small part of the heating curves could be evaluated when analyzing the inertia. Particularly, this is due to the transient response of the experimental setup in the first time ranges, which can therefore not be evaluated. Thus, for a reliable estimation of the system inertia, on the one hand, detailed modelling of the real constructional implementation is necessary, and on the other hand, the test stand setup needs to be modified in such a way that initial warm-up phases with lower transient behavior can also be utilized. Nevertheless, the approach already provided the possibility of carrying out an initial evaluation of the thermal behavior for the available scope of information and thus provided a substantially well-founded basis for decision making for the system developer without an in-depth design.

4. Conclusions and Outlook

In this paper, an overview of the validation of the method developed at the Institute for Automotive Engineering (ika) to evaluate transmission concepts is given. In the assessment of the loss modelling, it is shown that, even with the limited information available, a sufficiently good approximation of the measured losses could be given. The drag losses, which have been neglected, were only noticeable in areas of comparably low system performance. The evaluation of the thermal system properties based on this loss modelling was also able to provide a good approximation, particularly concerning the stationary end temperatures of the system. Deviations of a few degrees between the simulation model and the measurement were noted here. This is a promising result, especially considering the highly simplified housing within the simulation model, and thus provides a good supplement in identifying thermally critical system points. The evaluation of the thermal inertia, on the other hand, is afflicted with more significant deviations. In addition to the deviating housing design and the associated deviating thermal mass of the housing, deviations in the thermal masses of the shafts and gears due to the CAD design must also be considered as possible causes.

Recalculating the thermal model by scaling the loss components to the level of the measured losses led to a further improvement in the forecast for the final temperatures. The system inertia was also significantly improved by adjusting the thermal masses according to the CAD models. Thus, the function of the approach could be proven, but the use case of an early system evaluation does not allow such a system adaptation for the large number of systems considered. In summary, the loss calculation and the identification of the thermally critical components provide good results, and the modelling can therefore be considered validated for application within the 'BEV Goes eHighway–[BEE]'. Consequently, valuable predictions of the vehicle characteristics can be derived for a large number of powertrain configurations.

Author Contributions: Conceptualization, R.U. and T.K.; methodology, R.U. and T.K.; software, R.U., T.K., S.K. and G.W.; validation, R.U.; formal analysis, R.U., T.K., S.K. and G.W.; investigation, R.U. and T.K.; resources, R.U.; data curation, R.U.; writing—original draft preparation, R.U.; writing—review and editing, T.K. and G.W.; visualization, R.U., T.K., S.K. and G.W.; supervision, L.E.; project administration, R.U.; funding acquisition, L.E. All authors have read and agreed to the published version of the manuscript.

Funding: This research was funded by the Federal Ministry of Economics and Climate Action (BMWK) grant number [16EM5003] and grant number [01MY17002B].

Data Availability Statement: Not applicable.

Acknowledgments: The authors want to thank the Federal Ministry of Economics and Climate Action (BMWK) for the funding of the project 'BEV Goes eHighway—[BEE]' (16EM5003), which offers a comprehensive application scenario for the methodology developed at the Institute for Automotive Engineering (ika), as well as the funding of the project 'Concept ELV^2' (01MY17002B), which provided the data basis for the current application.

Conflicts of Interest: The Authors declare no conflict of interest.

References

1. European Parliament and Council. Regulation (EU) 2019/1242 of the european parliament and of the council. Presented at the CO_2 Emission Performance Standards for New Heavy-Duty Vehicles and Amending Regulations (EC) No 595/2009 and (EU) 2018/956 of the European Parliament and of the Council and Council Directive 96/53/EC, Brüssel, Belgium, 20 June 2019.
2. ACEA. *CO_2 Emissions from Heavy-Duty Vehicles: Preliminary CO_2 Baseline (Q3-Q4 2019) Estimate*; European Automobile Manufacturers Association: Brüssel, Belgium, 2020.
3. Rodríguez, F. *The Future of Vecto: CO_2 Certification of Advanced Heavy-Duty Vehicles in the European Union*; The International Council on Clean Transportation: Berlin, Germany, 2019.
4. Fiori, C.; Ahn, K.; Rakha, H.A. Power-based electric vehicle energy consumption model: Model development and validation. *Appl. Energy* **2016**, *168*, 257–268. [CrossRef]
5. Earl, T. Analysis of Long Haul Battery Electric Trucks in EU: Marketplace and Technology, Economic, Environmental, and Policy Perspectives. Presented at 8th Commercial Vehicle Workshop, Graz, Austria, 17–18 May 2018.
6. Wu, J.; Liang, J.; Ruan, J.; Zhang, N.; Walker, P.D. Efficiency comparison of electric vehicles powertrains with dual motor and single motor input. *Mech. Mach. Theory* **2018**, *128*, 569–585. [CrossRef]
7. Ahssan, M.R.; Ektesabi, M.M.; Gorji, S.A. Electric Vehicle with Multi-Speed Transmission: A Review on Performances and Complexities. *SAE Int. J. Alt. Power.* **2018**, *7*, 169–181. [CrossRef]
8. Gronwald, P.-O.; Kern, T.A. Traction Motor Cooling Systems: A Literature Review and Comparative Study. *IEEE Trans. Transp. Electrific.* **2021**, *7*, 2892–2913. [CrossRef]
9. Geiger, J. Wirkungsgrad und Wärmehaushalt von Zahnradgetriebem bei instationären Betriebsbedingungen. Ph.D. Dissertation, Technische Universität München, München, Germany, 2014.
10. Changenet, C.; Oviedo-Marlot, X.; Velex, P. Power Loss Predictions in Geared Transmissions Using Thermal Networks-Applications to a Six-Speed Manual Gearbox. *J. Mech. Des.* **2006**, *128*, 618–625. [CrossRef]
11. Höhn, B.-R.; Michaelis, K. Influence of oil temperature on gear failures. *Tribol. Int.* **2004**, *37*, 103–109. [CrossRef]

12. Kieninger, D. The Highly Integrated ModulED Drive Module–Holistic Concept Design. In Proceedings of the 20th International VDI Congress "Dritev", Bonn, Germany, 24–25 June 2020.
13. Kieninger, D.; Hemsen, J.; Köller, S.; Uerlich, R. Automated Design and Optimization of Transmissions for Electric Vehicles. *MTZ Worldw.* **2019**, *80*, 88–93. [CrossRef]
14. Hemsen, J.; Negri, T.; Trost, C.; Eckstein, L. Comparison of Permanent Magnet Rotor Designs for Different Vehicle Classes and Driving Scenarios: A Simulation Study. *SAE Int. J. Alt. Power.* **2021**, *10*, 207–226. [CrossRef]
15. Uerlich, R. Concept ELV^2 Design of an Electric Drive Axle for Heavy Distribution Traffic. In Proceedings of the 30th Aachen Colloquium Sustainable Mobility, Aachen, Germany, 4–6 October 2021.
16. Köller, S.; Uerlich, R.; Witham, G.; Eckstein, L. *Concept ELV^2—Development of an Electric Drive Axle for Heavy Commercial Vehicles, ELIV 2021*; VDI Verlag: Düsseldorf, Germany, 2021; 653-XVI.
17. Köller, S.; Schmitz, V. Systematic Synthesis and Multi-Criteria Evaluation of Transmission Topologies for Electric Vehicles. *Automot. Engine Technol.* **2022**, *7*, 65–79. [CrossRef]
18. *DIN ISO 6336:2019*; Calculation of Load Capacity of Spur and Helical Gears (ISO_6336:2019). Beuth Verlag GmbH: Geneva, Switzerland, 2019.
19. *DIN ISO 21771:2014-08*; Zahnräder_- Zylinderräder und Zylinderradpaare mit Evolventenverzahnung_- Begriffe und Geometrie (ISO_21771:2007). Beuth Verlag GmbH: Berlin, Germany, 2014.
20. Naunheimer, H.; Bertsche, B.; Ryborz, J.; Novak, W.; Fietkau, P. *Fahrzeuggetriebe*; Springer: Berlin/Heidelberg, Germany, 2019.
21. *ISO/TC 60/SC 2. ISO/TR 14179-2:2001*; Gears—Thermal Capacity—Part 2: Thermal Load-Carrying Capacity. Beuth Verlag: Berlin, Germany, 2001.
22. Blok, H. Recent Developments in Gear Tribology. In Proceedings of the Institution of Mechanical Engineers, London, UK, 1 September 1969.
23. Funck, G. Wärmeabführung bei getrieben unter quasistationären Betriebsbedingungen. Ph.D. Dissertation, Technische Universität München, München, Germany, 1985.
24. Uerlich, R.; Koch, T.; Theising, H.; Eckstein, L. Method for thermal evaluation of automotive gearbox packages taking into account load point-dependent oil distribution. *Automot. Engine Technol.* **2022**. [CrossRef]

World Electric Vehicle Journal

Article

Pathways to Carbon-Free Transport in Germany until 2050

Till Gnann [1,*], Daniel Speth [1], Michael Krail [1], Martin Wietschel [1,2] and Stella Oberle [3]

1 Fraunhofer Institute for Systems and Innovation Research ISI, Breslauer Str. 48, 76139 Karlsruhe, Germany; daniel.speth@isi.fraunhofer.de (D.S.); michael.krail@isi.fraunhofer.de (M.K.); martin.wietschel@isi.fraunhofer.de (M.W.)
2 Institute for Industrial Production (IIP) at the Karlsruhe Institute of Technology (KIT), Hertzstrasse 16, 76187 Karlsruhe, Germany
3 Fraunhofer Research Institution for Energy Infrastructures and Geothermal Systems IEG, Breslauer Str. 48, 76139 Karlsruhe, Germany; stella.oberle@ieg.fraunhofer.de
* Correspondence: till.gnann@isi.fraunhofer.de

Abstract: The transport sector has to be widely decarbonized by 2050 to reach the targets of the Paris Agreement. This can be performed with different drive trains and energy carriers. This paper explored four pathways to a carbon-free transport sector in Germany in 2050 with foci on electricity, hydrogen, synthetic methane, or liquid synthetic fuels. We used a transport demand model for future vehicle use and a simulation model for the determination of alternative fuel vehicle market shares. We found a large share of electric vehicles in all scenarios, even in the scenarios with a focus on other fuels. In all scenarios, the final energy consumption decreased significantly, most strongly when the focus was on electricity and almost one-third lower in primary energy demand compared with the other scenarios. A further decrease of energy demand is possible with an even faster adoption of electric vehicles, yet fuel cost then has to be even higher or electricity prices lower.

Keywords: transport; emissions; alternative fuel vehicles; electric vehicles; decarbonized; energy demand; electricity; transport demand model; simulation model

Citation: Gnann, T.; Speth, D.; Krail, M.; Wietschel, M.; Oberle, S. Pathways to Carbon-Free Transport in Germany until 2050. *World Electr. Veh. J.* **2022**, *13*, 136. https://doi.org/10.3390/wevj13080136

Academic Editor: Michael Fowler

Received: 30 June 2022
Accepted: 26 July 2022
Published: 28 July 2022

1. Motivation

The transport sector has to largely decrease its greenhouse gas emissions to contribute to the goals set in the Paris Agreement in 2015. For this reason, a number of countries have set goals to ban fossil-fueled cars from the road within the next two decades [1]. Since passenger cars contribute most to transport emissions in industrialized countries, this is certainly the most important transport sector to address, yet heavy-duty vehicles, ships, and aviation also need attention to completely decarbonize the transport sector.

Germany, as one important industrialized country, has manifested a target to be CO_2 emission free by 2045 in the German Federal Climate Change Act in 2021 [2]. To reach this target, energy generation, transport, and industry have to be almost emission-free by 2045 and transport has to cut its emission by 48% until 2030. Since there was hardly any progress over the last 30 years in emission reduction from transport, this requires tremendous effort and there are numerous policy support schemes under discussion and within the coalition treaty of the current German government that could initiate such a change [3,4].

In this paper, we proposed four scenarios that could achieve a decarbonized transport sector by 2050 in Germany. These four scenarios differ in the use of energy in the modes of transport: one scenario focuses on electrification, the second on a large use of hydrogen, the third on methane produced from electricity (Power-to-gas, PtG), and the last one on liquid synthetic fuels (Power-to-Liquid, PtL), i.e., gasoline and diesel produced from electricity via electrolysis and the Fischer–Tropsch process. In all scenarios, electricity generation is decarbonized by 2050 and all electricity-based fuels will be emission free by then. PtG and PtL fuels are not locally emission free, but a production of fuels with CO_2 from biomass or captured from the air permits a globally emission free use. These energy carriers are used whenever economically sensible in the scenarios. Such an analysis has been performed in other studies as well [5,6], yet here, we put special emphasis on one energy carrier per scenario to understand the demand for it. This paper is based on the results from ref. [7].

In the following Section 2, we briefly describe the methods and data used. In Section 3, results are shown and a discussion and conclusion are given in Section 4.

2. Methods and Data

For this analysis, we used two models that have been the basis for several publications [8–10]. The model ASTRA was used to determine the future development of transport demand in the different transport sectors. The model ALADIN was used to decide on the drive trains that will be used in future (see Figure 1 for model description and interaction).

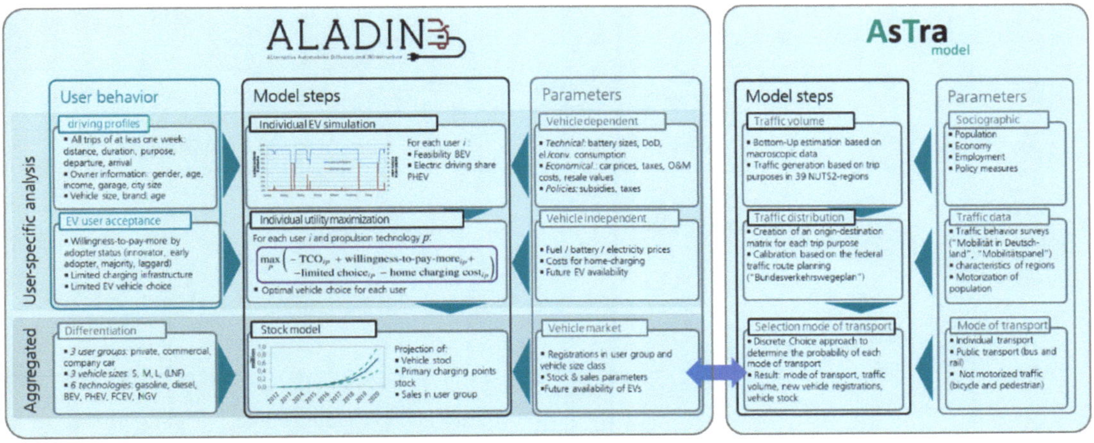

Figure 1. Models used and interaction for analysis.

The ASTRA model (ASsessment of TRAnsport Strategies) has been developed over more than 20 years in order to assess the impacts of transport policies on the transport system as well as on environmental indicators like GHG emissions [8]. Therefore, it assesses the future development of the national economy as well as its sectoral dynamics. On top, it includes a sophisticated population model that allows differentiating the population by attributes such as age, income, and employment status that enable an allocation of people into groups with similar mobility patterns. Based on these systems, ASTRA applies the classical 4-stage transport modeling approach (trip generation, trip distribution, modal split and assignment) to simulate the transport generation, its distribution, and the modal split both for passenger and freight transport. Therefore, it uses data generated from large mobility surveys such as the Mobility in Germany or the German Mobility Panel [11].

As ASTRA is implemented in VENSIM software based on System Dynamics methodology, it calculates the changes in the system every quarter year and allows considering feedback mechanisms such as the impact of road congestion on destination choice and modal split. Destination choice and modal split are mainly driven by changes in transport costs and transport times such that changes in drivetrain in vehicle fleets and thus increasing electrification of drivetrains induce changes in the transport related indicators like passenger- or vehicle-kilometers.

The model ALADIN (Alternative Automobiles Diffusion and Infrastructure) is a bottom-up simulation model that determines the market shares of drive trains based on individual driving profiles of vehicles. Driving largely determines the individual cost for each mode of transport. The individual total cost of ownership depending on weekly driving are calculated and accompanied by favoring and obstructing factors, e.g., for private passenger car users. Based on these individual decisions, the market shares of the different drive trains are determined for each year and transport mode.

This very detailed analysis can only be performed where individual driving profiles are publicly available, which is the case for passenger cars and heavy-duty vehicles in Germany. All developments of market diffusion and energy demand for the other modes of transport (bus, train, inland and overseas navigation, aviation) are added based on assumptions from literature. Although international navigation and aviation are not part of the country-specific CO_2 budgets, these are important to receive a full picture of the transport sector. Table 1 shows an overview of modes of transport, their importance based on the final energy demand in 2019 (before the COVID-19 pandemic), and the way these transport modes are analyzed.

Table 1. Overview of transport modes, their final energy demand in Germany in 2019, and the modeling approach.

Mode of Transport	Final Energy Demand in Germany in 2019 (PJ) [12]	Modeling of Future Market Diffusion
Passenger cars	1549	Individual buying decision based on TCO, favoring and hampering factors and vehicle and infrastructure availability
Heavy-duty vehicles	692	Individual buying decision based on TCO and vehicle and infrastructure availability
Trains	52	Literature based assumptions on future development
Buses	48	
Inland navigation	11	
Overseas navigation	57	
Aviation	435	

For passenger cars, we used about 7,000 individual vehicle driving profiles derived from the German Mobility Panel [11] and the REM2030 vehicle driving database [13]. These driving profiles contain all trips of vehicles of at least one week, and additional information on the driver socio-demographics and vehicle. They were shown to be representative for German car sales in ref. [14]. For heavy-duty vehicles, the annual mileages from 6000 trucks were analyzed that stem from truckscout24 and KiD2010 [15,16] and their representativeness was shown in ref. [10].

All detailed assumptions for vehicles, framework conditions, infrastructure, and energy carrier costs were given in ref. [7]. Here, we briefly show the differing energy, battery, and fuel cell prices in the scenarios (see Tables 2–6).

The buying decision in ALADIN uses end-user prices that contain all costs for generation, transport, and distribution, but also all taxes and other levies. Further, the energy prices in the transport sector do or will in future contain a cost for CO_2 that has to be considered. Here, we used 200 €/t in 2030 and 500 €/t in 2050. We can observe the additional taxes and levies, e.g., in the synfuels-focused scenario with an energy carrier cost of 0.122 €/kWh in 2050 (Table 2) and a final diesel price that is 100% synthetic in 2050 (Table 3) of 0.261 €/kWh (Table 4). Energy carrier prices are varied in the scenarios for further differentiation. In focus electricity, the household electricity price is continuously reduced up to 5 €ct/kWh in 2050 because of cost savings from load shifting. In focus synfuels, we reduce fuel prices by about 2 €ct/kWh in 2050 for gasoline and diesel. This change is performed equally in focus methane with CNG and LNG prices and the hydrogen price in focus hydrogen. In focus methane, the CNG and LNG price rose until 2040 due to a large share of conventional fuels and increased afterwards because of cheaper synthetic gas prices.

Battery and fuel cell prices are the important drivers in the buying decisions of alternative drive trains. We consider common, yet optimistic prices for battery price development in focus electricity and stagnating ones in the other scenarios (cf. Table 5). The same holds for fuel cell prices in scenarios focused on hydrogen with an optimistic development and higher price path in the other scenarios (see Table 6).

Table 2. Energy carrier cost for electricity-based fuels (€/kWh). All costs include transport and infrastructure cost.

Energy Carrier Cost	2020	2030	2040	2050
Power-to-hydrogen	0.285	0.220	0.150	0.120
Power-to-gas	0.300	0.195	0.160	0.122
Power-to-liquid	0.300	0.205	0.170	0.132

Table 3. Assumed share of synthetic fuels in transport.

Scenario	2020	2030	2040	2050
Focus—electricity	0%	10%	20%	50%
Focus—hydrogen	0%	10%	20%	50%
Focus—methane	0%	20%	50%	100%
Focus—synfuels	0%	20%	50%	100%

Table 4. Energy carrier prices (€/kWh).

Energy Carrier Price	2020	2030	2040	2050
Gasoline price	0.154	0.233	0.315	0.293
Diesel prices [1]	0.117	0.197	0.281	0.261
Hydrigen price [2]	0.285	0.285	0.282	0.235
CNG price [3]	0.088	0.190	0.273	0.257
LNG price [3]	0.097	0.212	0.317	0.304
Electricity price households [4]	0.329	0.321	0.313	0.311
Electricity price commercial [4]	0.226	0.217	0.210	0.208
Electricity price industrial [4]	0.130	0.131	0.136	0.135

[1] Lower in the synfuels-focused scenario, [2] Lower in the hydrogen-focused scenario, [3] Lower in the methane-focused scenario, [4] Lower in the electricity-focused scenario.

Table 5. Battery prices (€/kWh). Own assumptions based on refs. [17,18].

Scenario	EV Type	2020	2030	2040	2050
Focus—electricity	BEV	240	100	90	80
	PHEV	264	110	98	88
Focus—hydrogen	BEV	240	100	100	100
	PHEV	264	120	120	120
Focus—methane	BEV	240	120	120	120
	PHEV	264	132	132	132
Focus—synfuels	BEV	240	120	120	120
	PHEV	264	132	132	132

Table 6. Fuel cell prices (€/kW). Own assumptions based on refs. [17–20].

Scenario	2020	2030	2040	2050
Focus—electricity	234	80	66	55
Focus—hydrogen	234	78	62	50
Focus—methane	234	80	80	80
Focus—synfuels	234	80	80	80

3. Results

Let us now turn to results. Figure 2 shows the vehicle stock of passenger cars, light-duty trucks, medium-duty trucks, and heavy-duty trucks in rows 1–4 and the four scenarios in columns 1–4.

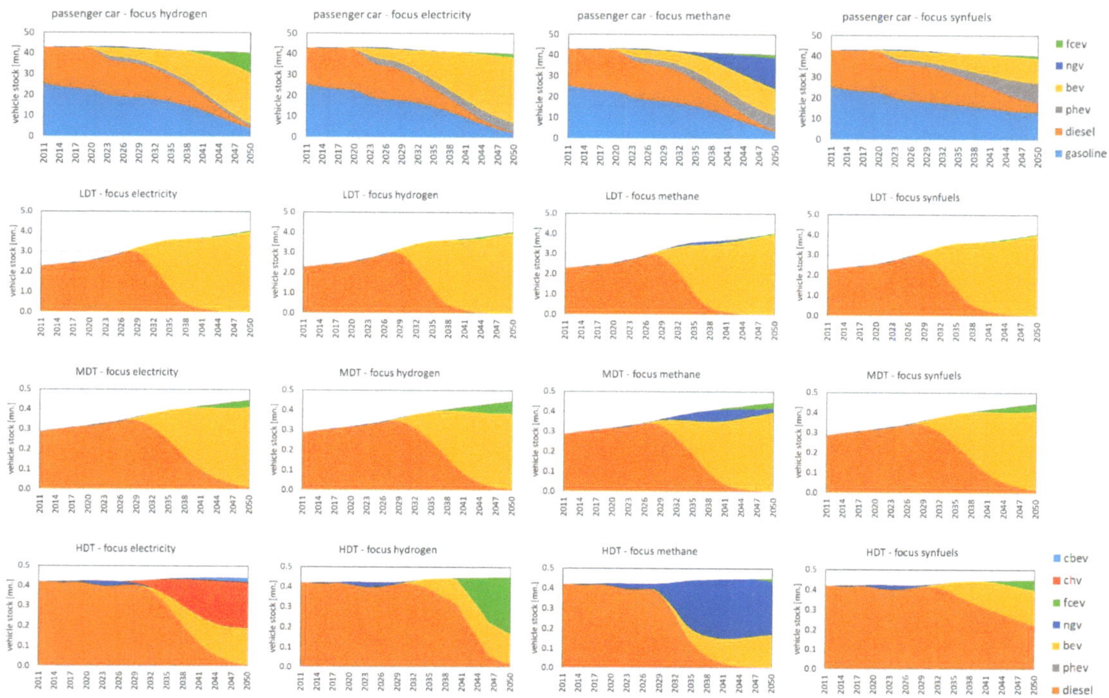

Figure 2. Vehicle stock in the four scenarios (**columns**) and vehicle size classes (**rows**). phev: plug-in hybrid electric vehicles: bev: battery electric vehicles; ngv: natural gas vehicles; fcev: fuel cell electric vehicles; chv: catenary hybrid electric vehicles; cbev: catenary battery electric vehicles.

For passenger cars, we find at least 25% and up to 70% BEV in 2050. The number of gasoline and diesel cars are up to 30% in the first three scenarios and some reach 50% in focus synfuels. Hydrogen and NGV only play a role if their fuel prices are very low. Thus, a quite large number of passenger cars will be electric vehicles while the rest depends on fuel prices for synthetic fuels and hydrogen.

It is less complicated for light-duty vehicles (up to 3.5 t), which will become electric in all scenarios. Medium-duty trucks (3.5–12 t) all contain a large number of battery electric trucks in 2050 in all scenarios (at least 50%), while all other transport modes depend on energy prices. This is similar for heavy-duty trucks where 30% are battery electric trucks and the other share are similar to medium-duty trucks. In focus electricity, the leftover vehicles are catenary hybrid vehicles; focus hydrogen, methane, and synfuels cover their long-distance trips with trucks powered by these fuels.

Thus, quite a large number of road transports will be powered directly with electricity while longer distances are covered with drive trains of the focus scenario.

Figure 3 contains exemplary results for the final energy demand in the transport sector in all four scenarios in 2030 and 2050 distinguished by energy carrier. In 2030, we find only some energy demand of alternative drive trains mainly stemming from the electricity demand of passenger cars. Heavy-duty vehicles can partly be fueled alternatively with electricity, hydrogen, or methane. There is a small energy demand for methane-powered vehicles in the methane-focused scenario.

In 2050, these differences are growing further. In focus electricity, earthbound transport and national aviation is carried out electrically, while the other sectors rely on liquid fuels due to technical restrictions. When counting on liquid fuels (focus PtL), the heavy-duty sector, national aviation, and parts of the railway system use synthetics or biofuels, resulting in a higher annual final energy demand. Hydrogen could play a role in the same transport sectors and partly also for long-distance driving passenger cars as could methane. The total final energy demand would be higher due to the lower efficiency of these drive trains.

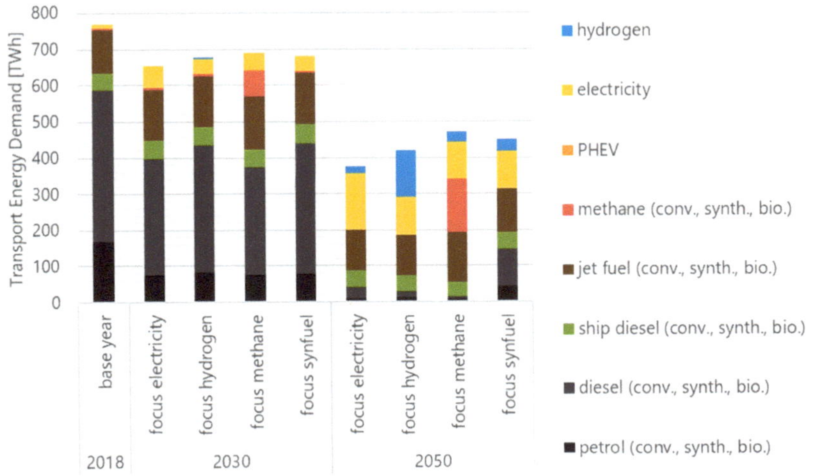

Figure 3. Final energy demand in 2030 and 2050 distinguished by energy carrier in the four scenarios.

Assuming that no biofuels are available for transport since they are needed in other sectors for a complete decarbonization, all fuels would have to be produced from electricity by 2050. If this was the case, the electricity demand for transport would be as shown in Figure 4. We can clearly observe that a large electrification would cause the lowest electricity demand for transport (~700 TWh) compared with the synfuels-focused or methane-focused scenario (~1000 TWh). However, the large shares of liquid or gaseous fuels in all scenarios showed the increasing importance of the aviation and navigation sectors that are not expected to be largely electrified by 2050.

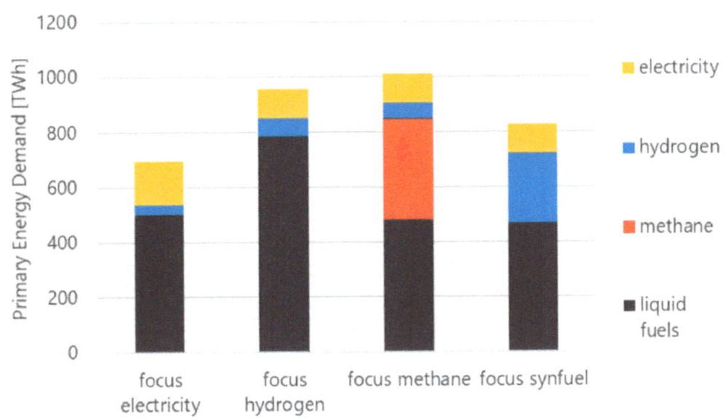

Figure 4. Electricity demand in 2050 distinguished by energy carrier type in the four scenarios.

To further bring down the share of conventional vehicles, we calculated sensitivities of energy prices on conventional vehicle market shares in 2050 in the elecricity-focused scenario. These are shown for diesel and gasoline vehicles for passenger cars in the first row and for light-, medium-, and heavy-duty vehicles in the second row of Figure 5.

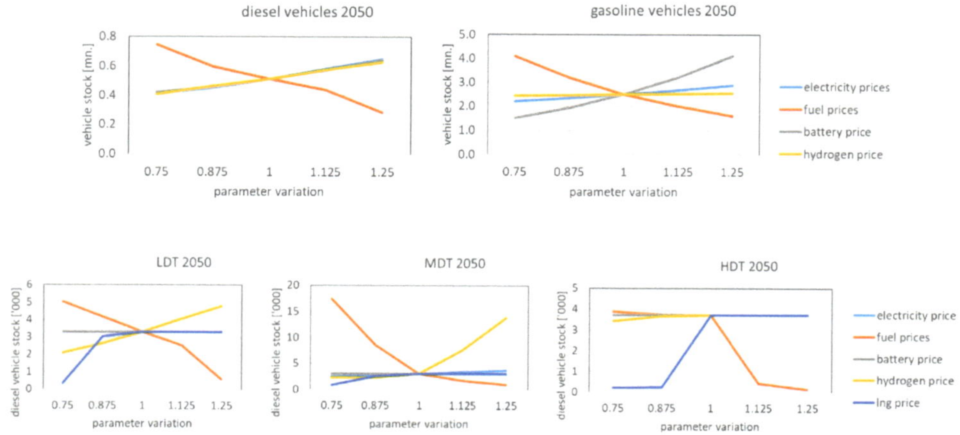

Figure 5. Sensitivities on conventional vehicle stock in 2050 in the electricity-focused scenario for changes in energy prices. Variation of energy prices by −25% to +25% shown on the x-axis and respective results for diesel or gasoline passenger cars (upper panel) and diesel vehicles in LDT, MDT, and HDT (lower panel). Changes for electricity price (blue), fuel (gasoline + diesel) prices (orange), battery price (gray), hydrogen price (yellow), and LNG price for trucks (dark blue).

We find the potential to further reduce gasoline and diesel passenger cars by almost half if we reduce electricity prices or increase synfuel prices by 25%. With the same changes, we could decrease the diesel truck stock in all size classes to almost zero in 2050. Thus, these changes can be considered in an even more ambitious scenario towards greenhouse gas neutrality in 2045.

4. Discussion and Conclusions

This paper described pathways for a complete decarbonization of transport in 2050 in Germany. These pathways are based on simulations with the models ASTRA and ALADIN and subject to a number of assumptions. Although the models have been used in multiple earlier publications, it is still discussible whether the approach is appropriate. Yet, based on literature reviews for market diffusion models in the transport sector, it is rather common [21–23]. Furthermore, the assumptions for all input parameters are subject to debate. All assumptions are based on literature and we performed a sensitivity analysis for the impact on model results for energy carrier prices to cope for this aspect.

What can we conclude from this analysis? We found high shares of electric vehicles in all scenarios and this will be the most energy efficient and economical option in the future. When batteries cause range limitations due to a lower energy density, the most cost effective energy carriers with higher ranges will be considered. In the scenarios, there can be some hydrogen vehicles if the hydrogen price is very low or also methane (or synfuels)-powered vehicles have a low price for methane (or synfuels). Yet, this comes at a price related to primary energy necessary for their production. We need about 300 TWh of additional electricity from renewables if converted to methane or synfuels and used in the transport sector. For comparison, the total renewable electricity generation in Germany was ~250 TWh in 2020, which shows the need for an efficient use of energy and direct electrification where possible. This will raise the question of importing these energy sources to Germany. Such import options do not exist at present and must first be built up with enormous investments. Furthermore, against the backdrop of the Russian war against Ukraine, questions of energy security then arise.

In any case, it is necessary to mention that these scenarios are very ambitious in assumptions and outcome and that the decarbonization of the transport sector will require large efforts.

Author Contributions: Conceptualization, T.G. and D.S.; methodology, T.G., D.S., M.K. and M.W.; software, T.G., D.S. and M.K.; validation, T.G., D.S. and M.K.; formal analysis, T.G., D.S. and M.K.; investigation, T.G., D.S., M.K., M.W. and S.O.; resources, M.W. and S.O.; data curation, T.G., D.S. and M.K.; writing—original draft preparation, T.G. and D.S.; writing—review and editing, T.G., D.S., M.K., M.W. and S.O.; visualization, T.G. and D.S.; supervision, M.W.; project administration, M.W. and S.O.; funding acquisition, M.W. All authors have read and agreed to the published version of the manuscript.

Funding: This research was funded by the Federal Ministry for Economic Affairs and Climate Protection (BMWi) in the projects "Long-term scenarios for the transformation of the energy system in Germany 3" and "MethQuest_MethSys-Production and use of methane from renewable sources in mobile and stationary applications; subproject: Energy system analysis for the production and use of methane from renewable sources" (FKZ 03EIV046A).

Institutional Review Board Statement: Not applicable.

Informed Consent Statement: Not applicable.

Conflicts of Interest: The authors declare no conflict of interest.

References

1. Plötz, P.; Axsen, J.; Funke, S.A.; Gnann, T. Designing car bans for sustainable transportation. *Nat. Sustain.* **2019**, *2*, 534–536. [CrossRef]
2. German Federal Ministry of Justice. Federal Climate Change Act. 2021. Available online: https://www.gesetze-im-internet.de/englisch_ksg/englisch_ksg.html (accessed on 21 July 2022).
3. German Environment Agency (UBA): Development and Goal Achievement of Greenhouse Gas Emissions in Germany in the Transport Sector According to Federal Climate Change Act. 2022. Available online: https://www.umweltbundesamt.de/bild/entwicklung-der-treibhausgasemissionen-des-verkehrs (accessed on 21 July 2022).
4. SPD. Bündnis 90/Die Grünen; FDP: Mehr Fortschritt wagen. Bündnis für Freiheit, Gerechtigkeit und Nachhaltigkeit. In *Koalitionsvertrag 2021–2025 zwischen SPD, Bündnis 90/Die Grünen und FDP*; Coalition Agreement: Berlin, Germany, 2021.
5. Blanck, R.; Kasten, P.; Hacker, F.; Mottschall, M. UBA: Treibhausgasneutraler Verkehr 2050: Ein Szenario zur zunehmenden Elektrifizierung und dem Einsatz stromerzeugter Kraftstoffe im Verkehr. In *Abschlussbericht im Auftrag des Umweltbundesamtes zum Forschungsvorhaben Verkehr 2050—Entwicklung von Parametern und Skizzierung eines Vereinfachten Energie und Emissionsszenarios*; Öko-Institut e.V: Berlin, Germany, 2013.
6. Schmied, M.; Wüthrich, P.; Zah, R.; Althaus, H.-J.; Friedl, C. UBA: Postfossile Energieversorgungsoptionen für einen treibhausgasneutralen Verkehr im Jahr 2050: Eine verkehrsträgerübergreifende Bewertung. In *Texte 30/2015, Studie im Auftrag des Bundesministeriums für Umwelt, Naturschutz, Bau und Reaktorsicherheit (UBA), Durchgeführt von INFRAS AG und Quantis*; Umweltbundesamt: Dessau, Germany, 2015.
7. Krail, M.; Speth, D.; Gnann, T.; Wietchel, M. Langfristszenarien für die Transformation des Energiesystems in Deutschland. *Treibhausgasneutrale Hauptszenarien Modul. Verk.* **2021**, *1*, 1–43.
8. Krail, M. System-Based Analysis of Income Distribution Impacts on Mobility Behaviour. Ph.D. Thesis, Universtity of Karlsruhe, Karlsruhe, Germany, 2009.
9. Gnann, T.; Plötz, P.; Wietschel, M. Can public slow charging accelerate plug-in electric vehicle sales? A simulation of charging infrastructure usage and its impact on plug-in electric vehicle sales for Germany. *Int. J. Sustain. Transp.* **2019**, *13*, 528–542. [CrossRef]
10. Wietschel, M.; Gnann, T.; Plötz, P.; Doll, C. Electric trolley trucks—A techno-economic assessment for Germany. *World Electr. Veh. J.* **2019**, *10*, 86. [CrossRef]
11. MOP. Mobilitätspanel Deutschland 1994–2010. Projektbearbeitung Durch das Institut für Verkehrswesen der Universität Karlsruhe (TH), Karlsruhe, Germany. Verteilt Durch die Clearingstelle Verkehr des DLR-Instituts für Verkehrsforschung. Available online: www.clearingstelle-verkehr.de (accessed on 22 June 2011).
12. *BMVI 2020: Verkehr in Zahlen 2020/2021*; Federal Ministry for Transport and Digital Infrastructure (BMVI); Zentrum für Luft-und Raumfahrt e.V. (DLR); Institut für Verkehrsforschung Deutsches Institut für Wirtschaftsforschung Berlin e.V. (DIW): Berlin, Germany, 2020.
13. Fraunhofer ISI. *REM2030 Driving Profiles Database V2015*; Fraunhofer Institute of Systems and Innovation Research ISI: Karlsruhe, Germany, 2015.
14. Gnann, T. *Market Diffusion of Plug-in Electric Vehicles and Their Charging Infrastructure*; Fraunhofer Verlag: Stuttgart, Germany, 2015.
15. Truckscout24: Sales Website for Used Commercial Vehicles. 2016. Available online: http://www.truckscout24.de (accessed on 13 February 2017).

16. KiD 2010: WVI, IVT, DLR, and KBA (2010). In *Kraftfahrzeugverkehr in Deutschland 2010 (KiD2010)*; WVI Prof. Dr. Wermuth Verkehrsforschung und Infrastrukturplanung GmbH, Braunschweig, IVT Institut für Angewandte Verkehrs-und Tourismusforschung e.V., Heilbronn, DLR Deutsches Zentrum für Luft-und Raumfahrt—Institut für Verkehrsforschung, Berlin; KBA Kraftfahrt-Bundesamt: Flensburg, Germany, 2010.
17. Lutsey, N. Cost Projection of State of the Art Lithium-Ion Batteries for Electric Vehicles Up to 2030. *Energies* **2017**, *10*, 1314. [CrossRef]
18. Zapf, M.; Pengg, H.; Bütler, T.; Bach, C.; Weindl, C. *Kosteneffiziente und Nachhaltige Automobile*; Springer Fachmedien Wiesbaden: Wiesbaden, Germany, 2019.
19. Hülsmann, F.; Mottschall, M.; Hacker, F.; Kasten, P. *Konventionelle und Alternative Fahrzeugtechnologien bei Pkw und Schweren Nutzfahrzeugen—Potenziale zur Minderung des Energieverbrauchs bis 2050*; Working Paper; Öko-Institut: Berlin, Germany, 2014.
20. Mottschall, M.; Kasten, P.; Kühnel, S.; Minnich, L. *Sensitivitäten zur Bewertung der Kosten Verschiedener Energieversorgungsoptionen des Verkehrs bis zum Jahr 2050—Abschlussbericht. Abschlussbericht*; Im Auftrag des Umweltbundesamtes; Dessau-Roßlau: Öko-Institut: Berlin, Germany, 2019.
21. Al-Alawi, B.M.; Bradley, T.H. Review of hybrid, plug-in hybrid, and electric vehicle market modeling Studies. *Renew. Sustain. Energy Rev.* **2013**, *21*, 190–203. [CrossRef]
22. Jochem, P.; Gómez Vilchez, J.J.; Ensslen, A.; Schäuble, J.; Fichtner, W. Methods for forecasting the market penetration of electric drivetrains in the passenger car market. *Transp. Rev.* **2018**, *38*, 322–348. [CrossRef]
23. Gnann, T.; Stephens, T.S.; Lin, Z.; Plötz, P.; Liu, C.; Brokate, J. What drives the market for plug-in electric vehicles?—A review of international PEV market diffusion models. *Renew. Sustain. Energy Rev.* **2018**, *93*, 158–164. [CrossRef]

World Electric Vehicle Journal

MDPI

Article

Development of the State of Warranty (SOW) for Electric Vehicles

Mikel Arrinda *, Denis Sánchez, Mikel Oyarbide, Haritz Macicior and Ander Zubiria

CIDETEC, Basque Research and Technology Alliance (BRTA), Po. Miramón 196, 20014 Donostia-San Sebastian, Spain; dsanchez@cidetec.es (D.S.); moyarbide@cidetec.es (M.O.); hmacicior@cidetec.es (H.M.); azubiria@cidetec.es (A.Z.)

* Correspondence: marrinda@cidetec.es

Abstract: There is an exponential increase in electric vehicles on the road that need a follow up in terms of warranty. The proposed state of warranty (SOW) is a metastate that qualitatively describes the warranty fulfillment level of an electric vehicle. All the relevant warranty information is synthesized in a single merit while maintaining the level of detail through the qualitative substates. The developed SOW is calculated with a rule-based logic of an expert system that evaluates the quantitative value of three substates: the remaining warranty, the remaining health and the remaining useful warranty. The SOW provides a synthesized and user-friendly description of the warranty fulfillment state while providing quantitative detailed information of the most relevant features of each of the different maintenance methodologies.

Keywords: diagnosis; prediction; electric vehicle; maintenance; off-board

Citation: Arrinda, M.; Sánchez, D.; Oyarbide, M.; Macicior, H.; Zubiria, A. Development of the State of Warranty (SOW) for Electric Vehicles. *World Electr. Veh. J.* **2022**, *13*, 135. https://doi.org/10.3390/wevj13080135

Academic Editor: Grzegorz Sierpiński

Received: 11 July 2022
Accepted: 26 July 2022
Published: 27 July 2022

1. Introduction

Electric vehicle sales are increasing steadily worldwide by 40% each year [1], while the number of cars with advanced warranty periods is increasing accordingly. The more electric vehicles are sold, the higher the repercussion of maintenance activities under warranty is for electric vehicle manufacturers. Inappropriate management of the warranty could lead to systematic replacement of the battery systems under warranty, which will end into undesired additional costs for the manufacturer. Consequently, many efforts are spent on this regard, specially on implementing a maintenance methodology (reactive, preventive and predictive) that better fits the needs of each original equipment manufacturer (OEM).

The simplest maintenance is reactive maintenance [2]. It consists of scheduling fixed maintenance activities and only acting in the case of vehicle failure. Therefore, interventions out of the scheduled plan would only be carried out if there is already a component that requires reparation. These interventions are scheduled immediately after failure notification (already broken). The most relevant features of reactive maintenance are the date of the beginning of life and the dates of scheduled maintenance events.

Nowadays, the most common maintenance is preventive maintenance [3]. It consists of diagnosing the actual state of the vehicle and acting once some damage thresholds are overcome. Maintenance activities are scheduled based on the state of the vehicle itself. Interventions are scheduled before the occurrence of a failure, but after a certain level of damage is reached. The interventions are scheduled in a short time period after receiving the notification (with serious damage but still working). The most relevant features are the warning thresholds of health indicator(s) and the observed/estimated state of health (SOH).

The maintenance with the highest interest from specialized enterprises is the predictive maintenance [4]. It consists of predicting failure before any real damage is inflicted. The evolution of health indicators of different elements of the vehicle are observed from the beginning of life. Recorded data are recurrently analyzed using aging models [5] and prognosis algorithms [6–8]. The results from those analyses are used to foresee and prevent failures in early states of damage. Hence, interventions can be scheduled before the occurrence of any failure and before observing any significant damage. Such interventions are scheduled in a long time period after receiving feedback (without any serious damage). The most relevant features are the expected lifespan (the prior) and the prognosed remaining useful life (RUL).

The actual maintenance schedules of electric vehicles are mostly carried out in a mix of reactive and preventive ways or in a mix of reactive, preventive and predictive ways. Maintenance activities of critical components such as the battery system are performed in a preventive or predictive way, while the remaining noncritical components are carried out in a reactive way. Consequently, the maintenance scheduling of electric vehicles must deal with a multicomponent, multimaintenance methodology scenario.

This scientific contribution tackles the need of easing such a multimaintenance methodology scenario through the study of the most critical element of an electric vehicle, the battery system. This study proposes a new qualitative metastate named the state of warranty (SOW). The SOW synthesizes the most relevant quantitative indicators (substates) of each of the maintenance methodologies in one merit. It allows for the simplification of the multimaintenance methodology into a global and unique maintenance methodology. Additionally, this SOW provides a milestone to develop a synthesized platform for multicomponent maintenance (a synthesized SOW of the electric vehicle determined by the estimated SOW of each of the components).

The paper is structured as follows. The proposal is explained in detail in Section 2. Then, the different tools required to determine the value of each of the quantitative indicators or substates used to estimate the SOW are described in Section 3: the SOH estimator, the prior and the RUL prognosis. After that, the results are shown and discussed in Section 4. Finally, the conclusions are drawn in Section 5.

2. State of Warranty

The needs of electric vehicle manufacturers to plan maintenance activities under the warranty period has motivated the development of a new state estimator: the state of warranty (SOW). The SOW is an off-board state that determines the warranty fulfillment level. This state provides a global understandable insight of the warranty state of an evaluated component while providing detailed and complex information about it if required.

The understandable insight is given by qualitatively determining the fulfillment level of the given warranty in a user-friendly color code. The detailed and complex information is given by quantifying in three substates related to the most relevant features of each of the different types of maintenance activities:

- The remaining warranty (RW) for the reactive maintenance.
- The remaining health (RH) for the preventive maintenance.
- The remaining useful warranty (RUW) for the predictive maintenance.

The qualitative value that defines the SOW is calculated with an expert system. This calculus is performed off-board. The rule-based logic applied by the expert system determines the SOW value based on the previously determined quantitative substates: the remaining warranty, the remaining health and the remaining useful warranty. That rule-based logic is disclosed in Table 1.

Table 1. SOW rule-based logic.

SOW	RW	RH	RUW	Severity Description
	>0.05	>0.75	>0.5	The warranty fulfilment level is correct.
	<0.05	>0.75	>0.5	The warranty fulfilment level is correct, but the end of warranty is close.

Table 1. *Cont.*

SOW	RW	RH	RUW	Severity Description
	-	>0.75	0.5 > RUW > 0.4	**ATTENTION**! Predicted 1 additional replacement. Early advice.
	-	0.75 > RH > 0.5	0.5 > RUW > 0.4	**ATTENTION**! Predicted 1 additional replacement. Middle advice.
	-	0.75 > RH > 0.5	0.4 > RUW > 0	**ATTENTION**! Predicted 1 additional replacement. Late advice.
	-	0.5 > RH > 0	0.4 > RUW > 0	**DANGER**! Predicted 1 additional replacement. Irreversible damages.
	-	-	0	**DANGER**! Predicted 2 additional replacements.
	-	0	-	**DEATH**! The battery has reached the EOL.
	¿?	¿?	¿?	**Undefined scenario**. Something unexpected is happening.

2.1. Remaining Warranty (RW)

The remaining warranty is the simplest substate. It is linked to reactive maintenance and refers to the warranty statement itself. It determines how much time of the warranty has been consumed and how much is still available. It is common to see warranties in form of kilometrage or lifespan.

This state aims at providing information about the closeness of the end of the provided warranty.

The value range of this state is defined with 0 and 1. A value equal to 1 represents the beginning of the warranty, i.e., the whole warranty is still available; 0 represents the end of the warranty, i.e., the warranty has been completely consumed.

This paper proposes to have four events of interest that describe this state. Two of those four events—first and fourth—correspond to the beginning and the end of warranty, respectively. Both determine the saturation levels of the proposed state since there is no interest on quantifying this warranty beyond those limits. The other two proposed events—the second and third events—correspond to warranty states where 5% and 1% of the total warranty remain available. These two values light up the advice state and alarm state, respectively.

Once the events are determined, how to fit them inside the 0 to 1 range is defined. For that aim, we proposed to use a first-order linear equation for the whole range.

Additionally, this paper proposes a color for each event, see Figure 1. The first event (RW = 1), the green color (in RGB logic [0,1,0]), represents the starting point of the warranty. The second event (RW = 0.05), the orange color (in RGB logic [1,0.832,0.212]), notifies that there is only 5% of the total warranty remaining. The third event (RW = 0.01), the red color (n RGB logic [1,0,0.153]), tells us that there is only 1% of the total warranty available. The fourth event (RW = 0), the black color (in RGB logic [0,0,0]), represents the end of the warranty period (all the warranty has already been consumed).

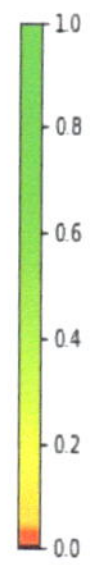

Figure 1. Proposed color map of the remaining warranty.

2.2. Remaining Health (RH)

The remaining health is linked with the most common maintenance: the preventive maintenance. It refers to the present SOH and quantifies its deviation from the prior and the closeness with the end of life (EOL). The aging behavior of a system is prone to suffer changes from design to real life. Changes could come from common modifications on the operating conditions (differences between the expected and the real ones), from a deficient description of the expected aging behavior (an inaccurate aging model will give as a result an inaccurate result) or form unexpected use/manipulation of the asset by the end user of the system.

This state has two objectives. Firstly, it places the present SOH in between the expectation and the EOL to rapidly check the SOH in a proper warranty context (the SOH estimation as itself has low information regarding the warranty unless it is placed in a proper context). Secondly, the remaining health shows the closeness to the appearance of a sudden failure (in this study, the EOL) so as to inform the vehicle manufacturer about the need for a preventive maintenance action.

The value range of this state is defined with 0 and 1. The 1 represents that the estimated SOH is equal or higher than the expected one at the present instant, and 0 represents that the EOL has been reached.

This paper proposes to have three events of interest that describe this state. Two of those three events—the first and third events—are the expected SOH (at the present instant) and the EOL, respectively. Both determine the saturation levels of the proposed state since there is no interest on quantifying it beyond those limits. In terms of diagnosis, if the present SOH is above the expected one, it means that the accumulated damage is lower than expected and, therefore, preventive actions are not required. At the same time, if the present SOH is below the EOL, it means that the warranty has not been fulfilled (a failure has already happened).

The second event of interest is the SOH value that triggers the alarms of the closeness of a sudden failure, the SOH threshold that forces preventive maintenance. The value of this second event is the instant when SOH value reaches the EOL plus 3%. This value is assumed to be big enough to give room for actions, but small enough not to light up hasty alarms.

Once the events are defined, the rest of the values of this 0 to 1 range need to be defined. We proposed to use a piecewise-defined function built with first order linear equations; see Equation (1):

$$\mathrm{RH}_k = \begin{cases} 1 & \mathrm{SOH}_k > \mathrm{SOH}_{expected\ k} \\ a_1 + \Delta\mathrm{SOH}_{1\ k} \cdot b_1 & \mathrm{SOH}_{\mathrm{event2}} > \mathrm{SOH}_k \geq \mathrm{SOH}_{expected\ k} \\ a_2 + \Delta\mathrm{SOH}_{2\ k} \cdot b_2 & \mathrm{EOL} > \mathrm{SOH}_k \geq \mathrm{SOH}_{\mathrm{event2}} \\ 0 & \mathrm{SOH}_k < \mathrm{EOL} \end{cases}, \tag{1}$$

where the RH_k is the remaining health at instant k; the SOH_k is the estimated SOH at instant k; the $\mathrm{SOH}_{expected\ k}$ is the prior at instant k; the $\mathrm{SOH}_{\mathrm{event2}}$ is the defined warning

SOH threshold (EOL + 3%); the $\Delta SOH_{1\ k}$ is the relative difference between the estimated SOH and the expected SOH, having as reference the difference between the expected SOH and the SOH_{event3} at instant k; the $\Delta SOH_{2\ k}$ is the relative difference between the estimated SOH and the EOL, having as reference the difference between the EOL and the SOH_{event2} at instant k; the a_x and b_x are the variables of the linear function in between the defined events, being x 1 and 2.

The proposed Equation (1) suffers from incongruency when the $SOH_{expected\ k} < SOH_{event2}$. It is only used from the beginning of life to the instant that the expectation reaches the second event. In this situation, the warranty is about to expire so the expectations are close to the failure threshold (the EOL). The remaining health loses its warning purpose since the prior already warns about it from the beginning of life. As result, the only purpose left for the remaining health is to contextualize the estimated SOH, which is carried out with a modified version of Equation (1), see Equation (2).

$$RH_k = \begin{cases} 1 & SOH_k > SOH_{expected\ k} \\ a_3 + \Delta SOH_{3\ k} \cdot b_3 & EOL > SOH_k \geq SOH_{expected\ k}, \\ 0 & SOH_k < EOL \end{cases} \quad (2)$$

where the $\Delta SOH_{3\ k}$ is the relative difference between the estimated SOH and the expected SOH having as reference the difference between the expected SOH and the EOL at instant k; the a_3 and b_3 are the variables of the linear function in between the EOL and the expected SOH.

This paper proposes a color for each event; see Figure 2A. The first event (RH = 1), the green color (in RGB logic [0,1,0]), represents that the estimated SOH is equal to the expected or higher. Everything goes well. The second event (RH = 0.5), the red color (in RGB logic [1,0,0.153]), represents that the EOL event is close (at 3% of SOH). An alert is lighted up and preventive actions are required not to suffer a failure of the vehicle before the warranty period. The third event (RH = 0), the black color (in RGB logic [0,0,0]), represents that the EOL has already been reached before the warranty period.

Figure 2. Proposed color map of the remaining health: (**A**) from the beginning of life until the expected SOH reaches the EOL + 3%; (**B**) from when the expected SOH is EOL +3% to the EOL.

For the life period that the expected SOH is lower than the one defined at the second event (EOL + 3%), an alternative proposal is made; see Figure 2B. In this period of the warranty, the second event has no sense (it has already been overcome), and it is taken from the color map. As consequence, the red color disappears and a color map of green and black is left.

2.3. Remaining Useful Warranty (RUW)

The remaining useful warranty is linked with the most interesting maintenance: the predictive maintenance. It requires the future SOH and refers to the remaining warranty from the present to the moment where the EOL threshold is reached; it refers to the RUL.

It quantifies the deviations between predicted and expected RUL. It gathers the initial expectations, calculated with the prior and the prognosed value, with the recorded SOH estimations. By comparing expectations and predictions, the remaining useful warranty represents the compliance of the stated warranty on the fulfilment of the application's operating specifications.

The objective of this state is to inform the vehicle manufacturer about the warranty state way before its nonfulfillment. Thanks to this, the remaining useful warranty gives enough room to the vehicle manufacturer to modify battery operation and achieve the designed warranty in absence of failure.

The value range of this state is defined with 0 and 1, with 1 being the fulfillment of the theoretical expectation and 0 the noncompliance of the warranty in half of the defined warranty period.

This paper proposes to have four events of interest that describe this state. The proposed first event is the theoretical remaining useful life based on the expectations described with 1. This event determines if the expectations are fulfilled or higher. We propose to neglect the states that represent higher compliance levels than those calculated prior since aging behavior is always expected to suffer negative deviations (higher degradation). Therefore, it is not likely to have states beyond this one.

The proposed second event is the RUL defined on the warranty. It usually consists of the value obtained by applying a reductive security margin to the theoretical RUL. Manufacturers used to add a security margin to their prior because of the uncertain level on the real application's operating conditions. In case it coincides with the theoretical RUL (no security margin is applied), the first and the second events would be the same. Therefore, there would be just one event that represents both the theoretical expectations and the defined warranty.

The proposed third event is the noncompliance of the warranty in 80% of the defined warranty period. It is represented by 0.4. This event tries to represent the state that lights up all the alerts. It becomes mandatory to act in order to correct the deviation suffered on the given warranty.

The proposed last event is the noncompliance of the warranty in half of the defined warranty period. This event defines the border where the battery replacement goes from 1 to 2 in order to keep the warranty. Remaining useful warranty values below this threshold could lead to a catastrophe to the battery manufacturer.

Once the events are defined, the rest of the values of this 0 to 1 range need to be defined. We proposed to use a piecewise-defined function built with first-order linear equations; see Equation (3).

$$\mathrm{RUW}_k = \begin{cases} 1 & \mathrm{lifespan}_k > \mathrm{lifespan}_{\mathrm{event1}} \\ a_1 + \mathrm{lifespan}_k \cdot b_1 & \mathrm{lifespan}_{\mathrm{event1}} > \mathrm{lifespan}_k \geqq \mathrm{lifespan}_{\mathrm{event2}} \\ a_2 + \mathrm{lifespan}_k \cdot b_2 & \mathrm{lifespan}_{\mathrm{event2}} > \mathrm{lifespan}_k \geqq \mathrm{lifespan}_{\mathrm{event4}} \\ 0 & \mathrm{lifespan}_k < \mathrm{lifespan}_{\mathrm{event4}} \end{cases}, \quad (3)$$

where the RUW_k is the remaining useful warranty at instant k; the $\mathrm{lifespan}_k$ is the predicted lifespan at instant k; the $\mathrm{lifespan}_{\mathrm{eventx}}$ is the lifespan value related with the event x, being x 1, 2, 3 and 4; the a_x and b_x are the variables of the linear function in between the defined events, where x is 1 and 2.

This paper proposes a color for each event; see Figure 3. The first event (RUW = 1), the green color (in RGB logic [0,1,0]), represents that the predicted lifespan is as the expected or higher. Everything goes well. The second event (RUW = 0.5), the orange color (in RGB logic [1,0.832,0.212]), means that the warranty lifespan and the predicted lifespan are the same. Since slight deviations can incur in the noncompliance of the warranty, a warning is set up. The third event (RUW = 0.4), the red color (n RGB logic [1,0,0.153]), represents that the warranty is not fulfilled 20% earlier than is defined on the warranty. An alert is lighted up. The fourth event (RUW = 0), the black color (in RGB logic [0,0,0]), represents the inflection point where the compliance of the warranty will come from 2 replacements of the whole battery system inside the warranty period. It represents a catastrophe.

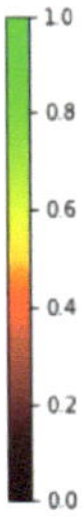

Figure 3. Proposed color map of the remaining useful warranty.

3. Required Tools

The SOW estimation can only be performed if some other elements are available along with the SOW estimator. The SOW cannot be estimated unless there is an on-board SOH estimator, the prior data are previously generated and an off-board RUL prognosis is already available.

3.1. SOH Estimator

The SOH is the state of a component that describes its health status. It is commonly used as the unique health indicator due to its level of synthesis. The SOH of a battery system is almost always represented by the relative dischargeable capacity at the present ($C_{present}$) in reference to the nominal capacity (C_n); see Equation (4):

$$SOH = \frac{C_{present}}{C_n}, \tag{4}$$

The SOH estimator of a battery system requires to estimate the dischargeable capacity at the present time instant. It can be directly estimated by undergoing a complete discharge process. Nonetheless, it is not common to undergo a complete discharge periodically in almost any real-life application of an electric vehicle. As consequence, indirect estimation methods are applied.

There are many different options available to conduct an indirect estimation of the dischargeable capacity of a battery system [9]. One of the most interesting methods consists of estimating the dischargeable capacity from the data generated in partial charges. This study has the advantage of the already integrated SOH estimator in an electric vehicle, see [10]. The obtained current profile of the partial charge is integrated in small time intervals so as to obtain the derivative profile of the capacity. These data are used along with a previously built model to obtain the SOH estimation.

The obtained on-board SOH estimations are used to directly estimate the remaining health. At the same time, the performed SOH estimations from the beginning of life (BOL) to the present are used to estimate the remaining useful warranty.

3.2. The Prior

The prior, or expectation, refers to the already obtained knowledge about the electric vehicle itself before it starts working; it is the knowledge available at the design phase of the electric vehicle. In this study, there are two different types of prior: the aging evolution and the warranty definition.

The prior related to the aging evolution consists of the data that describe the aging behavior. This knowledge is obtained by predicting the aging behavior at the most probable operating conditions from the BOL to the EOL. For that, there are four features that need to be defined.

Firstly, the most probable operating conditions are speculated. These operating conditions are established based on the application and battery characteristics. These characteristics are the following:

- The most probable power profile.
- The charge current rates.
- The operation limitations (often imposed by the battery management system (BMS)).
- The room environmental conditions.
- The voltage response of the battery system.
- The sizing characteristics of the battery.

Secondly, the EOL is specified. The EOL is an event of interest in the lifespan (in form of years or kilometrage) of the battery that describes the fulfillment of the application's operational specifications. It is common to use a fixed threshold such as the 80% of the SOH (20% of capacity decay) or the 200% of the ohmic resistance (100% of resistance increase). It is also possible to calculate the EOL based on the operational specifications (application oriented) [11].

Thirdly, the aging behavior of the battery at the most probable operating conditions is described. This is usually carried out with an aging model. There are many aging models available in the literature [12,13]. The accuracy of the model itself will lead to a lower or a higher uncertainty on the actual warranty definition (and more or fewer problems on the given warranty), but all of them are suitable options because they describe the aging behavior, i.e., what this study needs.

Finally, the SOH evolution from the BOL to the EOL considering the most probable operating conditions is predicted; see an example in Figure 4.

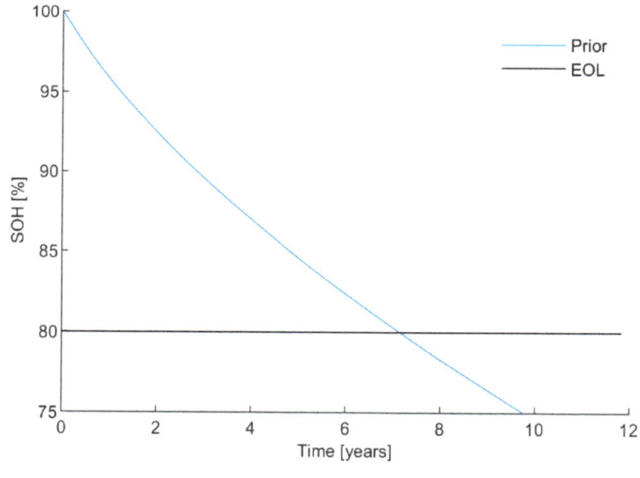

Figure 4. Example of the aging evolution of a battery system.

The prior related with the warranty definition consists of determining the lifespan that covers the warranty in terms of years and kilometrage. The manufacturers usually use the prior of the aging evolution to define these values. The theoretical lifespan is estimated by observing the moment the EOL is reached in the prior (the prior referring to the aging evolution). Then, the warranty lifespan is set up according to the uncertainty level that the calculated theoretical lifespan value has.

The prior knowledge related to the aging behavior is required to estimate the remaining health and remaining useful warranty. The prior knowledge related with the warranty definition, on the other hand, is used on the three substates: on the remaining warranty, on the remaining health and on the remaining useful warranty.

3.3. RUL Prognosis

The RUL represents the time or kilometers for which the electric vehicle can still be operative. The prediction of this merit as well as the uncertainty level on the given prediction, commonly defined as a probability density function (pdf), are fundamental elements on any predictive maintenance. The actions on estimating both (the prediction and its uncertainty) are performed by the prognosis.

Simplifying the RUL prognosis to the extreme, it can be defined as a trend search algorithm that is applied into the recorded SOH estimations. The obtained trend and its uncertainty are propagated until the EOL event in order to calculate the RUL and its uncertainty. The trend search algorithm can integrate prior knowledge of the trend itself (the prior referring the aging evolution) as a bias. This bias of the prognosis activity can greatly reduce the uncertainty on the RUL prediction if the trend underneath the data (the recorded SOH estimations) is similar to the trend on the bias. This study uses the prior that refers to the battery SOH evolution as the bias.

The RUL prognosis requires access to the recorded on-board SOH estimations. This is a must. Without these data, the RUL prognosis cannot be conducted, and therefore, any predictive maintenance is out of the scope. The RUL prognosis requires to have an SOH estimator on-board and a recorded device that provides the tools to have access to these data afterwards off-board.

Once data are accessible, a RUL prognosis algorithm can be applied to these data. There are plenty of different algorithms that can be used to predict the RUL [14]. The current study profits the already tested and validated prognosis algorithm based on a stochastic particle filter [7,8]; see example in Figure 5.

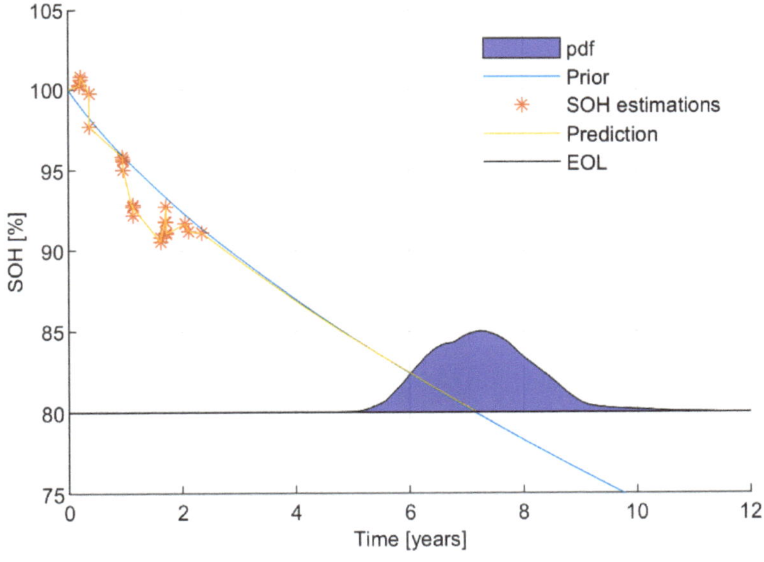

Figure 5. Example of a prognosis.

A stochastic particle filter is based on a Monte Carlo algorithm. It consists of a randomly generated finite number of particles to find the posterior distribution given a noise process and some partial observations. It allows for the filtering of the noise that hides the actual trend underneath the data. Once the trend is obtained and there are no more observations (SOH estimations), the trend and its distribution are propagated until the EOL is reached. As results, the RUL and its pdf are predicted [7].

The predicted off-board RUL values are used to estimate the remaining useful warranty.

4. Results and Discussion

The SOW estimation proposal has been applied into an artificially built example. Firstly, the example was characterized in order to generate the prior knowledge. The selected battery system was placed on a heavy-duty electric vehicle with a liquid-based thermal management system (the batteries are supposed to operate at 25 °C). This virtual electric vehicle travels a defined route (125 km with a specific orography) on a specified time schedule (320 days per year) during at least a specific lifespan (at least the 10 years of the imposed warranty). As result, the specifications of the operating conditions of the battery system were defined (the maximum required power rates based on the orography, the minimum dischargeable energy at the end of life of the battery and a minimum lifespan). Considering all this, the battery system was sized and the prior knowledge was generated. This virtual battery system was designed in five parts, containing 15 subparts each. The obtained aging evolution of the system under the designed operation conditions gives a lifespan of 12 years as a result (see the prior of Figure 6).

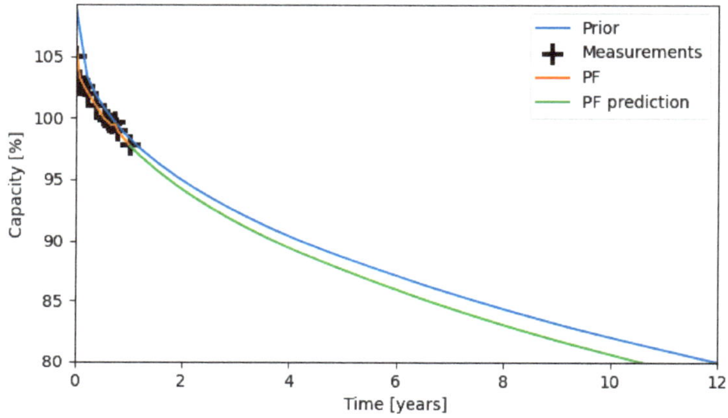

Figure 6. Aging evolution and recorded SOH estimations of the sized battery system for the electric bus.

After the design and sizing activity (and the prior generation), the designed battery system was constructed and operated based on the designed use case (all this was carried out in a virtual way and the data were artificially generated). The operation of the battery system provides on-board SOH estimations. One year of on-board SOH estimations were generated for each subpart; see the measurements of Figure 6. In order to simplify the example, it was assumed that all subparts have the same SOH evolution versus time.

Once SOH estimations of real operation of the battery were available (in this paper these data were artificially generated), off-board analysis such as the prognosis of the SOH evolution could be carried out. For this paper, a prognosis algorithm based on a particle filter [7] was developed in Python. The algorithm was applied to the artificially generated SOH estimations and the obtained prior from the sizing activity of the analyzed electric vehicle. The aging evolution from the observed first year onwards was then predicted (the worst case among all the subparts); see PF prediction in Figure 6. Based on the predicted aging evolution, the final lifespan was predicted to be 10.5 years (a RUL of 9.5 years).

Finally, the proposed approach to estimate the SOW was applied (once the prior knowledge, the SOH estimations and the RUL prognosis were available). The SOW estimation of every subpart was "green". The results of the estimated SOW are shown in Figure 7.

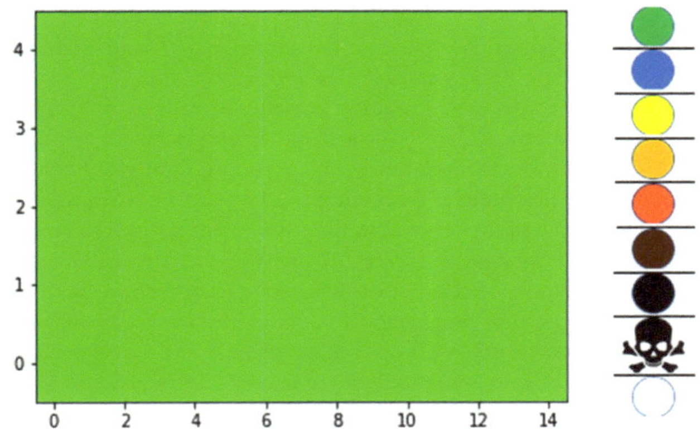

Figure 7. SOW estimation of each of the battery modules in the battery system.

The remaining warranty of the battery system after 1 year of operation is 0.9021.

The remaining health of each of the subparts is 0.9840. The SOH values of the different subparts are the same, so the remaining health of each of the subparts is also the same; see Figure 8 (left).

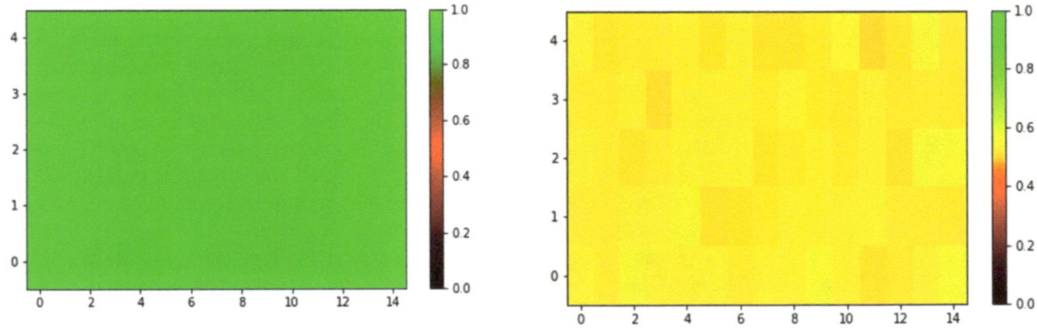

Figure 8. Remaining warranty (**left**) and remaining health (**right**) of each of the battery modules in the battery system.

The remaining useful warranty is shown in Figure 8 (right). Each subpart has a slightly different value due to the randomness of the applied prognosis algorithm. The obtained remaining useful warranty values go from 0.5153 to 0.5649.

The fact that the SOW is "green" indicates that the warranty goes well. If analyzing the substates in detail, the first thing that stands out is that both the remaining health and remaining warranty are high (above 0.9). The other aspect that stands out is that the remaining useful warranty is not the same for all the subparts as in the rest of the substates, even though the gathered data of each subpart are completely the same. This value changes from subpart to subpart.

Firstly, the remaining warranty is high since it passed just 1 year from the 10 years of operation that the battery system must accomplish. The battery system can be considered almost new, and so is indicated by this substate.

Secondly, the remaining health is high because of two reasons. The first reason is that the estimated SOH is only slightly lower than the expected SOH at the date of the estimations. The last SOH estimation is 97.8% and the expected SOH value at the date that this SOH estimation was performed is 98.2%. There is only 0.4% of difference between the SOH expectation and the SOH estimation. The second reason is that there is still a huge difference between the actual SOH and the EOL. The difference between the estimated SOH and the EOL is 17.8%. The high remaining health indicates that the experienced damage on the batteries is close to the expected one, and that in case predictions were far from optimistic, there would still be room for action.

Thirdly, the remaining useful warranty is not unique for all the subparts because of the characteristics of the applied prognosis algorithm. The particle filter is based on finite random simulations, which lead to slightly different results despite sharing the same data. The obtained remaining useful warranty values between 0.51 and 0.56 indicate that the predictions are worse than the expected one (below 1) but that the predictions still respect the imposed warranty (above 0.5). The results warn about the need to analyze the expectations and the prior. The expected operation conditions could not be the same as the real ones, thus generating a deviation between the estimated SOH and the expected SOH. Another cause could be that the prior of the aging evaluation itself is not correct, which would require an update of the used aging model and to act accordingly in order to fulfill the warranty satisfactorily.

To sum up, the results show that the warranty of 10 years is expected to be fulfilled if everything evolves as it happened during the observed first year. The worst case (the subpart with the lowest predicted lifespan) shows a predicted lifespan of 10.5 years, which is above the imposed warranty of 10 years.

5. Conclusions

This paper has proposed a metastate that qualitatively synthesizes the relevant information about the warranty fulfillment. For that, three substates have been defined and qualitatively quantified: the remaining warranty, the remaining health and the remaining useful warranty. The remaining warranty represents the relative passed time of the warranty with respect to the total warranty. The remaining health represents the relative health with respect to the expectations and EOL. The remaining useful warranty represents the relative RUL with respect to the warranty definition, the expectations and the event where the battery system has to be replaced twice. The qualitative values of these three substates (in a range of 1 to 0) are evaluated by an expert system to determine the SOW, which is represented in a color code.

The applicability of an SOW estimator is totally dependent to the availability of three elements: an on-board SOH estimator (and a way of accessing the generated data with this estimator), a prior definition of the most probable aging evolution (and the most probable lifespan of the system) and an off-board RUL prognosis algorithm. It is not possible to develop a SOW estimator unless these three elements are available. The prior knowledge generates the constrictions to relativize these substates. This is indispensable to obtain qualitative values. The SOH estimations are the health observations themselves. This is necessary to carry out any off-board analysis of the health, and therefore, of the warranty. The RUL prognosis provides the SOH predictions. This is indispensable to properly address the warranty. In this proposal, the uncertainty of the given prediction is neglected. The SOW and the remaining useful warranty are indicators and not final decision-making actions, and therefore, the integration of the uncertainty is considered to be necessary in a decision-making level rather than in a state estimation (diagnosis) level.

The results have shown some strong points of the proposed state. The SOW is able to show in a user-friendly manner if there are reasons to worry in terms of warranty fulfillment. Moreover, its three substates provide enough insight to understand completely how healthy the system is with respect to what it should be, and how different the final lifespan will be with respect to the expected one (and imposed one).

One of the most interesting uses of this state is the planification of maintenance activities. The SOW synthesizes the most relevant features of each of the different maintenance methodologies. Therefore, it can be used as the main indicator for maintenance scheduling and develop an improved predictive maintenance.

Another strong argument in favor of the SOW is the detection of failures and weak points of the system. The tracking of the fulfillment of the warranty provides the chance to apply data analysis, which could lead to detect defect components, errors on prior knowledge, weak points, harmful actuations, etc. To sum up, it can be used as the basis of a continuous improvement process of the system.

Last but not least, the use of SOW provides the basis to determine the optimal use of the batteries. The tracking of the SOW allows the final user to adjust the use conditions of the batteries to fit to their concept of optimal use. The use conditions of the batteries can be brought to the limits of the warranty, optimizing the total income from the batteries on the warranty period, or the use conditions can be modified to maximize the lifespan.

As for the future, we are currently researching the diagnosis assessments that will support the SOW estimator. Those assessments are based on the historic data of the operating conditions, and they will give insight in the reasons behind deviations of the expectations and/or prior knowledge.

Author Contributions: Conceptualization, M.A., D.S. and M.O.; investigation, M.A.; methodology, M.A.; software, M.A. and D.S.; writing—original draft preparation, M.A.; writing—review and editing, A.Z.; supervision, H.M.; project administration, A.Z. All authors have read and agreed to the published version of the manuscript.

Funding: This work has been financed within the initiative "ENERISLA 100% Renewable Isolated Energy Systems" (File: CER-20191002) by Centre for Industrial Technological Development (CDTI) under the Ministry of Science and Innovation in Spain.

Data Availability Statement: Not applicable.

Conflicts of Interest: The authors declare no conflict of interest.

References

1. International Energy Agency (IEA). Chapter 1: Trends in Electric Mobility. In *Global EV Outlook*; IEA Publications: Paris, France.
2. Wang, C.; Ghenniwa, H.; Shen, W. Distributed scheduling for reactive maintenance of complex systems. *IEEE Trans. Syst. Man Cybern. Part C Appl. Rev.* **2008**, *38*, 269–273. [CrossRef]
3. De Almeida, A.T.; Cavalcante, C.A.V.; Alencar, M.H.; Ferreira, R.J.P.; de Almeida-Filho, A.T.; Garcez, T.V. Preventive Maintenance Decisions. *Int. Ser. Oper. Res. Manag. Sci.* **2015**, *231*, 215–232. [CrossRef]
4. Calabrese, F.; Regattieri, A.; Bortolini, M.; Gamberi, M.; Pilati, F. Predictive maintenance: A novel framework for a data-driven, semi-supervised, and partially online prognostic health management application in industries. *Appl. Sci.* **2021**, *11*, 3380. [CrossRef]
5. Olmos, J.; Gandiaga, I.; Saez-De-Ibarra, A.; Larrea, X.; Nieva, T.; Aizpuru, I. Modelling the cycling degradation of Li-ion batteries: Chemistry influenced stress factors. *J. Energy Storage* **2021**, *40*, 102765. [CrossRef]
6. Richardson, R.R.; Osborne, M.A.; Howey, D.A. Gaussian process regression for forecasting battery state of health. *J. Power Sources* **2017**, *357*, 209–219. [CrossRef]
7. Arrinda, M.; Oyarbide, M.; Macicior, H.; Muxika, E. Comparison of stochastic capacity estimation tools applied on remaining useful life prognosis of lithium ion batteries. In Proceedings of the PHM Society European Conference, Utrecht, The Netherlands, 3–6 July 2018; Volume 4. [CrossRef]
8. Arrinda, M.; Oyarbide, M.; Macicior, H.; Muxika, E. Unified Evaluation Framework for Stochastic Algorithms Applied to Remaining Useful Life Prognosis Problems. *Batteries* **2021**, *7*, 35. [CrossRef]
9. Sui, X.; He, S.; Vilsen, S.B.; Meng, J.; Teodorescu, R.; Stroe, D.-I. A review of non-probabilistic machine learning-based state of health estimation techniques for Lithium-ion battery. *Appl. Energy* **2021**, *300*, 117346. [CrossRef]
10. Oyarbide, M.; Arrinda, M.; Sánchez, D.; Macicior, H.; McGahan, P.; Hoedemaekers, E.; Cendoya, I. Capacity and impedance estimation by analysing and modeling in real time incremental capacity curves. *Energies* **2020**, *13*, 4855. [CrossRef]
11. Arrinda, M.; Oyarbide, M.; Macicior, H.; Muxika, E.; Popp, H.; Jahn, M.; Ganev, B.; Cendoya, I. Application Dependent End-of-Life Threshold Definition Methodology for Batteries in Electric Vehicles. *Batteries* **2021**, *7*, 12. [CrossRef]
12. Lucu, M.; Martinez-Laserna, E.; Gandiaga, I.; Camblong, H. A critical review on self-adaptive Li-ion battery ageing models. *J. Power Sources* **2018**, *401*, 85–101. [CrossRef]
13. Arrinda, M.; Berecibar, M.; Oyarbide, M.; Macicior, H.; Muxika, E.; Messagie, M. Levelized cost of electricity calculation of the energy generation plant of a CO_2 neutral micro-grid. *Energy* **2020**, *208*, 118383. [CrossRef]
14. Hu, X.; Xu, L.; Lin, X.; Pecht, M. Battery Lifetime Prognostics. *Joule* **2020**, *4*, 310–346. [CrossRef]

MDPI AG
Grosspeteranlage 5
4052 Basel
Switzerland
Tel.: +41 61 683 77 34

World Electric Vehicle Journal Editorial Office
E-mail: wevj@mdpi.com
www.mdpi.com/journal/wevj

www.ingramcontent.com/pod-product-compliance
Lightning Source LLC
Chambersburg PA
CBHW041432210926
51850CB00013B/3409

* 9 7 8 3 7 2 5 8 3 9 4 1 4 *